MICROBIOLOGY

MICROBIOLOGY

Second Edition

EUGENE W. NESTER
University of Washington

C. EVANS ROBERTS
University of Washington

NANCY N. PEARSALL
University of Washington

BRIAN J. McCARTHY
University of California,
San Francisco

SAUNDERS COLLEGE
PHILADELPHIA

PART OPENING PHOTOGRAPHS

PART I: *Erwinia,* showing pili. (Courtesy of E.S. Boatman.)

PART II: Scanning electron micrograph of *Microcyclus marinus* showing doughnuts, horseshoes, and coils. (Courtesy of H. Raj.)

PART III: The bacteria that cause "Legionnaire's disease." (Courtesy of F. Schoenknecht, J. Lara, and S. Lara.)

PART IV: *Rhizobium* bacteria inside a soybean root nodule. (Courtesy of W. Brill; R. Ravelle, *Scientific American,* September 1976, p. 164.)

PART V: *Spirulina,* photosynthetic cyanobacteria used in the production of single-cell protein. (Courtesy of J.T. Staley.)

COVER: *Aspergillus fumigatus* culture. (Courtesy of F. Schoenknecht.)

Saunders College
West Washington Square
Philadelphia, PA 19105

Library of Congress Cataloging in Publication Data

Main entry under title:

Microbiology.

Includes bibliographies and index.
1. Microbiology. I. Nester, Eugene W.
QR41.2.M5 1978 576 77-10422
ISBN 0-03-019326-5

Printed in the United States of America
0 1 032 9 8 7 6 5

TO THE STUDENT

We have attempted to write a textbook with your interests, abilities, and needs clearly in mind. All of us have taught microbiology at the level at which this text has been written, so we have incorporated a number of teaching aids to help you understand and relate microbiology to material you have previously encountered. For example, new scientific terms are defined the first time they are used. You can also look up unfamiliar scientific words in the glossary, which gives the pronunciation of difficult words, including the names of organisms covered in the text.

You will find that topics discussed in more than one section of the book can be located by using either the cross references in the margin or the index. Major references for index entries are designated by boldface type.

We have divided the questions at the end of each chapter into two types: review questions, which you can answer directly from the material covered in the chapter, and thought questions, which require you to base your answers on material covered in the chapter and in other sources, either other chapters of the text or references given at the end of the chapter. The references we have included are interesting and informative supplementary readings and should be readily available to those of you using this text.

We sincerely hope that you will enjoy your study of microbiology and that you will find this book helpful in exploring an exciting life science.

E.W.N.
C.E.R.
N.N.P.
B.J.McC.

PREFACE

Why yet another microbiology text? The answer to this question comes from the authors' cumulative experience in teaching the introductory microbiology course at the University of Washington. This one-quarter course attracts a large percentage of the students preparing to enter health-related fields. Most of the students have had some exposure to chemistry and biology but are far from expert in handling these subjects. The course aims to teach basic microbiological concepts and the role of microorganisms in health and other applied areas. In teaching this course we have tried numerous texts or combinations of texts but have not found them satisfactory. They are either too detailed, presuppose too much chemical and biological knowledge, or neglect basic microbiology or the relationship of microbes to disease. The text we have written is designed to eliminate these problems. The gratifyingly wide acceptance of the first edition suggests that we have succeeded in large measure.

In this revised edition we have followed the same general approach of the first edition. First and foremost, we have attempted to present the fundamentals of microbiology so that their relevance to the applied material is clearly evident. In our experience, students inclined toward careers in the health sciences area have limited backgrounds in biology and chemistry and find discussions of such topics as microbial physiology, cell structure and function, and genetics not only difficult but not particularly important to their real interests. Thus they try to memorize facts merely for exams rather than to understand the material and apply it to the sections on applied microbiology. We have used several devices to interest students and help them interrelate material presented throughout the text. First, we have included numerous examples, particularly from familiar areas to illustrate abstract concepts. Second, we have employed cross-referencing in the margins to help the student recall material previously covered and be aware of relationships that may not have been obvious at first reading. Third, we have reduced the amount of minutiae by providing fewer facts than most texts. It is our belief that the understanding of a relatively few fundamentals is more important than the memorization of a host of facts, most of which will be forgotten soon after exams have been taken.

We have also tried to cover material to which every student has been exposed but which we often find is poorly understood or even forgotten. For this reason an early chapter covers the principles of biological chemistry. This chapter will serve as a review for some students and as reference material for others, but we predict it will provide some appreciation of the relationships of biology to chemistry for most students.

We have reduced the number of new terms in the text, for the student should not be expected to learn the language of the specialist. In particular, latinized names are used only in the genus and species names of microorganisms; disease and anatomical features are given their common names wherever possible, with the corresponding medical designation provided in the glossary. We have attempted to refer to microorganisms by the names being used in the laboratories on the forefront of research which we expect will be used by all microbiologists within the next few years.

The rapid advance of microbiology has required updating the second edition in a number of areas, notably in genetics, immunology, and medical virology. Moreover, the appearance in 1974 of the eighth edition of *Bergey's Manual of Determinative Bacteriology* has required that a number of scientific names be changed. We have also added two new chapters, "Bacteria of Medical Importance," to summarize data on certain organisms, and "Looking Ahead," to indicate some of the exciting possibilities in the future of microbiology. Certain material that was redundant or used only by a minority of instructors has been omitted.

We feel that the text is highly adaptable to different courses for students of varying interests. Thus the text has been written to include a number of core chapters appropriate for all courses and a number of additional chapters that are particularly appropriate for other courses. The "core" material consists of Chapters 1–10, 16, 19, and 20. Students in the health sciences area would probably read Chapters 17–33. Students concerned primarily with applied aspects of microbiology other than health-related topics might read Chapters 34–39. For students interested in an overall view of microbiology, Chapters 12–17, 32, 34–39 would be appropriate. Because all chapters have been cross-referenced, it is possible for the student to obtain relevant information on a particular subject without reading an entire chapter.

The chapters on microbial diseases represent a sufficiently radical departure from other textbooks to warrant some explanation. First, they are organized along an "organ-system" approach rather than by microbial classification, an approach that is more interesting to students although one that is more difficult to teach. Second, some elementary anatomical and physiological description of the area of the body involved is given to enhance understanding of host-microbial interactions. Third, diseases chosen for discussion are those which students are likely to be familiar with from the news media or prior experience. We have rejected the idea that only diseases of present-day suburban United States are important and have chosen examples according to their interest and the availability of sufficient data to illustrate biological principles. Finally, we have not attempted comprehensive coverage. These sections are not a handbook for paramedical personnel but rather exemplary material for all students taking an introductory course including those entering health-related fields. A laboratory manual and an instructor's manual accompany the text.

As before, we are greatly indebted to many students and teachers across the land for helpful criticisms and comments. Their comments have all been carefully considered.

We are especially grateful to the many members of our own department, notably Charles Evans, Neal Groman, Erling Ordal, John Sherris, James Staley, and Russell Weiser, who gave generously of their time and advice. Revision and additions to the art work were designed and executed by Iris Nichols. Kendall Getman and Jean Shindler of Holt, Rinehart and Winston shouldered much of the task of getting this edition into print.

We also want to thank the following persons for reading the manuscript and offering valuable suggestions for its improvement: Sylvia P. Brehm, Indiana State University; Ellis Brockman, Central Michigan University; Anthony Catena, San Francisco State University; Daniel P. Caylor, Palm Beach Junior College; Dennis E. Duggan, University of Florida; Cindy Erwin, City College of San Francisco; Marie Gilstrap, Highline Community College; Paul Hartman, Iowa State University; Alice Helm, University of Illinois; Eugene Hess, Los Angeles City College; Robert J. Janssen, University of Arizona; Maryann Luger, Pennsylvania State University; Eleanor K. Marr, Dutchess Community College; Ralph Mitchell, Harvard University; Anthony Mukkada, University of Cincinnati; P. J. VanDemark, Cornell University; Anne S. Vanderbilt, Purdue University; John M. Woodward, University of Tennessee.

We sincerely hope this book will be interesting and valuable to students and teachers of introductory microbiology. We encourage your comments and suggestions.

Seattle, Washington
June 1977

E. W. N.
C. E. R.
N. N. P.
B. J. Mc.

CONTENTS

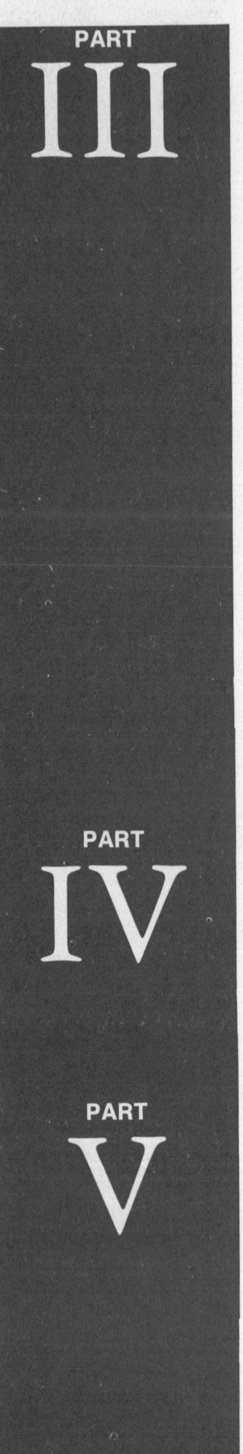

PART III

MICROBES AND HUMAN BEINGS 353

PART IV

MICROBES AND THE ENVIRONMENT 609

PART V

THE CONTINUING FRONTIER 689

MICROBIOLOGY

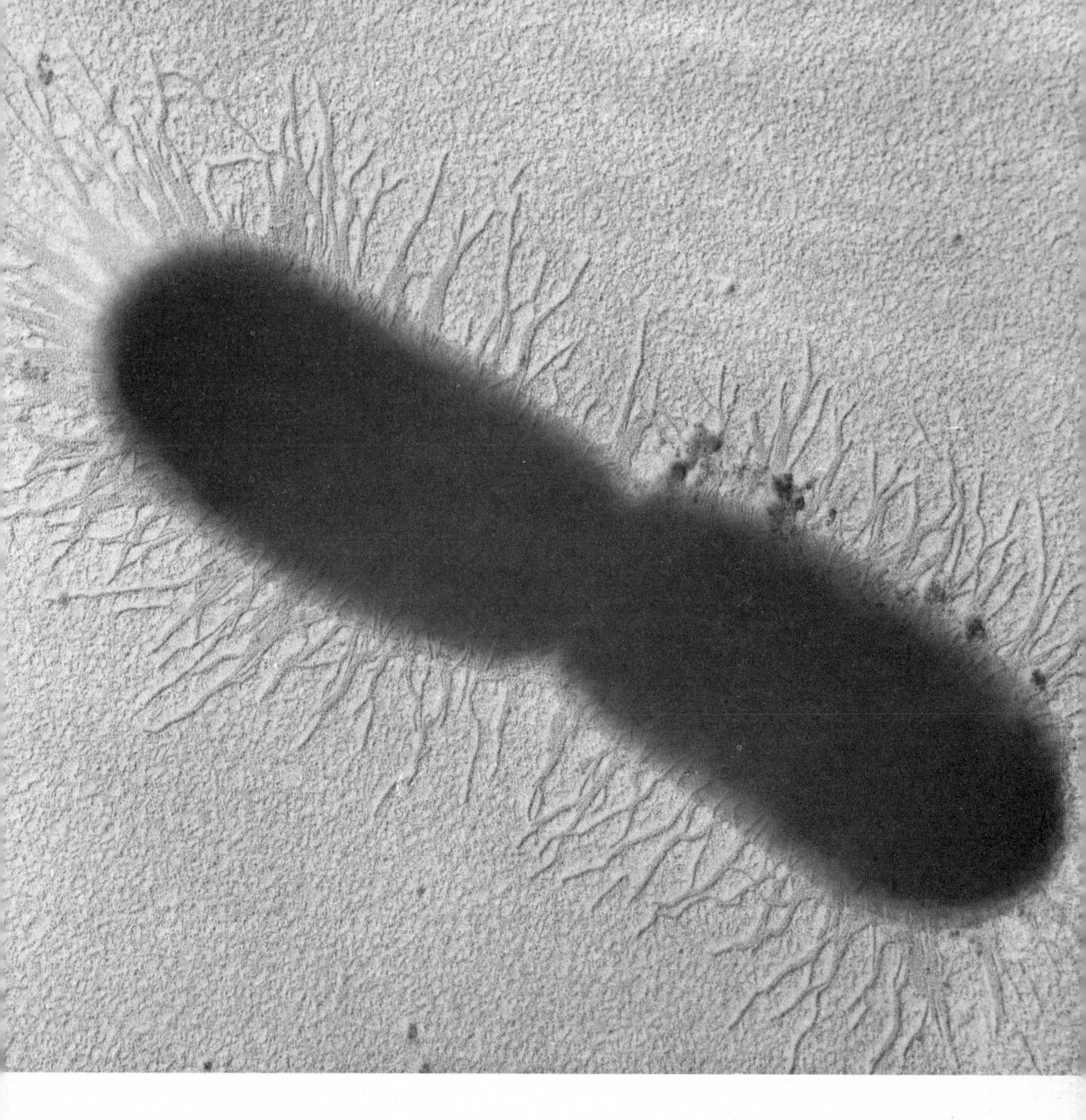

I FOUNDATIONS OF MICROBIAL LIFE

CHAPTER 1

MICROBIOLOGY IN THE BIOLOGICAL WORLD

DISCOVERY OF THE MICROBIAL WORLD

The birth of microbiology occurred in 1674 when Anton van Leeuwenhoek, an inquisitive Dutch drapery merchant, peered at a drop of lake water through a glass lens that he had carefully ground. What he observed through this simple magnifying glass was undoubtedly one of the most startling and amazing sights that man has ever beheld—the first glimpse of the world of microbes. As he recorded in a letter to the Royal Society of London, he saw:

> . . . very many little animalcules, whereof some were roundish, while others a bit bigger consisted of an oval. On these last, I saw two little legs near the head, and two little fins at the hind most end of the body. Others were somewhat longer than an oval, and these were very slow a-moving, and few in number. These animalcules had divers colours, some being whitish and transparent; others with green and very glittering little scales, others again were green in the middle, and before and behind white; others yet were ashen grey. And the motion of most of these animalcules in the water was so swift, and so various, upwards, downwards, and round about, that 'twas wonderful to see . . . .

This letter, one of his earliest to the Royal Society, became the first known description of *protozoa* and *algae.* It was followed by 150 more letters which spanned 50 years. Each of these remarkable letters was filled with descriptions of his important discoveries. Although van Leeuwenhoek had little formal education, his inquisitive nature led him to examine a wide variety of materials—river, well, and seawater as well as vinegar and pepper, to list only a few.

Van Leeuwenhoek left no descriptions of the apparatus with which he observed protozoa and bacteria. He kept "for himself alone" his best microscopes

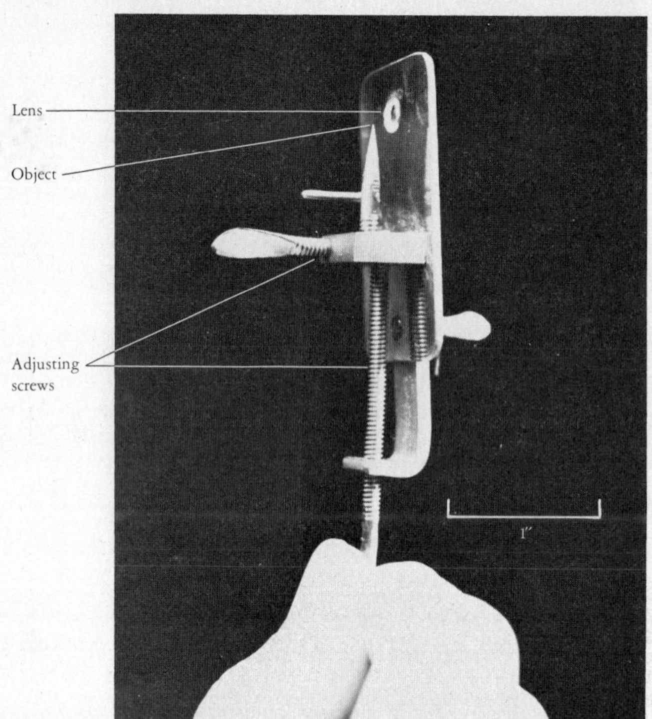

FIGURE 1-1
Model of a van Leeuwenhoek microscope. The original was made in 1673 and magnified almost 300 times. Stained bacteria between 1 and 2 μm in size can be seen with the original model. The object being viewed is brought into focus with the adjusting screws. This replica was made according to the directions given in *The American Biology Teacher,* 30:537 (1968). Note its small size. (Courtesy of J. P. Dalmasso.)

and his manner of observing the "animalcules." He never did divulge his method. His prized lenses and jealously guarded techniques enabled him to surpass all other microscopists for at least a century.

DESCRIPTION OF VAN LEEUWENHOEK MICROSCOPE

Van Leeuwenhoek left over 400 microscopes behind when he died; except for a few, all have long since mysteriously disappeared. We do, however, know something about his apparatus. The microscopes that he fashioned consisted of a spherical lens, meticulously ground and mounted between two metal plates (Figure 1-1). These instruments were not much more than powerful magnifying glasses. The material to be examined was placed on a blunt pin on one plate, and the specimen was brought into focus by moving the object relative to the lens with screws. Van Leeuwenhoek made a new microscope for every new specimen he wished to study. This accounts for the fact that he made several hundred instruments.

Darkfield microscopy, pages 49–51

Even though there is no question that van Leeuwenhoek was a master in the art of lensmaking, it is highly unlikely that he could have observed the detail and clarity in microorganisms that he reported unless he had unusually good methods of illuminating his specimens. It seems probable that he discovered the technique of darkfield microscopy, a technique in use today, in which faint objects appear as brightly lit against a dark background. Three hundred years after van Leeuwenhoek first saw his "animalcules," some of his observations were duplicated with a replica of one of his original microscopes, employing either sunlight or candlelight to provide darkfield illumination.

THE CONTROVERSY OVER SPONTANEOUS GENERATION

One of the most intriguing questions that the discovery of microbes immediately prompted concerned the origin of these microscopic forms. At the end of the seventeenth century the theory of spontaneous generation—that organisms visible to the naked eye can arise spontaneously from nonliving matter—had just been completely debunked. Francesco Redi, an Italian biologist and physician, showed conclusively that worms found on putrefying meat originated from the eggs of flies. If he surrounded the meat with a gauze fine enough to prevent flies from depositing their eggs, the worms no longer appeared.

Despite these findings the possibility that organic matter could spontaneously give rise to microscopic forms was quickly suggested. This hypothesis was not to be disproved for almost 200 years. Because gauze could not keep out microscopic forms, the general approach to determining whether microbes could arise spontaneously consisted of boiling the organic material in a vessel to kill all forms of life and then sealing the vessel to prevent the entrance of air. If the solution became turbid, microorganisms must be present, thus supporting the theory of spontaneous generation. However, the results of this kind of experiment were confusing since different investigators obtained different results.

In 1749 Needham, a Catholic priest and a capable scientist, reported that many different types of broth (infusions) gave rise to microorganisms even though they had been boiled and then sealed with corks. Needham's experiments, however, appeared to be refuted by the results of other, similar experiments published by Father Spallanzani in 1766. His experiments differed from Needham's in two significant ways: Spallanzani boiled the infusions for longer periods of time and then sealed the flasks by melting the glass neck of the flask. With this technique he demonstrated, by hundreds of separate experiments, that the infusions would remain free of microorganisms indefinitely. However, if the neck of the flask cracked, the broth rapidly became turbid. Spallanzani thus concluded that the microbes must have entered the infusion with the unsterilized air. The controversy, however, continued to rage. "Spontaneous generation" continued to be reported for the next 100 years by investigators who either did not boil their infusions long enough to kill highly resistant forms (endospores) of microorganisms or sealed their flasks imperfectly.

Experiments of Pasteur

Although Spallanzani's conclusions were correct, Needham remarked: "from the way he [Spallanzani] has treated and tortured his vegetable infusions, it is obvious that he has not only much weakened and maybe even destroyed the 'vegetative force' of the infused substances, but also that he has completely degraded . . . the small amount of air which was left in his vials. It is not surprising, thus, that his infusions did not show any sign of life. . . ." These objections were not convincingly answered for still another 100 years.

The two giants in science who did the most to lay the theory of spontaneous generation to rest once and for all were the French chemist Louis Pasteur, considered by many to be the father of modern microbiology, and the English physicist John Tyndall. The refutation published by Pasteur in 1861 was a masterpiece of logic. Indeed, the force of his logic was probably as important in swaying the scientific community to his side as were the results of his experiments themselves, many of which had already been performed by other investigators. First, he demonstrated that air was filled with microorganisms by filtering air through a cotton filter and then examining with a microscope the organisms which had been trapped by the cotton. Many of these trapped organisms were indistinguishable from those which had previously been observed in many infusions. Furthermore, if the cotton plug was dropped into a sterilized infusion, an increase in the number of these organisms promptly resulted. Perhaps Pasteur's most unique and dramatic experiment consisted in demonstrating that infusions, once sterilized, would remain sterile in specially constructed flasks open to the air. In these swan-necked flasks (Figure 1-2) organisms from the air would settle out in the bends and never reach the fluid. Only when the flasks were tipped would bacteria be able to enter the broth and grow. These experiments finally ended all claims that unheated air contained in itself a "vital force" sufficient for spontaneous generation.

PASTEUR REFUTES SPONTANEOUS GENERATION

Experiments of Tyndall

Although a large proportion of the scientific population, especially in France, was convinced by Pasteur's experiments, many respected scientists still doggedly held to the concept of spontaneous generation. This persistence undoubtedly stemmed in part from the fact that many scientists were unable to verify his results. The English physicist John Tyndall finally provided a logical explanation for the differences in experimental results among laboratories. Tyndall, an ardent admirer of Pasteur, was convinced of the validity of Pasteur's experiments, even though he was unable to obtain consistent results when he tried to repeat them. He then realized that different infusions required widely varying boiling times in order to be sterilized. Thus some materials could be sterilized by being boiled for 5 minutes; others, most notably hay infusions, could be boiled for 5 hours, and

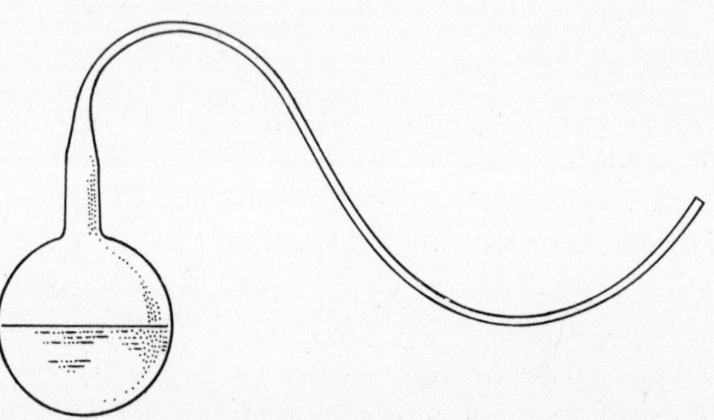

FIGURE 1-2
Swan-necked flask. Flasks of this type made by Pasteur have been sealed and remain sterile to the present day.

yet they still contained living organisms. Furthermore, if hay was present in the laboratory, it was virtually impossible to sterilize even the infusions that had previously been sterilized by boiling for 5 minutes. What was present in hay that caused these effects? Tyndall finally realized that heat-resistant forms of life were being brought into his laboratory on the hay. These were transferred to all infusions in his laboratory, thereby making everything difficult to sterilize. He concluded that some bacteria must be able to exist in two forms: a heat-labile form (*vegetative cell*) readily killed by boiling, and a heat-resistant form (*endospore*). In the same year (1876) a German botanist, Ferdinand Cohn, also described endospores and demonstrated their heat-resistant properties.

TYNDALL AND COHN DISCOVER ENDOSPORES

Endospore, pages 78–80

Tyndall then developed a procedure for destroying even the most resistant forms of bacteria. He boiled the infusion intermittently; the endospores developed into vegetative cells in the intervals between heatings and were then readily killed in the next round of heating. With this technique, named *tyndallization,* Tyndall could sterilize a hay infusion by boiling it for one minute, five separate times, even though continuous boiling for one hour would not kill all of the endospores. This discovery of the extreme heat resistance of spores explained the discrepancies between Pasteur's results and those of other investigators. Organisms which produce endospores are commonly found in the soil and would likely be present in hay infusions. Because Pasteur used only infusions prepared from sugar and yeast extract, his broth would not likely contain endospores.

Thus the concept of spontaneous generation, as envisioned by Needham and others, was disproved only about a century ago. This fact emphasizes the fantastic progress that has been made in our knowledge of microorganisms in the last 100 years. One of these microorganisms, *Escherichia coli,* is the most intensively studied organism in existence. A complete description of its living processes, in chemical terms, should be possible in the foreseeable future.

The rapid pace and importance of the studies on microorganisms can be gauged by the number of prizes that have been awarded to microbiologists. Fully one-third of the 67 Nobel Prizes in Physiology and Medicine ever awarded were given to people working in the areas of microbiology and immunology. Additional Nobel Prizes have been awarded to chemists studying microbial processes. Some of these advances, viewed in an historical context, will be discussed at the time the relevant facts of microbiology are considered.

Origin of Life

Although scientists do not believe that, under present-day conditions, life can originate from nonliving matter, it almost certainly did under the far different conditions which existed on earth when life first arose. Indeed, electron photomicrographs taken from the inside of a chert, a type of rock approximately 2 billion years old, reveal what appear to be the fossil remains of bacteria. But today, even the simplest microorganism is an extremely complex being and must have evolved over millions of years.

The studies of Pasteur and others demonstrated that microscopic and visible organisms are similar in at least one respect: they both arise only from preexisting organisms. To gain some insight into the role of microorganisms in the biological world, it will be enlightening to consider other similarities, and also differences, between the microorganisms and plants and animals.

CELL THEORY

Since most bacteria are single-celled organisms, an understanding of microbiology can best be gained by recognizing the basic significance of the *cell* to all life. About the time that van Leeuwenhoek was peering at "animalcules" through his simple microscopes, an English microscopist, Robert Hooke, was making significant observations on the structure of cork. He perceived that cork consisted of a "great many little boxes, separated out of one continued long pore, by certain Diaphragms. . . ." Hooke's "little boxes" are what we now call *cells.* Despite this significant discovery of the structure of living forms, however, cells were not studied intensively until the early nineteenth century.

THE CELL IS THE BASIC UNIT OF STRUCTURE AND FUNCTION

Before the 1830s most people viewed the entire organism as the fundamental unit of structure. But in 1838 a German botanist, Matthias Schleiden, and in 1839 a German zoologist, Theodor Schwann, published statements to the effect that all organisms were composed of cells and that cells were the fundamental units of life. They clearly recognized that these units were capable of carrying out all of the basic functions which were attributed to living organisms. Their hypothesis, called the *cell theory,* represents one of the most significant conceptual advances in biological thinking. Although Schleiden and Schwann were not the first to proclaim this concept, the clarity and logic of their statement made a great impact on the thinking of scientists. Previously the attention of scientists had focused on the apparent diversity of organisms; it now switched to their basic component parts, the cells, which share obvious similarities even in the most diverse forms of life. In the century and a half between the publication of the cell theory and today, the work of thousands of scientists in areas of zoology, botany, biochemistry, genetics, and microbiology has detailed just how basic these similarities are and how much variety nature can tolerate. The study of bacteria and other unicellular organisms has been especially important to an understanding of cell structure and its function.

CONCEPT OF BIOCHEMICAL UNITY AND CELLULAR DIVERSITY

ALL CELLS PERFORM THE SAME BASIC FUNCTIONS ALTHOUGH BY DIFFERENT MECHANISMS

Each cell, whether a free-living bacterium or a brain cell, which is part of a larger structure, has to deal with similar problems in order to live. The basic biochemical mechanisms for solving these problems are the same in all cells. However, there is extensive diversity in both the structures and the detailed mechanisms which different cells use in carrying on their life processes.

ALL CELLS MUST REPRODUCE THEMSELVES

The cells of all organisms, whether animals, plants, or unicellular life forms, must include *one* basic function in life—to reproduce exact copies of themselves. In order to replicate, a cell must contain genetic instructions for its exact duplication. In addition, replication requires that cells be able to synthesize the constituents of living matter from the much simpler foodstuffs available to them. Because these biosynthetic reactions require energy, all cells must also have mechanisms for deriving and utilizing the energy made available through the breakdown of foodstuffs.

One of the most striking findings from investigations of many cells (animal, plant, and unicellular) is the remarkable similarity in the basic mechanisms whereby these life processes are carried out. Representative cells from organisms of each group duplicate their genetic material in the same way, break down foodstuffs to gain energy through the same general sequence of biochemical reactions, and synthesize each of their cellular components from the same starting materials, using the same biochemical reactions. Several later chapters will be devoted to discussing these aspects of cell function.

Chapters 7 and 8

LEVELS OF ORGANIZATION

The science of microbiology can be considered in terms of a *hierarchical order of organization.* Each unit in this organization is composed of units lower in the order and in turn comprises part of a larger unit. Thus the levels of organization can be viewed as a trapezoid composed of horizontal slabs, one on top of another (Figure 1-3). The foundation, at least for the purposes of this course, is composed of *atoms.* The atoms interact to form *small molecules,* which in turn are joined together to form very large molecules, or *macromolecules.* The macromolecules

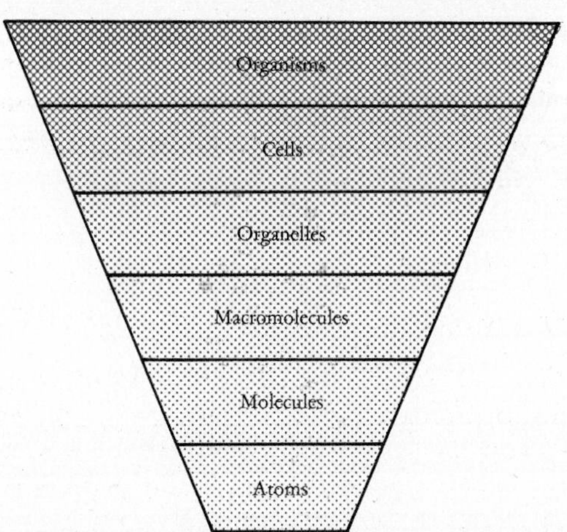

FIGURE 1-3
Levels of organization in the microbial world. Each level includes all levels beneath it. Neither tissues nor organs are present in members of the microbial world and are not included in this organizational scheme.

associate with other macromolecules to form structures which perform a specific function, the *organelles.* In turn, organelles and other cell structures are organized to form the *cell.* In the case of a unicellular organism the cell, of course, represents the entire organism, but in other cases the individual is composed of many cells.

The discussion will begin with the least complex levels, the atoms and molecules. Much of the rest of the text will be a consideration of the more complex levels.

Atoms of Life

C, H, O, and N COMPRISE 98% OF CELLS' WEIGHT

When cells taken from a diversity of organisms are analyzed for their chemical composition, an obvious uniformity in the kind and quantity of elements is apparent. Table 1-1 gives the elemental composition of a bacterial cell. Several interesting features stand out. First, cells are composed of only a small fraction of the total possible number of over 100 elements. In fact, four elements, carbon, hydrogen, oxygen, and nitrogen, comprise about 98.5 percent, by weight, of the atoms of cells. Phosphorus and sulfur together make up an additional 1 percent, and all of the remaining elements together account for less than 0.5 percent. This composition may differ in some specialized cells in which certain elements are associated with a specific structure or function. Cells in bone, for example, have a very high calcium content; diatom cells have a high silicon content. But in general, the basic chemical composition of living cells is remarkably constant.

TABLE 1-1
Elemental Composition of a Bacterial Cell

Element	*Percent of Cell Weight*	*Derived Macromolecules*
Carbon	13	All macromolecules
Hydrogen	10	All macromolecules
Oxygen	70	All macromolecules
Nitrogen	5.5	Proteins, nucleic acids
Phosphorus	0.8	Nucleic acids
Sulfur	0.3	Proteins
Potassium	0.3	
Magnesium	0.05	
Iron	0.01	
Calcium	0.01	
Manganese Cobalt Nickel Copper Zinc Molybdenum Vanadium Silicon Selenium Boron Sodium Chlorine	Total weight less than 0.01% of total	

CARBON ATOM HAS UNIQUE PROPERTIES

Carbon atoms form the backbone of most organic molecules in the cell, and life on this planet depends on the chemistry of carbon. Indeed, the chemistry of carbon is so unique and adapted to life on earth that all of the initial attempts to detect life on Mars in 1976 were based on the assumption that Martian life would also be based on the carbon atom. A significant feature of this element is its ability to form stable, covalent bonds with four other atoms. In many biologically important molecules, two of these atoms are also carbon atoms, thereby producing a skeleton, or chain, of carbon atoms. The other bonds may involve a variety of other elements, each conferring distinctive properties on the molecule.

Several groups of elements are excluded from living systems. These include the noble elements, such as argon and neon, which will not share electrons and therefore will not form bonds. A second, heterogeneous group contains the elements which are very large and have a great number of electron shells. Apparently the smaller-sized elements provide sufficient versatility in the numbers of bonds they can form to make it unnecessary for the cell to employ the larger-sized atoms.

Covalent bonds, pages 25–28

Molecular Composition

Carbon, hydrogen, oxygen, and nitrogen are found in all *important building blocks of cells* such as amino acids, purines, and pyrimidines, as well as in other molecules that play important roles in energy metabolism. Sulfur is found in fewer cellular constituents; these include certain amino acids, vitamins, and coenzymes.

Amino acids, pages 32–35

Purines and pyrimidines, pages 37–39

The most abundant molecule in all living cells is water, which usually makes up about 70 percent of the cell by weight. The chemistry of the cell is based on water as the solvent, and all the molecules within the cell would function quite differently if a material such as benzene were the solvent. All cells require an aqueous environment in which to grow, and the limitation of water restricts the growth of cells over large areas of the earth. The unique properties of water will be considered in Chapter 2.

The molecules within the cell which determine its chemical composition can be classified in terms of their broad functions (Table 1-2). One group, especially the inorganic ions, contributes to the *ionic environment.* Water, other inorganic, and some small organic molecules also contribute. Some small organic molecules, such as sugars, and other carbohydrates serve as fuel in the production of energy. Other small molecules, such as amino acids, purines, and pyrimidines as well as some sugars, serve as the subunits or precursors for macromolecules. The major macromolecules, the proteins and nucleic acids, are present in relatively small numbers of molecules.

Water, page 30

Macromolecular Composition

Most cells contain the same major macromolecules in approximately the same proportions (Table 1-2). Each kind of molecule performs essentially the same

TABLE 1-2
Chemical Composition of a Typical Bacterial Cell

Component	*Percent of Total Cell Weight*	*Approximate Number of Molecules per Cell*	*Number of Different Kinds*
H_2O	70	4×10^{10}	1
Inorganic ions	1	2.5×10^8	20
Carbohydrates and precursors	3	2×10^8	200
Precursors of proteins	0.4	3×10^7	100
Precursors of nucleic acids	0.4	1.2×10^7	200
Precursors of fats	2	2.5×10^7	50
Breakdown products of food molecules	0.2	1.5×10^7	200
Proteins	15	10^6	2000 to 3000
Nucleic acids			
DNA	1	1–4	1
RNA	6	5×10^5	1000

Enzymes, pages 143–149

DNA and RNA, pages 37–39

functions in all cells. For example, some *proteins* function as *enzymes,* the biological catalysts that facilitate the myriad reactions involved in energy metabolism and biosynthesis. Other proteins also form integral parts of certain cell structures. *Deoxyribonucleic acid* (DNA) is the macromolecule which carries the genetic information and specifies the structure and properties of the cell by determining the composition of its proteins. Another macromolecule, *ribonucleic acid* (RNA), plays a major role in converting the information coded in the DNA into the sequence of amino acids in the cell's proteins.

Amino acids formulas, page 33

Coenzymes, page 146

The fundamental unity of molecular composition in all forms of life is especially evident from an analysis of the chemical makeup of the two most important groups of biological macromolecules, proteins and nucleic acids. The same twenty amino acids appear in all cells as constituents of cell proteins. As a general rule, the same two purines and three pyrimidines make up the basic subunits of the DNA and RNA in all cells. A few smaller organic compounds are also ubiquitous, functioning universally in various crucial metabolic reactions. Thus *adenosine triphosphate* (ATP) serves as the storage and transfer form of energy in all cells. The *coenzymes,* which are synthesized from vitamins, function universally as agents for the transfer of atoms and electrons in cellular chemical reactions.

Cell Organelles

Macromolecules of the same or different types associate with each other in very specific patterns to form a variety of cell structures, the *organelles.* Each organelle has one or more functions in the life of the cell. Not all structures are found in all cells, and it is at the level of the structure and function of organelles that fundamental differences between cells can be recognized.

Basic Cell Types

TWO MAJOR CELL TYPES: PROCARYOTIC AND EUCARYOTIC

The assemblage of individual cellular structures into a smoothly functioning, autonomous unit—the *cell*—represents the next level of biological organization. It is at this level that basic differences between organisms become apparent. Two distinctly different cell types emerge: *procaryotic* and *eucaryotic* (Table 1-3). All organisms, both macroscopic and microscopic, are composed of eucaryotic cells, except for the bacteria (including the cyanobacteria, formerly called blue-green algae or blue-greens), which comprise only procaryotic cells. This division of cells into two types represents one of the most significant dichotomies in all of biology. Their differences are listed in Table 1-3. Although each of these differences is explored in more detail in Chapter 3, Table 1-3 illustrates the major point to be emphasized at this time—that the two cell types differ from each other in a great many respects. For one thing, the larger eucaryotic cells contain a number of structures that are absent in procaryotic cells, in particular the nuclear membrane and mitochondria. The physiological implications of these differences will be clearer as the functions of these structures are considered in Chapter 3.

The Organism

The next level of organization in the biological world relates to the physical and functional relationships between cells of *multicellular organisms.* If we consider these organisms as assemblages of cells, we can categorize the living world into groups according to the way in which these cells are organized to form the total being.

TABLE 1-3
Comparison of Eucaryotic and Procaryotic Cells

Features of Cells	*Procaryotic*	*Eucaryotic*
Size (diameter)	0.3–2 μm	2–20 μm
Genetic structure		
Chromosome number[a]	1–4 (all identical)	> 1 (all different)
Nuclear division by mitosis	–	+
Nuclear membrane	–	+
Cytoplasmic structures		
Cell wall	Unique chemical components	If present, composition varies between organisms
Mitochondria	–	+
Ribosomes	70S[b]	80S
Cytoplasmic streaming	–	+
Photosynthetic pigments associated with	Chromatophores	Chloroplasts

[a] The DNA of procaryotic cells functions as the chromosome of eucaryotic cells. However, its structure is far less complex than a true chromosome, and therefore it is often referred to as a nuclear body. In this text we employ the term chromosome, emphasizing that this defines its function and not its detailed structure.
[b] The size of the ribosome is indicated by the speed at which it sediments upon centrifugation (S value). The larger the S value, the heavier and larger the molecule.

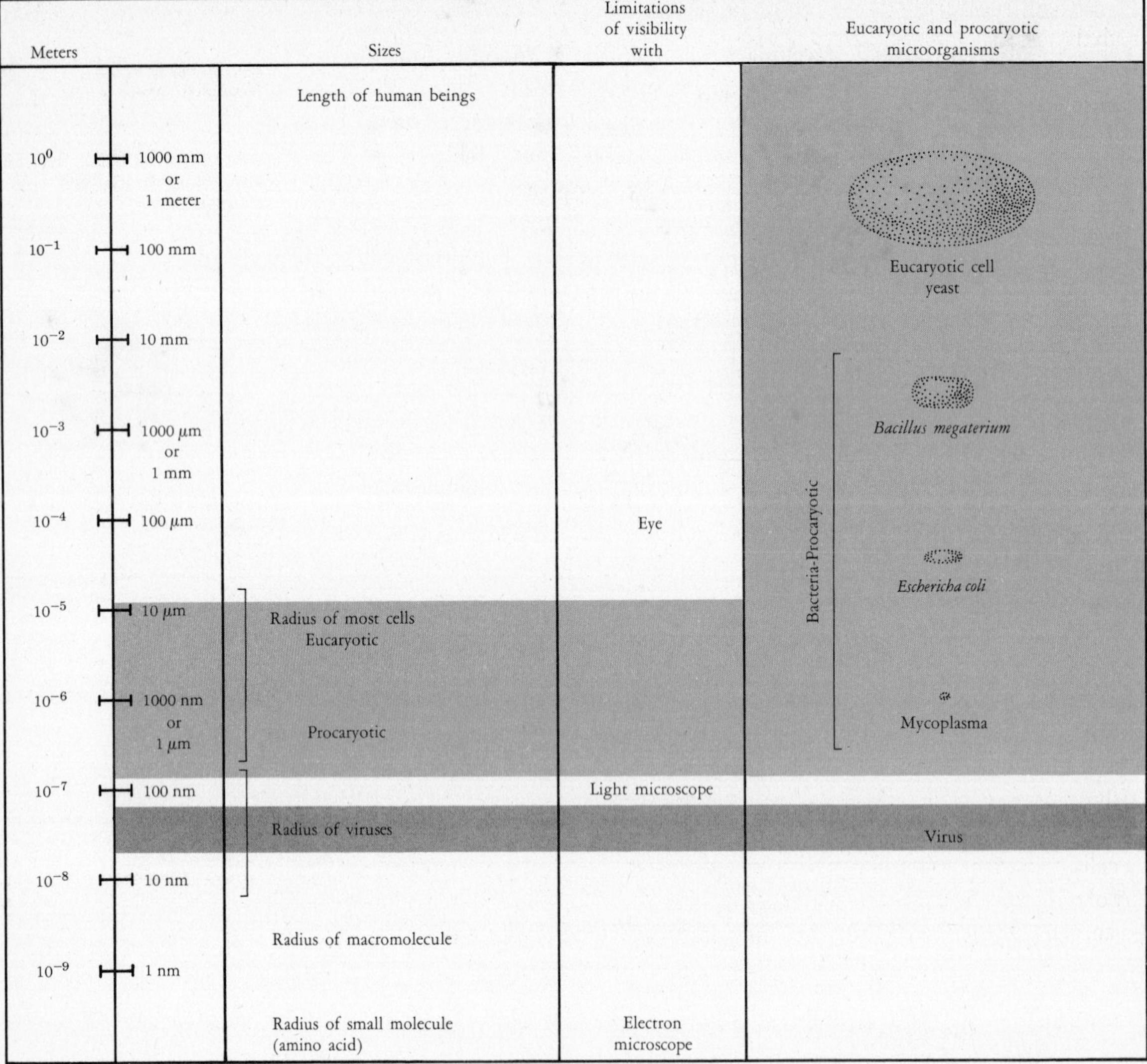

The basic unit of length is the meter (m), and all other units are fractions of a meter.

nanometer (nm) = 10^{-9} meter = .000000001 meter
micrometer (μm) = 10^{-6} meter = .000001 meter
millimeter (mm) = 10^{-3} meter = .001 meter
1 meter = 39.4 inches

These units of measurement correspond to units in an older but still widely used convention.

1 angstrom (Å) = 10^{-10} meter
1 micron (μ) = 10^{-6} meter

FIGURE 1-4 Measurements of molecules, cells, and organisms.

Single-celled organisms The simplest type of cell organization is the one in which the entire organism consists of a single cell. This organization predominates among the organisms we will study in this course. Single-celled (unicellular) organisms are almost invariably microscopic in size. Since such cells exist independently and are self-sufficient, each one must carry out all vital life functions. Although single-celled organisms are characterized by their small size, very profound differences may exist among them. Thus some unicellular microbes, such as protozoa and algae, possess the complex eucaryotic cell structure, whereas the bacteria, including the cyanobacteria, are of the simpler procaryotic type. As a result, there is a wide range in size as well as in internal organization among the unicellular organisms. Figure 1-4 shows the comparative sizes of some molecules, cells, and organisms, as well as the scales used for their measurement.

COMPLEXITY OF ORGANISM IS UNRELATED TO SIZE

Multicellular organisms Multicellularity is another type of cellular organization, in which organisms arise from single cells which increase in number but remain attached to one another in a characteristic way. Multicellular organisms can differ from one another in a number of fundamental ways. The most outstanding variation involves the level of *structural complexity* achieved within the organism. Because an organism is multicellular, it does not necessarily mean that its structures are complex. Many large multicellular organisms are composed of cells which are very similar to one another. In some cases the cells have become modified so that they now carry out a specific function. But in such cases these specialized cells function independently of one another, and not as one component of a group of identical cells which function as a unit. This concept can be illustrated by an organism known to most people, the mushroom (or toadstool). Mushrooms have several distinctive structures that are important in the life of the organism, such as the spore-bearing cells and the spores (Figure 1-5). Although the cells comprising these structures are specialized, the cells function *independently* of one another rather than cooperatively. In higher organisms, such as plants and animals, distinct regions of cells exist, all of which are essentially identical to

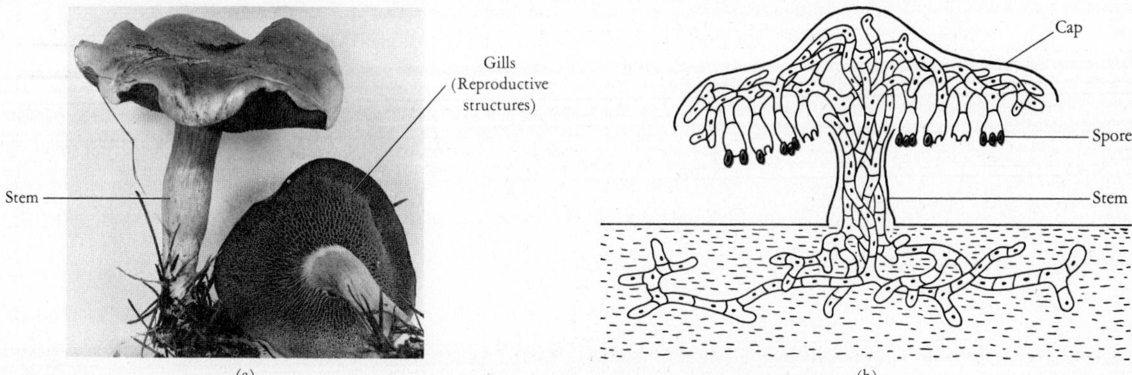

FIGURE 1-5 (a) Photograph of a common mushroom. (Courtesy of D. Stuntz.) (b) Diagram of a section through a mushroom. The entire stalk (stem) and cap are composed of mycelia. These mycelia differentiate to form the club-shaped structures in which reproductive spores are produced.

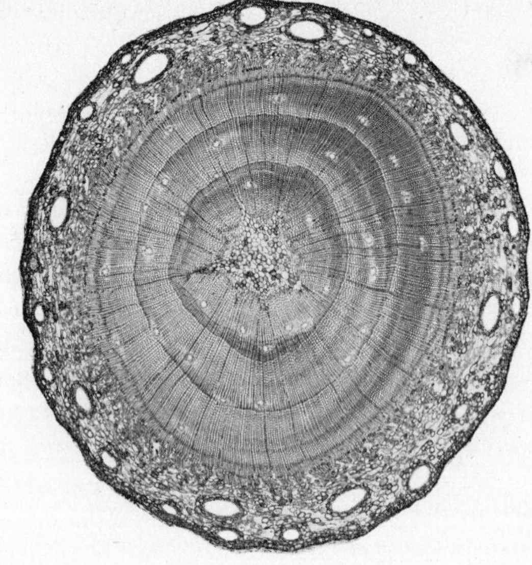

FIGURE 1-6
An organ. Cross section of a pine stem composed of many types of identical cells, each type grouped together to form a specialized functional unit, a tissue. The assemblage of tissues constitutes an organ. Some tissues are concerned with the transport of nutrients, others with water transport and still others provide structural integrity to the organ. (Courtesy of Carolina Biological Supply Company.)

one another but which differ in shape, structure, and function from cells in other regions. Each specialized group of cells is called a *tissue,* which performs a specific function *as a unit.* Different tissues in turn may combine into an even more complex structure, an *organ* (Figure 1-6). This complex type of organization contributes to a high degree of specialization. Each vital function of the organism is performed by a different organ, so that the life of the organism requires the proper functioning of each vital organ. These multicellular organisms represent the highest level of cell organization.

MEMBERS OF MICROBIAL WORLD DO NOT HAVE TISSUES

This increasing order of complexity of cells, tissues, and organs is characteristic of many plants and animals; *it is not typical of members of the microbial world.*

WHAT CONSTITUTES MEMBERSHIP IN THE MICROBIAL WORLD?

Being a member of the kingdom *Protista* constitutes membership in the microbial world. However, for thousands of years only two major groups of organisms had been observed. One group, classified as plants, was characterized by a type of nutrition which depended on sunlight for energy (photosynthetic), by rigid cell walls, and by being rooted in the soil and therefore stationary. The other group, animals, ingested its food, and its members were usually motile.

With the discovery of microorganisms, major difficulties in classification arose. Some forms were motile, ingested food, and were therefore classified as animals (protozoans). Others were nonmotile and photosynthetic and were naturally considered to be plants (algae). However, a large number of unicellular forms remained which had properties of both groups—being motile and photosynthetic or stationary and nonphotosynthetic. Perhaps most disturbing was the fact that in at least some single-celled plantlike organisms the environmental conditions under which the organism was grown determined whether or not it was photosynthetic.

Thus it was difficult to fit single-celled organisms into either the plant or the animal kingdom. Although many of them are not photosynthetic, some are. Also, many are motile, like animals, yet have rigid cell walls, like plants. The most reasonable solution to this dilemma seemed to be the recognition of insurmountable difficulties with the two-kingdom system and the creation of additional kingdoms.

PROTISTS INCLUDE BOTH EUCARYOTIC AND PROCARYOTIC CELL TYPES

Yet, while most taxonomists favor having more than two kingdoms, no single scheme of classification has met with universal approval. Perhaps the simplest and most satisfactory solution was advocated in 1866 by the German zoologist Haeckel, a student of Darwin. He proposed a third kingdom, the *Protista,* which would be on the same taxonomic level as the plant and animal kingdoms. The only basis for an organism being classified as a member of the Protista is its simple biological organization, characterized by a *lack of extensive tissue formation.* However, it should be recognized that some organisms have a *tendency* to form tissues, and it is difficult to classify these organisms with great certainty. Included in the kingdom Protista are *all bacteria, algae, fungi,* and *protozoa.* The group includes both microscopic, unicellular organisms and very large multicellular forms. It includes organisms that are procaryotic, the *procaryotic protists* (Table 1-4), as well as those of eucaryotic cell type, the *eucaryotic protists* (Table 1-5).

TABLE 1-4
Major Procaryotic Protists of the Microbial World

Bacteria	*Distinguishing Features*
Typical bacteria	Rigid cell walls; unicellular; multiply by binary fission; if motile, by flagella
Prosthecate and budding bacteria	May have unusual shapes, appendages (prosthecae); most prosthecate bacteria divide by budding; most budding bacteria are prosthecate
Mycelial bacteria	Often moldlike in appearance
Filamentous (sheathed bacteria)	Individual cells enclosed in a common sheath
Mycoplasma or pleuropneumonia-like organisms (PPLO)	Very heterogeneous group; properties distinct from bacteria; no cell wall and may not divide by binary fission; sterols in cytoplasmic membrane
Gliding bacteria	Organisms have flexible cell wall and may have a developmental cycle
Spirochetes	Helical cells with flexible cell walls; movement by axial filament
Rickettsia	Obligate intracellular parasites; have "leaky" cytoplasmic membrane
Chlamydia	Obligate intracellular parasites which lack some enzymes required for energy production
Cyanobacteria	Flexible cell wall which allows them to glide; gain energy by photosynthesis; reproduction primarily by binary fission

These groups will be covered in much greater detail in Chapter 12.

TABLE 1-5
Distinguishing Features of Eucaryotic Protists

Algae	*Fungi*	*Protozoa*
Gain energy from sunlight (photosynthetic)	Not photosynthetic Most multicellular, consisting of long filaments	Not photosythetic Unicellular Cells may be extremely complex

SOME MEMBERS OF MICROBIAL WORLD ARE NOT CELLS

Other members of the microbial world that we will consider are the *viruses* and *viroids* (Table 1-6). In actual fact, however, neither are cells and therefore are not organisms. A virus particle is actually a piece of genetic material, *either* DNA *or* RNA (not both) surrounded by a specialized protein coat. A viroid is a small piece of RNA which does not even have a protein coat. Both agents can only replicate inside living cells. Viroids have been shown to cause a number of plant diseases, and there is speculation that such agents may cause diseases in humans.

TABLE 1-6
Distinguishing Characteristics of Viruses and Viroids

Viruses	*Viroids*
Obligate intracellular parasites	Obligate intracellular parasites
Contain either DNA *or* RNA	Only RNA; no protein coat
Nucleic acid and protein duplicate independently	
Contain very few if any enzymes	No enzymes

Some photographs of members of each of these groups are given in Figures 1-7 through 1-11.

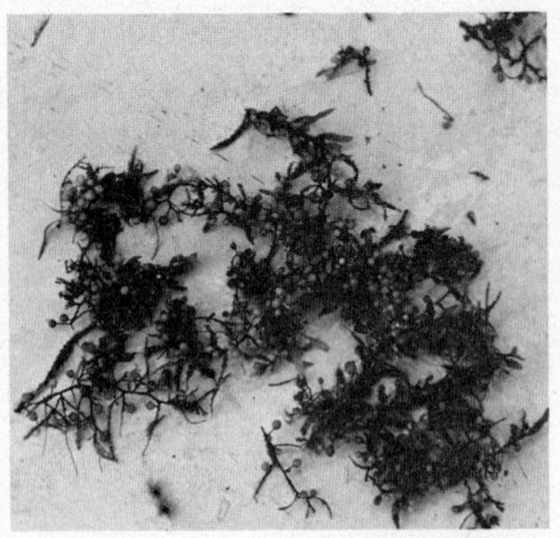

(a)

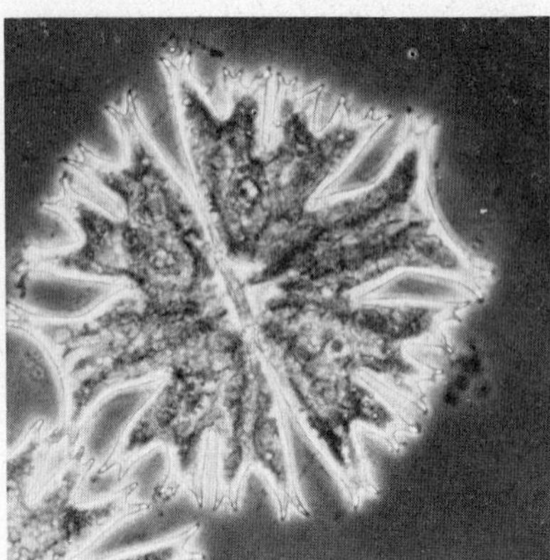

(b)

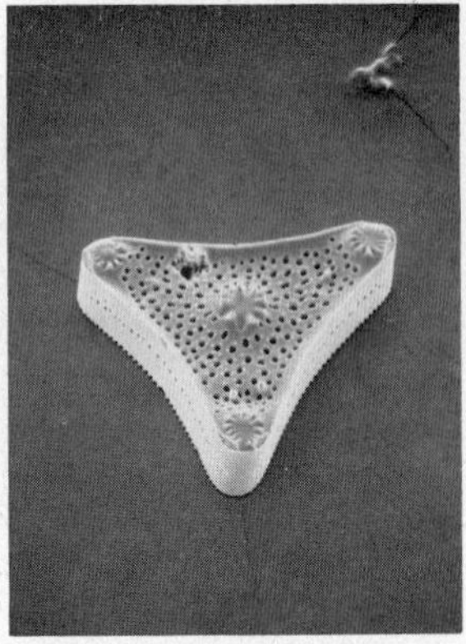

(c)

FIGURE 1-7 Algae. (a) *Sargassum,* a macroscopic alga one tenth its actual size. (Courtesy of J. T. Staley.) (b) *Micrasterias,* a green alga composed of two symmetrical halves. (×350) (Courtesy of J. T. Staley.) (c) Diatom organism which serves as food for aquatic animals, as viewed through a scanning electron microscope. (Courtesy of N. Hodgkin.)

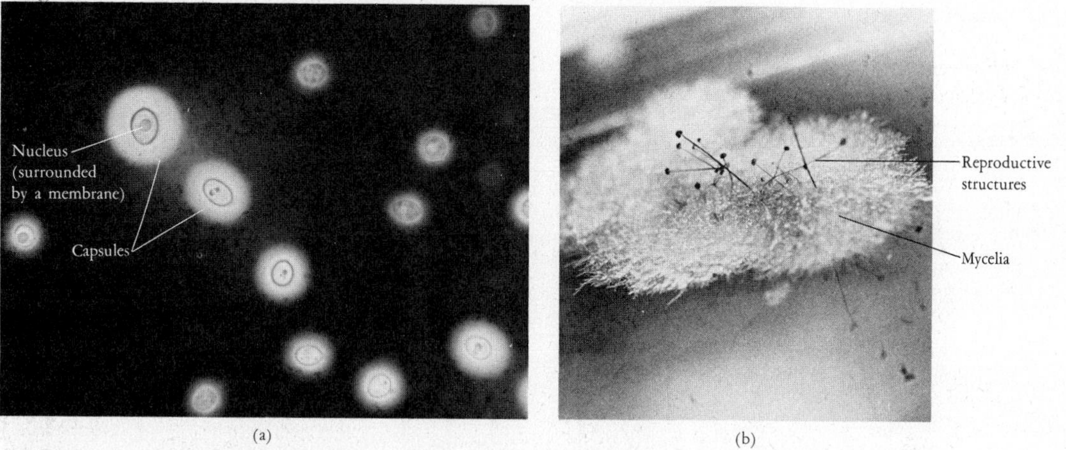

FIGURE 1-8 Fungi. (a) *Cryptococcus neoformans,* living cells of the yeast stained with India ink (negative stain) to reveal the large capsules which surround the cell. (×600) (b) *Aspergillus,* a typical mold whose dark reproductive structures rise above the mycelium. (×7) (Courtesy of J. P. Dalmasso.)

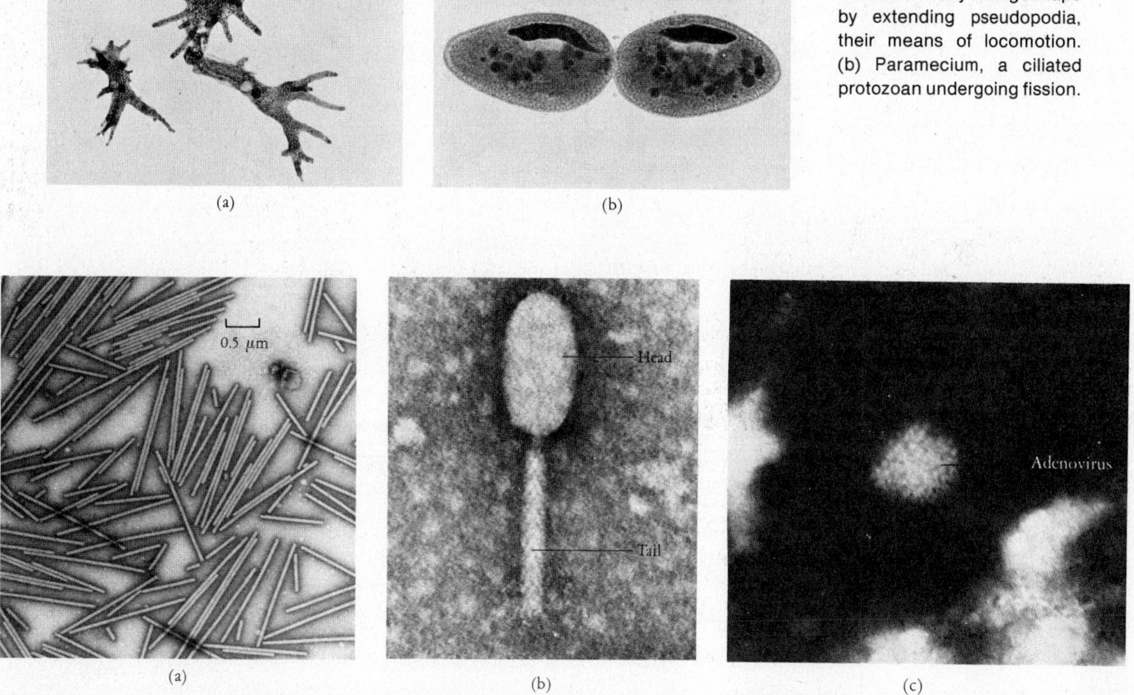

FIGURE 1-9
Protozoa. (a) *Amoeba,* whose cells continually change shape by extending pseudopodia, their means of locomotion. (b) Paramecium, a ciliated protozoan undergoing fission.

FIGURE 1-10 Viruses. (a) Tobacco mosaic virus that infects tobacco plants. A long, hollow protein coat surrounds a molecule of RNA. (Courtesy of E. S. Boatman.) (b) A bacterial virus (bacteriophage) that invades *Clostridium botulinum,* the causative agent of botulism. (×350,000) (Courtesy of E. S. Boatman.) (c) Adenovirus, a DNA-containing virus that causes respiratory infections in human beings and tumors in animals under certain conditions. (Courtesy of E. S. Boatman.)

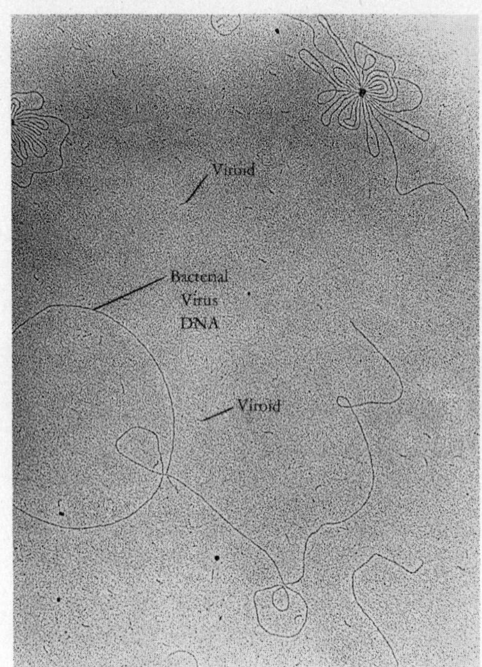

FIGURE 1-11
Viroid. A single-stranded, closed circular molecule of RNA which exists as a rodlike structure. Note how much smaller the viroid is than the molecules of bacteriophage DNA. The viroid consists of about 350 nucleotides. (Courtesy of T. Koller and J. S. Sogo, Swiss Federal Institute of Technology, Zurich.)

SIGNIFICANCE OF THE SIZE OF MICROORGANISMS

The smallness of microorganisms has a number of consequences. These relate primarily to the simplicity of the cell.

Simplicity of the Cell

The unicellular organisms range in cell volume from 0.01 to about 50,000 μm^3 (Figure 1-4). A few of the large single-celled organisms, such as the largest protozoans, can be seen without the aid of a microscope. The lower size limit of a microorganism is probably determined by the dimensions of the macromolecules it needs to carry out its life functions; the organism must be large enough to contain the equipment necessary to sustain life. Viruses are too small to contain all of the organelles necessary for life and therefore must utilize the machinery of the cells they invade in order to multiply. When not inside a cell they are incapable of performing life functions.

Significance of Size to Eucaryotic and Procaryotic Cells

The size of the smaller microorganisms also restricts the complexity of the cellular apparatus that they can contain. Since a single mitochondrion in a liver cell may be as large as an entire bacterial cell, the latter is obviously too small to contain a mitochondrion. The same holds true for many other organelles found in eucaryotic but not in procaryotic cells. As a general rule, as organisms increase in size, they also increase in complexity. Thus eucaryotic cells are generally larger than procaryotic cells (Figure 1-4, Table 1-3), and their increased size accommodates a number of organelles not present in procaryotic cells.

CELLS MUST BE LARGE ENOUGH TO CONTAIN ALL ORGANELLES NECESSARY FOR LIFE

The upper size limit of a single-celled organism is probably related to the ratio of its surface area to its cell volume. All the nutrients consumed by the cell must enter across the surface of the cell. Likewise, waste products must leave through the cell surface. As the size of the cell increases, the volume increases much more rapidly than the surface area: the volume increases as the *cube* of the cell radius, whereas the surface area increases only as the *square* of the radius. Thus when the cell's radius has increased three times, the volume has increased 27 (3^3) times, whereas the surface area has increased only 9 (3^2) times (Figure 1-12). With such a disparity between surface area and volume, the cell faces the potential problem of not having enough surface area, for the intake of nutrients and exit of waste materials, to supply the increased demands of the enlarged cell volume.

Eucaryotic cells, being larger, have solved this problem in a number of ways. Many such cells assume shapes which bring the internal parts of the cell close to the surface. Thus a large nerve cell has a long, threadlike shape. Also, the cytoplasmic streaming evident in many eucaryotic cells (but not in procaryotic) serves to stir up their cytoplasm, moving nutrients and waste products from one area to another. In addition, many cells, both procaryotic and eucaryotic, have their surface areas markedly enlarged as a result of extensive invaginations or convolutions of the cytoplasmic membrane through which nutrients and waste products pass.

INVAGINATIONS OF SURFACE INCREASE SURFACE AREA

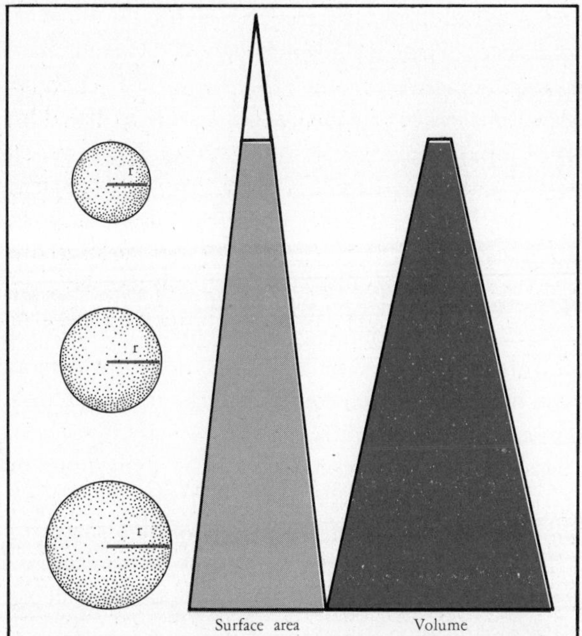

FIGURE 1-12
The relationship between surface area and volume of spheres. As the radius of a sphere increases, there is a much greater increase in the volume than in the surface area.

NOMENCLATURE OF ORGANISMS

ORGANISMS IDENTIFIED BY GENUS AND SPECIES NAMES

Nomenclature in biology refers to the system by which organisms are named. Virtually all organisms are named according to the *binomial system* of genus and species names devised by the Swedish botanist Carl von Linné, or Carolus Linnaeus, as he is more commonly called. (His name, like the names of the organisms he classified, became latinized.) A number of different species are included within the same genus. The first word in the name is the genus name, with the first letter always capitalized; the second word is the species name, which is not capitalized. Both words are always italicized. For example, *Escherichia coli,* commonly called the colon bacillus, is a member of the genus *Escherichia.* The genus name is commonly abbreviated with the first letter capitalized, that is, *E. coli.* Within a genus, members of the same species may differ from one another in minor ways. For example, one member may not be able to break down a foodstuff that another member can. This slight difference generally does not justify giving the organisms different species names, and so they are designated as *strains* or *varieties* of the same species. For example, two common strains of the colon bacillus are *E. coli* K-12, isolated from a patient at Stanford University Hospital, and *E. coli* ML, which is reputed to have been isolated from the bowel of a famous French scientist, whose initials are ML. Both strains are identical in their growth characteristics, but they differ in many details. The name and a description of all of the bacteria which are currently recognized are given in a reference text, *Bergey's Manual of Determinative Bacteriology.*

The precise identification of an organism requires both a genus and species name. However, many organisms are commonly referred to by less precise terms which serve to place the organism in a certain group. For example, the general term bacillus refers to rod-shaped bacteria; thus all bacilli (plural) are rod-shaped (cylindrical) organisms. These include members of the genus *Bacillus* as well as the genera *Clostridium, Lactobacillus,* and *Escherichia.* The general term bacillus is neither capitalized nor italicized and thus can be differentiated from the genus *Bacillus.*

SUMMARY

The combined work of a number of outstanding scientists established that the "animalcules" discovered by Anton van Leeuwenhoek 300 years ago originated from preexisting organisms which are ubiquitous in nature. Although the chemical elements and molecules, which comprise the basic unit of life—the cell—are uniform throughout the world, there is great diversity in the structures which carry out the basic functions of life. All cells can be divided into one of two distinct types: the simple cell type, called *procaryotic,* and the more complex type, called *eucaryotic.* The procaryotic cell lacks many of the structures of the eucaryotic. Procaryotic cells very often exist as single-celled organisms, whereas eucaryotic cells tend to be part of larger cellular aggregates, tissues, and organs, which in turn make up the organism.

All members of the microbial world are members of the kingdom Protista, a major grouping of organisms characterized by their lack of extensive tissue formation. All bacteria—the procaryotic protists, as well as algae, fungi, and protozoa, the eucaryotic protists—are included in this kingdom. The microbial world is also composed of subcellular agents, the viruses and viroids, which can only replicate inside living cells.

QUESTIONS

REVIEW

1. What major contribution did each of the following people make to the study of microbiology? Anton van Leeuwenhoek, Robert Hooke, Louis Pasteur, Ferdinand Cohn, John Tyndall, Father Spallanzani.
2. What are the major differences between eucaryotic and procaryotic cells?
3. What functions must cells perform in order to live?
4. Discuss the characterization of an organism which places it in the kingdom Protista. What was the justification for including macroscopic forms in this kingdom?
5. Why were rickettsia and chlamydia once classified as viruses? Why are they now classified as bacteria?

THOUGHT

1. Of the pioneers in microbiology considered in this chapter, who do you think made the greatest contribution? Why?
2. What properties of the carbon atom make it ideally suited for serving as the backbone of life? What other atom or atoms might you predict would have similar properties? Would you think it conceivable that another element might be more important than carbon if life is discovered on Mars? Why or why not?

FURTHER READING

BROCK, THOMAS, ed., *Milestones in Microbiology.* Washington, D.C.: American Society for Microbiology, reprinted 1975. A collection of historically significant papers in microbiology that cover the past 400 years. The editor briefly discusses the significance of each paper to the development of microbiology.

DOBELL, C., *Antony van Leeuwenhoek and His "Little Animals."* New York: Dover Press, 1960. A very readable and fascinating biography by an author who obviously greatly admired his subject. One of the best biographies ever written.

FRIEDEN, E., "The Chemical Elements of Life," *Scientific American* (July 1972).

GREEN, D. E., and R. F. GOLDBERGER, *Molecular Insights into the Living Process.* New York: Academic Press, 1967. A very lucid presentation of the principles underlying living processes. Chapter 1 is good supplemental reading.

KIRK, D., *Biology Today.* 2d ed. New York: Random House, 1975. This biology text covers the material of this chapter in a somewhat different way. Chapters 1–4 are especially relevant to this chapter.

LECHAVALIER, H. A., and M. SOLOTOROVSKY, *Three Centuries of Microbiology.* New York: McGraw-Hill, 1965. An historical account of the major developments in microbiology with many quotations from the original papers.

STANIER, R., E. A. ADELBERG, and J. INGRAHAM, *The Microbial World.* 4th ed. Englewood Cliffs, N.J.: Prentice-Hall, 1976. Probably the most highly regarded microbiology textbook, written for students with a prior background in chemistry and biology. It covers microbiology by emphasizing the role of the microbe. The Stanier book could serve as reference material for most of the topics covered in the present text and therefore will not be referred to at the end of each chapter.

CHAPTER 2

BIOCHEMISTRY OF THE MOLECULES OF LIFE

The understanding of various aspects of the life of microorganisms at the molecular level has provided an insight into the chemical basis of life. In order to achieve some understanding of microbiology at the chemical level, it is necessary to have some knowledge of certain fundamental facts derived from physics and chemistry, as well as microbiology.

The material in this chapter will serve as background as well as reference material for various aspects of microbiology covered throughout the text. For some, it may serve as a review of material already encountered. For others, it may be a first encounter with the chemistry of biological molecules. The discussion proceeds from the lowest level of organization—the atom—to the highly complex associations among macromolecules.

FORMATION OF MOLECULES: CHEMICAL BONDS

Most atoms can associate with other atoms to form molecules. The chemical bonds which hold atoms together to form molecules are of various types, each type characterized by its arrangement of electrons and by its strength.

Covalent Bonds

Except for carbon dioxide (CO_2) and carbon monoxide (CO), molecules that contain carbon (C) are called organic compounds. The atoms in organic molecules are held together by *covalent bonds,* formed by the *sharing* of electrons between two atoms, with each partner contributing electrons. The number of pos-

TABLE 2-1
Atomic Structure of Elements Commonly Found in the Living World

Element	*Symbol*	*Atomic Number (Total Number of Electrons)*	*Diagram of Outer Electron Shell*	*Number of Possible Covalent Bonds*
Hydrogen	H	1		1
Carbon	C	6		4
Nitrogen	N	7		3
Oxygen	O	8		2
Phosphorus	P	15		3
Sulfur	S	16		2

The number of electrons required to fill the outer shell determines the number of possible covalent bonds. The number of electrons in a completed outer shell varies depending on the distance of the shell from the nucleus.

sible bonds that each element can form depends on the number of electrons that the atom of the element requires to fill its outer electron shell. This information is given in Table 2-1 for the elements most important in the living world. The single most important atom, carbon, forms four covalent bonds (Figure 2-1). The dash or colon between two atoms represents two shared electrons (C—H or C:H). Note that electrons are sometimes shared in groups of four, called *double bonds* (for example, C═O). Once the outer shell of electrons is filled, the molecule is highly stable and is not capable of forming additional covalent bonds.

IMPORTANT FUNCTIONAL GROUPS INVOLVING COVALENT BONDS

The biologically important small organic molecules (discussed in later sections) can be classified into major groups, based on the type of *functional group* or groups they possess (Table 2-2). As Table 2-2 indicates, the most important covalent bonds in these molecules are those between carbon and carbon (C—C), car-

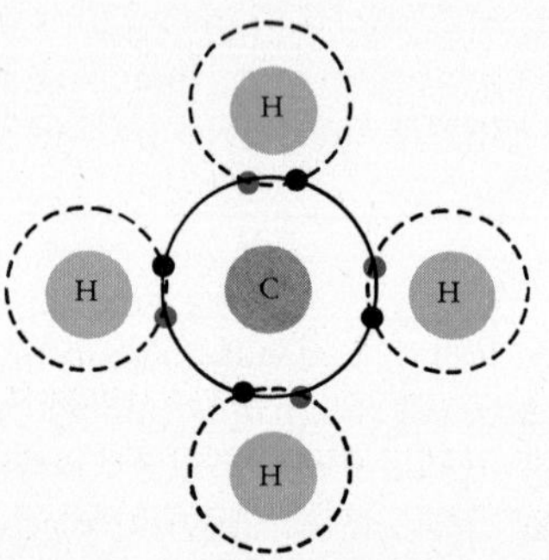

FIGURE 2-1
The carbon atom fills its outer electron shell by sharing a total of eight electrons with four H atoms.

TABLE 2-2
Classification of Small Molecules by Functional Groups

Group	*Name of Group*	*Example of Group in Biologically Important Molecule*		*Class of Molecule Found in*
$-\mathrm{CH_3}$ (—C(H)(H)—H)	Methyl	$\mathrm{CH_4}$ (H—C(H)(H)—H)	Methane	
—O—H	Hydroxyl	$\mathrm{CH_3{-}CH_2{-}OH}$	Ethanol (ethyl alcohol)	Alcohols
—C(=O)—O—H	Carboxyl	$\mathrm{CH_3{-}C(=O){-}O{-}H}$	Acetic acid	Acids
—N(H)—H	Amino	$\mathrm{H_2N{-}CH_2{-}C(=O){-}O{-}H}$	Glycine	Amines and amino acids
—C(=O)—	Keto	$\mathrm{CH_3{-}C(=O){-}CH_3}$	Acetone	Ketones
—C(=O)—H	Aldehyde	$\mathrm{CH_3{-}C(=O){-}H}$	Acetaldehyde	Aldehydes
—S—H	Sulfhydryl	$\mathrm{H{-}S{-}CH_2{-}CH(NH_2){-}C(=O){-}O{-}H}$	Cysteine	A few amino acids
—C(=O)— and —C(=O)—O—H	Ketone and Carboxyl	$\mathrm{CH_3{-}C(=O){-}C(=O){-}OH}$	Pyruvic acid	Keto acids
—O—H and —C(=O)—O—H	Hydroxyl and Carboxyl	$\mathrm{CH_3{-}CH(O{-}H){-}C(=O){-}OH}$	Lactic acid	Hydroxy acid

bon and hydrogen (C—H), carbon and oxygen (C—O), carbon and nitrogen (C—N), oxygen and hydrogen (O—H), and nitrogen and hydrogen (N—H). Some molecules of biological importance contain several functional groups. For example, every amino acid contains both an amino as well as a carboxyl group, hence the name amino acid. An acid which contains a keto group is often termed a keto acid; if the acid contains a hydroxyl group, it is termed a hydroxy acid. Unfortunately, many other compounds are commonly designated by trivial names which provide no clue as to what functional groups their molecules contain; moreover, some of these molecules may be referred to by more than one name. For example, inside the cell, salts of organic acids are generally present, rather than the acids themselves. For this reason "pyruvate," the name of such a salt, is used interchangeably with pyruvic acid, and "acetate" is commonly used for acetic acid.

Although all of the *bonds* described in Table 2-2 are covalent, covalent bonds between various atoms can differ in two ways which have consequences for biological systems. These differences are in their *strength* and in the *distribution of electrons* shared between the atoms.

DIFFERENCES IN STRENGTHS OF COVALENT BONDS

The relative strengths of the most common covalent bonds are given in Table 2-3. The stronger the bond the more difficult it is to break. Most covalent bonds are extremely strong and do not break unless large amounts of *energy*, generally as heat, are supplied. Since biological systems must function only within a narrow temperature range and cannot tolerate high temperatures, they utilize protein *catalysts*, the *enzymes*, which can break these covalent bonds at physiological temperatures. How enzymes break covalent bonds will be considered in Chapter 7.

DIFFERENCES IN ATTRACTION OF ELECTRONS: POLARITY

In addition to differences in strength of the covalent bonds between different atoms, covalent bonds also differ in the *distribution of shared electrons* between the atoms—the phenomenon of *polarity*. Polarity results from the fact that different atoms have different affinities for electrons. In covalent bonds between identical atoms such as H—H, the electrons have no tendency to associate with one atom more than with the other atom. This kind of *nonpolar* covalent bond also exists between different atoms, such as C—H, that have a similar attraction for electrons. In other cases, one of the atoms has a much greater affinity for electrons than does the other, and the electrons are *not* equally shared. The atom with the greater attraction for the electrons has a slight negative charge, leaving the other atom with a slight positive charge. Thus, in the O—H bond, the O has a much

TABLE 2-3
The Relative Strengths of Some Chemical Bonds

Name of Bond	*Relative Strength*
Covalent bonds (strong)	
H—H	1.0
C—C	.8
C—H	1.0
C—O	.8
C—N	.7
Hydrogen bonds (weak)	
H—O···O	
H—N···O	.02–.1
H—N···N	

TABLE 2-4
Polar and Nonpolar Covalent Bonds

Type of Covalent Bond	*Atoms Involved and Charge Distribution*	
Polar	$\overset{-}{O}$—$\overset{+}{H}$ $\overset{-}{N}$—$\overset{+}{H}$ $\overset{-}{O}$—$\overset{+}{C}$ $\overset{-}{N}$—$\overset{+}{C}$	The O and N atoms have a stronger attraction for electrons than C and H, so the O and N have a negative charge
Nonpolar	C—C C—H H—H	C and H have equal attractions for electrons, so there is an equivalent charge on each atom

greater affinity for electrons than H and will therefore have a negative charge, whereas the H atom will have a slight positive charge (Table 2-4). For this reason the water molecule is a highly polar molecule, having a slightly negative and two slightly positive charges at different locations (Figure 2-2).

Ionic Bonds

An extreme case of polarity is represented by *ionic bonds,* which are noncovalent. Electrons are attracted much more strongly by some atoms than by others, so that when both atoms are in close proximity the electrons completely leave one atom

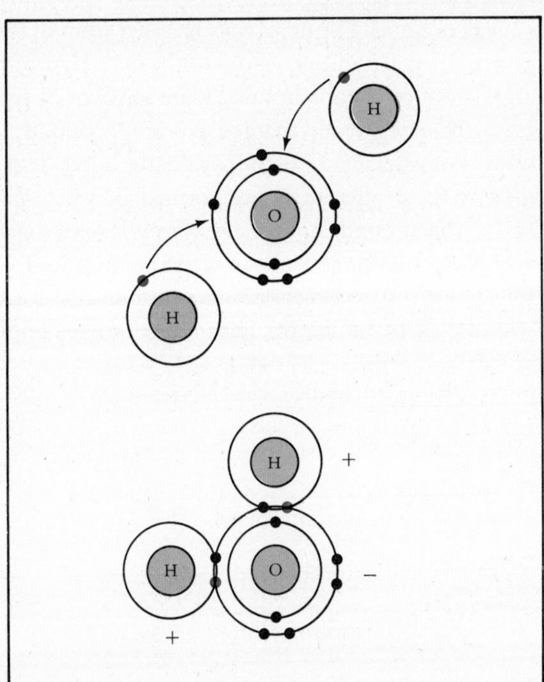

FIGURE 2-2
Formation of covalent bonds in water molecule. Oxygen has a greater attraction for the shared electron than do the hydrogen atoms. This results in the electrons being closer to the oxygen atom and conferring a negative charge on it. The hydrogen atoms in turn each have a positive charge.

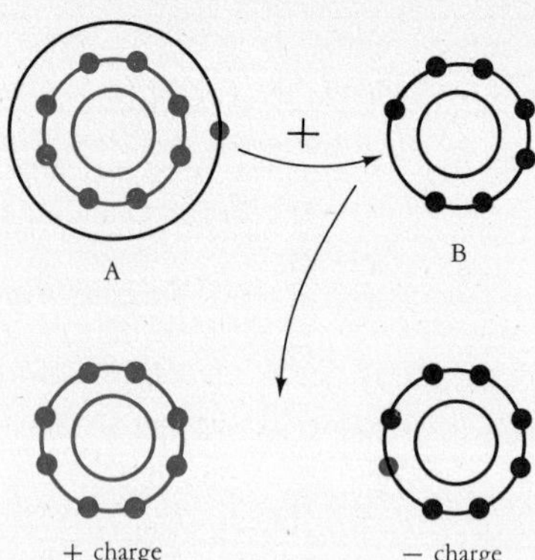

FIGURE 2-3
Ionic bond. Atom A gives up an electron to atom B. The result is that A acquires a positive charge and B a negative charge. Both atoms then have their outer electron shells filled.

and become incorporated into the outer electron shell of the other; in this situation the electrons are not shared between the two. The atom which gains the electrons becomes a negatively charged ion; the atom which gives up the electrons becomes a positively charged ion (Figure 2-3). Ionic bonds are the most common type of bonds in inorganic molecules.

Hydrogen Bonds

The phenomenon of polarity makes possible another type of bond, the *hydrogen bond.* This bond is formed when the hydrogen atom involved in a polar covalent bond (an atom which has a positive charge), interacts with the negative portion of another polar covalent bond. The negatively charged atom of the second bond attracts the positively charged hydrogen atom of the first. Hydrogen bonds commonly involve an H atom covalently bonded to either O or N (which gives the H a positive charge) and a negatively charged, covalently bound, acceptor atom. The acceptor atoms are generally N or O covalently bonded to H or C. One important example is the hydrogen bonds formed between water molecules (Figure 2-4). The hydrogen bonds hold the water molecules together to form a shifting latticework. This bonding results in water being a fluid, yet with a relatively high boiling point. Hydrogen bonding is responsible for the properties of many other molecules of biological importance, for example, DNA and protein.

HYDROGEN BONDS ARE WEAK BONDS

Several features distinguish hydrogen bonds from covalent bonds. The most important distinction is that *hydrogen bonds are much weaker than covalent bonds*

FIGURE 2-4
Water molecules held in a latticework, illustrating the hydrogen bonding between molecules.

```
|||OH|||OH|||OH
  H    H    H
|||OH   OH   OH
   H    H    H
|||OH|||OH|||OH
  H    H    H
```

(Table 2-3). If atoms have the option of forming either a covalent bond or a hydrogen bond, they will always form the stronger covalent bond.

Biological Significance of Weak Bonds

Weak bonds have several important ramifications in biology. They are constantly being made and broken at room temperature, thereby indicating that there is enough energy in the movement of electrons, even at room temperature, to break weak bonds. Since the average lifetime of a *single* weak bond is only a fraction of a second at room temperature, cells do not need a special mechanism (enzymes) to speed up the rate at which weak bonds are made and broken. This is in contrast to the strong covalent bonds.

MANY WEAK BONDS HOLD MOLECULES FIRMLY TOGETHER

Because of its weakness, a *single* hydrogen bond is unable to bind molecules together. However, a large enough number of such weak bonds can hold molecules together firmly. Thus weak bonds are of biological significance only when they occur in large numbers in a given molecule. The bonds which hold the two strands of the DNA molecule together are hydrogen bonds (Chapter 8).

WEAK BONDS ALLOW RECOGNITION BETWEEN MOLECULES

Another important feature of weak bonding forces is that they are effective only when the interacting surfaces are close to each other. Such close proximity is only possible when the atoms of two molecules form *complementary structures* such that a positive charge on the surface of one molecule is matched by a negative charge on the other. The interacting molecules must also have a lock-and-key relationship such that a protruding group of one molecule fits into a cavity of the other (Figure 2-5). Because of this requirement for complementarity, great specificity exists as to which molecules will lie next to each other in the cell. Most molecules in the cell can make good weak bonds with only a small number of other molecules.

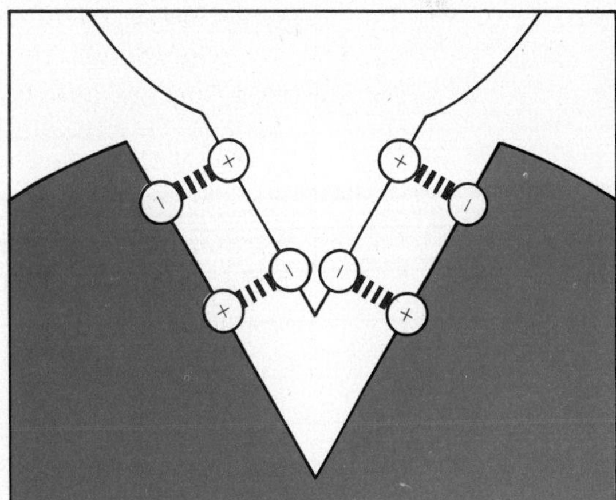

FIGURE 2-5 Lock-and-key fit of molecules that have complementary structures. The two molecules are complementary because the atoms on the "key" have the shape and charge to interact with atoms in the "lock." The interaction is manifest in the formation of many weak bonds which hold the complementary molecules loosely together.

IMPORTANT MOLECULES OF LIFE

Small Molecules

All cells contain a variety of small molecules and atoms, both organic and inorganic; many of the small molecules and atoms occur in the form of charged ions. About 1 percent of the total weight of a bacterial cell is composed of inorganic ions, principally Na^+, K^+, Mg^{+2}, Ca^{+2}, Fe^{+2}, Cl^-, PO_4^{-3}, and SO_4^{-2}. Many of the positively charged ions are required in very minute amounts for the functioning of certain enzymes. The phosphate ion plays a vital role in energy metabolism and is discussed in Chapter 7. The small organic molecules consist mainly of compounds that are being metabolized or have accumulated as a result of metabolism and those which serve as the building blocks of various macromolecules. Alcohols, aldehydes, and organic acids are the most frequent products of metabolism.

Macromolecules and Their Subunits: General Principles

Three major classes of biologically important macromolecules are considered: proteins, nucleic acids, and polysaccharides. The lipids, another class of molecules, not nearly as large as the macromolecules, also are briefly discussed.

As their name indicates, macromolecules are large molecules. Although each consists of a large number of atoms, generally more than 1000, all macromolecules have several structural features in common. The recognition and comprehension of these structural features make the structure of macromolecules relatively simple to understand.

All macromolecules are polymers—large molecules formed by the joining together of repeating small molecules, the *subunits.* Macromolecule synthesis occurs in two stages: first, the subunits are synthesized; second, they are linked together in a systematic fashion by covalent bonds.

The *individual* subunits of macromolecules contain two more hydrogen atoms and one more oxygen atom than are found after the subunits are linked together. This is because the joining together of two subunits involves a chemical reaction in which H_2O is removed. When the macromolecule is degraded into its subunits the reverse occurs, a water molecule is added. This type of reversible reaction, called a *hydrolytic reaction,* is common to the synthesis and degradation of a large number of molecules and only occurs through the action of specific enzymes.

The subunits of any single macromolecule have common functional groups. For example, amino acids are the subunits of proteins. Although there are more than twenty different amino acids, each one is characterized by a carboxyl group and an amino group in the same positions in its molecule.

Proteins as Macromolecules

Twenty different major amino acids are commonly found in proteins. Their formulas and names are given in Figure 2-6. All amino acids have several features in

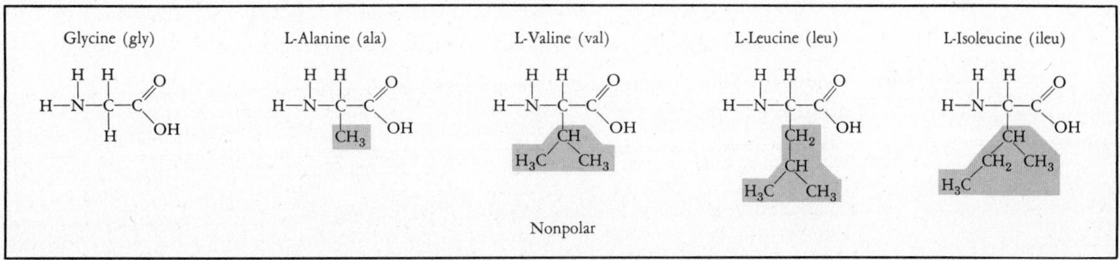

L-Serine (ser)
L-Threonine (thr)
Alcoholic

L-Tyrosine (tyr)
L-Phenylalanine (phe)
L-Tryptophan (try)
Aromatic

L-Aspartic acid (asp)
L-Glutamic acid (glu)
Carboxylic acids

L-Lysine (lys)
L-Arginine (arg)
L-Histidine (his)
Amine bases

L-Asparagine (asp NH_2)
L-Glutamine (glu NH_2)
Amides

L-Cysteine (cys)
L-Methionine (met)
Sulfur containing

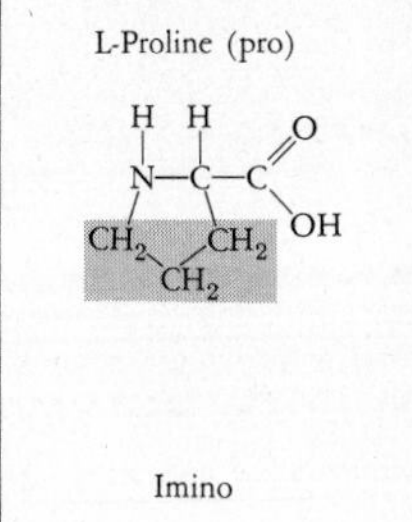

FIGURE 2-6 Amino acids. All amino acids have one feature in common: a carboxyl group and an amino group on the carbon atom next to the carboxyl group. Although the remainder of the molecule differs for each amino acid, they can be grouped because of certain features they have in common.

common, a terminal carboxyl group (—COOH) and amino group ($—NH_2$) bonded to the same carbon atom to which a side chain (shaded) that is characteristic for each amino acid is also bonded. The side chain, which gives individuality to each amino acid, may be a very simple group such as —H or $—CH_3$, a longer chain of carbon atoms, or one of various kinds of ring structures, such as the benzene ring found in tyrosine and phenylalanine. Because of certain similarities in their side chains, the amino acids can be subdivided into several groups (Figure 2-6). For convenience, the side chain is designated "R" when no specific amino acid is being considered. Some of these side chains may also contain carboxyl or amino groups giving negative or positive electrical charge properties to the amino acid. In a bacterial cell there are between 2000 and 3000 different proteins, each different in the arrangement of its amino acids.

Whenever a carbon atom is bonded to four different atoms, the combination can exist in two spatially different forms, one being the mirror image of the other. This situation exists in all amino acids (except glycine) since the C atom,

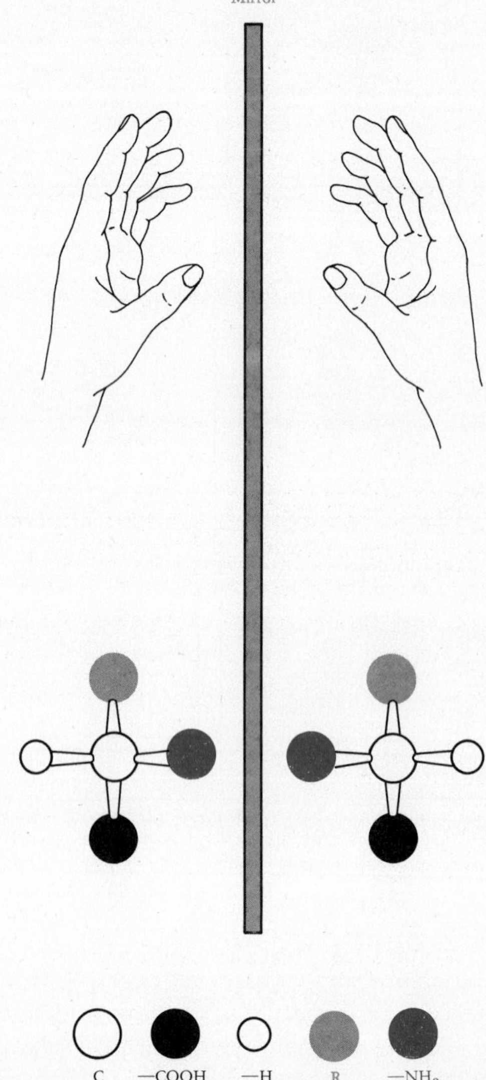

FIGURE 2-7
Mirror image forms of an amino acid. The joining of a carbon atom to four different groups leads to asymmetry in the molecule. Thus the molecule can exist in either the L- or D-form, one being the mirror image of the other. In no way can they be rotated in space to give two identical molecules.

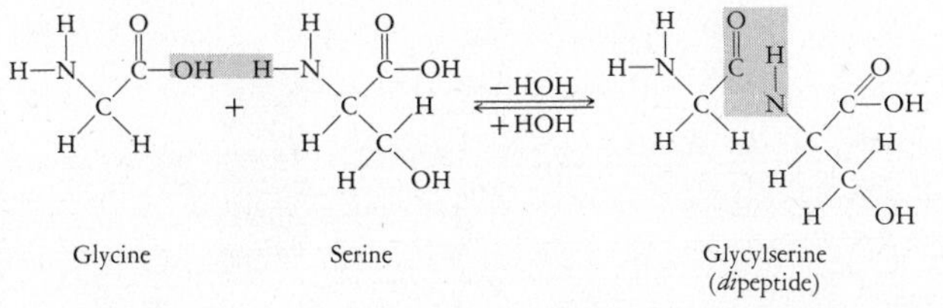

Glycine + Serine $\underset{+HOH}{\overset{-HOH}{\rightleftharpoons}}$ Glycylserine (*di*peptide)

FIGURE 2-8
Formation of a peptide bond between glycine and serine.

which is bonded to the —COOH and —NH_2 group, is also bonded to two other, different, groups. Thus this group of atoms can exist in either a "left-handed" or "right-handed" form known as the L-form or D-form, respectively (Figure 2-7). *Only* L-*amino acids occur in proteins,* and therefore they are designated the *natural amino acids.* D-amino acids are rare in nature, being found only in a few materials associated with bacteria—primarily cell walls and antibiotics.

The amino acids are bonded together by *peptide bonds,* a unique type of covalent linkage formed when the carboxyl group of one amino acid reacts with the amino group of the adjacent amino acid, with the release of water—a hydrolytic reaction (Figure 2-8). The chain of amino acids formed when a large number of amino acids are joined by peptide bonds is often referred to as a *polypeptide chain* (Figure 2-9).

A *protein* is defined as one or more polypeptide chains which are biologically functional. Some proteins consist of a single polypeptide chain; others are made up of a larger number (Figure 2-10). Proteins may vary in size from molecular weights of 6000 to many millions. The high molecular weight proteins are in fact *aggregates* of individual polypeptide chains which have associated with one another through weak bonding forces to form a highly complex, very large molecule. Some proteins are arranged in a very interesting fashion, and the entire structure is large enough to be visible through a high-powered microscope (Figure 2-11). Often, different proteins associate with one another to form even larger structures, called *multiprotein complexes* (Figure 2-10). The structure of any protein is described in several terms. The sequence of amino acids determines the *primary* structure. This is the most important aspect of its structure, because it largely determines the function as well as the three-dimensional structure of the protein molecule. Indeed, the substitution of one amino acid for another may completely destroy the function of the protein. Functional proteins with the same primary structure are identical in all other respects.

PRIMARY STRUCTURE DETERMINES THREE-DIMENSIONAL STRUCTURE

The three-dimensional structure of a protein is also important. A single polypeptide chain does not occur as an extended chain of amino acids. Rather it is

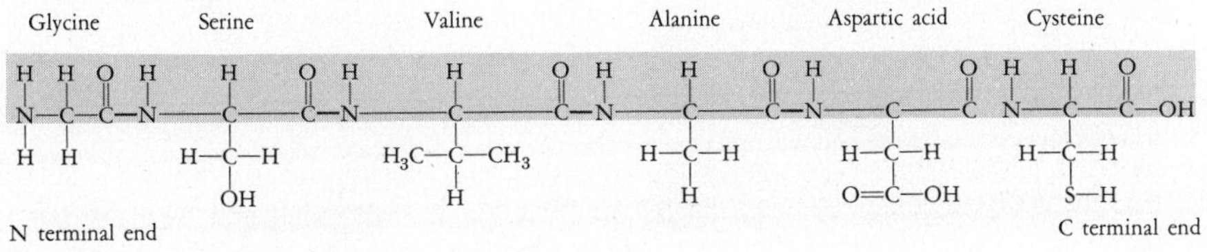

Glycine — Serine — Valine — Alanine — Aspartic acid — Cysteine

N terminal end … C terminal end

FIGURE 2-9 Polypeptide chain.

Amino acids — Polypeptide chains — Protein A — Protein B — Multiprotein complex

FIGURE 2-10 Steps in the buildup of multiprotein complexes.

folded and bent into its unique three-dimensional structure, attempting to achieve its most stable configuration. To achieve maximum stability, the molecule will bend in ways which allow weak bond formation between the side chains of appropriate amino acids. In addition, one covalent linkage is of great importance—the combination of the sulfur atoms in two cysteine amino acid subunits to form a covalent S—S bond. Most proteins tend to have a *helical* three-dimensional structure which can be visualized as a chain wound like a helix around a rigid rod (Figure 2-12). The nonpolar hydrocarbon side chains (H—C—H) are tucked inside the molecule and are not exposed to the aqueous environment. This regular structure is very stable primarily because the successive turns of the helix are held together by hydrogen bonds.

HELIX FORMATION COMMON IN PROTEIN STRUCTURE

The weak bonds which are involved in the three-dimensional structure are readily broken if the protein is heated (energy put into the system). Once this happens, the protein loses its three-dimensional structure and becomes *denatured*

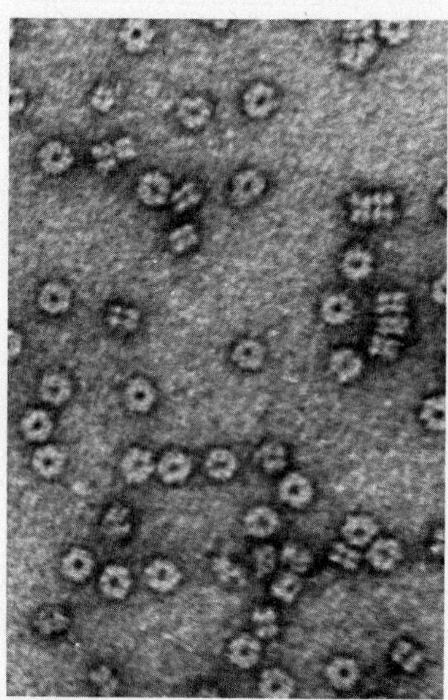

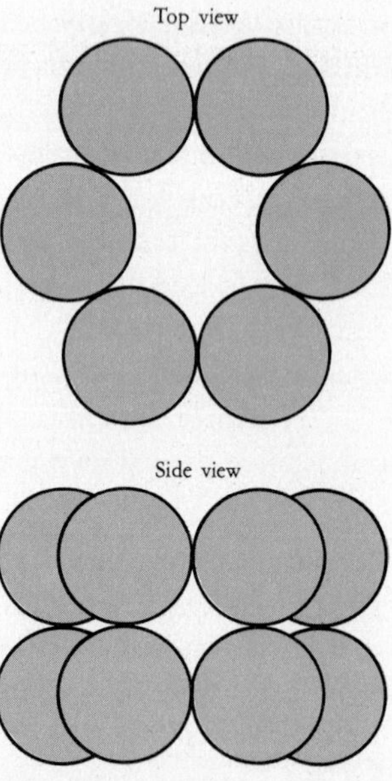

FIGURE 2-11
Electron micrograph and diagrammatic representation of the enzyme glutamine synthetase. The protein molecule is composed of twelve identical subunits forming two hexagons. The subunits bind together to form two closed circles, which are attached to each other. (Courtesy of D. Eisenberg.)

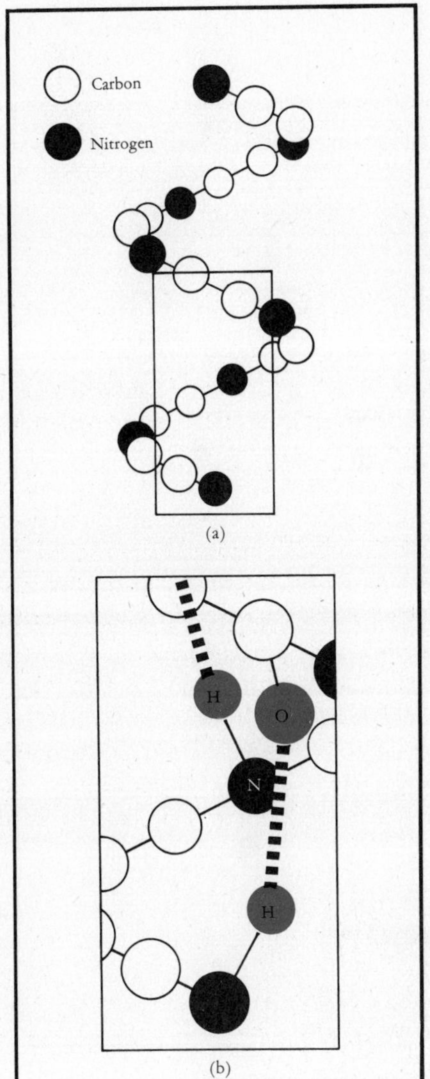

FIGURE 2-12
(a) Helical arrangement of polypeptide chains in protein. The turns in the helix are very regular and the regular arrangement is held together by hydrogen bonds between the O in the C—O group and the H in the N—H group. Although this helical arrangement, called the *alpha helix,* is the most common, other configurations are found in some proteins. (b) Enlarged section of the polypeptide chain above showing the hydrogen bond between oxygen and hydrogen.

(Figure 2-13). For example, boiling an egg results in a profound, readily observable denaturation of the protein (albumin) of the egg white.

Nucleic Acids as Macromolecules

SUBUNITS OF NUCLEIC ACIDS ARE MONONUCLEOTIDES

Nucleic acids—both ribonucleic (RNA) and deoxyribonucleic acid (DNA)—are linear polymers in which the subunits are called *nucleotides.* Each of these nucleotides is composed of three units: a nitrogen-containing ring compound, known as a base, covalently bonded to a 5-carbon sugar molecule, which in turn is bonded to a phosphate molecule. There are five different nitrogen-containing

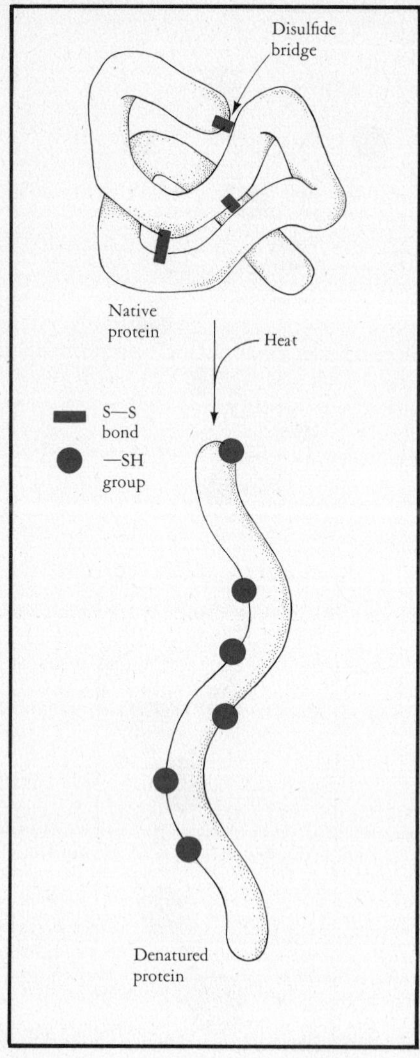

FIGURE 2-13
Heat denaturation of a protein molecule. The heating of proteins to high temperatures results in the disruption of the weak bonds as well as the S—S covalent bonds. The breaking of the S—S (disulfide) bonds results in the formation of S—H (sulfhydryl) groups.

bases, which can be divided according to their structure into two groups, *purines* and *pyrimidines* (Figure 2-14). The two types of nucleic acids, RNA and DNA, differ in the composition of their nucleotides. The nucleotides of RNA contain the sugar *ribose* and those of DNA the sugar *deoxyribose.* An additional difference is that the pyrimidine *uracil* in RNA is replaced by *thymine* in DNA. Both DNA and RNA contain the same two purines (adenine and guanine) and the pyrimidine cytosine. Thus each nucleic acid is composed of four particular kinds of nucleotide subunits (Figure 2-15). The subunits are joined together by a covalent bond which joins the phosphate of one nucleotide to the sugar of the adjacent nucleotide. This phosphate bridge joins the 3-carbon atom of one sugar to the 5-carbon atom of the other, so that a backbone composed of alternating sugar and

RNA AND DNA DIFFER IN CHEMICAL COMPOSITION

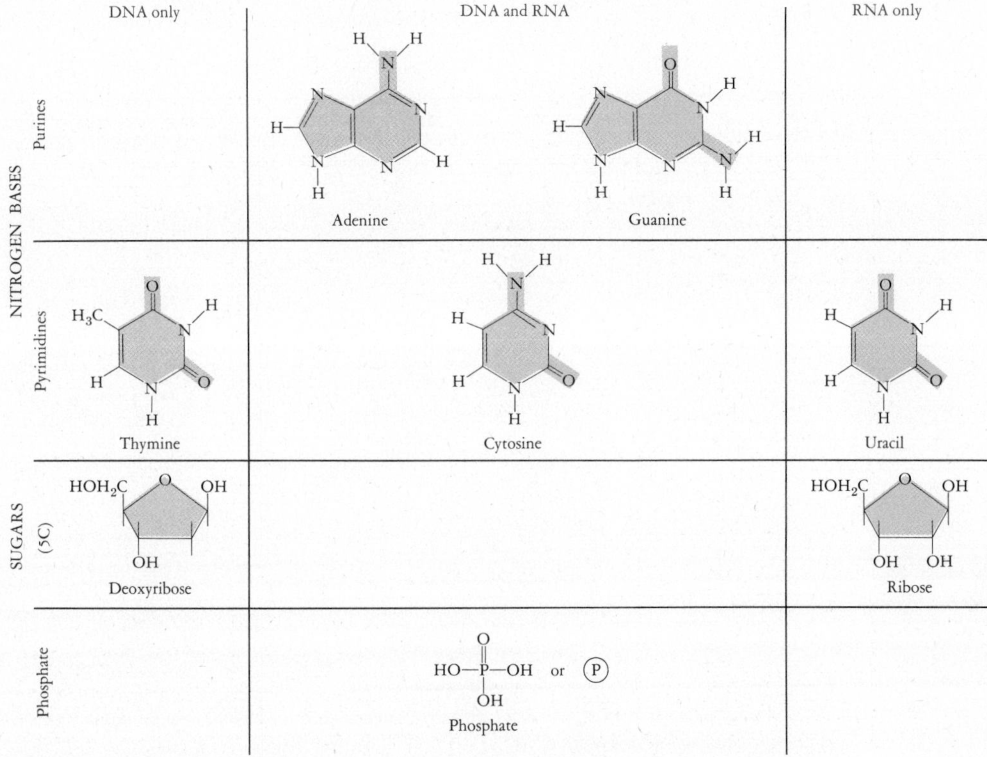

FIGURE 2-14 Components of RNA and DNA. Carbon atoms are not represented in the ring structures.

phosphate molecules is formed (Figures 2-16 and 2-17). In the reaction which joins the sugar and phosphate molecules, water is removed.

The size of the nucleic acid polymer varies depending on whether the macromolecule is RNA or DNA. A DNA molecule of a typical bacterial cell is composed of about 30 million nucleotides arranged in the form of a double-stranded helix. RNA is considerably smaller than DNA. The discussion of the structure of nucleic acids is continued in Chapter 8.

DNA carries the genetic information of the cell, coded in the sequence of purines and pyrimidines. This code must be translated into the sequence of amino acids that make up the cell's protein molecules, and RNA functions in this translation process in several ways. Three different types of RNA perform particular functions in the cell. Each type of RNA has a different size and slightly different three-dimensional structure from the others. The role of DNA and RNA in protein synthesis will be discussed in Chapter 8.

Polysaccharides as Macromolecules

Carbohydrates are compounds containing principally carbon, hydrogen, and oxygen atoms in a ratio of approximately 1:2:1. The carbohydrate macromole-

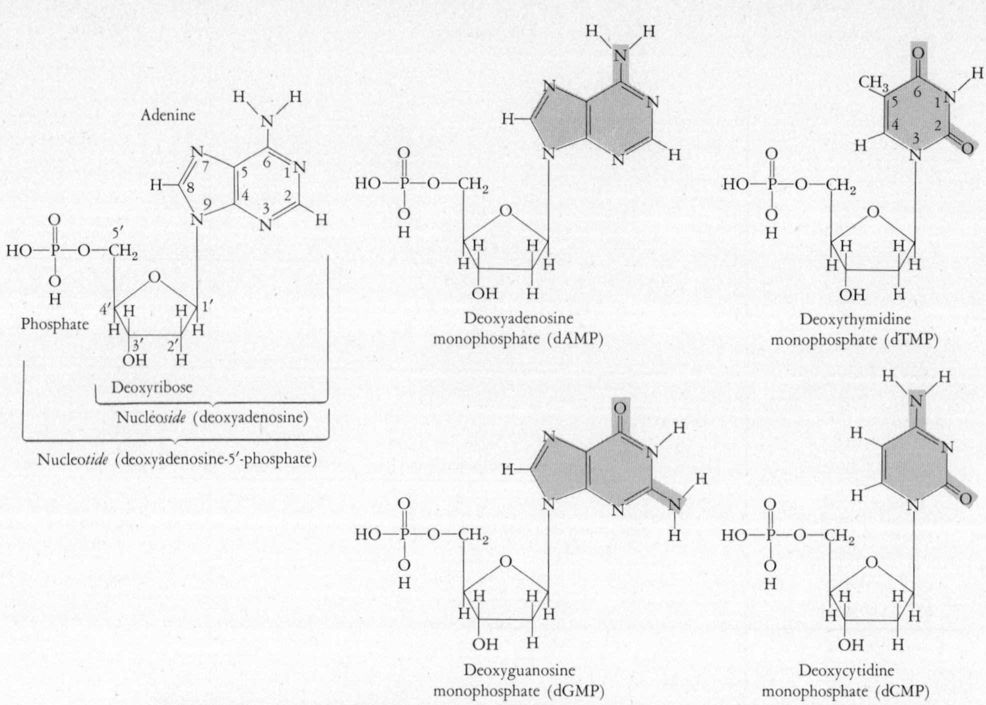

FIGURE 2-15 Subunits of DNA, the mononucleotides. Four mononucleotides are present in DNA, four in RNA. The nucleotides of RNA have the same structure except that ribose rather than deoxyribose and uracil in place of thymine are involved.

cules, the *polysaccharides,* are linear polymers of the subunit *monosaccharides. Disaccharides* are molecules consisting of two monosaccharides joined together. The term *sugar* is often applied to monosaccharides and disaccharides. The chief distinguishing feature of carbohydrates is that they contain a large number of alcohol groups (R—C—OH) which accounts for their ratio of carbon, hydrogen, and oxygen atoms. Carbohydrates also usually contain an aldehyde and, less commonly, a ketone group (Table 2-2). Some, the amino sugars, contain an amino group.

Monosaccharides are classified by the number of carbon atoms they contain. The most common in nature are the 5-carbon *pentoses* and the 6-carbon *hexoses.* The 5-carbon pentose sugars are represented by the sugars in nucleic acids, *ribose* and *deoxyribose* (Figure 2-14). Common hexoses are *glucose, galactose,* and *mannose.* All of these hexoses contain the same kind of atomic groups, but they differ in the spatial arrangements of the —H and —OH groups relative to the carbon atoms (Figure 2-18). For example, glucose and galactose have identical structures except for the arrangement of the H and OH attached to carbon 4. Mannose and glucose differ with respect to carbon 2. These small changes result in three distinct sugars with different physical and chemical properties.

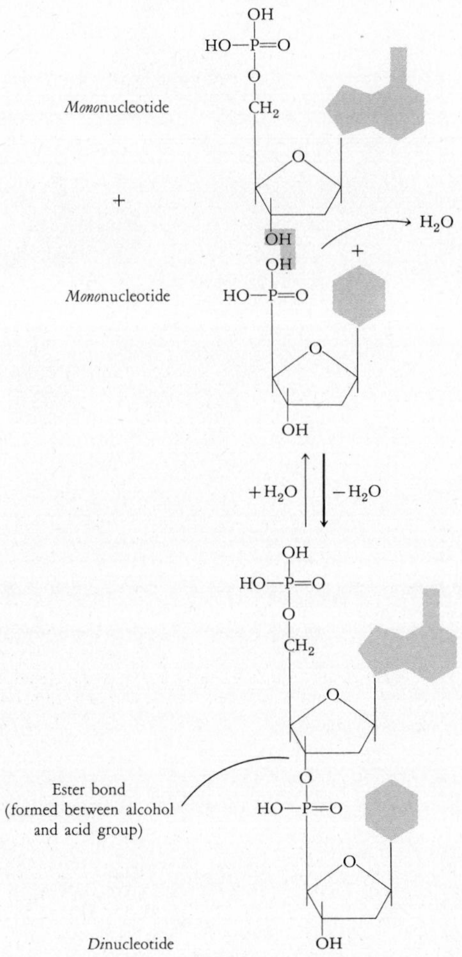

Mononucleotide

$$\text{HO—P(OH)(=O)—O—CH}_2\text{—(sugar)—OH}$$

+

Mononucleotide

$$\text{HO—P(OH)(=O)—(sugar)—OH}$$

$\rightarrow H_2O$ +

$+H_2O$ | $-H_2O$

Ester bond (formed between alcohol and acid group)

Dinucleotide

FIGURE 2-16
Formation of ester bond in joining together two nucleotides.

Sugars, like amino acids, have carbon atoms bonded to four different groups and so can also exist in two isomeric forms, D and L, which are mirror images of one another. By convention, if the —OH group on carbon 5 in hexoses (carbon 4 in pentoses) is written on the left, the sugar is L; if on the right, it is a member of the D-series. The majority of monosaccharides which occur in living organisms are of the D-configuration. This is the opposite of the situation with amino acids, in which the L-form is the biologically important form.

The most common *disaccharides* are the milk sugar, *lactose,* and the common table sugar, *sucrose.* Sucrose, derived commercially from sugar cane or sugar beets, is composed of one molecule of glucose and one of galactose. The linkage is formed in the reaction between the hydroxyl groups of two sugar units, with the loss of a molecule of water (Figure 2-19).

MONOSACCHARIDES ARE SUBUNITS OF POLYSACCHARIDES

The general structural feature of a polysaccharide is the repeating monosaccharide subunits, with D-glucose being the most frequent subunit. However, unlike proteins and nucleic acids, polysaccharides can have side chains branching

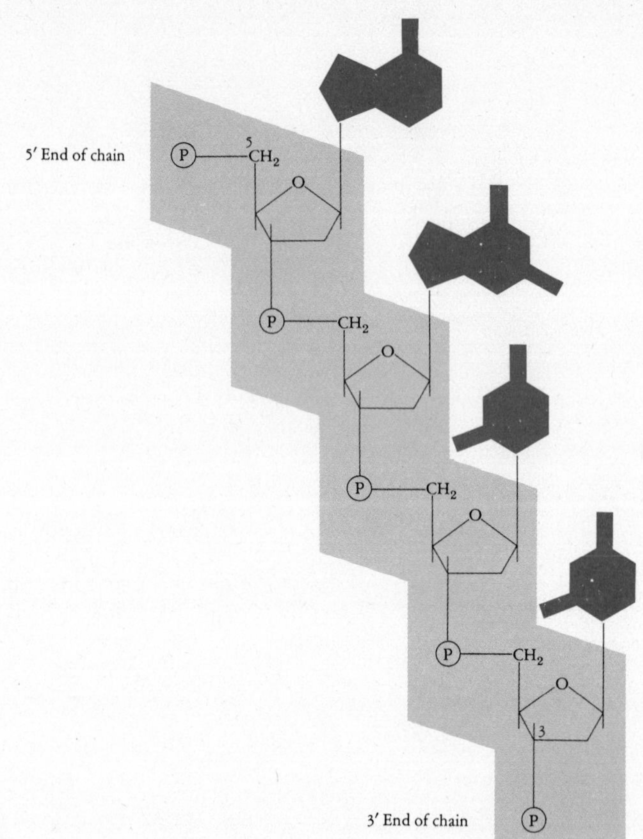

FIGURE 2-17
The repeating sugar-phosphate backbone of nucleic acids.

from the main, linear, chain. Polysaccharides are ubiquitous in nature and serve as the storage forms for carbon and energy, as well as being structural components of the cell. Cellulose, a polymer of glucose subunits, is the principal structural component of plant cell walls, and it is also synthesized by some bacteria. It is the

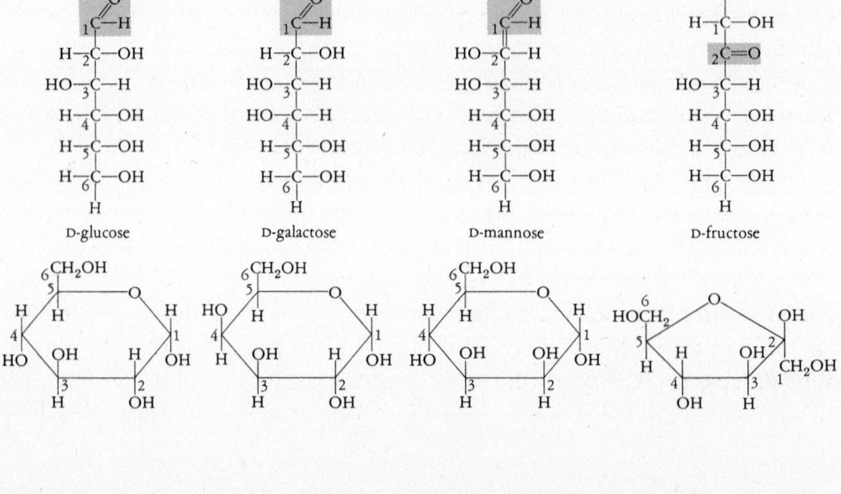

FIGURE 2-18
Formulas of some common sugars. The ring structures arise as a result of the reaction of the —OH group on carbon atom 5 with the aldehyde group on carbon atom 1, resulting in an O bridge between carbon atoms 1 and 5. Note that fructose is unusual in that it has a keto group on carbon atom 2.

D-glucose

D-galactose

D-mannose

D-fructose

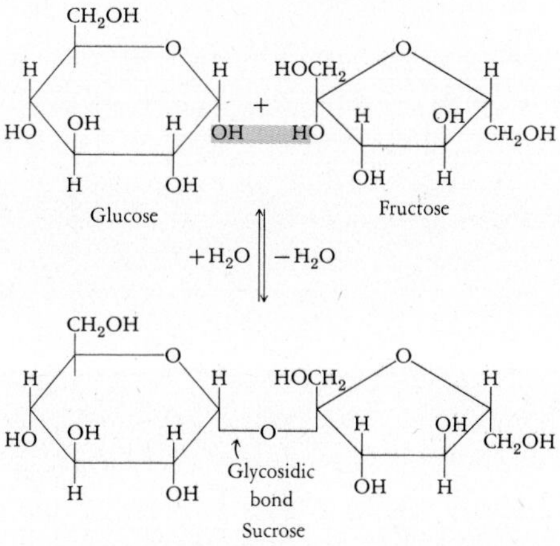

FIGURE 2-19
Formation of sucrose from glucose and fructose.

most abundant organic compound in the world. Glycogen, a carbohydrate storage product of animals and some bacteria, and dextran, which is also synthesized by bacteria, somewhat resemble cellulose, all three of these polymers being made up of glucose subunits, or residues. However, each is a distinct molecule. Polysaccharides which consist of the same subunits can differ in the following ways: (1) size of the polymer, (2) degree of branching of the chain, and (3) the carbon atoms involved in the glycosidic linkage. Thus a large variety of polysaccharides, made up of the same subunits, is possible.

Lipids as Macromolecules

LIPIDS ARE CHARACTERIZED BY THEIR INSOLUBILITY IN WATER

The *lipids* represent a very heterogeneous group of biologically important molecules, all of which have the property of being only slightly soluble in water but readily soluble in most organic solvents such as ether, benzene, and chloroform. The two general classes are the *simple* and the *compound lipids.* They are not macromolecules in the sense that the term has been used thus far, because they have molecular weights of no more than a few thousand. Furthermore, they have no well-defined subunits, but rather consist of a wide variety of substances of markedly different chemical structure.

FATS ARE ESTERS OF FATTY ACIDS AND GLYCEROL

The most common of the simple lipids are the *fats.* Fats are a combination of fatty acids and glycerol in which the bond forms between the *carboxyl group* of the acid and the *hydroxyl group* of the glycerol. The general formula for a fat is given in Figure 2-20. Note again the elimination of water in the formation of this bond. Glycerol has three hydroxyl groups which allow three fatty acid molecules to be bonded to it. These three fatty acids may be the same or different. There are many kinds of fatty acids, but the most common in nature are *palmitic* (16 carbon atoms), *stearic* (18 carbon atoms), and *oleic,* with 18 carbon atoms and some double bonds in the molecule. The fact that these molecules contain so many nonpolar groups, H—C—H, accounts for their insolubility in water. Water pulls

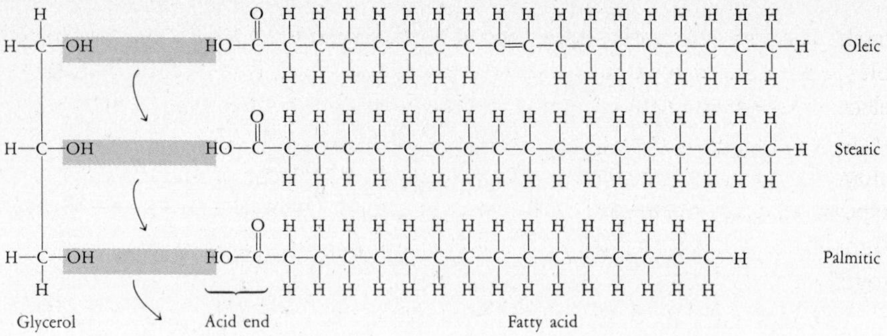

FIGURE 2-20
Formation and general formula of a fat. An ester bond is formed between the alcohol group of the glycerol and the acid group of the fatty acid.

apart polar compounds because of the strong attraction between the charged water molecules (Figure 2-2) and the polar compounds. Nonpolar compounds cannot form these attractions, so water does not pull them apart and they remain insoluble in water.

Another very important group of lipids is the *steroids,* all of which have the four-membered ring structure shown in Figure 2-21. These compounds are quite different from the lipids in chemical structure but are commonly classified as lipids because they have similar solubility properties. If an hydroxyl (—OH) group is attached to one of the rings, the steroid is termed a *sterol.* Other compounds in this general group of lipids are important hormones or vitamins.

MANY LIPIDS ARE ASSOCIATED WITH OTHER MACROMOLECULES

Compound lipids contain other elements, such as sulfur, nitrogen, or phosphorus, in addition to the carbon, hydrogen, and oxygen found in simple lipids. *Phospholipids* contain the fatty acids and glycerol and, in addition, a phosphate molecule bonded to a nitrogen-containing compound. Another group of compound lipids, the *lipoproteins,* are loose complexes of proteins and lipids held together by weak bonding forces. *Lipopolysaccharides* are molecules of lipids associated with polysaccharides through covalent bonds.

Lipids have several functions in living organisms. In microorganisms, phospholipids are found in the cytoplasmic membrane and in many cell walls. In human beings and other animals, fats act as the prime fuel reserve for metabolism, as well as being structural components of membranes.

SUMMARY

This chapter presents some fundamental chemical information required for understanding various facets of microbiology. Virtually all biological phenomena depend on the association of atoms and molecules. The forces or bonds involved

FIGURE 2-21
Cholesterol, a sterol. The carbon atoms in the ring structures are not shown.

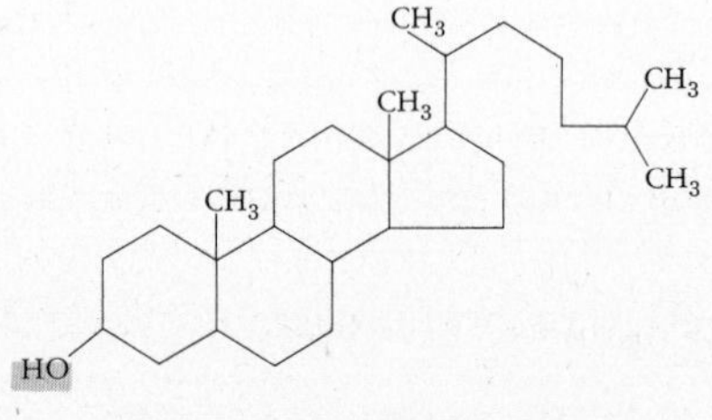

in the association of molecules can be of several types. One classification depends on the strength of the bonds. The covalent bond, which involves the sharing of electrons between atoms, is a very strong bond. Since some atoms attract electrons more avidly than others, the shared electrons in a covalent bond may be closer to one atom than to the other atom; this allows the formation of a weak type of bond, the hydrogen bond. Although a single weak bond has no biological significance, a large number are capable of holding very large molecules firmly together.

A large number of organic molecules play important roles in life processes. The small molecules can be grouped according to their functional group. Macromolecules range in size from 5000 to 1 billion in molecular weight. Their basic structures have been elucidated largely because all macromolecules are composed of repeating subunits (Table 2-5). These subunits are joined together by covalent bonds in a reaction that involves the splitting out of a molecule of water.

Proteins are composed of amino acids bonded together in a peptide linkage; proteins differ from each other in the sequence of the twenty kinds of amino acids which comprise the protein. The long protein chain assumes a three-dimensional, often helical, configuration which permits the maximum number of bonds to be formed between the amino acids in the chain.

The nucleic acids, RNA and DNA, are polymers of subunits, the nucleotides, which are composed of a molecule of a 5-carbon sugar—ribose or deoxyribose—a purine or pyrimidine, and a phosphate molecule. Covalent bonds between the acid groups of the phosphate molecule and an alcohol (H—C—OH) group on the sugar join the nucleotides. The two strands of DNA are held together by a large number of hydrogen bonds.

Polysaccharides comprise a number of molecules of varying sizes. The monosaccharides, especially glucose, bond together to form polysaccharides such as

TABLE 2-5
Structure and Function of Macromolecules

Name	*Subunit*	*Bond Joining Subunits*	*Atoms in Bond*	*Some Functions of Macromolecule*
Protein	Amino Acid	Peptide	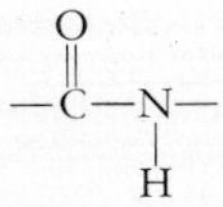 $-\overset{\overset{\displaystyle O}{\|\|}}{C}-\underset{\underset{\displaystyle H}{\|}}{N}-$	Catalysts; structural portion of cell organelles
Ribonucleic acid	Nucleotide	Diester	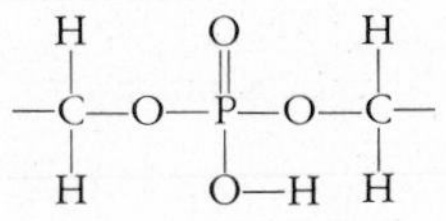 $-\overset{H}{\underset{H}{C}}-O-\overset{O}{\underset{O-H}{P}}-O-\overset{H}{\underset{H}{C}}-$	Various roles in protein synthesis
Deoxyribonucleic acid	Deoxynucleotide	Diester	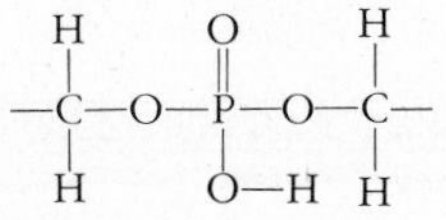 $-\overset{H}{\underset{H}{C}}-O-\overset{O}{\underset{O-H}{P}}-O-\overset{H}{\underset{H}{C}}-$	Carrier of genetic information
Polysaccharide	Monosaccharide	Glycosidic	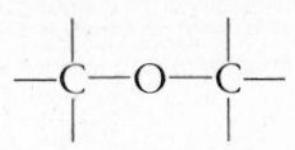 $-\overset{\|}{\underset{\|}{C}}-O-\overset{\|}{\underset{\|}{C}}-$	Structural component of plant cell wall; storage products

starch and cellulose. Variations in the bonds that join the monosaccharide molecules together, as well as the degree of branching, account for differences in molecules composed of the same monosaccharide subunits.

The lipids, too small to be considered macromolecules, are of two types, the simple and the compound. Simple lipids consist of glycerol bonded to three fatty acid molecules. They have unique biological properties because they are polar at the end which contains the carboxyl group of the fatty acid and nonpolar through the rest of the molecule. Compound lipids contain molecules that have either phosphorus or nitrogen in addition to fatty acids and glycerol. The steroids are another group of lipids, characterized by a four-membered ring structure.

QUESTIONS

REVIEW

1. Compare covalent and hydrogen bonds in terms of: (a) basis for bond formation, (b) biological significance, (c) strength, (d) compounds in which they are important.
2. A bacterium ingests the following radioactive compounds. In what macromolecules will the major elements in the compounds be found? (a) Nitrate (nitrogen), (b) glucose (carbon), (c) phosphate (phosphorus), (d) sulfate (sulfur).
3. Compare DNA, RNA and protein in terms of: (a) subunits, (b) bonding together of subunits, (c) components making up subunits.
4. What features do all macromolecules have in common? Relying on these features, would you consider lipids to be macromolecules?
5. What generalizations can you make about the chemical reactions in which the subunits of macromolecules are joined together?

THOUGHT

1. What are some of the properties of weak bonds that make them important in biological systems?
2. Proteins fold into very specific configurations. What feature of the molecule determines the three-dimensional configuration of the protein molecule?

FURTHER READING

The material covered in this chapter is covered much more extensively in any textbook of biochemistry. Two readable and up-to-date texts written for college juniors and seniors are the following:

LEHNINGER, A., *Biochemistry.* 2d ed. New York: Worth Publishing 1975.
STRYER, L., *Biochemistry.* San Francisco: Freeman Press, 1975.

Many textbooks of biology also cover this material. One such text, which is particularly well written, is:

KIRK, D., *Biology Today.* 2d ed. New York: Random House, 1975. Chapters 10 and 11 are relevant.

CHAPTER 3

FUNCTIONAL ANATOMY OF PROCARYOTIC AND EUCARYOTIC CELLS

Van Leeuwenhoek used only a single lens to view his "animalcules" and the magnification with this simple microscope was limited to approximately 300-fold. The details of cell structure cannot be observed at this magnification, so more detailed studies of the internal structure of cells had to await the development of a microscope that could magnify objects more than 1000 times. However, even at this magnification the finer structure of procaryotic cells cannot be elucidated, and it was not until the development of the electron microscope in the 1930s that the definitive studies on procaryotic cell structure could be done. The analysis of cell structure through microscopic techniques has been accompanied by a chemical analysis of the individual cell components that can be isolated in pure form. This combined chemical and microscopic approach has provided considerable insight into the function of the various structures of the cell and their role in the life of the organism.

MICROSCOPIC TECHNIQUES: INSTRUMENTS

Compound Microscope

The most commonly used instrument for observing any cells is the *light microscope,* so-named because it illuminates its object by visible light. There are many variations of this instrument. The light microscope used by van Leeuwenhoek, consisting of only a single magnifying lens, was called a *simple microscope.* The light microscopes most commonly used today have two sets of lenses, an objective and an ocular, and are called *compound microscopes.* The magnification achieved with such a microscope is the product of the magnification of each of the indi-

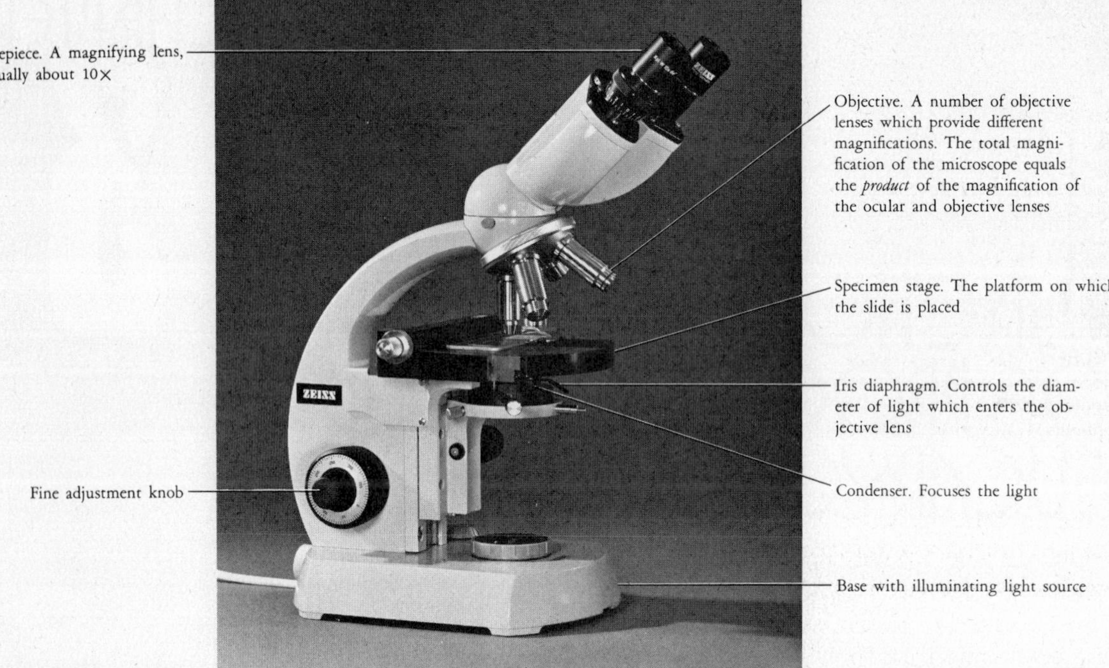
Eyepiece. A magnifying lens, usually about 10×
Objective. A number of objective lenses which provide different magnifications. The total magnification of the microscope equals the *product* of the magnification of the ocular and objective lenses
Specimen stage. The platform on which the slide is placed
Iris diaphragm. Controls the diameter of light which enters the objective lens
Fine adjustment knob
Condenser. Focuses the light
Base with illuminating light source
ZEISS
FIGURE 3-1 Compound microscope. (Courtesy of Carl Zeiss, Inc.)

vidual lenses. Most compound microscopes have a number of objective lenses, thereby making a number of different magnifications possible with the same instrument. A modern compound microscope, with its major components labeled, is shown in Figure 3-1. A very good compound microscope can magnify objects up to 2000 times.

RESOLVING POWER IS THE MOST IMPORTANT FEATURE OF A MICROSCOPE

The usefulness of a microscope depends not so much on the degree of magnification, but rather on the ability of the microscope to clearly separate or *resolve* two objects which are very close together. Indeed, the renowned English microscopist Robert Hooke was not as successful as van Leeuwenhoek in making out the details of microorganisms, even though the microscope that Hooke invented was a compound microscope with a greater magnification. This was because the lenses which Hooke employed suffered from serious optical defects which resulted in blurred images.

The *resolving power* of a microscope is defined as the minimum distance that can exist between two points such that the points are observed as separate entities. The resolving power determines how much detail can actually be seen. It depends on the quality of the lens, the magnification, and the preparation of the specimen under observation. The maximum resolving power of the best light microscope functioning under optimal conditions is 0.2 μm because the resolving power is limited by the wavelength of the light employed for illumination. The shorter the wavelength the greater the resolution that can be attained. Thus, microscopes have been developed which employ illumination of a shorter wavelength than visible light, and this technique has improved the resolving power (Figure 3-2).

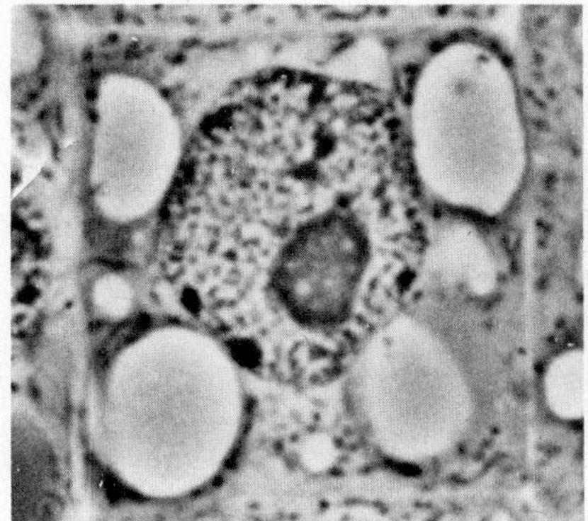
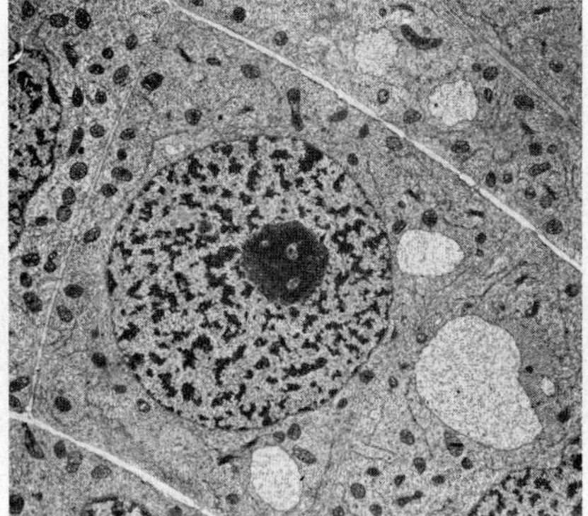

FIGURE 3-2 Comparison of the resolving power of a light microscope (left) and an electron microscope (right). The same preparation (onion root tip) was magnified 450 times. Note the difference in the degree of detail that can be seen at the same magnification. The magnification is the upper limit for the light microscope (phase contrast) but a lower limit for the electron microscope. (Courtesy of W. A. Jensen.)

In the electron microscope, a beam of electrons is analogous to the light source and its wavelength is only $\frac{1}{10,000}$ that of visible light. The unaided human eye has a resolving power of about 200 μm.

Phase Contrast Microscope

PHASE CONTRAST MICROSCOPE CAN REVEAL SOME STRUCTURES IN LIVING CELLS

Although bacteria are large enough to be seen with a light microscope when suspended in a drop of liquid, they are very difficult to see because they are transparent, usually colorless, and so tiny that they may even be difficult to find. To overcome the problem of transparency in all types of cells, a special type of light microscope, the *phase contrast microscope,* has been devised. This microscope, probably the one most commonly used in research laboratories for observing living microbes, is a variation of the compound microscope; it has special optical devices which increase the contrast between the microbes and the surrounding medium. Advantage is taken of the fact that cells are denser than the surrounding medium, and therefore illuminating light is slowed down more as it passes through the specimen as compared to the medium surrounding the specimen. The result is that even though the cells are not magnified to any greater extent, they nevertheless stand out from the background and are clearly visible. Thus it is possible to observe living organisms clearly and study their movements in the medium in which they are growing. Figures 3-3a and 3-3b are photomicrographs of the same organism, one viewed under the ordinary light microscope and the other under a phase contrast microscope. Note that even internal structures are discernible by phase contrast microscopy.

Darkfield Microscopy

Another commonly employed method for achieving a marked contrast between living organisms and the background is *darkfield microscopy.* In this technique light is directed toward the specimen at an angle, so that only light that is scat-

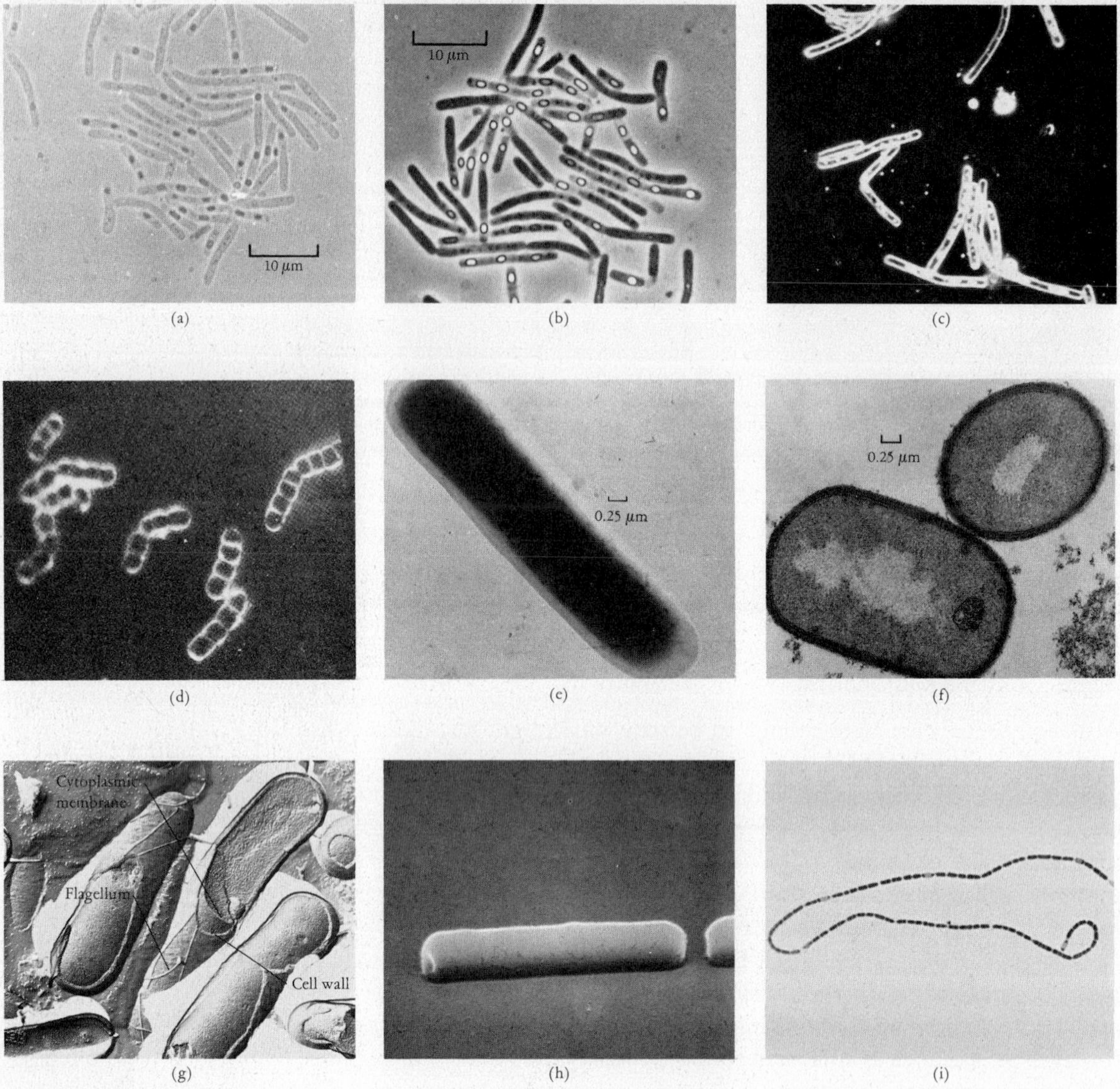

10 μm

(a)

10 μm

(b)

(c)

(d)

0.25 μm

(e)

0.25 μm

(f)

Cytoplasmic membrane

Flagellum

Cell wall

(g)

(h)

(i)

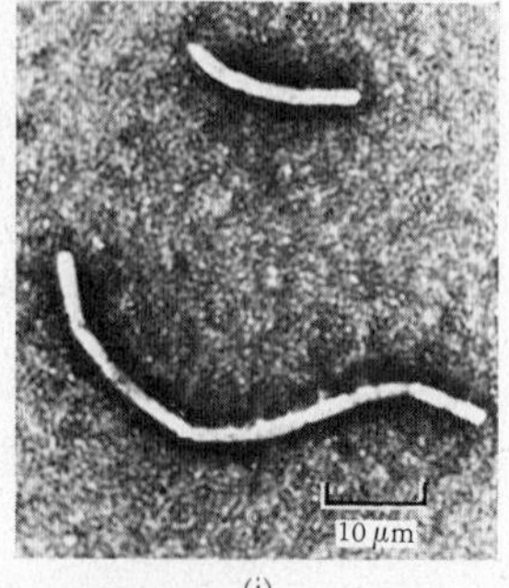

(j)

FIGURE 3-3 All of the photomicrographs shown here (except e, f, and g) are of *Bacillus megaterium*. They illustrate the details that are possible with each type of preparation and microscope. (a) Bright field (light) microscopy. The intracellular bodies are endospores. (Courtesy of J. T. Staley.) (b) Phase contrast microscopy. (Courtesy of J. T. Staley.) (c) Darkfield microscopy. (d) Fluorescence microscopy. The fluorescent material is bound to the cytoplasmic membrane. (Courtesy of B. A. Newton; B. A. Newton, *J. Gen. Microbiol.* 12:226, 1955.) (e) Electron photomicrograph of *Bacillus fastidiosus*. (Courtesy of E. Leadbetter; E. Leadbetter and S. Holt, *J. Gen. Microbiol.* 52:299, 1968.) (f) Thin section of *B. fastidiosus;* electron photomicrograph. (Courtesy of E. Leadbetter; E. Leadbetter and S. Holt, *J. Gen. Microbiol.* 52:299, 1968.) (g) Freeze-etched preparation of *Bacillus subtilis*. (Courtesy of S. Holt.) (h) Scanning electron photomicrograph. (Courtesy of N. Hodgkin.) (i) Simple, stained preparation, light microscopy. The light areas in some cells are endospores. (Courtesy of J. P. Dalmasso.) (j) Negative stain. (Courtesy of J. P. Dalmasso.)

tered by the specimen enters the objective lens and is visualized. The background is completely dark except for the objects being viewed, which are brilliantly illuminated (Figure 3-3c). This technique makes visible objects and cells that are invisible by ordinary light microscopy. When darkfield microscopy is used, *E. coli* is clearly visible at magnifications of only 100-fold. It is useful for observing very thin cells, in particular the organism causing syphilis, *Treponema pallidum,* which is barely visible using ordinary light microscopy. The advantage of using darkfield microscopy for observing living bacteria is that it permits the viewer to estimate their true size, shape, and mobility undistorted by fixing and staining procedures.

Fluorescence Microscopy

Another type of microscope of importance in laboratories concerned with identifying microorganisms is the *fluorescence microscope.* This microscope is used to visualize objects that fluoresce, that is, emit light when a light of a different wavelength impinges on the object. The fluorescence may be a natural property of the specimen being viewed, or it may result from a fluorescent compound being attached to a normally nonfluorescing material (Figure 3-3d). A common application of fluorescence microscopy is with the use of antibodies attached to a special fluorescent compound that appears yellow-green when exposed to near-ultraviolet light. A special lamp supplies the ultraviolet light, which reaches the object stained with the fluorescent-tagged antibodies. When stimulated by ultraviolet light, the fluorescent material emits a visible yellow-green light. Special filters must be used on the microscope to allow the visible light to pass, and to protect the eyes against the ultraviolet radiation. Techniques such as this one have numerous applications in medical microbiology.

Electron Microscope

The light microscope is only capable of defining the gross anatomy of the bacterial cell—its size, shape, and a few of its largest components. The light microscope was developed almost to the zenith of its resolving power over a century ago. In order to increase the resolving power significantly, a new type of microscope, the *electron microscope,* was constructed by Knoll and Hruska in Berlin in 1931. In this microscope, a beam of electrons, which is analogous to the light source, is focused by means of magnetic fields, which function as lenses. Some of the electrons pass through the specimen, others are scattered, and still others cannot pass. The electrons impinge on an electron-sensitive screen, creating an image which is determined by the ability of the electrons to pass through various parts of the object being viewed. A resolving power of 0.0003 μm (3 Å) can be achieved (about 600 times better than the light microscope) (Figure 3-3e). In order to increase the details of cell structure, investigators often slice the specimen to be viewed into very thin sections, commonly with a diamond knife, and observe these "thin sections" (each approximately 0.02 μm in thickness) (Figure 3-3f). In the past, specimens had to be fixed so that they were "bone dry" and extremely thin before they could be observed with the electron microscope. But since the fixation process may introduce a large number of artifacts into the preparation, one of the main jobs of electron microscopists is to interpret whether what they see is actually present in the living cell or merely an artifact that results from vigorous treatment in the fixation process.

Recently a technique (freeze-etching) has been devised which circumvents the need for chemical fixation and thereby presumably overcomes the problem of artifacts. The material is frozen and thin sections chipped off. The surface of a section is then coated with a thin layer of carbon, which is thin enough to be viewed with the electron microscope. The pictures from such freeze-etched preparations can be quite dramatic (Figure 3-3g).

A significant advance in microscopic techniques has been made in recent years with the development of a modified electron microscope, the *scanning electron microscope.* The material to be observed is coated with a thin film of metal. The electron beams scan back and forth over the surface and are reflected back into the viewing chamber. The most significant aspect of this technique is that relatively large specimens can be viewed and a dramatic three-dimensional figure is observed (Figure 3-3h).

MICROSCOPIC TECHNIQUES: STAINING

To overcome the difficulty of observing living, transparent, and often motile organisms, cells are frequently killed and then treated with one or more dyes that have a special affinity for one or more cellular components. Such treatment results in the entire organism, or part of it, achieving a marked contrast to the unstained background (Figure 3-3i). In the staining procedure a drop of liquid containing the organisms is placed on a glass slide and allowed to dry. The dried film is then "fixed" onto the slide, either with a chemical fixative or more commonly by passing it over a flame to coagulate the cell protein. The dyes, or stains, are then applied to the "fixed" organisms. A wide variety of stains and staining procedures are currently in use, each one having its own particular use. Some dyes will stain only a particular cell component. Other staining procedures will stain one group but not another group of organisms, thereby serving to divide organisms into groups based on their staining characteristics. Stains can be divided into two major groups based on their affinity for cell components. *Positive stains* have a strong affinity for one or more cell components and color these components when added to fixed cell preparations. *Negative stains* cannot penetrate the cell envelope and thus make the cell highly visible by providing a contrasting dark background. Negative stains are generally employed on living cells to demonstrate surface structures which are not stained well by positive stains (Figure 3-3j).

Simple Staining Procedures

In simple staining only a single dye is applied to the cells. This technique may be used to stain the entire cell or specific structures in the cell. The dye methylene blue is frequently applied to a fixed suspension of bacteria to stain entire cells a blue color without staining the background material. This basic dye binds primarily to the RNA and DNA of the cell, and little evidence of other internal structures is revealed. Acidic dyes, which include safranin, acid fuchsin, and congo red, stain basic compounds in the cell, primarily proteins carrying a basic charge. Sudan black is a dye that is very soluble in fat and is frequently used to identify the presence and location of fat droplets in bacteria.

Differential Staining Techniques

GRAM STAIN

It is possible to separate bacteria into groups, depending on their ability to take up and retain certain dyes. The most widely used staining procedure, capable of dividing virtually all bacteria into one of two groups, is the *Gram stain.* A Danish physician, Dr. Hans Christian Gram, working in a morgue in Berlin, developed this staining method in 1884 in order to distinguish the bacteria which cause pneumonia from the cell nuclei in infected mammalian tissues. Gram was disappointed with his method because not all bacteria retained the stain. But ironically, what he considered to be a defect in his staining method actually forms the basis of the most widely used characteristic test for bacteria. However, who actually first had the idea to use the Gram stain as a diagnostic tool to separate bacteria into two groups is unknown. The procedure involves the application of a basic purple dye, usually gentian or crystal violet (Figure 3-4), which stains all bacteria able to absorb this dye. Next, a dilute solution of iodine is added, which serves to decrease the solubility of the purple dye within the cell by combining with the dye to form a dye-iodine complex. An organic solvent, such as ethanol—which readily removes the purple dye-iodine complex from some but not other species of bacteria—is next added. A red stain is then applied. Those bacteria which are decolorized by the ethanol appear red or *Gram negative.* Those that retain the purple dye will appear purple or *Gram positive.* The reason why some bacteria retain the basic purple dye and others do not is related to the chemical structure of the cell wall. The cell wall of Gram-positive organisms is markedly different from that of Gram-negative organisms. Because the structure of the cell wall correlates with many other distinctive properties of the cell, whether an organism is Gram positive or Gram negative is correlated with these other properties of the cell.

ACID-FAST STAIN

Another differential staining procedure, the *acid-fast stain,* is used to characterize a small group of organisms which resist decolorization with an acidic solution of alcohol after being stained with a basic dye. Only a few groups of bacteria retain the basic dye—that is, are acid-fast—but because one acid-fast group causes tuberculosis, this stain has proven extremely valuable in diagnostic laboratories concerned with detecting these organisms.

Differential stains are also employed to stain specific cell structures such as flagella, endospores, and nuclear bodies. The staining procedure for each structure is different and takes advantage of the chemical composition and properties of the structure.

Because staining generally requires that the cells be heat-dried and killed, their

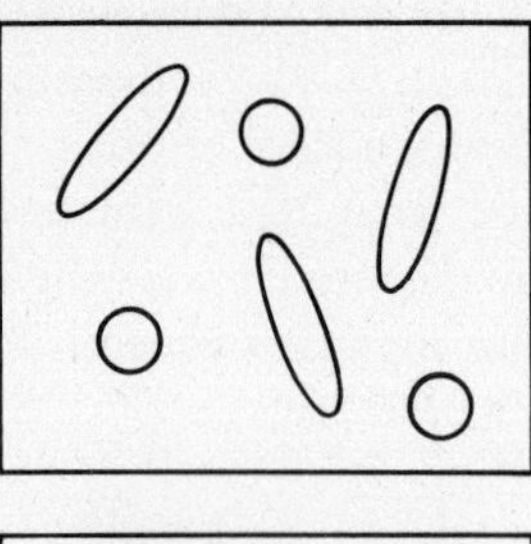
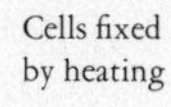
STEPS IN STAINING
APPEARANCE OF CELLS
Cells fixed
by heating
Shape of cells
becomes distorted
and cells shrink
in size

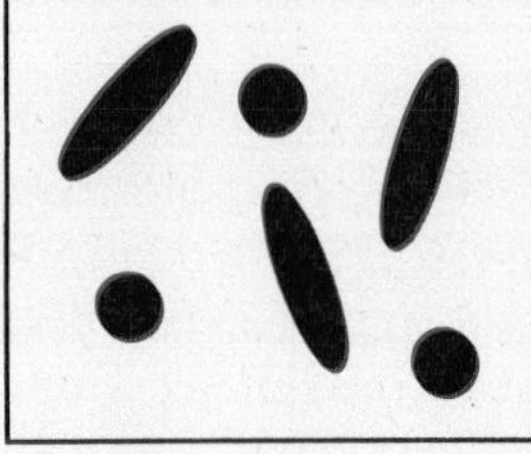
Basic dye, crystal
violet applied
All cells
stain purple

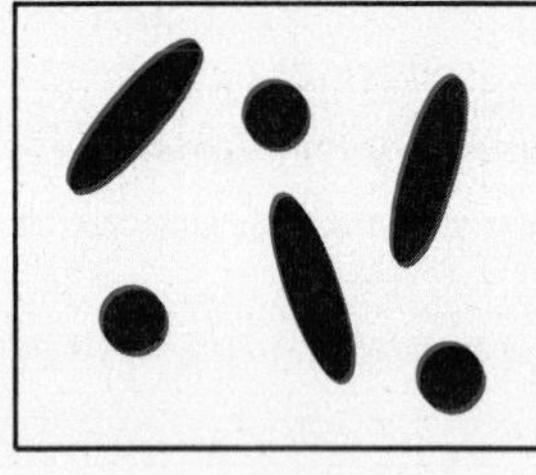
Addition of
iodine
All cells
remain purple

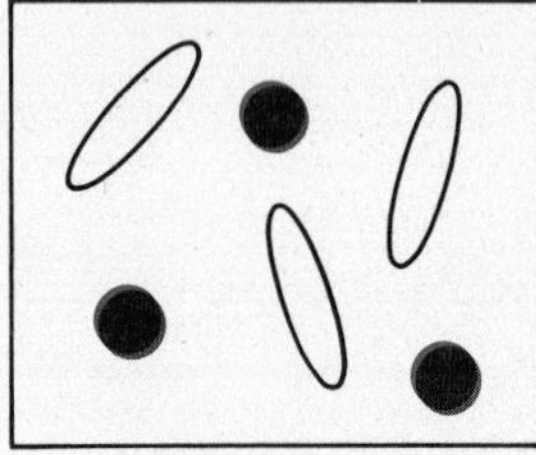
Addition of
alcohol
Gram-positive cells
remain purple;
Gram-negative cells
become colorless

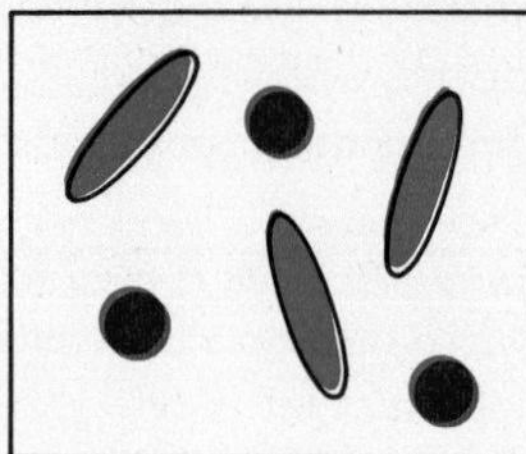
Addition of
counterstain,
often safranin
(red dye)
Gram-positive cells
remain purple;
Gram-negative cells
appear red

FIGURE 3-4
The steps in the Gram stain.

morphology can be easily distorted, producing artifacts. The fixation process tends to reduce the size of the cells, whereas the addition of dyes tends to increase their size. Indeed, flagella (organelles of locomotion) are so thin that they can be observed with the light microscope only if their size is increased by a specific staining procedure.

SHAPES OF BACTERIA

Typical bacteria have three shapes: spherical, comprising organisms which are commonly referred to as *cocci* (singular, coccus); cylindrical, comprising organisms commonly referred to as *rods* or *bacilli* (singular, bacillus); and helical, comprising organisms referred to as *spirilla* (singular, spirillum) (Figure 3-5). Most bacteria divide by binary fission to form two separate cells that are functionally distinct. However, these cells may not always separate from each other. This adherence of cells results in a characteristic arrangement which depends on the planes in which the bacteria divide (Figure 3-6). Those dividing in one plane form chains; those dividing in several planes at random appear as clusters. If division occurs in sequence in two or three planes perpendicular to one another, cubical packets result. Because such arrangements are characteristic properties of certain genera, they are useful aids in the microscopic identification of many bacteria. Thus certain cocci tend to form long chains; this is particularly characteristic of the genus *Streptococcus*. Others, especially members of the genus *Staphylococcus,* tend to occur in irregular clusters resembling bunches of grapes. The members of still other genera typically form packets of four, or cuboidal packets of eight or more cells.

FUNCTIONAL ANATOMY OF PROCARYOTIC CELLS (BACTERIA)

Because of the concern with bacteria in this text, the procaryotic cell type is emphasized. An electron micrograph and diagram of a common cylindrical bacterium is given in Figure 3-7. All typical bacteria have two characteristic features. They have a rigid cell wall and are motile (if they are motile) by means of flagella. These characteristics hold for most of the bacteria that will be considered in this text. However, atypical bacteria with other features, as well as eucaryotic protists, are important members of the microbial world and will also be covered to a limited extent.

The functional anatomy of procaryotic cells *must* provide the means:

1. To enclose the internal contents of the cell and separate it from the external medium
2. To store and replicate genetic information
3. To synthesize cellular components
4. To generate, store and utilize energy-rich compounds

In addition, *some* bacteria have the means for:

1. Cell movement
2. The transfer of genetic information
3. The storage of reserves of building blocks and energy

We will now consider in some detail the structures that are concerned with each of these functions.

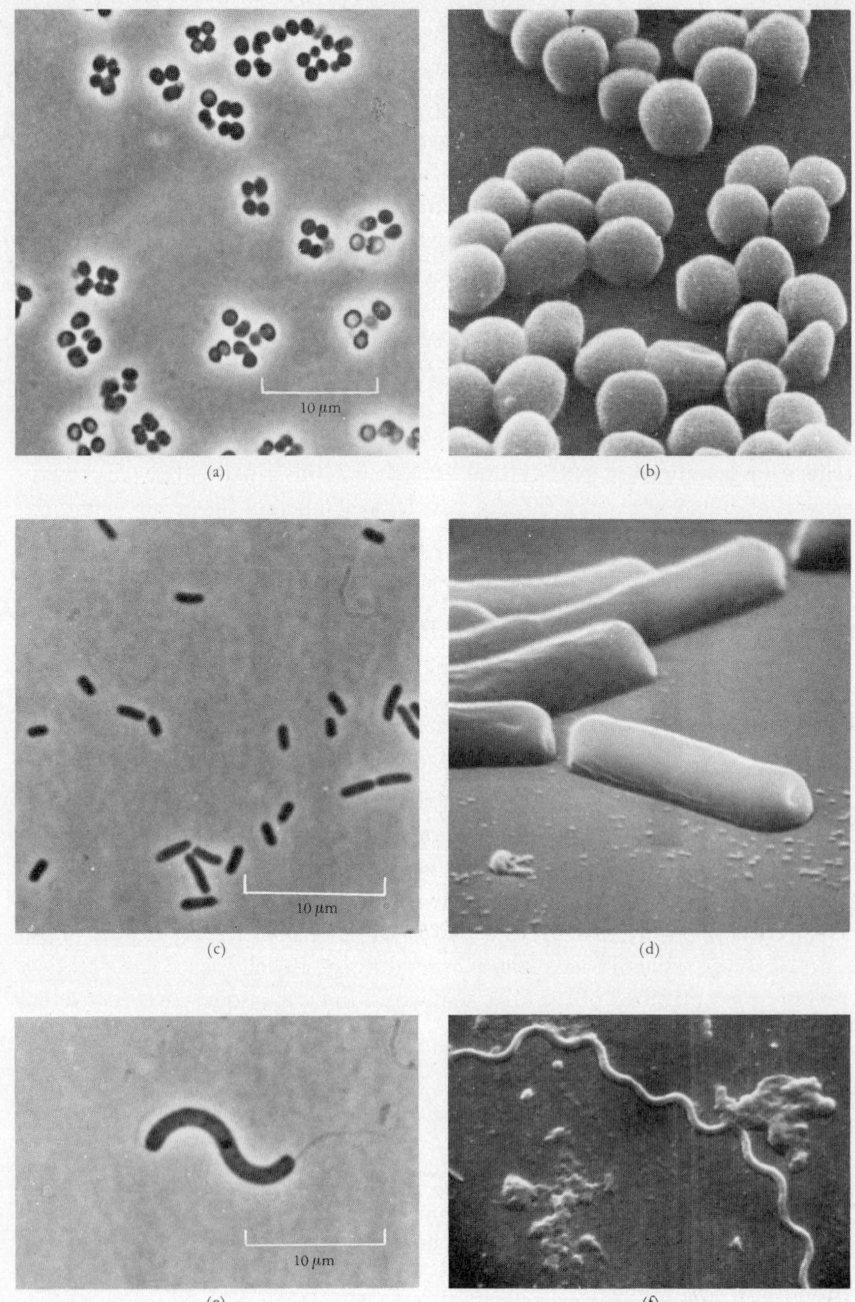

FIGURE 3-5 The three most common shapes of typical bacteria as viewed through a phase contrast microscope (left) and a scanning electron microscope (right). (Top) Spherical (coccus). (Center) Cylindrical or rod-shaped (bacilli). (Bottom) Spiral-shaped (spirilla). (a, c, and e courtesy of J. T. Staley and J. P. Dalmasso.) (b) (×30,000) (Courtesy of D. Greenwood; D. Greenwood and F. O'Grady, *Science* 163:1076, 1963. Copyright © 1963 by the American Association for the Advancement of Science.) (d) (×35,000) (Courtesy of N. Hodgkin.) (f) (Courtesy of A. Klainer; A. Klainer and I. Geis, *Agents of Bacterial Disease*. Hagerstown, Md.: Harper & Row, Publishers, 1973.)

FIGURE 3-6 Arrangements of cocci into characteristic patterns. (a) Chains (streptococci). (b) Clusters (staphylococci). (c) Cuboidal. (a, b, c courtesy of A. Klainer; A. Klainer and I. Geis, *Agents of Bacterial Disease*. Hagerstown, Md.: Harper & Row, 1973.)

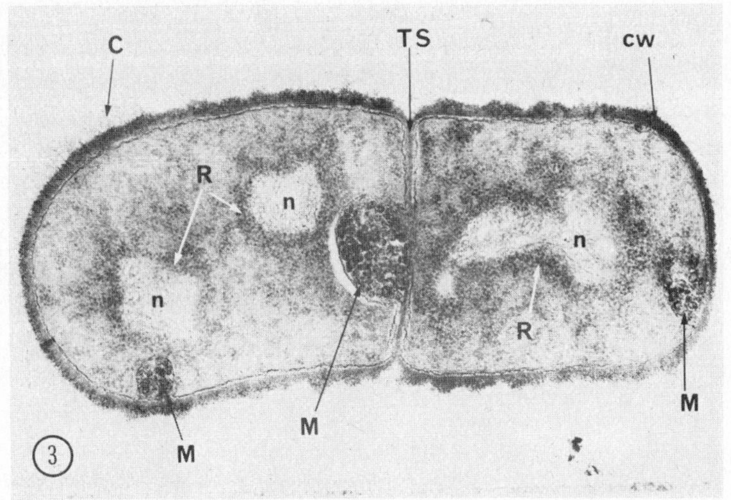

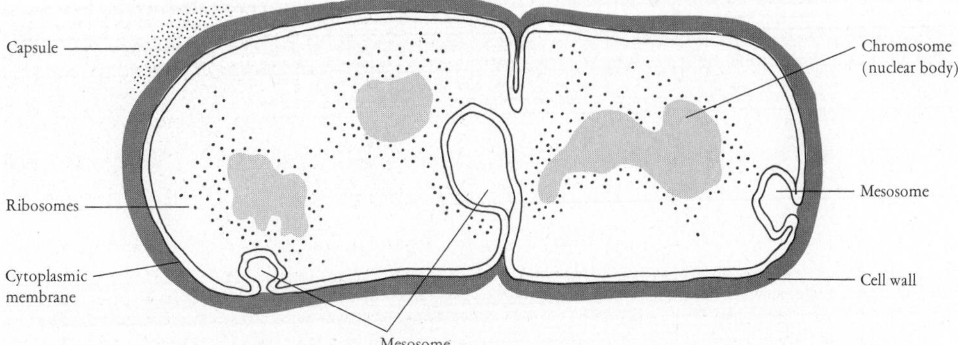

FIGURE 3-7 *Bacillus megaterium,* a representative procaryote, undergoing cell division. (Courtesy of D. G. Lundgren; D. J. Ellar, D. G. Lundgren, and R. Slepecky, *J. Bacteriol.* 94:1189, 1967.)

ENCLOSURE OF CYTOPLASM: THE CELL ENVELOPE

Most bacteria have two structures, the *cell wall* and the *cytoplasmic membrane,* that surround their cytoplasm, and some may have a third, the *capsule.* These layers are often referred to as the *cell envelope* (Figure 3-8). The discussion will proceed from the outermost (capsule) to the innermost layer (cytoplasmic membrane).

Capsule or Slime Layer

The capsule is a loose-fitting, gelatinous structure which surrounds some bacteria, and it is most readily demonstrated by negative staining. The capsular material is outlined as a light area against a darkened background (Figure 3-9a). Colonies of the bacteria that synthesize capsules are often relatively moist, glistening, and slimy in appearance when growing on solid medium (Figure 3-9b). Capsules are produced only by certain species of bacteria, and often only when these are grown under certain nutritional conditions.

Chemistry Capsules vary in their chemical composition. Some are composed of polysaccharide, while others consist of polypeptides of only one or two amino acids. These amino acids are generally of the unnatural, or D-configuration, as opposed to the natural or L-configuration always found in proteins.

Function The advantages a capsule confers are not evident for most bacteria, although it is definitely protective under certain situations for some species. Probably the best-studied capsule belongs to an organism causing bacterial pneumonia, *Streptococcus pneumoniae.* Some strains of this organism do not synthesize a capsule, and these strains are incapable of causing pneumonia. If it does not produce a capsule, the organism is quickly destroyed by the defenses of the infected animal. Apparently the phagocytic cells of the body, which engulf and destroy bacteria, have great difficulty engulfing the "smooth" encapsulated organisms, whereas the "rough" nonencapsulated organisms are quickly engulfed and destroyed. A bacterial capsule is also often essential for the development of dental caries. *Streptococcus mutans,* the major organism causing dental caries, must accumulate in large masses on the surfaces of teeth in order to cause dental decay. This attachment often requires a capsule composed of glucan and fructan, which is

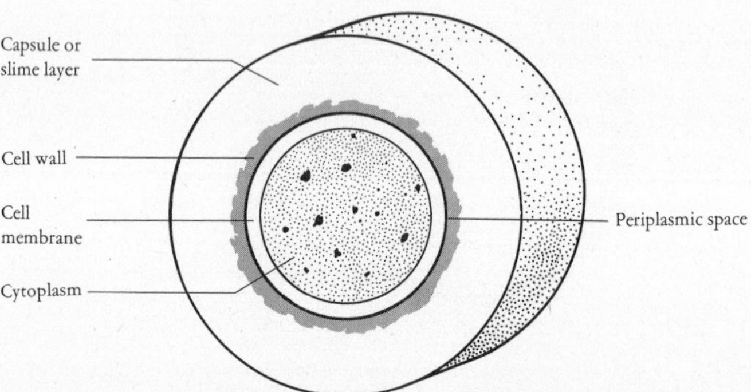

FIGURE 3-8
Diagrammatic representation of the surface layers surrounding the cytoplasm.

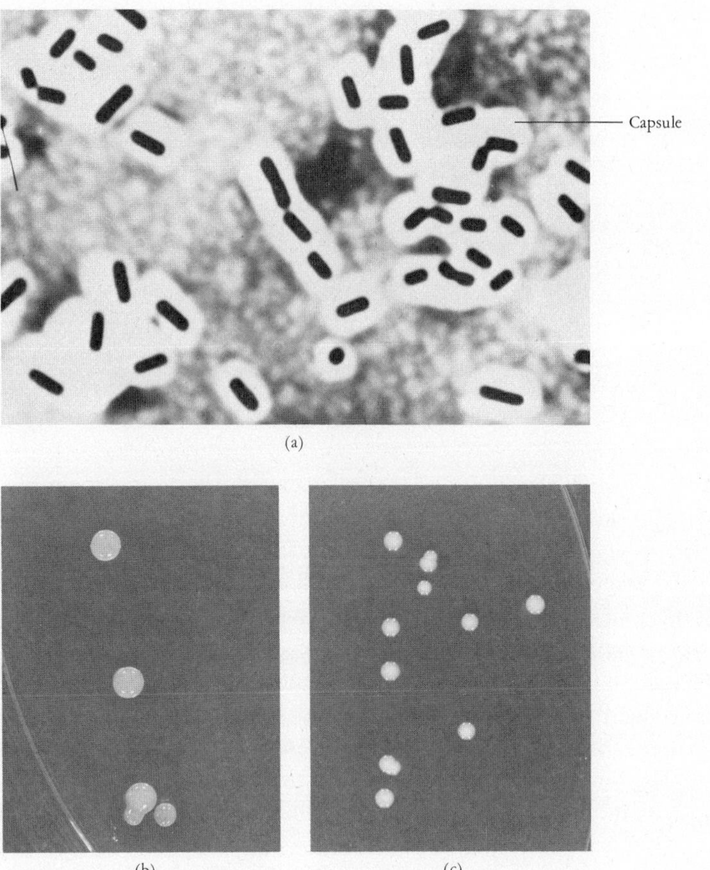

FIGURE 3-9
(a) Capsules as demonstrated by negative staining with India ink. The bacterial cytoplasm has been stained with a positive stain to increase the contrast. (Courtesy of V. Chambers.) (Below) Colonial morphology of *Streptococcus salivarius* growing on medium with sucrose (b) and without sucrose (c). This organism forms a capsule if sucrose is present in the medium and this capsule imparts a larger, gum drop appearance to the colonies. (b and c courtesy of J. P. Dalmasso.)

synthesized specifically from sucrose but not from other sugars. Although *S. mutans* can multiply in the absence of sucrose, this sugar, in the diet, allows the organism to synthesize a capsule, adhere to teeth, and cause decay.

Bacterial Cell Wall

The bacterial cell wall deserves special attention for several reasons: (1) it is composed of subunits found in no other place in nature, (2) the cell wall of certain organisms can produce symptoms of disease, (3) the site of action of some of the most effective antibiotic medications is the cell wall, and (4) differences in the chemical composition of the cell wall determine the Gram-staining characteristics of the cell.

The cell wall also determines the shape of the organism: cylindrical-shaped cells have a cylindrical cell wall; spherical organisms have a spherical cell wall. If the cell is broken apart, the cell wall does not collapse but maintains its rigid shape (Figure 3-10).

Chemistry The chemical structure of the bacterial cell wall is responsible for

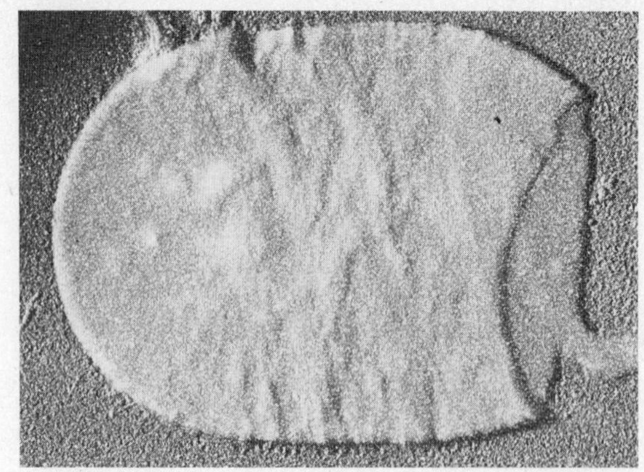

FIGURE 3-10
Rigidity of cell wall. (Courtesy of P. Gerhardt; J. H. Silvernale, H. L. Joswick, T. R. Corner, and P. Gerhardt, *J. Bacteriol.* 108:482, 1971.)

UNIQUE CHEMICAL STRUCTURE OF THE WALL

its rigid nature. The backbone of the cell wall is the macromolecule known as a *peptidoglycan,* consisting of two major subunits, amino sugars and amino acids.

The amino sugars, N-acetylmuramic acid and N-acetylglucosamine, are chemically related to glucose (Figure 3-11). These two subunits alternate to form a high molecular weight polymer. Although this polymer is found only in procaryotic cell walls, it bears a chemical relationship to the cellulose found in plant cell walls and to the chitin found in insect and crustacean exoskeletons and fungal cell walls. In a peptidoglycan, a chain of several amino acids is attached to each of the N-acetylmuramic acid molecules. Only a few of the twenty amino acids usually found in proteins occur in the peptidoglycan, and its exact composition varies in different bacterial species. One unusual amino acid, diaminopimelic acid, which is related to the amino acid lysine (Figure 3-12), is found only in certain bacterial cell walls. Many of the amino acids in the cell wall are of the D-configuration.

AMINO ACIDS CONNECT THE LINER POLYMERS TO ONE ANOTHER

Cell wall of Gram-positive bacteria: special features The cell wall of most Gram-positive bacteria consists of layer upon layer of peptidoglycan, with each layer connected to the one above and the one below through amino acid bridges. The resulting three-dimensional network, which is actually one large interconnected macromolecule, provides an unusually rigid, strong structure (Figure 3-13). In addition to its peptidoglycan portion, the cell wall of most Gram-positive

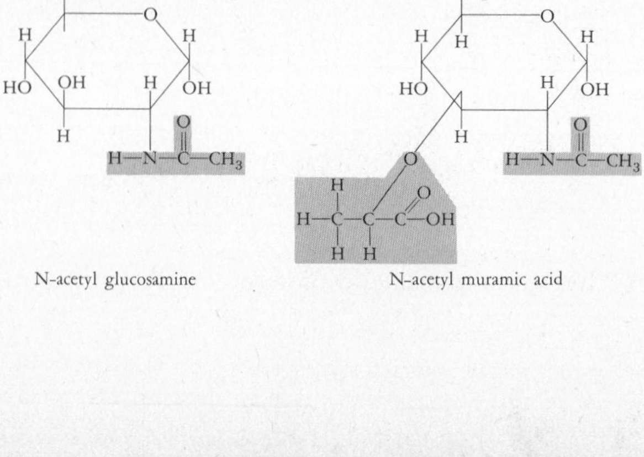

FIGURE 3-11
Chemical structure of N-acetyl glucosamine and N-acetyl muramic acid.

N-acetyl glucosamine

N-acetyl muramic acid

Diaminopimelic acid	L-lysine
O	
‖	H
C—OH	|
H—C—NH_2	H—C—NH_2
H—C—H	H—C—H
H—C—H	H—C—H
H—C—H	H—C—H
H—C—NH_2	H—C—NH_2
C—OH	C—OH
O⁄⁄	O⁄⁄

FIGURE 3-12
Chemical structure of an unusual amino acid, diaminopimelic acid, and its common relative L-lysine.

bacteria contains an additional component, the *teichoic acids.* These compounds have various structures, but a common teichoic acid has as its repeating subunit glycerol bonded to D-alanine and to phosphate (Figure 3-14). The teichoic acids are covalently bonded to the peptidoglycan layers.

Cell wall of Gram-negative organisms: additional features The wall of Gram-negative bacteria also contains a peptidoglycan layer, but it is much thinner than the multilayered structure found in Gram-positive bacteria. Consequently, the wall of Gram-negative cells is more easily broken by mechanical forces than is the wall of Gram-positive cells. Gram-negative organisms also have a layer of lipoprotein, lipopolysaccharide, and phospholipid which surrounds the peptidoglycan layer (Figure 3-15).

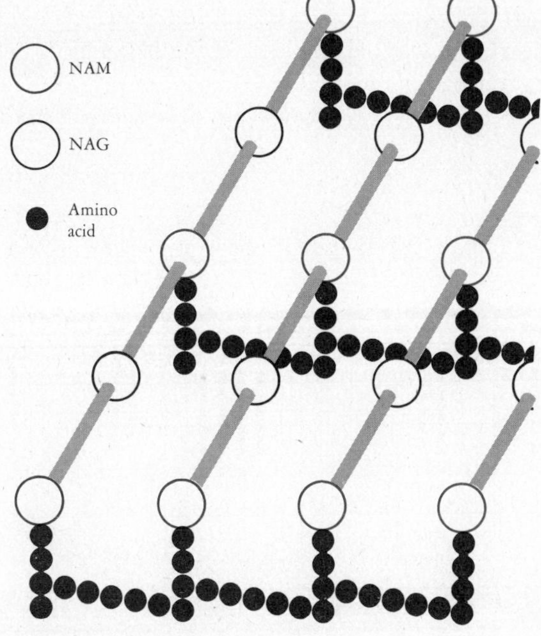

NAM = N-acetyl muramic acid
NAG = N-acetyl glucosamine

FIGURE 3-13
Representation of molecular structure of peptidoglycan portion of the cell wall of a Gram-positive organism.

```
                     H
                     |
                  H—C—H
                     |
D-alanine     H2N—C—H
                     |
                   O=C
                     |
        OH       H   O   H        OH
        |        |   |   |        |
  —O—P—O—C—C—C—O   —P—O—Peptidoglycan
        ||       |   |   |        ||
        O        H   H   H        O
     Phosphate      Glycerol

          REPEATING SUBUNIT
```

FIGURE 3-14
The repeating subunit of a common teichoic acid. Teichoic acids have different formulas; some contain 5- and 6-carbon alcohols instead of glycerol.

Lipopolysaccharide toxin, page 374

The lipopolysaccharide layer is of great interest because of its toxic properties. Many Gram-negative bacteria produce materials which are toxic for the host, thereby causing disease. The injection of the purified lipopolysaccharides isolated from many different Gram-negative bacteria causes symptoms characteristic of the infections caused by the whole organisms. The toxic component has now been shown to be a lipid fraction.

Function of cell wall The cell wall functions like a rigid girder framework on a dirigible to hold the cell together. Without the cell wall the cell would burst. Bacteria normally grow in dilute solutions, both in nature and in the laboratory. However, the bacterial cytoplasm is a very concentrated solution of inorganic salts, sugars, amino acids, and various other small molecules. There is a tendency for the *concentration* of particles (the number of molecules and ions per unit volume) to equalize on the inside and outside of the cell. Theoretically this equalization can occur by (1) the movement of particles out of the cell into the

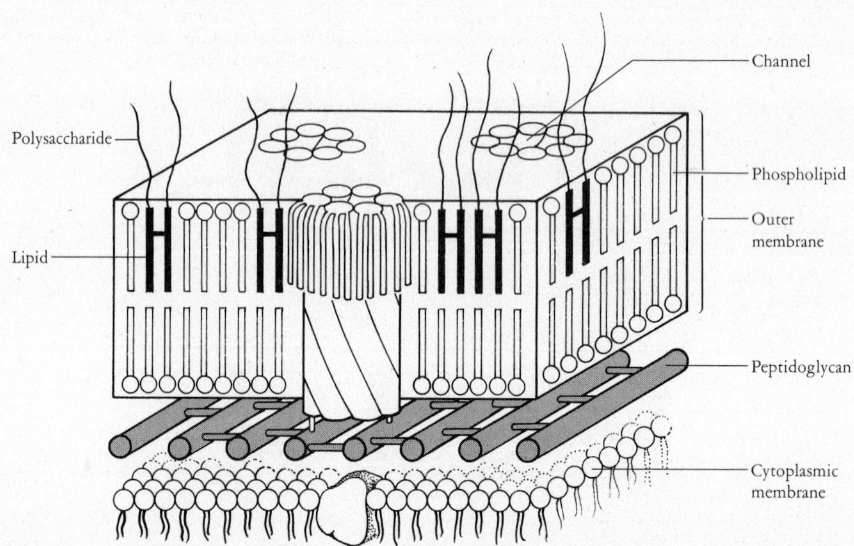

FIGURE 3-15
Representation of the cell wall of a Gram-negative organism.

medium or (2) the movement of water from the medium into the cell. The first mechanism cannot operate freely because the cytoplasmic membrane prevents the exit of many small molecular weight compounds. Water, however, can penetrate freely through the semipermeable membrane that allows some, but not other kinds of molecules, to pass. Therefore water flows from the medium into the cell in an attempt to equalize the concentration of particles on both sides of the membrane. This inflow of water exerts tremendous pressure, *osmotic pressure,* on all structures that enclose the cytoplasm, and as a general rule, unless the cell has a rigid cell wall which cannot expand, it simply balloons in size until it bursts. The osmotic pressure in some bacteria may be as much as 25 times the pressure of our atmosphere at sea level (1 atmosphere).

CELLS CAN EXIST WITHOUT A CELL WALL

The cell wall becomes less important if the bacterium is in an environment which has a high concentration of low molecular weight compounds. Not surprisingly, the only group of naturally occurring bacteria that have a cell wall lacking a rigid peptidoglycan layer is found in the Dead Sea (30% NaCl). In such an environment the tendency of water to enter the cell is matched by its tendency to leave the cell. The result is that there is no net movement of water into the cell, and under these conditions the cell can exist without a very rigid cell wall. If the salt concentration is lowered to 15%, however, the wall bursts.

Wall-deficient organisms It is possible to destroy the cell wall by treating the cell with *lysozyme,* an enzyme which cleaves the chemical bonds between the N-acetyl muramic acid and N-acetyl glucosamine subunits. Figure 3-16 illustrates the results of this treatment. In high salt or sugar solutions, the treated cell does not burst but rather becomes spherical, its most stable shape. The wall-less cell, a *protoplast,* can carry out metabolic processes.

Gram-positive cells generally lose *all* of their cell wall in lysozyme treatment. Gram-negative cells are also susceptible to lysozyme action, but only after a special treatment. Because Gram-negative cells contain layers other than the peptidoglycan, fragments of wall remain, although not enough to prevent the cell from becoming spherical. Such *wall-deficient* organisms are capable of limited multiplication; however, if the high salt solution around the cells is diluted, water enters the cell and it generally bursts.

Although the discussion thus far has emphasized the importance of the cell wall to bacteria in their natural environment, microorganisms which lack or contain very little cell wall material do occur in nature. The most notable are members of the genus *Mycoplasma.* These organisms tend to grow slowly and require a specially enriched medium, generally containing 20 percent serum, although many are capable of existing in dilute media. Some of the mycoplasma also have a requirement for *sterols* in the medium; these are then incorporated into the cytoplasmic membrane. It appears that sterols stabilize the cytoplasmic membrane by some unknown mechanism to prevent lysis. The lack of a rigid cell wall makes the mycoplasma extremely plastic in their shape (Figure 3-17).

WALL-DEFICIENT FORMS EXIST IN NATURE

Another group of wall-deficient organisms, L-*forms* (named after the Lister Institute in London where they were first described in 1935), superficially resemble the mycoplasma. These organisms are simply bacteria, both Gram positive

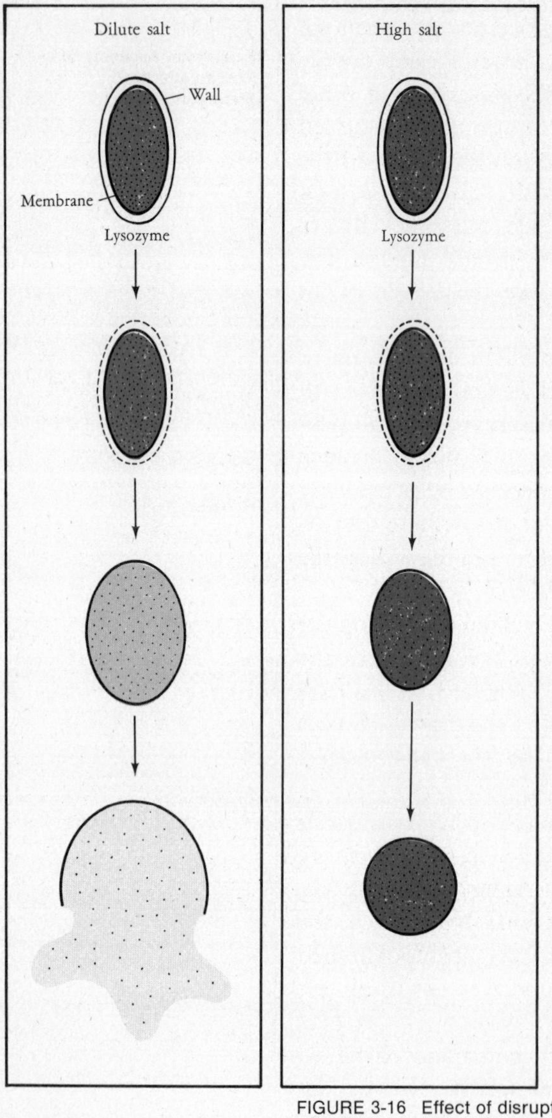

Effect on cell

Dilute salt medium	High salt medium
Lysozyme breaks bonds in muco-complex so that the wall develops gaps	Same as in dilute salt
Wall is lost completely. Water entering into the protoplast increases its size and it assumes a spherical shape	There is no net inflow of water but the protoplast begins to assume a spherical shape
Water continues to enter until the weak cytoplasmic membrane is no longer able to withstand the pressure; it breaks, releasing the cytoplasm into medium	The cell becomes spherical, the shape of all protoplasts

FIGURE 3-16 Effect of disruption of cell wall by lysozyme in a dilute salt and a high salt-containing medium.

and Gram negative, which have lost the ability to synthesize the peptidoglycan portion of their cell wall in either the normal quantity or quality. This loss can occur in the body of the host or with some forms spontaneously outside the host. In some of these organisms, the peptidoglycan layer is completely eliminated, in others a small amount is synthesized. Some L-forms derived from Gram-negative organisms cannot synthesize any peptidoglycan but can synthesize the lipopolysaccharide and lipoprotein layers. Others can synthesize the peptidoglycan layer, but it does not have the regular structure shown in Figure 3-13.

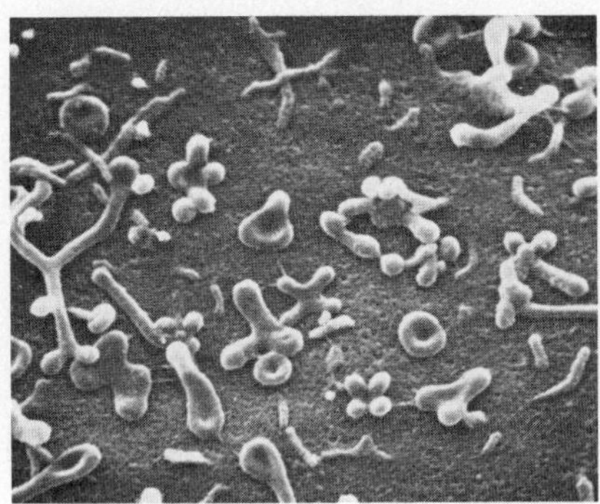

FIGURE 3-17
Mycoplasma pneumoniae, an organism that causes primary atypical pneumonia. (Courtesy of E. S. Boatman.)

PENICILLIN INHIBITS CELL WALL SYNTHESIS

Wall-deficient forms can also be artificially isolated by treating cells with penicillin or lysozyme and selecting the survivors. The cells capable of synthesizing normal amounts of the peptidoglycan layer are killed, but those cells which cannot are resistant to these agents which act on the peptidoglycan layer. If *all* of the peptidoglycan layer has been eliminated, then new layers of this wall component cannot be synthesized because new peptidoglycan layers must be synthesized on a preexisting layer. However, such peptidoglycan-deficient forms, which may still synthesize other wall components, can multiply in medium containing a high concentration of salt or sucrose. If the penicillin or lysozyme is removed, L-forms will either revert to the cell shape from which they were derived (revertable L-forms) or they will continue to grow as wall-less forms and never revert (stable L-forms). In order to multiply, some L-forms require a very concentrated medium to prevent the cell from bursting; others, however, are able to grow in dilute medium, perhaps because they can synthesize enough of their cell wall to retain their cytoplasmic contents. Although stable L-forms grossly resemble the mycoplasma, it now seems clear that these two groups are not closely related.

From the study of typical bacteria, L-forms, and mycoplasma it is clear that there is a gradation in the ability of organisms to survive in ordinary dilute nutrient medium without a normal cell wall. Most bacteria will lyse. Some (mycoplasma) are able to multiply slowly if the medium is fortified with additional materials such as blood serum and in some cases, sterols. Some L-forms are capable of multiplying in dilute medium. All wall-deficient organisms have several properties in common. They have no rigid cell shape, are relatively fragile, tend to multiply relatively slowly, and are resistant to those antibiotics which affect only cell wall synthesis.

Consequences of differences in cell wall composition Differences in cell wall composition between Gram-positive and Gram-negative bacteria probably account for their different staining characteristics. However, the staining dye is contained within the cell and not at the cell surface. The Gram-positive organism retains the purple dye after the cell is treated by iodine and washed with alcohol, whereas in Gram-negative cells the alcohol washes out the dye-iodine complex. Apparently the dye-iodine complex in the Gram-positive cell wall is relatively impermeable to alcohol, whereas in the Gram-negative cell wall it is not. This explanation may account for the observation that Gram-positive cells frequently become Gram-negative in old cultures, since it is known that enzymes which damage the cell wall are present in the cells in such old cultures. This damage alters the permeability of the cell wall to the dye-iodine complex, so that it is dissolved and then washed out by the alcohol. Furthermore, if the solvent (alcohol) is applied for too long a time in Gram staining, Gram-positive organisms will appear Gram-negative.

Penicillin kills bacteria by affecting the synthesis of the cell wall. Generally, but with notable exceptions, it is far more effective against Gram-positive than Gram-negative cells, apparently because penicillin has a more difficult time reaching its site of action in the latter. The layers of lipopolysaccharide and lipoprotein which surround the peptidoglycan layer in Gram-negative cells may prevent the penicillin from reaching its site of action in the peptidoglycan layer. The mode of action of penicillin will be covered more extensively in Chapter 33.

Cytoplasmic Membrane

Identification This membrane is the outer envelope in cells which lack a cell wall. In typical bacteria in dilute medium the osmotic pressure on the inside of the cell forces this thin, delicate, elastic membrane against the cell wall, although there is generally a space, the *periplasmic space,* between these two layers (Figure 3-8).

Chemical composition and structure The cytoplasmic membrane isolated from procaryotic cells contains approximately 60 percent protein and 40 percent lipid, much of which is in phospholipids.

The appearance of the cytoplasmic membrane, whether isolated from procaryotic or eucaryotic cells, is virtually identical when viewed in the electron microscope—two dark bands separated by a light band (Figure 3-7). This similarity has resulted in the term *unit membrane* being applied to these membranes, implying that all such membranes are quite similar in their molecular structure. Many of the properties of these membranes depend on the presence and structure of their phospholipid components, which have unique biological properties with respect to their solubility in water. They are polar at one end (which contains the phosphate group) and nonpolar throughout the rest of the molecule which consists largely of C—H atoms. Therefore, one end of the phospholipid molecule is insoluble in water, whereas the other, polar end is very soluble. This bimodal nature of many lipids results in their assuming a very ordered array when they are placed in water. All of the charged groups are on the outside of the phospholipids and the nonpolar groups are on the inside (Figure 3-18). Aqueous solutions are present on both sides of cell membranes, with polar groups oriented toward each side. However, the fact that membranes play so many different physiological roles suggests that in addition to this basic plan, differences must exist in membranes to allow them to perform different functions. These differences probably exist in the different protein molecules which are embedded in the phospholipid bilayer (Figure 3-18).

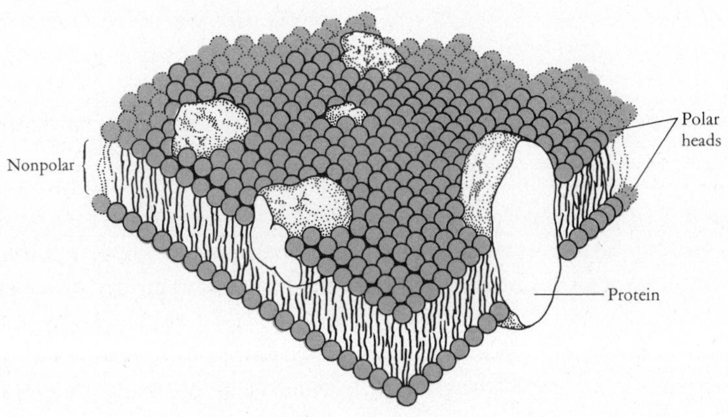

FIGURE 3-18
A representation of the lipid bilayer with proteins in the cytoplasmic membrane. (Modified from S. J. Singer and G. L. Nicolson, *Science* 175:720, 1972. Copyright © 1972 by the American Association for the Advancement of Science.)

Functions The cytoplasmic membrane of bacteria performs several functions vital to the life of the cell.

A large number of enzymes concerned with the degradation of foodstuffs and the production of energy are associated with the membrane. Infoldings of the membrane are especially noticeable in bacteria that are capable of carrying out photosynthesis or generating energy at a very high rate (Figure 3-19). These infoldings apparently provide the organism with a greatly enlarged surface area which allows for additional enzyme molecules of photosynthesis and energy production.

The membrane is semipermeable. Generally, only low molecular weight materials (molecular weight no greater than several hundred) can penetrate to the inside of the cell. All molecules which enter bacteria and are metabolized must be polar since they must be soluble in water. Water pulls apart these polar compounds because of their strong hydrogen bonding attraction to both water molecules and other polar or *hydrophilic* (water-loving) compounds, and in this way dissolves them. Nonpolar compounds, such as benzene, are water-insoluble because they cannot form hydrogen bonds with water.

Compounds enter the bacterial cytoplasm by one of two distinct proc-

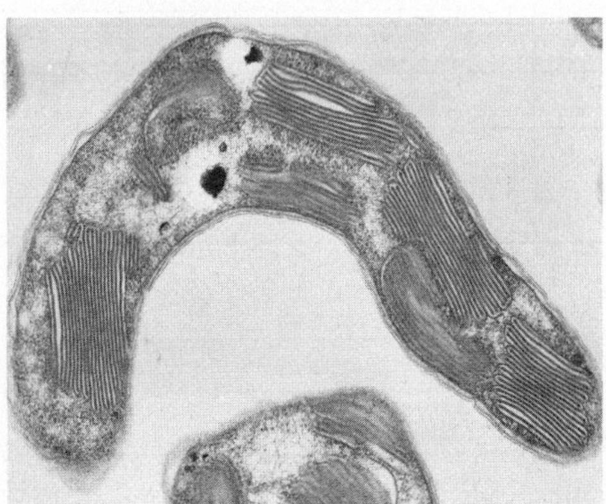

FIGURE 3-19
Extensive infoldings of the cytoplasmic membrane in the photosynthetic bacterium, *Ectothiorhodospira mobilis* (×86,200). (Courtesy of S. Watson; photograph taken by J. Waterbury.)

esses—*passive diffusion* or *active transport.* In passive diffusion the molecules flow freely into and out of the cell without the cell expending energy. Diffusion will occur until the concentration of a particular molecule is the same inside as outside the cell (note that the concentration of *each* specific molecule is the same, not the sum total of *all* molecules). In active transport the cell expends energy to transport molecules into and out of the cell. In general, the cell transports more molecules in than it transports out, with the net result being an accumulation of molecules inside the cell (Figure 3-20). In active transport the concentration of a particular nutrient may be more than a thousand times greater inside the cell than it is in the environment. The transport system, called a *permease,* is located in the membrane. A permease probably consists of a number of enzymes, associated with the protein portion of the membrane, which catalyze a sequential series of reactions, some of which require energy. Generally a separate permease exists for each nutrient. For example, one permease transports glucose and another lactose. In some cases the same permease may be involved in the transport of several compounds which have a similar chemical structure.

PERMEASES CONCENTRATE NUTRIENTS INTRACELLULARLY

The ability to concentrate nutrients allows the bacterial cell to maintain a relatively constant intracellular environment, even though the external environment changes drastically. Since bacteria live in environments in which many of the nutrients are present in extremely low concentrations, free diffusion would not be sufficient to provide the intracellular concentration of nutrients required for rapid growth. Thus bacteria have evolved a mechanism by which they can actively transport and accumulate the majority of nutrients.

CELLS CAN UTILIZE HIGH MOLECULAR WEIGHT COMPOUNDS

Bacteria also have the means for utilizing the high molecular weight compounds which are often present in the environment but which cannot penetrate the cell wall or cytoplasmic membrane. The subunits of these polymers are potential nutrients. Most bacteria contain enzymes which degrade the large molecules into their subunits, which can then be transported into the cells, generally through the action of specific permeases. Thus proteins are broken down to amino acids by proteolytic enzymes or proteases. Other enzymes break down polysaccharides into their sugar subunits. Many enzymes concerned with the break-

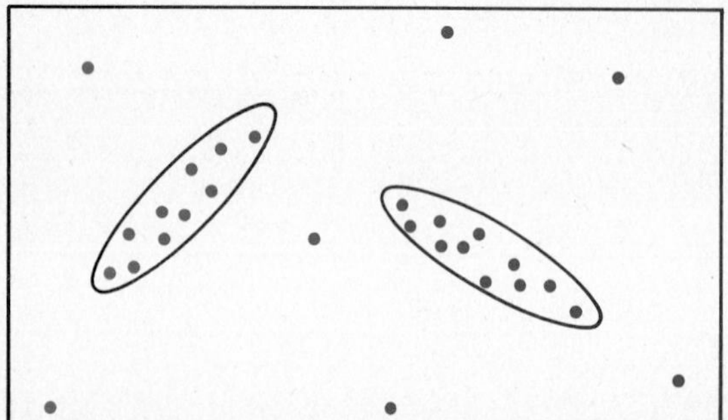

Active transport: because of the process, the internal milieu of a cell is generally much different than the external milieu of the cell's surroundings

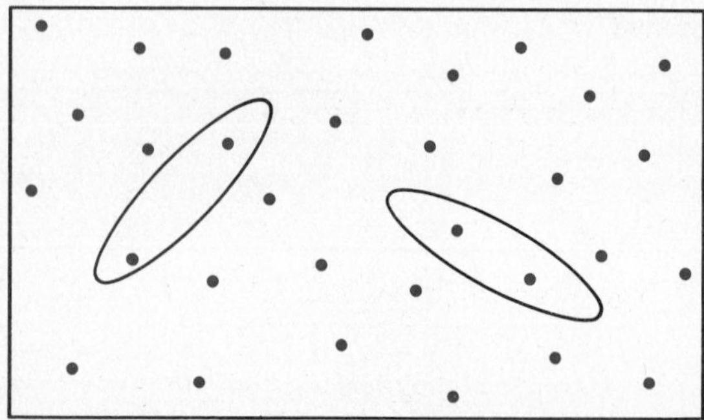

Passive diffusion: the internal concentration and external concentration are identical

FIGURE 3-20 Active transport and passive diffusion.

down of large molecules are excreted by the cell into the medium, where they act. These *extracellular enzymes* are found outside the cell in some bacteria and are located in the *periplasmic space* between the cell wall and the cytoplasmic membrane in other bacteria. Thus any large molecules which pass through the cell wall through the pores (Figure 3-15) will come in contact with these enzymes, be broken down into smaller molecules, and be transported through the cytoplasmic membrane via the permeases (Figure 3-21).

Mesosomes

MESOSOMES INCREASE SURFACE AREA

In Gram-positive bacteria the cytoplasmic membrane may form localized folds in the cytoplasm, called *mesosomes* (Figure 3-7). These membranous structures are not apparent in most Gram-negative bacteria for reasons that are not clear. Mesosomes greatly increase the surface area of the membrane and therefore may increase the ability of cells to concentrate nutrients. In addition, they seem to be located near the center of the cell, at the site at which the cross-wall forms when the cell divides. Since the DNA of the nuclear body appears to be attached to the mesosomes, the mesosomes may play a role in cross-wall formation and in the partitioning of the DNA of the parent cell into each of the two daughter cells.

STORAGE OF GENETIC INFORMATION

Chromosomes

The major structure in which the genetic information of procaryotic cells is stored is the *chromosome,* located in the *nucleoplasm* (Figure 3-7). In procaryotic cells, the chromosome is *not* surrounded by a nuclear membrane. Each chromosome is composed of a single long molecule of DNA, which when extended to its full length is about 1 millimeter long, approximately 1000 times longer than the entire bacterium (Figure 3-22). The acidic DNA is probably coated with basic protein inside the cell, but this protein is easily removed when the cell walls are broken. The long chromosome is tightly packed into a volume of about 10 percent of the volume of the cell and exists as a closed circle in all bacteria in which it has been carefully studied. Its intracellular structure probably resembles the structure shown in Figure 3-23.

Plasmids

Plasmids,
pages 209–211

Additional genetic information in many bacteria is carried in *plasmids,* a form of *extrachromosomal* DNA, which is not a part of the chromosome. These fragments of DNA, which are approximately 1 percent of the chromosome in size, often play an important role in the transfer of genetic material between bacteria. The genetic information that determines whether a bacterial cell is resistant or sensitive to certain antibiotics often resides in plasmids.

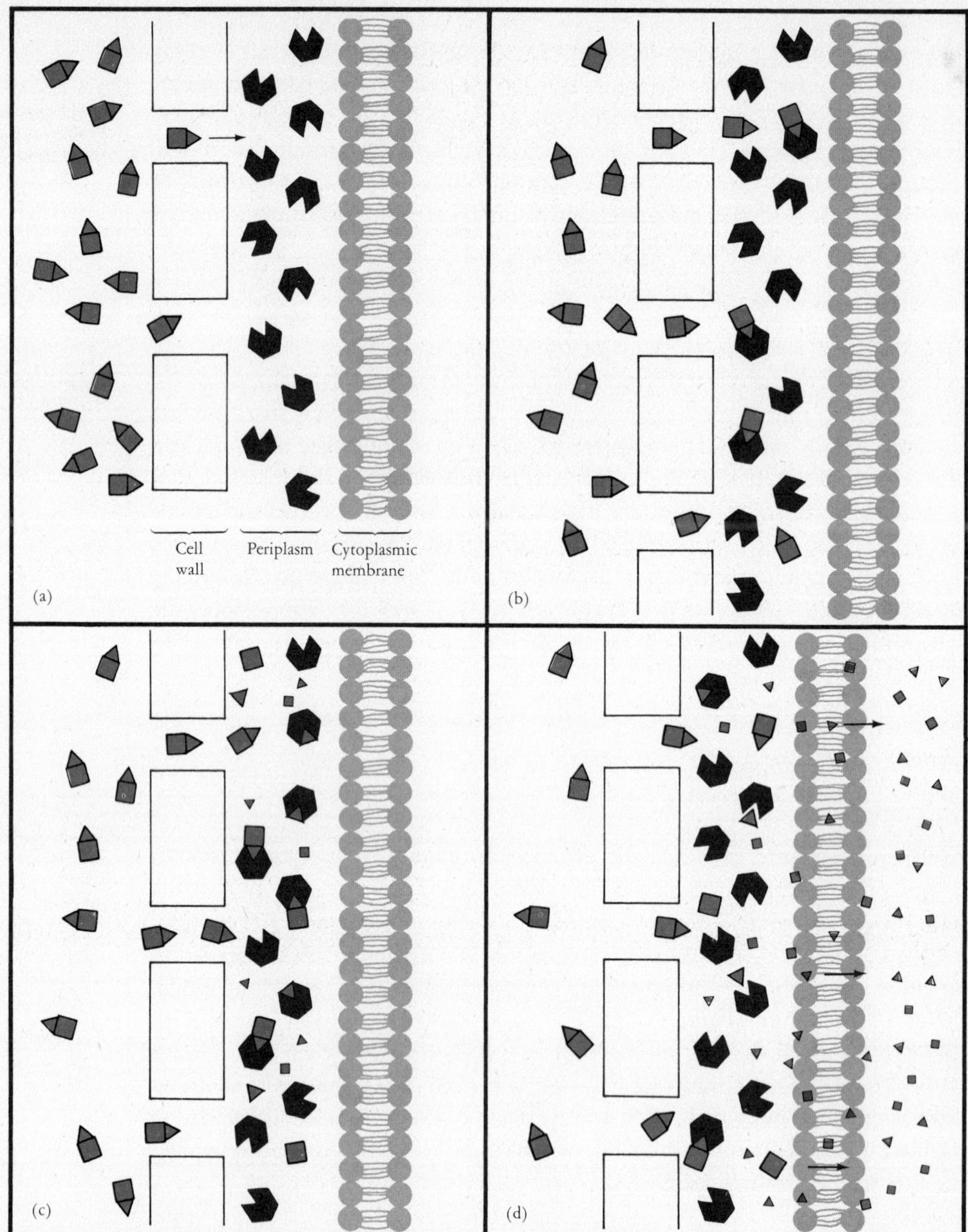

FIGURE 3-21 The role of enzymes in the periplasmic space in breaking down large, impermeable molecules into small, permeable subunits.

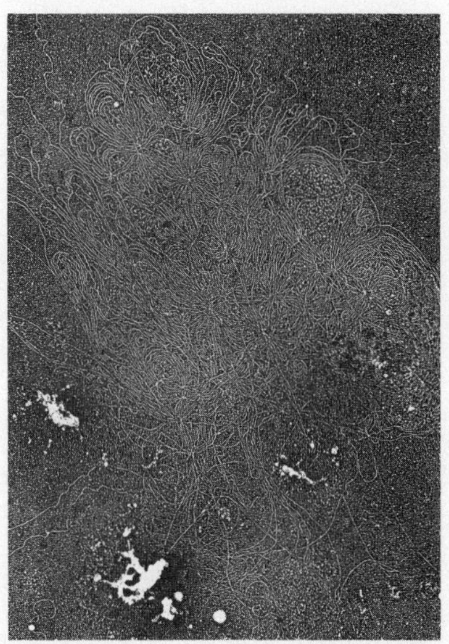

FIGURE 3-22
The single chromosome of *Micrococcus lysodeikticus* as it appears in the electron microscope after the cell has been gently lysed. Inside the cell DNA may not have any free ends. (×12,000) (Courtesy of A. K. Kleinschmidt.)

SYNTHESIS OF PROTEIN

Ribosomes

Protein constitutes over 50 percent of the dry weight of a typical bacterial cell, and almost 90 percent of all of the energy that the cell utilizes in synthesizing cell components is used for the synthesis of proteins. In the bacterial cell the most conspicuous structures devoted to the synthesis of proteins are the ribosomes (Figure 3-7). These structures are the *workbenches* on which amino acids are joined together to form proteins. Up to 15,000 of these small granules are found in the cytoplasm, the exact number depending on how rapidly the cell is synthesizing protein. The greater the rate of protein synthesis the greater the number of ribosomes. Each granule is composed of two parts, and each part in turn is composed of two different macromolecules, protein and RNA, often called *ribosomal protein* and *ribosomal RNA* to distinguish them from other types of protein and RNA which also play a role in the synthesis of proteins (Figure 3-24). Many kinds of ribosomal proteins form a part of the ribosome structure.

RIBOSOMES FUNCTION AS WORKBENCHES FOR PROTEIN SYNTHESIS

Ribosomes are characterized by their sedimentation properties when centri-

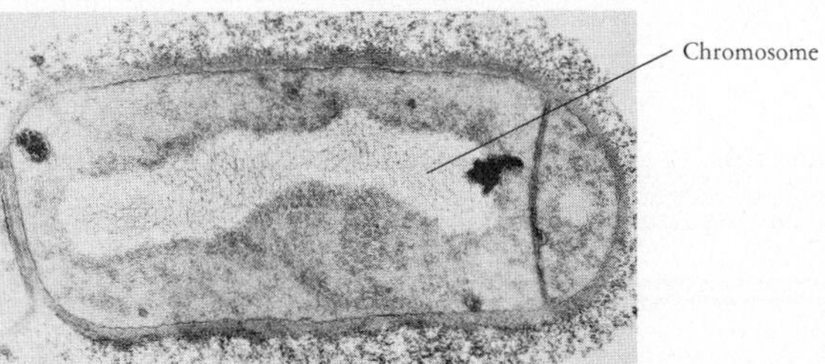

FIGURE 3-23
Chromosome of *Bacillus megaterium*. The granular appearance of the DNA is readily observable. (×40,000) (Courtesy of D. G. Lundgren; D. J. Ellar, D. G. Lundgren, and R. A. Slepecky, *J. Bacteriol.* 94:1189, 1967.)

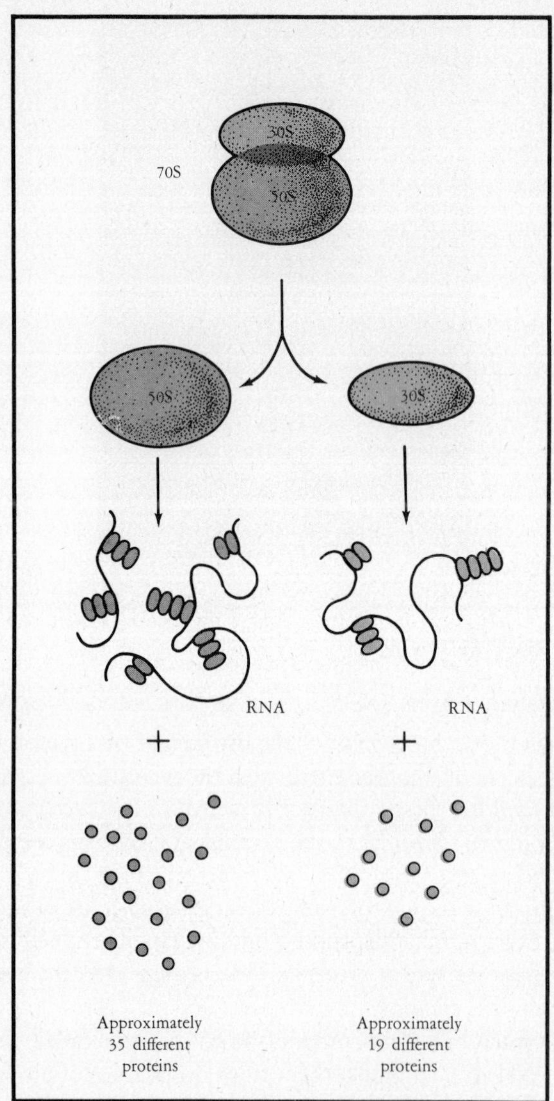

FIGURE 3-24
The procaryotic ribosome and its constituent parts. Under appropriate conditions of incubation, ribosomes dissociate into their component parts. First, the 70S ribosome dissociates into two sub-units, 30S and the larger 50S. These subunits in turn will break down into their component parts, ribosomal RNA and structural proteins. It is these structural proteins that bind antibiotics which interfere with protein synthesis. The ribosomes from eucaryotic cells are larger (80S) and contain different structural proteins. The dissociation of ribosomes into their two 50S and 30S subunits is important in the process of protein synthesis (Chapter 8).

fuged at very high speeds in an ultracentrifuge. The faster they sediment the greater their density. Bacterial ribosomes are termed 70S ribosomes, the S referring to a unit of sedimentation, which honors the Swedish physical chemist Svedberg, who played a dominant role in the development of the ultracentrifuge.

GENERATION OF ENERGY

Cytoplasmic Membrane

Most cells obtain their energy by degrading foodstuffs. In procaryotic cells a large number of the enzymes involved in the generation of energy are attached to the

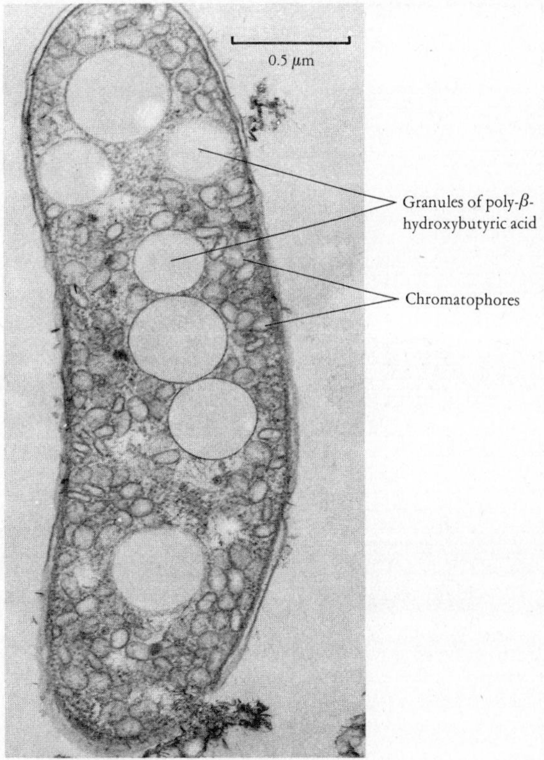

FIGURE 3-25
A thin section of a photosynthetic bacterium, *Rhodospirillum rubrum*. The cytoplasm is filled with chromatophores and inclusion granules. (Courtesy of E. S. Boatman.)

cytoplasmic membrane (membrane-bound enzymes). Most other enzymes are located in the cytoplasm (soluble enzymes).

The photosynthetic bacteria have pigments as well as enzymes which are concerned with converting the energy of light into chemical energy. These pigments and enzymes are usually localized on invaginations of the cytoplasmic membrane which extend into the cytoplasm. The name given to these membranous structures is *chromatophores* (Figure 3-25).

CELL MOVEMENT

Flagella

FLAGELLA PROPEL BACTERIA

Structure The flagella (singular, flagellum) are long protein filaments, helical in microscopic appearance, which are responsible for the motility of most bacteria. The bacterial flagellum is composed of three parts: the *filament,* the *hook,* and the *basal body* (Figure 3-26). The filament is composed of several chains of protein that are twisted together into a helix with a hollow core. The hook is attached to the near end of the filament, and the basal body is attached to the hook, anchoring the flagellum to the cell wall and cytoplasmic membrane.

Bacteria can have different arrangements of flagella (Figure 3-27). Some

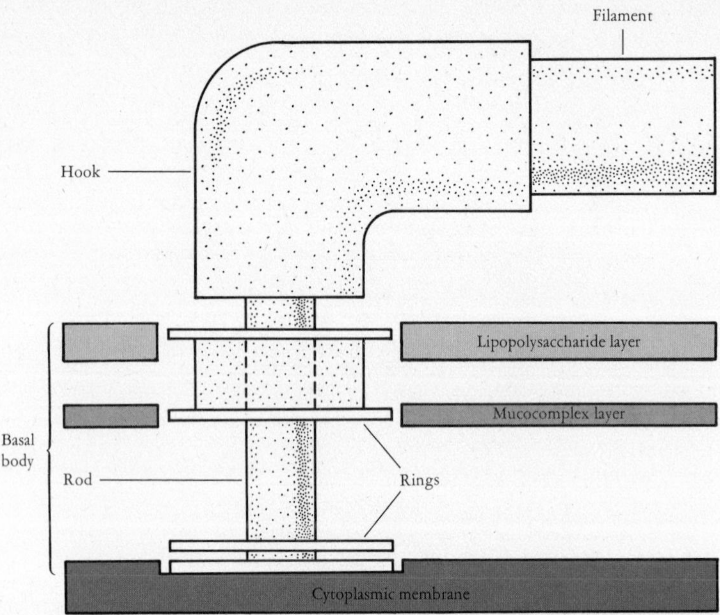

FIGURE 3-26 Model of a bacterial flagellum of *Escherichia coli*. The integration of the basal body of the flagellum into the cell wall and cytoplasmic membrane is shown. This basal body is composed of a number of rings which surround a rod. The filament extends from the cell. (Modified from M. L. DePamphilis and J. Adler, *J. Bacteriol.* 105:395, 1971.)

have a single flagellum at one end of the cell; some have a tuft. Other bacteria have a flagellum at either end, and still others have their flagella inserted at one or many points along the side of the cell. These arrangements help characterize the species and are often useful in identifying an organism.

Mechanisms of locomotion That flagella are the organelles of locomotion is clear from the fact that when flagella are sheared off by treating the cells gently in a blender, the cells remain alive but are no longer motile. If these nonmotile cells are incubated in a medium in which they can synthesize their flagella, the cells

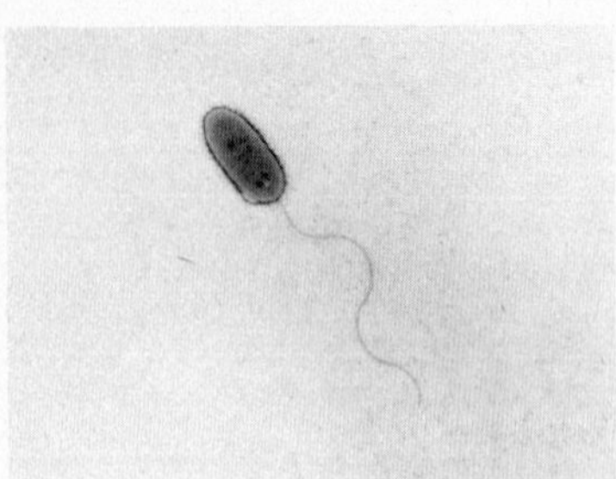
(a)

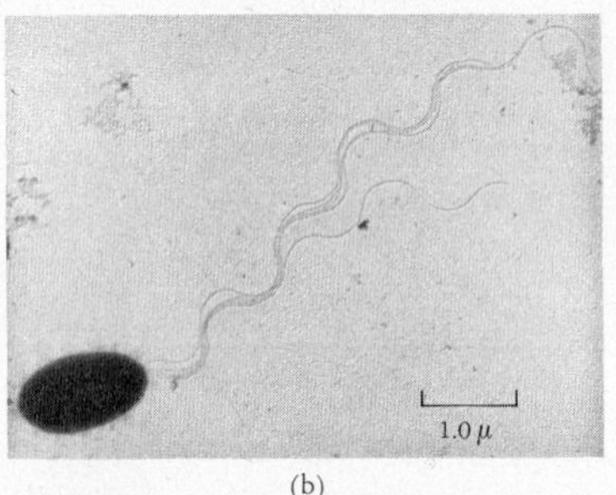

(b)

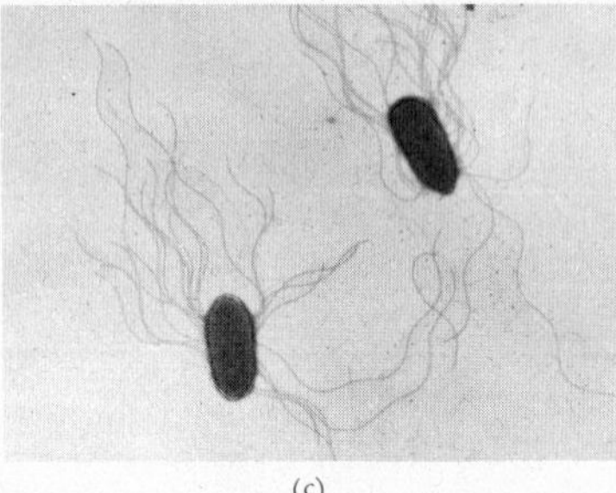
(c)

FIGURE 3-27 Arrangements of flagella. (Left) Single polar flagellum, *Pseudomonas aeruginosa* (polar flagellation). (×5250) (Courtesy of V. Chambers.) (Center) Tuft at one end (polar flagellation—lophotrichous). (Courtesy of E. S. Boatman.) (Right) Flagella inserted throughout *Proteus mirabilis* (peritrichous flagellation). (×3500) (Courtesy of V. Chambers.)

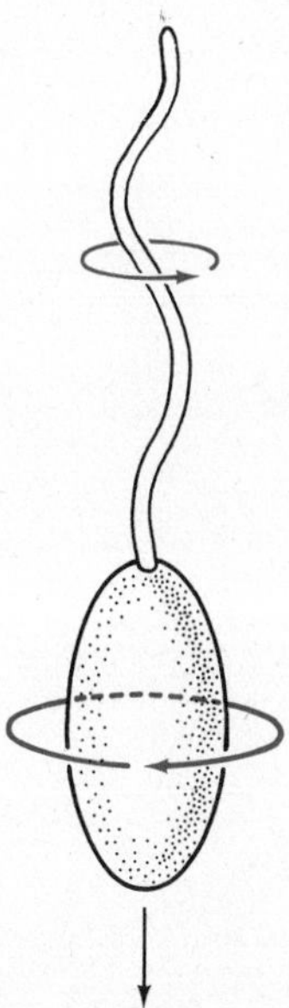

FIGURE 3-28
Flagellum propelling a cell. The flagellum and cell rotate in opposite directions.

regain their motility simultaneously with regaining their flagella. The flagella propel the cell through a liquid by pushing the cell, much like a propeller pushes a ship through water. Thus the flagella are rotors that spin around their long axis (Figure 3-28). The nature of the motor that drives flagella is not clear. But what work the flagella must do! To a bacterium, water has the viscosity of molasses.

Cell movement requires coordination in the motion of all flagella of a cell. If each flagellum were to push the cell in a different direction, the cell would not move in any single direction but would rather thrash about. In a series of elegant experiments it has been shown that in cells with multiple flagella, all the flagella can form a single bundle and thereby act as a coordinated unit. However, this coordination occurs only if the flagella are rotating in a counterclockwise direction. When the flagella rotate in a clockwise direction, the bundle flies apart, and all coordination between individual flagella is lost; the cell no longer moves in any direction, but merely thrashes about or tumbles (Figure 3–29).

Chemotaxis In a homogeneous solution bacteria move in one direction for a second or so, tumble for a fraction of a second, and then move again in a different direction. Movement is random. If, however, nutrients are available in the environment, or repellants are in the vicinity, then bacteria move in the direction of the nutrients (positive chemotaxis) or away from the harmful chemicals (negative chemotaxis). This movement is not random. The cells are able to detect the presence of nutrients, and this information is transmitted to the flagella so that they rotate in a counterclockwise direction and push the cell. Once the cell no longer detects the attractant, the flagellar bundle flies apart, and the cell tumbles. Thus the movement of the cell is determined by the frequency of these tumbles. As long as the cell detects the positive attractant, tumbling will be less likely, and the cell will move slowly but surely in the

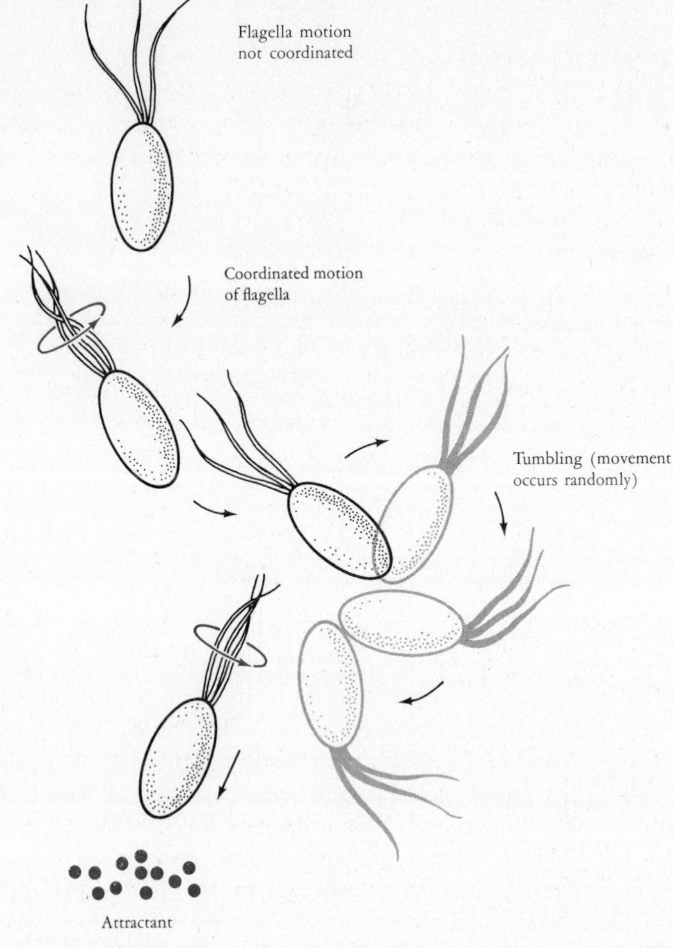

FIGURE 3-29
The cell moves in the direction of the attractant by lengthening the time that the flagellar bundle remains intact.

direction of the attractant (Figure 3–29). The opposite situation holds for repellants. Repellants increase the frequency of tumbling, and so the organism tends to move away from the repellant.

What is the mechanism by which bacteria sense the presence of chemicals? The entire story has not yet been unraveled, but it is clear that each cell has receptors which bind the chemical attractant or repellant. These receptors are often the same proteins in the cytoplasmic membrane that are associated with the activity of permeases. How this information is passed on to the flagella remains unanswered. However, from the data that have been collected thus far, it appears that the basic components required for behavior in higher organisms, namely, sensory receptors, as well as a system for processing and transmitting sensory information to the structures which produce movement, are all present in the bacterial cell. For these reasons the study of bacterial behavior represents one of the most exciting research areas in biology today—neurobiology at the molecular level.

In addition to responding to chemicals, some bacteria can respond to a gradient in light (phototaxis) and others may respond to the concentration of oxygen (aerotaxis). Organisms that require oxygen for growth will move in the direction of oxygen, and those bacteria that grow only in the absence of oxygen tend to be repelled by oxygen. Recently, certain motile bacteria have been shown to respond to the earth's negative field and move in a northerly direction (magnetotaxis).

Flagella synthesis How do flagella increase in length? Apparently the protein subunits that comprise flagella are synthesized at the base of the flagellum and move to the tip via the hollow core. Flagella in a single species have a characteristic length, and once this length is reached, the synthesis of the subunits ceases.

ORGANELLES OF ATTACHMENT

Pili

Many Gram-negative bacteria (but apparently only one Gram-postive species so far observed) possess hundreds of hairlike appendages called *pili* which are considerably shorter and thinner than flagella (Figure 3-30). Like flagella, pili consist of protein subunits wound around one another, generating a hollow core. They can also be removed without any loss of viability if the cells are agitated in a blender.

Function Pili can be divided into a number of types based on their function. One group, called the *F,* or *sex,* pili, are concerned with attaching two cells together prior to the transfer of DNA from one cell to the other. Other pili may serve to keep bacteria near the surface of liquid or where oxygen is most available. The ability of strains of *Neisseria gonorrhoeae* to cause gonorrhea correlates with piliation of the strain. The cells with pili attach more readily to many types of eucaryotic cells, and this attachment appears to be necessary for disease because nonpiliated strains fail to cause gonorrhea.

DNA transfer, pages 203–209

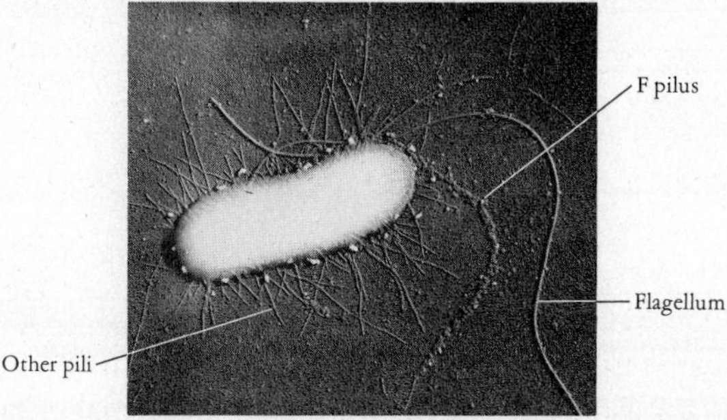

FIGURE 3-30
Pili of *Escherichia coli.* The variety of pili can be judged from the differences in their lengths. The F pilus can be readily identified because certain bacterial viruses (Figure 1-10b) specifically adsorb to it. Note that the flagellum is considerably longer than any of the pili. (×11,980) (Courtesy of C. Brinton, Jr.)

STORAGE MATERIALS

β-hydroxybutyric acid, see Figure 3-25, page 73.

Many bacteria form and store granules in their cytoplasm in the form of high molecular weight polymers. These granules, which serve as reserve food supplies, can be seen with special stains and are generally large enough to be readily detected by light microscopy. One common granular inclusion is *glycogen,* which acts as a storage form of both carbon and energy. Another storage form of both carbon and energy is the polymer of *β-hydroxybutyric acid.* Phosphate can be stored in the form of granular inclusions of polymeric phosphate, called *volutin.* These stain red with certain blue dyes, so they are also called *metachromatic granules.* The cell avoids the problem of increasing the osmotic pressure inside the cell by converting numerous small molecules into a few large molecules.

The synthesis of inclusion bodies depends on the environment and the organism. If the environment contains an excess of a nutrient, the cell will often convert the nutrient into a macromolecular form and store it until it is needed. As a general rule, only one kind of reserve material is stored by a particular species.

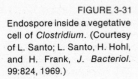

ENDOSPORES

A very distinctive structure, the *endospore* (sometimes simply called a "spore"), is often seen inside certain species of Gram-positive bacilli (Figure 3-31) or free in

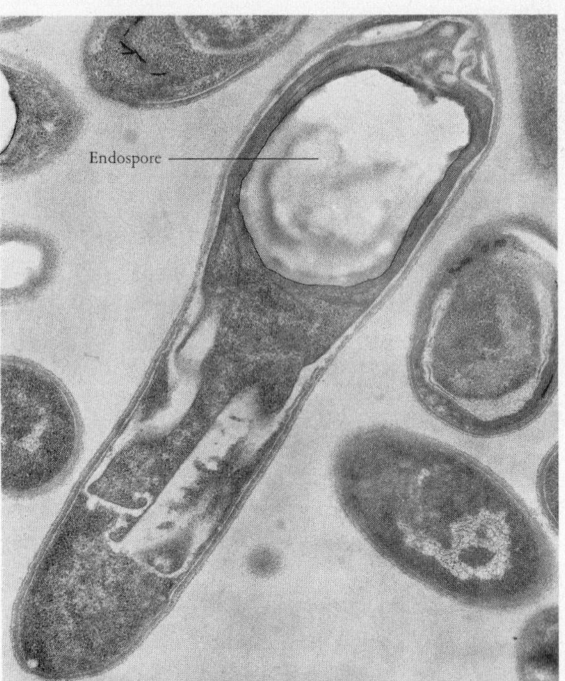

FIGURE 3-31
Endospore inside a vegetative cell of *Clostridium*. (Courtesy of L. Santo; L. Santo, H. Hohl, and H. Frank, *J. Bacteriol.* 99:824, 1969.)

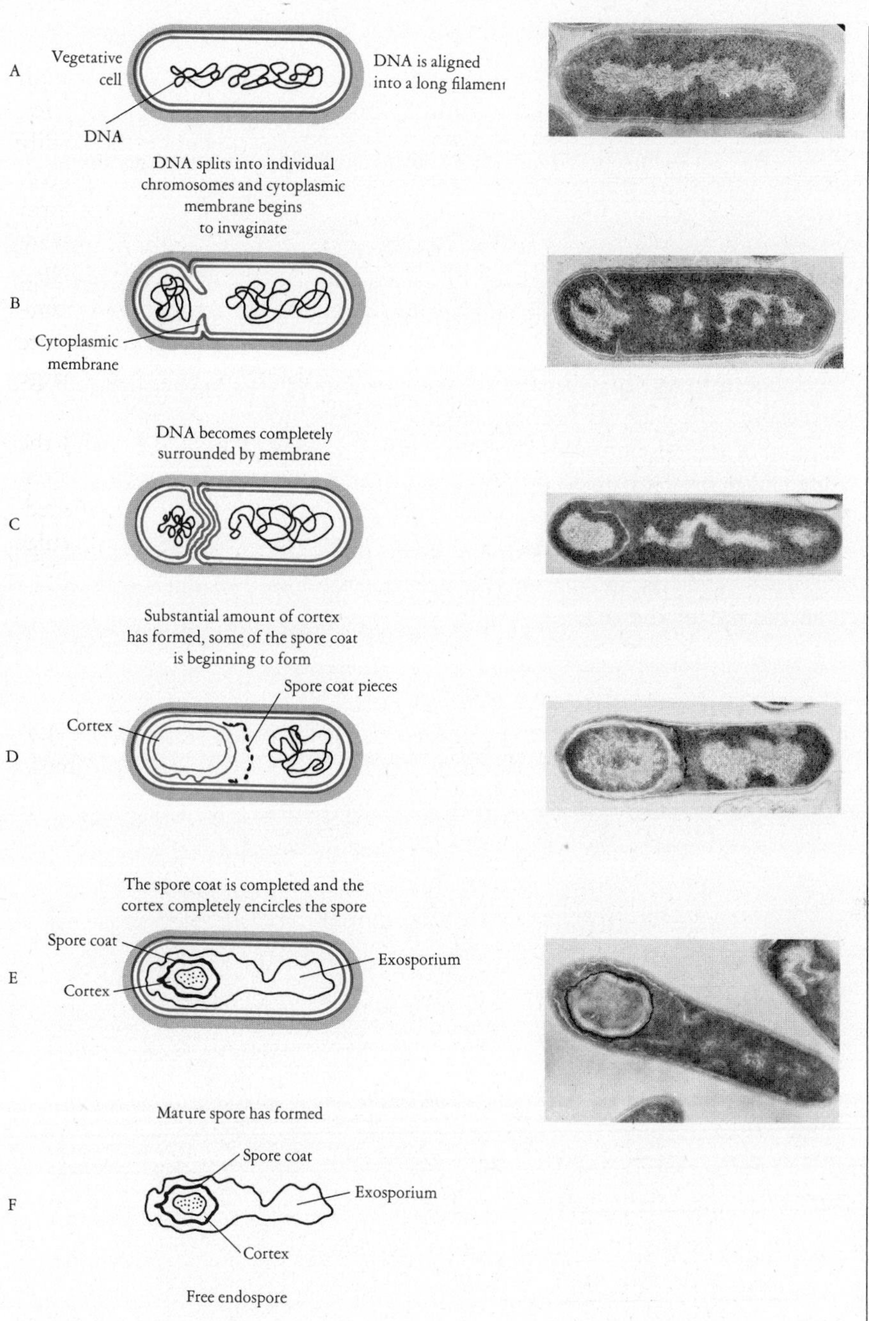
A
Vegetative cell
DNA
DNA is aligned into a long filament
DNA splits into individual chromosomes and cytoplasmic membrane begins to invaginate
B
Cytoplasmic membrane
DNA becomes completely surrounded by membrane
C
Substantial amount of cortex has formed, some of the spore coat is beginning to form
Spore coat pieces
Cortex
D
The spore coat is completed and the cortex completely encircles the spore
Spore coat
E
Cortex
Exosporium
Mature spore has formed
Spore coat
F
Exosporium
Cortex
Free endospore

FIGURE 3-32
Major steps in sporulation as observed through the electron microscope. The time from the first to the last picture was 16 hr. (Photos courtesy of L. Santo; L. Santo, H. Hohl, and H. Frank, *J. Bacteriol.* 99:824, 1969.)

the medium. This structure is a *unique cell type* that is formed within the vegetative cell when this cell is growing under suboptimal nutritional conditions. The process of sporulation involves a complex, highly ordered sequence of morphological changes called *sporogenesis* (Figure 3-32). This process represents an example in procaryotes of cellular differentiation which involves very

Regulation of sporulation, pages 232–233

sophisticated regulation of cellular activities. Some aspects of this regulation will be considered in Chapter 10.

Endospores can remain as endospores for long periods of time (reportedly up to 150,000 years), but under certain conditions an endospore can go through a sequence of events in which it is transformed back into a vegetative cell:

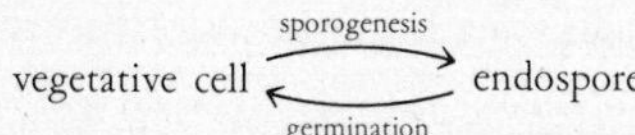

This is the process of *germination.* Since one vegetative cell gives rise to one endospore, germination is obviously not a means of reproduction.

ENDOSPORE A RESISTANT, NONMETABOLIZING CELL

The most distinctive feature of the endospore is that it has virtually no metabolic activity. It is usually resistant to a variety of agents that kill the vegetative cell, such as heating, drying, freezing, chemicals, and radiation. Because of its resistance, and the fact that some bacteria produce deadly toxins form endospores, it is especially important that foods which might contain these spores be processed adequately.

Endospores are readily detected under the light microscope because they are highly refractive and do not take up stains through their thick spore coat as does the rest of the cell.

STRUCTURES OF EUCARYOTIC CELLS

An electron micrograph and diagram of a eucaryotic cell is shown in Figure 3-33. It should be apparent from a comparison of this figure with Figure 3-7

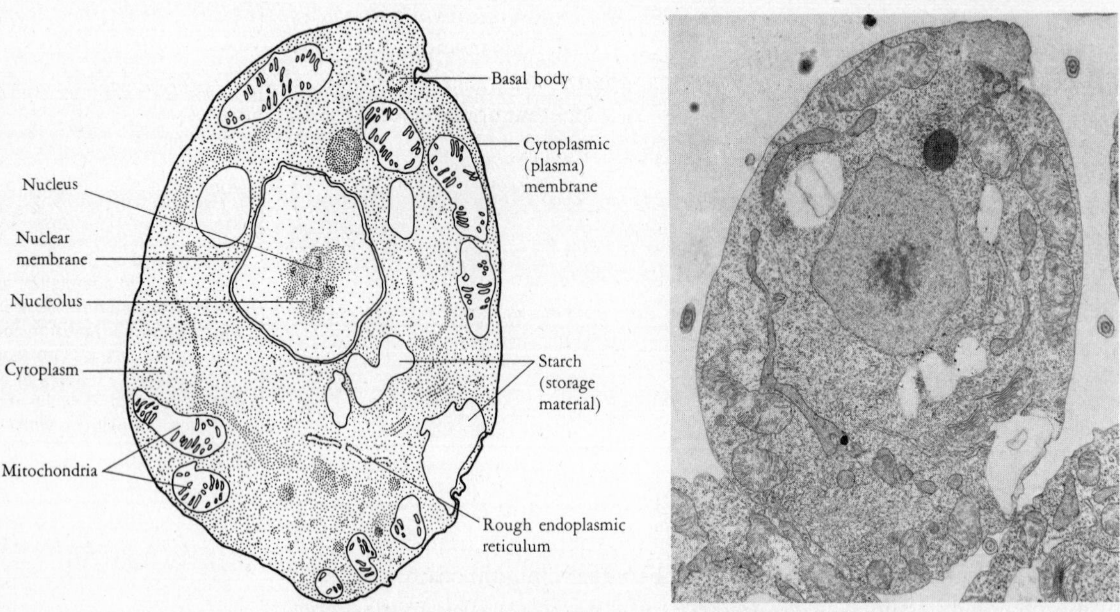

FIGURE 3-33 A eucaryotic cell, the protozoan *Polytomella agilis*. The outer structure is the cytoplasmic membrane because it does not have a rigid cell wall. (Photo courtesy of J. Moore. Reprinted with permission of the Society of Protozoologists from *J. Protozool.* 17:672, 1970.)

that the eucaryotic cell is structurally far more complex than the procaryotic cell. Another major difference between the two cell types is that the cytoplasmic organelles in eucaryotic cells consist of, or are bounded by, unit membranes. In procaryotic cells internal membranes are not nearly as common, and the organelles are not bounded by unit membranes. However, because both cell types must perform essentially the same functions in order to live, the simpler structures of the procaryotic cell are apparently fully capable of performing many of the same functions that are carried out by complex structures in the eucaryotic cell.

Cell Wall

The chemical structure of the cell walls of eucaryotes varies, but in general they are far simpler than bacterial cell walls. Most algae and fungi have rigid cell walls which contain cellulose. Other fungi have walls similarly composed of repeating subunits of glucose, but since these subunits are bonded together differently from their bonding in cellulose, the resulting polymer is given another name, *glucan.* Fungal walls also often contain chitin, a polymer consisting of repeating units of N-acetyl glucosamine. In contrast, protozoa do not have a cell wall even though they do have a definite, specific shape. What confers this structural rigidity is not clear. The cells of many higher animals, such as mammals, also lack a rigid cell wall.

N-acetyl glucosamine, page 60

Cytoplasmic Membrane

The cytoplasmic membrane, commonly termed the plasma membrane, is the outer covering in eucaryotic cells that lack a cell wall. This membrane must be much stronger than its counterpart in procaryotic cells. The membrane is very extensively arrayed and forms a series of internal membranes termed the *endoplasmic reticulum.* Ribosomes are attached to the endoplasmic reticulum to form the *rough endoplasmic reticulum.* Other internal membranes enclose a wide variety of organelles which are found in eucaryotic but not procaryotic cells.

The chemical composition and structure of the eucaryotic cytoplasmic membrane is grossly similar to its procaryotic counterpart, namely, a bilayer of phospholipids, with different proteins inserted into the bilayer. Carbohydrates may be attached to the protein layer. Sterols, which are present in the mycoplasma that lack a cell wall, are also commonly present in eucaryotic membranes, and they probably contribute to the ability of these membranes to resist osmotic lysis.

Nucleus: Storage of Genetic Information

The eucaryotic nucleus is made up of a number of chromosomes and is bounded by a nuclear membrane, which is actually a part of the cytoplasmic membrane (Figure 3-33).

There appear to be pores in the nuclear membrane so that areas of the nucleus are actually in contact with the cytoplasm. The chromosomes consist of DNA to which several kinds of basic proteins, the *histones,* are tightly bound. This nucleoprotein structure is called *chromatin* (Figure 3–34). Acidic proteins are also associated with the basic proteins, and both kinds are involved in the functioning of DNA as the carrier of genetic information. The nucleus also contains the *nucleolus,* which plays an important role in the synthesis of the RNA that comprises ribosomes (Figure 3–33).

The quantity of DNA is much greater in the eucaryotic than in the typical bacterial cell, ranging from approximately three times greater in a yeast cell to 700 times greater in a human cell. For both the eucaryotic and procaryotic members of the microbial world, there is a direct correlation between the haploid amount of DNA per cell and the complexity of the cell.

Mitochondria and Chloroplasts: Generation of Energy

Mitochondria are highly complex structures about the size of a bacterial cell which are enclosed by both an outer and inner membrane (Figure 3–35). The enzymes concerned with the generation of energy are localized on the inner membrane.

CHLOROPLASTS AND MITOCHONDRIA MAY BE REMNANTS OF BACTERIA

Mitochondria also contain DNA, ribosomes, and the other materials necessary for protein synthesis. They do not contain enough DNA to code for all of the information necessary for their own synthesis, so some of this information must be coded by nuclear genes. However, mitochondria do elongate and divide in a fashion similar to bacteria. This observation, coupled with the observation that the ribosomes in mitochondria are 70S (like those of bacteria rather than the 80S ribosomes of eucaryotes), has given rise to the controversial theory that mitochondria have evolved from a procaryotic cell. By this theory the procaryotic cell presumably entered into a symbiotic relationship

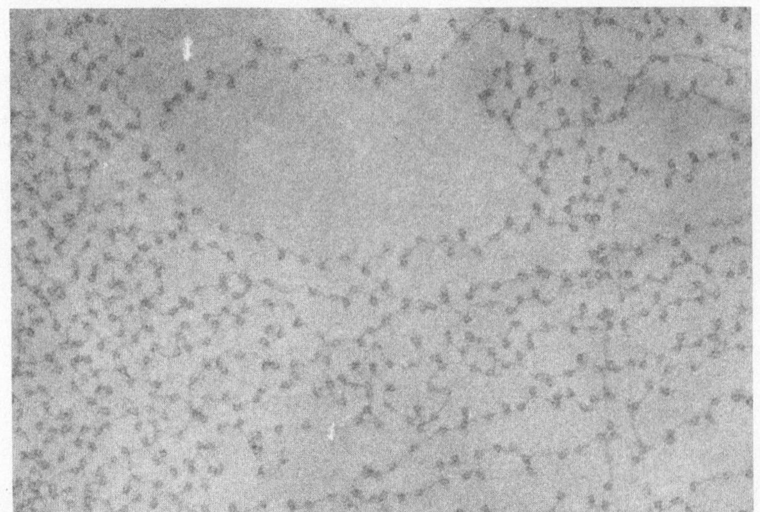

FIGURE 3-34
Chromatin fibers streaming out of a nucleus of a red blood cell of a chicken. The threads as well as the "beads" on the threads consist of DNA and histone protein. These fibers and beads are packed tightly into the nucleus. The beads are 70 ångstroms in diameter; the threads are 140 ångstroms in length. (×143,200) (Courtesy of A. L. Olins; A. L. Olins and D. E. Olins, *Science* 183: 330, 1974. Copyright © 1974 by the American Association for the Advancement of Science.)

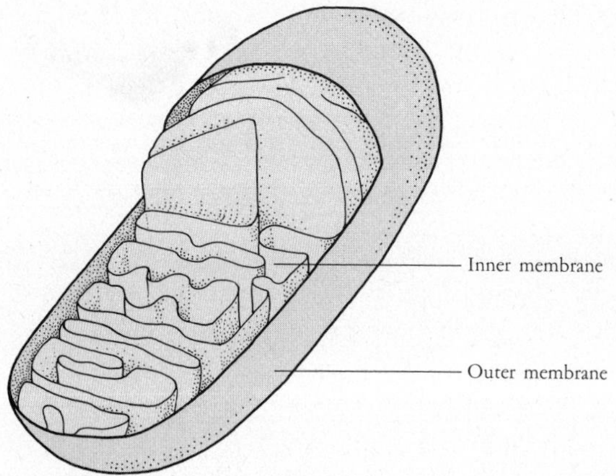

FIGURE 3-35
Diagrammatic representation of a mitochondrion (see Figure 3-33). The enzymes are located in two positions. Some are in the fluid region inside and others are attached to the inner membrane, which because of its convolutions has an extraordinarily large surface area.

with a larger cell that had engulfed it and then evolved into the eucaryotic structures we observe today. Such a theory is strengthened by the observation that some eucaryotic cells living today have procaryotic cells living inside them.

Another structure, the *chloroplast,* is also concerned with the generation of energy in organisms that can utilize light energy through the process of photosynthesis. This unit-membrane bounded structure contains the pigments and enzymes required for photosynthesis. It resembles the mitochondrion, being enclosed by an outer membrane and composed of a large number of closely packed internal membranes. Chloroplasts also divide from preexisting chloroplasts and contain procaryotic-size ribosomes, as well as DNA and enzymes for protein synthesis. They closely resemble cyanobacteria and in one case have been shown to possess remnants of a bacterial cell wall.

Flagella and Cilia: Movement

Many motile eucaryotes are propelled either by flagella or cilia. The flagella of eucaryotes are far more complex than those of procaryotes. Nine pairs of hollow protein fibers or *microtubules* are arranged around a central pair (Figure 3-36), and the entire structure is enclosed in a membranous sheath. These fibers

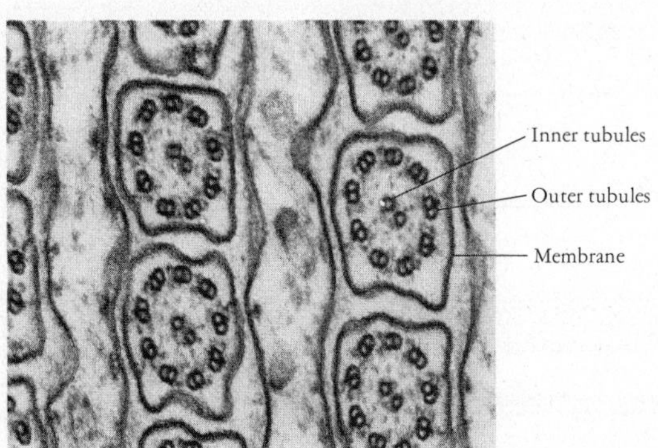

FIGURE 3-36
Cross section of eucaryotic flagella form the protozoa *Trichonympha*. Each flagellum consists of nine pairs of outer tubules, two inner tubules (9 + 2 arrangement), and an enclosing membrane. (×65,000) (Courtesy of A. W. Grimstone.)

originate from a basal body within the cytoplasm. The structure of cilia, including those of some cells of the human body, is identical to the structure of eucaryotic flagella; cilia can be considered to be short flagella.

SUMMARY

The elucidation of the detailed anatomy of microorganisms has been accomplished primarily through observations with the electron microscope, an instrument which can resolve objects only 0.0003 μm apart.

The following table (Table 3-1) summarizes important information about the structures in eucaryotic and procaryotic cells. The structures in the procaryotic cells are divided into two groups, typical structures and those found only in some species. Eucaryotic cells have several structures that are completely absent from procaryotic cells and other structures that are common to both cell types. However, these latter structures, such as flagella, ribosomes, and chromosomes, are more complex (or larger) in eucaryotic cells. In eucaryotic cells internal organelles are commonly unit membrane-bounded structures, while in procaryotes these organelles are not bounded by such a membrane.

TABLE 3-1
Major Structures of Procaryotic and Eucaryotic Cells

		Structure	*Function*	*Chemical Composition*	*Distinguishing Features*
PROCARYOTIC	Cell envelope	Capsule	Protective	Varies; polysaccharide or polypeptides	Synthesized only under certain nutritional conditions by some bacteria
PROCARYOTIC	Cell envelope	Cell wall	Encloses cytoplasm; shapes cell	Peptidoglycan mainly	Structure varies in Gram-positive and Gram-negative cells
PROCARYOTIC	Cell envelope	Cytoplasmic membrane	Metabolic activity; active transport	Lipoprotein	Unit membrane structure common to all membranes
PROCARYOTIC		Ribosome	Workbench for protein synthesis	Protein; ribonucleic acid (RNA)	Size varies in eucaryotic and procaryotic cells
PROCARYOTIC		Chromosome	Carrier of genetic information	Deoxyribonucleic acid (DNA)	Circular, double stranded DNA molecule
PROCARYOTIC		Flagella	Motility	Protein	Complex structure, consisting of several parts; propels cell by pushing
EUCARYOTIC		Cell wall	Provides rigidity and gives cell shape	Varies; chitin, cellulose	Chemical structure much simpler than in procaryotes
EUCARYOTIC		Cytoplasmic (plasma) membrane	Semipermeable	Lipoprotein	Outer covering in most eucaryotic cells; often contain sterols portions associated with ribosomes
EUCARYOTIC		Nucleus chromosomes	Carrier of genetic information	DNA	Bounded by nuclear membrane associated with protein (histones) to form chromatin
EUCARYOTIC		Mitochondria	Energy generation	Protein and lipoprotein	Contain all structures necessary for protein synthesis

A few bacteria form an endospore, a unique cell type that developes from the vegetative bacterial cell by a sequence of morphological changes. This cell type is metabolically inert and is extremely resistant to high temperatures and drying. Because of its unique resistance properties it can survive, for long periods of time, adverse conditions that would kill ordinary vegetative cells.

QUESTIONS

REVIEW

1. What structures enclose bacteria? What function does each serve, and what is the chemical composition of each?
2. Compare, in terms of chemical composition and appearance, the structures in eucaryotic and procaryotic cells responsible for (a) generation of energy and (b) storage of genetic information.
3. Compare the chemical composition of cell walls in Gram-positive and Gram-negative cells, and discuss the significance of this difference to (a) the Gram stain and (b) the mode of action of penicillin.
4. Discuss the response of bacterial cells to lysozyme if the cells are in (a) a dilute solution and (b) a high concentration of salt.
5. Discuss chemotaxis toward an attractant in terms of the motion of the bacterial cell.

THOUGHT

1. What features might a cell which was intermediate between a eucaryote and procaryote have?
2. What kinds of experiment would demonstrate that capsules are only formed under certain nutritional conditions?

FURTHER READING

ADLER, J., "The Sensing of Chemicals by Bacteria," *Scientific American* (April 1976).

BERG, H., "How Bacteria Swim," *Scientific American* (August 1975).

CAPALDI, R., "A Dynamic Model of Cell Membranes," *Scientific American* (March 1974).

EVERHART, T., and T. HAYES, "The Scanning Electron Microscope," *Scientific American* (January 1972).

FOX, C. F., "The Structure of Cell Membranes," *Scientific American* (February 1972).

GOODENOUGH, U., and R. LEVINE, "The Genetic Activity of Mitochondria and Chloroplasts," *Scientific American* (November 1970).

LURIA, S., "Colicins and the Energetics of Cell Membranes," *Scientific American* (December 1975).

MARGULIS, L., "Symbosis and Evolution," *Scientific American* (August 1971).

SATIR, P., "How Cilia Move," *Scientific American* (October 1974).

SHARON, N., "The Bacterial Cell Wall," *Scientific American* (May 1969).

SINGER, S. J., and A. L. NICHOLSON, "The Fluid Membrane Model of the Structure of Cell Membranes," *Science* 175:720 (1972).

STOECKENIUS, W., "The Purple Membrane of Salt-Loving Bacteria," *Scientific American* (June 1976).

CHAPTER 4

DYNAMICS OF MICROBIAL GROWTH UNDER LABORATORY CONDITIONS

PURE CULTURE METHODS

The variety of organisms of different sizes and shapes found in most environments in nature represents a *mixed population* or *mixed culture.* Today we recognize that a mixed culture of bacteria represents different genera and species which have become adapted to a particular environment. In the 1860s, however, a different interpretation was favored by many bacteriologists—that bacteria have the capacity to change their shape and size, and perhaps even their function. That bacteria having different shapes represent different species was demonstrated by separating all the progeny of a single bacterial cell from all other bacteria to obtain a *pure culture.* The concept of a pure culture and the means for achieving it in the laboratory represent one of the most significant advances in bacteriology. Without it, the field would have floundered badly. With it, the "Golden Age of Bacteriology" developed.

KOCH PIONEERS PURE CULTURE METHODS

The person who contributed the most to pure culture techniques was Robert Koch, a German physician, who succeeded in combining a medical practice with a remarkably successful and productive research career for which he received a Nobel Prize in 1905. Koch was primarily interested in isolating and identifying disease-causing bacteria. However, early in his career in the late 1870s he realized that it was necessary to have simple methods for obtaining pure cultures of bacteria. Koch recognized that the isolation of pure cultures required a solid medium on which a single, isolated cell could multiply in a limited area. *The population of cells which arise from such a single bacterial cell in one spot is called a* colony. About one million cells are required for a colony to be easily visible to the naked eye.

Koch initially experimented with growing bacteria on the cut surfaces of potatoes, but he found that a lack of nutrients in this medium prevented growth of some bacteria. To overcome this difficulty, Koch realized that it would be advantageous to be able to solidify any liquid nutrient medium, and he conceived the idea of adding gelatin (the same material used in gelatin desserts) as a hardening agent. Since the gelatin-containing nutrient medium is liquid at temperatures above 28°C, it was only necessary to heat the medium above this temperature, pour it into a sterile container, and allow it to cool and harden. Once hardened, a loopful of bacteria could be drawn lightly over the surface so as to deposit single bacterial cells at intervals. Each of these single cells divides, and after a day or two enough cells are present to form a visible colony. This technique, referred to as the "streak plate" method, is the simplest and most commonly used technique for isolating the progeny of single bacteria (Figure 4-1). A gelatinized nutrient poses certain problems, however. The medium must be incubated at temperatures below 28°C if it is to remain solid. Furthermore, gelatin itself can be degraded by many microorganisms, resulting in liquefaction of the medium. Thus its use is severely restricted.

The perfect hardening agent came out of a household kitchen. Frau Hesse, the New Jersey–born wife of one of Koch's associates, suggested agar, a solidifying agent that her mother had used to harden jelly. Agar is a polysaccharide extracted from certain marine algae. In contrast to gelatin, a 1.5 percent agar gel must be heated above 95°C before it liquefies, and it is therefore solid over the entire range of temperatures at which bacteria grow. Once melted it can be cooled to about 40°C before it soldifies. Nutrients which might be destroyed at higher temperatures can be added aseptically before the agar hardens. Another advantage of agar is that very few bacteria can degrade it. In all these intervening years no better hardening agent has ever been found.

Source of agar, page 287

Another method for obtaining isolated single colonies, which also takes

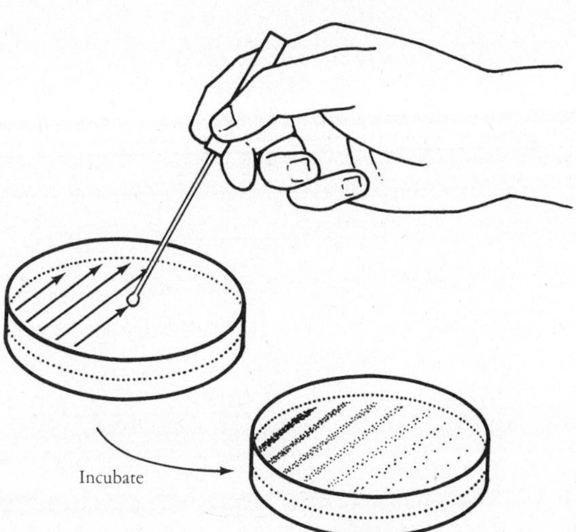

FIGURE 4-1
Streak plate method. Fewer and fewer bacteria are deposited on the surface of the agar as the loop is stroked across it.

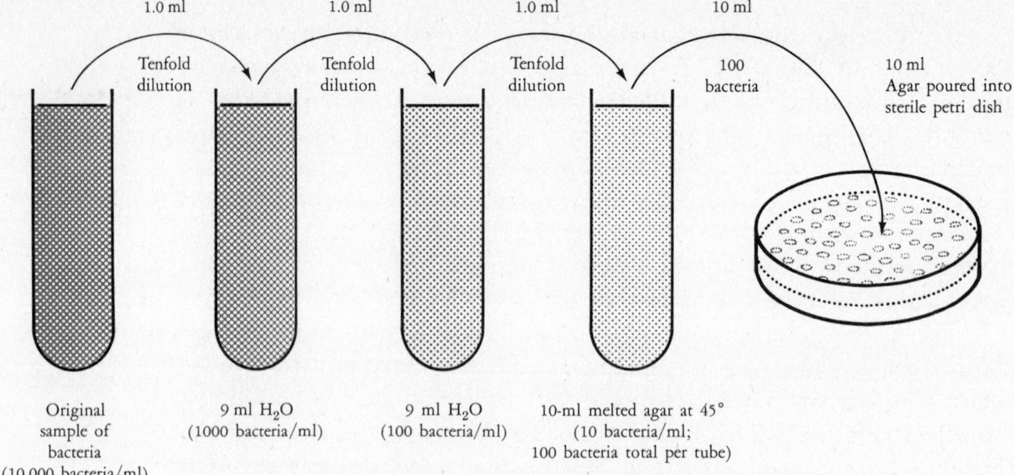

FIGURE 4-2 Pour plate method of achieving isolated colonies.

advantage of the properties of agar, is the *pour plate method* (Figure 4-2). One merely mixes an appropriate number of bacteria with melted agar at a temperature of about 45°C and then pours the agar containing the bacteria into a sterile dish. The agar solidifies and traps the organisms in a defined space. The plate can then be incubated at an appropriate temperature, and the single cells will multiply and form colonies. This technique is possible because most bacteria are not killed by a short exposure to the low temperature to which agar can be cooled after being melted and still remain liquid (45°C). If the original sample contains a very large number of organisms so that isolated, well-separated colonies cannot be obtained, it becomes necessary to dilute the sample with water or another suitable diluent before mixing it with the agar.

Another major technical advance made in Koch's laboratory was the development of a two-part glass dish which could be readily sterilized and maintained in a sterile condition. This dish served as a convenient container during the incubation of the bacteria streaked on the surface of the medium. The dish remains unmodified to the present day, and it retains the name of its inventor, Petri (Figure 4-2).

The significance of the pure culture technique to the development of microbiology cannot be overestimated. This method proved crucial for the rapid isolation of pathogenic (disease-causing) as well as other bacteria. Within 20 years after the development of pure culture methods, the causative agents of most of the major bacterial diseases of mankind were isolated and characterized.

ORGANISMS REPRODUCE EXACT COPIES OF THEMSELVES

Furthermore, the concept that bacteria did indeed breed true was incontrovertibly established. Koch recognized that bacteria of different shapes and sizes could be considered different species. The fact that there were so many distinguishable forms with so many different morphological and metabolic properties had a tremendous impact on the development of the embryonic science of bacteriology.

Scientists working with microorganisms in the laboratory are very careful to insure that they are working with pure cultures. Otherwise, it would not be possible to obtain consistent, interpretable results. However, it is also true that the understanding of the interactions of different organisms with their natural environment will require the use of mixed populations.

MEASUREMENT OF CELL GROWTH

Definition of Growth

Growth is the orderly increase in the quantity of all components of the bacterial cell. After a bacterial cell has almost doubled in size and in amount of each of its components, it divides into two daughter cells—the process of *binary fission.* The time required for one cell to divide into two is the *doubling* or *generation time.* Consequently, the growth of bacteria is measured in terms of an increase in the number of cells, rather than a significant increase in the size of a single organism. Thus, with bacteria, the growth of a population is measured, whereas with a multicellular organism growth is reflected primarily by an increase in the size of a single organism.

Measurements of Increase in Cell Mass

It is essential to measure accurately the growth of a *bacterial population,* since only in this way can the effects of agents which inhibit or stimulate bacterial growth be assayed. Two parameters of growth can be measured, namely, the increase in the *mass* of the population and the increase in *cell number.* Each technique has certain advantages and disadvantages, and different procedures may provide different information. It is important to be fully aware of the information each measurement provides as well as its limitations. Although measurements of cell mass and cell number generally yield comparable figures, this is not always the case. For example, although doubling of the mass of a culture is generally accompanied by a doubling in cell number, under certain conditions an organism may accumulate reserve material or may synthesize capsular material without dividing. Furthermore, the size of cells, and therefore their mass, varies according to their rate of multiplication: the faster they multiply the greater the mass of each cell.

CELL MASS CAN BE DETERMINED BY WEIGHING CELLS

Cell mass can be measured in several ways. In one technique a measured volume of the culture is dried in an oven and then weighed. Another method involves the determination of one component of the cells, most commonly protein. The major problem with these methods for the study of unicellular organisms is their lack of sensitivity. Each bacterial cell weighs about 50×10^{-13} grams. To be accurate, then, this technique requires the processing of large volumes of media, ranging from 100 ml upward.

CELL MASS CAN BE DETERMINED BY LIGHT SCATTERING

The most convenient and simplest method of determining the mass of cells is based on the fact that small particles, such as bacteria, scatter light which is passed through the cell suspension. The amount of light-scattering is proportional to the mass of cells present, and because the bacterial cells are relatively uniform in size (mass), the number of cells can be measured from the amount of light that reaches a sensing device after passing through the suspension (Figure 4-3). This method is far more sensitive than weighing the cells. However, it too has its limitations. A medium containing a million bacterial cells (10^6) per milliliter is perfectly clear, and one containing about 10 million cells (10^7) per milliliter has only a barely discernible turbidity. Thus, although a turbid solution suggests the presence of bacteria, a clear solution does not guarantee their absence. The apparent absence of large numbers of bacteria, inferred from a medium that is crystal clear, may have serious consequences in certain situations.

Bacteria in clear solutions, pages 121–123

The technique of light-scattering is also subject to certain errors. Because the light-scattering is proportional to cell mass, larger cells will scatter light more than smaller cells. Furthermore, because the mass of a cell depends on the conditions under which it is growing, the same number of cells growing under different conditions will scatter light to different degrees. Thus it is very

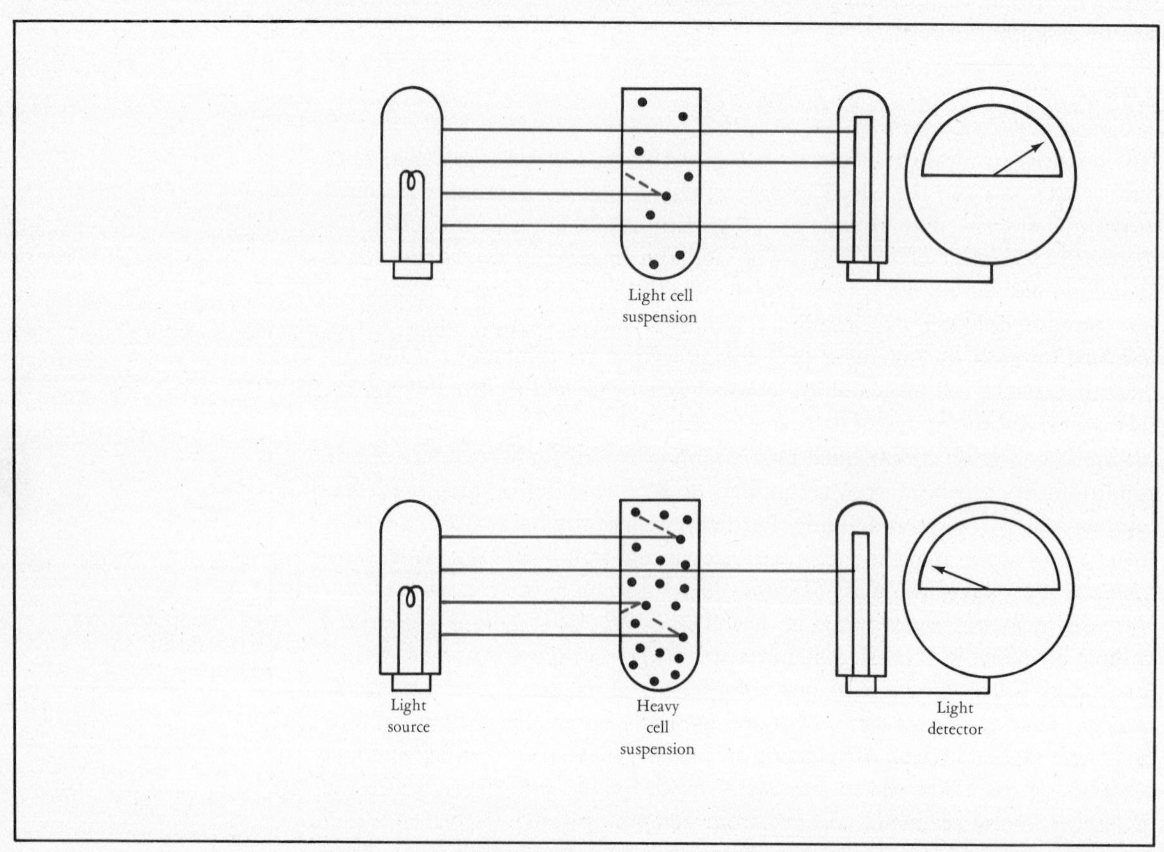

FIGURE 4-3 Measurement of bacterial mass by transmission of light. The amount of light which impinges on the light detector is proportional to the number of bacteria in the suspension.

important to relate cell number to cell mass under defined conditions of cell growth and species of bacteria.

MEASURE OF CELL NUMBERS

Two common laboratory techniques for directly measuring cell number are the *plate count* and the *direct microscopic count.*

Plate count The plate count method is based on the fact that a single living bacterium deposited on a solid nutrient medium will multiply to form a visible colony. The number of colonies that appear after a suitable period of incubation thus represents the number of living or *viable* cells in the original suspension. Because it is difficult and inaccurate to count more than about 300 colonies per petri dish, it is generally necessary to dilute the original culture before pipetting a known volume of this onto the solid medium. If the cells do not separate from each other following cell division (a common situation, for example, with the chain-forming streptococci), then the viable cell count measures the colony-forming units rather than individual cells. The viable cell method measures only living cells, but it is extremely sensitive since the presence of only a few cells per milliliter can be readily detected. However, it is slow because at least a one-day incubation is generally required before a colony is readily visible.

Direct microscopic count A much more rapid method of determining cell number involves counting the number of bacteria in an accurately measured volume of liquid with a microscope. There are special slides called *counting chambers* that hold such a known volume of liquid. These counting chambers can be viewed with the microscope, and the number of bacteria contained in the known volume of liquid can be accurately counted. However, about 10 million bacteria per milliliter are required in order to gain an accurate count, and this method does not distinguish living from dead bacteria.

FACTORS INFLUENCING MICROBIAL GROWTH

General Considerations

A variety of environmental factors influences the growth of microorganisms. These can be divided into two broad categories: *physical environment* and *nutritional factors.* The physical environment includes temperature, pH, and osmotic pressure. The nutritional environment comprises sources of energy and constituents for synthesis of cell components.

As a group, microorganisms will grow under an immense range of environmental conditions. The organisms which can adapt to a wide range of conditions are widely distributed in nature, whereas other organisms, which can multiply only under a limited range of environmental conditions, are found only in restricted natural habitats. No single species is able to multiply over the entire range of conditions found in nature. As expected, most species live under those conditions most commonly found on the earth's surface, but some organisms can be isolated from extreme environments. Such organisms have generally become adapted to their particular environment and do not grow well if cultivated under ordinary conditions.

Shortly after an organism is put into culture the requirements for its growth in the laboratory often change; alternatively, once the organism has been maintained in the laboratory for a long period of time, it may no longer survive in the habitat from which it was orginally isolated. For example, in order to grow when it is isolated initially, *Brucella abortus,* the bacterium responsible for undulant fever in humans, requires 5–10 percent CO_2 in its environment. However, after a few transfers on medium while in the presence of CO_2, this requirement is lost. An example of the second situation relates to the famous strain of *Escherichia coli* K-12. This strain, originally isolated from a human intestinal tract, is no longer capable of growing in that habitat.

PHYSICAL FACTORS INFLUENCING MICROBIAL GROWTH

Temperature

Temperature and enzyme activity, page 147

Most bacteria grow over a temperature range of approximately 30°C, with each species having a well-defined upper and lower limit. As a general rule, the optimum temperature for growth of a species is close to the upper limit of its range. Growth slows as the temperature approaches the lower limit. Above the upper limit there is a very sharp drop in the speed with which cells grow (Figure 4-4). This reflects the fact that temperature affects primarily the enzymes

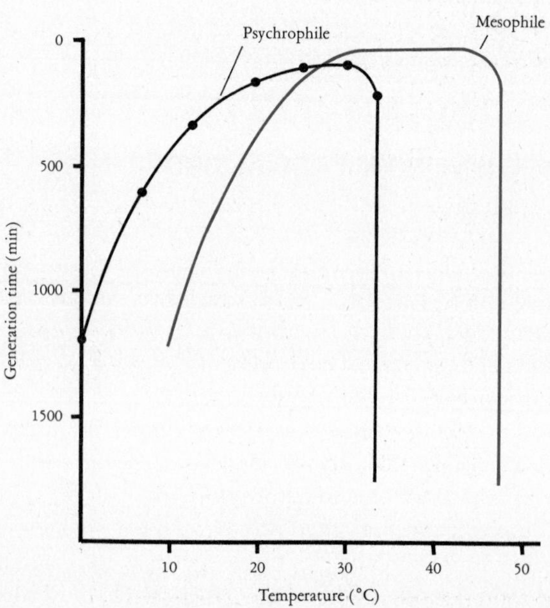

FIGURE 4-4
Effect of temperature on the generation time of a typical mesophile and a psychrophile. The generation time decreases as the temperature is raised until a temperature is reached at which enzymes start to denature. Cell growth slows down precipitously at this temperature.

of the cell. The rise in temperature increases enzyme activity, making the cells grow faster. If the temperature rises too high, enzymes critical to the life of the cell are denatured, and the cells grow more slowly or die.

Bacteria are customarily divided into three groups according to their *optimal* growth temperatures—the temperatures at which they have the *shortest* generation time (Figure 4-5). *Psychrophiles* grow best between -5°C and 20°C. Most actually grow slowly near the lower end of the range and have their optimum near the upper limit. *Mesophiles,* which constitute the majority of

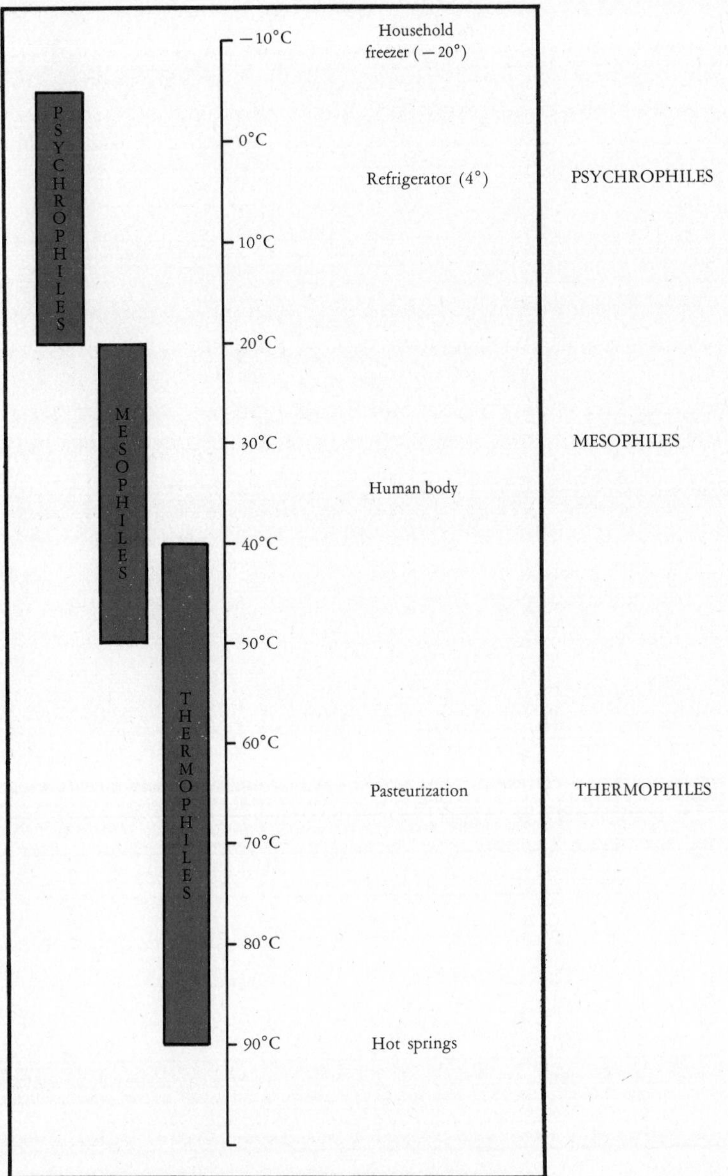

FIGURE 4-5
Temperature growth range of microorganisms. Organisms in each category cannot grow over the entire range for the group.

bacteria, grow best within the range of from 20°C to about 50°C. Most disease-causing bacteria adapted to growth in the human body have their optimum temperature between 35°C and 40°C. Those found in the soil, a colder environment, generally have a lower optimum, close to 30°C. The third group, the *thermophiles,* have their optimum temperature between 50°C and 60°C. It has recently been shown that some thermophilic bacteria isolated from hot springs can actually grow at temperatures above 90°C. As a general rule, protein molecules from thermophiles differ from those from the mesophiles in not being denatured at high temperatures. The thermostability of these proteins of thermophilic organisms is intrinsic and rests on the composition and sequence of amino acids in the proteins. The sequence of amino acids results in the formation of strong bonds—as, for example, covalent disulfide bridges, as well as many hydrogen and other weak bonds—which stabilize the structure of the proteins. The ribosomes from thermophiles are also thermostable, apparently because of their proteins.

Although it is convenient to classify organisms into three groups according to their growth temperatures, in actual practice there is no sharp dividing line which separates the temperatures at which organisms in each group will grow (Figure 4–5). Furthermore, not every organism can grow over the entire range indicated for its group. The upper limit of growth appears to be related to the acidity of the environment. Thus in hot springs the upper temperature limit at which bacteria are found decreases as the springs become more acid. In highly acid water the upper limit of growth is 80°C; in neutral or alkaline springs bacteria are found at temperatures over 90°C.

Since the microorganisms that cause food spoilage are in most cases mesophiles, storage of food at low temperatures, approximately 15°C, is a commonly used method of preserving food (cellar storage). Fruits, vegetables, and cheeses are stored at such low temperatures to retard their spoilage. Refrigerators, which maintain temperatures around 4°C, have become a major tool in preserving foods. However, it is important to recognize that organisms can multiply at refrigerator temperatures, and this can cause problems in storage of food and other materials, such as blood, which are nutritious enough to support microbial growth. To overcome this problem, freezing has often been employed as a means for long-term storage of those items that can withstand freezing.

pH

The pH is a measure of the degree of acidity or alkalinity of a solution (Figure 4-6). Most bacteria grow best at a neutral pH of 7, although they can tolerate ranges from pH 5 (acidic) to pH 8 (basic). Yeasts and molds grow best in an acid medium. Some bacteria can tolerate very high acid concentrations, with some of those that form vinegar (acetic acid) actually able to live in 1N acid. The genus *Thiobacillus,* which accumulates sulfuric acid from its metabolic processes, can grow at pH values of 1 and even below. The pH

H^+ *Ion Concentration* (*moles per liter*)		*pH*		*Common Materials with pH Indicated*
1.0	10^0	0	↑ Increasing acidity	
0.1	10^{-1}	1		
0.01	10^{-2}	2		Human gastric contents
0.001	10^{-3}	3		Ginger ale and wines
0.0001	10^{-4}	4		Tomatoes, orange juice
0.00001	10^{-5}	5		Beans
0.000001	10^{-6}	6		Milk begins to taste sour
0.0000001	10^{-7}	7	Neutral	Pure water, human saliva
	10^{-8}	8	Increasing alkalinity ↓	Seawater
	10^{-9}	9		
	10^{-10}	10		Soap solutions
	10^{-11}	11		Household ammonia
	10^{-12}	12		
	10^{-13}	13		
	10^{-14}	14		

FIGURE 4-6
The pH scale is defined as the negative logarithm of the H^+ ion concentration. The use of logarithms is convenient since it expresses in a single number what would originally require many 0's to express. Note that there is a tenfold variation in the H^+ ion concentration between the two successive figures in the pH scale.

inside the cell must be considerably higher than this, and the ability of this organism to grow depends on the cell's ability to keep H^+ (acid) ions out of the cell. How the cell accomplishes this remarkable feat is not known. Many bacteria produce enough acid as a by-product of the breakdown of foodstuffs to inhibit their own growth, and buffers are generally added to media to maintain certain pH levels. Buffers are often supplied to the cell as a mixture of two salts of phosphoric acid, the sodium phosphates Na_2HPO_4 and NaH_2PO_4. These salts resist pH changes because they have the ability to combine chemically with the H^+ ions of strong acids and the OH^- ions of alkaline solutions according to the following equations:

MECHANISM OF BUFFERING ACTION

$$(1)\quad H^+ \text{ (strong acid)} + HPO_4^{2-} \longrightarrow H_2PO_4^- \text{ (weak acid)}$$
$$(2)\quad OH^- \text{ (strong base)} + H_2PO_4^- \longrightarrow HPO_4^{2-} \text{ (weak acid)} + HOH$$

The buffers maintain solutions at a pH of neutrality because in reaction (1) a strong acid is converted into a weak acid and in reaction (2) the OH^- combines with H^+ to form HOH, which is neutral.

Osmotic Pressure

The osmotic pressure of a medium depends on the concentration of dissolved substances in solution. Such a substance commonly encountered in nature is sodium chloride (common salt). Most bacteria can grow over a broad range of salinity because the cell is capable of maintaining a relatively constant internal salt concentration through the activity of specific permeases. If the salt concentrations on the outside of a cell become too high, however, water is lost from the cell, and cell growth is inhibited. This is the basis of food preservation by salting, used in such foods as beef jerky, bacon, salt pork, and anchovies.

Permeases, page 68

Besides the preservative effect of salts, many foods such as jams, jellies, honey, preserves, and sweetened condensed milk have a high sugar content which leads to the same result. The process of *pickling* combines salt with a very acid environment and is quite effective in controlling the growth of most microorganisms. However, some organisms, called *halophiles,* actually require high salt concentrations, at least in part to maintain the integrity of their cell wall. Indeed, some of these organisms may only grow in environments in which the salt concentration approaches saturation (above 30 percent). Halophiles isolated from the Dead Sea (salt concentration of 29 percent) contain a very high internal salt concentration which is required for the functioning of many of their enzymes as well as their ribosomes. Marine organisms require the salt concentration found in their natural environment (about 3.5 percent) in order to multiply. If the concentration varies much from this level, they may die.

NUTRITIONAL FACTORS INFLUENCING MICROBIAL GROWTH

For an organism to grow, it must not only be compatible with its physical environment, but foodstuffs must be available which can serve as raw material for the synthesis of cell components and also as a source of energy.

Spectrum of Nutritional Types

Organisms can be divided into two large groups based on the foodstuffs providing the carbon that they convert into their cell constituents. *Heterotrophic* organisms utilize *organic compounds* as sources of carbon for synthesis as well as sources for energy. A variety of organic compounds can serve as energy sources. In fact many of the compounds that serve as sources of carbon for biosynthetic processes can also serve as sources of energy. *Autotrophic* organisms utilize *carbon dioxide* as the *major* source of carbon for the synthesis of their cell components. Generally this group gains its energy from either the sun (through photosynthesis) or by metabolizing *inorganic compounds.* A number of different inorganic compounds can serve as sources of energy, and this fact points up the nutritional diversity of the microbial world. Such simple substances as hydrogen gas (H_2), ammonia (NH_3), reduced iron (Fe^{2+}), nitrite (NO_2^-), and hydrogen sulfide (H_2S) can serve as sources of energy for different microorganisms.

Inorganic compounds in metabolism, Table 7-2, page 153, and pages 159–160

Although heterotrophs can utilize a variety of organic compounds as their source of carbon for biosynthetic needs, they are most commonly cultured in the laboratory on media containing glucose. Members of the genus *Pseudomonas* are amazingly versatile. Certain members of this genus are able to degrade over 90 different organic compounds. Most organic molecules, regardless of complexity, can be degraded by at least one species of microorganism which has the enzymes required for the degradation. This versatility does not hold for any single species of microorganism, but rather for the microbial world as a whole. Any single species is capable of degrading only a limited number of organic compounds, somewhere, however, exists a microorganism capable of degrading virtually any particular organic molecule. In some cases the molecule may be broken down into smaller molecules, which can then be converted to the subunits of macromolecules. In other cases the compound is degraded and provides energy, rather than being a source of carbon. It is becoming increasingly apparent, however, that although most biologically generated compounds in the environment can be broken down by microorganisms, a large number of man-made products cannot be degraded by any known organism.

Mineral Requirements

In addition to a source of carbon, all organisms require sources of each of the other elements found in cell constituents—primarily nitrogen, sulfur, and phosphorus. Most microorganisms, both heterotrophs and autotrophs, can utilize inorganic salts as a source of each of these elements.

Microorganisms obtain nitrogen in a variety of forms for subsequent incorporation into amino acids, purines, and pyrimidines. Most organisms cannot utilize atmospheric nitrogen and require nitrogen provided in the form of an ammonium or nitrate salt, such as ammonium chloride (NH_4Cl), sodium nitrate ($NaNO_3$), or even more complex molecules.

The element sulfur is often provided by a sulfate salt, such as ammonium sulfate $(NH_4)_2SO_4$, and is in turn converted into the sulfur part of sulfur-containing amino acids, coenzymes, and other cellular constituents.

Phosphorus is present in nucleic acids and phospholipids, as well as in the key molecule of energy metabolism, ATP. It is generally supplied to the cell as a salt of phosphoric acid, such as the potassium phosphates K_2HPO_4 and KH_2PO_4.

Formulas for amino acids, page 33

ATP, page 145

Oxygen Requirements

The highly reactive gas oxygen is an essential nutrient for many bacteria, but for other organisms it is lethal. Between these two extremes are bacteria that require oxygen, but at concentrations of about 5 percent, which is lower than that found in the atmosphere (20 percent). Thus there exists an entire range of oxygen requirements, and bacteria are classified according to these requirements.

Oxygen and metabolism, pages 156–157

1. *Obligate (strict) aerobes.* These organisms have an absolute requirement for oxygen. They grow best if they are grown in flasks which are continuously shaken in order to keep oxygen always available.
2. *Obligate (strict) anaerobes.* These organisms cannot utilize oxygen. Some members of this group are actually killed by traces of oxygen, while others are not killed but cannot grow in its presence.
3. *Facultative anaerobes.* These organisms can utilize oxygen if it is present but can grow in its absence. Growth is generally more rapid with oxygen.
4. *Microaerophilic organisms.* These represent an ill-defined group of organisms, which have the common feature of requiring oxygen, but only in concentrations less (2–10 percent) than the concentration in air (20 percent).

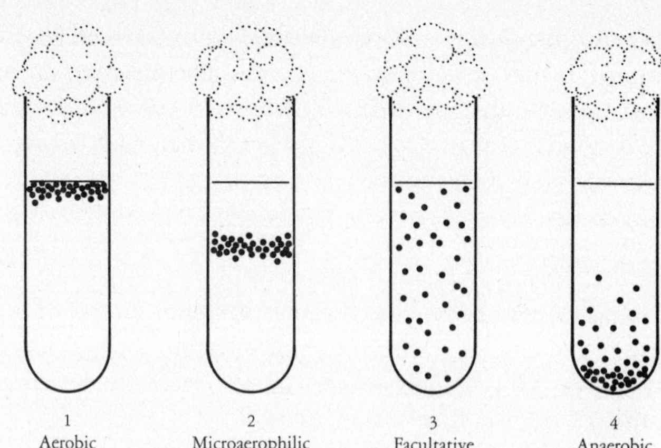

FIGURE 4-7
Shake tubes that demonstrate the oxygen requirements of different organisms.

The response of organisms to oxygen is dramatically illustrated by the growth positions of different organisms in "shake tubes" (Figure 4–7). The bacteria, which are shaken and dispersed throughout the entire tube before the nutrient agar solidifies, are trapped in place by the solidification of the agar. The top of the tube is highly aerobic, the bottom anaerobic, with gradations of oxygen between these extremes.

The broad spectrum in response of cells to oxygen reflects two considerations—whether the metabolic pathways that the cell possesses for degrading foodstuffs require oxygen, and whether the cell possesses the enzymes to dispose of the highly toxic compounds produced from oxygen by cell metabolism. These considerations will be covered in Chapter 7.

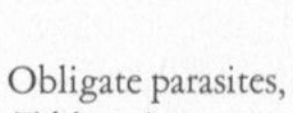
Obligate parasites,
Table 1-4, page 17

Growth Factors

RELATIONSHIP BETWEEN GROWTH REQUIREMENTS AND ENZYME CONSTITUTION

Despite their similarity in chemical composition, microorganisms display a wide spectrum in their growth requirements; these reflect the enzymatic capabilities of the cell. At one extreme are the obligate intracellular parasites, such as the members of the genera *Rickettsia* and *Chlamydia,* which lack some of the enzymes or structures necessary for an independent existence. At the other extreme are the autotrophs, such as the photosynthetic bacteria, which have enzymes that allow them to multiply with very simple nutrients—sunlight and inorganic salts. Heterotrophs do not have all of the enzymes present in autotrophs and therefore require those organic compounds which they cannot synthesize themselves. However, all three groups have many enzymes in common. Comparative biochemical studies on a wide variety of organisms have made it apparent that most organisms can carry out a fundamental core of enzymatic reactions. Many heterotrophs (and a few autotrophs) do not have all of the enzymes necessary for synthesizing some of the small molecules required for growth. These small organic molecules, called *growth factors* (such as amino acids, vitamins, purines, and pyrimidines, to name a few), must therefore be provided in the medium. The more enzymes for the biosynthesis of small molecules which an organism lacks, the more factors are required for its growth. The need for a large number of growth factors is illustrated by the group called the lactic acid bacteria, to which must be supplied as much as 95 percent of the subunits which comprise their cell material.

ISOLATION AND IDENTIFICATION OF VITAMINS AS BACTERIAL GROWTH FACTORS

The microorganisms' requirements for growth factors were successfully exploited to isolate several vitamins long before anything was known about these vitamins' chemical structure and metabolic function. A number of such vitamins had been recognized as unknown growth factors for certain microorganisms; their presence in extremely small amounts (10^{-12} gram) in "rich," complex media could be detected by their ability to stimulate the growth of these microorganisms. Thus, Vitamin B_{12} was first isolated by purifying a substance that served as a growth factor for a species of *Lactobacillus.*

The technique whereby microorganisms are used to assay the presence of any material is termed *microbiological assay.* The exact quantity of any growth factor can be assayed by using a strain of microorganism which requires the growth factor. For example, the lactic acid bacteria as a group require a large number of amino acids, purines, pyrimidines, and vitamins for growth. If any one of these growth factors is completely left out of the medium and an unknown amount then added, the final level of cell growth is directly related to the amount of the growth factor present in the medium (Figure 4-8). In a microbiological assay, it is important to note that the *rate* of growth of the organism is identical in a medium with a limiting amount of the required nutrient to its growth rate in a medium with an excess amount. The concentration inside the cell is essentially the same in both cases because permeases can concentrate low levels of the nutrient until all of the nutrient in the medium has become exhausted.

Microbiological assays are currently being used in the diagnosis of certain diseases, such as phenylketonuria. This severe human mental disease comes from the lack of an enzyme which converts phenylalanine to tyrosine. In persons with phenylketonuria the concentration of phenylalanine increases dramatically. Its presence can be easily detected if bacteria which require phenylalanine grow when the whole blood of such a person is added to a medium that is lacking phenylalanine.

CULTIVATION OF BACTERIA IN THE LABORATORY

Basic Media

Microbiologists use two basic types of media for cultivating bacteria. One contains known amounts of pure chemical compounds; it is therefore a chemically defined or *synthetic medium.* The other type is a "rich" complex nutrient medium containing poorly defined materials, such as digests of ground meat, plants, and fish.

The composition of a synthetic medium commonly used to grow the bacterium *Escherichia coli* is given in Table 4-1. This species can obviously synthesize all of its cellular constituents from glucose and a few inorganic salts. Glucose also serves as the species' energy source. This glucose-salts medium can serve as a basic

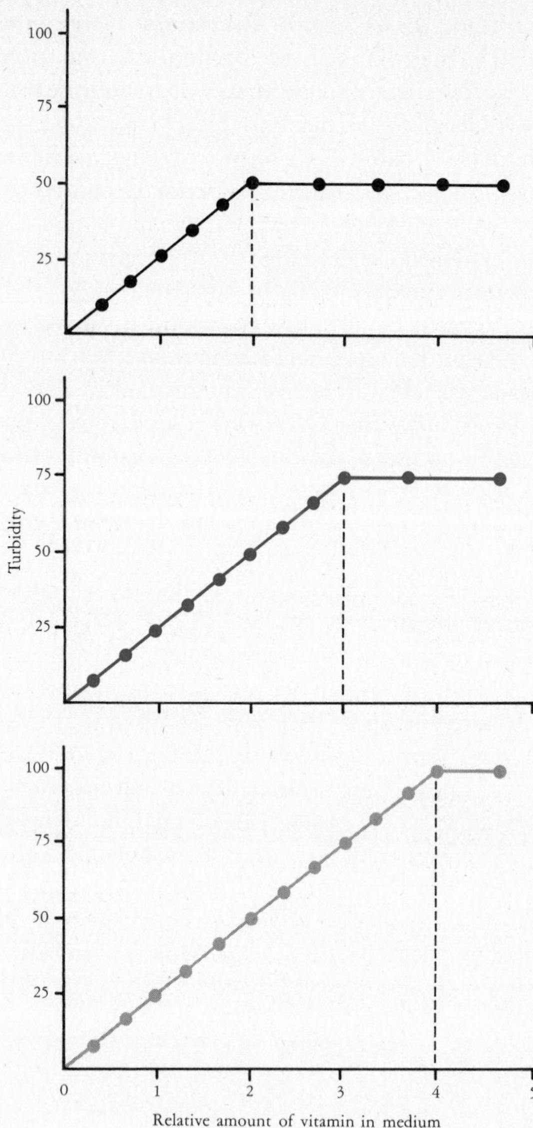

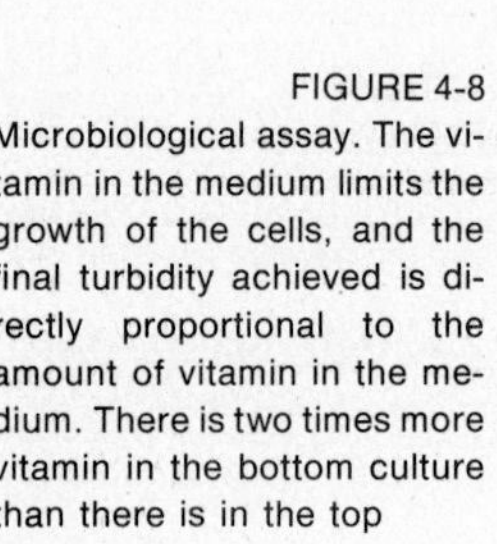
FIGURE 4-8
Microbiological assay. The vitamin in the medium limits the growth of the cells, and the final turbidity achieved is directly proportional to the amount of vitamin in the medium. There is two times more vitamin in the bottom culture than there is in the top

Permeases, page 68

medium to which various growth factors can be added to support the growth of more fastidious organisms. For example, some members of the neisseria group (which includes the causative agents of gonorrhea) will grow if the supplements listed in Table 4-2 are added to a glucose-salts medium. Other bacteria may have different nutritional requirements and will not grow even in this fortified medium. It often becomes necessary to add undefined nutrients, such as yeast and meat extracts, to permit growth of some organisms. As a group, the pathogenic organisms are particularly exacting in their nutritional requirements. Their fastidious nature probably results from their having become adapted to a particular

TABLE 4-1
Synthetic Medium for Growth of *E. coli:* Glucose–Inorganic Salts

Ingredient	*Grams per Liter of Medium*
Glucose	5
Dipotassium phosphate (K_2HPO_4)	7
Monopotassium phosphate (KH_2PO_4)	2
Magnesium sulfate ($MgSO_4$)	0.08
Ammonium sulfate $(NH_4)_2SO_4$	1.0
Water	1,000 ml

1. The glucose is generally sterilized separately from the salts, because heating the glucose with the salts converts glucose to materials which may be toxic to cell growth.
2. For the preparation of solid medium, 1.5 percent agar (final concentration) is generally added.

environment—the animal body—which contains a variety of "rich" nutrients. They have therefore been able to dispense with many of the enzymes involved in the synthesis of these nutrients.

It is generally much more convenient to incubate cultures in a "rich" medium than it is to painstakingly determine the exact nutrient requirements of various species step by step. This is especially true in laboratories in which a variety of different organisms are grown. For this reason clinical diagnostic bacteriology laboratories rely almost solely on undefined media.

A typical rich medium nutrient broth is shown in Table 4-3. The peptones in the medium are breakdown products of plant, fish, and meat protein and consist primarily of peptides of various sizes. Beef extract is a water extract of lean meat and provides vitamins, minerals, and other nutrients. To make the medium

TABLE 4-2
Synthetic Medium for Growth of *Neisseria*

Component	*Component*
NaCl	Uracil
K_2SO_4	Hypoxanthine
$MgCl_2$	
NH_4Cl	Spermine
K_2HPO_4	Hemin
KH_2PO_4	Nitrilotriethanol
$CaCl_2$	Polyvinyl alcohol
$Fe(NO_3)_3$	
Amino acids:	Sodium lactate
20 different types	Glycerine
	Oxalacetate
Vitamins:	Sodium acetate
7 different types	Glucose

Source: Adapted from Catlin, B.W. *Journal of Infectious Diseases* 128:178 (1973).

TABLE 4-3
Rich Undefined Medium–Nutrient Broth

Ingredient	*Amount per Liter*
Peptone (0.5%)	5.0
Meat extract	3.0
Distilled water	1000 ml

richer, yeast extract may be added. Yeast extract, the broth of autolyzed yeast cells, serves as an excellent source of vitamins.

PROBLEMS IN DEVISING A UNIVERSAL MEDIUM

It is theoretically possible to devise a single medium with the proper nutrients to support the growth of any bacterium, no matter how unique its growth requirements. In practice, however, no such medium exists, for not only must one contend with providing growth factors, but many materials in rich media are toxic to various bacteria. Pathogenic microorganisms seem to be especially sensitive. Other material must then be added to neutralize the toxins. Some organisms, notably some mycoplasma, grow only in media containing a high concentration of serum (20 percent). Its major function is to supply serum albumin, which binds small molecules such as fatty acids, metal ions, and detergents, all of which are toxic to many cells.

Haemophilus, pages 437–438

Members of the genus *Haemophilus* and certain other fastidious bacteria are generally grown on chocolate agar, a medium enriched with blood and then heated sufficiently to release the hemin from the denatured hemoglobin of the blood (giving a chocolate brown color to the medium). The heating also denatures an enzyme, present in blood, which destroys a growth factor required by many bacteria. The manufacture of hundreds of different types of media is the primary concern of several companies. Each of these media has been specially formulated to permit luxuriant growth of one or several groups of organisms.

Even with the availability of these rich media, however, some microorganisms still have never been cultivated in the laboratory in the absence of living cells. These include the rickettsiae and the spirochete causing syphilis, *Treponema pallidum*. One large group of organisms which present special problems for cultivation are the anaerobes.

Treponema pallidum, page 441

Cultivation Methods for Anaerobes

Since their growth is inhibited to varying degrees by free oxygen, special techniques are required for cultivating anaerobes. Two general methods are available. One involves incubating cultures of these bacteria in jars from which the oxygen has been removed. One common type of anaerobic jar contains a disposable hydrogen generator and a catalyst which combines the hydrogen with oxygen at room temperature. Thus no free ozygen is available to inhibit growth.

Another technique is to add to media various chemicals which reduce the level of oxygen by chemically combining with it. These compounds include sodium thioglycollate, cysteine, and ascorbic acid. The medium is often boiled first to drive out the air before inoculation; the surface can then be covered with sterile vaseline and paraffin to limit the reentrance of oxygen.

The cultivation of anaerobic organisms presents a great challenge to the microbiologist. It is often difficult to provide strict anaerobes with suitable conditions quickly enough to keep them alive once they have been removed from their natural source. Unquestionably, as techniques for culturing anaerobes are improved, many more such organisms will be isolated from habitats in which they had previously appeared to be absent.

From this discussion of the growth requirements for both aerobes and anaerobes, it should be clear that the inability to demonstrate the presence of an organism does not mean that the organism is not there. It may merely reflect the fact that the proper nutritional and environmental conditions of the organism have not been met.

Enrichment Cultures

POSSIBLE TO PROMOTE GROWTH OF SELECTED ORGANISMS

An enrichment culture enhances the growth of one particular organism in a mixed population and thereby promotes its dominance. In practice this method allows one to isolate in pure culture an organism from natural sources. The technique is based on the well-documented concept that in any environment the best adapted organism will come to the fore. The first consideration in setting up an enrichment culture is selecting the sample from which the organism is to be isolated. It should be selected from an area in nature in which the organism most likely occurs. The sample is next inoculated into a medium that contains nutrients favoring as much as possible the growth of the desired organism but not the other cells in the sample. The medium is then incubated under conditions which also favor the growth of the desired organism. By judicious culture methods, a skilled microbiologist can sometimes manipulate the growth of certain species so dramatically that a single species will predominate. It is then relatively simple to plate the predominant organism on agar and select a single colony as a pure culture.

The isolation of an organism capable of degrading crude oil in sea water illustrates a practical use of enrichment culture techniques. Crude oil was added to unsterilized sea water containing phosphate and ammonium sulfate to provide phosphate, nitrogen and sulfur. After incubation for a week the oil became evenly dispersed. Bacteria were growing on the oil, using it as their sole source of added carbon and energy. When this culture was then plated on agar containing oil, colonies of eight different types of bacteria arose, which could be readily isolated and cultivated on nutrient agar. One of these isolates appears to be useful in preventing the major cause of oil pollution, the discharge from oil tankers of seawater which serves as ballast and contains about 1 percent of the oil left over from the ships' last oil cargoes. It has been estimated that oil tankers dump about 1 million tons of oil into the oceans annually from this source alone.

If these oil-dispersing bacteria are added to the oil-containing seawater in the hold of a ship, together with urea (a cheap source of nitrogen) and potassium phosphate, the bacteria will within a week convert the oil to a form which is no longer waxy and which quickly breaks down in the ocean. Not only will there then be no oil slick, but the ship's owner will be spared the expense of cleaning his holds by tedious and expensive mechanical means. As an added bonus, the crop of bacteria can be harvested to serve as a source of high quality animal food (see Chapter 39).

POSSIBLE TO SELECTIVELY INHIBIT THE GROWTH OF UNDESIRED ORGANISMS

One common difficult problem is the selection of a very fastidious organism when most organisms in the environment are less demanding. One solution involves using *selective inhibitors,* which can be added to media to discourage the growth of some organisms in the population (the organism being sought must be resistant to the inhibitor). For example, one diagnostic procedure for identifying typhoid fever involves isolating the causative organism, *Salmonella typhi,* from the stool, a habitat containing a large number of organisms related to *Salmonella.* The stool specimens are first incubated in selenite or tetrathionate broth, both of which inhibit the growth of the flora normally found in the stools far more than they inhibit multiplication of *Salmonella.* Many other examples involving other pathogens will be considered later.

Salmonella, pages 435–436

Once an organism becomes the dominant species, it may be tentatively identified by its pattern of growth on *indicator or differential agar medium.* These media contain a component which is changed in a recognizable and somewhat unique way. In cases of suspected typhoid fever clinical microbiologists in some laboratories plate the organisms on bismuth sulfite agar. *Salmonella typhi* forms hydrogen sulfide, which results in the formation of black colonies. Blood agar is a common indicator medium used in clinical microbiology laboratories. Some bacteria produce enzymes, hemolysins, which break down blood. Colonies of the organism causing "strep" throat are readily detected on blood agar plates by their ability to destroy red blood cells and produce a clear zone around the colony (beta hemolysis) (Figure 4-9). Certain dyes added to media will change color if a medium becomes acidic. A wide variety of other materials can serve as diagnostic aids when incorporated into media.

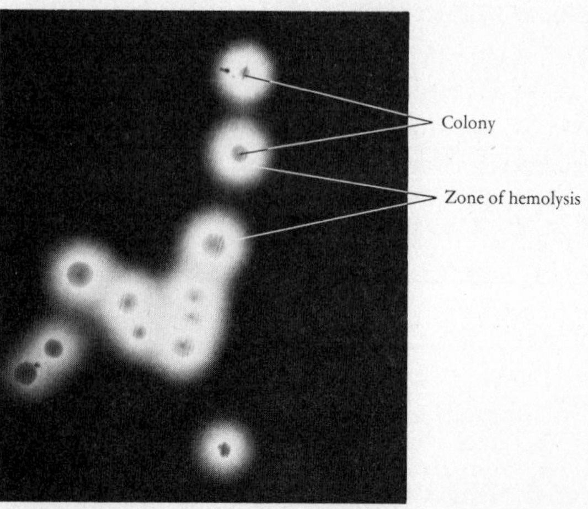

FIGURE 4-9
The colony excretes a substance which lyses the red blood cells (sheep) in the medium, resulting in complete clearing. (Courtesy of J. P. Dalmasso.)

CELL DIVISION

Division of one cell into two occurs after a cell has duplicated its components. The process of cell division can be separated into several stages (Figure 4-10). First, the cytoplasmic membrane grows into the center of the cell, forming a *transverse septum.* Then the cell wall on both sides grows inward toward the center of the cell, forming a cross wall thicker than the ordinary bacterial cell wall. Once the cell wall forms this completed septum, dividing the cell in two, cell separation is initiated by cleavage along the length of this thickened cell wall, each half forming part of the wall of each daughter cell.

Although the overall process of cell division can be described in these simple terms, the regulation of numerous cellular processes that must be coordinated in the process of cell division is not well understood.

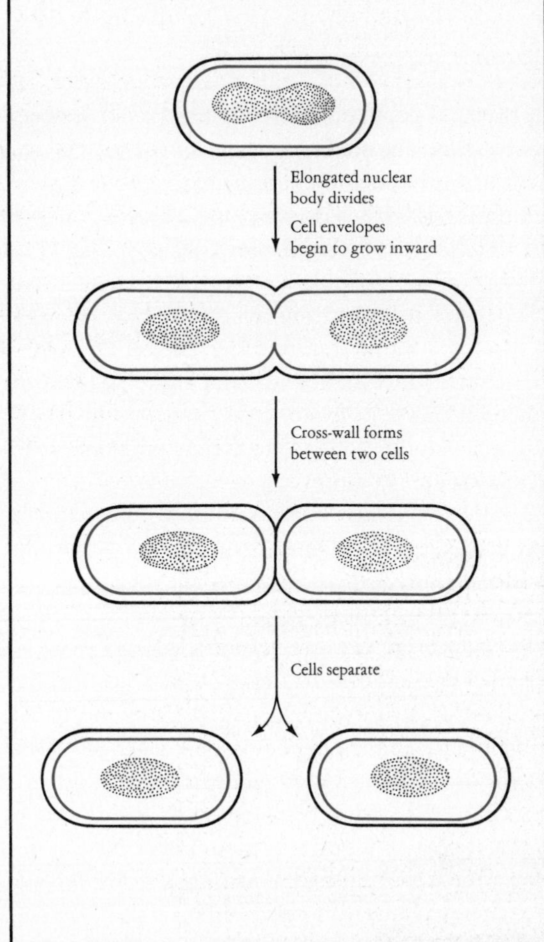
Elongated nuclear body divides
Cell envelopes begin to grow inward
Cross-wall forms between two cells
Cells separate

FIGURE 4-10
Diagrammatic representation of cell division.

Cell Wall Synthesis

Before dividing, the cell must duplicate its cellular components. In some species of bacteria, DNA may be synthesized throughout the life of the cell before it divides. Other species appear to synthesize DNA only during a portion of the lifespan of the cell. The size of the cell also enlarges before division, which also entails the synthesis of new cell wall material. One question that has been studied is: Where does new cell wall synthesis take place—at the ends, center, or throughout the old cell wall? The answer apparently depends on the organism. In *Streptococcus* and a species of *Bacillus*—both Gram-positive organisms—the new wall is synthesized only at the sites of septum formation, that is, near the center. In several Gram-negative organisms newly synthesized material is intercalated between layers of old cell wall (Figure 4-11). In this case cell wall synthesis occurs at various sites, in contrast to the limited area of synthesis in the two Gram-positive organisms. The reasons for this difference are not understood.

DYNAMICS OF POPULATION GROWTH

EXPONENTIAL INCREASE IN CELL NUMBER CHARACTERIZES BACTERIAL MULTIPLICATION

The growth of a bacterial population can be measured in terms of an increase in the number of viable cells. The doubling or *generation time* conveniently expresses the rate of growth of the population. The increase in cell number in a growing bacterial population is *logarithmic* or *exponential,* because each cell gives rise to two cells, each of which in turn gives rise to two cells—a total of four—and so on. When the number of cells in a population is graphed against time of incubation, a straight line is generated if the ordinate (Y-, or vertical, axis) representing number of cells is a logarithmic scale and the abscissa (X-, or horizontal, axis) representing time of incubation is an arithmetic scale. The straight line indicates that there is an identical percentage of increase in the number of cells during any constant time interval (Figure 4-12). Thus if the number of cells increases from one to four in 1 hour (a generation time of 30 minutes), there will be another fourfold increase in the population after an additional hour and yet another fourfold increase after each succeeding hour. Note, however, that the increase in the *number* of cells varies within the three intervals. The total number of cells depends on the number of cells present initially and the number of cell generations. Since many bacteria can divide every 30 minutes (and a few species even divide every 10 minutes), the number of these cells in a population reaches high levels in a relatively short time. Thus it is very convenient and sometimes absolutely necessary to plot the number of cells in the population on a logarithmic scale.

CALCULATIONS OF CELL GROWTH

For cells in the logarithmic phase of growth, the following simple growth equation expresses the relationship between the original number of cells (zero time), the number of cells in a population at a later time, and the number of divisions (or generations) the cells have undergone. If any two values are known, the third can be easily calculated from the equation $b = B2^n$ where $B =$ number of cells at zero time, $b =$ number of cells at any given later time, and $n =$ number of cell generations.

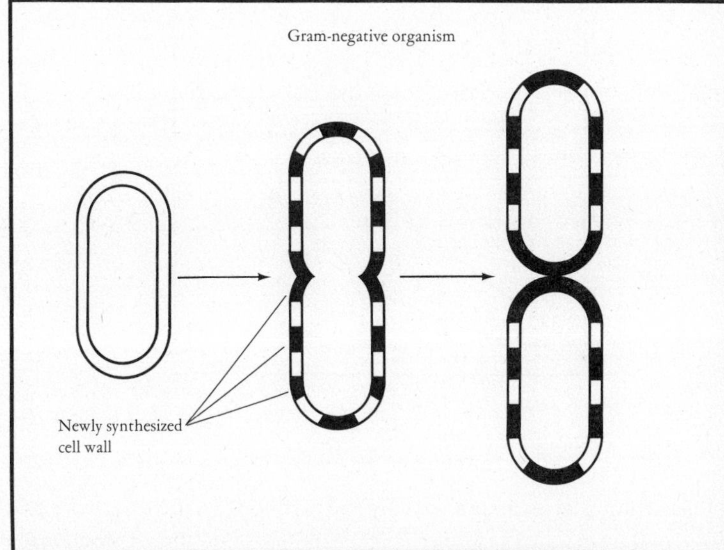

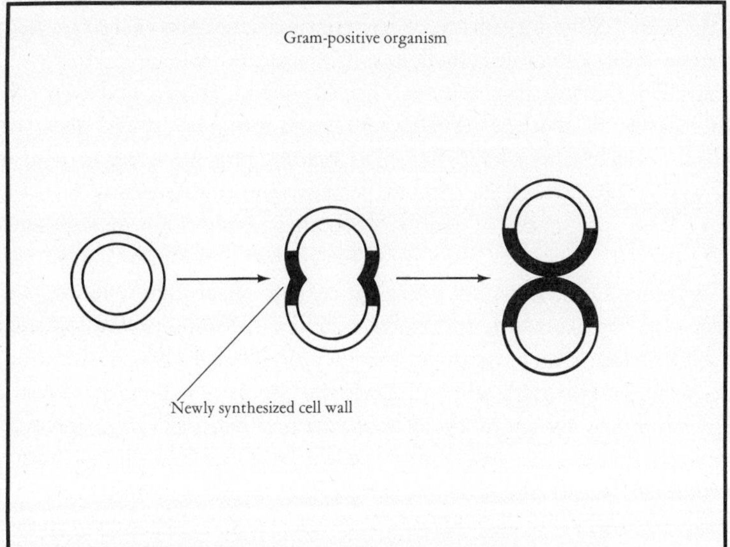

FIGURE 4-11
Diagrammatic representation of cell wall addition to growing Gram-positive and Gram-negative cells.

Phases of Bacterial Growth

FOUR STAGES IN GROWTH OF A BACTERIAL CULTURE

Thus far we have only been concerned with bacteria that are multiplying. However, bacteria are not always increasing in number even when they are in a medium that can support their growth. If bacteria are transferred from one medium into another, the population undergoes a characteristic and predictable sequence in its rate of increase in cell numbers (Figure 4-13). There are four readily distinguishable phases to this sequence: (1) the *lag phase,* in which there is no increase in the number of viable cells; (2) the *log phase* in which the cell population in-

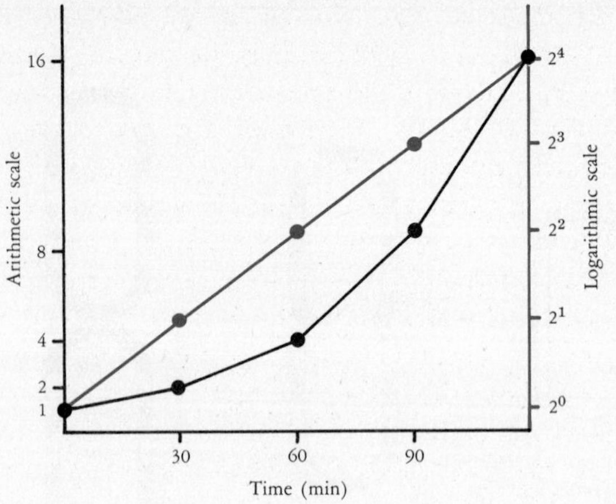

FIGURE 4-12
Relationship of cell number to number of generations plotted on arithmetic and logarithmic scales.

creases logarithmically or exponentially with time; (3) the *stationary phase,* in which the total number of viable cells remains constant; and (4) the *death phase,* which is characterized by an exponential decrease in the number of viable cells.

LAG PHASE: "TOOLING UP"

Lag phase When a culture of bacteria is diluted and placed in fresh medium, the number of viable cells does not increase immediately. However, each cell grows by increasing its cell mass. The cells enlarge in size, and there is extensive macromolecule synthesis. This period is a "tooling up" stage: growth without cell division during which the cell is synthesizing and accumulating molecules it will require for cell division. Especially important are molecules of ATP which represent the supply of energy, ribosomes, required for protein synthesis, and any enzymes which may be required for growth.

Log phase During the log phase the cells divide at their maximum rate. A convenient way of expressing the growth rate of a culture is in terms of the number of cell divisions which occur per hour. A culture with a generation time of 30 minutes will have a growth rate of 2; a generation time of 60 minutes will give a growth rate of one. The larger the growth rate number the faster will the cells be multiplying. The increase in cell number does not occur in discrete jumps but rather gradually with time, thus indicating that individual cells are at different

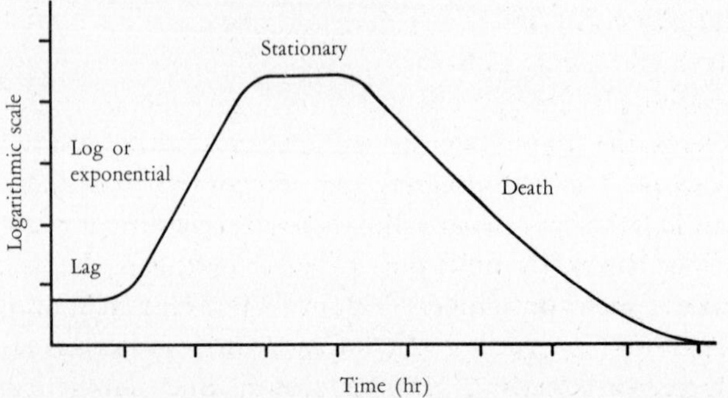

FIGURE 4-13
Bacterial growth curve. The duration of each phase depends on the environment and the particular organism.

stages in their division cycle: some have just divided, others have enlarged to the size at which they are about ready to divide, still others are at intermediate stages in the division cycle. Thus, cells in the population divide nonsynchronously.

STATIONARY PHASE: TOTAL NUMBER OF CELLS REMAINS CONSTANT

Stationary phase The size a cell population reaches before it stops growing is controlled primarily by the environment but also depends on the genetic constitution of the organism. The generation time of many bacteria can be as short as 20 minutes. With this generation time the population starting with a single cell will reach 10^9 cells in 10 hours. Since the volume of an average bacterial cell is 1 μm^3, the total cell volume (size of each cell times the number of cells) of a culture which continued growing exponentially for 45 hours would be greater than the volume of the earth. Fortunately, bacterial cultures stop growing once they have reached a population density of approximately 10^9 cells per milliliter.

Two parameters characterize the exponential growth phase of any culture: the growth rate and the duration of the period of exponential growth. The faster cells multiply, the faster the growth rate, and in terms of the graph the steeper the line or greater the slope of the line representing exponential growth.

The growth rate is influenced by all of the environmental conditions previously mentioned in this chapter. Obviously the optimal conditions for growth will vary for each particular organism. One important generalization that can be made is that the *growth rate of a bacterium is proportional to its rate of energy metabolism.* All of the environmental and nutritional factors which influence the growth rate do so primarily by influencing the rate of ATP production by the cell.

The reason populations cease increasing in number has traditionally been attributed to one of three possible mechanisms: physical overcrowding, limitation of a required nutrient, or accumulation of toxic products during growth. Physical crowding is not a reasonable explanation, since cells in colonies can grow rapidly on solid medium in which crowding is much more severe than in liquid medium. It has generally been possible to explain the stoppage of cell growth in all cases investigated by the exhaustion of a nutrient or accumulation of toxic products.

Microbiological assay, page 99

If cells growing in a synthetic medium exhaust one of the ingredients essential for growth, growth ceases. Bacteria can exhaust a required nutrient even when growing in a rich nutrient medium. The material most commonly exhausted by aerobic organisms is oxygen. This gas is not very soluble in water, and once the population reaches a high cell density there are enough cells at the air-liquid interface to utilize all of it immediately. As a result, the cells beneath the surface of the liquid cannot grow.

Cells growing under anaerobic conditions generally run out of a usable energy supply when the population reaches high cell densities. They also produce a variety of waste products in their metabolism, many of which are toxic.

As a general rule toxic materials produced by cells growing under anaerobic conditions limit the final level that the population can attain under otherwise optimal growth conditions. This explains why some facultative anaerobes (organisms capable of living under aerobic or anaerobic conditions) will invariably grow to a higher cell density if they are grown under aerobic conditions. For reasons that will be discussed in Chapter 7, not as much toxic material is likely to accumulate under aerobic conditions.

The stationary phase is reached when the total number of *viable* cells in the population no longer increases. This constancy may result from one of two different events, depending on the particular organism and nutritional conditions. Every cell in the population may stop multiplying, and none die. Alternatively, there may be a balance between cell division and cell death such that for every cell that divides another cell dies. In both cases the total number of *viable* cells in the population remains constant. In the second case, however, the *total* number of cells, both living and dead, would increase.

Acid formation in metabolism, Figure 7-11, page 155

The duration of the stationary phase varies, depending on the species of organism and environmental conditions. Some organisms remain in this phase for only a few hours; others for several days. Generally if toxic products have accumulated in the exponential phase, the cells begin to die rapidly. Some cells lyse very quickly once they stop multiplying because of the action of a lytic enzyme present in the cell wall. This same enzyme may be concerned with the extension of cell walls during cell division.

Death phase A decrease in the total number of viable cells in the population signals the onset of the death phase. A cell is dead if it is no longer capable of multiplying. In this case a graph of viable cells on a logarithmic plot versus time on an arithmetic plot normally yields a straight line (Figure 4-13). The same mathematical treatment given to the kinetics of growth in the logarithmic phase holds true here. But now there is an exponential decrease in the number of viable cells rather than an exponential increase. Thus if there is a tenfold decrease in the number of viable cells during the first hour of the death phase, an additional tenfold decrease is observed by the end of the next hour.

DEATH PHASE: EXPONENTIAL DECREASE IN THE NUMBER OF VIABLE CELLS

The vast majority of organisms in a culture are similar enough that death in a population may be treated as a statistical phenomenon. Since each organism is equally likely to die, which cells actually do die in any time period depends entirely on chance. The likelihood of a cell dying remains constant during most of the death phase and so the cells continue to die at a constant rate. However, as soon as the vast majority of the cells have died, the death curve flattens out (Figure 4-13). This indicates that some cells in the population are more resistant than others to dying. Why a few cells in the population are more resistant than all other cells is not understood. This resistance is not transferred to daughter cells, however, since survivors that are diluted into fresh medium undergo the same stages of growth and die at the same rate as the original population.

An inherently more resistant form of the cell, the *endospore,* may be present in cultures of some bacteria during the death phase. These spores, which develop from vegetative cells during the stationary phase, would certainly be more resistant to dying in the death phase than vegetative cells.

If a population of cells in the death phase is diluted and inoculated into fresh medium, the cells must go through a lag phase before their log phase. Since cells in the death phase are not primed for rapid multiplication, they will spend an extended period of time in "tooling up" before they can multiply. In contrast, cells transferred from a log phase stage into fresh medium do not exhibit a lag phase unless the two media differ in composition.

Age of Culture versus Age of Individual Cells

A bacterial culture is said to be "old" when it is no longer in the log phase of growth. However, any cells which have just finished dividing in the old culture would be young. Cells which divide by binary fission do not have a single life cycle that includes youth, senescence, and death. Accordingly, the concept of age has an entirely different meaning as far as typical bacteria are concerned.

GROWTH OF A BACTERIAL COLONY

DIFFERENT PARTS OF THE SAME COLONY ARE IN DIFFERENT GROWTH PHASES

The growth and development of a bacterial colony involve many of the features of bacterial growth discussed in this chapter. A period of time elapses before a single cell on a solid medium starts to divide. The initial multiplication of cells may be exponential. However, after a very few divisions, the close proximity of cells to one another results in very crowded conditions. Further development of the entire colony is a very complex process. The crowding results in competition for available nutrients, which must diffuse from the medium into the cells. The cells on the periphery of the colony are favorably located to obtain oxygen from the air as well as nutrients diffusing into the colony. A gradient of nutrients is thus set up from the outside to the center of the colony. The center tends to be anaerobic, with only a limited supply of nutrients available. Thus the colony will enlarge through multiplication of cells at its periphery.

The toxic end products of growth also influence the development of the colony. Toxic end products, such as acids, must diffuse from the colony into the medium, or they may kill cells in their immediate vicinity. Thus, within a single colony, there are cells which may be in the log, stationary, or death phase, depending upon their location. Cells at the edge of the colony may be growing exponentially, whereas those in the center may be in the death phase.

SUMMARY

This chapter has focused on the principles of microbial growth in the laboratory. Bacteria can be grown in a variety of media, each of which must provide sources of energy as well as the elements carbon, nitrogen, phosphorus, and sulfur in forms that can be used by the organism for the synthesis of cell material. An unusually broad spectrum of nutrients can serve these purposes for different microorganisms. At one end of the spectrum are the autotrophs—organisms which derive their energy from sunlight or from the metabolism of inorganic compounds and their carbon from carbon dioxide. At the other end are obligate intracellular parasites—bacteria that cannot multiply unless they are inside living cells. Intermediate forms, the heterotrophs, represent the vast majority of bacteria and gain carbon for the biosynthesis of their cellular components from organic sources. Heterotrophs gain their energy by metabolizing organic materials. Many organisms can use inorganic sources of nutrients such as nitrates, phosphates, and sulfates for the synthesis of proteins and nucleic acids, although some cells require certain preformed subunits, called growth factors. Oxygen is a required nutrient for some cells, a toxic element for others, and still others can grow either in its presence or absence. The complexity of nutrient requirements is inversely related to an organism's biosynthetic capabilities. Also important to the growth of bacteria are several environmental factors, including the temperature, pH of the medium, and the osmotic pressure. Any single species can grow only within a limited range of conditions. However, as a group, microorganisms can grow over a very broad range. The growth of bacteria in the laboratory follows a predictable pattern, with four readily identifiable phases: lag phase, log phase, stationary phase, and death. During these phases the culture undergoes a period of active metabolism without cell multiplication (lag), logarithmic increase in the number of viable cells (log), no further increase in the number of viable cells (stationary), and exponential decrease in the number of viable cells (death).

QUESTIONS

REVIEW

1. Describe the various methods of assaying bacterial *mass.* Which method would you use if you wanted to determine the percentage of viable cells in a culture? The total number of cells?
2. Discuss the relationship between the number of enzymes in a cell and the nutritional requirements of the cell.
3. Discuss the classification of bacteria according to their response to oxygen.
4. Discuss the classification of bacteria according to their optimum temperature for growth.
5. What are the phases of bacterial growth, and what macromolecules are being synthesized during each?

THOUGHT

1. You are attempting to isolate a halophilic aerobic spore forming bacteria which can utilize N_2 gas and CO_2. How would you proceed to isolate such an organism?
2. The bacteria you are culturing in a nutrient medium stop growing when they have reached a cell density of 10^8 cells per milliliter. What possible reasons might you give to explain the stoppage of growth? What experiments would you do to determine which of your explanations is correct?

FURTHER READING

BAILEY, R., and E. SCOTT, *Diagnostic Microbiology.* 4th ed. St. Louis, Mo.: C.V. Mosby, 1974. The last chapter of this book contains a concise section on the formulas and preparation of culture media. It is probably the single best reference book for such information. Another chapter gives formulas and procedures for staining.

Difco Manual of Dehydrated Culture Media and Reagents for Microbiological and Clinical Laboratory Procedures. 9th ed. Detroit: Difco Laboratories, 1971. Difco, one of the leading manufacturers of bacteriological media, has published a book which describes the composition of all of the media that it sells as well as some of the salient features of these media. Many references are given on how the media were formulated and what they can accomplish. To keep the book up to date, the company publishes supplementary literature on new products that it introduces.

CHAPTER 5

DYNAMICS OF MICROBIAL GROWTH IN NATURE

The last chapter was concerned with some principles of microbial growth as studied in the laboratory. The growth of microorganisms in nature is much more complex. Instead of growing as *pure cultures* under defined cultural conditions, organisms in nature exist as parts of *communities* under poorly defined environmental conditions that are often changing. In the laboratory the investigator provides the nutrients necessary to ensure the optimal growth of the culture. In nature food supplies are generally limiting, growth rates tend to be slower, and intense competition develops between populations. Organisms may be highly dependent on the presence of other specific organisms. In the laboratory organisms are often grown in liquid culture, whereas in nature most microorganisms grow in association with solid surfaces—on soil particles, on sediment particles in marine and fresh water, and on plants and animals. In this chapter are emphasized some *principles* which govern the growth and development of populations in nature as well as the relationships between organisms and their environment. Specific examples of these relationships will be considered throughout the text, which is in large measure concerned with the growth of organisms in their natural environment.

PRINCIPLES OF MICROBIAL ECOLOGY

All the living organisms of a given area—the community—interact with the nonliving environment of that area to form an ecological system or *ecosystem,* the unit of biological organization one step above the community. Many major ecosystems exist in the world, each with its particular physical environment and spectrum of living organisms. These ecosystems include seas, streams and rivers, lakes and ponds, marshes, deserts, tundras, grasslands, and various types of forests. Each ecosystem is populated with microorganisms that may be unique to the particular ecosystem. *Ecology* is the study of the relationship of organisms to each other and to their environment. The *function* that an organism performs in its particular *ecosystem* is called its *ecological niche.*

Cycles in Nature

CARBON CYCLE

In order for a major ecosystem to sustain the life characteristic of that particular area, all of the chemical elements that make up living cells must be continuously recycled. A fixed and limited amount of these elements exists on the earth and in the atmosphere. These elements cannot be utilized for the synthesis of cell components, however, unless they are in a specific form. Only if the elements are continuously recycled from nonutilizable to utilizable forms will they be sufficient to support the growth of new organisms. Thus recycling processes for nitrogen, phosphorous, sulfur, oxygen, and carbon are essential. As an example of how recycling processes function, the carbon cycle is diagrammed in Figure 5-1. Other cycles are discussed in Chapter 34.

NUMEROUS ORGANISMS PARTICIPATE IN THE CARBON CYCLE

Typically, plants are eaten by animals which die and decompose through the action of numerous microorganisms. A number of microorganisms are capable of degrading the complex cell constituents to inorganic substances, such as nitrates, sulfates, and phosphates. The cycle begins with the fixation of atmospheric CO_2 into organic compounds by green plants through the process of photosynthesis. Some of the carbohydrate synthesized is utilized by the plant to obtain energy, with the release of CO_2 through the leaves or roots by a process called *respiration,* which is in many ways the opposite of photosynthesis. Plants provide the animal world with organic material. When animals respire they also release CO_2. When animals and plants die, they are ultimately decomposed by heterotrophic microorganisms in the soil, with the production of more microbial cells and certain decomposition products. These chemical transformations of carbon require the activity of organisms in three ecological niches: *producers, consumers,* and *decomposers.* The producers convert CO_2 into organic material. The consumers eat organic matter; the decomposers digest and convert dead plant and animal cell components into small molecules utilizable by both the consumers and producers.

In order for recycling to operate, every major ecosystem requires members of all three functional groups to be present for each chemical element used in cell structures. The loss of any group would break the cycle. In the carbon cycle the producers are algae, cyanobacteria, and other bacteria capable of utilizing the carbon of CO_2 as their sole carbon source. Protozoa are typical consumers, pursuing and eating algae, bacteria, or other smaller protozoa. Most members of the microbial world are able to decompose organic material.

In addition to a supply of chemical elements for the synthesis of cell components, all ecosystems must have access to sources of utilizable energy. In con-

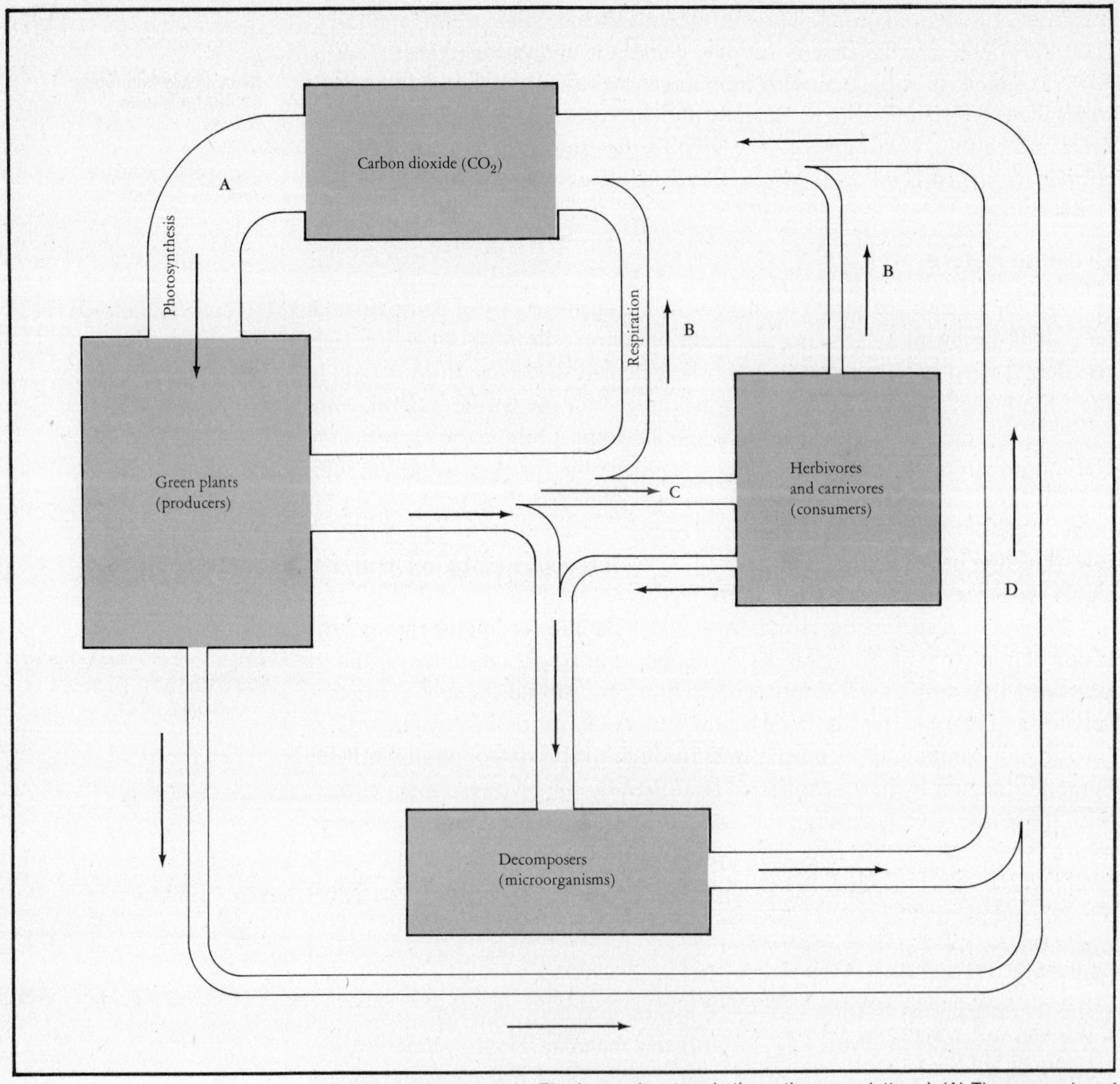

FIGURE 5-1 Carbon cycle. The four major steps in the pathway are lettered: (A) The conversion of inorganic carbon (CO_2) into organic forms by green plants. (B) The conversion of organic carbon into CO_2 through metabolic processes of plants and animals. (C) The eating of green plants by animals. (D) The conversion of organic forms of carbon into CO_2 through decomposition of dead plants and animals by microorganisms. The width of the tubing in the pathway gives an indication of the relative amount of material channeled through that pathway.

ELEMENTS ARE RECYCLED; ENERGY IS NOT

trast to the recycling of chemical elements, energy cannot be reutilized. Thus there is a continual loss of energy from biological systems. To compensate for this outflow there must be a continual inflow. The fundamental energy being continually added to all ecosystems is solar or light energy, but before this is useful, it must be converted into chemical bond energy. This conversion, mediated by chlorophyll-containing plants and microorganisms, makes the energy available to the consumers and decomposers. The requirement for solar energy explains why life does not exist everywhere in and on the earth. Solar energy is confined primarily to a thin shell in the upper levels of the soil and water and to the lower levels of the atmosphere. It does not penetrate to deep soil, rocks, or the depths of the oceans. Any organisms present in these zones must depend for their food on organisms that have access to light.

Populations in Ecosystems

ABILITY TO MULTIPLY IS CRUCIAL TO SURVIVAL OF THE SPECIES

The region of the earth inhabited by living organisms is called the *biosphere,* and the *ecosystems* are its major subunits. Within the total biosphere a wide diversity exists both in the number of different types or species of organisms inhabiting particular ecosystems and in the total quantity of cell material present in any given area. What determines whether or not a particular organism will be found in a particular ecosystem? The fundamental requisite for all organisms in any given environment is that they be capable of multiplying at least sometime during the year. If not, they will soon disappear. In an environment conducive to life, such as the top few inches of fertile soil, there are likely to be a large variety of different species as well as a large quantity of protoplasmic mass. Table 5-1 shows the relative abundance of microorganisms in fertile soil. The bacteria are represented by many different species, and it has been estimated that the top 6 inches of fertile soil may contain more than 2 tons of bacteria and fungi per acre. In harsh environments, exposed to extremes in temperature, moisture, or other environmental conditions, relatively few species are capable of multiplying, and therefore of existing.

The external environment has a more profound effect on conditions inside microbial cells than it does on cells in higher organisms. This is especially obvious in the case of temperature. Multicellular organisms contain sensitive controls in specialized organs; these maintain reasonably constant internal conditions despite wide fluctuations in the external environment. In microorganisms the external and internal environment may be similar, although even bacteria have the ability to preferentially concentrate essential foodstuffs and to keep out some potentially harmful chemicals.

Permease, page 68

TABLE 5-1
Number of Organisms at Different Depths in the Soil (Organisms per Gram of Soil)

Depth (cm)	*Bacteria except Actinomycetes*[a]	*Actinomycetes*	*Algae*	*Fungi*
3–8	9,750,000	2,080,000	25,000	119,000
35–40	570,000	49,000	500	14,000
135–145	1,400	0	0	3,000

[a] Actinomycetes are branching filamentous bacteria that are very plentiful in the soil.
Source: Data taken from M. Alexander, *Introduction to Soil Microbiology,* New York: John Wiley & Sons, 1961.

Actinomycetes, page 260

The true environment of a microorganism is often difficult to identify and measure. The environment immediately surrounding the microbial cell, the *microenvironment,* is most relevant to the cell. The more readily measured gross environment, or *macroenvironment,* has probably far less effect on the cell. To the bacterial cell living within a crumb of soil, for example, the environment inside the crumb is more important than the environment outside it. And not only may the two environments differ radically from each other, but a variety of different environments may actually be found within a single crumb.

Competition for Survival

Although many different species may be capable of living in the same microenvironment, it has been observed repeatedly that few species—sometimes only one—actually occupy a specific microhabitat. To return to the previous example, a large number of different species are found within the crumb of soil. If the crumb is carefully dissected into squares of about 70 μm on each side, only a single species will be found living as a microcolony within any one of these squares. Thus the heterogeneous variety of species in the total crumb disappears when we consider each of its individual microenvironments. The species best adapted to live in a particular microenvironment is apparently the one that occupies it to the exclusion of all others. Thus the organism best fitted to its environment is the one that endures.

TWO FACTORS IMPORTANT FOR SURVIVAL: ABILITY TO MULTIPLY RAPIDLY AND ABILITY TO WITHSTAND ADVERSE CONDITIONS

The fitness of an organism, as measured by its ability to compete successfully for its ecological niche or for its microhabitat, is generally related to the *rate* at which the organism multiplies, as well as to its ability to withstand adverse conditions of the environment. If cells of two different species are introduced into a microhabitat which is suitable for both, the organism which multiplies faster will soon become dominant while the other species may disappear completely. The complete takeover by one of the species is especially likely when the two species have similar requirements for cell growth and one of these requirements is in limited supply. Thus, with a limited supply of food, the organism which multiplies faster will yield the greater cell population, thus utilizing most of the limited food supply. The more slowly growing species is then deprived of more and more food and may soon be eliminated. In environments such as the soil a species will only be successful if it is capable of withstanding periods of dryness and starvation. Many soil organisms can develop into forms such as endospores and cysts which are resistant to such adverse conditions.

Perhaps nowhere in the living world is competition more fierce and the results of competition more quickly evident than among microorganisms. This is especially true of different organisms growing in the same liquid culture. Because cells are often able to divide at least once every several hours, any small difference in their generation times will result in a very large difference in the total number of cells of each species which exist after a relatively short time. For example, if 10 cells each of organisms A and B are growing in the same natural environment, under conditions in which the generation time of A is 100 minutes and that of B is 99 minutes, and if the number of cells reaches about 10^9, there will be 1.35

more cells of B than of A. If the cells are then occasionally diluted so that they can continue to multiply, the ratio of B to A will increase to the point where there are very few A cells left.

One or more of these considerations probably explains the observation that the bacterial composition of the human intestinal flora remains quite stable with time. Despite swallowing of large populations of strains closely related to strains already present, the new strains have great difficulty in becoming established. The best evidence available suggests that the resident population competes very successfully for a limited food supply. Furthermore, the resident population may produce toxic metabolites which tend to inhibit the growth of organisms not closely related to the indigenous population. These factors may represent a natural defense mechanism providing protection against growth of intestinal pathogens.

In some cases, especially evident in colonies growing in fixed positions, the cells only hamper one another's growth by competing for the same nutrients. Strict competition can be observed when bacterial colonies grow very close to one another on agar plates. The cells in one colony compete with those of the neighboring colony. The result is that colonies close to one another are much smaller than those which are well separated. (Figure 5-2).

Interdependence between Populations

Many populations' existence depends on the presence of other populations. For example, anaerobic organisms may exist in association with facultative anaerobes which remove traces of oxygen, thus creating anaerobic conditions. The interrelationships between organisms in a medical context are covered more fully in Chapter 19.

Changes in Microbial Populations

Environments do not remain constant, and changes in an environment bring about changes in its population. The best adapted organism several inches beneath the surface of an untilled field probably fails to remain the best adapted if

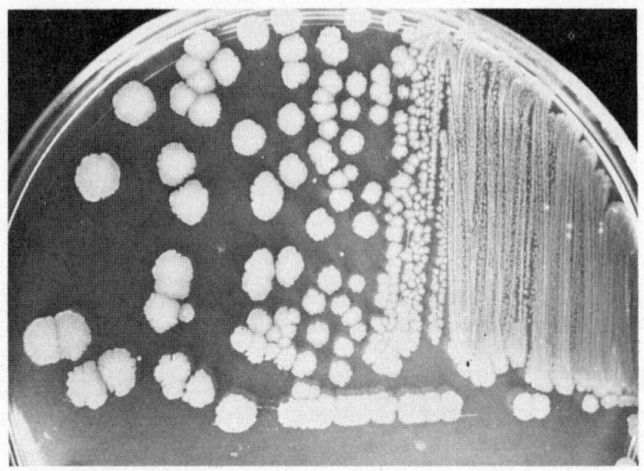

FIGURE 5-2
Competition for nutrients among bacteria. The closer the cells are to one another, the smaller the colonies. (Courtesy of J. P. Dalmasso.)

the field is plowed up, fertilized, and irrigated. If the organism is to survive under these new conditions, it must in some way adapt to the changed environment. Cells can adapt in a number of ways. The same cells may be able to synthesize new enzymes which help them cope with the new environment. As a general rule, cells will only synthesize the enzymes that they absolutely require for growth. To synthesize others is a waste of both amino acids and the energy required for synthesis. Under changed environmental conditions, however, additional or different enzymes may become important. The cell will then synthesize these enzymes, and it may stop synthesizing other enzymes which have lost their usefulness in the changed environment. Frequently the *preexisting* population changes. Thus a *mutant* within a preexisting population may be especially well adapted to the changed environment, whereas the mutant was in the minority in the old environment. It may then become the dominant organism in the new environment.

A BETTER ADAPTED MUTANT MAY BECOME DOMINANT

The emergence of antibiotic-resistant organisms in hospital environments illustrates this type of adaptation. Antibiotic-resistant mutants always occur as a very small proportion of the total cell population. In the absence of antibiotics they generally multiply more slowly than the antibiotic-sensitive cells and are therefore *selected against.* In the presence of antibiotics, however, the sensitive cells are killed, leaving only the antibiotic-resistant cells to multiply. Accordingly, an antibiotic resistant population replaces the sensitive population. The dramatic increase in antibiotic-resistant cells in hospitals since the introduction of antibiotics is in part the result of this phenomenon.

A NEW SPECIES MAY REPLACE A PREEXISTING SPECIES

Another possible alteration in the population in response to a changed environment is the replacement of the *dominant species* by *another species,* originally present only as a minor species in the ecosystem. Thus, although the latter species was not able to compete well with other organisms in its old environment, its new environment is much more favorable to its multiplication, and it now becomes the dominant organism. One example of such a situation can be demonstrated on the skin. Gram-positive species of bacteria are the dominant organisms in the armpit (axilla), although small numbers of Gram-negative organisms are also present. The Gram-positive organisms excrete enzymes that degrade the greasy secretions of the glands; this produces fatty acids which inhibit the growth of the Gram-negative cells and give rise to "body odor." If a deodorant containing an agent which kills Gram-positive organisms is applied, the population gradually shifts, principally to a Gram-negative flora of about the same cell density as the original Gram-positive flora. The Gram-positive flora returns as soon as the deodorant use is discontinued.

In addition to changes in the environment that result from external sources, the growth and metabolism of organisms themselves may change the environment dramatically. Foodstuffs may become depleted, and a variety of waste products, many of which are toxic, may accumulate. In some environments the changing conditions bring about a highly ordered and predictable *succession* of organisms: first, one species of organism, then another, then a third becomes dominant. An example of such a succession is that which occurs in unpasteurized milk. Such milk contains a variety of different organisms, including various species of bacteria, yeasts, and molds, which are derived mainly from the immediate environment around the cow. Initially the dominant organism is a bacterium (*Streptococcus lactis*) that breaks down the milk sugar lactose to lactic acid. The acid inhibits the growth of most other organisms in the milk, and eventually enough acid is produced to even inhibit the further growth of the *Streptococcus.* The acid also sours the milk and *denatures* (curdles) the proteins in it. However, other bacterial species (*Lactobacillus casei* and *Lactobacillus bulgaricus*) are capable of multiplying in this highly acidic environment. These species break down the rest of the lactose and form even more acid until their growth is also inhibited. Yeasts and molds, which grow very well in this highly acidic environment, now become the dominant group and convert the lactic acid into nonacidic products. Because all of the milk sugar has already been broken down by the bacteria, the major genera of bacteria, *Streptococcus* and *Lactobacillus,* cannot resume multiplication because they lack this food. Food, however, is still available in the form of protein (casein), which can be utilized for energy by members of the genus *Bacillus,* which excrete proteolytic enzymes that digest the protein. This breakdown of protein, known as *putrefaction,* results in a product which is completely clear and very odorous. The milk thus goes through a succession of changes with time, first souring and finally putrefying. These changes are brought about by organisms which themselves have undergone a succession as dominant populations.

If milk is *pasteurized* after being drawn from the cow, a markedly different succession may occur. The pasteurization kills most of the nonendospore-forming organisms (streptococci, lactobacilli, yeasts, and molds) and the heat-resistant spores of the bacilli may predominate. These organisms break down the milk proteins, resulting in putrefaction. The milk does not sour, it rots (putrefies).

Quantitative Aspects of Growth of Population in Nature

SIZE OF POPULATIONS MAY REMAIN CONSTANT THROUGH BALANCE OF CELL MULTIPLICATION AND DEATH

The number and types of microorganisms in many ecosystems remain relatively constant over a reasonably long period of time as long as the environment remains relatively constant. The types of organisms that are present are those that have adapted best to the ecosystem. Population size is generally determined by the outcome of the competition for limiting nutrients because in natural environments the food supply is generally the limiting factor. However, although the number of cells of each species remains relatively constant, the cells are not inert. Generally there is a large turnover of the population; some cells may be killed and decompose, or are devoured, while others multiply to take their place, keeping the total number constant. Actually the population is in the log phase of growth, but the limited amount of food available limits the size that the population can reach. Furthermore, the environmental conditions are seldom optimal for the growth of each microorganism, so its generation time is much longer than it would be under optimal conditions.

Growth of Bacteria in a Hospital Environment

A natural habitat which illustrates a number of interesting and unusual aspects of bacterial growth is the reservoir of distilled water in mist therapy units commonly used by hospitals to combat lung diseases. In one published case approximately 10^7 viable cells of *Pseudomonas aeruginosa* were found per milliliter of the distilled water in these units. When these cells were diluted into sterile distilled water at a cell density of approximately 10^3 per milliliter and the culture incubated, there were approximately 10^7 cells per milliliter after 48 hours. The cell density remained at this level for at least 42 days.

Such experimental results demonstrate in a practical sense some points about the growth of bacteria in nature. It is clear that the *Pseudomonas* in this case were able to grow in a medium which contained a very low level of nutrients. It is not clear what these nutrients were, but they are likely to have been organic compounds that were absorbed either from the air or from inadequately cleaned units. The fact that bacteria can grow in such dilute medium indicates that they are able to effectively concentrate nutrients inside themselves.

Species of *Pseudomonas* are commonly found in such apparently hostile environments because members of this genus can utilize a wide variety of organic materials as their sources of carbon and energy. The limitation in nutrients, however, apparently prevents the cell density from increasing much above 10^7 cells per milliliter. The cell concentration is not high enough to result in a noticeably turbid solution, especially because the cells are only one-fourth their normal size, as would be expected from their slow generation time. However, their concentration is high enough to represent a definite menace if the contaminated water is sprayed into the lungs of a weakened patient. The generation time of the cells growing in the mist therapy units was about 4.5 hours, which is not the fastest at which the organism is capable of growing, but still fast enough to reach high levels in a relatively short time. Their growth was probably slower than the optimum because of the lower-than-optimal growth temperature.

In this situation a dangerous organism was able to multiply without detection (no visible turbidity) in an inhospitable environment, distilled water. Such organisms can then be transferred to other locations in the hospital via the airborne route, fomites, or personnel. It is not surprising that *Pseudomonas aeruginosa* is an important causative agent of hospital-acquired infections.

SUMMARY

Microorganisms are ubiquitous in nature and are found in areas which will support no other forms of life. The organisms living in a particular environment have become adapted to the physical and nutritional aspects of that environment. Their interactions with other organisms may include competition, generally for limiting sources of food, and interdependence on one another. Organisms fill ecological niches in any ecosystem and may be classified as producers, consumers, or decomposers. The combined operation of these groups results in the recycling of the chemical elements found in biological materials. This recycling is essential to the maintenance of life.

The organismal content of an ecosystem remains relatively constant as long as the environment does not change. However, although the numbers of each group of microorganisms do not change significantly, there is a continual turnover of organisms, some dying, others multiplying. The growth of microorganisms in nature is generally considerably slower than it is in the laboratory even though the organisms are in the log phase of growth.

QUESTIONS

REVIEW

1. Give the four major steps in the carbon cycle, and give the chemical form that the carbon is in at each stage.
2. List the major ecological niches that must be filled in nature, and give examples of organisms that fill each niche.
3. What is the function of members of each ecological niche?
4. Describe the succession of events in pasteurized and unpasteurized milk.
5. Compare the recycling of chemical elements and energy, and the consequences of this recycling or lack of it.

THOUGHT

1. What qualities must members of the genus *Pseudomonas* have that allow them to grow in presumably distilled water?
2. By what experimental approach could you decide between adaptation of the whole population and mutation in accounting for changes in microbial populations in natural environments?

FURTHER READING

ALEXANDER, M.J., *Microbial Ecology.* New York: John Wiley & Sons, 1971. A scholarly, lucid, and somewhat detailed account of microbial ecology.

BROCK, T., *Principles of Microbial Ecology.* Englewood Cliffs, N.J.: Prentice-Hall, 1966. The first textbook on this subject by a leader in the field.

CHET, I., and R. MITCHELL, "Ecological Aspects of Microbial Chemotactic Behavior," *Annual Review of Microbiology* 30:221–239 (1976). A recent thorough review on the ecological significance of chemotaxis in bacteria, slime molds, fungi, and algae.

FAVERO, M., N. PETERSON, L. CARSON, W. BOND, and C. HINDMAN, "Gram-Negative Water Bacteria in Hemodialysis Systems," *Health Laboratory Science* 12:321–334 (1975). A well-written, easily understandable review of the conditions under which water bacteria can grow and the problems to which they can give rise.

JANICK, J., C. NOLLER, and C. RHYKERD, "The Cycles of Plant and Animal Nutrition," *Scientific American* (September 1976).

MARPLES, M., "Life on the Human Skin," *Scientific American* (January 1969).

CHAPTER 6

CONTROL OF MICROBIAL GROWTH

To a large extent, routine control of undesirable microbes is achieved by their mechanical removal (washing and scrubbing), by drying, and by exposure to the competitive action of other microbes. However, it is frequently important to use special methods for decontaminating materials and keeping them free of unwanted microbes. Such technology plays a role, for example, in hospitals, in the food and beverage industry, and in space exploration. Some of the major approaches to accomplishing microbial control are discussed here.

DEFINITIONS

Sterilization

Spontaneous generation, page 6

Removing or killing *all* microbes on an object or in any material is called *sterilization.* "Killing" means making microbes unable to reproduce *even under the most favorable growth conditions.* Unhappily, history is full of episodes in which microbes that had apparently been killed were able to grow later on when conditions became favorable. For example, in the argument over spontaneous generation, Pasteur showed that urine heated to a temperature sufficiently high to kill bacteria would remain "sterile" if protected from airborne contamination. However, one of his detractors demonstrated bacterial growth in such urine upon addition of sterile alkali. Only later was it determined that bacterial spores had survived heating of the urine but were incapable of germinating until the normal acidity of the urine was neutralized by the added alkali.

Disinfection

Disinfection is the reduction or elimination of pathogenic microbes in or on materials so that they are no longer a hazard. Unlike sterilization, disinfection implies that some living microbes may persist. The term *disinfectant* is generally used for chemical agents employed to disinfect inanimate objects. In the United States the term *antiseptic* is generally used to indicate a nontoxic disinfectant suitable for use on animal tissues. *Decontamination* is often used interchangeably with disinfection, but it implies a broader role including inactivation or removal of microbial toxins as well as of living microbial pathogens themselves.

Germicide

An agent capable of killing microbes rapidly is a *germicide.* Special terms (such as bactericide, fungicide, and viricide) are used to indicate killing action against specific microbial groups. Related terms (such as bacteriostatic and fungistatic) indicate that an antimicrobial agent is primarily inhibitory in its action, preventing growth without substantial killing. Growth can resume if the inhibitory agent is neutralized or diluted. It is important to recognize that a germicidal agent against one species of microbe may be only inhibitory against another, and that the nature of its action may change depending on its concentration, pH, temperature, and other factors.

PRINCIPLES OF MICROBIAL KILLING

There have evolved certain generalizations concerning the killing of bacteria, and presumably these apply to other microbes as well. They may be summarized as follows.

Only a fraction of the microbes present in or on a given material die during any given time interval. As pointed out in Chapter 4, death of microbes usually follows a logarithmic curve, which becomes a straight-line graph if the logarithm of viable microbes is plotted against time. Such a curve means that a constant fraction of the surviving organisms is being killed per unit time. Thus, if 60 percent of the organisms are killed in the first 2 minutes by a given treatment, then 60 percent of those surviving the first 2 minutes will be killed during the next 2 minutes. Curves of this kind never quite reach "zero organisms." To choose another example, if the concentration of organisms in a fluid was 100 per milliliter to begin with and 90 percent were killed every minute, we would have after 3 minutes the seemingly strange result of one-tenth of an organism per milliliter. This result does not seem so peculiar, however, if we interpret it (as we should) as indicating one organism per 10 milliliters, or, to be more precise, as indicating the odds of finding a living organism in any milliliter to be one in ten. Sterility can thus be considered to be the state where the probability of finding even a single viable organism in the total volume involved, or of finding such an organism on the object we are dealing with, is extremely low. Obviously, the larger the volume the greater the probability of finding a living organism.

Death phase, page 110

LOGARITHMIC DEATH CURVE

The time it takes to achieve sterility depends on the number of organisms present on a material at the beginning. Thus sterilization of a heavily contaminated substance would require one unit increase in time for every tenfold increase in the number of organisms (Figure 6-1). This is one reason why prevention of contamination and mechanical methods of removing microbes (such as scrubbing, mopping, and washing) are important preliminaries to the successful use of disinfecting or sterilizing agents.

Not all microorganisms have the same susceptibility to sterilizing agents. There is variation among and within species. There is also variation with growth phase, most organisms being at peak susceptibility when in the logarithmic phase of growth. Spores of microorganisms are generally more resistant than the vegetative forms. Fungal spores are not as resistant as bacterial endospores, although many of those responsible for food spoilage easily resist pasteurization. Finally, some viruses (such as the hepatitis virus) are killed relatively slowly by heat and disinfectants.

HEAT AS AN ANTIMICROBIAL AGENT

Use of heat for microbial control long antedates the discovery of the microbial world and remains to this day the most generally satisfactory control method. Heat is fast, reliable, and cheap, and it does not introduce toxic substances into the material being treated. Use of heat is limited, however, to substances that can withstand the temperatures required for microbial killing and to objects small enough to fit into a suitable container. Cooking of food is a familiar example of

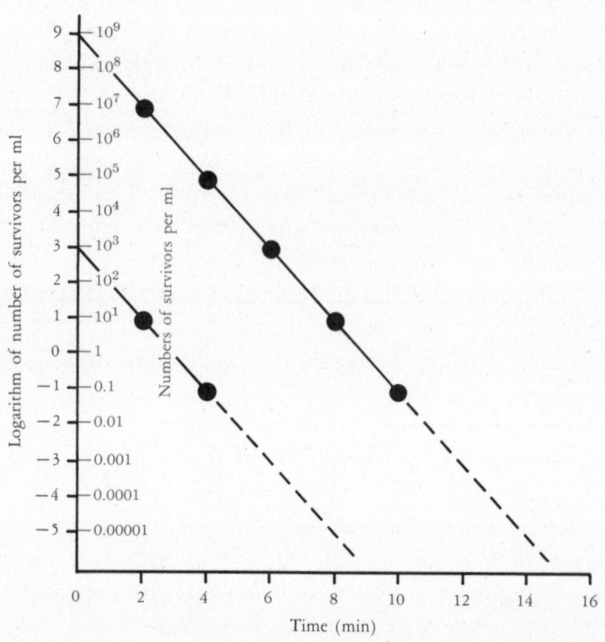

FIGURE 6-1
A bacterial killing curve. Notice that for each tenfold increase in the number of organisms at the start, it takes one additional unit of time to achieve satisfactory killing (in this example, 6 more minutes to kill 10^9 organisms than to kill 10^3 organisms).

the use of heat for microbial killing. However, a "well-done" appearance does not always guarantee that a food is safe to eat. Temperatures sufficient to kill the microbes in question must be maintained for an adequate time *throughout* the material being heated, and not just on its surface.

Dry Heat

Heating is frequently used to kill microorganisms in materials other than food. Sterilization of laboratory equipment and glassware is a familiar application, and similar techniques are also used on landing modules in space exploration. For example, the Viking spacecraft that landed on Mars was sterilized by heating it between 111°C and 180°C for 1½ to 2 hours with static air. Circulating hot air sterilizes in half the time of static air because of more efficient transfer of heat to laboratory glassware, and generally dry heat requires more time than wet heat to kill microbes.

Boiling

The use of boiling to prevent disease extends back at least to the time of Aristotle, who advised Alexander the Great to have his armies boil their drinking water. Under ordinary circumstances, with concentrations of microorganisms of less than 1 million per milliliter, suspensions of vegetative cells and eucaryotic spores can be sterilized in boiling water at 100°C in about 10 minutes. Most viruses also succumb, although large concentrations of hepatitis virus and scrapie agent may possibly survive boiling for longer times. Some bacterial endospores, including certain strains of *Clostridium perfringens* and *Clostridium botulinum,* survive boiling for hours.

Scrapie, page 350

When heat is applied to water at sea level, the temperature rises until it reaches 100°C (212°F). An additional amount of heat energy (heat of vaporization) is then taken up by the water without a change in temperature, whereupon boiling begins to occur. The temperature does not go above 100°C, and at high altitudes water even boils several degrees lower than this because of the reduced atmospheric pressure. Alkali is sometimes added to boiling water to increase its killing power against bacterial endospores. Even with such methods, boiling cannot be relied upon to kill the more resistant strains of spores. It is therefore not a reliable sterilizing technique.

Pressure Cooking and Autoclaves

The first pressure cookers were probably used in the early 1800s to produce canned food for Napoleon's armies. These devices simply enclosed the vessel containing heated water so that pressures above those of the atomosphere might be attained. A suitable safety valve was included to prevent an explosion. Water and steam at temperatures above 100°C could thus be used, killing even the most heat-resistant microbes in a few minutes.

The modern autoclave is, in principle, a sophisticated pressure cooker with mechanisms for regulating the steam pressure and ensuring complete evacuation of air from the chamber (Figure 6-2). The presence of air would allow an increase in pressure within the chamber without a corresponding increase in temperature. Thus it is important always to check the temperature as well as the pressure, because the latter is of little or no significance in killing microbes. Superheated steam (dry steam) is likewise unsatisfactory and kills at slow rates, similar to those of hot air. The pressures and temperatures inside an autoclave should correspond, as illustrated in Figure 6-3. The conditions usually employed are 15 pounds pressure above atmospheric pressure and 121°C (250°F).

IMPORTANCE OF WET HEAT

The main reason for the great effectiveness of the autoclave as compared to the hot air oven relates to the fact that heat coagulation of proteins, including essential enzymes, occurs more readily under moist conditions than in those which are dry. Release of the latent heat of vaporization by steam condensing on organisms may play a role in the coagulation of proteins and rapid killing. On the other hand, dry proteins are resistant to denaturation, and the killing of organisms with dry heat occurs by unknown mechanisms. Killing appears to take place almost equally well whether or not oxygen is present, indicating that oxidation of microbial substances by atmospheric oxygen is not the primary cause of death.

The autoclave is the simplest and most consistently effective means of sterilizing most objects. The following are some practical aspects of its use.

1. *Air trapping.* The importance of removing air from the chamber has already been mentioned. Because steam displaces air downward in the

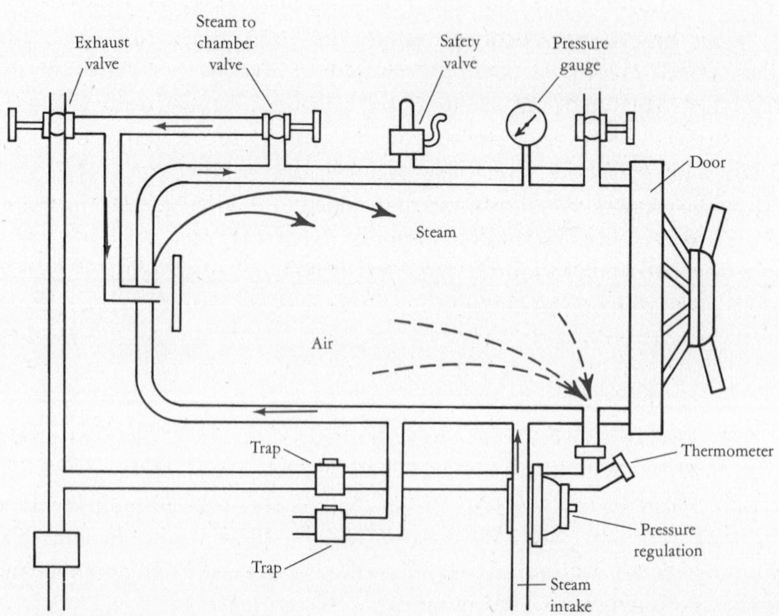

FIGURE 6-2
Steam-jacketed autoclave. Entering steam displaces air downward and out through a port in the bottom of the chamber. Objects should be placed in the autoclave so as to avoid trapping air.

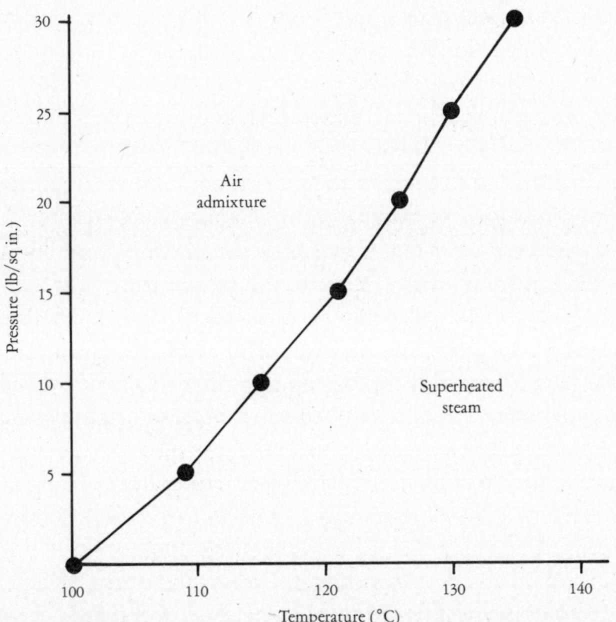

FIGURE 6-3
Relationship between temperature and pressure of steam. Operating conditions falling to the left or right of the curve are unsatisfactory for autoclaving.

chamber, long thin containers should not be inserted into the autoclave in an upright position.

2. *Heat penetration.* A cold object placed in an autoclave takes time—over and above the time at which the chamber temperature reaches 121°C—to warm up. For example, a 1-liter container of fluid must be autoclaved for 25 minutes after the chamber reaches 121°C, but a 4-liter container requires an hour for sterilization.
3. *Contact with steam.* Dry microorganisms protected from contact with steam (for example, within oil, occlusive wrappings, or containers of talcum powder) cannot reliably be killed by autoclaving.
4. *Heat indicators.* Test tubes and tapes containing a heat-sensitive chemical indicator are often used with objects when they are autoclaved. The indicator changes color during the autoclaving and thus gives a visual means of checking whether the objects have been subjected to adequate heat. A changed indicator does not, of course, always indicate that the object is sterile, because heating may not have been uniform, and even wrapped objects may have organisms reintroduced by insects, water, air, and so on.

USE OF AUTOCLAVES

Tubes or envelopes containing large numbers of heat-resistant spores of the nonpathogenic bacterium *Bacillus stearothermophilus,* which act as biological indicators, are also frequently included with packs of materials being autoclaved. Death of these spores indicates adequate killing at the point where the tube or envelope was placed, preferably near the center of the pack.

5. *Elevated boiling points under pressure.* Fluids in the autoclave are prevented from boiling, even though well above their normal boiling point, by the elevated pressure. If at the end of the period of autoclaving, valves are opened to release this pressure, these fluids will immediately boil—and may even explode their containers. The pressure must be maintained to prevent them from boiling until temperatures have dropped below the boiling point at atmospheric pressure.
6. *Deleterious effects of 121°C heat on some materials.* Most autoclaves have a mechanism allowing operation at temperatures lower than 121°C. The use of lower temperatures is satisfactory for materials that can be tested for successful sterilization before being put into use. For example, certain heat-sensitive bacteriological media are sterilized at 115°C. This temperature is usually effective because heat-resistant microbial contaminants are rarely present in the media. Moreover, samples of such autoclaved media are tested for sterility before being used.
7. *Prevention of recontamination.* Objects to be autoclaved are usually wrapped in paper or cloth to allow penetration of steam during sterilization and to prevent recontamination thereafter. When wet, these coverings are readily permeable to bacteria and should therefore be allowed to dry before removal from the autoclave chamber.

Pasteurization

In the 1800s the French wine industry was simultaneously threatened by foreign competition and by a microbial invasion from within. Good wine is the result of properly controlled conversion of fruit sugars to ethyl alcohol by yeast. Bacteria and molds from the environment may become established in the wine, performing unfortunate degradations of the alcohol and producing various ill-flavored metabolites. Pasteur found that with moderate heating of wine, at just the right conditions of time and temperature, these spoilage microbes could be killed without significantly changing the taste of the wine. He verified his findings by supplying heated and unheated wines to ships of the French Navy, which consistently found the heated wines to be superior. This process of controlled heating at temperatures below boiling (now called *pasteurization*) helped save the French wine industry and is widely used today for ridding milk and other foods of disease-causing bacteria. Certain items of hospital equipment, such as oxygen masks, which may not withstand the higher temperatures of the autoclave, can also be pasteurized. The temperatures and times used vary according to the organisms present and the heat-stability of the material. Pasteurization is not equivalent to sterilization. However, pasteurization causes a substantial reduction in the numbers of microbes present and thereby retards spoilage in foodstuffs.

Tyndallization, page 7

As indicated in Chapter 1, repeated controlled heating at relatively low temperatures (tyndallization) can sterilize, because it will kill even spore-forming bacteria, provided the suspending medium enables their spores to germinate. Recommended procedures for tyndallization specify heating at 100°C for 30 minutes on three successive days or at 60°C for 1 hour on five successive days. The material is usually incubated at 37°C following each period of heating.

FILTRATION

IMPORTANCE OF STRUCTURE AND COMPOSITION OF FILTERS

Filters capable of removing bacteria from fluids were devised during the last decade of the nineteenth century. They were made from metallic compounds of silica, including porcelain, diatomaceous earth, or asbestos, partially fused by intense heat. Filters of this type are still used today. Their action is often thought to be that of a microscopic sieve, holding back microbes while letting the suspending fluid pour through the small holes in the sieve. In fact the passages that run through such filters are very tortuous, some with large open areas and many others blind. The diameter of the passage is often considerably larger than that of the microbes they retain, and trapping of microbes by electrical forces may be involved because the walls of the filter passages are electrically charged.

From these considerations it is easy to see that the effectiveness of such microbial filters varies not only with their pore size but also with the chemical nature of the suspending fluid and the amount of pressure used to transfer it across the filter. These filters are often used satisfactorily to remove microbial agents larger than viruses from aqueous fluids. They are relatively inexpensive and unlikely to clog. They are not suitable for some other purposes, however, because of the relatively large amount of filtrate they absorb, the possible introduction of metallic ions into the filtrate, or the adsorption of biological substances such as enzymes from the filtrate.

MEMBRANE FILTERS

In recent years membrane filters composed of compounds, such as cellulose acetate, have been widely used in the laboratory and in industry. Beer may now be filtered by this means rather than being pasteurized. The filters are paper thin and are produced with graded pore sizes extending below the dimensions of the smallest known viruses. They are relatively inert chemically and adsorb very little of the suspending fluid or its biologically important constituents. Furthermore, they become semitransparent when immersion oil is applied, thus allowing microscopic inspection for any microbes that adhere to their surface. Their chief disadvantages are that they are expensive and they clog easily.

Filters also figure in plans for space exploration, being used to sterilize heat-susceptible fluids in spacecraft. Because scientists want to avoid introducing terrestrial microbes to other planets, the International Committee on Space Research has specified that when a vehicle is landed on another planet, the probability of its carrying a single microbe must be less than 1 in 10,000. As indicated previously, heat is a highly reliable method of ridding the metallic surfaces of space vehicles of living microbes. But what about the reliability of the filters needed to free heat-labile fluids of microbes? Studies comparing different kinds of filters have indicated that the older types—diatomaceous earth, porcelain, and asbestos—were not sufficiently reliable. Only membrane filters gave satisfactory results for use in sterilizing heat-susceptible fluids in space vehicles. Even so, it is very difficult to guarantee the complete absence of every microorganism, although we can usually state with some precision the probability of a microorganism escaping the procedure designed to remove it.

Except for the smallest pore size membranes, filters will not sterilize fluids containing viruses. Bacteriophages and other viruses readily pass through most bacteriologic filters and, if present in the original suspension, will be present in the filtrate.

CHEMICAL AGENTS

Most chemical methods are unreliable for sterilization, and experience has repeatedly shown that it is dangerous to assume that they are effective for this purpose. Chemical agents do, however, have value as disinfectants and preservatives. Table 6-1 gives some of the antimicrobial uses of chemical agents.

In chemical disinfection the best results are obtained when initial microbial counts are low and when the surfaces to be disinfected are clean and free of interfering substances, such as protein. A very large number of chemical disinfecting agents are in use. The major groups of these chemical agents and their activities under ordinary conditions of use are shown in Table 6-2. Heat generally enhances the action of these disinfectants. Chemical disinfectants work very slowly against bacterial endospores, against the bacteria that cause tuberculosis, and against some viruses. The hepatitis viruses represent a special problem, because they require extended treatment with viricidal agents (Table 6-3). Obviously, the choice of chemical and how it is used depends on the nature of the contaminant.

LIMITATIONS OF DISINFECTANTS

The ingredients of soaps, medicines, deodorants, contact lens solutions, foodstuffs, and many other everyday items often include an antimicrobial agent. Meats may contain potassium nitrate, and apple cider may have sodium benzoate added to it; heparin (a substance used to prevent blood clots) may contain a phenol derivative; contact lens solutions include a preservative; and a leather belt

TABLE 6-1
The Principal Uses of Germicidal Chemicals

Chemical Agent	*Consumer Products*[a]	*Cooling Water*	*Food Products*	*Paint*	*Paper*	*Wood Preservatives*
Halogens, halogen compounds	x	x		x	x	x
Phenolics	x		x	x		x
Quaternary ammonium compounds	x	x			x	
Arsenicals	x					x
Mercurials	x			x		
Other metallic compounds	x	x	x	x		

[a] Consumer products include cosmetics, toiletries, disinfectants, proprietary drugs, soaps and other cleansers, swimming pool algicides, waxes, and polishes.

Source: C. H. Kline and Company.

TABLE 6-2
The Major Groups of Chemical Agents Useful in Disinfection

Type of Disinfectant	*Level of Activity*[a]
Formaldehyde	
Formaldehyde in alcohol	High
Alkaline glutaraldehyde	High
Formalin	Intermediate
Halogens	
Iodine in alcohol or water	Intermediate
Iodophores	Low
Chlorine, chlorine compounds	Intermediate
Phenolics	Intermediate
Alcohols	
Ethyl	Intermediate
Isopropyl	Low
Quaternary ammonium compounds	Low
Mercurials	Low

[a] High: -cidal activity against vegetative bacteria, bacterial endospores, fungi, and human viruses except hepatitis.
Intermediate: -cidal activity against vegetative bacteria, fungi, and human viruses except hepatitis; activity inadequate against endospores.
Low: -cidal activity against most vegetative bacteria and fungi; activity inadequate against tubercle bacilli, endospores, and viruses.

may be treated with a mercury compound plus one or more phenol derivatives. The purpose of these agents is to prevent or retard spoilage by microbes which are inevitably introduced from the environment. Unfortunately, they do not always work, and toxic effects or hypersensitivity states sometimes occur in persons repeatedly and unwittingly exposed to them.

Alcohols

Ethyl and isopropyl alcohol are rapidly effective in killing vegetative bacteria and fungi but are of little value against spores and some medically important viruses. In the past some doctors, dentists, and other medical personnel prepared their instruments by soaking them in alcohol. This sometimes resulted in serious illness and death because of the failure of alcohol to kill hepatitis viruses on the

TABLE 6-3
Chemical Agents Useful against Hepatitis Viruses[a]

Agent	*Time*
Sodium hypochlorite 0.5–1%	30 minutes
Formalin 40%	12 hours
Alkaline glutaraldehyde 2%	10 hours

[a] For use only when heat or ethylene oxide gas sterilization are impossible.

instruments. Alcohols act by coagulating essential proteins, so they are only effective when water is present. The final concentration of alcohol in the material being disinfected should be about 75 percent by volume for optimal results. Because of the difficulty in hydrating spores, however, they are frequently quite resistant to alcohols. Valuable as disinfecting agents by themselves, alcohols can also enhance the activity of other chemical agents, such as iodine, chlorhexidine, and quaternary ammonium compounds. Use of alcohols is, of course, limited to materials resistant to their solvent action.

Chlorine

Chlorine is widely used in disinfection in the form of free gas, in chlorine-releasing organic compounds such as chloramines, and as hypochlorite ion. Chlorine gas reacts with water to produce hypochlorite, which is thought to act on microbes by oxidation of essential proteins. Household bleaches (such as Clorox) consist of about 5.25 percent sodium hypochlorite solution, and three tablespoons of such preparations in a gallon of water results in a solution of about 600 parts per million (ppm) of available chlorine. This is several hundred times the amount lethal to most pathogenic microbes, but these high concentrations are usually necessary to speed action and also because organic material, which is often present, greatly interferes with the disinfectant activity. Use of chlorine disinfectants is limited to materials resistant to the corrosive action of these compounds. Certain kinds of rubber and certain metals, for example, are broken down by chlorine.

Iodine

Iodine, like chlorine, is very active against most microbial species. Its mode of action, however, is thought to relate to iodination of the amino acid tyrosine present in the enzymes and structural proteins of microorganisms. Like several other disinfectants, iodine has enhanced activity when dissolved in alcohol to produce a tincture. However, simple swabbing of the skin with tincture of iodine does not kill all the spores of *Clostridium perfringens* when they are experimentally applied, probably because organic material interferes with the action of the iodine. Compounds of iodine with surface-active agents (iodophores) have been very popular because they do not irritate the skin or sting as much as tincture of iodine when applied to wounds, and they are not as likely to stain. Careful evaluation, however, shows that these benefits are accompanied by a substantial reduction of antimicrobial activity. Tincture of iodine and the iodophores are widely used as antiseptics. Their use is contraindicated in the occasional individual with allergy to iodine.

Formaldehyde

Eight-percent formaldehyde is an extremely active disinfectant, killing most forms of microbial life in minutes. Lower concentrations, however, are only slowly active against some bacterial endospores and some viruses. Failure to recognize this has caused several near disasters from the use of formaldehyde to kill living agents in vaccines. Solutions of formaldehyde also have limited use because of their irritating vapors, and in recent years other aldehydes (such as glutaraldehyde) have been used instead. Like formaldehyde, these appear to act by combining with amino groups in microbial proteins.

Phenolics

Phenolics are a very large group of compounds more or less chemically related to phenol (carbolic acid). The group includes cresols and xylenols (Figure 6-4). In high concentration (from 5 to 10 percent) they kill most bacteria, including the tubercle bacillus (*Mycobacterium tuberculosis*). Five-percent phenol can also be used against all the major groups of viruses, including picornaviruses, but some of the more commonly used derivatives of phenol lack sufficient antiviral activity. Lower concentrations of phenol (or its derivatives) may be ineffective. In fact some Gram-positive cocci and Gram-negative rod-shaped bacteria not only grow in the presence of 0.1 percent phenol but are actually able to utilize it. The major advantage of phenolics is the ability of most of them to remain active when mixed with soaps and detergents. An exception to this rule occurs with some of the halogenated xylenols and cresols, which are inactivated by organic materials. Phenolics are thought to act mainly against the cytoplasmic membrane of microorganisms. Their mode of action against viruses probably involves the lipid outer coat in most cases.

Hexachlorophene

One phenol derivative, hexachlorophene, deserves special mention. It was once widely used in household hand soaps, deodorants, and the antiseptic scrub of surgeons. It has substantial bacteriostatic and bactericidal activity against *Staphylococcus aureus* and tends to be retained by the skin, so that its antimicrobial activity increases with repeated use. However, it is not effective against many Gram-negative bacteria. Use of hexachlorophene has now been restricted, since infants repeatedly immersed in soaps containing this disinfectant have absorbed quantities which produced brain damage in test animals. Hexachlorophene, however, is of proven value in preventing nursery staphylococcal infections and with its judi-

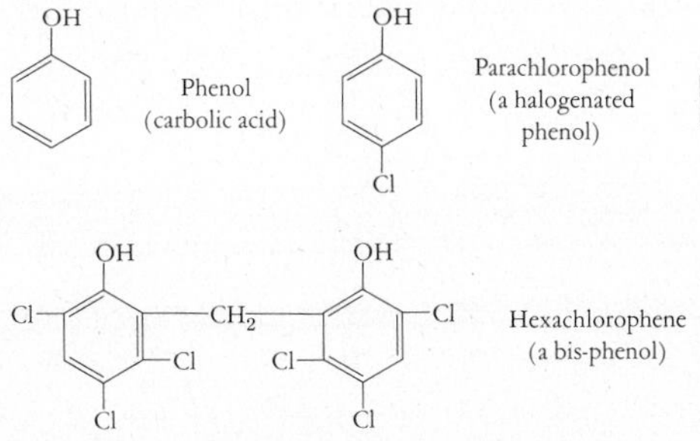

OH
Phenol
(carbolic acid)

OH
Cl
Parachlorophenol
(a halogenated
phenol)

CH_3
OH
Metacresol
(a cresol)

OH OH
Cl— —CH_2— —Cl
—Cl Cl—
Cl Cl
Hexachlorophene
(a bis-phenol)

FIGURE 6-4
Phenolic disinfectants.

cious use the risk of brain damage in infants would appear to be too small to justify its abandonment. Immersion in hexachorophene baths is now avoided, as is repeated use of hexachlorophene in conditions that facilitate its absorption, such as extensive areas of broken skin.

Quaternary Ammonium Compounds

These compounds may be thought of as derivatives of ammonium chloride in which the hydrogen atoms have been replaced by more or less complex organic radicals (Figure 6-5). They represent a larger group of compounds called surface-active agents, which reduce the surface tension of liquids. In contrast to other surface-active agents (such as household detergents) quaternary ammonium compounds are cationic, which means that their active groups have a positive electrical charge. Like the phenolics, they act against many vegetative bacteria (attacking mainly the cytoplasmic membrane), and they attack some viruses in a way which is not yet clear. Quaternary ammonium compounds are widely used for purposes of disinfection and preservation, such as prevention of bacterial growth in aqueous medications. Such uses have sometimes led to tragic infections in human beings because some bacteria, notably species of *Pseudomonas,* are able to resist the quaternary ammonium compounds and sometimes grow in their presence. In other instances there has been failure to recognize that these substances are easily inactivated by soaps, detergents, and by organic materials such as gauze.

Metallic Compounds

Mercury compounds are largely bacteriostatic. They have been in use as disinfectants and antiseptics for many years. Formerly, they were thought to have strong antimicrobial properties, including the ability to kill spores. Because of this, as recently as two decades ago many people were infected by medical instruments that had been soaked in mercury compounds and were erroneously thought to be sterile. Mercury acts by binding reversibly to sulfhydryl (—SH) groups in microbial proteins. Trade names include Mercurochrome and Merthiolate.

Compounds of mercury, tin, arsenic, copper, and other metals are widely used as preservatives and to prevent microbial growth in recirculating cooling water and industrial processes. Their extensive use has resulted in serious pollution of natural waters (see Chapter 35) and has led to strict controls on some of

FIGURE 6-5
Quaternary ammonium compounds.

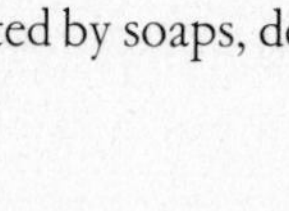

$H-\overset{+}{N}H_2-H \cdot Cl^-$ Ammonium chloride

$C_6H_5-CH_2-\overset{+}{N}(CH_3)_2-C_nH_{2n+1}{}^* \cdot Cl^-$ Benzalkonium chloride (a quaternary ammonium disinfectant)

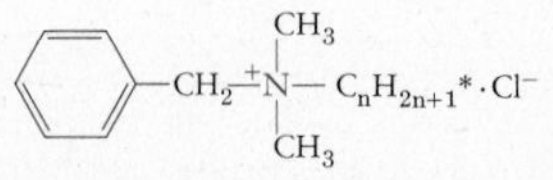

* n ranges from 8 to 18

these compounds. As an example, mercurials may no longer be employed in pulp mills for controlling slime-producing microbes.

Gaseous Agents

Fumigation has been employed as a weapon against infection at least as far back as ancient Greece, where burning sulfur was apparently used for purification. Formaldehyde gas has likewise had many years of use and is still recommended for decontaminating air filters on the cabinets used for culturing *Mycobacterium tuberculosis.* Heating the powdery polymer of formaldehyde (paraformaldehyde) can be a safe and effective way of generating the gas if concentration and relative humidity are controlled to prevent explosions and repolymerization. Formaldehyde generated in this manner can not only kill the endospores of *Clostridium botulinum* but can also effectively neutralize the powerful nervous system toxin produced by this organism.

Ethylene oxide is the most effective and useful gaseous antimicrobial agent. It penetrates well into fabrics and equipment (in contrast to formaldehyde) and is one of the very few chemical agents that can be relied on for sterilization. Ethylene oxide readily reacts with amino, carboxyl, sulfhydryl, and other chemical groups of organic substances. Because it is explosive, it must be mixed with some inert gas such as carbon dioxide or freon. The effectiveness of ethylene oxide also depends on temperature and relative humidity, which must be carefully controlled. Its use is therefore restricted to a special chamber where temperature, pressure, and relative humidity can be regulated. In practice, 4 or more hours are generally required to sterilize, and an additional period of from 4 to 24 hours is needed to allow release of absorbed gas, depending on the object being treated. One must allow absorbed ethylene oxide to dissipate because of its irritating effect on tissues or the undesired persistence of its antimicrobial effect. Disposable plastic dishes, syringes, and other heat-sensitive items used in laboratories or for medical purposes are commonly sterilized with ethylene oxide.

RADIATION

Electromagnetic radiations can be thought of as waves having energy but no mass. Examples are X-rays, gamma rays, ultraviolet, and visible light. Electromagnetic rays all travel at the speed of light, and the energy they possess is proportional to the frequency of the radiation (the number of oscillations of the wave per second). Thus electromagnetic radiations of short wavelength, such as gamma rays, have a lot more killing power than those with long wavelength, such as visible light. The spectrum of electromagnetic radiations ranges from electric waves (of very low frequency) through radio waves, heat rays, visible light, ultraviolet light, X-rays, and gamma rays to the very high frequency secondary cosmic rays, with wavelengths of only 10^{-11} centimeters. Gamma rays and ultraviolet radiations have proved to be valuable as tools for microbial control.

Gamma Rays

Gamma rays are examples of ionizing radiations which cause biological damage by producing hyperreactive ions and other molecular forms when they give up their energy to a microbe. Gamma radiation from the radioisotope cobalt 60 has proved to be of practical value in the control of medically important microorganisms. Gamma irradiation, for example, is used in killing pathogens such as *Salmonella* in food products. Such applications are analogous to the use of heat in the pasteurization process, with complete sterilization being impractical because of undesirable changes in color, flavor, or consistency with higher doses of radiation. A number of biological materials (such as penicillin) and numerous disposable plastic items (such as hypodermic syringes) can be effectively sterilized with high dose gamma irradiation without altering the material.

Various substances and conditions alter the susceptibility of microbes to ionizing radiations. Oxygen and certain chemicals may decrease microbial resistance, while vitamin C, ethyl alcohol, and glycerol may increase resistance. Protective chemicals are, of course, of great interest in view of modern-day radiation hazards.

Bacterial endospores are the most radiation-resistant microbial forms, whereas the Gram-negative rod-shaped bacteria, such as species of *Salmonella* and *Pseudomonas,* are among the most sensitive. Some bacteria which do not form spores are peculiarly radiation-resistant, however. *Micrococcus radiodurans* and some laboratory derived mutants, for example, have enzymes that can repair moderate degrees of radiation damage.

Ultraviolet radiation

Certain wavelengths of radiation are much more effective antimicrobial agents than those immediately shorter or longer. The most important example of this is the band of wavelengths from 200 to 310 nanometers—the ultraviolet zone. This electromagnetic zone of enhanced killing can be explained by the fact that it includes the wavelengths optimally absorbed by nucleic acids with resulting damage to their structure and function. The absorbed energy causes this damage by producing changes in purines and pyrimidines so that their normal function in nucleic acid is impaired. However, in microbial populations which have apparently been killed by ultraviolet radiation, some members will recover if irradiated with longer wavelengths of light. Furthermore, recovery of damaged viruses may also occur if they infect cells possessing repair enzymes.

In practice, ultraviolet light with satisfactory germicidal properties can be produced by passing an electric current through vaporized mercury in a special glass tube similar to a fluorescent light bulb. Most of the light produced has a wavelength of 253.7 nm, close to the killing optimum of 265 nm. The antimicrobial effect falls off almost in a linear fashion very close to the bulb, and faster, in proportion to the square of the distance, as one moves farther from the bulb. Ultraviolet rays penetrate very poorly, so that a film of grease on the bulb may

LIMITATIONS OF ULTRAVIOLET IRRADIATION

markedly reduce effective microbial killing, as will extraneous materials covering the microbes. Therefore ultraviolet lamps are of greatest value against exposed microbes in air or on clean surfaces and at close range.

Acidity, growth phase of the microbe, and the presence of spores definitely alter the effectiveness of ultraviolet irradiation, while the effect of moisture is variable. Most types of glass and plastic effectively screen out ultraviolet radiation. It has for years been assumed that the microbial killing effect of sunlight is due entirely to ultraviolet rays. However, much of the sun's ultraviolet radiation is absorbed by the atmosphere, and the bactericidal effect of sunlight on earth is primarily due to another mechanism, *photo-oxidation.* In photo-oxidation light energy of wavelengths longer than ultraviolet is absorbed by microbes, producing lethal oxidation in the presence of atmospheric oxygen. By contrast, microbial killing by ultraviolet irradiation does not require the presence of oxygen.

COMPARISON OF AGENTS USED AGAINST MICROBES

Different means of killing microbes are compared in the laboratory by counting the number of surviving organisms after varying time intervals or by counting the survivors at a fixed time. The count must be achieved after *neutralizing or removing the killing agent.* Different dilutions are inoculated on appropriate media, and the number of survivors able to grow into a visible colony is determined. Curves can then be plotted and "D" or "Z" values determined. The D value, or "decimal reduction time," is the time it takes for 90 percent of the organisms to be killed at a given temperature, while the Z value represents the temperature required to obtain 90-percent killing in a given time period. These values, rather than the time required for complete killing, are used in comparing different methods of microbial killing. The complete-killing time varies widely due to the shape of the killing curves and is therefore of little use in scientific studies.

EVALUATION OF DISINFECTANTS

Disinfectants may be evaluated by comparing the ratio of their antimicrobial activity to that of pure phenol. The larger the resulting value—known as the *phenol coefficient*—the more active the agent under those conditions. The phenol coefficient has general applicability but may be highly misleading since differences in killing rates depend on the species of microorganism, stage of growth, nature of the killing agent, suspending medium, pH, salt concentration, and many other factors. For these reasons any agents being compared for possible use in a given situation (such as disinfection of a hospital floor or decontamination of a spaceship) *must be evaluated under the same conditions in which they will actually be used.* Widely conflicting information on the value of disinfecting agents continually arises because various and often inappropriate conditions of testing are used. For example, in counting surviving microbes there may be failure to employ a medium that neutralizes the disinfectant being tested. If this happens, traces of disinfectant carried over to the counting medium may prevent growth of the microbes and falsely indicate that they had been killed. In the United States proper testing procedures are defined in a manual of the Association of Official Analytical Chemists (AOAC tests). Disinfectants shipped in interstate commerce must have federal approval and are subject to evaluation in laboratories of the Environmental Protection Agency.

SUMMARY

Sterilization means killing or removing all viable organisms, while disinfection means reducing or eliminating organisms that might represent a hazard. Heating is one of the most practical methods of controlling unwanted microbes and is also the most reliable. Heat is most effective when applied under moist conditions. Autoclaving and tyndallization can bring about sterilization, while pasteurization and boiling can reduce or eliminate unwanted microbes. Filtration can be used to sterilize fluids, although with less reliability than heating. In using filters it is important to know their approximate pore size, response to solvents and filtration pressure, and their affinity for biologically active substances, such as enzymes. Most chemical agents are unreliable for sterilizing but have wide use as disinfectants. The major classes of chemical disinfectants include alcohols, chlorine and iodine, formaldehyde, phenolic substances, and quaternary ammonium compounds. Gaseous agents used in sterilization include formaldehyde vapor and ethylene oxide. A variety of chemical agents may be used as preservatives to prevent growth of destructive organisms in foodstuffs, leather goods, medications, and other materials. Electromagnetic radiations are useful both for sterilizing and for reducing the numbers of potentially harmful microbes.

Germicidal agents generally cause a logarithmic decrease in numbers of susceptible microbes, necessitating time to achieve sterilization or satisfactory reduction in numbers of viable organisms. Microbes vary in their susceptibility depending on pH, temperature, growth phase, strain, and species differences.

Laboratory comparisons of germicidal agents may be made by analyzing their plotted curves of microbial killing. The most meaningful of germicide comparisons are made under conditions of their actual use.

QUESTIONS

REVIEW

1. Explain the superiority of moist heat as a sterilizing agent.
2. Explain why conditions of temperature and pressure other than those given in Figure 6-3 may be unsatisfactory for autoclaving.
3. Why do doctors and dentists no longer use alcohol to sterilize their instruments?
4. Discuss the limitations of ultraviolet light for sterilization and disinfection.
5. Why is it that laboratory comparisons of disinfectants may not always reflect their relative value in actual practice?

THOUGHT

1. What happens to preservatives, such as those used in paints, leather, paper and other products, as these products deteriorate?
2. Give reasons why an object that has been autoclaved might not be sterile.

FURTHER READING

LAWRENCE, C. A., and S. S. BLOCK, *Disinfection, Sterilization and Preservation.* Philadelphia: Lea and Febiger, 1968. A large and comprehensive reference book.

PERKINS, J. J., ed., *Principles and Methods of Sterilization in Health Sciences.* Springfield, Ill.: Charles C. Thomas, 1969. Practical applications relating to the medical field.

PHILLIPS, G. B., and W. S. MILLER, eds., *Industrial Sterilization.* Durham, N.C.: Duke University Press, 1972. Papers presented at an international symposium on this subject in 1972, covering heat, gas, and radiation primarily.

SPAULDING, E. H., and D. H. M. GRÖSCHEL, "Hospital Disinfectants and Antiseptics," *Manual of Clinical Microbiology.* Ed. E. H. Lennette et al. Washington, D.C.: American Society for Microbiology, 1974. Clear, concise, and up-to-date discussion of these agents.

TAYLOR, G. R., "Space Microbiology," *Annual Review of Microbiology* 28:121–137 (1974).

CHAPTER 7

METABOLISM

The growth of all organisms requires the synthesis of cytoplasm. This synthesis usually involves the breakdown of foodstuffs both to provide compounds which are converted to cell material and to supply energy. This energy is used to drive the large number of biosynthetic reactions by which the cell converts relatively simple molecules to the supramolecular structures that make up its cytoplasm. One fantastic aspect of this entire operation is the unbelievable speed with which cell components are synthesized. For example, a single *Escherichia coli* cell dividing every hour synthesizes 4000 molecules of lipid, almost 1000 protein molecules (each containing about 300 amino acids), and 4 molecules of RNA per second.

All of this is made possible through the action of hundreds of different *enzymes*—the biological catalysts. They allow the cell to carry out its reactions at relatively low temperatures. In a chemical laboratory these same reactions cannot be duplicated without employing considerably higher temperatures. Before we discuss the biology of energy generation and biosynthesis, it will be helpful to consider the properties and mechanisms of enzymes.

CHEMICAL KINETICS

Enzymes are organic catalysts that speed up a chemical reaction without being used up or changed by the reaction. Thus, if the conversion of a substance to a product is measured at varying intervals, data such as those presented in Figure 7-1 are observed. In the absence of the enzyme, the *substrate* (the material on which the enzyme acts) remains essentially unchanged. In the presence of the enzyme, however, the substrate is converted to a product.

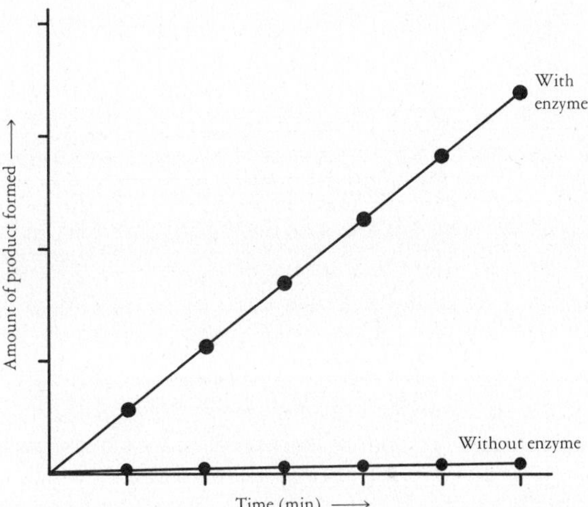

FIGURE 7-1
Effect of enzyme addition on product formation.

How does an enzyme act? An understanding of enzyme action requires some simple, yet essential knowledge of the kinetics of chemical reactions. Much of this discussion relates to information given in Chapter 2, involving the formation of chemical bonds. *All reactions, whether or not they are catalyzed by an enzyme, represent a rearrangement of molecules.* Any rearrangement requires the breaking of old bonds and the formation of new ones. Much of this discussion can relate to a specific example, the formation of water (H—O—H) from oxygen (O—O) and hydrogen (H—H). When both gases are brought together, they will not react. If, however, some form of energy, such as a spark, is added to the system, a violent explosion erupts with the release of energy in the form of heat and light. Water (HOH) is formed. The reaction can be summarized by the equation:

Covalent chemical bonds, pages 25–26

$$2\text{H—H} + \text{O—O} \longrightarrow 2\text{H—O—H} + \text{energy}$$

Note that new bonds are formed, O—H in place of H—H and O—O. Two aspects of this reaction are considered: First, why must energy be put *into* the system before a reaction occurs? Second, why is energy released in the reaction? Understanding these two features aids in understanding enzymatic reactions.

ENZYMES

Enzymes and Activation Energy

When oxygen and hydrogen gas are mixed, the gaseous molecules collide with one another and with the walls of the container. Each molecule has a certain level of energy, which is its *kinetic energy,* or *energy of motion.* At room temperature, however, this energy is not sufficient to allow the hydrogen atoms to break the strong covalent O—O bonds or to permit the oxygen atoms to push apart the H—H bonds. No new bonds are formed because atoms can only recombine with new partners after they have been partially separated from their original partners.

As the temperature increases, however, the hydrogen and oxygen molecules move more rapidly, and some collisions have sufficient energy to break apart the H—H and O—O bonds and to form O—H bonds. An electric spark initiates the reaction, but once started, the heat energy given off by the reaction of a few molecules is sufficient to raise the kinetic energy of other molecules. A chain reaction is thus set in motion, leading rapidly to an explosion.

The reason why energy must first be put into the system can be explained by the analogy described in Figure 7-2. In this figure the hydrogen and oxygen molecules are represented as balls located near the top of a hill. Potentially, the molecules have considerable kinetic energy. This energy can only be expressed, however, after they have rolled down the hill. But they are prevented from rolling because they are trapped in an "energy depression." If enough energy is supplied so that the molecules can surmount this hurdle, they will spontaneously roll down and generate kinetic energy. The amount of kinetic energy the molecules generate is much greater than the amount needed to bring them out of their energy depression. Indeed, a reaction will occur *only* if the products roll *downhill,* that is, *release* kinetic energy, which is designated by the symbol $\Delta G(-)$. The amount of energy required to lift the molecules out of the energy depression is called the *activation energy. This is the point at which enzymes act.* Enzymes *lower* the activation energy hurdle to values that can be surmounted by the kinetic energy of the original molecules at room temperature. Thus the enzyme-catalyzed reaction proceeds without any additional heat. The activation energy barrier is enor-

ENZYMES LOWER ACTIVATION ENERGY

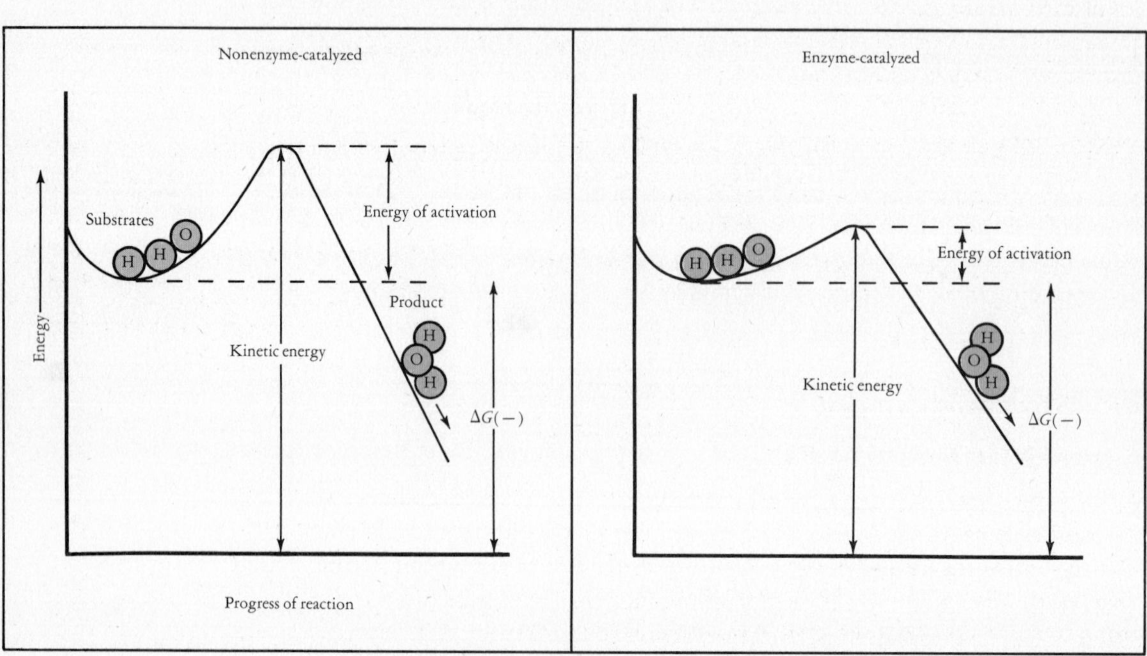

FIGURE 7-2 Energy barrier in chemical reactions. The enzyme lowers the energy of activation enough so that the energy in the molecules of H and O is sufficient to raise them to the top of the hill.

mously important to life. Without it there would be a continuous rearrangement of all atoms and molecules, and life could not exist. It would indeed be difficult to live in a world in which wood and coal could burst into flames spontaneously!

Even though an enzyme is present, many reactions of biological importance will proceed only if additional energy is supplied. These are the reactions that would run *uphill.* In such reactions, which generally involve the biosynthesis of metabolites, a "downhill," energy-yielding reaction $\Delta G(-)$ is *coupled* to the uphill energy-requiring reaction $\Delta G(+)$ in such a way that a net loss of energy $\Delta G(-)$ results. The reaction will then proceed since the overall reaction now runs downhill. The coupling of an energy-requiring reaction to an energy-yielding reaction occurs through a compound that is a component of the energy-yielding as well as the energy-requiring reaction. This compound is frequently *adenosine triphosphate* (ATP) (Figure 7-3), a compound that contains chemical energy in the form of high-energy bonds. When these high-energy bonds are broken, they release large amounts of energy (Figure 7-4), which can then be used by the energy-requiring reactions of the cell to allow them to run downhill.

ENZYMES CATALYZE ONLY REACTIONS THAT WILL PROCEED SLOWLY IN THEIR ABSENCE

Just exactly how enzymes lower the activation energy is still not well understood. Part of their action can be explained by the fact that when an enzyme combines with its substrate through weak bonding forces, it places a stress on those chemical bonds in the substrate which will be broken. This stress weakens these bonds in the substrate enough to permit them to break and re-form new bonds. To function, the substrate must bind to a specific portion of the enzyme, the *active* or *catalytic site,* to yield an enzyme-substrate complex. Next, the products of the reaction are released, leaving the enzyme unchanged. The substrates that can bind to any one enzyme are limited since a complementary relationship between the substrate and enzyme is required. This "lock-and-key" arrangement is diagramed in Figure 7-5. This complementary specificity requires essentially that every reaction in the cell be catalyzed by a different enzyme.

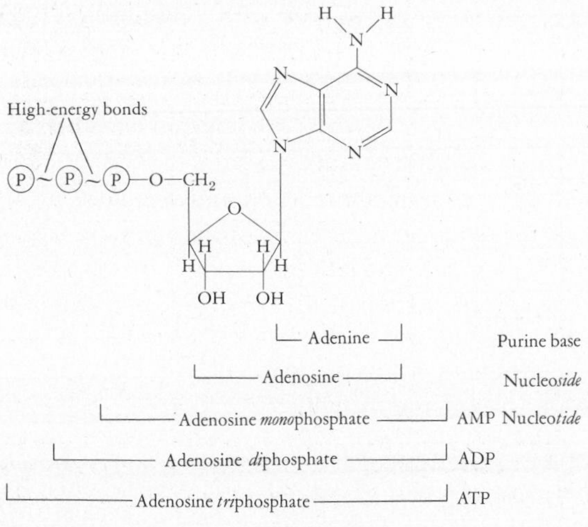

FIGURE 7-3
Formula of adenosine triphosphate (ATP). The two high-energy bonds are denoted by a wavy line.

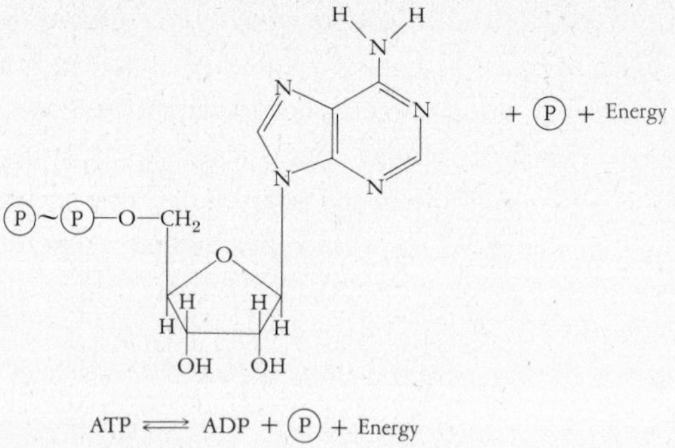

$$ATP \rightleftharpoons ADP + (P) + Energy$$

FIGURE 7-4
Energy is liberated when the high-energy phosphate bond of ATP is broken to yield ADP and inorganic phosphate.

Nomenclature of Enzymes

Unfortunately, enzymes have not always been named according to a systematic rule. However, most enzymes end with the suffix *-ase* and are commonly named after the substrate on which they act or the type of reaction that they catalyze. For example, an enzyme that degrades cellulose is termed a *cellulase;* the enzyme that degrades DNA is a *deoxyribonuclease.* The enzyme that polymerizes mononucleotides into an RNA polymer is called RNA *polymerase.*

Structure of Enzymes

All enzymes are proteins. Each enzyme has a unique sequence of amino acids that determines its three-dimensional structure and, therefore, the substrate to which it will fit and the reaction it will catalyze.

Some enzymes contain a nonprotein portion that is essential for their functioning. If this nonprotein portion is an organic molecule that is readily separated from the enzyme, it is called a *coenzyme.* Coenzymes generally function in the transfer of small molecules from one protein to another. As an enzyme binds a substrate molecule, a coenzyme associates with it, picks up a small molecule or atom (such as a hydrogen atom) from the substrate, dissociates from the enzyme, and transfers the small molecule to another acceptor molecule (Figure 7-6). The coenzymes are not specific; they can associate with a number of different enzymes. However, each coenzyme is able to transfer only one type of small molecule or atom.

Compared to enzymes, coenzymes are comparatively small molecules themselves and are synthesized from vitamins by a series of enzymatic steps (Table 7-1). Since coenzymes, like enzymes, are recycled and therefore are needed only in minute quantities, the required dietary level of the vitamins, whether for man or microorganism, is very small. The lack of a vitamin that is converted to a coenzyme results in the nonfunctioning of all of the different enzymes that require the coenzyme in order to function.

EFFECT OF ENVIRONMENT ON GROWTH REFLECTS EFFECT OF ENZYMES ON THE CELL

Factors Influencing Enzyme Activity

The growth of any organism depends on the functioning of most of the enzymes that it contains. The environment exerts its influence on cell growth primarily by

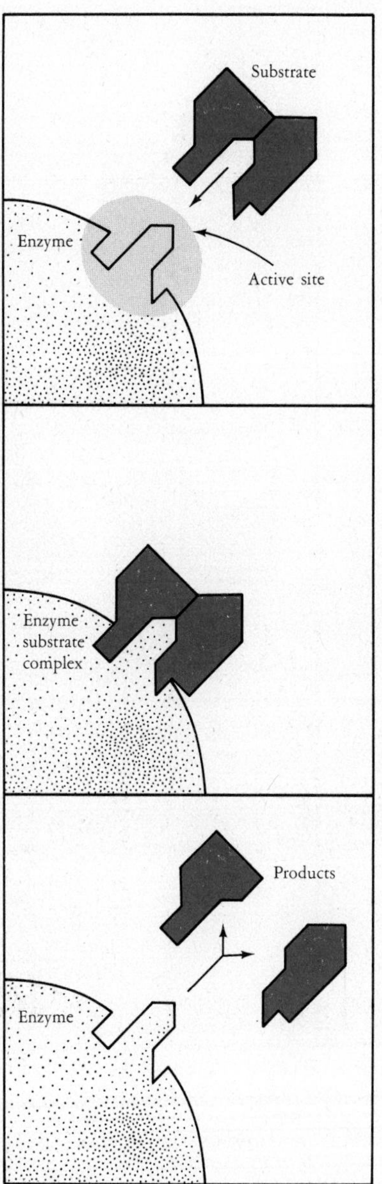

FIGURE 7-5
Enzyme catalysis. The requirement for a lock-and-key relationship between enzyme and substrate.

influencing the activity of enzymes. The optimum temperature, pH, and salt concentration for growth of any cell represents the best average for the functioning of all of its cellular enzymes.

In general, increasing the temperature promotes more rapid bacterial growth. This is true because an increase in temperature increases the energy of motion of molecules and the rates (speed) of reactions. However, if the temperatures become too high, the enzymes are denatured and the organism dies.

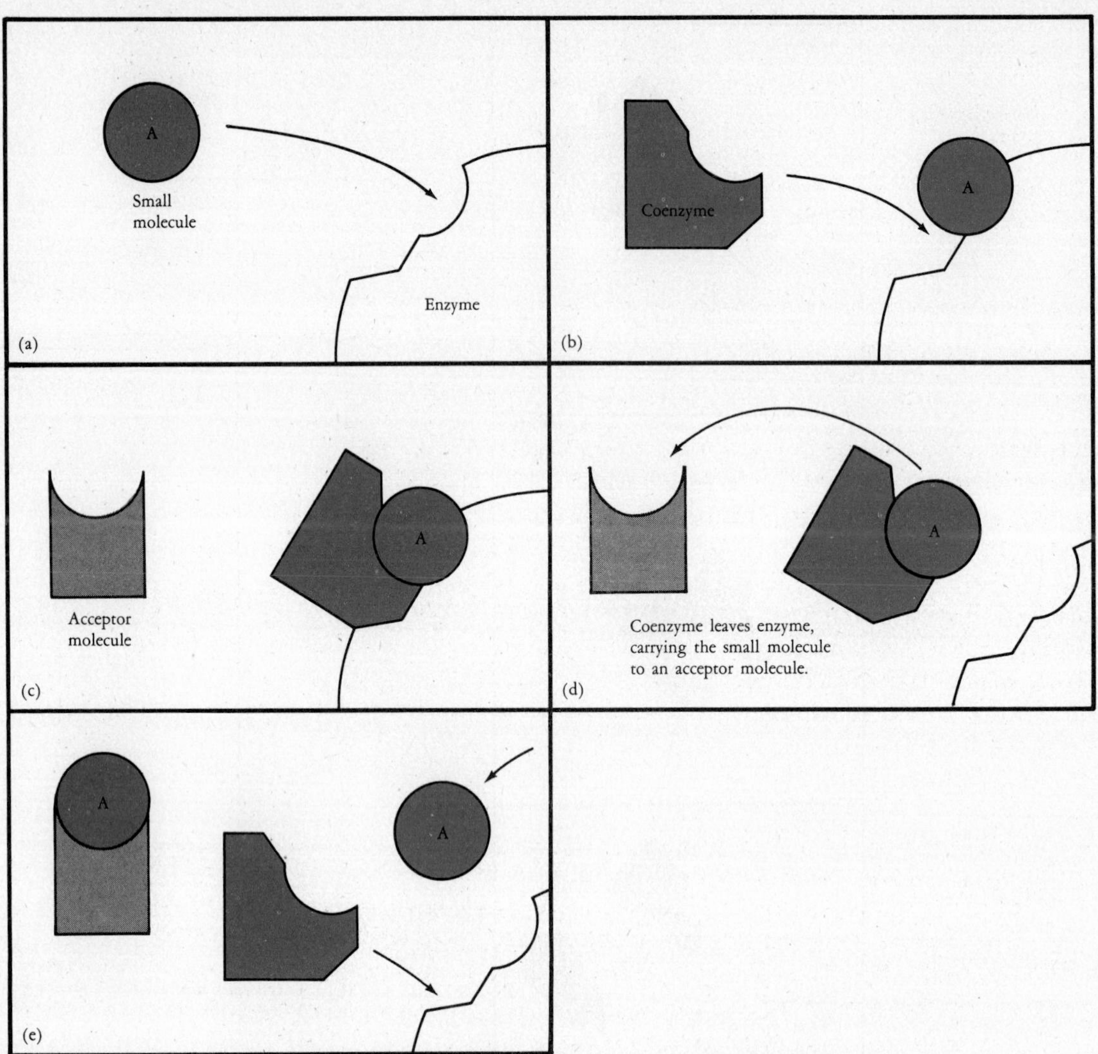

FIGURE 7-6 Enzyme-coenzyme interaction. The coenzyme carries small molecules from an enzyme to the acceptor molecule, which may be a protein or another coenzyme.

Effect of Inhibitors on Enzyme Activity

The most potent poisons known exert their effects by inhibiting enzyme function. Thus cyanide, arsenic, mercury, and the nerve gases all combine with and prevent the functioning of enzymes necessary for survival. The inhibition of enzyme function also plays an important role in the treatment of bacterial infections. The classic example is provided by the sulfa drugs. Sulfanilamide has a structure very similar to paraaminobenzoic acid (PABA) (Figure 7-7). The latter compound is an essential nutrient that is enzymatically converted to a required vitamin, folic acid. The structures of PABA and sulfanilamide are so similar that

MODE OF ACTION OF SULFA DRUGS

TABLE 7-1
Coenzymes

Name of Coenzyme	*Vitamin from Which It Is Derived*	*Entity Transferred*
Nicotinamide adenine dinucleotide (NAD)	Niacin	Hydrogen atoms
Flavin adenine dinucleotide (FAD)	Riboflavin	Hydrogen atoms
Coenzyme A	Pantothenic acid	Acetyl group
Thiamine pyrophosphate	Thiamine	Aldehydes
Pyridoxal phosphate	Pyridoxine	Amino group
Tetrahydrofolic acid	Folic acid	1-Carbon group

an enzyme involved in converting PABA to folic acid often combines with the drug rather than with PABA. Thus the synthesis of folic acid (and consequently bacterial growth) is inhibited. This is an example of *competitive inhibition,* since the drug competes with PABA for the active site of the enzyme (Figure 7-8). The higher the ratio of sulfanilamide molecules to PABA molecules the greater the inhibition. Sulfa drugs are selective against bacteria and are therefore of value in the treatment of certain diseases because man does not synthesize folic acid from PABA, whereas many bacteria do. Sulfa drugs are discussed further in Chapter 33.

Sulfa drugs, page 588

ENERGY METABOLISM

Energy metabolism can be summarized as follows: the original nutrient material contains a great deal of energy in the bonds between the atoms which compose it. However, this energy is spread throughout the molecule and is not in a readily usable form. The pathways of energy metabolism involve a series of enzymatic reactions in which the diffuse nonusable energy present in the many low-energy bonds is first concentrated in a few high energy bonds. Then the energy is transferred to ADP to form ATP, the readily utilizable energy currency of all cells. These energy conversions require a series of reactions in which the energy is slowly but surely concentrated into high-energy bonds. We shall now consider these pathways in more detail.

When a few yeast cells of the species *Saccharomyces cerevisiae* are placed in a glucose-salts medium in a flask and incubated for several days, a cursory glance at

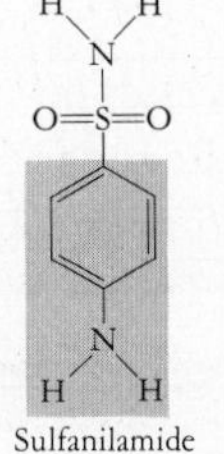

PABA

FIGURE 7-7
Structure of sulfanilamide (sulfa) and para-amino-benzoic acid (PABA).

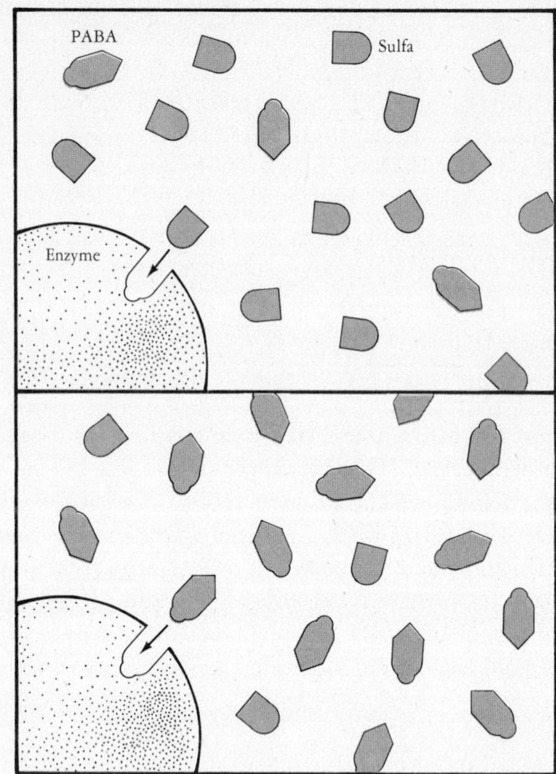

FIGURE 7-8
Competitive inhibition of folic acid synthesis by sulfa. The sulfa molecule occupies the same site on the enzyme as the PABA molecule and therefore competes with PABA for this site. Whether the enzyme will be occupied by a PABA or sulfa molecule depends on the proportion of these two molecules. Both the PABA and sulfa are converted to different products by the enzyme.

the flask and a smell of its contents are enough to convince anyone that profound changes have taken place. Whereas the contents of the flask were perfectly clear at the beginning of the incubation, they are now turbid, with bubbles of gas at the surface. In addition, the odor of ethyl alcohol emanates from the previously odorless liquid. The increase in turbidity and the formation of ethyl alcohol and gas reflect the entire spectrum of metabolic activity of the yeast cells. The cells have transformed some of the glucose and inorganic salts into more yeast cells, which increase the turbidity of the medium. In addition, they have converted some of the glucose into ethyl alcohol and CO_2. The cell extracts energy from this degradation of glucose. The following equation summarizes the metabolism:

ENERGY METABOLISM RESULTS IN PRODUCTION OF ENERGY AND SMALL MOLECULES FOR SYNTHESIS OF CYTOPLASM

$$\underset{\text{glucose}}{C_6H_{12}O_6} \longrightarrow \underset{\text{ethanol}}{2CH_3CH_2OH} + \underset{\text{carbon dioxide}}{2CO_2} + \text{yeast cells}$$

Other organisms degrade glucose to products other than ethyl alcohol and CO_2. But in all instances the degradation results in the release of energy $\Delta G(-)$ which the cell traps as chemical energy in the form of ATP. This ATP is then used as a source of energy for the synthesis of more yeast cells. In this process the glucose is metabolized to small molecules that are converted first to subunits, then to macromolecules, and finally to cells.

Glycolysis

There is a tremendous variety of compounds which cells can degrade. Such compounds include both organic and inorganic materials. Of all organic compounds the one most commonly used in the laboratory as a source of energy is glucose. This sugar can be degraded along a number of different pathways; the most common is called *glycolysis,* the pathway that yeast employ to degrade glucose. The glycolytic pathway is diagramed in abbreviated form in Figure 7-9 and in more complete form in Appendix I. Figure 7-9 and Appendix I illustrate that in a series of reactions, one glucose molecule is degraded to two molecules of pyruvic acid. The rearrangement of the atoms in the molecules in the course of this breakdown concentrates the energy previously dispersed throughout the glucose molecule. The difference between the bond energy in glucose and that in pyruvic acid is 52,000 calories per mole, and the cell can trap much of this difference in energy content in the form of high-energy bonds in ATP. This amounts to a total of four high-energy phosphate bonds created in ATP for each molecule of glucose metabolized. However, as Figure 7-9 indicates, two high-energy bonds are consumed at the beginning of the pathway, so there is actually a net gain of only two high-energy bonds. Thus pyruvic acid still contains a great deal of energy which can be tapped for further ATP synthesis. The tapping of this energy involves *oxidation-reduction* reaction.

Oxidation-Reduction Reactions

COENZYMES TRANSFER HYDROGEN ATOMS FROM ONE COMPOUND TO ANOTHER

In addition to the two molecules of pyruvic acid and the two molecules (net) of ATP produced from each molecule of glucose that is degraded in glycolysis, one other product is produced—two pairs of *hydrogen atoms.* These hydrogen atoms are removed from compounds that are being metabolized through the glycolytic pathway. Each of the hydrogen atoms consists of one *electron* and a *hydrogen ion* (H^+). The *removal* of either an electron or an entire H atom from a compound results in the compound becoming oxidized, and the removal is therefore termed an *oxidation.* The biological importance of oxidation reactions is that the energy decrease of a molecule is associated primarily with such reactions.

In oxidation reactions the electrons and hydrogen ions removed from a molecule do not remain free in solution but combine immediately with another compound. The compound from which they came becomes *oxidized;* the compound that combines with the electrons and hydrogen atoms becomes *reduced.* These two reactions (which occur virtually simultaneously) are collectively termed *oxidation-reduction reactions.* Compounds that initially contain a high proportion of hydrogen atoms, such as glucose, are highly reduced compounds; those that contain very few, if any, hydrogen atoms, such as CO_2, are highly oxidized compounds. Reduced compounds contain much more energy than oxidized compounds, and a compound that contains many hydrogen atoms can be recognized as a potential foodstuff with much energy. Thus energy metabolism involves the degradation of foodstuffs by a series of reactions in which highly

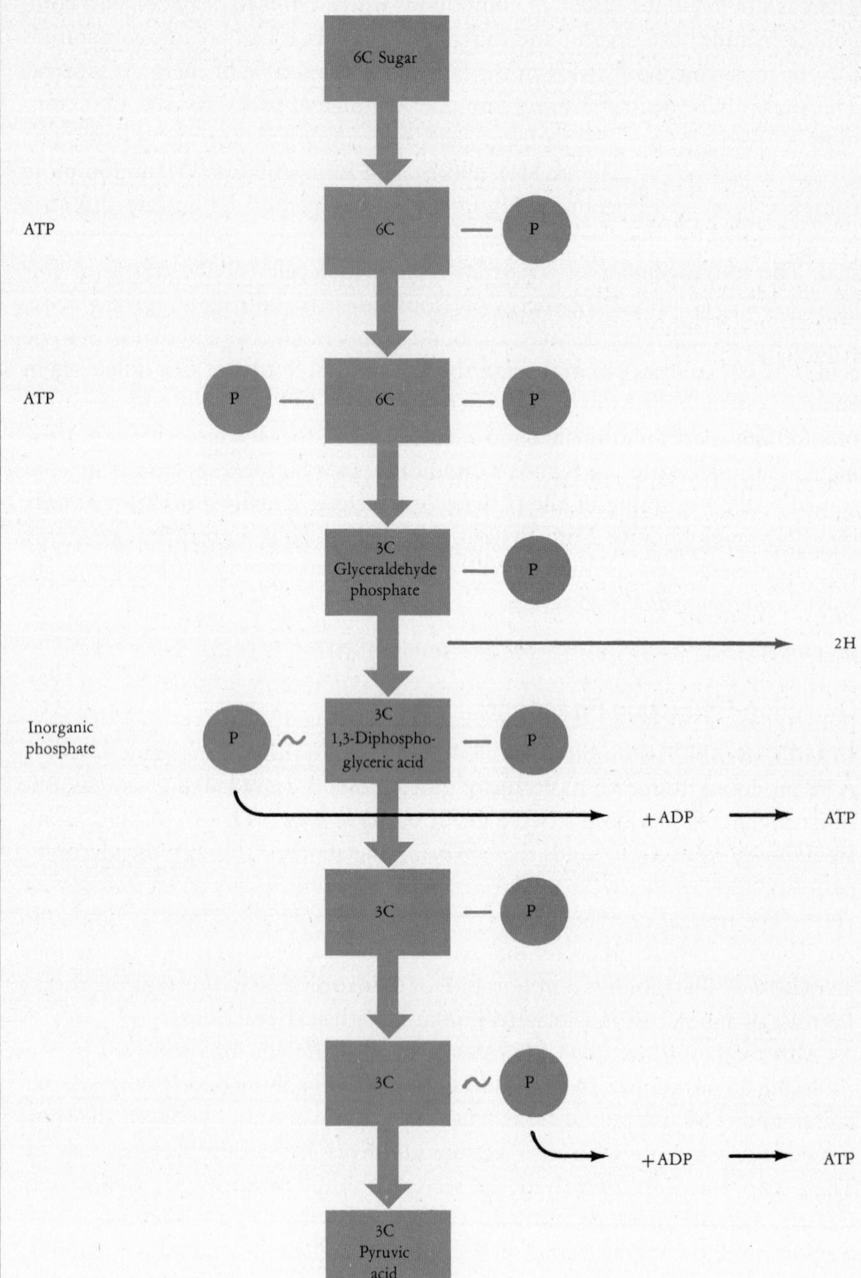

FIGURE 7-9
Glycolysis. This pathway involves the transformation of a 6-carbon sugar to two 3-carbon molecules. In the course of this transformation energy is extracted and transferred in the form of high-energy phosphate molecules to form ATP. In addition, 2H are removed. These atoms can react with pyruvic acid, also generated, to form a variety of products (shown in Figure 7-11).

reduced compounds become increasingly oxidized. The energy a compound gives up as it is oxidized is largely captured in the high-energy bonds of ATP.

The hydrogen atoms (electrons and hydrogen ions) that are given up in an oxidation reaction are removed by enzymes that have an associated coenzyme. This coenzyme, which is most commonly a derivative of the vitamin niacin, is nicotinamide-adenine-dinucleotide (NAD). When NAD picks up and carries hydrogen atoms, it becomes reduced and is designated NAD_{red}. Because there is only a limited supply of this coenzyme in the cell compared to the large amounts of glucose that are broken down, NAD_{red} must transfer its hydrogen atoms to another compound, becoming NAD_{ox} (the same as NAD), in order to be able to pick up more hydrogen atoms from other molecules. The acceptor compound for the hydrogen atoms varies, depending on the particular organism and the environment in which the glucose is being metabolized. However, the acceptor chosen for the hydrogen atoms determines the materials that will be synthesized from pyruvic acid.

Metabolism of Pyruvic Acid

PYRUVIC ACID CAN BE CONVERTED TO A VARIETY OF DIFFERENT COMPOUNDS

The metabolism of pyruvic acid can take several different directions, each leading to different end products. Since each reaction requires the participation of a specific enzyme, which all organisms may not have, the end products depend to some degree on the species of microorganism. In all of these reactions the hydrogen atoms are taken from one organic compound in the glycolytic pathway and transferred to another. There are a number of types of compounds which can serve as the hydrogen acceptor, and each pathway is named after the type of compound that serves as the hydrogen or electron acceptor (Table 7-2). If the final hydrogen acceptor is an *organic* compound, the entire reaction sequence is called a *fermentation.* A very important feature of all fermentation reactions is that most of the energy of the glucose molecule remains untapped. Of the 686,000 calories of bond energy in each mole of glucose, approximately 634,000 calories still re-

TABLE 7-2
Hydrogen or Electron Acceptors in Energy Metabolism

Metabolic Process	*Conditions of Growth*	*Hydrogen or Electron Acceptor*	*Reduced Product Formed*
		Inorganic compound, that is:	
Respiration	Aerobic	O_2	H_2O
	Anaerobic	NO_3^-	NO_2^-, N_2O, N_2
		$SO_4^=$	H_2S
		CO_2	CH_4
		Organic compound, that is:	
Fermentation	Aerobic or anaerobic	$CH_3-C(OH)(H)-C(=O)-OH$	$CH_3-C(=O)-C(=O)-OH$

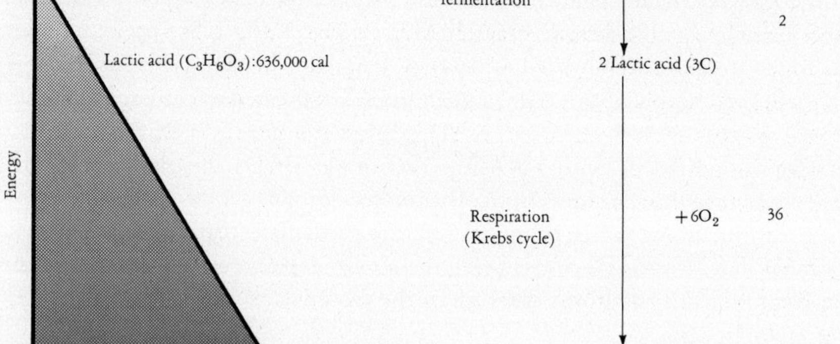

FIGURE 7-10
Energy gain in breakdown of glucose.

main in the products of fermentation (Figure 7-10). Fermentations can proceed under anaerobic conditions, and they are named after the final product formed in the fermentation.

Lactic acid fermentation. In this fermentation two molecules of NAD_{red} reduce the two molecules of pyruvic acid formed in glycolysis to two molecules of lactic acid. NAD_{ox} is regenerated.

$$2CH_3{-}\overset{\overset{\large O}{\|}}{C}{-}\overset{\overset{\large O}{\|}}{C}{-}OH + 2NAD_{red} \rightleftharpoons 2CH_3{-}\underset{\underset{\large H}{|}}{\overset{\overset{\large OH}{|}}{C}}{-}\overset{\overset{\large O}{\|}}{C}{-}OH + 2NAD_{ox}$$

pyruvic acid — lactic acid

This fermentation is important in both the spoilage and food production of a number of food products. Depending on the microorganisms present, fermentation of milk results in a variety of foodstuffs. These include yogurt, buttermilk, and cheeses. Sauerkraut results from the lactic acid fermentation of fresh cabbage and pickles from a similar fermentation of cucumbers.

Alcoholic fermentation. This reaction is the conversion of pyruvic acid to acetaldehyde and CO_2.

$$2CH_3{-}\overset{\overset{\large O}{\|}}{C}{-}\overset{\overset{\large O}{\|}}{C}{-}OH \rightleftharpoons 2CH_3{-}\overset{\overset{\large O}{\|}}{C}{-}H + 2CO_2$$

pyruvic acid — acetaldehyde

The acetaldehyde then becomes reduced to ethyl alcohol by another enzyme.

$$2CH_3{-}\overset{\overset{\large O}{\|}}{C}{-}H + 2NAD_{red} \rightleftharpoons 2CH_3{-}\underset{\underset{\large H}{|}}{\overset{\overset{\large OH}{|}}{C}}{-}H + 2NAD_{ox}$$

acetaldehyde — ethyl alcohol

These two reactions are carried out by yeast (*Saccharomyces*) and form the basis for wine and beer production.

Other products of fermentation. Pyruvic acid can be converted to a wide variety of other products of fermentation. Some of the final products of these reactions are given in Figure 7-11. In all of these cases the metabolism of pyruvic acid involves the same principles. The NAD_{red} must give up its hydrogen atoms to an organic molecule so that it can participate further in glucose metabolism. Pyruvic acid can be converted to a variety of products which are further reduced by accepting the hydrogen atoms from NAD_{red}.

Applied Aspects of the Metabolism of Pyruvic Acid

The metabolism of pyruvic acid has important consequences for many areas of microbiology. To some degree the products of this metabolism may help identify particular organisms, because the ability to carry out any one of these metabolic reactions is specific to a particular group of organisms. One very important tool for identifying disease-causing organisms, therefore, involves determining their products of metabolism.

Many of the metabolic products of microorganisms have great commercial value. Needless to say, the ability of yeast to degrade a wide variety of grains with the production of ethyl alcohol has had profound human effects in all parts of the world.

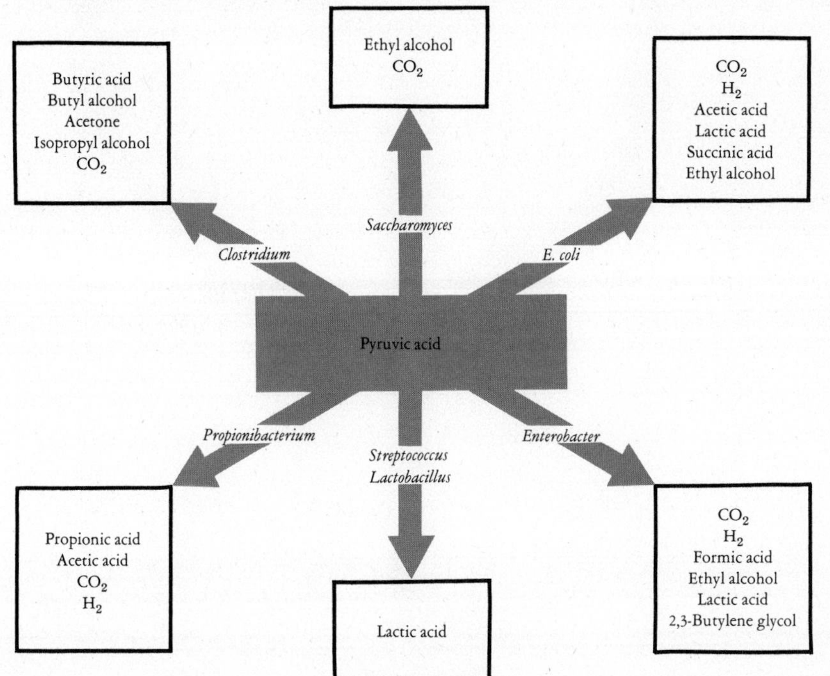

FIGURE 7-11
Metabolic conversions of pyruvic acid. This "key" intermediate in metabolism can be converted to a variety of end products, depending on the organism and the electron acceptors available.

Tricarboxylic Acid or TCA Cycle

ENERGY IN GLUCOSE MOLECULE REMAINS LARGELY UNTAPPED AFTER GLYCOLYSIS

Because fermentation extracts only a very small fraction of the total energy available in glucose, the remaining energy can be tapped if the pyruvic acid formed is oxidized to compounds with less bond energy—CO_2 and H_2O. If this occurs, all of the energy in the glucose molecule is liberated. Before entering this oxidative pathway, pyruvic acid is first converted to acetic acid, which is then oxidized, via the cycle, to CO_2 and H_2O (Figure 7-12 and Appendix I). This pathway of oxidation of acetic acid is sometimes called either the *Krebs cycle,* after the German biochemist working in England who pioneered in its elucidation, or the

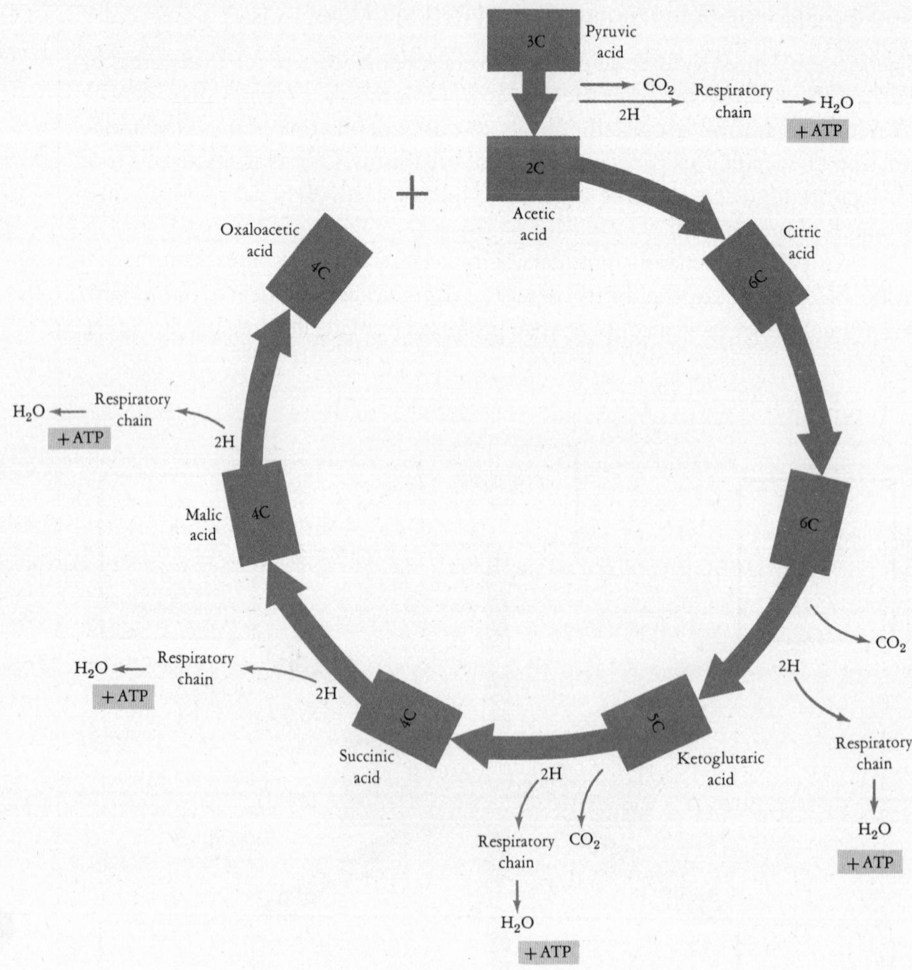

SUMMARY 3C (Pyruvic acid) + $O_2 \longrightarrow$ 2C (Acetic acid) + 3ATP + CO_2

2C (Acetic acid) + $2O_2 \longrightarrow 2CO_2$ + 12ATP

FIGURE 7-12 Abbreviated form of Krebs cycle. In every turn of this cycle the 2C compound (acetic acid) is converted to two molecules of CO_2. The 4C compound which combines with the 2C compound to initiate the cycle is always regenerated.

tricarboxylic acid cycle (TCA), because the first compound formed in the cycle, citric acid, has three carboxyl (—COOH) groups.

HYDROGEN IONS AND ELECTRONS PASSED FROM ONE MEMBER OF RESPIRATORY CHAIN TO ANOTHER

Respiration is the name given to the sequence of reactions in which an *inorganic* compound is the final electron and hydrogen acceptor (Table 7-2). The large amounts of energy gained in respiration result from the transfer of electrons through the *electron transport chain* (Figure 7-13). Each member of the chain, beginning with NAD, alternately becomes reduced and then oxidized as it first picks up and then gives away the electrons to the next member of the chain. The hydrogen ions formed (H^+) are simultaneously passed along the chain. The final acceptor compound for the electrons and H^+ is usually oxygen, resulting in the formation of water.

GENERATION OF ATP

Exactly how the bond energy of ATP is generated as a result of the passage of electrons and hydrogen ions along the electron transport chain is not known. When they are first removed from compounds in the TCA cycle, the electrons must have a high level of energy. As the electrons are passed along the chain, they lose energy. The energy that they lose is transferred into the high-energy phosphate bonds of ATP by a method which as yet can only be speculated on. The

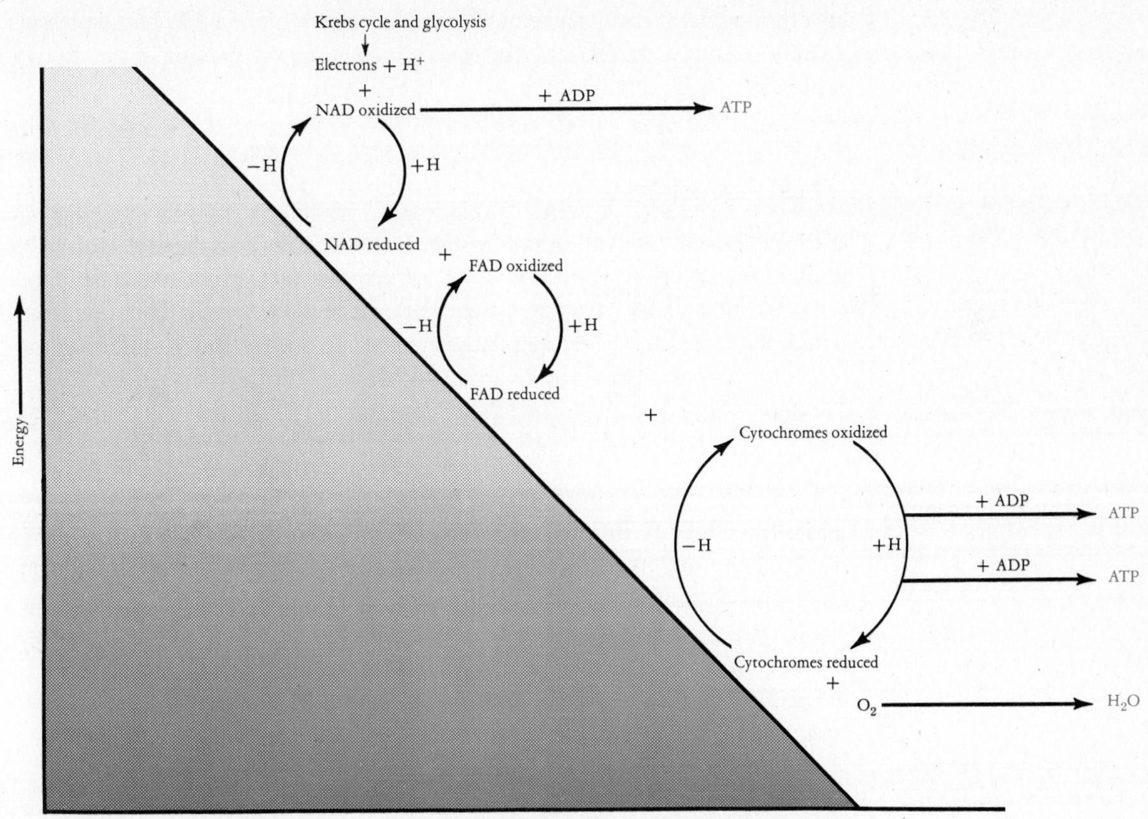

FIGURE 7-13 Diagrammatic representation of the formation of ATP molecules as the electrons and H atoms roll downhill from a high to a low energy level.

generation of ATP by this process of electron transport is referred to as *oxidative phosphorylation.*

Energy Balance Sheet for Metabolism of Glucose

The energy balance sheet for the metabolism of glucose can now be summarized. Figure 7-10 illustrates the tremendous energy bonus a cell gains by oxidizing pyruvic acid to CO_2 and H_2O (respiration) as compared to the energy that it extracts when pyruvic acid or another organic compound is the hydrogen acceptor (fermentation). Oxidative phosphorylation generates the greatest amount of energy available from glucose.

Not all of the energy of the glucose molecule is captured as chemical energy in ATP, because no process is 100 percent efficient. Some of the chemical energy is wasted as heat energy. The cell, however, is remarkably efficient in capturing energy from the metabolism of glucose in the high-energy bonds of ATP. Approximately 38 moles of ATP (net) are produced from the oxidation of 1 mole of glucose to CO_2 and H_2O. Since one high-energy bond of ATP yields approximately 7000 calories when it is broken, a total of 38×7000 or 266,000 calories are captured in the form of chemical energy by each mole of ATP. This represents an almost 40 percent efficiency in conserving the energy present in the glucose molecule.

OTHER PATHWAYS

We have discussed only the major pathways by which bacteria degrade glucose. A number of other pathways exist which are very important to some bacteria, because not all bacteria have a glycolytic pathway and TCA cycle. Like the glycolytic and TCA cycle, these other pathways provide both energy and starting materials for the synthesis of the components of cytoplasm.

TYPES OF WORK

Basically, cells perform three types of work: (1) chemical work, which involves the work required to synthesize chemical bonds; (2) work of transport and concentration of molecules from the outside to the inside of the cell; and (3) mechanical work, such as the movement of flagella. When any kind of work is done, the high-energy phosphate bonds of ATP are hydrolysed, yielding a large amount of energy. The energy released in this reaction drives those reactions that require energy.

ECONOMIES OF FERMENTATION VERSUS RESPIRATION

In order to obtain the same amount of energy, a cell growing under anaerobic conditions must degrade about 20 times more glucose than a cell growing under aerobic conditions. Because the amount of work that a cell can perform depends on its supply of ATP, a cell growing under aerobic conditions can synthesize far more cell material per unit time (and therefore multiply more rapidly) than the same cell growing under anaerobic conditions. This is especially true if the cell has a limited supply of its nutrient energy source. In the laboratory the investigator generally supplies organisms with an excess of foodstuffs, but in nature the nutrient supply generally limits the number of cells in any particular location. Anaerobic conditions may also limit microbial growth because organic acids, commonly the end products of anaerobic metabolism, are often toxic to cells.

Limited food supply, page 118

BREAKDOWN OF COMPOUNDS OTHER THAN GLUCOSE

COMPOUND METABOLIZED TO INTERMEDIATE COMPOUNDS IN GLYCOLYTIC PATHWAY

Many compounds in addition to glucose can be degraded to provide energy. Indeed, if enzymes are available to catalyze the degradation, anything that has chemical bond energy can be degraded and its energy converted into the phosphate bond energy of ATP. Most naturally occurring compounds are degradable by some microorganism, although the pathways by which these compounds are degraded are in most cases unknown. It is certainly true, however, that degradation of organic compounds will often generate ATP—most likely through oxidation-reduction reactions.

Most organic compounds that are degraded, including some amino acids, lipids, and carbohydrates, are converted into the intermediates of either the glycolytic pathway or the TCA cycle and thus enter the energy-yielding sequence of the cell's reactions.

ANAEROBIC RESPIRATION

Respiration has been defined as an oxidation in which the terminal electron (or hydrogen) acceptor is an inorganic compound (Table 7-2). The most common such acceptor is oxygen, and therefore respiration usually occurs only under aerobic conditions. If oxygen is not available, however, some organisms can utilize other inorganic compounds as electron acceptors. The most common of these other acceptors are nitrate (NO_3^-), sulfate (SO_4^{2-}), and carbon dioxide (CO_2). This type of metabolism is called *anaerobic respiration,* because it occurs in the absence of oxygen but *does* involve an inorganic terminal electron acceptor—part of the definition of respiration.

UTILIZATION OF INORGANIC COMPOUNDS AS ENERGY SOURCES

Many microorganisms can oxidize a variety of inorganic compounds to gain energy. Table 7-3 presents the relevant information about some of these groups of organisms. A number of these organisms and their metabolic processes are considered elsewhere in the text (Chapters 12, 34, and 36). Many of the organisms

TABLE 7-3
Energy Metabolism of Some Aerobic Bacteria Which Derive Energy from Inorganic Sources

Common Name of Organism	*Source of Energy*	*Oxidation Reaction (Energy Yielding)*	*Important Features of Group*	*Common Genera in Group*
Hydrogen bacteria	H_2 gas	$H_2 + \frac{1}{2}O_2 \rightarrow H_2O$	Can also use simple organic compounds for energy	*Hydrogenomonas*
Sulfur bacteria (nonphotosynthetic)	H_2S	$H_2S + \frac{1}{2}O_2 \rightarrow H_2O + S$ $S + 1\frac{1}{2}O_2 + H_2O \rightarrow H_2SO_4$	Some organisms of this goup can live at pH of O; *Beggiatoa* and *Thiothrix* can only use H_2S as energy source	*Thiobacillus* *Beggiatoa* *Thiothrix*
Iron bacteria	Reduced iron (Fe^{2+})	$2\,Fe^{2+} + \frac{1}{2}O_2 + H_2O \rightarrow 2\,Fe^{3+} + 2\,OH^-$	Question as to whether *Sphaerotilus* will oxidize Fe^{2+}, although iron oxide is present in sheaths	*Sphaerotilus* *Gallionella*
Nitrifying bacteria	NH_3 HNO_2	$NH_3 + 1\frac{1}{2}O_2 \rightarrow HNO_2 + H_2O$ $HNO_2 + \frac{1}{2}O_2 \rightarrow HNO_3$	Important in nitrogen cycle Important in nitrogen cycle	*Nitrosomonas* *Nitrobacter*

that gain energy from inorganic oxidations can also gain their energy by oxidizing organic compounds.

ENZYME DIFFERENCES BETWEEN AEROBES AND ANAEROBES

ANAEROBES LACK CERTAIN ENZYMES OF RESPIRATORY CHAIN

All bacteria have enzyme systems that can react with oxygen. One of the end products of such reactions is H_2O, as we have already discussed. However, in other cases toxic compounds, such as hydrogen peroxide, are often formed, and sometimes an especially toxic form of oxygen called *superoxide* (O_2^-). The bacteria for which oxygen is not toxic (aerobes and facultative anaerobes) have enzymes that degrade these toxic byproducts of oxygen metabolism. H_2O_2 (hydrogen peroxide) is degraded to H_2O and O_2 by catalase or peroxidase enzymes. The enzyme superoxide dismutase, found only in aerobic organisms, is responsible for degrading superoxide.

Since strict anaerobes cannot utilize oxygen as the final electron acceptor, they lack many of the enzymes concerned with oxygen utilization. These include many of the electron transport chain enzymes, through which electrons are passed to oxygen (the cytochromes), as well as many enzymes of the TCA cycle, which are not utilized in anaerobic energy metabolism.

PHOTOSYNTHESIS

The discussion thus far has focused on the mechanisms which bacteria use to gain their energy by converting the chemical energy present in the bonds of either inorganic or organic molecules into ATP. The bacteria that get their energy in this way are the most common. Some organisms, however, have the ability to convert the radiant or light energy absorbed by chlorophyll into useful chemical energy in the form of ATP molecules, through the process of *photosynthesis.* This chemical energy is then available to the organism itself as well as to heterotrophic organisms which can degrade the organic material in photosynthetic organisms as a source of energy.

PHOTOSYNTHESIS REQUIRES ENERGY AND SOURCE OF REDUCING POWER

The overall reaction of photosynthesis carried out by green plants, eucaryotic algae, and cyanobacteria is summarized in the equation:

$$6CO_2 + 6H_2O \xrightarrow{\text{light}} C_6H_{12}O_6 + 6O_2$$

These organisms utilize CO_2 and H_2O, from which, in the presence of light, they synthesize a carbohydrate and release oxygen. The carbohydrate ($C_6H_{12}O_6$) represents not only the starch that these organisms synthesize but is also essential to all structures that their cells synthesize as they grow. This reaction therefore summarizes all of the biosynthetic reactions of the cell. It is the *reverse* of the reactions by which glucose is oxidized completely via glycolysis and the TCA cycle:

PHOTOSYNTHESIS IS REVERSE OF RESPIRATION

$$C_6H_{12}O_6 + 6O_2 \longrightarrow 6CO_2 + 6H_2O + \text{energy}$$

Because energy is gained when glucose is broken down, energy must be provided when it is synthesized from its constituents. Because hydrogen atoms are removed in the oxidation of carbohydrates, hydrogen atoms must be supplied for their synthesis.

Recent experiments have employed radioactive *isotopes* as tracers to follow the fate of each of the atoms in CO_2 and H_2O during photosynthesis. These experiments have conclusively demonstrated that the carbohydrate is synthesized from CO_2 and the hydrogen atoms of water, while the O_2 originates from the oxygen atom of water. The process of photosynthesis is more accurately shown by this balanced equation:

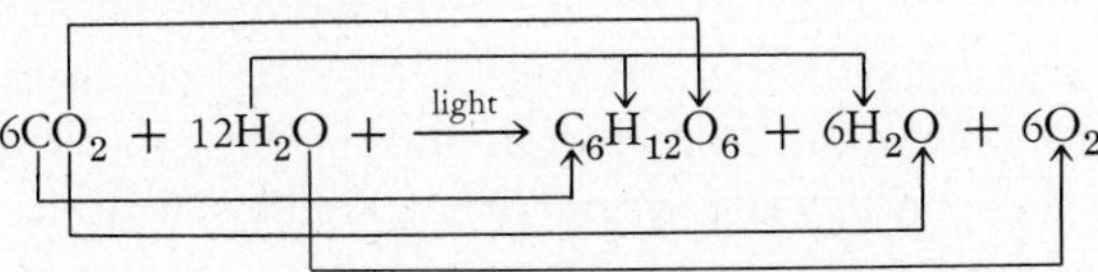

The source of the hydrogen atoms used for reducing CO_2 to $C_6H_{12}O_6$ is, therefore, water. The equation for photosynthesis in all bacteria except the cyanobacteria differs from the above equation. Whereas cyanobacteria and plants utilize H_2O as their source of hydrogen atoms for the reduction of CO_2, other photosynthetic bacteria never use the hydrogen atoms from water, but rather utilize other reduced compounds, either inorganic or organic, depending on the particular organism, to reduce CO_2. As a result, these bacteria never liberate oxygen, whereas plants, algae and cyanobacteria always do. Some groups of bacteria utilize a reduced inorganic sulfur compound (H_2S) in place of water. In this case the hydrogen in the $C_6H_{12}O_6$ comes from H_2S, and sulfur, rather than oxygen, is released. Other bacteria use reduced organic compounds as hydrogen donors. The *general* equation for photosynthesis, covering all bacteria, algae and green plants, may be restated as:

Photosynthetic bacteria, pages 272–273

$$CO_2 + H_2X \xrightarrow{\text{light}} C_6H_{12}O_6 + X$$

LIGHT REACTIONS INVOLVE CHLOROPHYLL; DARK REACTIONS INVOLVE ENZYMES

The overall process of photosynthesis can be divided into *two* series of interrelated reactions: the "light" reactions, which occur *only* in the presence of light, and the "dark" reactions, which do not require light. The light reactions are unique to photosynthetic organisms. In these reactions the green pigment *chlorophyll* absorbs light energy. Electrons in the chlorophyll increase their energy by gaining the energy of light. These energized electrons are transferred along an electron transport chain, and the high-energy phosphate bonds of ATP are generated. Some of the electrons and hydrogen ions in photosynthesis also serve to reduce small organic compounds, which are formed from CO_2 and H_2O in the reactions in which these two substances are converted to $C_6H_{12}O_6$. Thus the absorption of light by chlorophyll generates both chemical energy (ATP) and reducing power (H atoms).

Once light has generated ATP and reducing power, the formation of carbohydrate in photosynthesis can occur in the dark. Such dark reactions involve the incorporation of CO_2 into an organic molecule and then, by additional enzymatic steps, the conversion of this molecule to carbohydrate. Nor are these reactions unique to photosynthetic organisms; they also occur in heterotrophic organisms as well as in autotrophic organisms, which gain their energy by oxidizing inorganic compounds.

BIOSYNTHETIC METABOLISM

Amino acids formulas, page 33
Purines and pyrimidines, page 39

Thus far, the discussion has concentrated on the reactions by which cells obtain energy. Because glucose or some other energy source is also often the only source of carbon available to cells, it must also serve as the initial source of carbon for the synthesis of all organic molecules within these cells. Many intermediate compounds of the glycolytic pathway and TCA cycle serve as starting compounds for the synthesis of amino acids, purines and pyrimidines, and lipids. If the intermediate is channeled along one pathway, it will be converted into a component of the cell's structure; if it is metabolized along another pathway, it will serve to generate ATP. The synthesis of any amino acid requires the participation of a number of enzymes, each catalyzing one reaction in the synthesis of the particular amino acid. Figure 7-14 illustrates in abbreviated form the various intermediates of glucose degradation which serve as starting materials for the synthesis of some subunits of macromolecules.

As pointed out previously, anaerobes do not have an intact TCA cycle because they have no need for one for the generation of their energy. However, because intermediates of this pathway serve as substrates for the synthesis of amino acids, it is perhaps not surprising that most of this pathway exists in *all* bacteria that can synthesize these amino acids. Anaerobes synthesize oxaloacetic acid, the first step in the pathway, by combining CO_2 with pyruvic acid (Figure 7-14). Thus pyruvic acid can enter the cycle either directly or after being converted to acetic acid.

The same enzymes may be involved in synthesizing several amino acids that

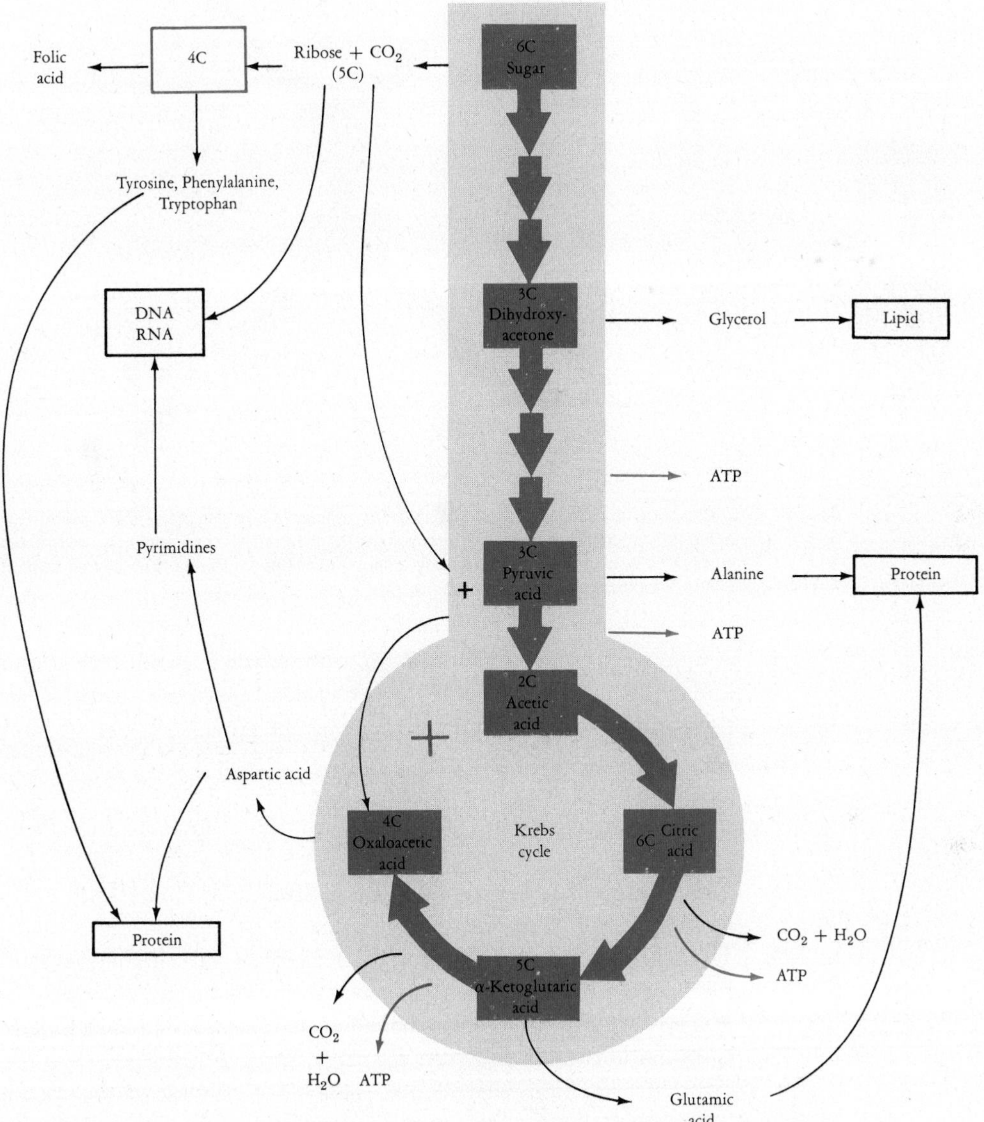

FIGURE 7-14 Summary of compounds originating in the glycolytic pathway and Krebs cycle leading to biosynthetic products. This diagram illustrates that most reactions involved in the degradation of a compound also serve to synthesize the beginning materials for the synthesis of macromolecules.

are members of the same family. This is illustrated in the pathway leading to the synthesis of the three amino acids tyrosine, phenylalanine, and tryptophan (Figure 7-15), all of which have a benzene ring as part of their structure. This same pathway is involved in the synthesis of several vitamins. The sequence of reactions illustrates features common to a large number of biosynthetic pathways.

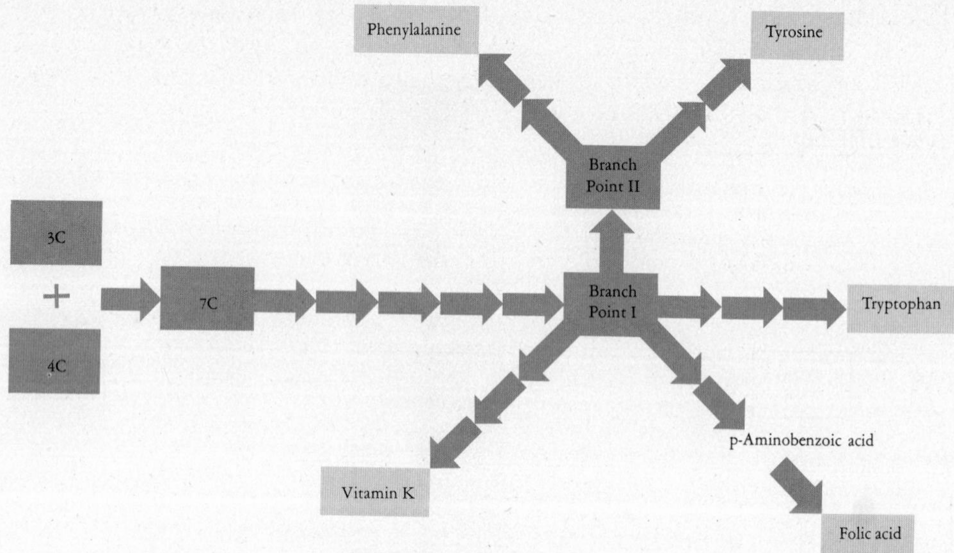

FIGURE 7-15 Biosynthetic pathway of aromatic acid biosynthesis. A complex biosynthetic pathway in which three different amino acids and several vitamins are synthesized. (The number of arrows do not reflect the number of steps actually involved.) The compound at Branch Point I is chorismic acid, at Branch Point II the compound is prephenic acid.

1. A number of enzymes are common to the synthesis of all three amino acids and several vitamins. If any one of these enzymes is defective, all three amino acids and two vitamins must be provided in the medium for the defective cells to grow.
2. Energy must be supplied at several steps in the pathway. The high-energy compounds that take part are phosphoenolpyruvic acid (PEP) and ATP.
3. There are several branch points in the pathway at which a compound is metabolized along several different paths.
4. The cell must regulate the flow of metabolites into the biosynthetic pathway commensurate with the availability of its energy supply. It makes no sense for the cell to channel metabolites used in biosynthetic pathways into these pathways if energy is not available to drive the many such reactions that require energy. For example, if the energy sources of the cell are limited, the cell will utilize phosphoenolpyruvic acid as a source of cellular energy rather than use it in the aromatic acid biosynthetic pathway. The regulatory mechanisms which determine whether a metabolite will be converted into a subunit for macromolecule synthesis or degraded further for energy are also considered in Chapter 10.

Utilization of Ammonium Salts

One essential feature of amino acid biosynthesis is the introduction of nitrogen to form the amino (—NH_2) group. A number of enzymatic reactions catalyze the incorporation of ammonia (or the ammonium ion NH_4^+) into an organic molecule. A key reaction links energy metabolism to biosynthesis. In this reaction α-ketoglutaric acid, a 5-carbon intermediate compound in the TCA cycle (Figure 7-11 and Appendix I), will react with NH_4^+ and be converted to the amino acid, glutamic acid:

$$\underset{\text{ammonium ion}}{NH_4^+} + \underset{\alpha\text{-ketoglutaric acid}}{HO{-}\overset{O}{\overset{\|}{C}}{-}\underset{H}{\overset{H}{C}}{-}\underset{H}{\overset{H}{C}}{-}\overset{O}{\overset{\|}{C}}{-}\overset{O}{\overset{\|}{C}}{-}O{-}H} + NAD_{red} \rightleftharpoons$$

$$\underset{\text{glutamic acid}}{HO{-}\overset{O}{\overset{\|}{C}}{-}\underset{H}{\overset{H}{C}}{-}\underset{H}{\overset{H}{C}}{-}\underset{H}{\overset{NH_2}{C}}{-}\overset{O}{\overset{/\!/}{C}}{-}OH} + NAD_{ox}$$

Because this reaction is reversible, glutamic acid can also be readily converted to α-ketoglutaric acid. The latter compound enters the TCA cycle and is transformed to CO_2 and H_2O with the production of energy (Figure 7-11).

Transamination Reactions

Glutamic acid plays a very important role in the synthesis of most other amino acids. It donates its amino group to a precursor of the amino acid being synthesized, always an α-keto acid. This converts the α-keto acid to the amino acid, and simultaneously the glutamic acid is converted to α-ketoglutaric acid. The synthesis of most amino acids involves such a transamination reaction. This transamination reaction can be expressed as follows:

$$\underset{\text{glutamic acid}}{HO{-}\overset{O}{\overset{\|}{C}}{-}\underset{H}{\overset{H}{C}}{-}\underset{H}{\overset{H}{C}}{-}\underset{H}{\overset{NH_2}{C}}{-}\overset{O}{\overset{/\!/}{C}}{-}OH} + \underset{\alpha\text{-keto acid}}{R{-}\overset{O}{\overset{\|}{C}}{-}\overset{O}{\overset{\|}{C}}{-}OH} \rightleftharpoons$$

$$\underset{\text{amino acid}}{R{-}\underset{H}{\overset{NH_2}{C}}{-}\overset{O}{\overset{/\!/}{C}}{-}OH} + \underset{\alpha\text{-ketoglutaric acid}}{HO{-}\overset{O}{\overset{\|}{C}}{-}\underset{H}{\overset{H}{C}}{-}\underset{H}{\overset{H}{C}}{-}\overset{O}{\overset{\|}{C}}{-}\overset{O}{\overset{\|}{C}}{-}OH}$$

R stands for any of a variety of carbon side chains, depending on the amino acid to be synthesized. For example, pyruvic acid ($CH_3{-}\overset{O}{\overset{\|}{C}}{-}\overset{O}{\overset{\|}{C}}{-}OH$) can be

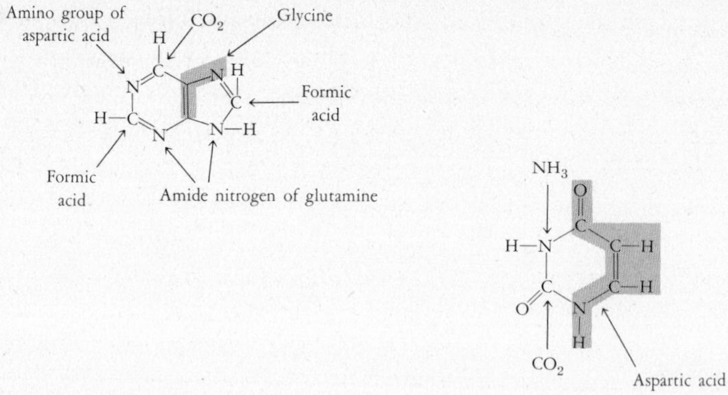

FIGURE 7-16
Origin of atoms for the biosynthesis of a purine molecule (top) and pyrimidine molecule (bottom).

converted to the amino acid alanine which has a carbon side chain of CH_3 ($R{=}CH_3$) by transamination:

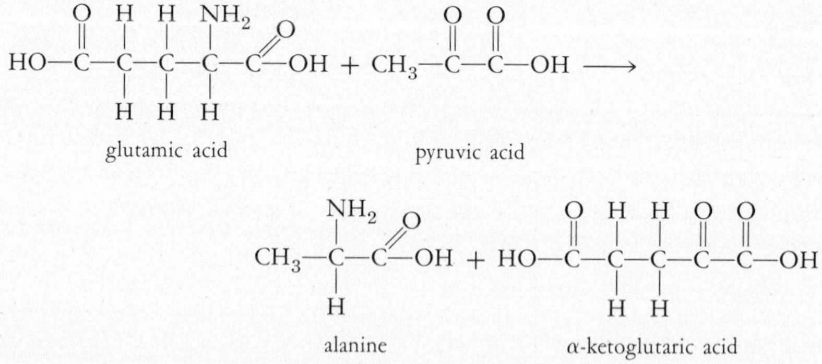

$$\underset{\text{glutamic acid}}{HOOC{-}CH_2{-}CH_2{-}CH(NH_2){-}COOH} + \underset{\text{pyruvic acid}}{CH_3{-}CO{-}COOH} \longrightarrow$$

$$\underset{\text{alanine}}{CH_3{-}CH(NH_2){-}COOH} + \underset{\alpha\text{-ketoglutaric acid}}{HOOC{-}CH_2{-}CH_2{-}CO{-}COOH}$$

Synthesis of Purines and Pyrimidines

The synthesis of these components of nucleic acids illustrates how compounds can be built up by the stepwise addition of a number of simpler molecules. The carbon and nitrogen atoms originate from a variety of sources (Figure 7-16). The synthesis of purines and pyrimidines depends on the availability of the amino acids and other molecules that are component parts of these nucleic acid precursors, as well as about a dozen different enzymes, each one catalyzing one step in the pathway. The synthesis of purines and pyrimidines illustrates the interrelationship between all biosynthetic processes in the cell.

SUMMARY

The metabolic transformations that cells carry out are dedicated to increasing the quantity of all cellular components. Since the synthesis of these components requires energy, the cell must also derive energy from foodstuffs. In many instances the foodstuff, often a sugar, serves as both a source of the carbon skeletons from which are synthesized all of the cellular components as well as the material from which is derived cellular energy. The requirement for energy involves the conversion of diffuse chemical bond energy into the concentrated high-energy bonds of

ATP, or, in the case of photosynthetic organisms, the conversion of light energy into the chemical bond energy of ATP. ATP serves as the energy currency in all cells. Heterotrophs obtain energy by oxidizing reduced organic compounds through a series of enzymatic reactions. The energy lost between the starting material and end products is largely captured in the form of high-energy phosphate bonds. One of the most common pathways for the breakdown of carbohydrates in all forms of life is the glycolytic pathway, which functions under both aerobic and anaerobic conditions. For every molecule of glucose metabolized, the products of glycolysis are two molecules of pyruvic acid, two molecules of ATP (net), and two molecules of NAD_{red}. Pyruvic acid can be further metabolized in a variety of ways which yield a variety of products, including ethanol, CO_2, lactic acid, acetic acid, and several other acids and alcohols. Under aerobic conditions, and provided the cell has the necessary enzymes, the acetic acid may be metabolized through the TCA cycle. If glucose is oxidized only to organic compounds, most of the energy available in the glucose molecule remains untapped. However, if it is oxidized completely to CO_2 and H_2O, then about 40 percent of the energy originally present in the glucose molecule is trapped by ATP. The ATP is generated in the course of transfer of electrons along the electron transport chain.

The pathways of glucose degradation provide the starting compounds from which many small molecules are synthesized. Many steps in these biosynthetic reactions require energy, and the ATP generated in the breakdown of sugar drives these reactions.

QUESTIONS

REVIEW

1. Under what conditions of oxygenation would a yeast cell: (a) multiply to the highest cell yield? (b) produce the most alcohol? (c) have the shortest generation time? Explain your reasoning for each answer.
2. Give some of the products that can be synthesized from pyruvic acid, and name the genera that can synthesize these end products.
3. Discuss the major properties of enzymes, and explain how they function.
4. Discuss the reasons why some cells are killed by oxygen and others are not.

THOUGHT

1. Some organisms grow *equally well* in the presence or absence of oxygen. How can this be explained?
2. Why do diseases of vitamin deficiency have widespread and far ranging effects on the body?
3. Why is it important that an organism have certain enzymes of the TCA cycle even though the organism is an anaerobe?

FURTHER READING

GREEN, D., and R. GOLDBERGER, *Molecular Insights into the Living Process.* New York: Academic Press, 1967. An extremely well-written and understandable presentation of metabolic principles, both degradative and biosynthetic. Chapters 4–8 are especially relevant to this chapter.

KIRK, D., *Biology Today.* 2d ed. New York: Random House, 1975. Chapters 12, 14, and 15 present the material covered in this chapter, especially material on the catalytic activity of enzymes and eucaryotic metabolic processes, in much greater detail than is given in the present Chapter 7.

LEHNINGER, A., *Bioenergetics.* 2d ed. Menlo Park, Calif.: W. A. Benjamin, 1971. Probably the best single volume on the generation, storage, and utilization of energy in eucaryotic and procaryotic cells. For students who have had some chemistry.

INFORMATIONAL MACROMOLECULES: FUNCTION AND SYNTHESIS

In Chapter 7 the mechanisms by which cells obtain energy and synthesize small molecules were explored. In this chapter the biosynthesis of several macromolecules—RNA, DNA, and protein—is considered. These macromolecules have several features in common. First, each embodies information that is contained in the sequence of the subunits that constitute the macromolecule. The biological function of these macromolecules depends on the exactness of this sequence. Second, the synthesis of each of these macromolecules requires *templates*—other macromolecules whose structures provide specific surfaces which serve as patterns for the synthesis of the macromolecules. Third, recognition of purine and pyrimidine base sequences through *complementarity* plays a vital role in the synthesis of all three classes of macromolecules. Fourth, the synthetic reactions require energy in order to proceed, and a high-energy bond is broken as each subunit is bonded to its neighbor to form the macromolecule.

CHEMISTRY OF DNA AND RNA

Subunits of nucleic acid, page 39

DNA usually does not occur as a long, single strand. Rather it occurs as a *double-stranded helical structure* (Figure 8-1). The two strands of this helix are held together by hydrogen bonds between complementary surfaces of the purine molecule (adenine) and the pyrimidine molecule (thymine). Also, because the purine guanine is complementary to the pyrimidine cytosine it will form hydrogen bonds with cytosine. This requirement for complementarity explains why, in all double-stranded molecules of DNA, the number of molecules of thymine equals that of adenine, while the number of molecules of cytosine equals that of gua-

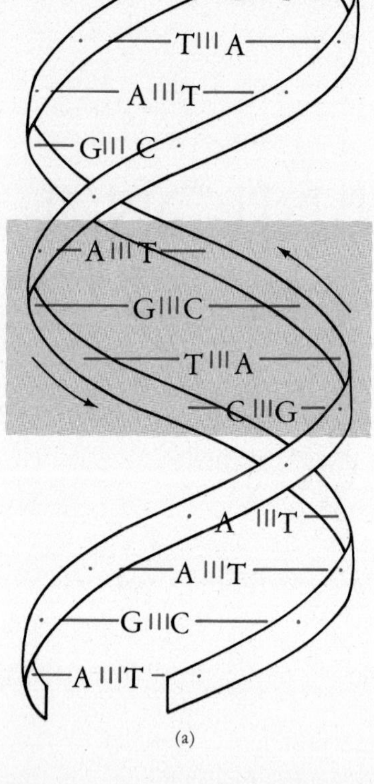

(a)

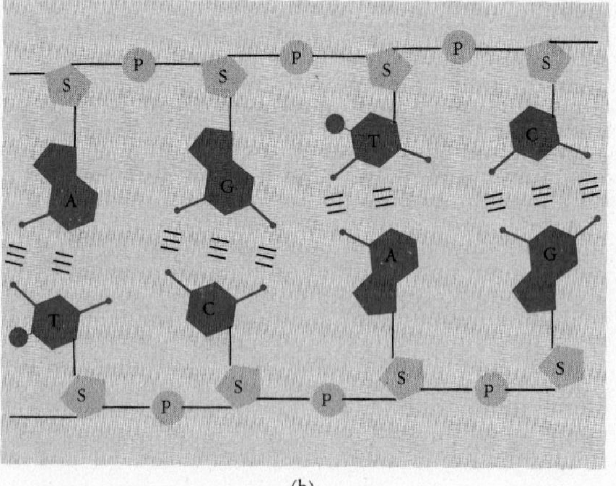

(b)

FIGURE 8-1
(a) Helical structure of DNA and (b) its appearance when not twisted. The helical DNA molecule can be looked on as a spiral staircase, in which the sugar-phosphate repeating sequences form the two railings and the purine and pyrimidine bases represent the stairs. The railings move in opposite directions, one going up, the other down. If the helix is untwisted, the molecule assumes the shape of a ladder. The fact that the two strands run in opposite directions is most readily seen in this untwisted form. The two strands are held together by hydrogen bonding.

Hydrogen bonds, page 30

nine. The elucidation of the structure of DNA in the 1950s by the Englishman Dr. Francis Crick and his colleague, the American Dr. James Watson, represents one of the most fundamental contributions ever made in molecular biology. In 1962 they received the Nobel Prize for these studies. The hydrogen bonding of this double-stranded structure again demonstrates how a sufficiently large number of weak bonds can confer great stability on a molecule. The complementarity of the purines and pyrimidines is important for another reason: the sequence of purines and pyrimidines on one strand dictates the sequence on the other strand. This has significance in the synthesis of new strands of DNA in the course of cell growth and division. Like DNA, RNA consists of a sequence of nucleotides, but unlike DNA it usually exists as a single stranded structure.

Concept of Information Storage and Transfer

CENTRAL DOGMA OF MOLECULAR BIOLOGY

Biological macromolecules can contain information specified by the arrangement of the subunits making up the *polymer.* For example, the four different subunit-bases in DNA may be thought of as a four-letter alphabet in which the sequential arrangement spells out a message. In all normal cells DNA is the ultimate genetic language in which all the genetic information of the cell is written. On the other hand, proteins contain twenty different kinds of amino acid subunits, the arrangement of which constitutes another language written in a completely different alphabet.

Two basically different processes are involved in information storage in cells. In one process the information is duplicated in an identical molecule which is passed on to the daughter cell. An example of this process is the replication of DNA. The second process involves the transfer of information to another similar, but not identical molecule, and the further transfer of this information into other kinds of macromolecules containing different subunits. The second process generally involves two stages, *transcription,* the transfer of information from one language into a *similar language,* and *translation,* the transfer from the similar language into a *second language.* An example of this process is the transfer of information from DNA to RNA (transcription) and then from RNA to protein (translation). This transfer of information is often referred to as the *central dogma* of molecular biology (Figure 8-2). It was once believed to be unidirectional, from DNA to RNA to protein. It has been shown, however, that certain viruses that can cause cancer in animals can transfer information from RNA to DNA using an enzyme with the descriptive name *reverse transcriptase.*

Reverse transcriptase, pages 359–361

The synthesis of *any* polymer which is synthesized on a template, whether it be DNA or RNA, follows the same general pattern. This sequence of steps is

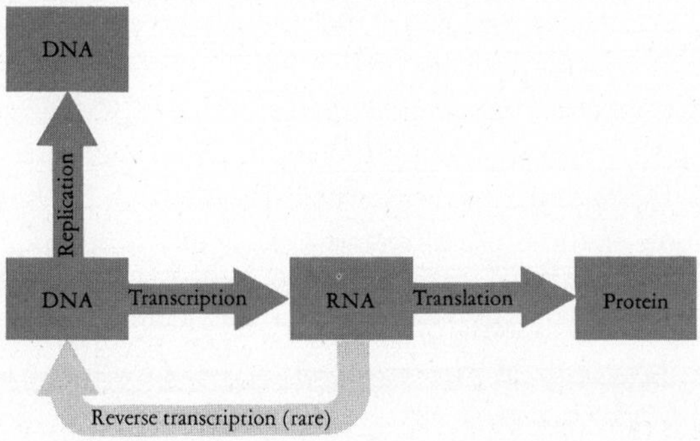

DNA
Replication
DNA Transcription RNA Translation Protein
Reverse transcription (rare)

FIGURE 8-2
Flow of information in biological systems. Reverse transcription occurs in a few tumor virus systems.

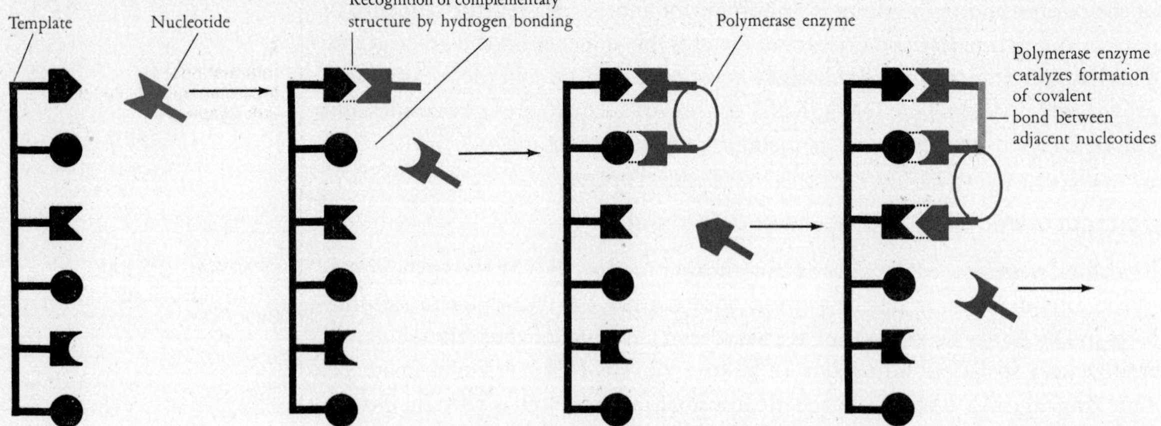

FIGURE 8-3 Sequence in template-directed nucleic acid synthesis. This general reaction sequence diagrams the synthesis of DNA and RNA.

diagrammatically represented in Figure 8-3. We shall first consider DNA and then RNA synthesis.

Replication of DNA

The bacterial *chromosome* is divided into a large number of genes, each *gene* being the sequence of purine and pyrimidine bases that specifies the sequence of amino acids that will comprise one polypeptide chain.

In the process of DNA replication of a chromosome which is usually circular in bacteria, the two strands separate at the site at which the new strands are to be synthesized to form a Y-shaped structure referred to as the *replication fork.* This fork then progresses around the entire circular DNA molecule. As it moves, two new strands of DNA are synthesized, each one having a base sequence complementary to one of the original strands of the DNA (Figure 8-4). There are specific sites on the chromosome at which replication is initiated. Sometimes synthesis proceeds in one direction; in other cases it proceeds in two. The strands are not synthesized as a single chain; instead, short segments are synthesized and then joined together (Figure 8-5). The *initiation* of DNA replication first requires the synthesis of a short chain of RNA that is complementary to the DNA. The DNA that is then synthesized is joined to this short piece of RNA. The DNA polymerase then digests the RNA and replaces it with DNA (Figure 8-5). When the replication fork has moved along the entire DNA molecule, two complete double-stranded DNA molecules are formed. Because each double-stranded molecule contains one of the original strands and one new strand, this process of replication is termed *semiconservative.* This type of replication results in only two of the daughter cells in a bacterial population having the same DNA strands that were present in their original parent cell, no matter how many cells eventually descend from this cell (Figure 8-6).

Base sequence complementarity, page 31

SYNTHESIS OF DNA FIRST REQUIRES SYNTHESIS OF RNA

DNA SYNTHESIS IS SEMICONSERVATIVE

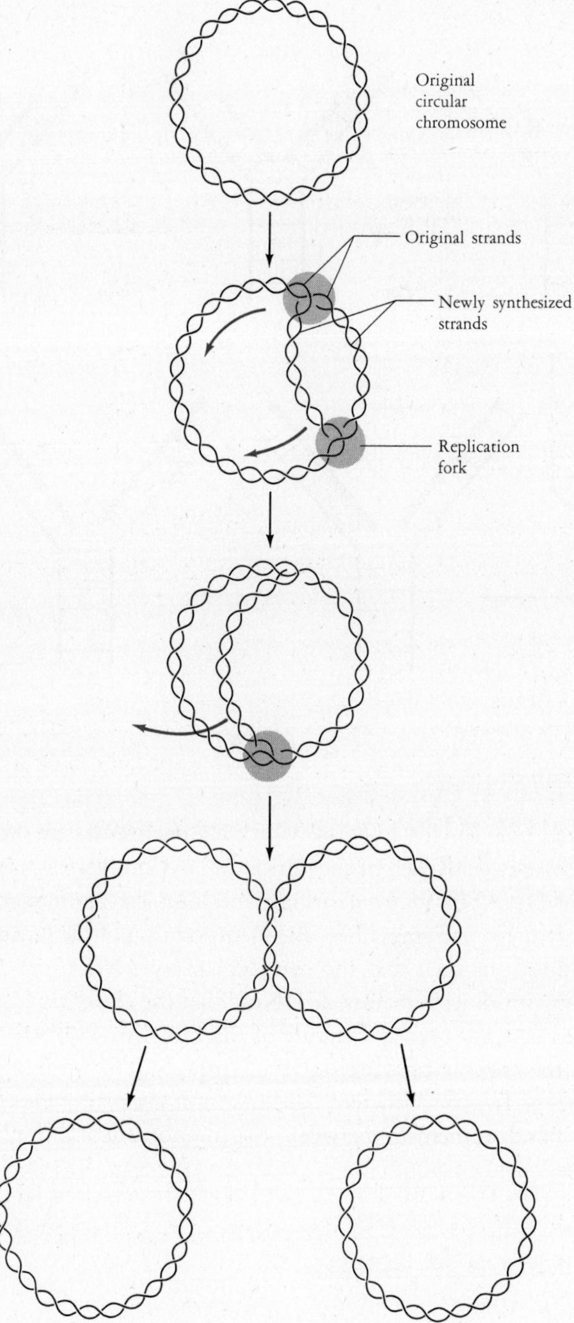

FIGURE 8-4
The replication of a circular chromosome in bacteria.

A number of components required for DNA synthesis must be present at the replication fork. These include the two enzymes DNA polymerase and RNA polymerase, which join the substrate molecules to form DNA and RNA, respec-

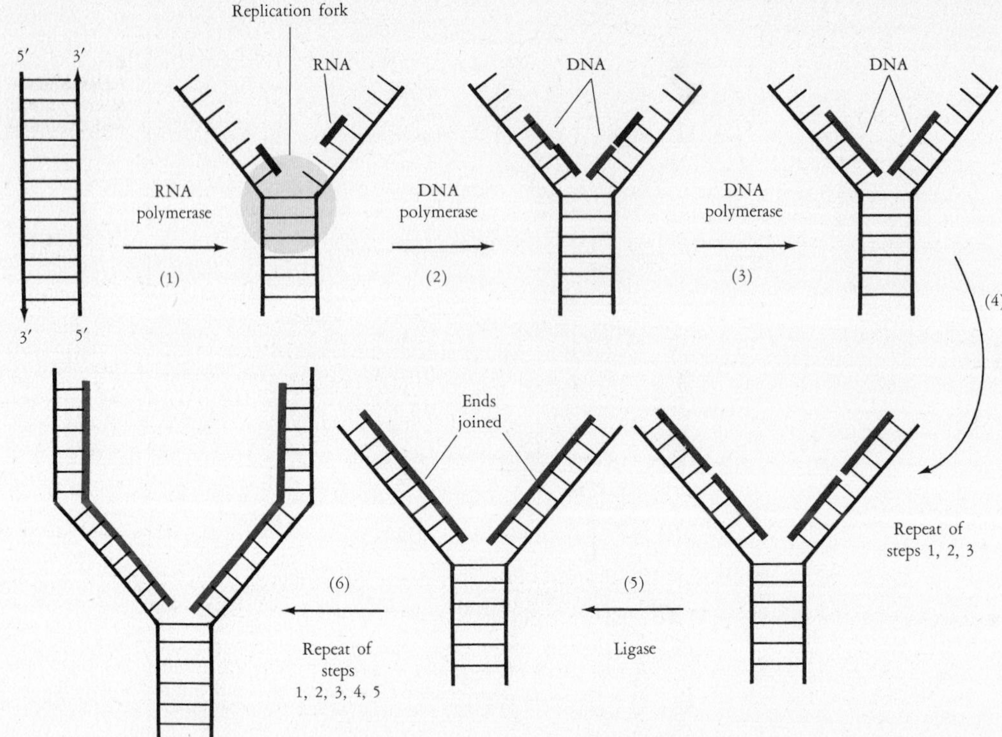

FIGURE 8-5 The details of chromosome replication in bacteria. The process requires not only DNA synthesis but also RNA synthesis.

tively, the *substrates* for the reaction, which are *deoxynucleoside triphosphates* and *nucleoside triphosphates* of each of the purine and pyrimidine bases, and a *template,* consisting of DNA. In addition, an enzyme joins the short segments (Figure 8-5). The polymerase enzymes direct the incorporation of the purines and pyrimidines into positions such that the complementary relationship between these purines and pyrimidines is maintained. Note that the structure of the DNA that is synthesized depends on the structure of the DNA that serves as the template and not on the enzyme carrying out the synthesis. The breaking of the high-energy bonds of the four different substrate molecules provides the energy required for bond formation between the mononucleotide subunits of the nucleic acids.

DNA TEMPLATE DICTATES THE BASE SEQUENCE OF THE NEWLY SYNTHESIZED DNA

THE EXPRESSION OF GENES

Transcription

All of the genetic information of a cell resides in the sequence of purines and pyrimidines that make up the genes in its DNA. The transcription process is the first step in gene expression. The transcription of the message encoded in the specific sequence of purine and pyrimidine bases in the DNA proceeds by an enzymatic

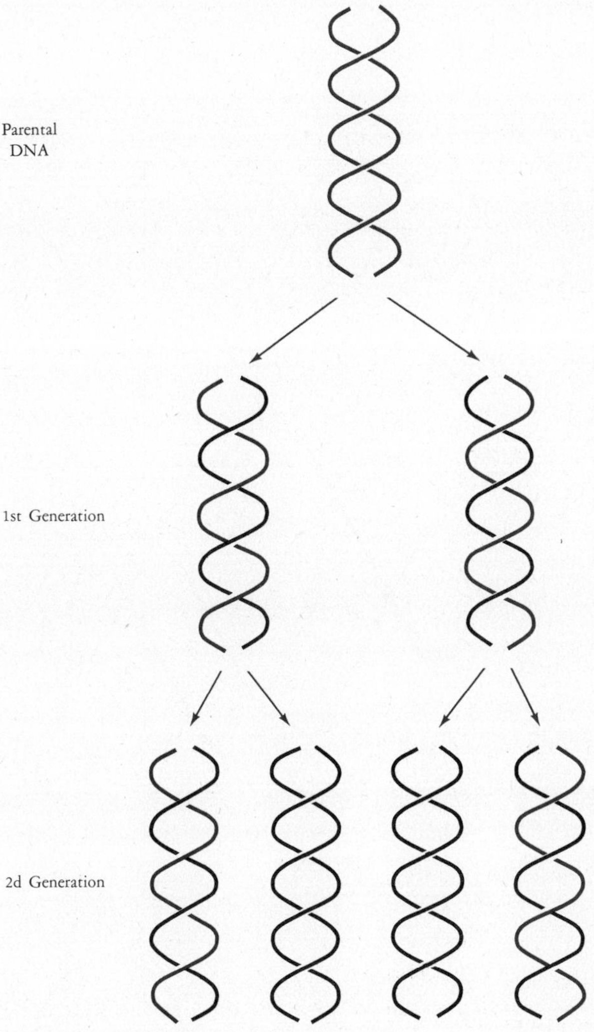

FIGURE 8-6
Semiconservative replication of DNA. The parental DNA is found in two progeny cells *only,* no matter how many generations occur.

process that is analogous to the process of DNA replication. The enzyme *RNA polymerase* binds to the DNA close to the beginning of a gene, moves along the gene, and synthesizes a single-stranded RNA molecule which is complementary to *one* of the strands of the DNA. The RNA molecule synthesized is termed *messenger RNA* (mRNA) and corresponds in length to one gene (Figure 8-7). Genes vary in size, but a reasonable average estimate of their length is 1000 nucleotides. The mRNA which is transcribed from a gene of this length is also 1000 nucleotides.

Translation

PROTEIN SYNTHESIS

The sequence of purines and pyrimidines in mRNA is translated into a sequence of amino acids in a protein (polypeptide) molecule whose structure is coded by

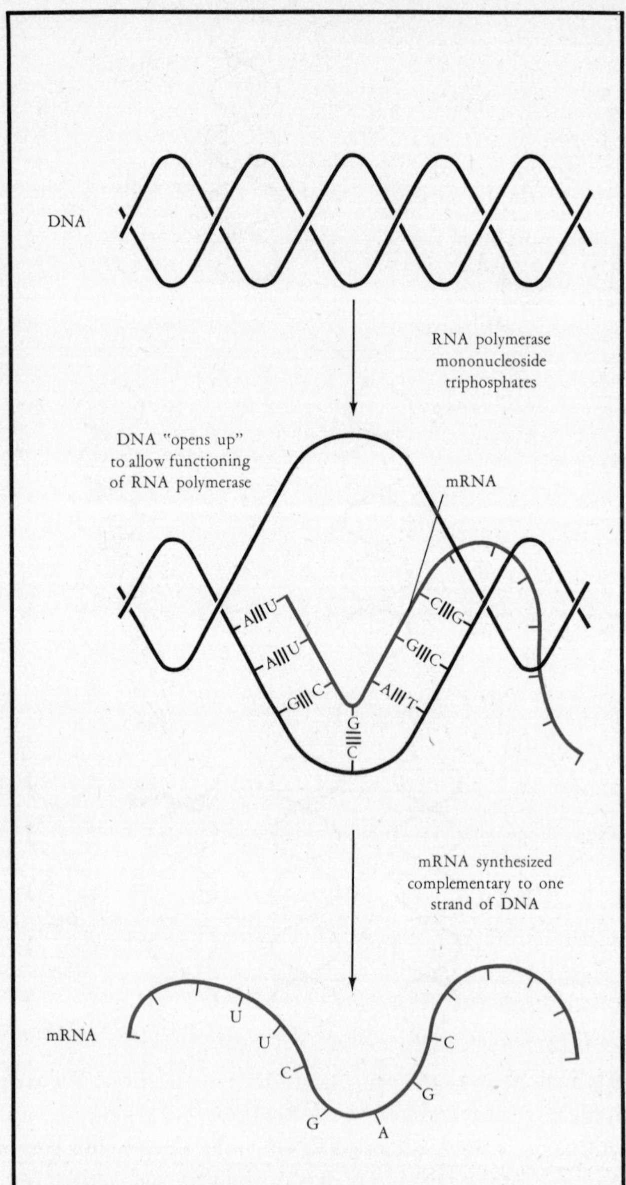

FIGURE 8-7
Gene transcription.

EACH AMINO ACID BINDS TO A SPECIFIC tRNA MOLECULE

the transcribed gene. This process is the most elaborate biosynthetic mechanism known. It involves several different enzymes, protein "factors," and two additional types of RNA, *transfer RNA* (tRNA) and *ribosomal RNA* (rRNA). About 90 percent of the total energy expended by a growing bacterial cell is dedicated to protein synthesis. The major aspects of this process include the following (Figure 8-8).

Step 1. The twenty different amino acid molecules are "activated" by the

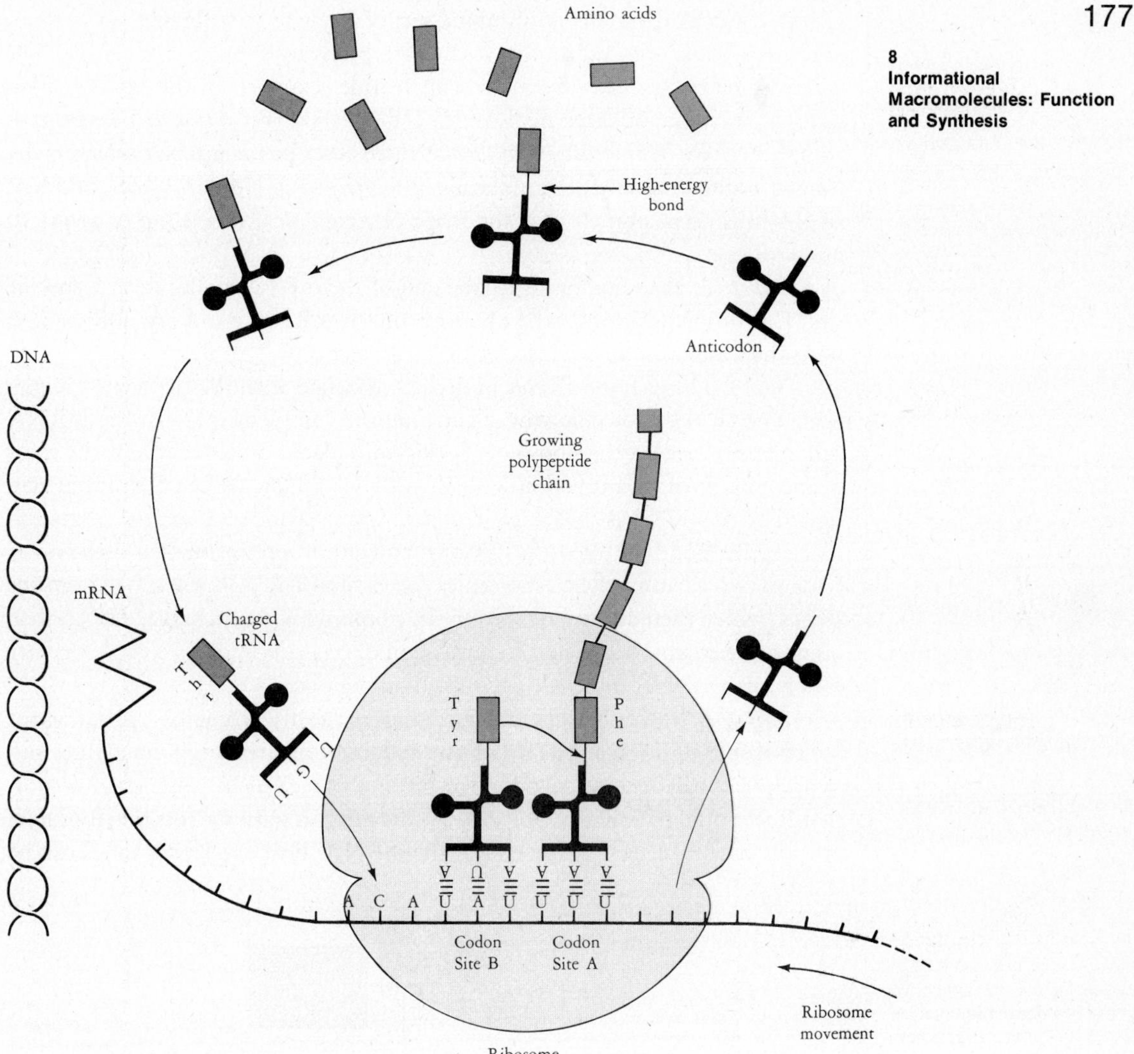

FIGURE 8-8 The process of protein synthesis.

formation, at the expense of high-energy phosphate bonds in ATP, of a high-energy bond between the amino acid carrier molecule, which is tRNA, and the amino acid. The tRNA carrying the amino acid is thus "charged"; when it is not carrying an amino acid, it is "uncharged." The activation of each different amino acid requires a different "activating enzyme." A unique class of tRNA molecules exists for each amino acid, and in the case of some amino acids as many as five different tRNA molecules are capable of binding to and carrying the same amino acid. Each tRNA molecule functions as an *adaptor* between the mRNA for a protein and the amino acids that will make up the protein. In order to function as an adaptor between the nucleotide bases of mRNA and amino acids, each

tRNA IS AN ADAPTOR BETWEEN AMINO ACIDS AND NUCLEIC ACIDS

tRNA molecule contains two unique sets of nucleotide sequences. One set of sequences determines the amino acid which binds to the tRNA molecule. The other sequence is *complementary* to a nucleotide sequence on the mRNA. This latter *site* consists of three nucleotides and is called the *anticodon* because it is complementary to the *codon*—the set of three bases in the mRNA which codes for one amino acid. A tRNA molecule is diagramed in Figure 8-9. All species of tRNA thus far examined have the shape of a cloverleaf consisting of about 70 nucleotides.

Ribosomes, pages 71–72

Step 2. A ribosome binds to one end of the mRNA molecule at a specific site. This initial position of the ribosome on the mRNA is important because it starts the translation of the mRNA in the correct codon reading frame.

Step 3. The charged tRNA molecules associate with the ribosome (Figure 8-8). The tRNA molecule whose anticodon is complementary to the mRNA codon positioned on the ribosome "locks" into place on that codon by the complementary fit of its anticodon with the mRNA codon. When the amino acid carried by the tRNA is thus fixed in position, an enzyme catalyzes the formation of a peptide bond between the carboxyl and amino groups of two such amino acids on two adjacent tRNA molecules. The amino acid is then released from the tRNA, which then diffuses away from the ribosome. This uncharged tRNA then binds another amino acid and its anticodon once again finds its complementary codon in an mRNA molecule positioned on the ribosome.

mRNA MOVES ALONG RIBOSOME

Step 4. A "rachet"-type mechanism moves the mRNA across the surface of the ribosome through a distance of one codon, or one reading frame (three nucleotides), so that the next codon (or reading frame) of the mRNA is now in position to bind to a tRNA molecule. As the mRNA moves across the ribosome, other ribosomes attach to the end of the mRNA. Five or six ribosomes can be

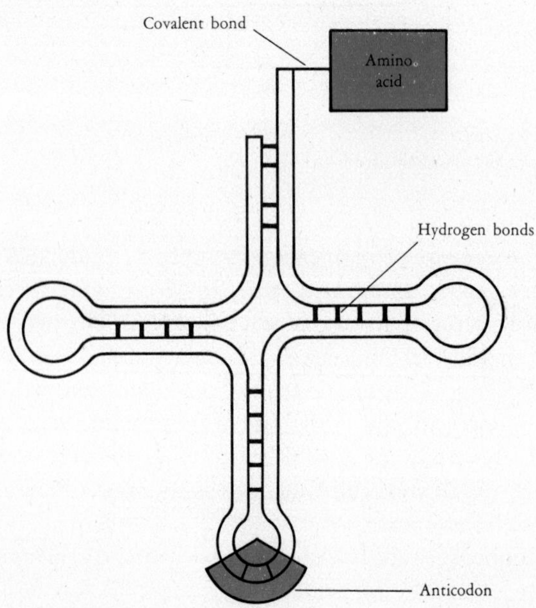

FIGURE 8-9
Molecule of tRNA. All tRNA molecules have an anticodon site at one end of the molecule and a site which binds a specific amino acid at another end. There are at least 20 different kinds of tRNA molecules, each with a different anticodon and the ability to bind a different amino acid. Note the hydrogen bonds which give the molecule its characteristic shape.

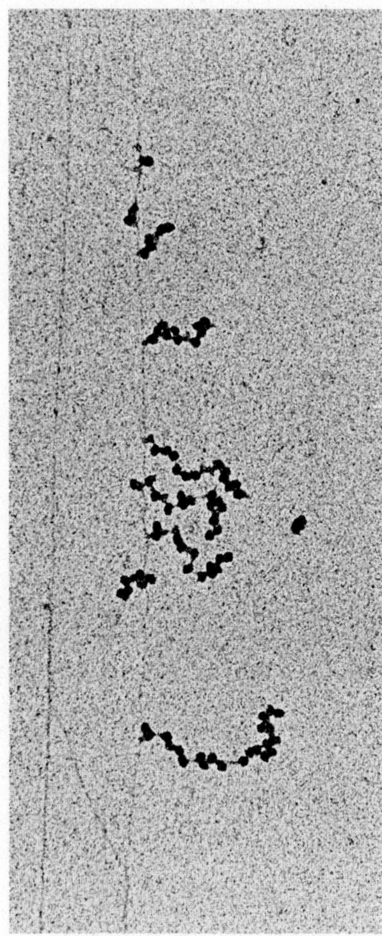

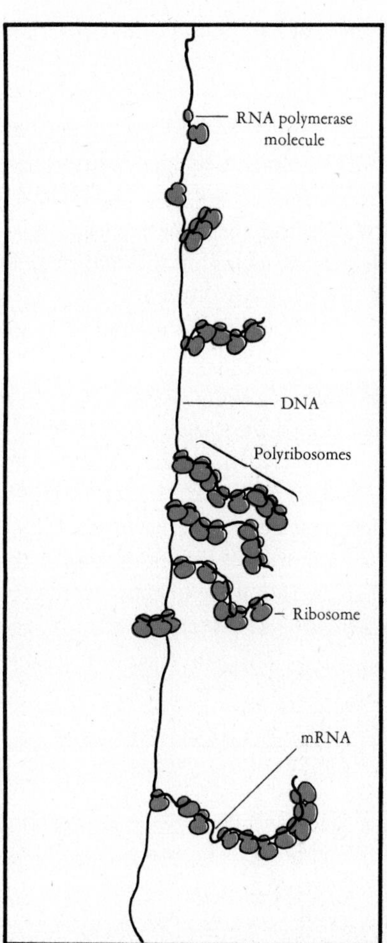

FIGURE 8-10
Genes in action. DNA is being transcribed into mRNA by RNA polymerase. The mRNA is bound to the ribosomes, thereby creating the polyribosomes. This figure illustrates how closely integrated the processes of transcription and translation are. (Photo courtesy of O. L. Miller, Jr., and B. A. Hamkalo; O. L. Miller, Jr., B. A. Hamkalo, and C. A. Thomas, Jr., *Science* 169:392, 1970.)

attached at different positions to the same mRNA molecule to form a *polyribosome* (Figure 8-10).

Step 5. As the peptide chain grows by the addition of amino acids, a typical polypeptide chain requires that steps 1–4 must be repeated 300 times for the synthesis of the complete chain. A complete polypeptide chain is synthesized on each ribosome, beginning when the ribosome attaches to the beginning (one end) of the mRNA molecule and being completed when the ribosome reaches a specific "stop" signal. At this "stop" signal the polypeptide chain dissociates from the last tRNA molecule, the mRNA continues to move across the surface of the ribosome, and a different polypeptide chain is synthesized by repeating steps 1–5.

POLYPEPTIDE COMPLETED AS RIBOSOME COMES TO END OF mRNA MOLECULE

THE GENETIC CODE

The term *genetic code* describes the relationship between the sequence of purines and pyrimidine bases in the DNA and mRNA molecules and the sequence of amino acids in the proteins these molecules encode. A *codon* consists of three

TRIPLET CODE: 3 BASES CODE FOR 1 AMINO ACID

purine or pyrimidine mononucleotides and specifies one amino acid in a polypeptide chain. There are 64 possible triplet combinations but only 20 amino acids. Therefore several nucleotide triplets may code for the same amino acid, but no individual triplet codon specifies more than one amino acid. In one of the most significant achievements of modern biology, the catalog of code words for each amino acid was worked out (Figure 8-11). The person mainly responsible for cracking the genetic code was Dr. Marshall Nirenberg, an American biochemist working at the National Institutes of Health. In 1968 he shared the Nobel Prize for this work.

EACH CODON SPECIFIES A SINGLE AMINO ACID EXCEPT FOR 3 CODONS WHICH SERVE AS PUNCTUATION MARKS

Three of the 64 triplets do not code for any of the 20 amino acids, and these codons are called *nonsense* codons. Yet, these codons are very important, because they serve to signal the end of the protein chain. The three nonsense codons are recognized by specific proteins which release the completed polypeptide chain from the tRNA molecule. Returning to the analogy between human languages and the genetic code, such termination signals correspond to punctuation marks. If there was no punctuation at all in the genetic code, all of the proteins coded by a single mRNA molecule would be connected to one another. It is also crucial that the initiation of transcription begin at the correct codon and at the correct nucleotide *within* that codon. A sequence of bases may be read in three distinct ways depending on the point in the reading frame at which reading is initiated

FIGURE 8-11
Genetic code. The dictionary of the genetic code words. The codons are read from left to right (5′ → 3′). The three nonsense codons (designated as three different colors) serve as punctuation marks between genes, and no tRNA recognizes them. Note that there are many codons which specify the same amino acid. The significance of this feature of the code is not known.

	Middle Letter							
	U		C		A		G	
First Letter U	UUU	Phe	UCU	Ser	UAU	Tyr	UGU	Cys
	UUC	Phe	UCC	Ser	UAC	Tyr	UGC	Cys
	UUA	Leu	UCA	Ser	UAA	Ochre	UGA	Umber
	UUG	Leu	UCG	Ser	UAG	Amber	UGG	Trp
C	CUU	Leu	CCU	Pro	CAU	His	CGU	Arg
	CUC	Leu	CCC	Pro	CAC	His	CGC	Arg
	CUA	Leu	CCA	Pro	CAA	Gln	CGA	Arg
	CUG	Leu	CCG	Pro	CAG	Gln	CGG	Arg
A	AUU	Ile	ACU	Thr	AAU	Asn	AGU	Ser
	AUC	Ile	ACC	Thr	AAC	Asn	AGC	Ser
	AUA	Ile	ACA	Thr	AAA	Lys	AGA	Arg
	AUG	Met	ACG	Thr	AAG	Lys	AGG	Arg
G	GUU	Val	GCU	Ala	GAU	Asp	GGU	Gly
	GUC	Val	GCC	Ala	GAC	Asp	GGC	Gly
	GUA	Val	GCA	Ala	GAA	Glu	GGA	Gly
	GUG	Val	GCG	Ala	GAG	Glu	GGG	Gly

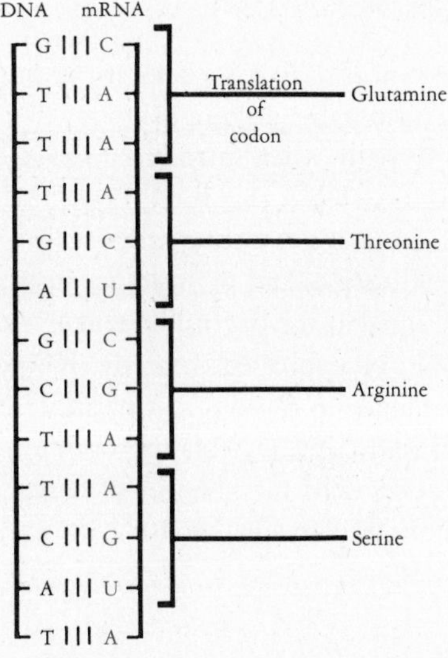

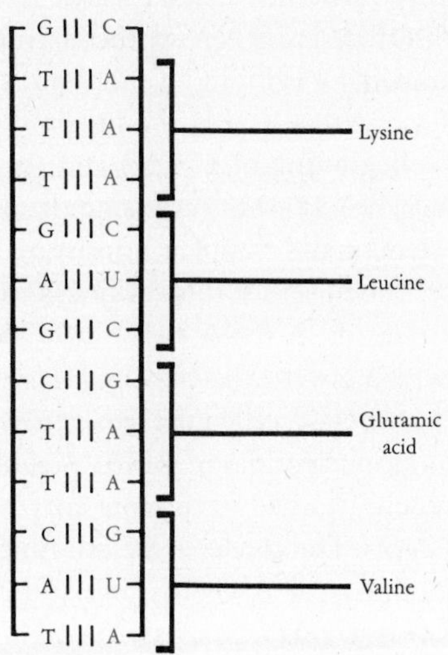

FIGURE 8-12
Importance of initiating translation of mRNA in the correct frame. A completely different amino acid sequence is generated depending on the frame in which translation is initiated.

(Figure 8-12). Thus, depending on the site of initiation in the first codon, a completely different set of amino acids can be specified. This emphasizes the importance of precise initiation of transcription of DNA and translation of mRNA. Once the initial site of transcription and translation is set, the reading frame is maintained by the sequential reading of three nucleotides at a time (Figure 8-12).

The nature of the genetic code also demands that the sequence of codons in a gene corresponds directly to the order of amino acids in the protein coded for by that gene. This expectation has been experimentally verified in several systems. This finding also indicates that both transcription and translation proceed in an orderly, linear fashion from one end of the gene to the other.

THE ARTIFICIAL GENE

In 1976 Dr. H. G. Khorana and his associates at the Massachusetts Institute of Technology reported a formidable accomplishment, the *chemical* synthesis of a gene which functions normally when introduced into a cell of *E. coli.* The gene codes for the synthesis of the tRNA that binds tyrosine. In order to have it function as a normal gene, Khorana not only synthesized a DNA molecule containing the exact sequence of purines and pyrimidines in the gene, but he also synthesized a short "start" signal at the beginning and a "stop" signal at the end of the gene so that it would be expressed correctly. Khorana hopes to study the effect of altering the sequence of purines and pyrimidines in the start and stop signals on the control of transcription. This approach might provide insight into how certain genes in cancer cells may be out of control. The control of gene expression is covered more fully in Chapter 10.

RNA POLYMERASE

RNA POLYMERASE HAS SEVERAL FUNCTIONS

This enzyme is structurally very complex and serves several functions. It apparently is involved in both DNA and RNA synthesis. In transcription, RNA polymerase is concerned not only with joining mononucleotide subunits to form the RNA polymer, but also with selecting the site on the DNA at which RNA synthesis is *initiated* (the beginning of a gene) and the site at which synthesis *terminates* (the end of the gene). It is not surprising that an enzyme which fulfills all of these functions is structurally complex, consisting of several subunits. But for all of the information known about this enzyme, a mystery still remains as to how a sequence of amino acids in RNA polymerase recognizes in DNA a sequence of nucleotides which says to the enzyme "start transcribing here." The major portion of RNA polymerase, called the *core enzyme,* catalyzes the formation of bonds between the mononucleotides of RNA. A subunit of the enzyme, the *sigma factor,* which is loosely attached to the core enzyme, determines the site at which RNA synthesis begins. The core enzyme can function without the sigma subunit, but it has no specificity as to where it initiates RNA synthesis. The discovery of the function of the sigma factor has created much excitement in biology because it has many implications for regulation of enzyme synthesis and cellular differentiation.

RNA Polymerase and Control

In all cells at all times certain functions are necessary whereas others are superfluous. Cells only express *a part* of their genetic information. This is obvious when we consider that all of the cells in any human body contain the same genetic information. In terms of enzymatic functions certain enzymes are vital in some circumstances but unnecessary in others. One important control over which genes are expressed, and to what extent, is exerted on the *rates* at which enzymes are synthesized. This differential synthesis of enzymes is controlled largely through the *initiation* of gene transcription. Some control over the *initiation* of transcription is exerted by the RNA polymerase enzyme itself, particularly by the sigma factor which controls the site on the DNA at which RNA synthesis is initiated. Additional control mechanisms are considered in Chapter 10.

SUMMARY

The informational macromolecules DNA, RNA, and protein contain information in the sequence of the subunits composing their molecules. The information in DNA—the ultimate information—is passed on to succeeding generations through the process of DNA replication. This is a semiconservative replication mechanism in which new strands of DNA, which are complementary to each of the original strands, are synthesized. The base sequence of DNA also codes for the sequence of amino acids in the corresponding proteins. The relationship between the various functional components of DNA involved in information transfer is diagramed in Figure 8-13. The mechanism of protein synthesis involves the transcription of the genetic message into mRNA, which binds to several ribosomes. Transfer RNA, carrying an amino acid, recognizes a codon of the mRNA, and the anticodon of the tRNA binds complementarily with this mRNA codon. Peptide bonds form between adjacent amino acids on the tRNA molecules, and the completed protein peels off the ribosome. The enzyme RNA

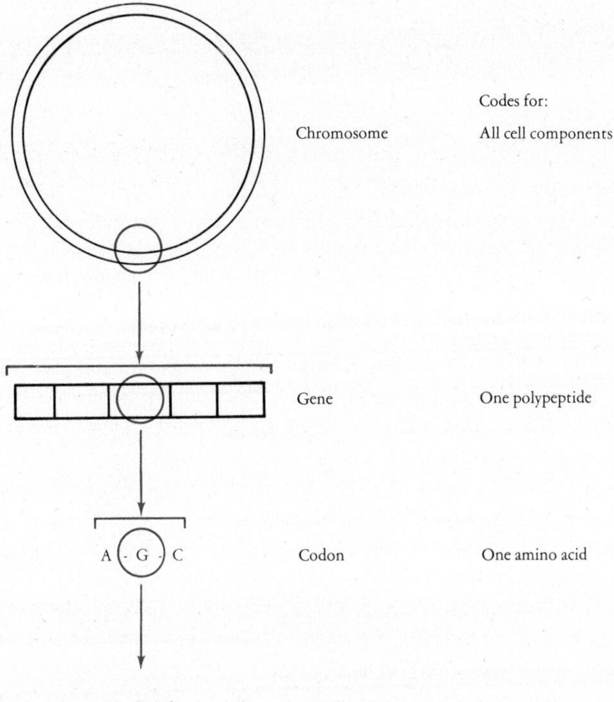

FIGURE 8-13
Levels of coding DNA in the cell.

polymerase, responsible for the synthesis of mRNA, determines the site at which the transcription of each mRNA molecule is initiated. It is this enzyme which probably determines in large part which genes are expressed and which are not in any particular situation.

The most complex, and inherently most interesting, enzyme in this entire process is RNA polymerase. It consists of a core enzyme and several subunits. Together the enzyme not only polymerizes the substrates of RNA into a macromolecule, but the subunits of RNA polymerase also determine the site at which RNA synthesis is initiated and the site at which the synthesis of a molecule of mRNA terminates. The recognition of specific sites on DNA molecules by such proteins points out that complementarity is not only important in recognition between nucleic acid molecules but is of very great importance in protein-nucleic acid interactions.

QUESTIONS

REVIEW

1. Consider the stages in protein synthesis in which complementarity between purine and pyrimidine bases plays a role.
2. How does transfer RNA serve as an adaptor in protein synthesis?
3. What components must be present in a test tube in order for DNA synthesis to take place? Which component determines the sequence of purine and pyrimidine bases that will be in the product?
4. Define transcription and translation and relate them to nucleic acid and protein synthesis.
5. What is the flow of information in biological systems, and what is each stage called?

THOUGHT

1. Why does it make sense to the bacterial cell for mRNA to function for a few minutes before it is degraded?
2. The faster bacterial cells multiply the more chromosomes they contain. Why does this make sense?

FURTHER READING

CAIRNS, J., "The Bacterial Chromosome," *Scientific American* (January 1966).

CLARK, B. F. C., and K. MARCKER, "How Proteins Start," *Scientific American* (January 1968).

CRICK, F. H. C., "The Genetic Code: III," *Scientific American* (October 1966).

GORINI, L., "Antibiotics and the Genetic Code," *Scientific American* (April 1966).

HOLLEY, R., "The Nucleotide Sequence of a Nucleic Acid," *Scientific American* (February 1966).

KORNBERG, A., "The Synthesis of DNA," *Scientific American* (October 1968).

———, *DNA Synthesis.* San Francisco: W. H. Freeman, 1974.

MILLER, O., "The Visualization of Genes in Action," *Scientific American* (March 1973).

The Molecular Basis of Life. San Francisco: W. H. Freeman, 1968. A collection of offprints of articles originally appearing in *Scientific American,* including most of those listed here.

NIRENBERG, M., "The Genetic Code: II," *Scientific American* (March 1963).

NOMURA, M., "Ribosomes," *Scientific American* (October 1969).

WATSON, J. D., *Molecular Biology of the Gene.* 3d ed. New York: W. A. Benjamin, 1975. This book presents the fundamentals of molecular biology with unusual clarity. The insight and perception that the author brings to the discussion of biological phenomena make this book a classic. It was written for an undergraduate course in molecular biology and biochemistry.

———, *The Double Helix.* New York: New American Library (Signet Books), 1969. The fast moving story of the discovery of the structure of DNA told by one of the participants.

YANOFSKY, C., "Gene Structure and Protein Structure," *Scientific American* (May 1967).

CHAPTER 9

MICROBIAL GENETICS

Bacteria are excellent for genetic studies. They are small and therefore require very little laboratory space. Bacteria also grow and divide quickly, allowing many experiments to be performed within a short time span. Under appropriate conditions billions of cells can be screened for a single characteristic by simple techniques that require little time and expense. Furthermore, all progeny of a single cell are essentially identical. For these reasons the study of microbial genetics has made fundamental contributions to understanding microbiological phenomena as well as to many problems central to all of biology. Very recently the useful genetic properties of bacteria have been exploited as a tool for identifying potential cancer-causing agents in the environment.

FUNCTION AND CELLULAR ORGANIZATION OF DNA

The tremendous diversity in the living world stems from two major factors. The first of these is the unique genetic information each organism possesses. For example, bacteria give rise to bacteria and nothing else: bacteria do not have the genetic information required to become a rabbit or a tree. The other factor is the influence on organisms of the different environments with which they come in contact. Two organisms with the same information in their DNA may demonstrate remarkably different qualities if they are exposed to two different environments. For instance, a facultative anaerobe growing in a glucose-salts medium will form different end products depending on whether it is grown in the presence or absence of air. In this case the environment is controlling the *functioning of enzymes.* In addition, the environment can regulate *gene functions.* The sum of the genetic constitution of an organism is its *genotype.* The characteristics of an organism that are expressed within a given environment make up the *phenotype* of the organism. The genotype represents the potential of the organism; the phenotype describes the actual state of the organism. Some of the mechanisms by which the environment acts on the genotype of the bacterial cell are discussed in Chapter 10.

NUCLEOTIDE SEQUENCE IN DNA DETERMINES AMINO ACID SEQUENCE IN POLYPEPTIDE CHAIN

The genotype of a cell is determined solely by the genetic information contained in the cell's chromosome or *genome.* The genome is divided into genes, each gene coding for the arrangement of the amino acids in one polypeptide chain.

Because an average polypeptide chain contains approximately 300 amino acids, and three bases in the DNA serve as the codon for each amino acid, the average gene contains over 900 nucleotide *base pairs* (both strands of the double helix are included). The molecular weight of a gene is approximately 600,000 or 6×10^5 (900×700, the molecular weight of a nucleotide pair), and the molecular weight of the entire bacterial chromosome is approximately 2×10^9. The chromosome therefore contains enough DNA to code for approximately 3000 different proteins, or $(2 \times 10^9)/(6 \times 10^5)$. Because only about 1000 different enzymes have been identified thus far however, it is apparent that much of the DNA either has functions other than coding for enzymes or that many enzymes still await discovery. Both are probably true.

All properties of a cell depend on the proteins that the cell possesses. The substitution of even one amino acid for another among the several hundred in the protein molecule may cause the protein to be nonfunctional. Thus any change in the sequence of purines and pyrimidines in one of the genes in the cell's DNA produces a change in the sequence of amino acids coded by that gene. Such a change in the nucleotide sequence of DNA is called a *mutation.*

GENE MUTATIONS

Although gene mutations occur at a low frequency, they can arise in a number of ways. The most common mechanism of mutation involves the process of DNA synthesis. At the time of duplication the strands of the double helix separate, and two new strands are synthesized, each one complementary to one of the original strands. In rare instances, however, an incorrect purine or pyrimidine is incorporated into the DNA during DNA replication (Figure 9-1). Mutations in a gene that codes for an enzyme concerned with inserting the correct purines and pyrimidines into their positions in nucleic acid itself (DNA polymerase) can increase the frequency of mutation in all other genes from ten-to a thousandfold. Apparently the defective enzyme is unable to insert the correct bases with the precision and accuracy of the normal enzyme.

DNA replication, pages 172–174

MOST MUTATIONS OCCUR DURING REPLICATION OF DNA

Mutations can also result from the *deletion* of either a purine or pyrimidine nucleotide or an entire section of a gene. Nucleotides can also be *added* within a gene, resulting in a change in the gene's nucleotide sequence and thus in a muta-

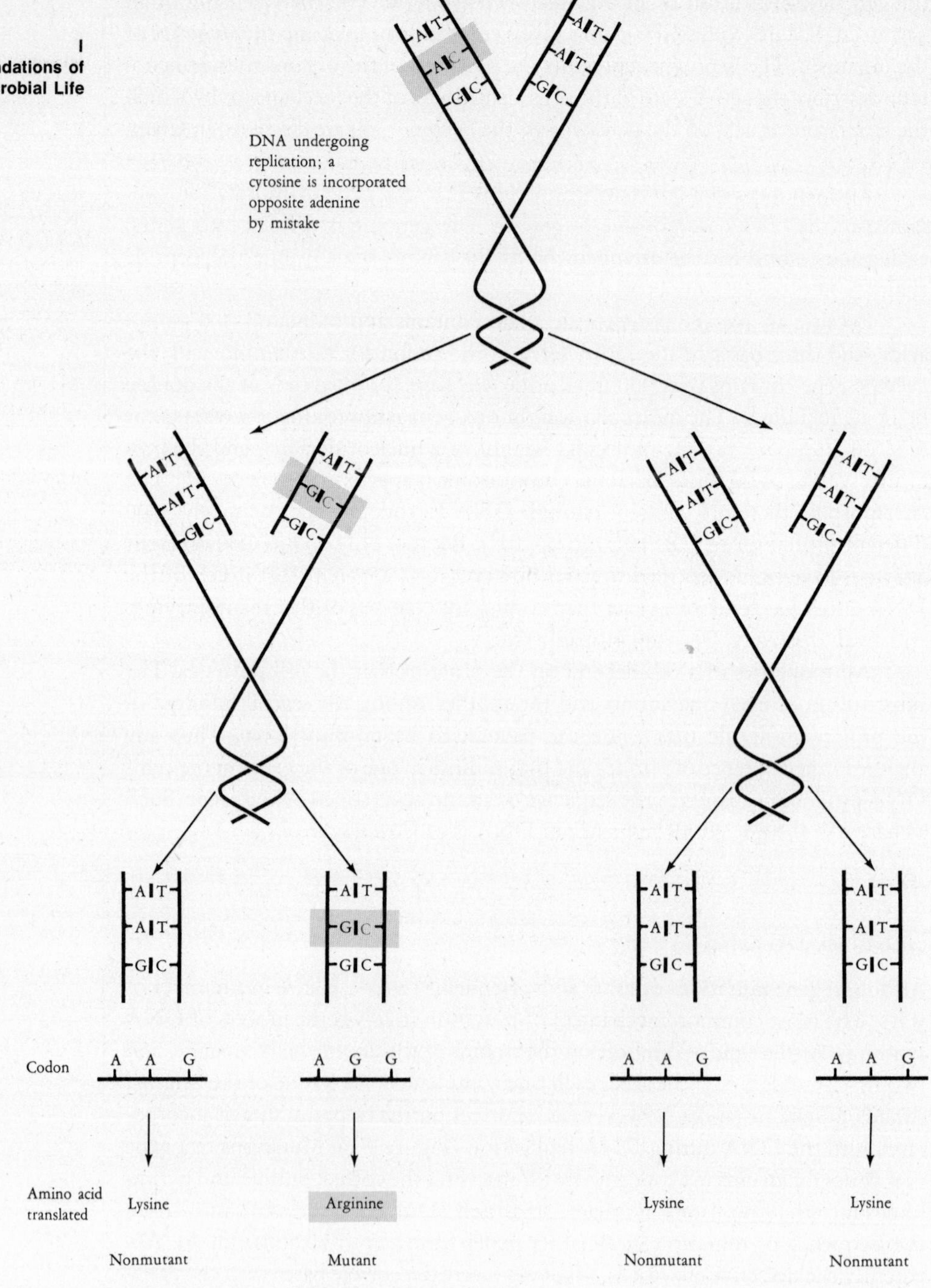

FIGURE 9-1 The generation of a mutant organism as a result of the incorporation of a pyrimidine base (cytosine) in place of thymine in DNA replication.

tion. The addition or deletion of a purine or pyrimidine nucleotide produces a shift in the "reading frame" of the DNA when it is transcribed into mRNA (Figure 9-2). Note that such an addition or deletion of nucleotides determines which amino acids are incorporated at all sites beyond the original site at which the purine or pyrimidine base was inserted or deleted. These mutations are thus more severe than those that result from the substitution of one purine or pyrimidine base for another.

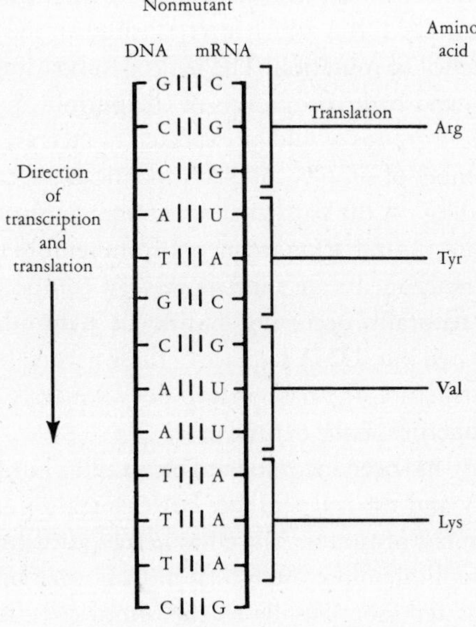

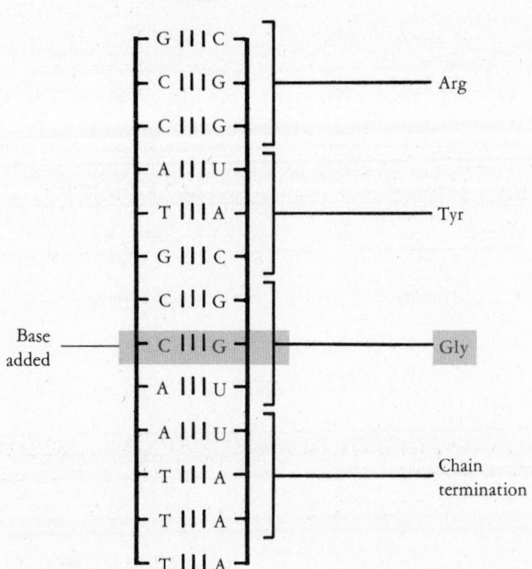

FIGURE 9-2
Production of mutation as a result of base addition. The addition of a nucleotide in the DNA results in a "frame shift" in the transcription of the DNA and a new triplet code word which is translated as a new amino acid. The deletion of a nucleotide would have essentially the same effect. The protein chain terminates when certain triplets are encountered in the DNA.

Because the frequency with which naturally occurring or *spontaneous mutants* arise in the population is extremely low, geneticists trying to isolate mutants often treat cultures with agents which increase the frequency of their occurrence. Such an agent is called a *mutagenic agent* or *mutagen.*

Mutagenic Agents

MODIFICATION OF HYDROGEN-BONDING PROPERTIES OF BASES RESULTS IN MUTATIONS

Chemical mutagens. Because mutation occurs spontaneously when one purine or pyrimidine base does not pair with its complementary base during the duplication of DNA, any chemical treatment that can alter the hydrogen-bonding properties of a base in the DNA which is responsible for the complementarity will increase the frequency of mutation. The hydrogen-bonding properties are a function of the amino and hydroxyl groups of the purines and pyrimidines, so any modification in these groups would be expected to increase the likelihood of mutations. A large number of simple, common chemicals are capable of modifying DNA and, depending on the particular agent and conditions of treatment, can increase the spontaneous mutation frequency from tenfold to a thousandfold or more. Other mutagenic agents are the *base analogs,* compounds which structurally resemble the naturally occurring purine or pyrimidine bases closely enough to be incorporated into DNA in place of these natural bases (Figure 9-3). Although similar in structure, the base analog does not have exactly the same hydrogen-bonding properties as the natural base.

Base sequence complementarity, page 31

Once a base analog has been incorporated into a cell's DNA, the DNA may not duplicate normally and the cell will die. For this reason, base analogs have been used in the treatment of tumors as well as herpes virus infections. Because tumor cells generally replicate more often than most normal body cells, the mutagenic agent, incorporated more rapidly by the tumor cells, is selectively toxic for the tumor to some extent. Indeed, all mutagenic agents are powerful killers. For every cell that is mutated by a mutagenic agent, several hundred are probably killed.

Normal nitrogenous base | *Analog*

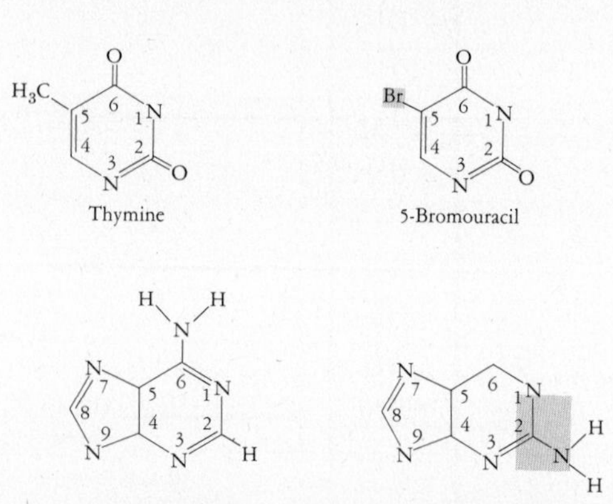

Thymine | 5-Bromouracil

Adenine | 2-Amino purine

FIGURE 9-3
Common base analogs and the normal bases they replace in DNA. The hydrogen atoms bonded to the carbon and nitrogen atoms in the ring are not shown.

MUTAGENIC EFFECT OF UV LIGHT

Radiation. Certain types of radiation, *X-rays and ultraviolet light* (UV) in particular, are also powerful mutagens. Although the mutagenic effect of UV radiation was first recognized almost 50 years ago, only within the past fifteen years has its mode of action become understood. This radiation causes the formation of a covalent bond between adjacent thymine molecules on the same strand of DNA (intrastrand bonding), resulting in thymine-thymine *dimer* formation (Figure 9-4). The covalent bond distorts the DNA strand sufficiently so that the

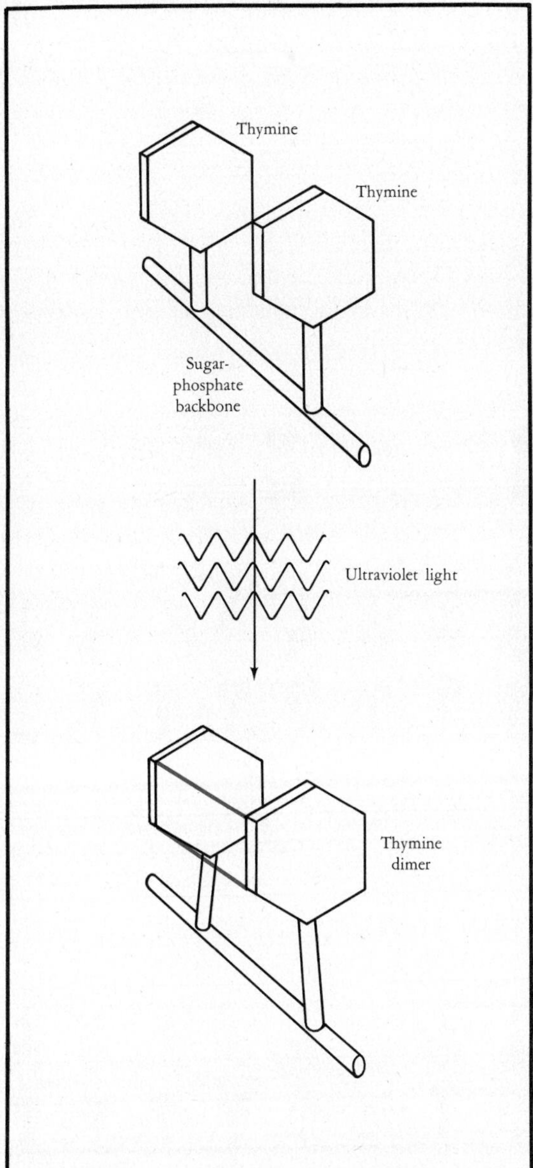

FIGURE 9-4
Formation of a thymine-thymine dimer as a result of ultraviolet irradiation.

hydrogen-bonding properties of the purines and pyrimidines near the dimer are altered. The result is that incorrect purines and pyrimidines may be inserted into a new strand of DNA as it forms along the altered strand. Since UV is a component of sunlight, cells are often exposed to this mutagenic and killing agent.

UV repair. To combat the effects of UV light many species of bacteria have enzymes that can repair the damage it causes. There is an enzyme that in the presence of visible light is able to break the covalent bond joining the thymine bases, a phenomenon called *light repair* (Figure 9-5). To prevent this repair, cultures treated with UV light to induce mutants are kept in the dark.

BACTERIA HAVE MECHANISMS TO REVERSE DAMAGE TO DNA

Some bacteria also have both enzymes that cut out, or *excise,* the damaged single strand of DNA and other enzymes that then repair the resulting break by synthesizing a strand complementary to the undamaged strand. Because visible light is not required for the action of these enzymes, the term *dark repair* describes this process (Figure 9-6).

It is a well established fact that people who are exposed to the sun for long periods of time have a much higher incidence of skin cancer compared to people who are not. Apparently humans also have a dark repair mechanism to repair UV light damage to their DNA. Some people who have a defective dark repair enzyme have also been identified. The incidence of skin cancer in these people is markedly elevated.

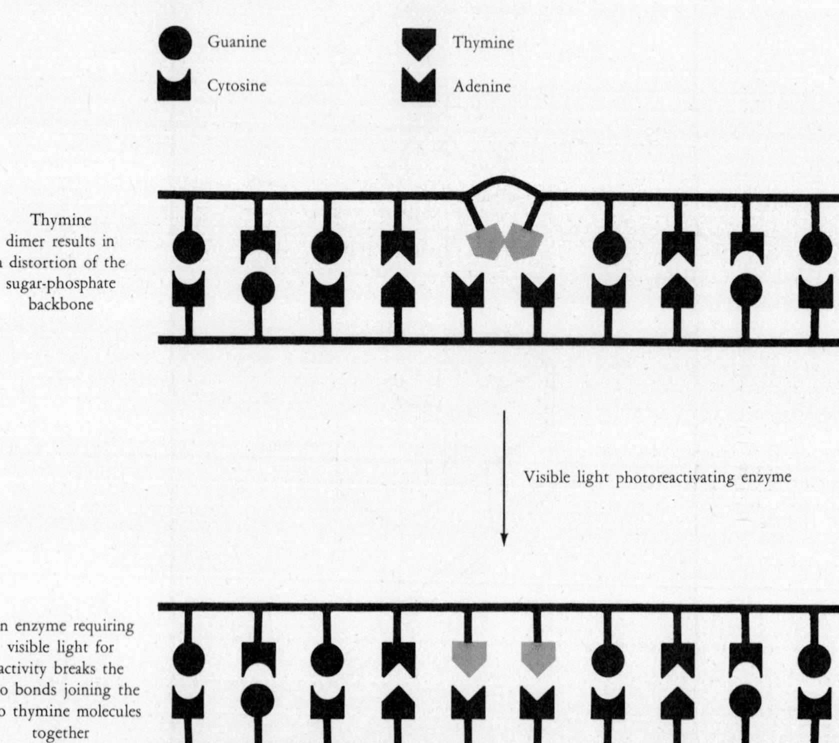

FIGURE 9-5 Light repair of thymine-thymine dimer.

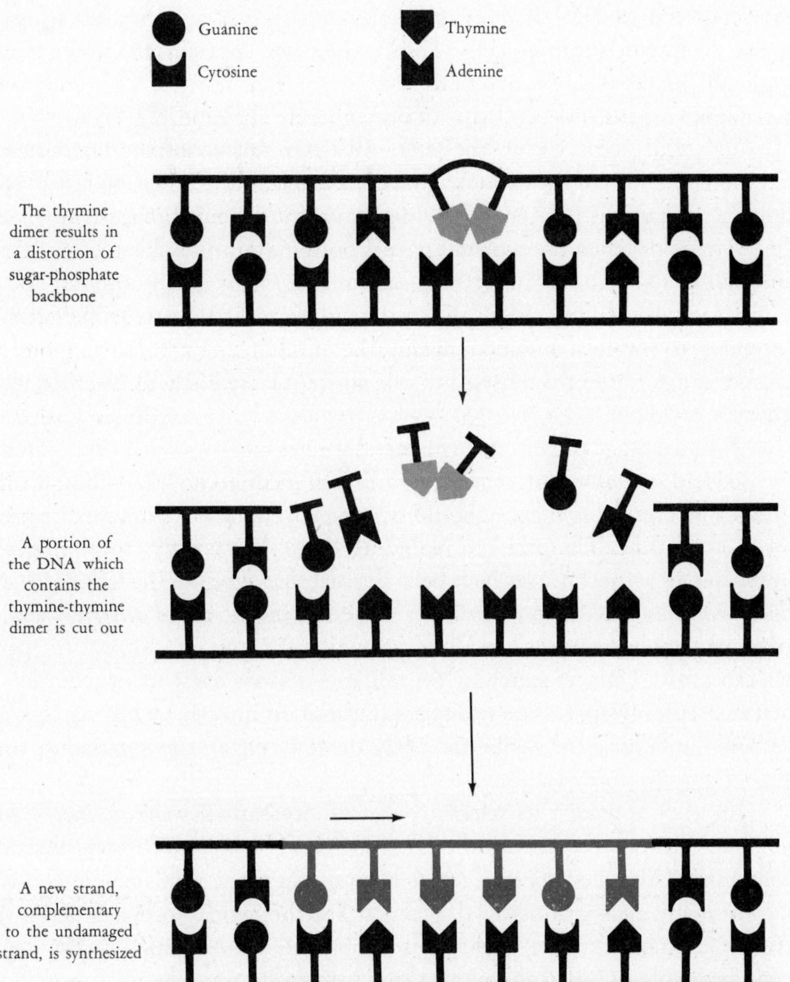
Guanine
Cytosine
Thymine
Adenine
The thymine dimer results in a distortion of sugar-phosphate backbone
A portion of the DNA which contains the thymine-thymine dimer is cut out
A new strand, complementary to the undamaged strand, is synthesized

FIGURE 9-6
Dark repair of a thymine-thymine dimer.

Although mutagenic agents generally affect a specific purine or pyrimidine in a defined way, they also affect all genes. No mutagen has yet been found which selectively mutates a specific gene. Furthermore, any agent that is capable of disrupting hydrogen bonds or removing amino groups is a potential mutagen, so it is not surprising that many compounds in the environment are mutagenic.

Rate of Mutation

MUTATIONS ARE RARE AND OCCUR INDEPENDENTLY OF ONE ANOTHER

The *rate of mutation* is defined as the probability at which a particular gene is likely to mutate each time the cell divides. It is generally expressed in the form of a negative exponent per cell division. Thus, if there is one chance in a million that a gene will mutate when the cell divides, the mutation rate for that gene is 10^{-6} per cell division. Of course, the mutant will be recognized only if it is readily distinguished from the parent cells. Because the mutation rate of any single gene varies between 10^{-3} to 10^{-9} per cell division, each *gene* will, on the average, be mutant in one in a million cells. Thus the concept that all cells arising from a single cell are identical is not strictly true.

Genes mutate independently of one another. Therefore the chance that two mutations will occur within the same cell is the *product* of the single rates of mutation (the sum of the exponents). For example, if the mutation rate to streptomycin resistance is 10^{-6} per cell division and that to penicillin resistance is 10^{-8} per cell division, then the probability that both mutations will occur within the same cell is $10^{-6} \times 10^{-8}$, or 10^{-14}—an infinitesimally small number. For this reason it is common practice in the treatment of certain diseases (tuberculosis in particular) to employ combined therapy, the simultaneous administration of two or more drugs. Any cells resistant to one antibiotic are likely to be killed by the other.

The advantage of performing genetic studies with organisms that can multiply to yield several billion cells per milliliter of medium now becomes apparent. Mutants in every single gene should be represented in a population which fits into a small tube. The problem is finding them. Fortunately, some unusually simple but clever techniques have been devised that simplify the task of locating this proverbial "needle in a haystack." One technique, called *direct selection,* involves plating bacteria on a medium on which the mutant, but not the parent cell, can grow. Thus, in searching for mutants that are resistant to an antibiotic, such as streptomycin, it is easy to select this mutant directly by plating cells on a medium containing the antibiotic. Only the few cells in the population which are resistant will form a colony.

AN INGENIOUS TECHNIQUE: REPLICA PLATING

Another approach to selecting mutant organisms is the *indirect selection technique,* illustrated by the ingenious *replica plating technique.* This was devised by Joshua and Esther Lederberg, a research team that made fundamental contributions in many areas of microbial genetics. For this work, Joshua Lederberg received the Nobel Prize in Medicine in 1958. The replica plating technique has been invaluable in isolating mutants that specifically *require* one or more nutrients. Because many bacteria, *Escherichia coli* in particular, can grow on a glucose-salts medium, it is possible to isolate a wide range of such nutritional mutants. Microorganisms that require particular growth factors are called *auxotrophs;* the organism is then said to be auxotrophic for the growth factor. The parent cell, which requires no organic metabolites other than glucose, is called a *prototroph.* The major problem with isolating the rare auxotroph is that a medium that supports growth of the auxotroph also supports growth of its parent, the prototroph. Thus, finding the rare auxotroph in the midst of a very large number of prototrophs had always been very tedious until the Lederbergs devised the technique of replica plating.

Replica plating involves the transfer of a part of every colony, using a piece of sterile velveteen that contains thousands of tiny bristles, from the "master" petri plate containing a rich medium (on which all cells will grow) to unsupplemented glucose-salts–containing plates on which auxotrophic cells will *not* grow (Figure 9-7). It then becomes a simple matter to identify the colony on the master plate that does not appear on the unsupplemented medium. The nutritional

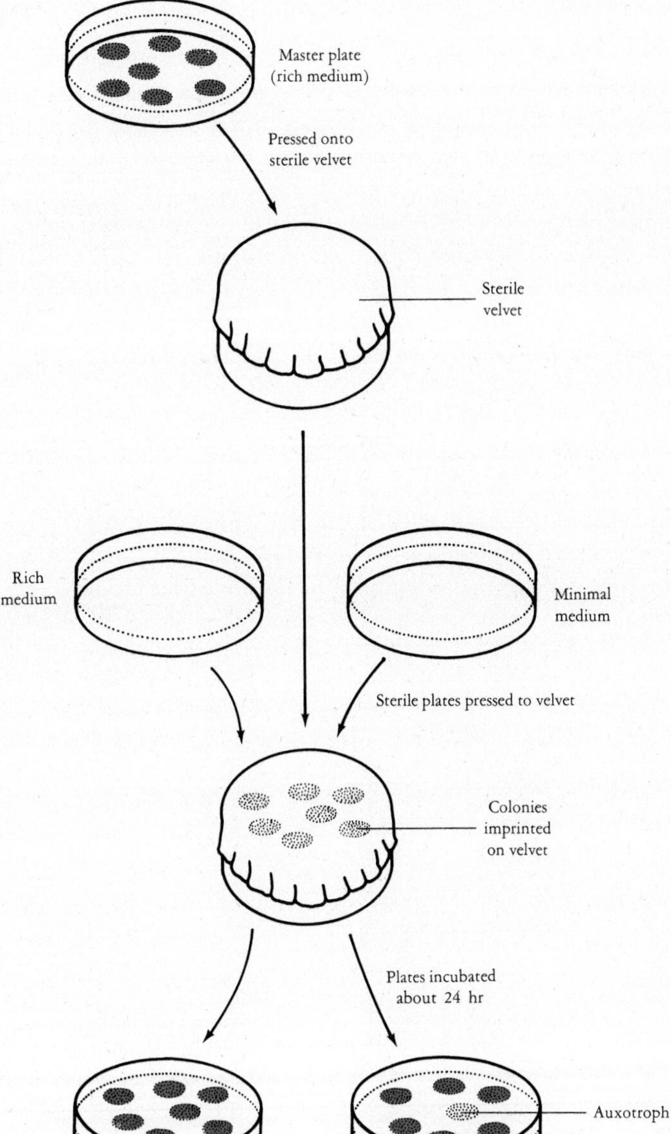

FIGURE 9-7
Replica plating technique. A bit of each colony is imprinted on the velveteen, which is then pressed to both rich and glucose-salts media. Any colony originally growing on the rich medium which has a requirement for any growth factor will not grow on the glucose-salts medium.

requirement of the auxotroph can be identified by replica plating colonies on the master plate to a medium containing each of the individual growth factors. The auxotroph will grow on the plate containing the proper growth factor. Before the invention of replica plating, individual bacterial colonies had to be inoculated into the unsupplemented medium to determine whether or not each would grow.

Conditional Mutants

SOME MUTANTS ARE NORMALLY LETHAL

All of the mutants discussed thus far will grow if the metabolite which they cannot synthesize is added to the medium. However, there exists a very large class of potential mutants that is defective in the enzymes required for the joining of subunits to form macromolecules and for the association of macromolecules to form organelles. Ordinarily, these mutants would never arise because their macromolecular end products cannot be supplied to overcome their defect. The mutation would therefore be lethal because the mutant cell could not divide. Such mutants cannot be isolated by using only those techniques that have been discussed thus far. To isolate these mutants, advantage is taken of the fact that mutant proteins that have a different amino acid sequence than the nonmutant parent protein are often not as stable to heat as the nonmutant protein. Advantage has been taken of this to devise simple but clever techniques for isolating these mutants. One approach is that which appears in Figure 9-8. The culture is treated with a mutagen, and the cells are plated out. The plates are then incubated at an elevated temperature (usually 40°C for *E. coli*). The cells that grow are marked, and the plate is then incubated at a temperature about 10° lower. Any new colonies that appear must contain a temperature-sensitive gene product. The next step is to determine the vital process in which the mutant gene is involved and finally

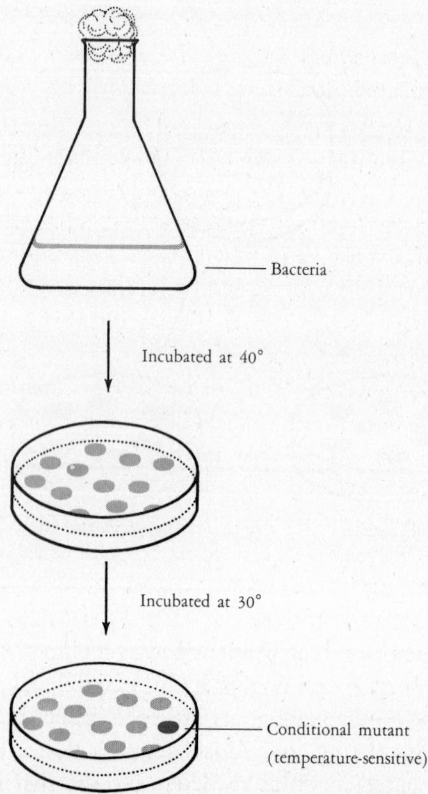

FIGURE 9-8
Isolation of temperature-sensitive mutants. This procedure selects for mutants that only grow at low temperatures. Cold-sensitive mutants that only grow at elevated temperatures have also been isolated.

the enzyme process involved. Those mutants which grow only if the environmental conditions are proper are called *conditional mutants.* Such mutants have been isolated for most of the steps in DNA, RNA, and protein synthesis. The precise identification of the mutant gene products in conditional mutants has allowed a genetic dissection of many processes vital to the life of the cell.

Expression of Mutations

Procaryotic cells contain one or several *identical* copies of each gene per cell and are thus *haploid.* In contrast, each chromosome in eucaryotic cells occurs in pairs, and each member of the pair differs from its partner in the genetic information it contains. These cells are thus *diploid.* In procaryotic cells any mutation in the cell's DNA will be expressed very quickly. Thus a mutation that results in the synthesis of a nonfunctional version of an enzyme necessary for growth causes the cell to require a growth factor, and the cell becomes auxotrophic. However, if there is more than one gene coding for this enzyme per procaryotic cell, the mutation is not expressed immediately because the nonmutant gene can synthesize enough of the functional enzyme to allow growth of the cell in the absence of the growth factor.

Testing for Chemical Carcinogens

EFFECT OF A CHEMICAL ON DNA CAN BE ASSESSED BY SCORING FOR REVERTANTS

There is growing evidence that a substantial proportion of cancers are caused by chemicals in the environment. However, to test one suspected compound for its cancer-producing or carcinogenic potential in animals takes from 2 to 3 years and costs about $100,000. Recently a simple screening test for potential carcinogens was developed by Dr. Bruce Ames at the University of California in Berkeley. This test takes only 3 days and costs about $200. It takes advantage of the ease, speed, sensitivity, and low cost of working with bacteria. The test is based on the fact that a change in the DNA of a mutant can result in a cell that resembles the original, nonmutant cell. In such a case the mutant *reverts.* The rationale of the test is based on the reasonable premise that mutagenesis and cancer induction both result from alteration of the DNA of a cell. Experimentally, over 90 percent of the carcinogens tested are mutagens. The Ames test simply measures the rate of reversion of histidine auxotrophs of *Salmonella* to prototrophy in both the presence and absence of the chemical being tested. If the chemical is mutagenic and *by implication* carcinogenic, it will increase the reversion rate (Figure 9-9). The test also gives some idea of how powerful a mutagen—and therefore how potentially hazardous—a chemical is by the *number* of revertants that arise. The more powerful the mutagen the greater the number of revertants. The vast amount of information available about microbial genetics has allowed Dr. Ames to continually improve the test. Thus there are now being employed bacterial mutants that lack the dark repair system and therefore cannot repair damaged DNA readily. Furthermore, these histidine-requiring tester strains also have mutations in the lipopolysaccharide layer of their cell wall which increase their permeability to the

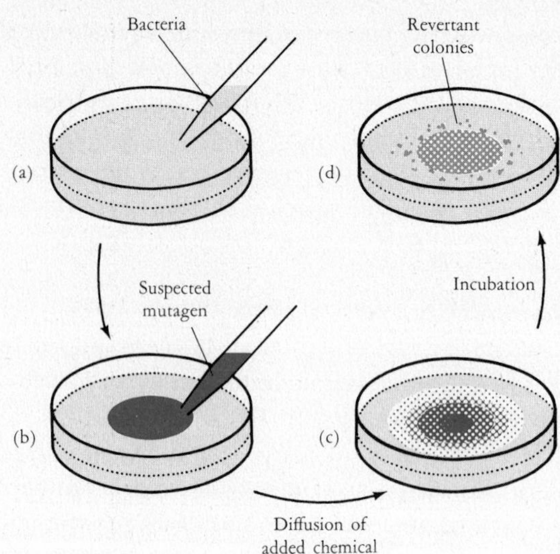

FIGURE 9-9
The Ames test for detecting possible mutagens.

compounds being tested. Of course, additional testing must be done on those mutagenic agents that are presumably carcinogenic in order to verify that they are indeed carcinogenic in animals. Unfortunately, results are not available on what percentage of mutagens are carcinogens, but there seems little doubt that the Ames test is useful as a rapid screening test for indicating those compounds that are potentially hazardous. Dr. Ames himself has tested over 300 chemicals for their mutagenic capabilities; some of these are listed in Table 9-1.

MECHANISM OF GENETIC EXCHANGE

The fact that bacterial mutants arise and breed true was the first indication that bacteria must have something analogous to the genes of higher organisms. The

TABLE 9-1
Testing of Chemicals for Mutagenic and Carcinogenic Activity

Product or Chemical	*Area of Use*	*Mutagenic*	*Carcinogenic in Animals*
Vinyl chloride	Chemical Industry	+	+
Captan	Fungicide	+	+
Dieldrin	Insecticide	0	+
Sevin	Insecticide	0	0
Cigarette smoke condensate	Pleasure	+	+
Phenobarbital	Drug	0	Weak
Nicotine	Tobacco	0	0
Caffeine	Coffee	0	Now being tested
Atropine sulfate	Medicine	0	Now being tested
Medium Golden Brown No. 612	Hair dye	+	Not tested
Moon Haze No. 32	Hair dye	+	Not tested

isolation of mutants made it possible to determine whether or not *sexual* or *genetic recombination* occurred in bacteria. Sexual recombination results in the formation of a cell that has properties of both parents. Therefore it is possible to look for sexual recombination only if there are available strains of bacteria which differ from one another in stable, easily recognized characteristics. The procedure used by Joshua Lederberg, then a young medical student at Columbia, and Edward Tatum, a professor at Yale, in looking for recombination in *E. coli* illustrates the general approach to recognizing genetic recombination (Figure 9-10). A strain of *E. coli* which required threonine, leucine, and thiamine (abbreviated thr^-, leu^-, thi^-) but could synthesize all other requirements for growth was mixed with a strain requiring phenylalanine, cystine, and biotin (phe^-, cys^-, and bio^-). After allowing the bacteria time to recombine, the mixture was plated on a glucose-salts medium without additional nutrients. On this medium only cells able to synthesize the six growth requirements can grow. Such recombinants must contain genes from each of the parental strains. In this experiment about 100 colonies grew when approximately 10^8 cells of each type were mixed to-

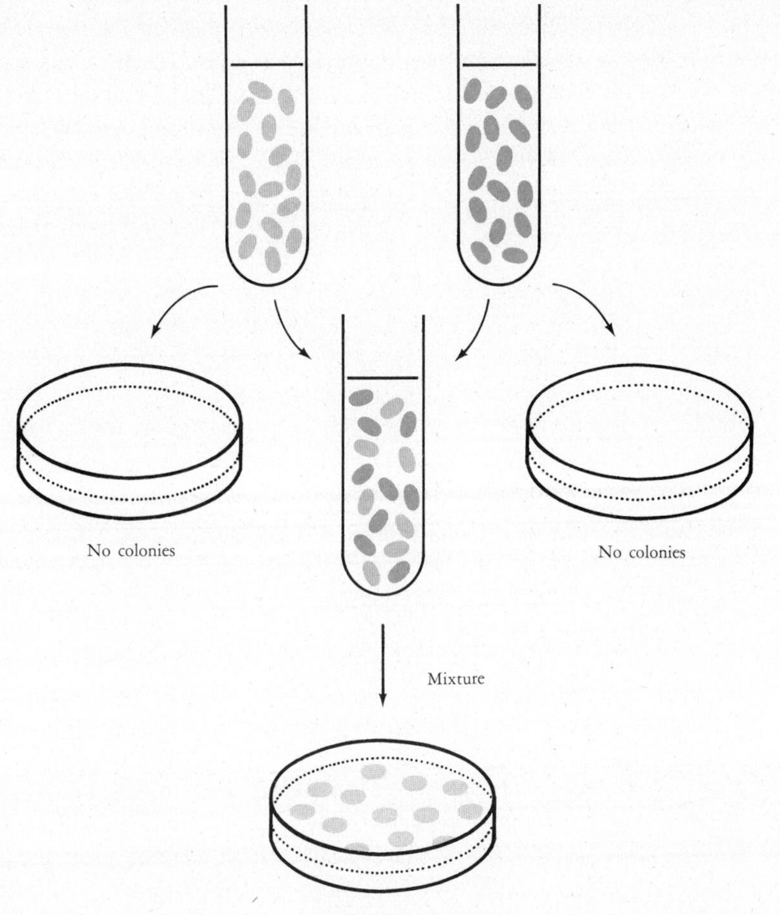

FIGURE 9-10
General experimental approach for detecting gene transfer in bacteria.

gether and plated. When the cells of each parent were plated on the glucose-salts medium separately, no colonies appeared. This was not surprising since the three mutations would have had to revert to prototrophy in their original cells in order for the cells to grow on the glucose-salts medium. The chance of this happening is about 1 in 10^{18} ($10^6 \times 10^6 \times 10^6$)! Therefore Lederberg and Tatum concluded that sexual recombination had indeed occurred.

The mechanism by which genes are transferred between cells can be divided into three categories. However, once the DNA has been transferred, the fundamental mechanisms of the recombination process are very similar for all three processes. The three processes are the following:

THREE MECHANISMS OF GENE TRANSFER IN BACTERIA

1. *DNA mediated transformation.* Genes are transferred as "naked" DNA.
2. *Conjugation.* Genes are transferred between cells which are in contact with one another.
3. *Transduction.* Bacterial genes are transferred by a bacterial virus.

Before discussing the specifics of each process, it is useful to discuss their basic similarities. The cells do *not* fuse, but in all three mechanisms only a part of the DNA is transferred from one cell, the *donor cell,* to another cell, the *recipient cell.* Once inside the recipient cell the donor DNA is positioned alongside the recipient cell's DNA so that homologous genes are adjacent. One or more enzymes then act on the recipient DNA, causing breaks and subsequent excision of a piece of the recipient DNA. Next, the donor DNA is *integrated* into the recipient chromosome in place of this excised DNA by a process termed *breakage and reunion.* The details of this process are not yet understood, but it is likely that the same enzymes or enzymes similar in function to those involved in the repair of UV damage in DNA are also involved in the recombination process (Figure 9-11). The end result is that the chromosome of the recombinant cell contains DNA derived from both the donor and recipient cells. The piece or pieces of DNA which are excised from the recipient chromosome are probably broken down by DNA-degrading enzymes. Recombination is precise; there is no increase or decrease in the number of nucleotides in the recombinant, only a precise substitution of some donor for some recipient nucleotides.

To recognize that genetic exchange in a particular part of the chromosome has occurred, it is necessary to distinguish the recombinant from the parent cells. Each of the three processes of exchange is a rare event generally affecting no more than 1 percent of a cell population. It is therefore advantageous to be able to select *directly* for the recombinants by plating on a medium on which only the recombinants will grow. Since several billion cells can be plated on the agar contained in a single petri dish, it is possible to detect rare recombinants.

DNA-mediated Transformation

The isolation and characterization of DNA as the transforming principle provided the first definitive evidence that DNA comprises the genetic material of cells. The study of transformation dates back to the late 1920s when an English

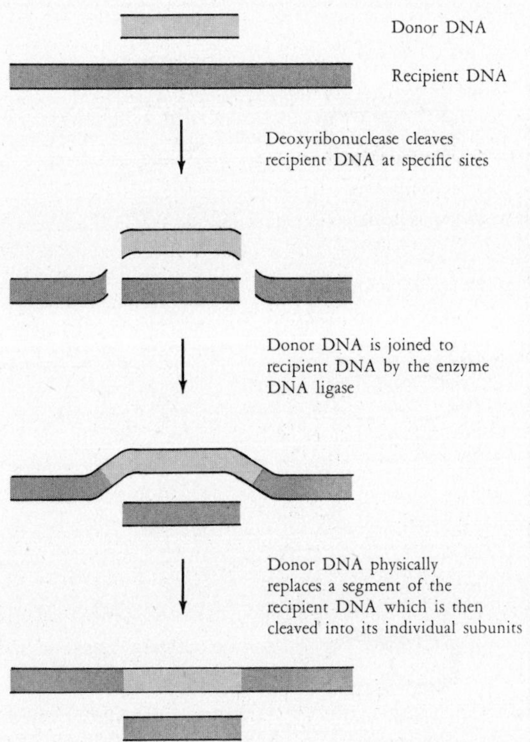
Donor DNA
Recipient DNA
Deoxyribonuclease cleaves recipient DNA at specific sites
Donor DNA is joined to recipient DNA by the enzyme DNA ligase
Donor DNA physically replaces a segment of the recipient DNA which is then cleaved into its individual subunits

FIGURE 9-11
General mechanism by which donor DNA becomes integrated into the recipient genome.

physician, Griffith, first observed the process in an organism that causes bacterial pneumonia, *Streptococcus pneumoniae.* This organism, as usually isolated in nature, has a polysaccharide capsule. If only a few of these encapsulated cells are injected into a mouse, the animal dies of infection. Mutants devoid of a capsule can be isolated in the laboratory, but these organisms are readily killed by the host animal and therefore do not cause disease. Griffith inoculated one group of mice with nonencapsulated cells, a second group with heat-killed encapsulated cells, and a third with both living, nonencapsulated cells and heat-killed, encapsulated cells (Figure 9-12). As expected, the mice in the first two groups were unaffected. Much to Griffith's surprise, however, mice in the last group developed infection and died. Furthermore, he was able to isolate living encapsulated cells from the dead mice. It was shown later that when encapsulated cells were ground up and added to nonencapsulated cells in a test tube, living encapsulated cells would arise. But it was not until 1943 that a group of investigators at the Rockefeller Institute (Drs. Oswald Avery, Colin MacLeod, and Maclyn McCarty) identified as DNA the active material which brought about this transformation of nonencapsulated to encapsulated cells. It was this identification that proved that DNA was the genetic material of cells. A few years earlier, two geneticists at Stanford University, George Beadle and Edward Tatum, had demonstrated that the auxotrophic mutants in a fungus resulted from a mutation in a gene. The demonstration that DNA is the genetic material supplied convincing evidence that DNA must function by determining the specificity of enzymes.

HISTORY OF DNA-MEDIATED TRANSFORMATION

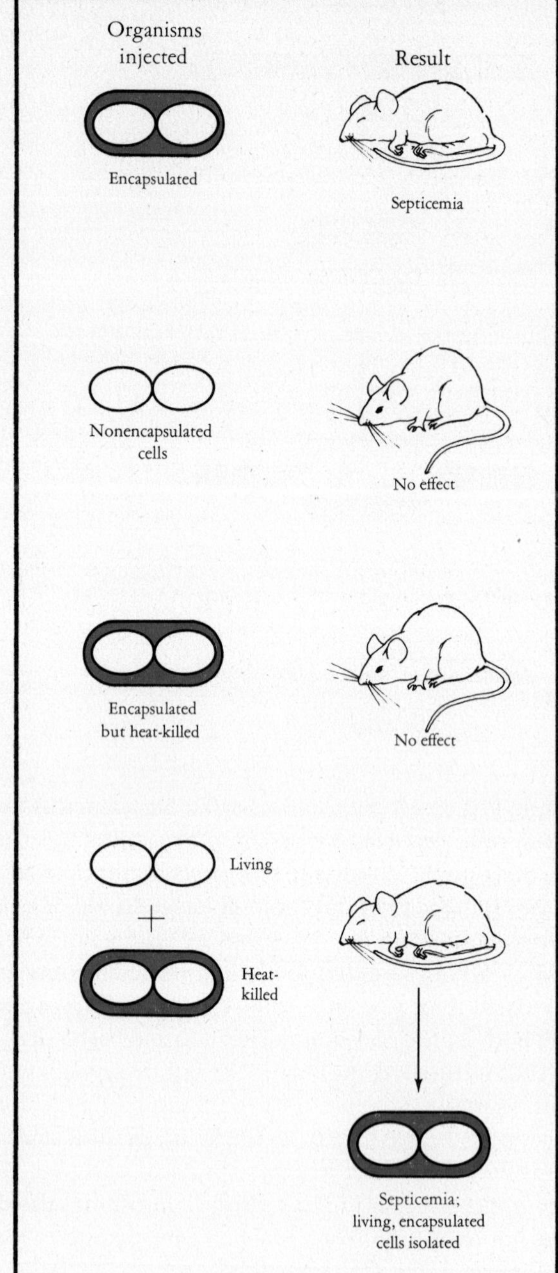

FIGURE 9-12
DNA-mediated transformation in pneumococci. The effects of injecting various preparations of pneumococci into mice.

In hindsight the observations made by Griffith in 1928 can now be explained in terms of mutation, transformation, and selection. The heat-killed encapsulated cells carried the genetic information for synthesizing the capsule; the mutant, nonencapsulated strain had a defective gene so that it could not synthesize the capsule. Even though the encapsulated cells were killed, their DNA remained intact and could be transferred to the nonencapsulated recipients. Those cells which integrated the donor genes concerned with capsule formation were now capable of synthesizing the capsule. These cells, although initially few in number, were then resistant to the body defense cells of the mouse and multiplied. The nonencapsulated cells were rapidly and *selectively* killed by the host animal, so that the bacterial population shifted in favor of the encapsulated cells. When these cells reached a high enough concentration, the mouse died of the infection.

Since the original experiments with the pneumococcus (*S. pneumoniae*), transformation has been observed in other organisms, including *Bacillus subtilis, Neisseria meningitidis, Haemophilus influenzae, Escherichia coli,* and certain strains of streptococci and staphylococci. Investigations of transformation in these organisms have revealed certain common features (Figure 9-13). When the donor cells are broken, the long, circular molecule of DNA is fragmented into about 100 pieces; each piece consisting of about 20 genes. Therefore 20 genes can be transferred simultaneously.

One of the most amazing features of the transformation process is that such a very large piece of DNA, with the consistency of a coiled door spring, passes through the bacterial cell wall and membrane. The recipient cells must be grown under rigidly controlled conditions to become *competent* to take up this large fragment of DNA. What makes a cell competent is not known, but it seems likely that the cell wall may have to be modified so that the DNA can pass through.

DNA transformation provides a useful system for studying the effects of a variety of physical and chemical treatments on the biological functions of DNA. Thus it is possible to isolate and purify DNA from donor cells and then determine whether treatment with a certain chemical has any effect on the ability of the DNA to transform recipient cells. This technique was used by Lederberg to demonstrate that chlorine compounds used in swimming pools can damage DNA. Mutagenic chemicals, such as nitrous acid, quickly destroy the transforming ability of a portion of the DNA. Mutations also arise in the recipient cells that have integrated the mutated genes.

Conjugation

This process differs from transformation in that cell-to-cell contact is required. This need can be shown experimentally by the following technique: if two different auxotrophic mutants are placed on either side of a filter through which bacteria cannot pass, inside a U tube, recombination does not occur. However, if the filter is removed, allowing cell-to-cell contact, the DNA of the cells does recombine (Figure 9-14).

F^+. Two types of cells exist in any population of *E. coli* in which conjugation occurs. One is termed the F^+ (or male) cell, the other the F^- (or female) cell.

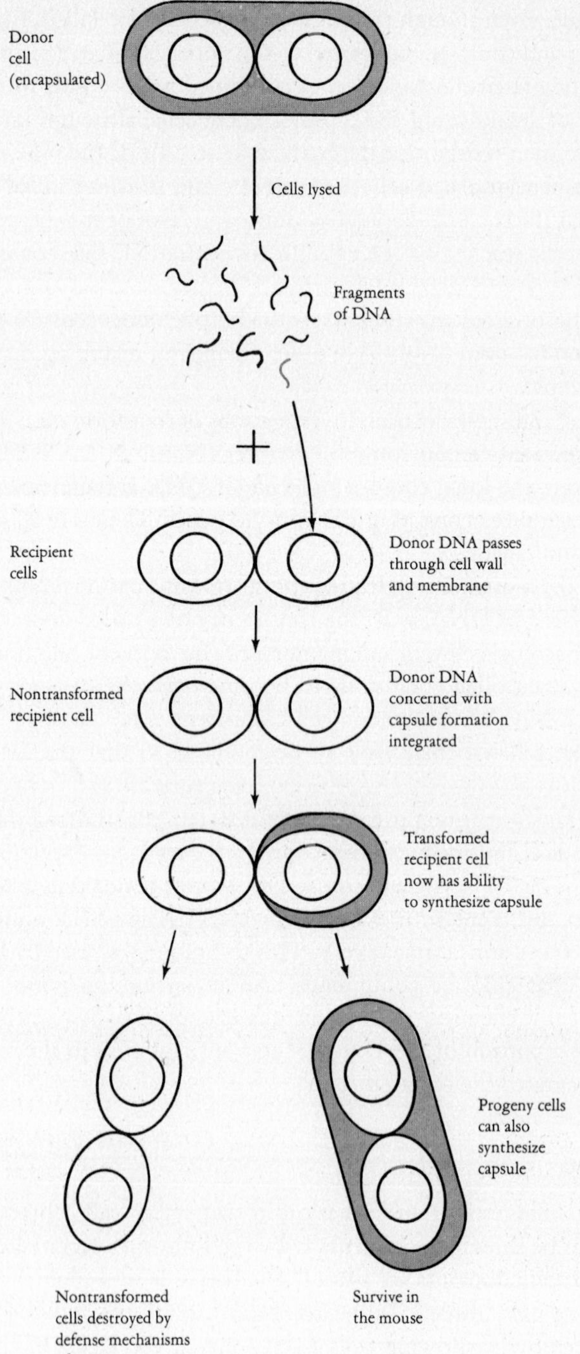
Donor
cell
(encapsulated)
Cells lysed
Fragments
of DNA
Recipient
cells
Donor DNA passes
through cell wall
and membrane
Nontransformed
recipient cell
Donor DNA
concerned with
capsule formation
integrated
Transformed
recipient cell
now has ability
to synthesize capsule
Progeny cells
can also
synthesize
capsule
Nontransformed
cells destroyed by
defense mechanisms
of mouse
Survive in
the mouse

FIGURE 9-13
General features of DNA-mediated transformation.

Extra chromosomal DNA, page 69

The F^+ cell has a few pieces of DNA in its cytoplasm which are not a part of the chromosome (*extrachromosomal DNA*) but exist as double-stranded closed cir-

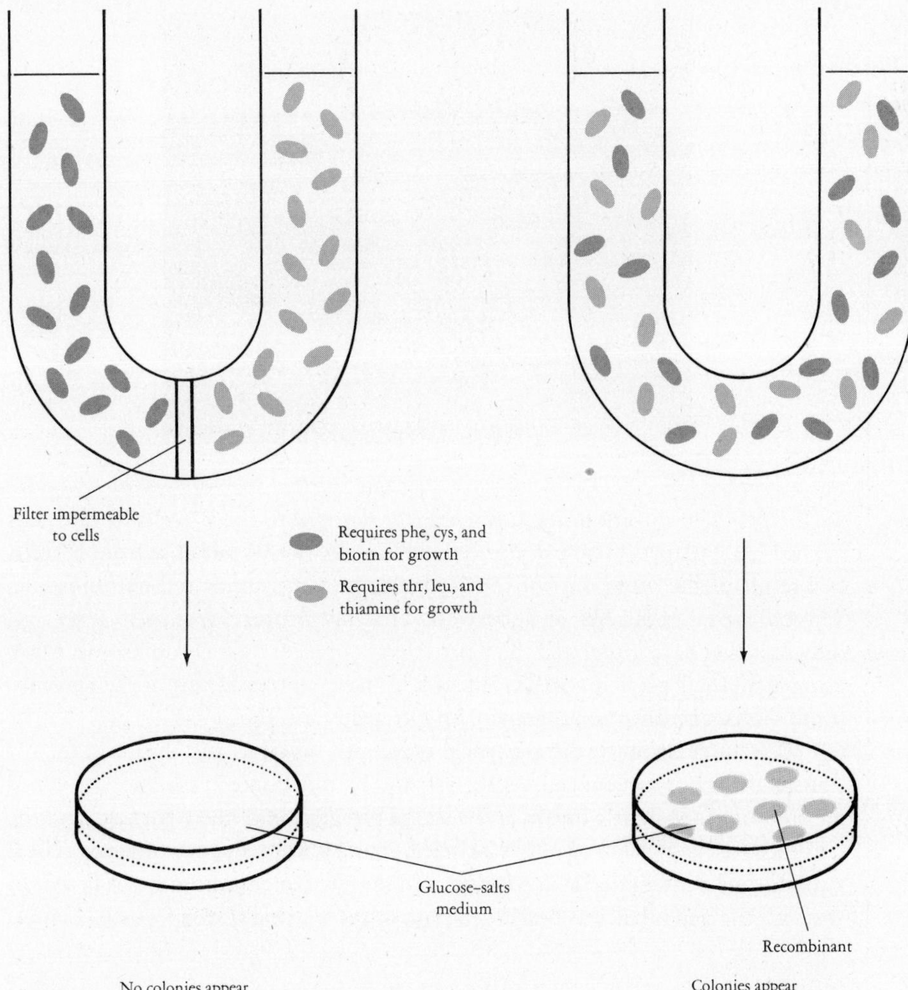

FIGURE 9-14 Demonstration of requirement for cell-to-cell contact for recombination in conjugation. To grow on the glucose-salts medium, the bacteria must be able to synthesize all amino acids and vitamins.

cular DNA molecules (Figure 9-15). These *F particles* are absent in the female. Each particle, approximately 1 percent the size of the chromosome, multiplies independently of the chromosome. Because this extrachromosomal DNA carries genetic information, the F^+ cell synthesizes proteins that the F^- cell does not, the most distinguishing proteins being the *sex pili.*

Pili,
page 77

When a population of F^+ and F^- cells is mixed together, the F^+ cells attach to the F^- cells by means of these sex pili (Figure 9-16). Within minutes one of the F particles from the male cell enters the recipient female, *perhaps* passing through the hollow pilus. Because all F^- cells in the population receive the F particle, the entire culture quickly becomes F^+. In this process the chromosome is *not transferred,* and recombinants do not arise (Figure 9-17a).

ONLY THE F PARTICLE IS TRANSFERRED IF DONOR CELL IS F^+

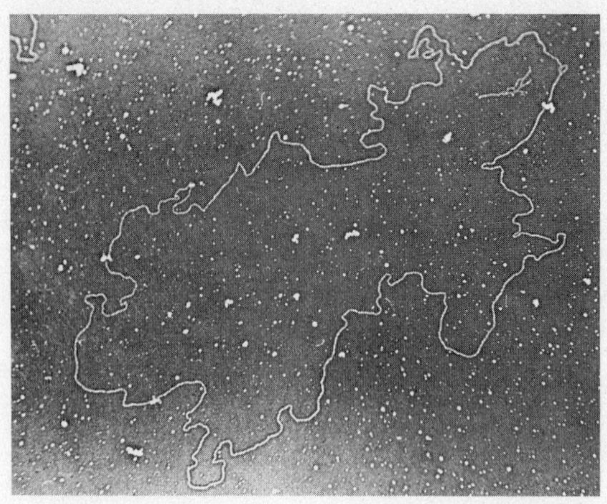

FIGURE 9-15
Extrachromosomal DNA. This DNA exists as a closed circular molecule. (Courtesy of R. Saiki; photo by T. Currier.)

Hfr. The chromosome, however, *is transferred* to the F^- cells by rare cells in the F^+ population. These donor cells are usually not F^+ but arise from F^+ cells as a result of the *integration* of the F particle into the donor cell chromosome. This cell type is called *Hfr,* an abbreviation for *high frequency of recombination,* and only cells that have integrated the F particle can transfer their chromosome. Once integrated, the F particle remains integrated and replicates as part of the chromosome. Consequently, the progeny of an Hfr cell is also Hfr.

The integration of the F particle apparently *mobilizes* the chromosome for transfer into the recipient cell. As the Hfr and F^- cells come in contact, the donor chromosome apparently breaks at the site of integration of the F particle, and the chromosome is transferred not as a circle but as a linear sequence of genes. The F particle can be integrated at any location in the circular chromosome of *E. coli.* In transfer, the gene that was *next* to the site of integration of the F particle enters

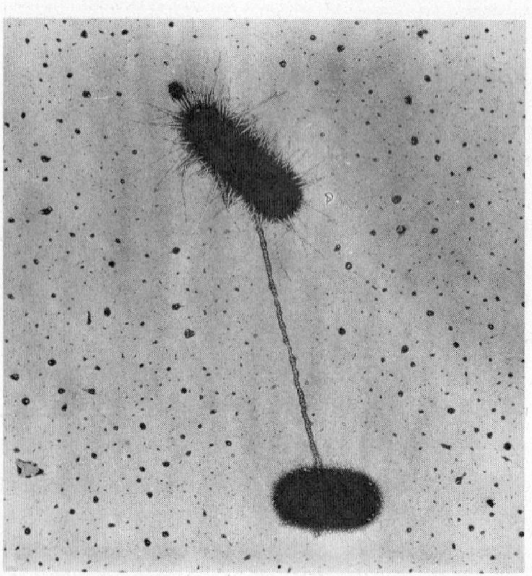

FIGURE 9-16
Sex or F pilus holding together a donor and a recipient cell of *Escherichia coli* during DNA transfer. The "dots" on the pilus are bacterial viruses that have adsorbed to the pilus. Some investigators believe that the DNA is transferred through the pilus so that the pilus should be considered an organelle of transport. (Courtesy of C. Brinton, Jr., and J. Carnahan.)

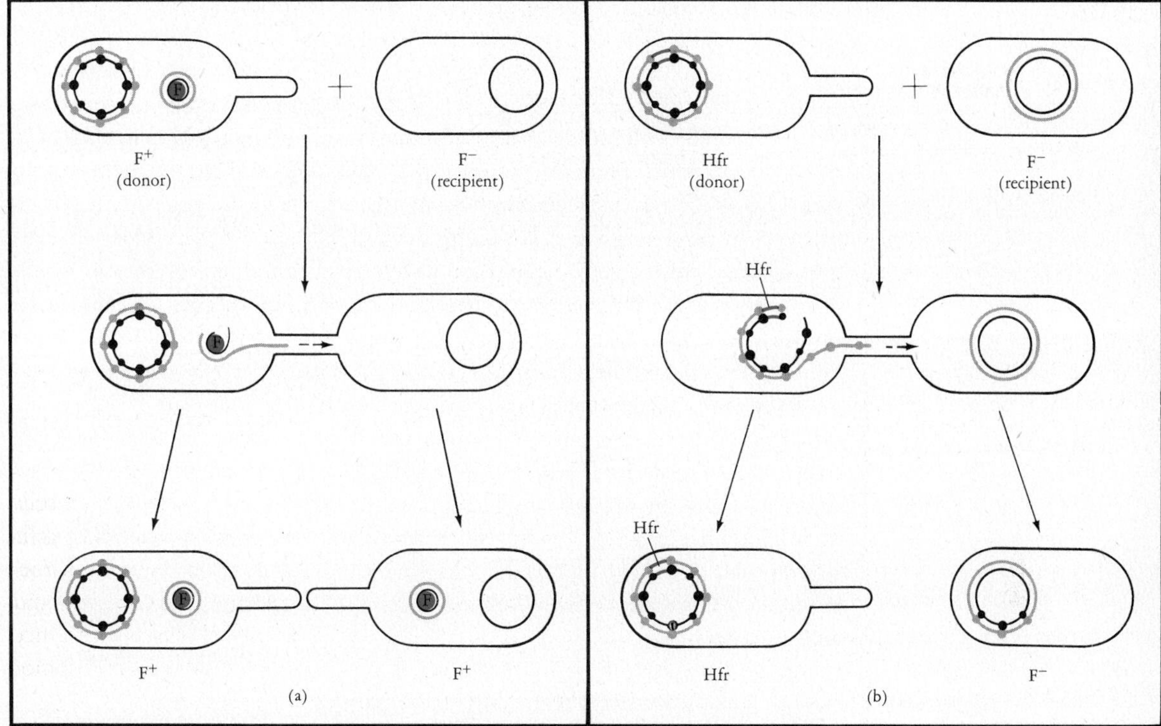

FIGURE 9-17 (a) Transfer of the F particle from the donor to the recipient cell results in the recipient cell becoming F$^+$. Because only one strand of the DNA of the F particle is transferred, the complementary strands of DNA are synthesized in both the donor and recipient cells. Therefore the donor cell also remains F$^+$. (b) Transfer of a portion of the chromosome of the donor cell. Because the Hfr fragment remains in the donor cell, the recipient cell remains F$^-$.

the recipient cell first, the *origin,* and the integrated F segment always enters last (Figure 9-17b). Because the F particle can become integrated at many different sites on the chromosome, there is a large number of different Hfr strains, each one identified by the genes that are transferred first.

In contrast to the rapid transfer of the F particle into F$^-$ strains, the transfer of the entire chromosome of Hfr strains into F$^-$ cells requires about 90 minutes. The rate of transfer is constant during the entire process. That is, 30 minutes after conjugation exactly one-third of the donor chromosome will be transferred into the recipient cell; after 45 minutes, one-half. The pilus holding the two cells together is not strong; in most cases it breaks long before the entire chromosome is transferred. Thus only the genes close to the origin are generally transferred, and the vast majority of the recipient cells remain F$^-$.

F′. Two different types of DNA can be transferred in conjugation: the extrachromosomal F particle if the donor is F$^+$ and a portion (or, more rarely, all) of the chromosome if the donor is an Hfr strain. Another type of extrachromosomal particle which some strains are capable of transferring combines the F particle and a segment of the chromosome. This extrachromosomal particle is termed F′ (F prime), and the donor strains which transfer these particles are termed F′ strains. These strains arise in the following manner: just as the F particle can integrate into the chromosome, in rare instances it can detach from the chromosome and once again become extrachromosomal. This detachment of the F particle often carries some of the chromosomal DNA that is in close proximity to the integrated F particle (Figure 9-18). Such cells contain an extrachromosomal F particle with several chromosomal genes attached. This particle has the property of the F particle of being rapidly and efficiently transferred to F^- cells and so some chromosomal genes can also be transferred rapidly to all F^- cells. However, because the recipient cell already has the chromosomal genes carried on the F′ particle, the recipient cell will be *diploid* for these few genes. The F′ particle remains extrachromosomal, except in rare cases when it becomes integrated into the chromosome.

How can a donor cell remain alive if it donates its chromosome to the recipient cell? The donor DNA strands separate, and one strand is transferred to the recipient while the other strand remains behind. The complementary strands to each are then synthesized as the one strand enters the recipient cell. The same situation holds for the transfer of other genetic elements from the donor to the recipient cell. Therefore the F^+ cell still remains F^+ even though it has transferred an F particle.

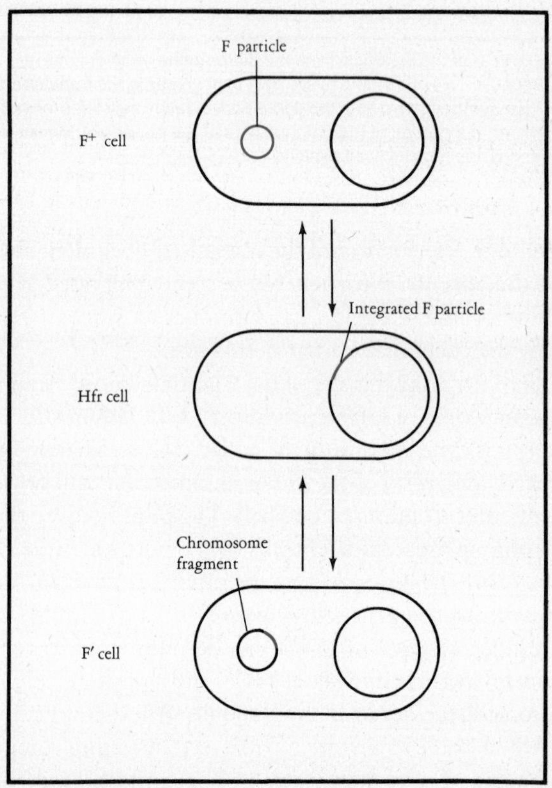

FIGURE 9-18
Relationship of F^+, Hfr, and F′ cells.

The two people most responsible for unraveling the relationship between the F particle and Hfr strains were the two French scientists, Jacques Monod and François Jacob. They articulated a new concept for biologists—the realization that the same fragment of DNA can exist in one of two forms, either extrachromosomally or integrated into the chromosome; a genetic element with these properties is called an *episome.* For these as well as other fundamental studies on the regulation of gene expression Monod and Jacob received the Nobel Prize in 1965.

The relationships that exist between the extrachromosomal particles concerned with gene transfer and the donor chromosome are illustrated in Figure 9-18. The F factor represents only one example of what appears to be a general phenomenon among the bacteria. There are remarkable similarities between the F factor and certain bacterial viruses which can exist in the same two states as fragmentary DNA.

PLASMIDS

PLASMIDS CONSIST OF TWO PARTS: GENES THAT CODE FOR VARIOUS TRAITS AND GENES THAT CODE FOR PLASMID TRANSFER

In addition to F particles, other extrachromosomal genetic elements occur in the cytoplasm but do *not* become integrated into chromosomes. These are called *plasmids.* Plasmids code for a variety of different functions and are widely dispersed throughout various genera of bacteria (Table 9-2). As a general rule, the traits coded by plasmid genes provide additional capabilities to the cell and are not required for its life. One of the most important groups of plasmids are the *resistance* or *R factors.* Superficially, they have many features of the F′ particles. They are extrachromosomal genetic elements and are composed of two parts, a *resistance transfer factor* (RTF), concerned with the transfer of the plasmid, and multiple genes concerned with drug resistance (*R genes*) (Figure 9-19). For example, a strain of bacteria carrying the R factor may be resistant to four widely used antibacterial medicines: sulfanilamide, streptomycin, chloramphenicol, and tetracycline. Other bacterial strains may be resistant to some, but not all of these drugs, while still others may be resistant to more drugs than these four agents. The unusual and distinguishing feature of the R factor is that it can be rapidly transferred to antibiotic-sensitive cells in a population by conjugation, thereby conferring resistance on these recipient cells. Furthermore, this transfer can occur not

TABLE 9-2
Some Plasmid-coded Traits

Trait	*Organism*
Antibiotic resistance	*E. coli, Salmonella, Shigella, Neisseria*
Colicin production	*E. coli*
Enterotoxin production	*E. coli*
Camphor degradation	*Pseudomonas*
Tumor induction in plants	*Agrobacterium tumefaciens*
Oil degradation	*Pseudomonas*

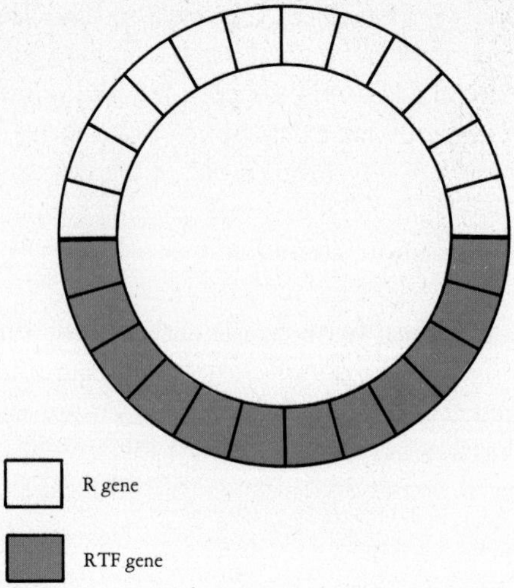

R gene

RTF gene

FIGURE 9-19
Diagrammatic representation of R factors. There are many individual genes concerned with resistance (R genes). The R genes are transferred only if they are attached to the RTF genes.

only between strains in the same species but also between one species and other closely related organisms, including members of the genera *Shigella, Salmonella, Escherichia, Yersinia, Klebsiella, Serratia,* and *Proteus.* The R factor can also be transferred to members of some other genera which are not closely related, such as the genera *Vibrio* and *Pseudomonas.* Thus, if cells in any one of these genera contain the R factor, it can often be rapidly transferred to other cells in any of these other genera by cell-to-cell contact.

GENES CAN JUMP FROM ONE DNA MOLECULE TO ANOTHER

Transposons. In today's world the widespread use of antibiotics has strongly selected for mutants that are antibiotic-resistant. This selection process, however, does not explain how and where R factors originated. Some new insights into the origin of certain R genes have recently arisen. It is now clear that certain R genes which determine resistance to different antibiotics are capable of migrating from one plasmid to another in the same cell as well as being capable of jumping from a plasmid to the chromosome of the cell (Figure 9-20). These small pieces of DNA, not much larger than a gene, are called *transposons.* They cannot replicate by themselves but must be integrated into a plasmid or chromosome in order to replicate. Thus, not only are there entire genetic units, the episomes, which can

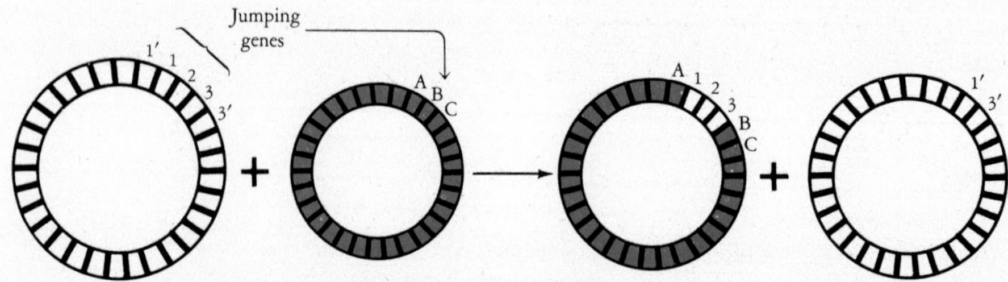

FIGURE 9-20 Transposon. This segment of DNA has the ability to "jump" from one piece of DNA to another, where it becomes integrated.

exist in a number of states, but small pieces of plasmids the size of a single gene, the transposons, can migrate between molecules of DNA of grossly different base composition. By this mechanism chromosomes and extrachromosomal genetic elements of widely different genera of bacteria may contain some of the same genes. Indeed, it has now been shown that all of the R genes that determine resistance to penicillin are very similar, even though they occur in different genera and species.

It is abundantly clear that R factors represent a serious threat to antibiotic therapy. Thus an ever-increasing number of drug-resistant strains are continuously emerging in the bacterial population, both from mutation and selection of antibiotic resistant cells as well as from the transfer of drug-resistance genes.

Although the discussion thus far has emphasized plasmids and episomes as the sites of genes specifying resistance to antibiotics, many chromosomal genes also confer resistance to the same antibiotics. In many cases the mechanism of resistance to the same antibiotic differs depending upon whether the resistance gene is chromosomal or extrachromosomal.

In addition to the F and R factors, other extrachromosomal elements of DNA are important in some bacteria. Among the most interesting are plasmids called *bacteriocin factors.* These are closed circular pieces of DNA that carry the genetic information for the synthesis of the *bacteriocins,* proteins which kill bacteria. There is a large number of bacteriocins which are highly specific in the bacteria they kill. Bacteriocins produced by one species kill only other strains of the same or related species. Their mechanism of killing varies depending on the particular bacteriocin. One group appears to have its primary effect on the cytoplasmic membrane; another on the ribosome. Bacteriocins have proved useful for distinguishing between certain strains of the same species of bacteria in clinical diagnostic laboratories.

Use of bacteriocins in diagnostic bacteriology, page 248

Although many bacteria, both Gram positive and Gram negative, synthesize bacteriocins, the most intensively studied have been some produced by *E. coli*—the *colicins.* Some of the colicinogenic factors are episomes; others are plasmids. Like the F particle, these factors, if they are integrated into a chromosome, can promote chromosome transfer; in the extrachromosomal state they are rapidly transferred to recipient cells.

Transduction

In this mechanism of genetic exchange, bacterial DNA is carried within a bacterial virus, called a *bacteriophage* (or *phage*), from the donor to the recipient cell. In the most common type of transduction, the infection of a bacterial cell by the phage results in the fragmentation of the bacterial chromosome into about 100 pieces. One of these fragments may then be accidentally packaged into a phage as the phage replicates. Once the phage is released from the infected cell, it may infect another cell, thereby transferring the bacterial genes. This process, diagramed in Figure 9-21, is also considered in Chapter 16, in which the life cycle of viruses is discussed in greater detail.

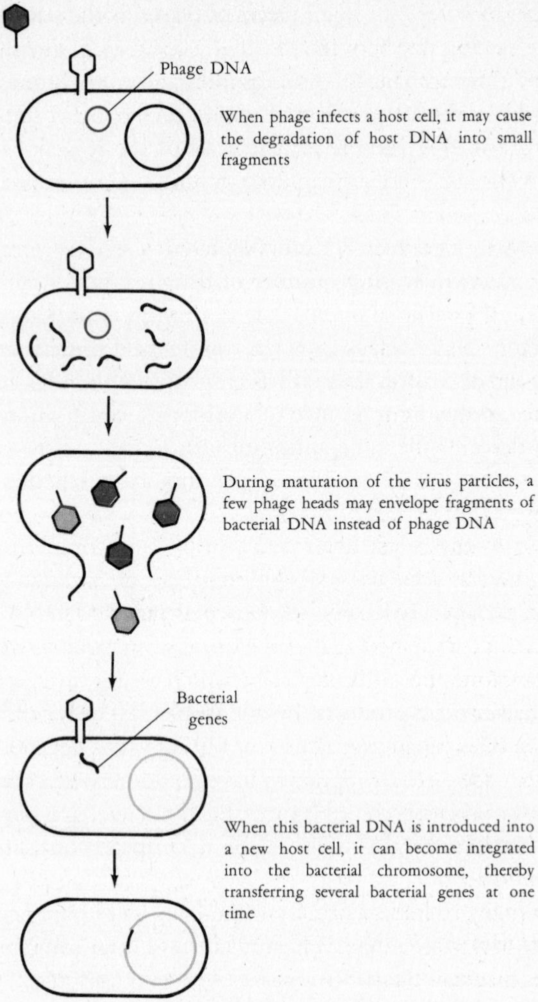

Phage DNA

When phage infects a host cell, it may cause the degradation of host DNA into small fragments

During maturation of the virus particles, a few phage heads may envelope fragments of bacterial DNA instead of phage DNA

Bacterial genes

When this bacterial DNA is introduced into a new host cell, it can become integrated into the bacterial chromosome, thereby transferring several bacterial genes at one time

FIGURE 9-21
Generalized transduction by a bacterial virus (bacteriophage). Only certain viruses have the capability of transferring DNA from one organism to another.

RESTRICTION AND MODIFICATION OF DNA

Restriction of DNA

BACTERIA USUALLY EXCLUDE THE ENTRANCE OF FOREIGN DNA

In general, bacteria will take up DNA by transformation, conjugation, or transduction *only* if the DNA is synthesized in the same species or variety of bacteria that it is entering. Bacteria can recognize whether entering DNA is foreign, and if it is, enzymes present in the cell will degrade it. These enzymes, called *restriction nucleases* because they restrict the entrance of nucleic acids, have a very unusual ability. They recognize specific sequences of nucleotides in an alien DNA and cleave the DNA only at these sites. One of the most intensively studied restriction nucleases is an enzyme isolated from *E. coli* called Eco R1 (an abbreviation of *E. coli*). Its action is typical of a large number of restriction nucleases.

The Eco R1 nuclease recognizes a sequence of six purine and pyrimidine bases and cleaves at the two sites indicated in Figure 9-22. The DNA molecule, originally a circle, breaks apart to yield a linear molecule which contains two overlapping ends. In the cell this linear molecule of DNA is probably chopped up further by other DNA degrading enzymes and is destroyed. Thus the Eco R1 enzyme protects the cell from entering foreign DNA. However, if a circular plasmid molecule is cleaved in a test tube with purified Eco R1 nuclease, interesting possibilities for *genetic engineering* arise. Note that the two ends of the cleaved DNA molecule are *complementary* to one another, so that they can associate with each other by hydrogen bonding. If the same test tube contains different DNA molecules which have also been treated with the Eco R1 nuclease, they will have the same complementary ends and thus will also be able to hydrogen bond with the ends of one another (Figure 9-22). It is thus a simple matter to join two different molecules of DNA.

In order to produce a functional DNA molecule, however, several additional requirements must be fulfilled. First, the molecule must be transported into a cell. Second, once inside it must be capable of replicating itself. Third, it must be capable of expressing its genes so that its presence can be recognized. For these reasons, it is convenient to join a DNA molecule being studied with a plasmid that contains genes concerned with drug resistance. This hybrid molecule can then be introduced into cells of *E. coli* (mutant in the restriction enzyme) by DNA transformation. Once inside, the hybrid plasmid replicates and the transformant cells can be easily identified by their resistance to the antibiotics. As the plasmid replicates, the piece inserted into the plasmid also replicates. Thus by this simple technology (Figure 9-22) it is theoretically possible to introduce any segment of DNA from an animal, a plant cell, or another bacterium, into *E. coli,* and have it replicate to produce an infinitely large number of identical molecules of DNA, as well as the products of these molecules.

The ability of any gene to replicate in *E. coli,* whether it be from an animal, a plant cell, or another bacterium, presents tremendous possibilities. Conceivably it may be possible to develop strains of *E. coli* that will synthesize antibiotics and hormones such as insulin. However, there is also considerable concern that combining eucaryotic and procaryotic genes and introducing them into bacteria may result in hazardous gene combinations which might pose a serious threat to life on earth.

Modification of DNA

BACTERIAL ENZYMES CHEMICALLY ALTER THE CELL'S DNA

If bacteria contain restriction nucleases, why don't they destroy their own DNA? The answer is that bacteria contain another enzyme, a *modification enzyme,* which slightly changes the chemical composition of certain purine and pyrimidine bases of their own DNA so that the restriction nuclease cannot recognize these bases and therefore doesn't cleave the DNA. These changes, such as the addition of methyl groups to cytosine, do not alter the hydrogen bonding characteristics of the bases and so do not induce mutations.

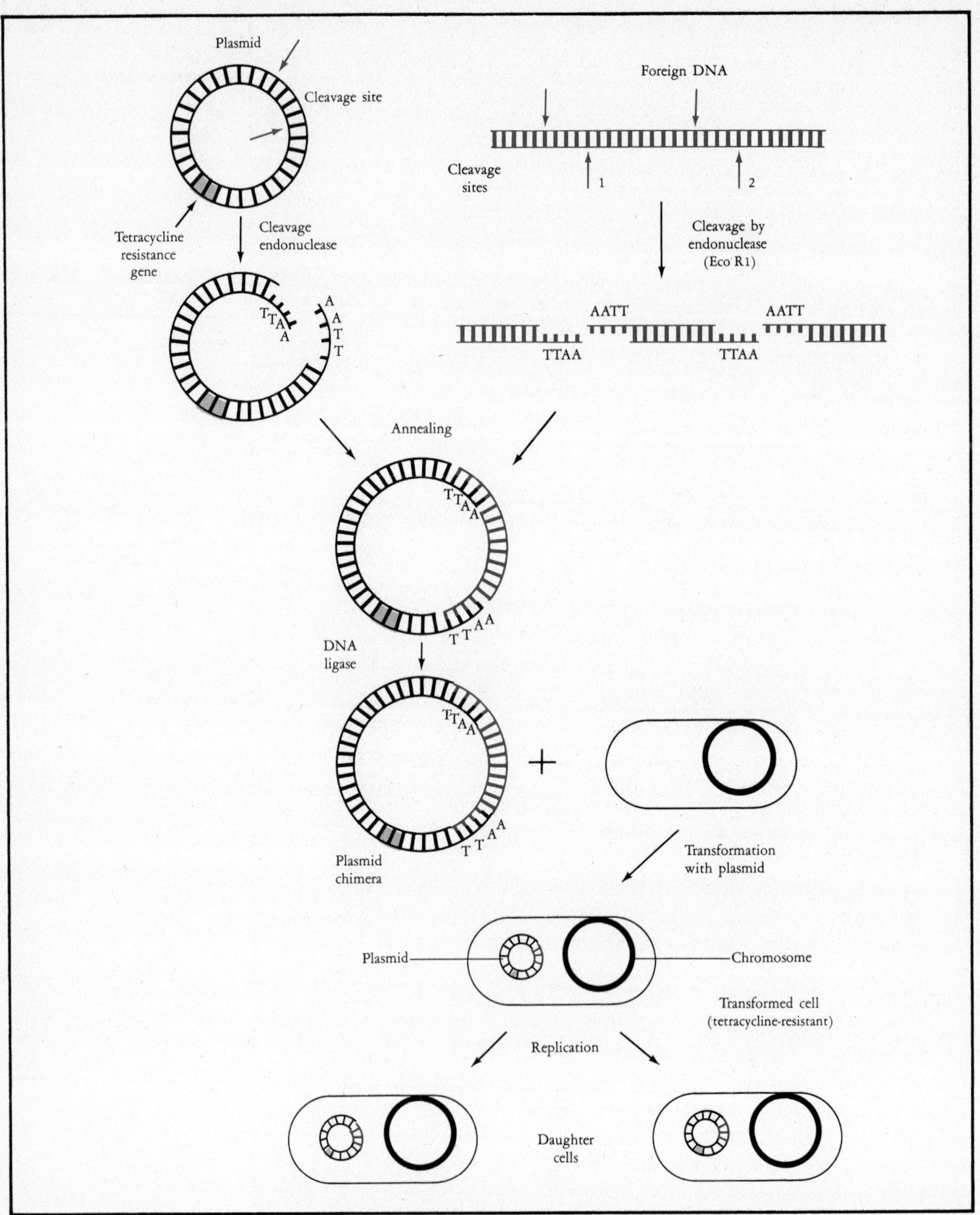

FIGURE 9-22 The scheme for forming recombinant DNA and then cloning it. The plasmid has the appearance of the DNA molecule shown in Figure 9-15.

DNA-mediated transformation and conjugation have been described in only a relatively few species of bacteria (Table 9-3). Thus conjugation has been shown to occur in only about a dozen species of mostly Gram-negative bacteria. Transformation has been demonstrated in the laboratory in about a dozen species, both Gram positive and Gram negative. Transduction is probably the most widespread mechanism of genetic exchange, having been demonstrated in virtually all species in which it has been actively sought. The fact that conjugation and transformation have been found in relatively few species does not necessarily mean that these processes do not occur in other species. It may only mean that other species have not yet been investigated, or that if they have, the proper conditions for demonstrating the exchange have not yet been found.

The role gene transfer plays in nature is not always apparent; nevertheless, it is easy to imagine how gene transfer and recombination can serve a very useful function in microorganisms. The major source of variation within a bacterial species results from mutation. Because bacteria are haploid, the mutation quickly becomes evident. Because of the rapid multiplication of bacteria by binary fission, selection within the population also operates very rapidly. For example, in a hospital environment in which a large variety of antibiotics are constantly present, a mutation to streptomycin resistance will confer a selective advantage on the mutant organism. The antibiotic-sensitive organisms will be killed by the antibiotic, and the population will change very quickly from a population in which the resistant cells are in the minority to one in which the resistant organisms predominate. However, only a *single* gene will undergo *mutation* at any one time. In contrast, a *considerable number* of genes can be *transferred* at the same time; therefore, *groups* of genes can be tried out in new environments and in different combinations with genes of the recipient cells. Transposons provide an additional flexibility to microorganisms. They make available a greater number of genetic backgrounds on which the selection process can operate. Probably the best example of the importance of gene transfer to bacterial survival is the transfer of the R genes. This transfer must be important in nature because strains carrying these factors are extremely widespread in any country in which antibiotics are commonly used. However, the restriction and modification systems found in most

TABLE 9-3
Some of the Species in Which Gene Transfer Has Been Observed

Transformation	*Transduction*	*Conjugation*
Bacillus subtilis	*Escherichia coli*	*E. coli*
Streptococcus pneumoniae	*Salmonella typhimurium*	*S. typhimurium*
Neisseria meningitidis	*Bacillus subtilis*	*P. aeruginosa*
Haemophilus influenzae	*Pseudomonas aeruginosa*	*Neisseria gonorrhoeae*
Rhizobium meliloti	*Staphylococcus aureus*	
Micrococcus radiodurans		
Staphylococcus aureus		
Escherichia coli		

bacteria serve to insure that *major* alterations in their genetic constitution do not occur.

SIGNIFICANCE OF GENE TRANSFER TO THE MICROBIAL GENETICIST

GENES CAN BE MAPPED ON THE CHROMOSOME

An analysis of gene transfer provides the means whereby genes can be located or mapped on the bacterial chromosome. The basic idea behind all mapping theory is this: the closer two genes are on the chromosome, the more likely they will be transferred together from the donor to the recipient cell; the farther apart they are, the less the chance that they will be transferred together. In DNA-mediated transformation the bacterial chromosome of the donor fragments, and only those genes that are close enough to each other to be on the same DNA fragment, will be transferred to the same recipient cell. In conjugation the frequency with which genes are transferred into the recipient cell is directly related to their distance from the gene which enters the recipient first. It is actually possible to time the entrance of genes into the recipient cell. The distance between two donor genes is inversely proportional to the number of minutes that separate the entrance of the two genes into the recipient cell.

By measuring the frequency with which genes are transferred either together or separately, it is possible to map a large number of genes on the chromosomes of a variety of organisms which undergo genetic exchange. The most intensively mapped chromosome is that of *E. coli;* the locations of over 150 of its genes have been determined. Unfortunately, genetic maps have been constructed mainly for those organisms that undergo genetic recombination, so we have no information on the location of genes in the vast majority of microorganisms.

From the data that have been collected thus far, several interesting facts have emerged about the map locations of bacterial genes. Organisms which are closely related, such as *E. coli* and *Salmonella typhimurium,* have very similar genetic maps; the genes which specify the same functions are located at approximately the same location on the chromosome. In unrelated organisms, such as *E. coli* and *B. subtilis,* the genetic maps are quite different. In all bacteria, genes which code for enzymes in the same pathway are often located very close together. This phenomenon is so prevalent in bacteria that it cannot have occurred by chance but must have been strongly selected for in the course of evolution. The selective advantage that such gene clustering confers on the cell can only be speculated on at this time. One popular notion is that the function of many genes can be efficiently controlled if they are located close to one another. The means of this control is considered in Chapter 10.

SEXUAL RECOMBINATION IN EUCARYOTES

The processes of transformation, transduction, and conjugation all display the fundamental features of genetic recombination: namely, the union of the ge-

IN PROCARYOTES ONLY A PORTION OF THE CHROMOSOME IS TRANSFERRED; IN EUCARYOTES NUCLEI FUSE

netic material of two individuals to form the genetic material of a third individual. One feature of all three of these processes that makes them unique in the biological world is that all three entail only DNA of the donor cell—and in most cases only a small portion of the entire chromosome—entering the recipient cell. Because the entering DNA is usually quickly incorporated into the recipient cell, replacing a segment of the recipient cell's DNA, this recombinant cell remains haploid. In eucaryotic organisms an actual *fusion* of the haploid nuclei of the two parental cells takes place, rather than a replacement of recipient DNA by donor DNA. A diploid cell is thus formed. The two cells that fuse are called *gametes,* and the product of their fusion is a *zygote.* The diploid zygote has two copies of each chromosome (2n); the haploid gametes only one (1n). Each species of eucaryotic organism has a characteristic number of chromosomes; in humans there are 23 pairs (total 46). In order for the organism to reproduce sexually (that is, for its gametes to undergo fusion to form a zygote), the number of chromosomes must first be reduced from 2n to 1n. Thus in humans, both the egg and sperm have the reduced (haploid) number of 23 chromosomes each. The process by which this reduction in chromosome number occurs is called *meiosis* (Figure 9-23a). This process involves a number of steps, the first being a *duplication* of the DNA by the process of *mitosis.* In this process each nucleus is duplicated, giving rise to two identical nuclei (Figure 9-23b). The process of mitosis always precedes the division of one eucaryotic cell into two when such a cell divides asexually by fission. In meiosis, after the DNA has been duplicated, a series of steps occur in which the daughter cells undergo two divisions, resulting in each cell having the 1n number of chromosomes. The processes of mitosis and meiosis share many features in common but also differ in a number of respects. In particular, in meiosis homologous chromosomes line up along their entire length (*synapse*) whereas this does not occur in mitosis.

The stages in the life cycle of eucaryotic protists at which meiosis occurs differ, depending on the organism. In some organisms the life cycle of the organism is characterized primarily by the diploid phase. In other organisms the zygote undergoes meiosis very quickly after it is formed so that the organism is haploid during most of its life cycle. The processes of meiosis and mitosis are fundamentally the same in all organisms and both sexes.

One eucaryote that has proven very useful in genetic studies is the single-celled alga *Chlamydomonas.* This organism reproduces both sexually and asexually (Figure 9-24). In the sexual phase of its reproduction two different haploid mating types (+ and −) fuse to form the diploid zygote, which is encysted in a thick cell wall. The zygote undergoes a period of maturation (6 days duration) and then undergoes meiosis, resulting in the formation of four haploid vegetative cells, two + and two −. Each of these cells can divide by mitosis to form clones of identical haploid cells (asexual reproduction). Under certain conditions, however, the vegetative cells undergo a process of differentiation and turn into gametes, these being of the same mating type as the vegetative cells from which they were derived. The gametes then fuse and the life cycle is repeated. In the case of this alga, the haploid phase is the dominant one in the life of the organism.

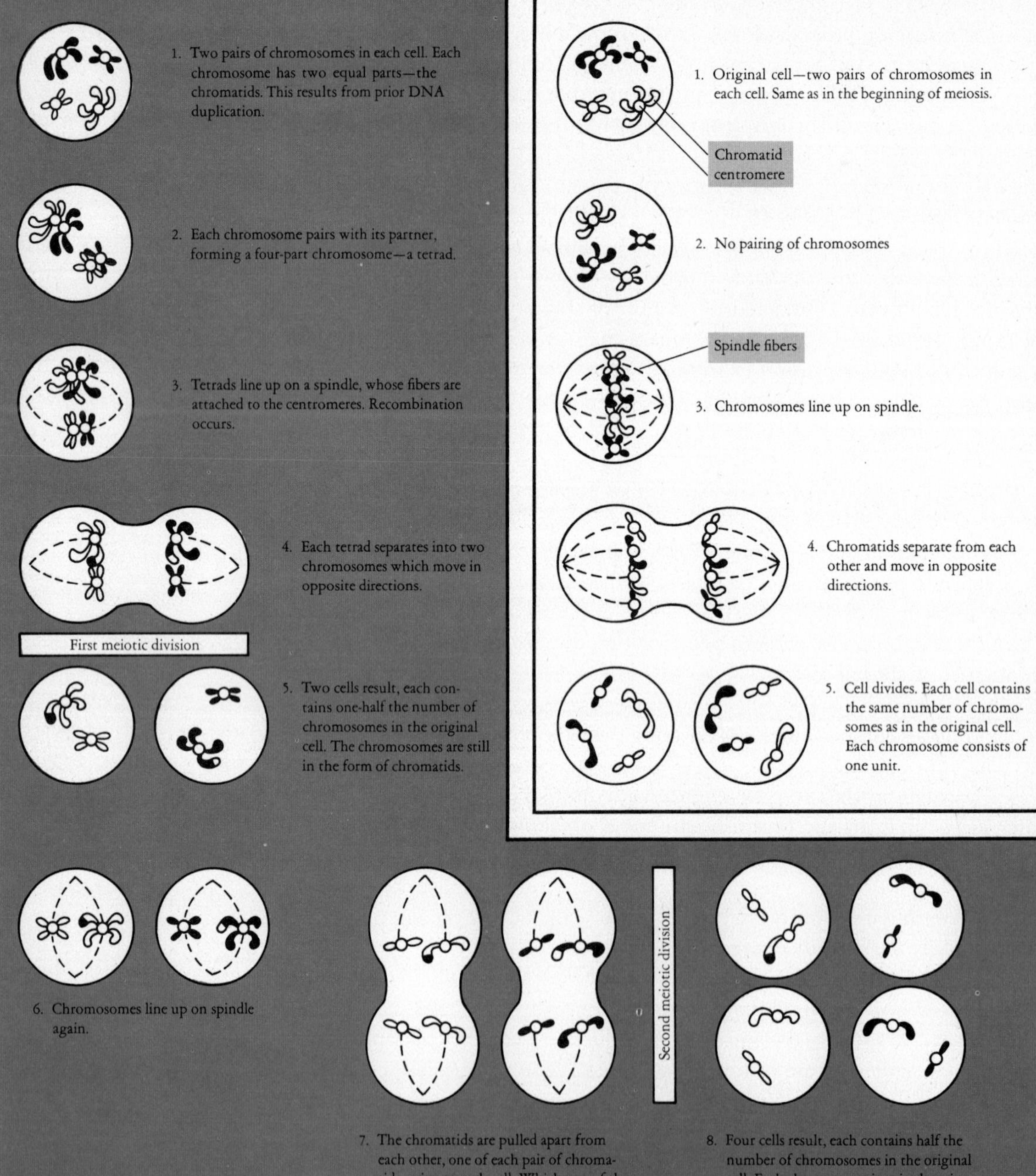

A Meiosis
B Mitosis
1. Two pairs of chromosomes in each cell. Each chromosome has two equal parts—the chromatids. This results from prior DNA duplication.
2. Each chromosome pairs with its partner, forming a four-part chromosome—a tetrad.
3. Tetrads line up on a spindle, whose fibers are attached to the centromeres. Recombination occurs.
4. Each tetrad separates into two chromosomes which move in opposite directions.
First meiotic division
5. Two cells result, each contains one-half the number of chromosomes in the original cell. The chromosomes are still in the form of chromatids.
6. Chromosomes line up on spindle again.
7. The chromatids are pulled apart from each other, one of each pair of chromatids going to each cell. Which one of the chromatid pairs goes to which cell is random.
Second meiotic division
8. Four cells result, each contains half the number of chromosomes in the original cell. Each chromosome is a single unit.
1. Original cell—two pairs of chromosomes in each cell. Same as in the beginning of meiosis.
Chromatid
centromere
2. No pairing of chromosomes
Spindle fibers
3. Chromosomes line up on spindle.
4. Chromatids separate from each other and move in opposite directions.
5. Cell divides. Each cell contains the same number of chromosomes as in the original cell. Each chromosome consists of one unit.

FIGURE 9-23 (a) Meiosis and (b) mitosis. In meiosis, one diploid cell gives rise to four haploid cells. In mitosis, one cell divides into two cells, each having the same DNA content as the original cell.

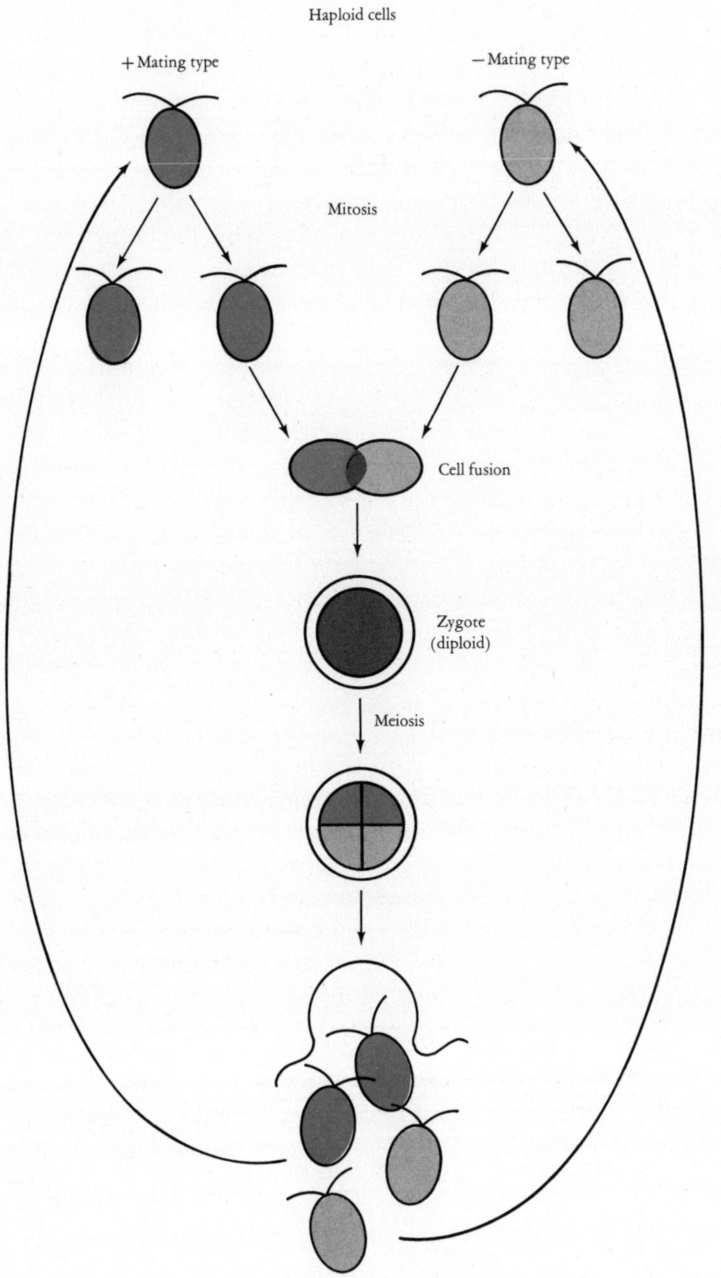

FIGURE 9-24
Life cycle of *Chlamydomonas*. This single-celled eucaryotic protist reproduces sexually by cell fusion and asexually by fission.

In *Chlamydomonas* chemical differences exist between the gametes, so that a + gamete will mate only with a − gamete. In other organisms all gametes are identical, and any gamete will fuse with any other gamete.

SUMMARY

Variations can arise in a single microbial cell as a result of two phenomena: changes in the nucleotide sequence of its DNA (mutations) and changes in the environment that determine which genes will be transcribed and which enzymes will function. Changes in the nucleotide sequence of a DNA molecule generally arise when the improper base is inserted into the growing DNA chain as the DNA is being replicated. These variations are rare but can be increased by mutagenic agents, which may increase their frequency by over a thousandfold. For mutations in which a missing end product cannot be supplied, special conditions must be employed to allow the mutants to grow.

Three mechanisms of genetic exchange have been demonstrated in bacteria: transformation, in which "naked" donor DNA is taken up by the recipient cell; transduction, in which the donor DNA is carried from the donor to the recipient cell inside a bacteriophage; and conjugation, in which the donor DNA passes from the donor to the recipient cell during cell-to-cell contact. The three mechanisms of gene transfer have basic features in common. In all three the donor DNA enters the recipient cell, pairs with the homologous region of the recipient chromosome, and is then integrated. A portion of the recipient chromosome is thus replaced by the donor chromosome. In DNA transformation highly purified DNA can pass through the cell wall and membrane of the competent cell. In conjugation there is unidirectional transfer of DNA from the Hfr to the F^- cell. If the donor cell is F^+, only the F particle is transferred. If the donor cell is of an Hfr strain, the chromosome, or, more commonly, a portion of it, is transferred into the F^- cell. All processes of genetic exchange occur at a relatively low frequency, so it is necessary to plate the recipient cultures on selective media to detect recombinants. Genetic exchange does not occur between different organisms because specific endonucleases degrade foreign DNA as it enters the cell.

In eucaryotes sexual recombination results from the fusion of haploid nuclei to form a diploid zygote. Many eucaryotic organisms multiply asexually as diploids, with their nuclei dividing by the process of mitosis. Other eucaryotes reduce their chromosome complement to 1n immediately after formation of the zygote and multiply asexually.

QUESTIONS

REVIEW

1. What are the three mechanisms by which DNA can be transferred from one bacterial cell to another? What are the distinguishing features of each mechanism?
2. What are the major differences between light repair and dark repair?
3. A base change occurs in an mRNA molecule of a cell, and the cell undergoes two divisions. How many cells would contain the "mutant" mRNA?
4. What kind of mutation would be likely to be the most severe? Why?

5. If an investigator was attempting to map genes that were located long distances from one another, what mechanism of genetic exchange would he be most likely to use? Why?

THOUGHT

1. Many mutations are silent in that changes in the DNA do not result in changes in the protein coded by the mutant gene. Why is this true?
2. What argument can be made that genetic recombination in bacteria has great significance even though it is a rare event?
3. Why is it reasonable to find that plasmids code only for functions that are dispensable to the cell?

FURTHER READING

BROWN, D., "The Isolation of Genes," *Scientific American* (August 1973).

CLOWES, R., "The Molecule of Infectious Drug Resistance," *Scientific American* (April 1973).

COHEN, S., "The Manipulation of Genes," *Scientific American* (July 1975).

CURTISS, R., III, "Genetic Manipulation of Microorganisms: Potential Benefits and Biohazards," *Annual Review of Microbiology* 30:507–533 (1976). An up-to-date review of this topic by an individual very much involved in laboratory research. The emphasis is on the consequences of genetic engineering.

FALKOW, S., *Infectious Multiple Drug Resistance.* London: Pion Limited, 1975. An advanced text which covers in considerable detail most aspects of plasmids. Written by one of the major research investigators in this field.

GROBSTEIN, C., "The Recombinant DNA Debate," *Scientific American* (July 1977).

TOMASZ, A., "Cellular Factors in Genetic Transformation," *Scientific American* (June 1958).

WOLLMAN, E., and F. JACOB, "Sexuality in Bacteria," *Scientific American* (July 1956).

CHAPTER 10

REGULATION OF CELLULAR ACTIVITIES

THE ENVIRONMENT AND CONTROL SYSTEMS

ENERGY SUPPLY OFTEN LIMITING IN NATURE

To survive in any environment, microorganisms must often reproduce more rapidly than any other organism in the same environment. Because the supply of available energy is generally the limiting factor in bacterial growth in nature, it is crucial that the bacterial cell synthesize the maximum amount of cell material from a limited supply of energy. And because biosynthetic reactions often require energy, cells have developed very elaborate mechanisms for the control of their biosynthetic pathways. Specifically, cells shut down these biosynthetic pathways when the products of the pathways are not needed. Likewise, cells do not synthesize degradative enzymes unless the substrates that these enzymes degrade are present in the environment.

The control mechanisms that microorganisms have evolved take into account two fundamental principles related to their environment. First, if the organism can utilize the end products of biosynthetic reactions, which may be available to the cell from the environment, energy can be conserved. This is why cells *shut down biosynthetic pathways* when the products of these pathways are available to them. Second, because the environment can change drastically, often in a matter of minutes, such control mechanisms must be *reversible.* Consider the situation with *E. coli:* for over 100 million years this organism has inhabited the gut of mammals, where it reaches concentrations of 10^6 cells per milliliter. Obviously, it is very well adapted to this particular niche in nature. In this same niche, however, it is subjected to alternating periods of feast and famine. For a limited time, following a meal, the *E. coli* cells prosper, wallowing in a medium composed of amino acids, vitamins, purines, pyrimidines, and other end products of biosynthetic pathways. The cells actively take up these metabolites through their permease systems, which require a minimal expenditure of energy. Simultaneously, the cells shut down their own biosynthetic pathways, conserving their energy in order to channel it into the rapid synthesis of macromolecules. Under these conditions the cells divide at their most rapid rate. However, famine follows the feast. Between meals, which may be many days in the case of some mammals, the end products of metabolism are not available. Therefore, the cells' biosynthetic pathways must be activated to synthesize the different components of macromolecules. Cell division then markedly slows down because energy is drained off for the synthesis of these small component molecules. Thus, whereas cells can divide several times an hour in an environment composed of rich nutrients, they may divide only once every 24 hours in a famished mammalian gut.

Permease, page 68

In the following discussions we shall consider the control of the synthesis of only one end product, the amino acid tryptophan (Figure 10-1). However, the *principles* for the regulation of this biosynthetic pathway apply to all amino acids, purines, pyrimidines, vitamins, and other metabolites.

MECHANISMS OF CONTROL OF ENZYMATIC ACTIVITY

TWO CONTROL MECHANISMS: CONTROL OF ENZYME ACTIVITY AND CONTROL OF ENZYME SYNTHESIS

Cells control the synthesis of metabolites by two mechanisms. In both cases the end product of a pathway inhibits its own synthesis. Thus any tryptophan in the environment will enter the cell and shut off the synthesis of tryptophan. This inhibition occurs in two entirely *independent* ways by *two different mechanisms.* In the first of these, the end product of a pathway *inhibits the activity* of a preformed enzyme of the pathway involved in its synthesis. This mechanism is called *feedback* or *end product inhibition.* In the second mechanism, the end product of a pathway *inhibits the synthesis of the enzymes* of the pathway involved in its synthesis. This is called *end product repression.* The fundamental contributions to the elucidation of these control mechanisms were largely made by Monod and Jacob.

Feedback Inhibition

Enzymes, pages 143–149

In addition to the *active* or *catalytic* site which all enzymes have, the *first* enzyme of most biosynthetic pathways has an additional site which combines with the end product of the pathway. This additional site is called the *allosteric site,* and the end product is called a *feedback inhibitor.* The combination of the allosteric site of the enzyme with the end product of the pathway through weak bonding forces

END PRODUCT COMBINES REVERSIBLY WITH FIRST ENZYME AND INACTIVATES IT

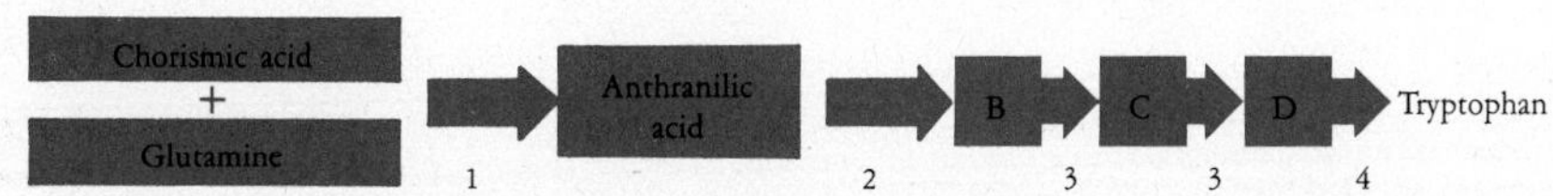

FIGURE 10-1 Pathway of tryptophan biosynthesis. The numbers indicate the enzymes and the letters the intermediate compounds in the pathway. Note that the same protein catalyzes two sequential reactions in the pathway. "Double-headed" enzymes are unusual.

Tryptophan biosynthesis, Figure 7-15, pages 163–164

changes the conformation or shape of the enzyme molecule so that the enzyme is no longer capable of catalyzing its usual reaction (Figure 10-2). By this mechanism the end product prevents the formation of the product of the first enzyme step of the pathway, thereby effectively shutting down the entire pathway. Consider again the biosynthesis of tryptophan, which involves five enzymatic steps (Figure 10-1). Any tryptophan in the medium combines with the first enzyme and changes this enzyme's shape. As a result, the product of the first enzymatic reaction is not synthesized. Therefore, because this unsynthesized product is also the substrate of the second enzyme, the pathway is shut down instantaneously. Note that only the *first* enzyme in the pathway has evolved an allosteric site, but this single site is sufficient to shut down the entire pathway.

The importance of feedback inhibition in the regulation of end product synthesis is illustrated by observing what happens when feedback control does not function. Mutants can be isolated in which tryptophan cannot change the shape of the first enzyme of its biosynthesis although the catalytic site on this enzyme is normal. Such mutants are called *feedback-resistant.* They invariably overproduce and actually excrete large amounts of tryptophan into the medium.

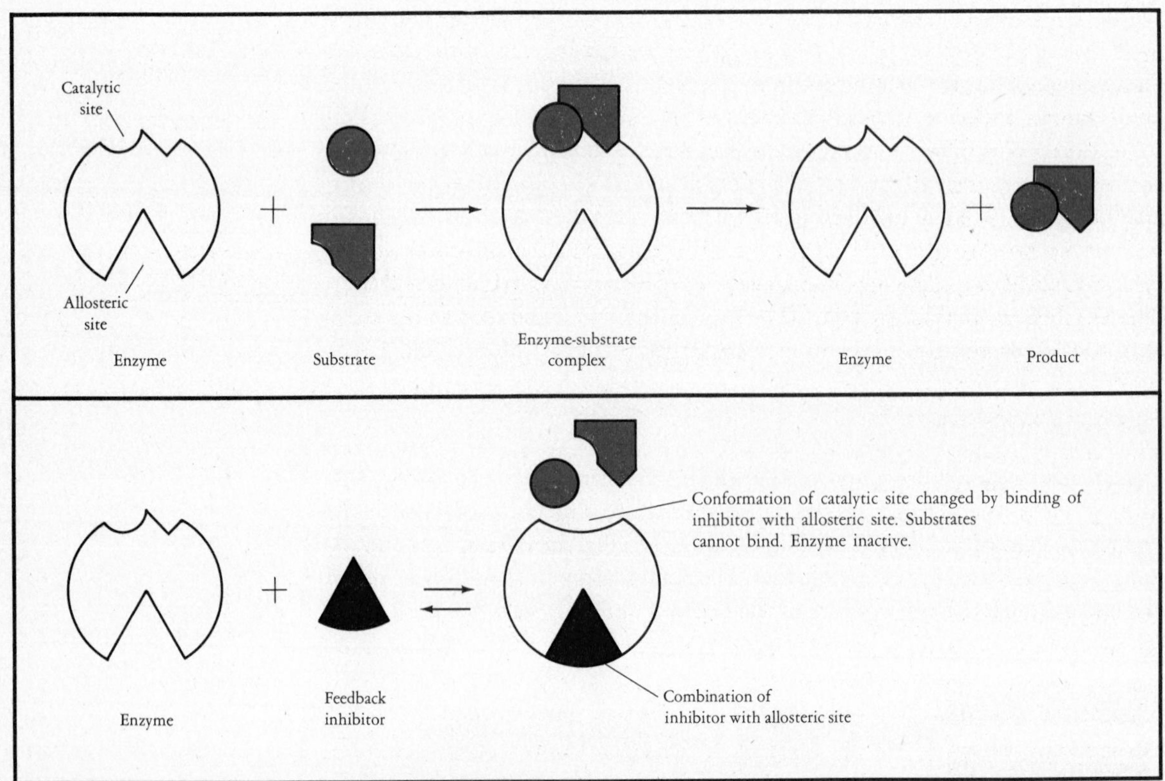

FIGURE 10-2 Effect of feedback inhibitor on catalytic activity of enzyme. The combination is reversible, and if the level of inhibitor drops, the reaction goes in the reverse direction, so the enzyme becomes functional again.

This phenomenon is readily detected by growing a few colonies of the feedback-resistant mutant on a medium lacking tryptophan but seeded with a very large number of cells of a tryptophan-requiring mutant. The excretion of the tryptophan by the feedback-resistant mutant allows the tryptophan-requiring cells to grow around the tryptophan resistant colony, giving rise to a halo of background cells.

A wide variety of feedback-resistant mutants, which overproduce and excrete metabolites, have been isolated. Such mutants are useful to industrial concerns interested in isolating these metabolites on a commercial scale.

End Product Repression

Although feedback inhibition effectively shuts off the synthesis of the end product of a pathway, it still allows some waste of energy and carbon because feedback inhibition has no effect on the *synthesis of the enzymes* of the pathway. Therefore, another mechanism, *endproduct repression,* comes into play. In this mechanism the end product represses or prevents the synthesis of the enzymes concerned with the synthesis of that particular end product. Returning to the example of tryptophan synthesis, the presence of tryptophan in the medium inhibits the synthesis of *all* five enzymes of this pathway. This inhibition results in a saving of the many molecules of ATP which must be expended for the synthesis of these enzymes. However, because all of the enzyme molecules which already exist in the cell are still functional, this regulatory mechanism is slow to act in regulation. Only when the number of enzyme molecules per cell decreases as a result of cell division does this mechanism operate. Note that there is great specificity involved. Tryptophan shuts down *only* the tryptophan pathway.

END PRODUCTS CONTROL THE RATE OF ENZYME SYNTHESIS

The mechanism by which an end product represses the synthesis of the enzymes involved in its creation is only partially understood at the molecular level. Indeed, it appears that the details of this mechanism vary for different amino acids, although the general aspects of the mechanism are reasonably clear. In the case of tryptophan biosynthesis the genes that code for the five required enzymes are linked to one another on the chromosome in *E. coli* (Figure 10-3). In addition, another gene, called a *regulatory gene,* codes for a repressor protein that is concerned with regulating the *functioning* of these five genes. This protein has the unusual ability to recognize and bind to a specific nucleotide sequence at the extreme end of the sequence of tryptophan genes. This region, the *operator region,* is adjacent to the site at which RNA polymerase binds the *promoter* region and initiates transcription of the genes of tryptophan synthesis (Figure 10-4a). The binding of the repressor protein to the operator region prevents the RNA polymerase enzyme molecule from binding and *initiating* transcription. Therefore none of the enzymes concerned with tryptophan biosynthesis are synthesized. What role does tryptophan play in this story? The repressor protein must be *activated* by combining with tryptophan before it can bind to the operator region (Figure 10-4b); tryptophan is thus called a *co-repressor.* How the repressor protein is activated is not clear. It seems likely, however, that the repressor protein is an

RNA polymerase, page 182

Transcription, Figure 8-7, pages 174–176

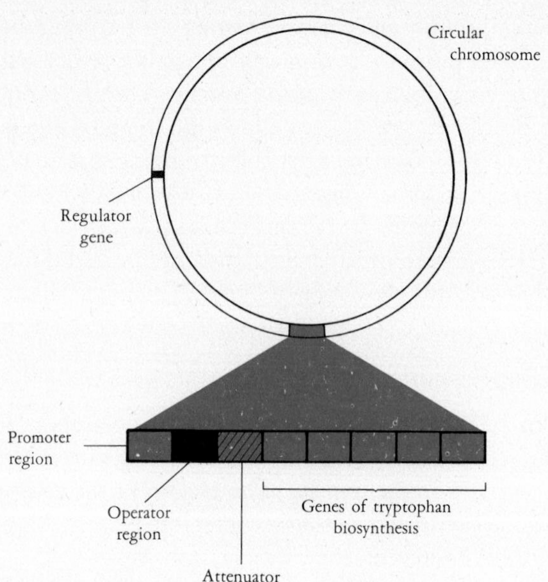

FIGURE 10-3
Genes of tryptophan biosynthesis. All of the genes that code for the enzymes of tryptophan biosynthesis are closely linked on the chromosome. The regulatory gene that codes for a repressor protein is not located close to these genes. The operator region and the promoter region are contiguous and next to the genes of tryptophan biosynthesis.

allosteric protein, so it can exist in two different shapes. In one shape it does not bind to the DNA. However, if the end product (tryptophan) binds to the repressor protein and changes its shape, it becomes able to bind to the DNA.

REPRESSION PREVENTS INITIATION OF TRANSCRIPTION

If tryptophan is removed from the medium, tryptophan dissociates from the repressor, and the repressor detaches from the DNA. The RNA polymerase is then able to bind to the promoter gene, initiating transcription. This regulatory mechanism is therefore reversible.

ONCE TRANSCRIPTION IS INITIATED, IT MUST CONTINUE PAST THE ATTENTUATOR BARRIER

There is another site for the control of enzyme synthesis. Between the site at which transcription is *initiated* and the site of the first structural gene of tryptophan synthesis, there is an *attenuator region,* a barrier which the RNA polymerase enzyme must traverse if it is to transcribe the genes of tryptophan synthesis. In the presence of tryptophan, most RNA polymerase molecules fall off of the DNA in this region and therefore do not transcribe the tryptophan genes. In the *absence* of tryptophan, the RNA polymerase molecules remain attached to the DNA and are able to successfully traverse the attenuator region. Thus tryptophan controls the synthesis of its own enzymes in two independent ways: by controlling the *initiation* of transcription at the promoter region and by controlling the *continuation* of transcription at the operator region.

Enzyme Induction

In biosynthetic pathways the presence of the end product *turns off* gene expression. Metabolites can also turn on inactive genes so that they are transcribed. The enzymes synthesized as a result of genes being turned on are called *inducible enzymes,* and the process involved is called *enzyme induction.* The low molecular weight compound that activates gene transcription is called the *inducer.* Inducers are usually compounds that cells are *potentially* capable of utilizing as a source of

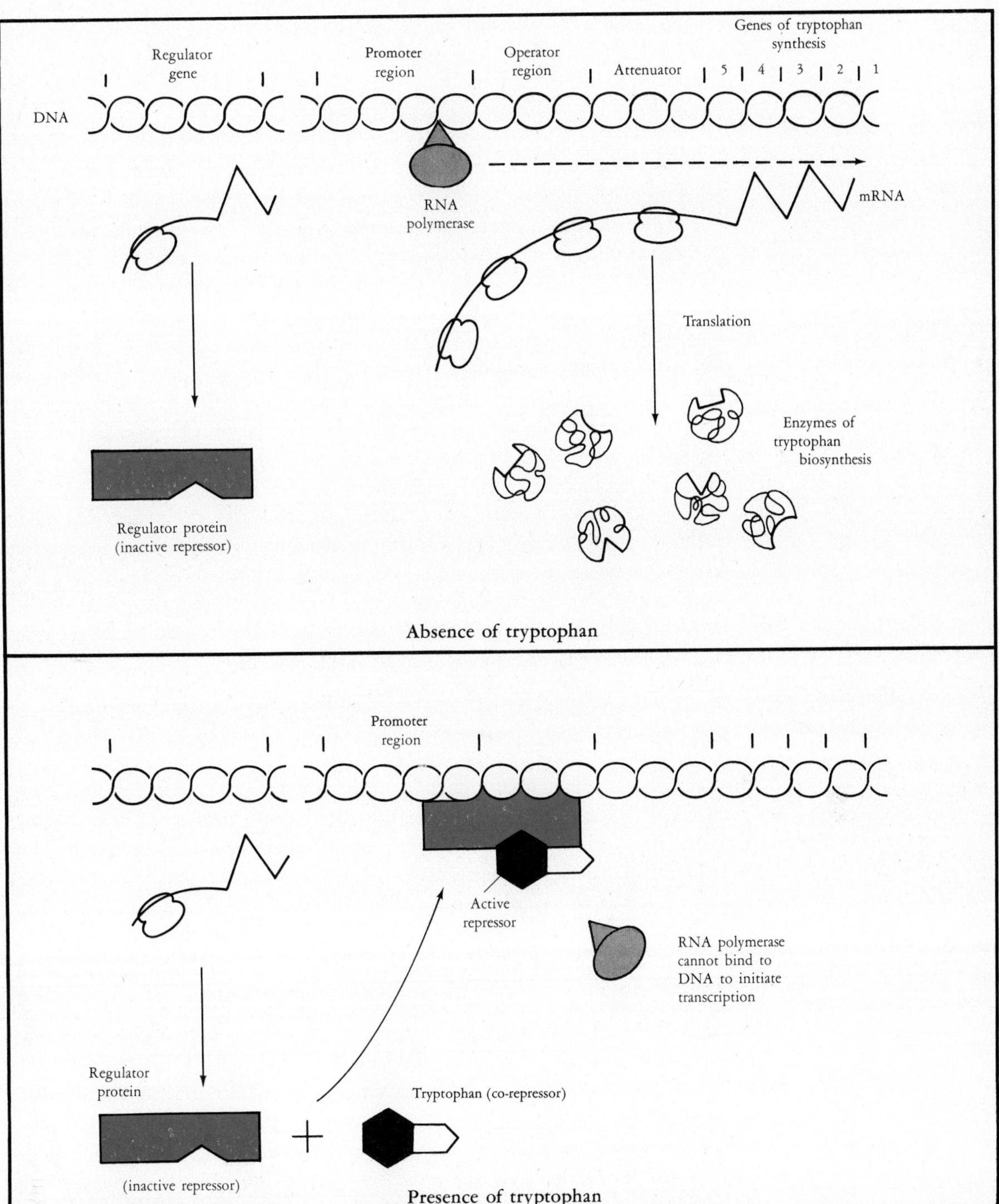

FIGURE 10-4 Diagrammatic representation of the regulation of synthesis of enzymes of tryptophan biosynthesis. Major control is mediated through the frequency of initiations of RNA synthesis at the promoter region, controlled by the repressor binding to the operator region. Minor control is mediated by the frequency with which the RNA polymerase molecules traverse the attenuator region. The control at the attenuator region involves charged tryptophan transfer RNA, but the details are not known.

energy, carbon, or both, but which are not always present in the cells' environment. Until the compound is actually present, the cells will not synthesize the enzymes required for its breakdown. In this way the cells do not waste energy and carbon by synthesizing unneeded enzymes.

Lactose, page 41

The most thoroughly studied case of induced enzyme synthesis involves the enzymes of lactose degradation in *E. coli.* This system serves as a model for most other inducible systems. Lactose, the most abundant sugar in milk, is a disaccharide of glucose and galactose. It induces the synthesis of two enzymes required for its breakdown: a permease which actively transports the sugar into the cell and an enzyme (β-galactosidase) which degrades the sugar into glucose and galactose.

The mechanism of induction (Figures 10-5a and 10-5b) is basically similar to that of enzyme repression except that an attenuator region is not involved in this system. The same genetic elements are involved: a regulatory gene which synthesizes a repressor protein, an operator region which binds the repressor protein, and the two genes of the operon which code for the permease and β-galactosidase enzymes. However, there is one major difference. In the case of inducible enzymes, the repressor protein is active only in the *absence* of the inducer; the inducer *inactivates* the repressor protein. Accordingly, in the presence of the inducer, the repressor does not bind to the operator region, and the genes of the lactose operon are therefore transcribed. The inducer is also able to inactivate any repressor bound to the operator, causing it to dissociate from the operator. Only in the absence of lactose does the repressor bind to the operator, with the result that the genes of lactose metabolism are not transcribed.

Constitutive Enzymes

Many enzymes are neither inducible nor repressible but remain at the same level no matter what nutrients are present in the growth medium. These enzymes, synthesized continuously at the same rate, are called *constitutive* and generally include those that the cell needs under all conditions of growth. The enzymes of glycolysis are one class of constitutive enzymes.

Regulation of Enzyme Synthesis in Other Pathways

The control of lactose degradation that has been described serves as the model for the control of a large number of inducible enzyme systems. However, it is now apparent that many other inducible systems are controlled in ways slightly different than the lactose system. In some systems a regulatory protein is absolutely essential for the initiation of transcription. In other systems the regulatory proteins are enzymes that function in pathways other than those which they control. It is now clear that bacteria have evolved a variety of control mechanisms whose function appears superficially to be similar in all cases: the *control of initiation of gene transcription.* But the molecular mechanisms that the cells use to achieve this control varies in different pathways. Compared with inducible systems, the

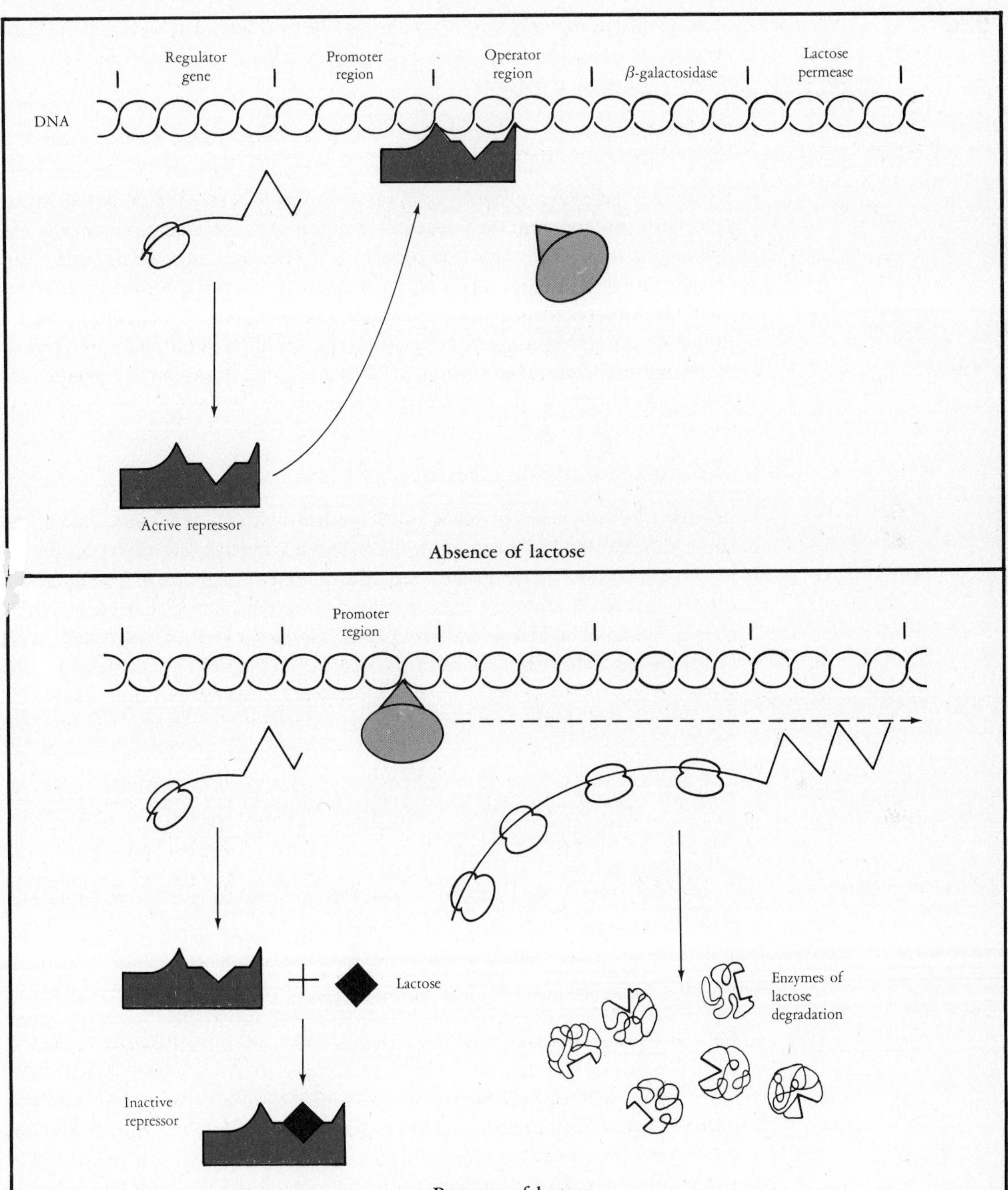

FIGURE 10-5 Induction of a degradative enzyme system. Note that the same genetic elements take part as in the regulation of a biosynthetic pathway except an attenuator region does not appear to be present.

molecular basis of control of many of these biosynthetic pathways is poorly understood, but it is clear that many of the different biosynthetic pathways are controlled in slightly different ways.

Messenger RNA Instability

A feature of mRNA also contributes to the ability of bacteria to adapt quickly to a changing environment. Most bacterial mRNA molecules are metabolically unstable, so that after they function for about two minutes, they are degraded into their subunits, which are then used to synthesize new mRNA molecules. Thus every few minutes there is a complete replacement of the templates for the cell's proteins, and new messages can be synthesized without waiting for the cell to dilute out its content of preexisting mRNA molecules through cell division. The precise details of how this degradation occurs are not known.

Gene Clusters and Regulation

Because all of the genes involved in tryptophan synthesis are closely linked, the entire cluster is transcribed as a unit to create one large mRNA molecule (Figure 10-4a). Therefore a single operator region at one end can control the synthesis of this single mRNA molecule and consequently the synthesis of all enzymes of the tryptophan pathway. Gene clustering thus provides a very efficient mechanism for regulating the synthesis of *all* of the enzymes of a pathway as a unit. The term *operon* defines such a cluster of genes which are controlled as a unit. However, genes which code for biosynthetic pathways are not always clustered on the chromosome. In cases where they are not, there must be an operator region next to each gene that codes for a particular enzyme of the biosynthetic pathway. In such cases, however, there is still only one repressor protein synthesized. Operons are very common in bacteria but rarely occur in eucaryotic cells. The reason for this difference is not clear.

A summary of the mechanisms that the end product employs to control its own synthesis is diagramed in Figure 10-6.

Catabolite Repression or the Glucose Effect

Because glucose is broken down by enzymes which are constitutive, whereas the breakdown of lactose requires the induction of two enzymes, what happens when the cell is grown in a medium containing limiting amounts of *both* glucose and lactose? Does the cell metabolize both sugars simultaneously or sequentially? The answer to this question has revealed a whole new aspect of biosynthetic control in procaryotes. In *E. coli* glucose is always metabolized first. Only after the glucose is completely utilized is lactose degraded. Thus cells growing in a medium containing both lactose and glucose exhibit a two-stage or diauxic growth curve (Figure 10-7).

In the diauxic growth curve, the first period of growth occurs at the expense

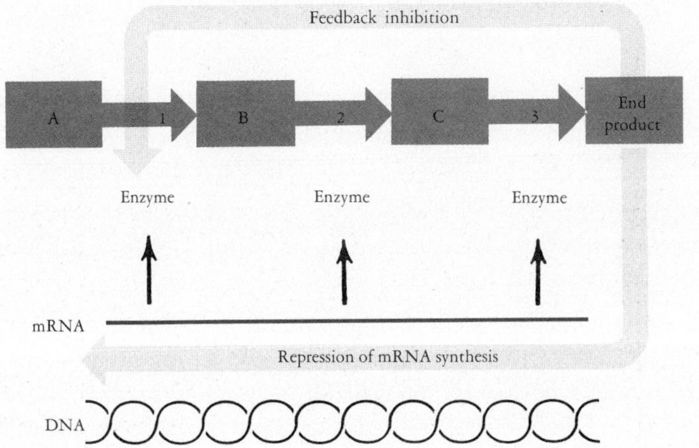

FIGURE 10-6
Effect of end product on control of its own synthesis. The end product feedback inhibits the activity of the *first* enzyme of the pathway and prevents the synthesis of *all* enzymes in the pathway.

of glucose, and the enzymes for lactose metabolism are not induced. After all of the glucose has been degraded, a short period of time elapses during which the enzymes of lactose synthesis are induced, and growth then resumes at the expense of lactose. Glucose represses the synthesis of a very large number of inducible enzymes in many different bacteria. This general repression, called *catabolite repression* or the *glucose effect,* is not completely understood, but some interesting features of it have been uncovered. The addition of glucose to a growing culture results in an almost immediate *release* of an important nucleotide from the cell, 3′,5′-cyclic AMP (Figure 10-8). The significance of this nucleotide was not previously appreciated in *E. coli,* although it had been known for a number of years that in mammalian systems many hormones functioned by regulating the synthesis of cyclic AMP. This compound is in turn directly responsible for the effects of many hormones. This effect of cyclic AMP appears to be related to the fact that

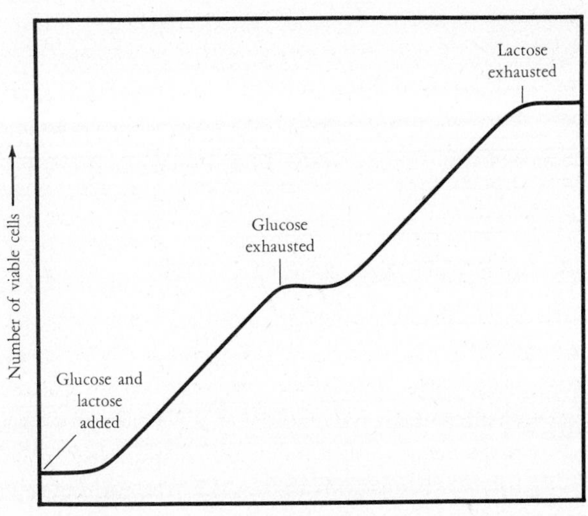

FIGURE 10-7
Diauxic growth curve.

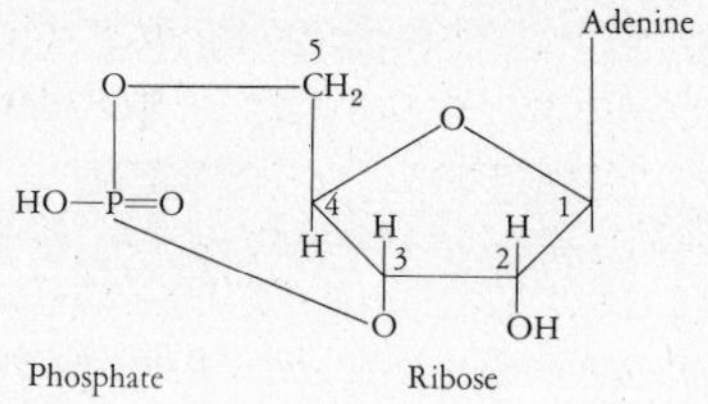

FIGURE 10-8
Cyclic AMP. The phosphate molecule is bound to carbon atoms 3 and 5, thereby giving the molecule its name.

certain important enzymes seem to function only in the presence of cyclic AMP. It is therefore noteworthy that cyclic AMP is also an important component of regulatory processes in bacteria and other procaryotes. In bacteria this nucleotide is required for the *initiation* of transcription of a large number of inducible enzyme systems, most of which are involved in energy metabolism.

Glucose inhibits enzyme induction by causing the loss of cyclic AMP from cells by some unknown mechanism. Furthermore, mutants that lack the enzyme necessary for the synthesis of cyclic AMP cannot synthesize most inducible enzymes unless cyclic AMP is added to the culture in which they are growing. However, the role of cyclic AMP in induced enzyme synthesis is even more complicated. This nucleotide does not function alone, instead, it requires a protein to which it must bind to carry out its function of promoting selective transcription. The protein which binds to the cyclic AMP also apparently binds to the DNA that is to be transcribed, to facilitate transcription. Thus, to promote gene transcription, not only must lactose be present to inactivate the repressor protein, but cyclic AMP must also be available to bind to the protein which binds to the DNA. Dr. Earl Sutherland, the American biochemist who isolated, identified, and recognized the general significance of cyclic AMP in both animals and bacteria, was awarded the Nobel Prize in Medicine in 1971.

CYCLIC AMP BINDS TO A PROTEIN WHICH BINDS TO DNA

Catabolite repression, the control of enzyme synthesis by readily metabolized sources of energy, serves a very useful function in bacteria. It requires the cell to utilize the best available source of energy. Glucose is generally the most commonly utilizable source, thus it inhibits indirectly the synthesis of enzymes which metabolize poorer sources of energy, such as lactose. As predicted, glucose is *not* a catabolite repressor in species that cannot metabolize this sugar easily.

REGULATION OF BACTERIAL SPORULATION

Endospore and sporulation, pages 78–80

In Chapter 3 the composition, properties, and development of a bacterial *endospore* were described. The progressive physiological and morphological changes that are involved in the conversion of a *vegetative cell* to the dormant endospore requires the participation of all the regulatory mechanisms we have discussed, acting in a specific sequence to turn some genes on and other genes off.

The cytological transformations involved in the formation of the mature endospore are accompanied by profound changes in the chemical and enzymatic machinery of the vegetative cell. Thus many enzymes which are not detected in the vegetative cell appear during certain stages of sporulation. In contrast, many enzyme systems which are characteristic of the vegetative cell are present at very low levels or even completely absent in the spore. In addition, a variety of new structural components that are not present in the vegetative cell appears in the spore. All of these facts indicate that a considerable amount of genetic information concerned specifically with the process of sporulation is not expressed during vegetative growth. Perhaps fifty genes concerned specifically with sporulation are located at widely separated sites on the chromosome of a bacterial cell.

What controls which of the cell's genes are transcribed? The answer relates to the mechanisms by which cells undergo differentiation, one of the central questions in biology. This answer is not entirely known, but recent investigations have provided some clues. First, it appears that each step in the process of sporulation requires the participation of the preceding step. For example, if an early spore enzyme is defective, the development of the spore will cease at that stage, and none of the enzymes or spore structures of later stages of development will be synthesized. Second, it has been shown that the RNA polymerase of the vegetative cell is modified in the course of sporulation. Thus it appears that in a sporulating cell there occurs a modification of this polymerase that renders it unable to transcribe genes associated with vegetative growth but able to transcribe genes associated with sporulation. This could account for the fact that in sporulation certain genes associated with vegetative growth are turned off and genes associated with sporulation are turned on.

Catabolite repression by glucose may also be responsible for repressing the synthesis of a variety of the enzymes involved in sporulation. To sporulate, bacteria must be growing at either a very slow rate or have stopped growing altogether. Because glucose starvation results in sporulation, it seems possible that glucose may function as a catabolite repressor for at least some spore-forming enzymes. Nitrogen starvation also results in the conversion of vegetative to sporulating cells. Rapid growth and sporulation seem to be mutually exclusive aspects of cell development. Perhaps nature has employed this elaborate control system, involving starvation for required nutrients, to ensure that sporulation will occur when conditions are no longer favorable for vegetative growth.

SUMMARY

Bacteria control the flow of metabolites through synthetic pathways by two basic mechanisms: end product inhibition and end product repression. In the first of these mechanisms the end product *inhibits the activity* of a preformed enzyme (generally the first) of a biosynthetic pathway. This results in an *immediate* shutdown of the pathway. The second control mechanism involves the control of enzyme synthesis by control of the initiation of gene transcription. The end product *activates* a repressor protein which binds to an operator gene, thereby preventing transcription of *all* the genes involved in the synthesis of the particu-

TABLE 10-1
Features of Inducible and Repressible Enzyme Systems

Features	*Inducible Enzyme Systems*	*Repressible Enzyme Systems*
Name of controlling metabolite	Inducer	Repressor
Effect of small metabolite	Inactivates repressor	Activates repressor
Effect of cyclic AMP	In many, stimulates transcription by RNA polymerase	No known effect
Pathways affected	Degradative	Biosynthetic

lar end product. In both cases the product of a pathway controls *only* its own pathway. In addition to these regulatory mechanisms for the control of biosynthetic pathways, cells have a mechanism to control the synthesis of degradative enzymes. This mechanism is basically similar to end product repression, except that in enzyme induction the substrate *inactivates* an active repressor protein. Thus the substrate prevents the repressor from combining with the operator region, and the genes that encode the inducible enzyme are transcribed. These control mechanisms are specific in that only a single pathway is controlled by an inducer or repressor. Glucose, however, can prevent the induction of most of the inducible enzymes concerned with the generation of energy. Indeed, if cells are growing on glucose, most of the inducible enzymes concerned with energy generation will not by synthesized even in the presence of their substrate. Glucose apparently exerts its effect by controlling the level of cyclic AMP in the cell. This nucleotide, as well as a protein which binds it, are required for the initiation of transcription of most inducible enzymes. These salient features of the regulation of enzyme synthesis are summarized in Table 10-1.

The development of an endospore from the vegetative cell represents a process of differentiation in bacteria. In this process some genes are turned off and others are turned on, some enzymes are activated, others inactivated, resulting in profound changes in the morphological, physiological, and biochemical properties of the resulting endospore.

QUESTIONS

REVIEW

1. Tryptophan is added to a culture of *E. coli* growing in a glucose-salts medium. Explain the effect that tryptophan exerts on the *synthesis* of the enzymes concerned with tryptophan synthesis. Discuss the mechanism by which tryptophan exerts this effect.
2. Regarding Question 1, what effect does tryptophan have on the activity of *preformed* enzymes of tryptophan synthesis? Which enzymes are involved?
3. Which effect takes place more rapidly—the shut-off of tryptophan synthesis by feedback inhibition, or end product repression?
4. Compare the role of repressor molecules in enzyme induction and end product repression.
5. Explain the role of cyclic AMP in catabolite repression.

1. Why it is essential that the effects that end products exert on their own synthesis be reversible?
2. What would be the phenotype of a cell that was unable to synthesize cyclic AMP?
3. How would you recognize a mutant that synthesized a defective repressor protein of tryptophan biosynthesis?

FURTHER READING

HANSON, R. S., "Gene Expression during Bacterial Sporogenesis," *Microbiology—1975*. Ed. D. Schlessinger. Washington, D.C.: American Society for Microbiology, 1975, pp. 475–483. A brief review of some of the current speculations on how genes concerned with sporulation are temporally expressed.

KOSHLAND, D., "Protein Shape and Biological Control," *Scientific American* (October 1973).

MANIATIS, T., and M. PTASHNE, "A DNA Operator-Repressor System," *Scientific American* (January 1976).

PASTAN, I., "Cyclic AMP," *Scientific American* (August 1972).

WATSON, J. C., *Molecular Biology of the Gene.* 3d ed. New York: W. A. Benjamin, 1975. Chapter 14 is a beautifully written, well-illustrated chapter that covers the major aspects of control of enzyme synthesis and function.

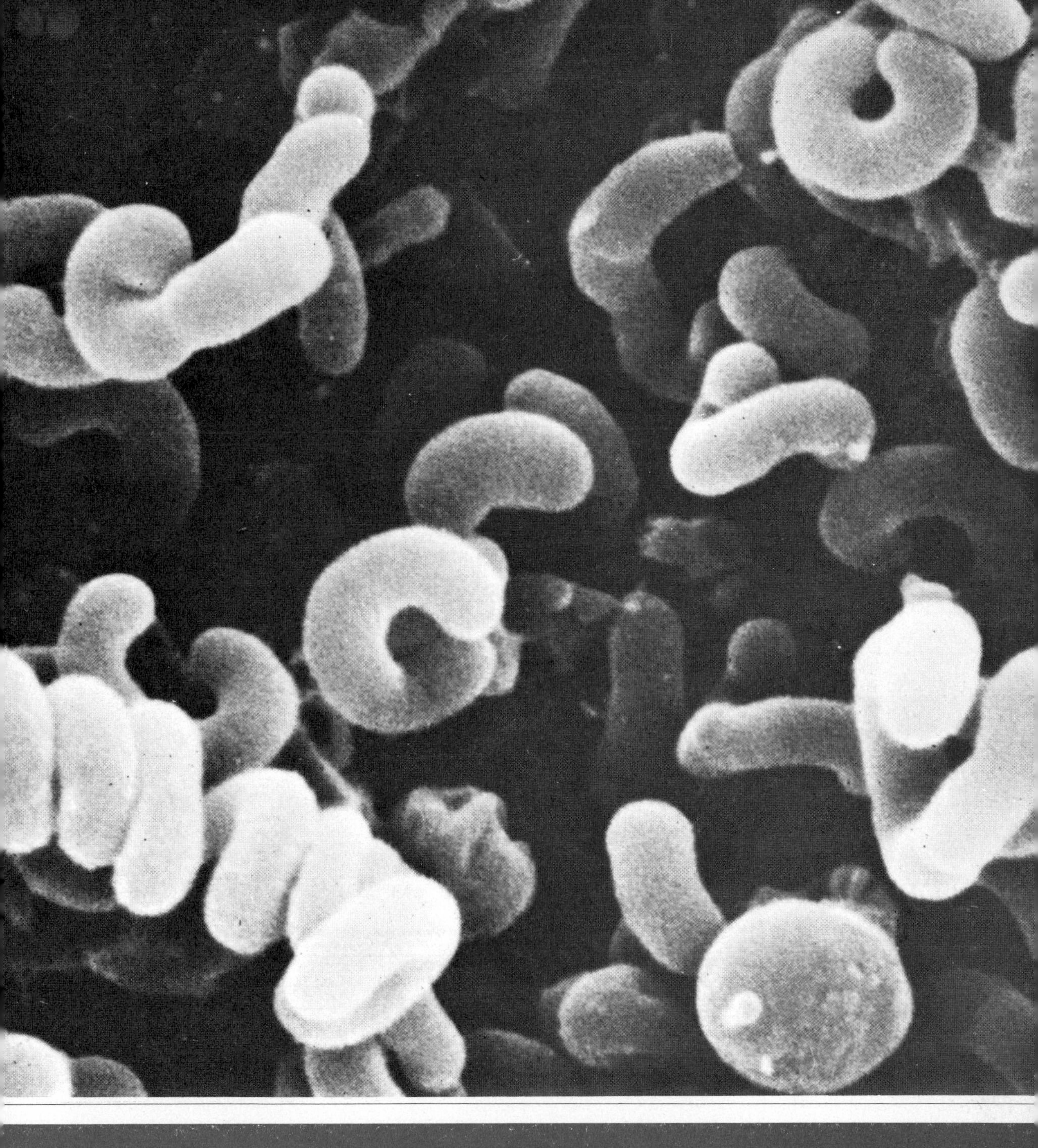

II THE MICROBIAL WORLD

CLASSIFICATION OF MICROORGANISMS

PROBLEMS IN CLASSIFICATION OF BACTERIA

The taxonomy of microorganisms is difficult and complex because the number of different organisms is immense and because the various groups of microbes are very different from one another. In fact some, like Protozoa, fall within the province of zoology while others, such as algae, are dealt with by botanists. In such cases the microorganisms are classified according to conventions developed for animals or higher plants which are not necessarily appropriate for unicellular organisms.

CONCEPT OF SPECIES

The purpose of all classification schemes is to group organisms with similar properties and to distinguish those which are different. There are several basic differences between bacterial taxonomy and the taxonomy of higher plants and animals. In both taxonomic schemes the species is the basic taxonomic unit. However, the definition of a species is different for the higher organisms than for the bacteria. For the higher organisms a species may be fairly precisely defined as an *inbreeding group having a limited geographic distribution.* This results in individuals having distinct morphological characters that distinguish them from individuals of other species. These same criteria cannot normally be applied to bacteria, because their morphological characters are limited and sexual reproduction among bacteria is seldom obligatory, appearing to be confined to a relatively few genera. Therefore, among bacteria, a species can be defined only in terms of a population of cells, or *clone,* all derived from a single cell. The members of such a clone are virtually indistinguishable from one another but do differ from those of a clone of another species, generally in several features. Thus the major problem in the taxonomy of bacteria becomes one of trying to decide how different two clones must be for their members to be classified as different species. As more and more sophisticated methods evolve for looking at microbes, unsuspected differences and similarities become apparent. For one thing, as more organisms are studied in detail, the divisions between species tend to become blurred. However, just as for higher organisms, several species of bacteria are grouped into a genus, and several genera are included within a family. At present only a few families of bacteria are recognized. Families in turn are grouped into orders; many orders constitute a class.

BACTERIA CLASSIFIED ACCORDING TO THEIR COMMON PROPERTIES

The taxonomic classification scheme for bacteria turns out to be one of convenience, consisting of descriptions of organisms in terms of morphology, staining characteristics, and nutritional and metabolic properties, among others. For bacteria these descriptions are contained in a reference text, *Bergey's Manual of Determinative Bacteriology,* the "Bible" of bacterial taxonomy. This manual, which is now in its eighth edition, groups organisms according to their shared properties, so it is not surprising that each edition is revised extensively as ideas about the relative importance of different shared properties change and as new methods of examining bacteria are developed. Because *Bergey's Manual* is a dictionary of descriptions of generally accepted species of bacteria, it is useful for bringing uniformity into the nomenclature of bacteria. However, its taxonomy does not necessarily represent a scheme based on evolutionary relationships.

Until simple techniques, which can show the genetic relationships between organisms, are devised, classification schemes will be subject to the changing practices of taxonomists. Fortunately, several modern approaches to taxonomy provide hope that evolutionary relationships between organisms can be established more precisely than has thus far been possible. Some of these modern techniques are now briefly considered.

NUMERICAL TAXONOMY

This approach attempts to use indirect methods, relying on phenotypic characteristics to estimate the genetic similarity of organisms. In numerical taxonomy the investigator runs a very large number of tests, each describing whether a certain character is present or absent in an organism. The proponents of numerical taxonomy hold that although some characteristics, such as shape and form, are more basic descriptions of an organism than an arbitrarily chosen set of biochemical characteristics, if *enough* characters are examined, there is no need for weighting different characteristics. Thus numerical taxonomy yields results that are unbiased by subjective judgments.

The final result of a classification by numerical taxonomy is expressed in terms of a *similarity coefficient,* defined as the percentage of the total number of characters tested that is held in common between two strains of microorganisms (Figure 11-1). In calculating the similarity coefficient, it is important not to include any characters which are negative or absent in both organisms, for the lack of a character should not imply relationship. For example, the absence of flagella in two organisms need not imply similarity, although flagellar presence and absence in two strains that are otherwise related are useful facts. It is also important

Strain Number	1	2	3	4	5	6	7
1	100						
2	5	100					
3	10	95	100				
4	0	90	95	100			
5	80	15	35	15	100		
6	70	25	40	10	80	100	
7	95	10	20	10	90	75	100

(a)

		A				*B*		
	Strain Number	1	7	5	6	3	2	4
A	1	100						
	7	95	100					
	5	80	90	100				
	6	70	75	80	100			
B	3	10	20	35	40	100		
	2	5	10	15	25	95	100	
	4	0	10	15	10	95	90	100

(b)

FIGURE 11-1 Similarity matrices. Seven strains of organisms were tested in 100 different ways, each test resulting in a positive or negative result. Each strain is then compared to the other strains by determining the similarity coefficient, the percentage of the total characters tested which are held in common in each of the strains. In (a) the strains are arranged randomly; in (b) they are arranged so that the organisms with similar matrices are grouped together. Note that the seven organisms fall into two unrelated groups. (Modified from J. Mandelstam and K. McQuillen, *Biochemistry of Bacterial Growth,* Oxford: Blackwell Scientific Publications, Ltd, 1968. By permission of the publisher.)

not to score characters which have a common basis. For example, for organisms that break down glucose through the TCA cycle, the pH of the growth medium does not become highly acidic, and this is true for most such organisms. When a large number of characters are examined in a large number of strains, a computer program is used to calculate similarity coefficients for each pair of organisms and to construct a similarity matrix. On the basis of this matrix, it is possible to arrange the strains in a hierarchy such that those that have more than a 90 percent similarity are termed a single species, while other more distinct groups are classified into different species and perhaps even different genera.

Numerical taxonomy has been used with some success with several groups of bacteria. For some groups of organisms the agreement of this technique with more classical methods or with results based on genetic and molecular methodology is quite impressive, whereas for others the agreement is poor.

MOLECULAR APPROACHES TO TAXONOMY

Recent developments in methods of classification are based on attempts to compare more directly the information contained in the DNA of different groups of organisms. Because the genetic information of a cell is encoded in its DNA, the relatedness of two organisms is directly related to the gross composition as well as to the sequence of purines and pyrimidines (base sequence) in their DNA. As the base sequence of DNA diverged, organisms diverged in evolution. Different proteins were then synthesized, and this resulted in organisms that no longer could mate with each other. Thus modern approaches to an evolutionary taxonomy have often involved (1) one or more of the comparisons of base sequences in DNA described in the next section, (2) the ability of two organisms to mate, or (3) a comparison of the amino acid composition of a specific protein or group of proteins in different organisms.

The Composition of DNA

DNA structure, Figure 8-1, pages 169–171

The structure of double-stranded DNA is such that the proportion of adenine (A) equals the proportion of thymine (T), and the proportion of guanine (G) equals that of cytosine (C). The relative proportions of the four bases are usually expressed as the percent of guanine plus cytosine (GC). This is calculated from the equation

$$\frac{\text{moles G} + \text{moles C}}{\text{moles G} + \text{moles C} + \text{moles A} + \text{moles T}} \times 100$$

DESCRIPTION OF BACTERIA INCLUDES THEIR GC CONTENT

The relative proportion of AT and GC base pairs varies widely among different organisms. In fact these variations in base composition are of considerable value for classification. *Escherichia coli* DNA has 50 percent GC while *Bacillus subtilis* DNA has 40 percent GC. These numbers mean that the two DNAs contain 50 percent and 60 percent AT base pairs, respectively.

Knowledge of the base composition of DNA has revealed some very interesting facts. The base composition of DNA is extremely variable, ranging from about 22 percent GC to about 78 percent GC. However, *organisms known to be related by other criteria have DNA base compositions which are similar or identical.* Thus if the base ratios of two organisms are widely dissimilar, that is, they have a difference of more than 10 percent in their GC, they cannot be closely related. However, it is important to realize that a *similarity* of base composition does not necessarily imply a relationship. This is because many sequences of bases are possible, even though the proportions of these bases may be identical. Thus one should not conclude that human beings are closely related to *B. subtilis* because both have 40 percent GC in their DNA. Obviously, the correspondence between the two 40 percent GC DNAs is fortuitous, tells nothing about the base sequence, and provides no real clue for classification.

Hybridization of DNA

When double-stranded DNA is heated, the two complementary strands dissociate or *denature.* If this mixture of single strands is then incubated for periods of time at the appropriate temperature, the strands can reassociate to form a double-stranded molecule whose properties are very similar to those of the original double-stranded DNA molecule (Figure 11-2). Because the reassociation reaction

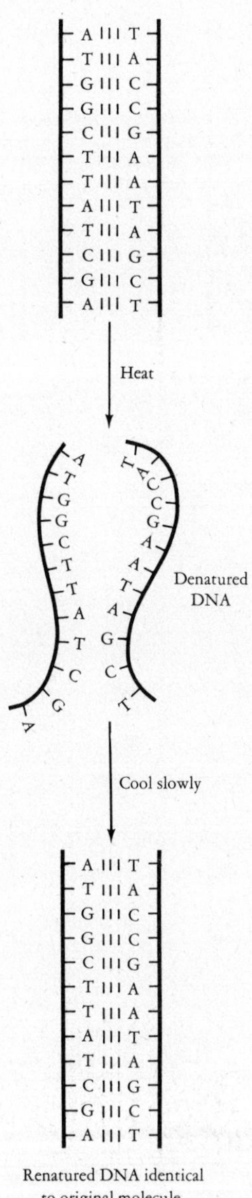

FIGURE 11-2
Dissociation and reassociation of double-stranded DNA.

HYBRIDIZATION MEASURES THE SIMILARITY OF BASE SEQUENCE IN DNAs OF RELATED ORGANISMS

requires that the two strands of DNA be complementary to re-form the double-stranded helix, the same procedure can be employed to assess the extent of similarity of base sequence in the DNAs of two related organisms. In this case single-stranded DNA from the two different organisms is incubated together. If the two organisms are closely related, the sequences of purine and pyrimidine bases in their DNA will be similar, and the single strands will associate to form a double-stranded molecule (Figure 11-3). Alternatively, one of the two strands may be mRNA rather than DNA, because the base sequence of an mRNA molecule is

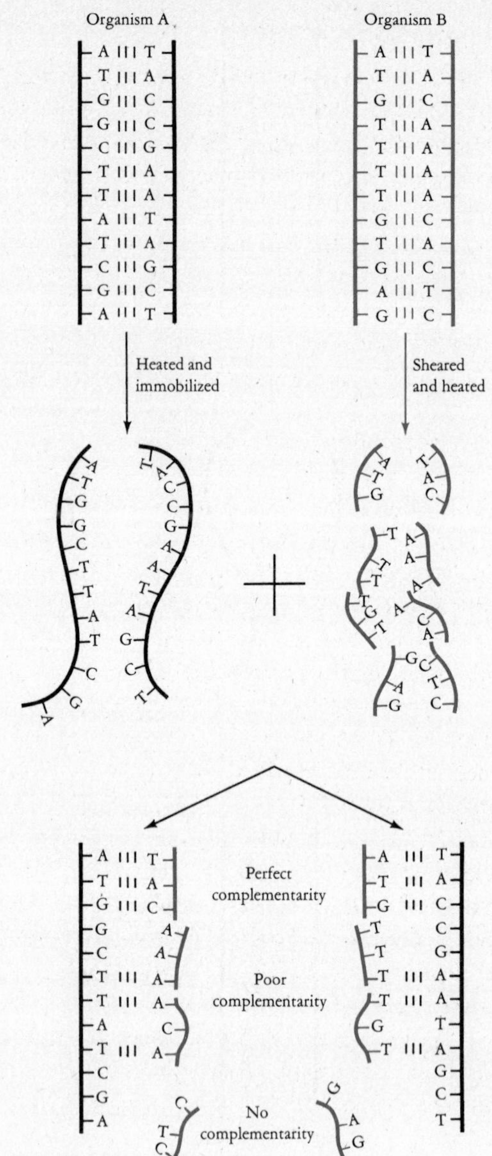

FIGURE 11-3
Measuring relatedness between organisms by the similarity in their DNA base sequences. Whether the DNA of organism A will hybridize with organism B depends on the DNA base sequences being complementary to one another. The single-stranded DNA can be readily separated from double-stranded DNA.

complementary to that of one of the two strands of DNA and identical to that of the other. This approach, referred to as *nucleic acid hybridization,* is a direct and powerful method for assessing relatedness between organisms.

At the present time nucleic acid hybridization has been applied as a means of classifying many groups of bacteria. Some groups of eucaryotic protists, such as fungi, have also been examined in this way. The method is also applicable to plants and animals.

In those groups of bacteria in which genetic exchange has been demonstrated, relationships may be explored by what amounts to *biological hybridization.* This approach, like that used in the taxonomy of higher plants, involves the formation of hybrids between two species. In various groups of bacteria, transduction, transformation, or conjugation—whichever is appropriate—is used as a taxonomic index. In principle each of these types of genetic exchange, although mediated by different transfer processes, depends on the integration of a piece of DNA originating from a donor organism into the chromosome of the recipient organism. Because this integration requires that the DNA of the donor and recipient be complementary to each other, the *efficiency* of the process is a measure of genetic relatedness.

Chromosome transfer among different strains by conjugation has been studied in members of the same bacterial group. Hybrids may be obtained within each of the genera *Escherichia, Shigella,* and *Salmonella.* These three genera appear to be closely related by DNA hybridization. However, no such biological hybridization is obtainable between *Escherichia* or *Salmonella* and *Enterobacter, Serratia,* or *Proteus,* even though these genera are detectably related by DNA hybridization. This example emphasizes the fact that *considerably higher degrees of base sequence homology are necessary for recombination to form biological hybrids than are required for DNA hybridization in the test tube.*

Techniques involving DNA-mediated chromosome transformation are also now being developed, and these should help identify certain bacterial species quickly and accurately in hospital laboratories. These tests are based on the fact that for chromosomal DNA to be successfully donated, the donor bacteria must be of the same species or even variety as the recipient cells. Therefore it is necessary only to lyse the donor cells and add the crude lysate to the recipient test strain plated on a selective medium on which only transformants will grow. If the donor and recipient strains belong to the same species, colonies of transformants will arise. This general technique has been used successfully to distinguish between certain closely related species of Gram-negative cocci, some of whose members cause gonorrhea and meningitis. The major difficulty with the technique is that it never eliminates the possibility that members of closely related species may give a low but positive level of transformants. It may therefore be necessary to run further tests to insure that the original diagnosis is correct. There are two advantages to this type of analysis: (1) the bacteria do not have to be purified before the test is run, and (2) less than 100 donor cells is sufficient to yield positive results.

Neisseria,
page 433

Proteins as Tools for Molecular Taxonomy

A single gene can be described either in terms of its base sequence or in terms of the amino acid sequence of the protein molecule for which it codes. Thus a comparison of the amino acid sequences of corresponding proteins from two organisms provides insight into the relationship between the genes that code for this type of protein. Although this approach is precise, it is limited to a small portion of the total genetic information of the cell, in contrast to the overall estimation of DNA homology that can be made by DNA hybridization. Furthermore, although the complete amino acid sequences have been determined for many proteins, few of these are of microbial origin.

DIAGNOSTIC MICROBIOLOGY

METHODS OF IDENTIFYING BACTERIA

The major goal of a clinical diagnostic laboratory is to quickly identify, as an aid to diagnosing a disease, the agents which are isolated from patients. The recognition of an organism among a relatively limited number of possibilities is much more important than genetic interrelatedness. Thus, under clinical laboratory circumstances, an organism is characterized in as many ways as is practically possible. Based on the results of these tests, the organism is identified as to genus and perhaps species. It is important to recognize that the number of tests that can reasonably be done under such circumstances represents the gene products of only a very small fraction of the total genome of the organism. The more tests that are done the more accurate the identification. The major ways of characterizing organisms, especially in diagnostic microbiology, will now be considered.

Colonial Characteristics

Appearance (size, shape, color, degree of opacity), consistency (buttery, gumlike, easily fractured), and effect on medium (pitting, adherent, color-changing) often are part of the first assessment of an unknown species and determine which additional identification procedures need be undertaken. For example, a 2-mm-in-diameter, cream-colored, buttery colony that produces a small amount of clearing of blood agar medium is likely to be *S. aureus,* and it usually requires only one or two specific additional tests to confirm its identification. Colonial characteristics may depend greatly on medium, duration, temperature, and atmosphere of incubation, and these factors must be taken into account.

Microscopic Analysis

Cell shape, page 55

Shape of cells. Although the number of general shapes of microorganisms is small relative to the enormous number of species, shape may be used to identify a microbe within narrow limits. However, morphology of many species varies greatly with the type of medium employed, and the stage of growth and these factors must be taken into account.

Cell arrangement, Figure 3-6, pages 55, 57

Cell arrangement. These arrangements can often be quite useful diagnostically in distinguishing between organisms that have other properties that are similar. Thus *Streptococcus* growing in broth culture produces chains, while *Staphylococcus* characteristically shows a "bunch of grapes" arrangement.

Inclusions, spores, and capsules. Some of these structures are almost diagnostic for certain species. For example, the terminal endospore of *Clostridium tetani* helps distinguish this species from other Gram-positive, rod-shaped bacteria.

Staining. The Gram stain and acid-fast stains are important stains for dividing bacteria into large groups. It is important to realize that cultures tend to change from Gram-positive to Gram-negative as the cultures reach the stationary phase of growth.

Gram stain, Figure 3-4, pages 53–54

Motility. The determination of whether an organism is motile or not may be helpful in its identification. The characteristics of movement are often distinctive. A bacterium with unipolar flagella moves rapidly with a single straight course while organisms with peritrichous flagella tend to move in a zig-zag pattern.

Flagella arrangement, Figure 3-27, page 74

Biochemical Tests

Many ingenious tests identify specific biochemical capabilities of microbial cells. These tests include the detection of breakdown products of sugars and other metabolites such as acids and various gases, as well as the ability to grow on certain nutrient sources of carbon and energy. A partial listing of some of the more important of these tests is given in Table 11-1.

Requirements for Growth

Nutritional requirements. Some organisms (obligate intracellular parasites) grow only inside of living cells whereas most microorganisms are able to grow on

TABLE 11-1
Some Biochemical Tests for Identifying Bacteria

Name of Test	*Reaction*	*Observed Reaction*
Voges-Proskauer	Metabolism of glucose to 2,3-butylene glycol	Red color develops upon addition of chemicals which detect acetoin, a precursor of 2,3-butylene glycol
Gelatinase	Enzyme breakdown of gelatin to polypeptides	Solid gelatin converted to liquid
Hydrogen sulfide production	Enzymes convert the S in sulfur containing amino acids to H_2S	Black precipitate forms in the presence of iron salts
Phenylalanine deaminase	Enzyme removes the amino group from the amino acid phenylalanine	Phenylpyruvic acid reacts with ferric chloride to give a green color
Lysine decarboxylase	Carboxyl group removed from the amino acid lysine	Medium becomes more alkaline, and a pH indicator changes color
Indole	Amino group removed from the amino acid tryptophan	The product, indole, changes the color of a chemical reagent to a red color

medium in the absence of living cells. Many microbes of medical importance, however, have unique growth requirements. Thus *Haemophilus influenzae* must be supplied NAD and hemin as growth factors. Many microbes require increased levels of carbon dioxide.

O_2 requirements, Figure 4-7, pages 97–98

Requirements for O_2. Microbes have all degrees of requirement for oxygen. Some organisms have an absolute requirement, others get along without or tolerate it, and still others are killed by it. The identification of this requirement provides important diagnostic information.

Immunological techniques, pages 390–397

Use of antisera. Antibodies can often detect subtle chemical variations in microbial structures, and can even detect toxins or enzymes which are not structural components of the microorganism. Such antisera may help distinguish one species from another, as well as distinguish strains within a species. Capsules, cell walls, and flagella are commonly identified by their ability to elicit the formation of specific antibodies which will then react with them in agglutination, precipitation, or flocculation reactions. All strains of a species possessing the same antigens compose a *serotype.* Over 1000 serotypes of *Salmonella* can be distinguished based on cell wall, capsular, and flagellar antigens.

Other Biological Reagents

Bacteriocins, page 211

Bacteriophage and bacteriocins. Besides using antisera, differences in the *surface* composition of closely related microbes can be detected using bacteriophage and bacteriocins. Both agents require specific receptors on the cell surface in order to attach and kill the cell. Because these substances are active only against a narrow range of organisms, this method is useful in characterizing different strains of the same species.

Antimicrobial agents. Antimicrobial agents such as dyes, chemicals, and antibiotics are also useful in distinguishing both between strains of the same species as well as between similar species. For example, optochin (a derivative of quinine) and bile salts are very toxic to *Streptococcus pneumoniae,* but the closely related viridans streptococci are quite resistant to both. Pathogenic organisms, isolated from patients, are often tested with several antibiotics to determine their pattern of susceptibility to these drugs prior to treating the patient. These susceptibility patterns have been used to provide a quick means of identifying epidemic strains of *Staphylococcus aureus.*

The exact procedure that a diagnostic clinical laboratory will use to identify an organism depends on many factors. Some of these include the capabilities of the laboratory, the condition of the patient, the symptoms of the disease, and the time required for the test.

SUMMARY

Although there are problems in defining a microbial species, a working definition is a group of strains, clones, or cultures that has many properties in common yet is easily distinguished from other species. The traditional schemes of bacterial classification, in which organisms that share common properties are grouped together, are convenient. However, because the members of each group also share properties with members of other groups, the decision as to which properties take precedence in classification is quite arbitrary. Although traditional classification schemes are useful for providing a description of each species, such taxonomic schemes are not based on evolutionary relationships.

Probably the most accurate method for assessing the relatedness of organisms is based on the nature of their DNA. The DNAs of closely related organisms have similar base compositions. A more quantitative estimate of their relatedness can be gained by determining the degree of base sequence homology in their DNA or RNA. This can be accomplished in vitro by studying the degree of reassociation of single-stranded DNA from each of the two organisms or by determining the efficiency of the organisms' DNAs in undergoing genetic recombination with one another.

In the clinical diagnostic laboratory, microorganisms are identified by characterizing them mainly in regard to their morphology and their staining and biochemical characteristics.

QUESTIONS

REVIEW

1. Discuss the reasons why bacteria, as compared to higher forms of life, present special problems in classification.
2. What are some features of bacteria that are important in identifying them?
3. Explain the reason why nucleic acid hybridization serves to indicate relatedness between organisms.
4. Explain the theory behind classification by numerical taxonomy.
5. What conclusions can we draw on the relatedness of organisms that have the same overall base composition?

THOUGHT

1. Why may plasmids be transferred between organisms of different species even though the chromosomes of the organisms will not undergo genetic recombination?
2. Do you think that the ability of an organism to grow in air is more important in its classification than the ability of the organism to grow on lactose as a carbon source? Why or why not?

FURTHER READING

BUCHANAN, R. E., and N. E. GIBBONS, eds., *Bergey's Manual of Determinative Bacteriology.* 8th ed. Baltimore: Williams and Wilkins, 1974. This manual is the major book to consult when trying to identify an unknown bacterium. It describes each species as part of a hierarchical classification scheme. The introductory chapter, "A Place for Bacteria in the Living World," discusses the problem of bacterial classification and is well worth reading.

DAYHOFF, M. O., "Computer Analysis of Protein Evolution," *Scientific American* (July 1969).

DICKERSON, R., "The Structure and History of an Ancient Protein," *Scientific American* (April 1972).

LENNETTE, E., E. SPAULDING, and J. TRUANT, eds., *Manual of Clinical Microbiology.* 2d ed. Washington, D.C.: American Society for Microbiology, 1974. The most up-to-date and useful book for identifying disease-causing organisms. The book, written by 125 authorities, provides practicing clinical laboratory microbiologists with the most recent techniques of collecting and examining clinical specimens, as well as for identifying the suspected organism. This is a reference that students in the allied health programs should know about.

MACFADDIN, J. F., *Biochemical Tests for Identification of Medical Bacteria.* Baltimore: Williams and Wilkins, 1976. A manual that provides considerable information on the tests used to identify bacteria. Includes the principles of the various tests, their purposes, the biochemistry involved, the media and reagents employed, as well as interpretations of and references to the tests. This is an invaluable reference.

SOKAL, R., "Numerical Taxonomy," *Scientific American* (December 1966).

SPIEGELMAN, S., "Hybrid Nucleic Acids," *Scientific American* (May 1964).

CHAPTER 12

BACTERIA

Bacteria are ubiquitous, inhabiting virtually every ecological niche capable of supporting life. The eighth edition of *Bergey's Manual of Determinative Bacteriology* lists only 1576 recognized bacterial species falling into 245 genera, but hundreds of others are also mentioned which are of uncertain taxonomic position. In addition, more than 200 distinct groups of cyanobacteria have been identified. It has been proposed that all the procaryotic microorganisms (that is, the bacteria) should be classified within a new kingdom, *Procaryotae.*

In this chapter the bacteria are divided into three groups, depending on their principal energy source. Members of one group obtain energy from chemical reactions involving organic materials. A second group utilizes inorganic chemicals. The members of the remaining group trap energy from sunlight. Readers are referred to Appendix II for the more detailed classification used in *Bergey's Manual.*

Procaryotic cell type, page 13

BACTERIA USING ORGANIC CHEMICALS FOR ENERGY

Bacteria that require organic substances as energy sources represent a vast and well-studied group with a wide spectrum of metabolic capabilities. A large number of these species are important in food, agricultural, veterinary, and medical microbiology, and some of these will be discussed in detail in later chapters. As with the other two major groups of bacteria, those using organic chemicals for energy include both unicellular and multicellular types of various sizes and shapes. For convenience in discussion we will divide the bacteria using organic chemicals into eight principal morphological types: (1) unicellular rods and

TABLE 12-1
Morphological Types of Bacteria Using Organic Compounds for Energy

Type	*Characteristics*
Unicellular rods and cocci	More or less cylindrical or spherical; occur as single cells or loosely adherent chains or clumps
Endospore-forming	Generally resemble the above except for the formation of endospores under certain growth conditions; most are Gram-positive bacilli, but coccoid and Gram-negative species are known
Budding or prosthecate	Usually asymmetrical reproduction; prosthecae contain cytoplasm; in some species attach to larger objects
Spiral	Helical or wave-shaped, single or multicelled, with rigid or flexible cell walls
Branching filamentous	Multicelled branching filaments; some fragment readily into single cells at some stage of growth; some produce endospores or conidia
Nonbranching filamentous	Multicelled filaments; many show gliding motility; some are enclosed by a sheath
Myxobacteria	Usually unicellular rods that aggregate and produce fruiting bodies, often with specialized resting cells
Mycoplasmas	Lack a cell wall, are easily deformed; include the smallest cells known

cocci, (2) endospore-forming, (3) budding or prosthecate, (4) spiral, (5) branching filamentous, (6) nonbranching filamentous, (7) myxobacteria, and (8) mycoplasmas (Table 12-1).

Unicellular Rods and Cocci

Gram stain, page 53

Anaerobic respiration, page 159

Pseudomonads. Unicellular rods and cocci that use organic materials for energy (Table 12-2) include the pseudomonads, aerobic Gram-negative rods which are motile by polar flagella and often produce nonphotosynthetic pigments. Although most are strict aerobes, a few can grow anaerobically in the presence of nitrate, which substitutes for oxygen as the final electron acceptor, an example of anaerobic respiration. Many species of pseudomonads are free-living and harmless, while others can produce disease in plants and animals. The genus *Pseudomonas* contains species with remarkably diverse biochemical capabilities. Compounds used by members of this genus include unusual sugars, amino acids, and far more complex compounds such as those containing aromatic rings. Thus

TABLE 12-2
Examples of Simple Unicellular Bacteria Using Organic Compounds for Energy

Group	*Characteristics*
Pseudomonads	Respiratory Gram-negative rods; can metabolize a wide diversity of organic compounds
Azotobacteria	Respiratory Gram-negative rods; fix atmospheric nitrogen; form resting cells called *cysts*
Bdellovibrios	Respiratory Gram-negative curved rods; prey on other bacteria
Enterobacteria	Gram-negative rods; facultative anaerobes that employ respiratory and fermentative metabolism; the nature of flagella, cell wall, and capsules and biochemical tests distinguish numerous species; *E. coli* is an example
Vibrios and related	Gram-negative rods, often curved, with polar flagella; some are luminescent
Obligately parasitic	Under natural conditions require a higher organism host to supply nutrients; some can only live inside host cells, most are Gram negative
Lactic acid bacteria	Gram-positive rods and cocci that require rich growth media and produce lactic acid as a major end product of carbohydrate fermentation; streptococci and lactobacilli are important subgroups
Arthrobacters	Respiratory pleomorphic, Gram-positive rods that have a tendency to stain Gram negatively in actively growing cultures; spherical cells formed as culture ages

Pseudomonas species play an important role in the degradation of many man-made and natural compounds that are refractory to breakdown by most other microorganisms. The ability to carry out some of these degradations depends on plasmids.

Azotobacteria. Another group of unicellular Gram-negative organisms, the azotobacteria, is unique in being able to fix atmospheric nitrogen under aerobic conditions. Nitrogenase enzymes, which are needed for nitrogen fixation, are ordinarily sensitive to traces of oxygen. However, azotobacteria have exceedingly high respiratory rates, and their metabolism probably consumes oxygen that might otherwise inhibit their nitrogenase enzyme. Members of the genus *Azotobacter* live in alkaline soil. They are able to form a type of resting cell called a *cyst*. Each cyst is formed through division and shortening of a vegetative cell, followed by the elaboration of a thick protective wall which surrounds the cell (Figure 12-1). Thus the azotobacteria represent one of the few groups of unicellular bacteria able to produce a resting cell. Cysts are resistant to drying and ultraviolet irradiation but are not highly resistant to heating; they differ from endospores in both their formation and their lesser degree of resistance to deleterious agents. Furthermore, whereas the formation of endospores involves a sequential series of steps involving the regulation of the expression of a large number of genes, the

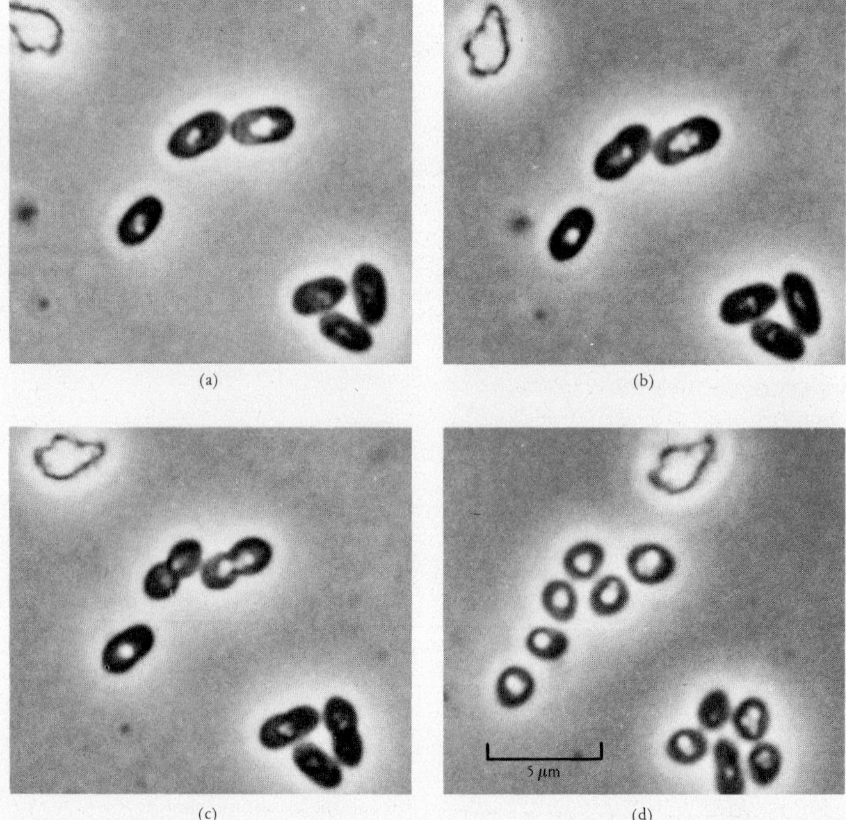

FIGURE 12-1
Development of cysts of *Azotobacter* (a) 30 min after inoculation, (b) $2\frac{1}{2}$ hr after inoculation, (c) $3\frac{1}{2}$ hr after inoculation, (d) 8 hr after inoculation. (H. L. Sadoff, E. Berke, and B. Loperfido, *J. Bacteriol.* 105:184, 1971.)

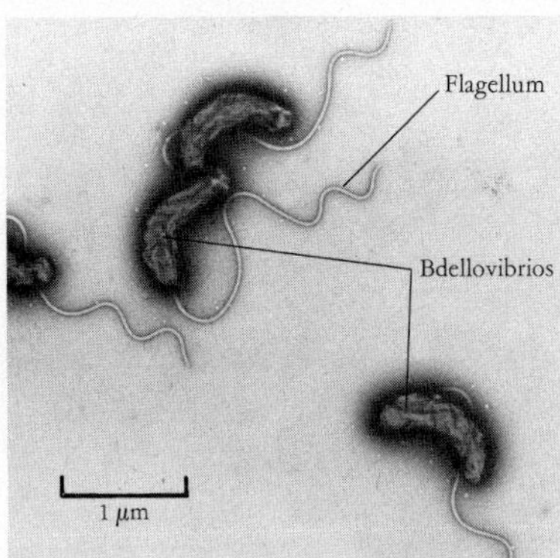

FIGURE 12-2
Bdellovibrio bacteriovorus. Note the small size and single thick flagellum. (S. C. Rittenberg, *J. Bacteriol.* 109:432, 1972.)

formation of cysts primarily involves changes in the cell wall. The enzyme content of the cyst is very similar to that of the cell from which it arises.

Bdellovibrios. Yet another very interesting aerobic group, the bdellovibrios, preys on other bacteria. Bdellovibrios in many instances require a specific host bacterium as their only food source. They are also unique in appearance, being tiny Gram-negative, curved rods possessing a long, sheathed flagellum at one end of the cell. The generic name *Bdellovibrio* describes the organism's behavior and shape: they attach to other bacterial cells (bdello, from the Greek word for "leech") and have a comma shape like bacteria of the genus *Vibrio. Bdellovibrio bacteriovorus* (Figure 12-2), the most intensively studied species of this group, is approximately 0.25 μm wide and 1 μm long. *Bdellovibrio* attacks by striking at high velocity so that its nonflagellated end attaches to the host cell. The *Bdellovibrio* then penetrates through a hole made in the host cell wall by the use of digestive enzymes and a spinning motion. The flagellum is left behind. The *Bdellovibrio* then lodges between the host's cell wall and cytoplasmic membrane, growing in length at the expense of the cellular contents of the host. Bdellovibrios utilize components of the host after they are only partially degraded, thereby saving the energy that would be needed for complete resynthesis. Following division, the *Bdellovibrio* progeny become motile and leave to find new hosts and repeat the growth and reproduction cycle (Figure 12-3). Some bdellovibrio strains are able to grow heterotrophically in the absence of host cells.

Heterotrophy, page 96

One surprising fact about *Bdellovibrio* is that it was discovered only as recently as 1962, even though this type of organism can now be easily isolated from soil, sewage, and natural waters. Viruses, which are not even visible with the light microscope, were discovered in 1892 and rickettsiae in 1909. How did *Bdellovibrio* escape the detection of so many scientists using sophisticated microscopic techniques for all those years? The major reason for this oversight is probably that there was no a priori reason to believe that such a creature existed. Although many people had undoubtedly observed this organism while examining the microbial flora in soil, there was no inkling that it was anything more than a typical, albeit small, bacterial cell.

Facultative anaerobes, page 97

Enterobacteria. The enterobacteria are facultatively anaerobic, Gram-nega-

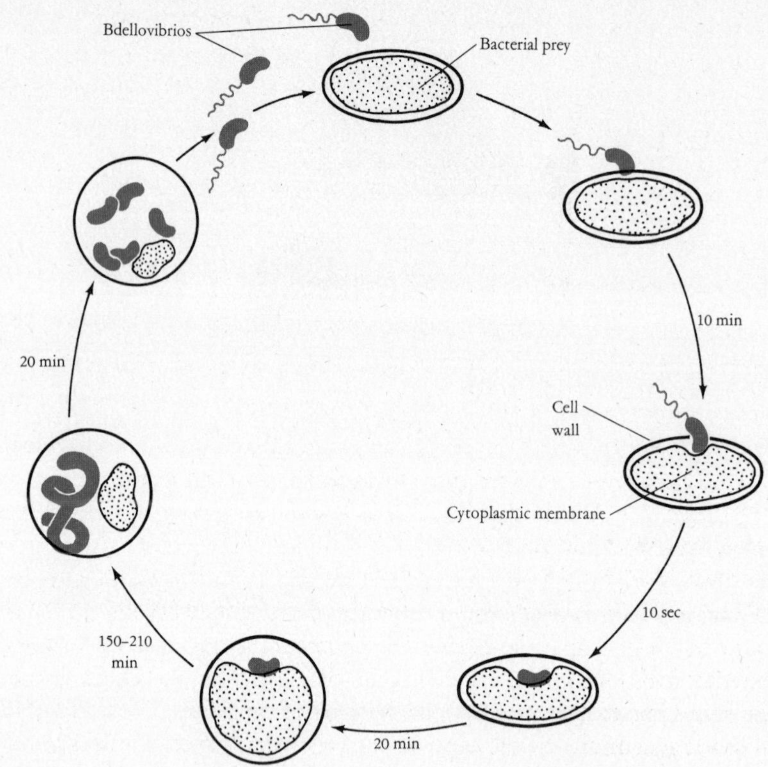

FIGURE 12-3
Life cycle of *Bdellovibrio bacteriovorus*. Note that the bdellovibrio multiplies between the cell wall and cytoplasmic membrane of the bacterial cell upon which it preys. (Adapted from an illustration courtesy of S. C. Rittenberg.)

tive rods. They include a large number of medically important species, as well as plant pathogens and saprophytes. *Escherichia coli* is the most extensively studied member of this group.

The different enterobacteria are distinguished partly on the basis of biochemical tests (Table 12-3) and partly by antigenic structure. The general structure of an enterobacterial cell is shown in stylized form in Figure 12-4. Note that

TABLE 12-3
Use of Biochemical Tests to Identify Species of Enterobacteria[a]

Bacterium	*Indole*	*Methyl Red*	*Voges-Proskauer*	*Citrate Utilization*	*Lactose Fermentation*	*Urease*
Escherichia coli	+	+	−	−	+	−
Proteus mirabilis	−	+	−	+	−	+
Enterobacter aerogenes	−	−	+	+	+	−
Serratia marcescens	−	−	+	+	−	−

+ = positive test; − = negative test.
[a] In practice numerous additional biochemical tests may be employed. Reactions shown here are the usual results; individual strains may vary slightly.

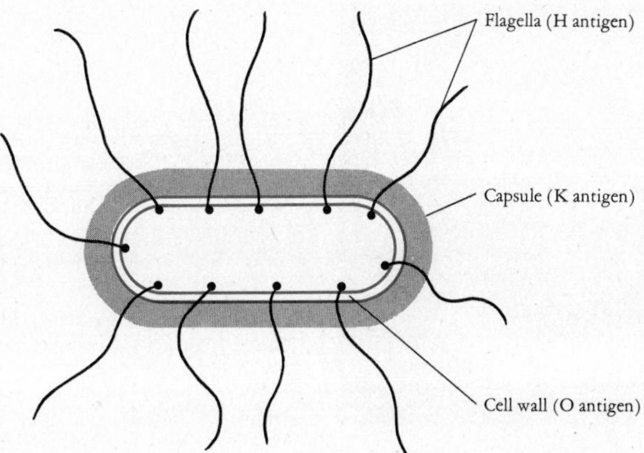

FIGURE 12-4
Schematic drawing of a typical enterobacterial cell (the cytoplasmic membrane is not shown).

the typical organism has three main antigenic components (these are antigens because they stimulate production of specific antibodies when the organisms are injected into an animal; see Chapter 20). The three types of antigens are chemically distinct. The cell wall antigen (called the "O" antigen) is a lipoprotein-polysaccharide complex, the flagellar antigen (called the "H" antigen) is protein, and the capsular antigen (called the "K" antigen) is polysaccharide. Each of these substances varies in chemical composition not only from species to species but even from strain to strain within a species. Identifying the specific kinds of O, H, and K antigens of an organism can therefore be very helpful in identification of the species or strain.

Vibrios and related bacteria. Vibrios and related bacteria comprise another large group of Gram-negative facultative anaerobes. Some cause diseases of animals and human beings. Both curved (Figure 12-5) and straight rods occur

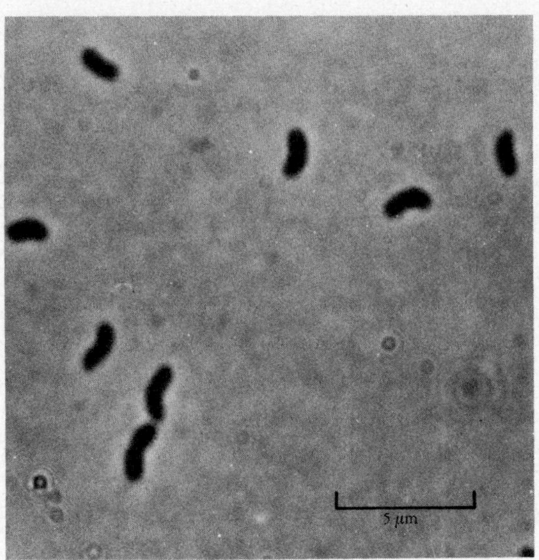

FIGURE 12-5
Vibrio cholerae, a curved bacterium, which causes cholera. (Phase contrast photomicrograph courtesy of J. T. Staley and J. P. Dalmasso.)

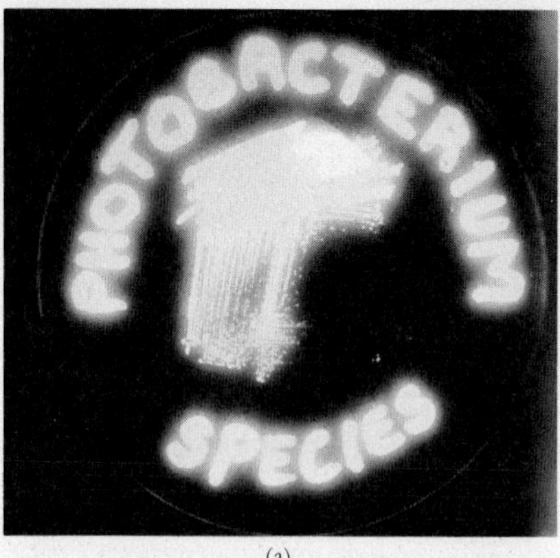

(a)

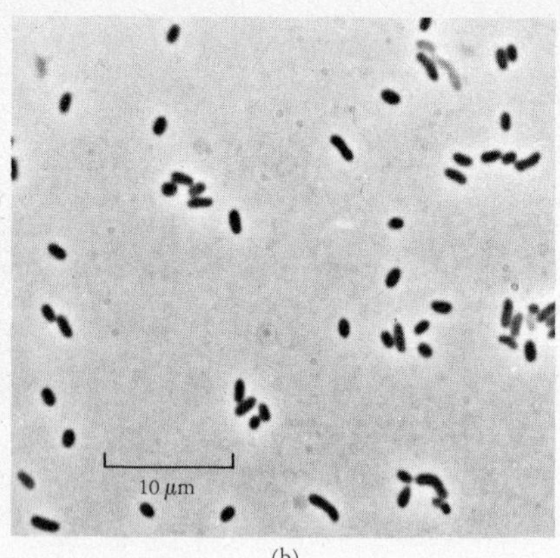

(b)

FIGURE 12-6
Photobacterium species. (a) This photograph was taken using light produced by the bacteria themselves. (b) Phase photomicrograph of cells. (Courtesy of J. T. Staley and J. P. Dalmasso.)

within this group, and some species are luminescent (Figure 12-6). The latter are represented by species of *Photobacterium,* salt-requiring marine organisms that produce light in the presence of oxygen. The reaction is mediated by the enzyme luciferase. Photobacteria live in specialized pouches on certain fish and are thus responsible for these fish's luminescent spots.

Parasitic bacteria. Many species of unicellular bacteria utilize the tissues or body fluids of higher organisms and are therefore parasites. The parasitic mode of existence offers advantages to a microorganism because the host provides substances that would otherwise have to be synthesized by the microorganism. Once the problem of ensuring its transfer from host to host is solved, the parasitic bacterium has no further need of the biochemical machinery for synthesizing these substances; therefore, as the parasitic relationship evolves, this excess machinery is dropped. Such is the case, for example, with *Haemophilus influenzae,* which lacks the ability to synthesize NAD and hemin and therefore requires these components of blood to grow on laboratory media. Other species have even mastered the problem of growing within the cytoplasm of host cells, although retaining the ability to grow without, while still others *require* the intracellular environment in order to reproduce. Examples of obligate intracellular bacteria include the medically important species of *Chlamydia* and *Rickettsia* (Chapter 32). At a simpler level other obligate intracellular bacteria infect protozoa. Some of these bacteria infecting protozoa fail to cause apparent harm to their hosts and even approach the end stage of parasitic evolution in conferring survival advantage to the host. This has caused considerable speculation about what the ultimate fate of such an *endosymbiont* might be in eucaryotic hosts. There is speculation, for example, that the chloroplasts of algae represent the remnants of ancient procaryotic photosynthetic endosymbionts, and indeed, in one species of red algae, the chloroplasts have traces of procaryotic cell wall material.

NAD, page 153

Hemin, page 102

Lactic acid bacteria. The *lactic acid bacteria* comprise an important group of Gram-positive unicellular bacteria that utilize organic energy sources. They commonly occur loosely associated in chains of bacteria, require rich growth media, and produce lactic acid as a major end product of their metabolism of sugars. The lactic acid bacteria include both the streptococci and the lactobacilli. Lactobacilli are nonsporeforming, rod-shaped bacteria, common among the microbial flora of the mucous membranes of human beings and especially abundant in the vagina during the childbearing years. These bacteria have virtually no disease-producing ability, although they may contribute to dental caries under certain conditions. Lactobacilli tend to prefer acidic and relatively anaerobic conditions. In addition to their animal habitats they are commonly present in decomposing plant material and in milk and other dairy products. They are important in such processes as pickling and cheese-making.

Arthrobacters. The nonsporeforming, Gram-positive rods of the genus *Arthrobacter* represent an important group of soil bacteria. Strains of some species of *Arthrobacter* are largely responsible for the degradation of insecticides and herbicides in the soil. During growth members of this genus may undergo transitions in which the shape of the cells is markedly altered. In the stationary phase the cells are essentially spherical, but when such a culture is diluted into fresh medium, the spherical cells elongate and form the irregular rods characteristic of coryneform (club-shaped) bacteria. The rods once again revert to spherical cells when the culture reaches the stationary phase (Figure 12-7). The molecular basis for these transitions probably lies with variations in the structure of the cell wall, perhaps in the degree of crosslinking of peptidoglycan layers. Such changes in the cell wall may also account for the unusual property of cells of *Arthrobacter* to stain Gram negatively in young cultures, whereas the spherical cells of old cultures are Gram positive. This is the opposite reaction to that which most species undergo as a culture ages.

Stationary phase, page 109

Peptidoglycan, page 60

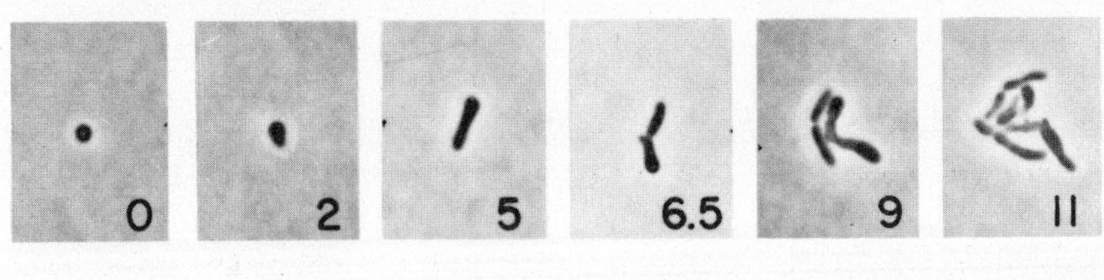

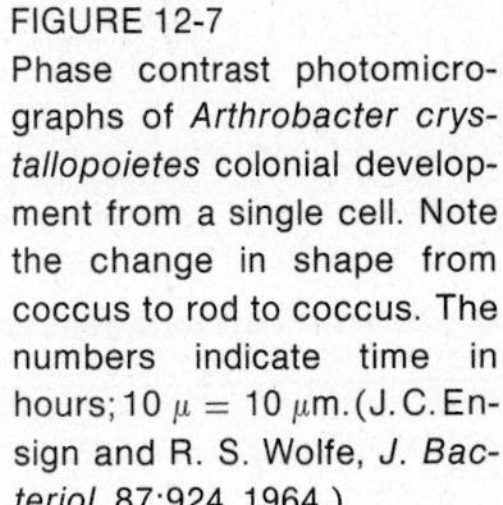

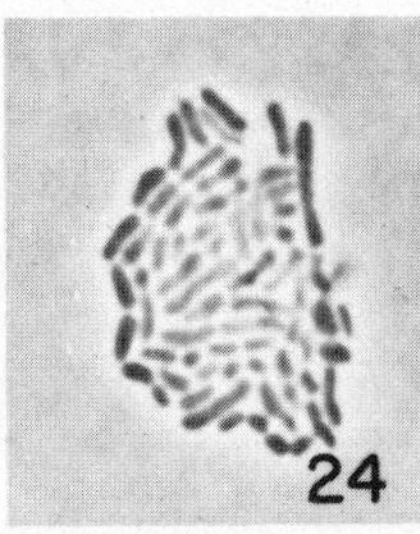

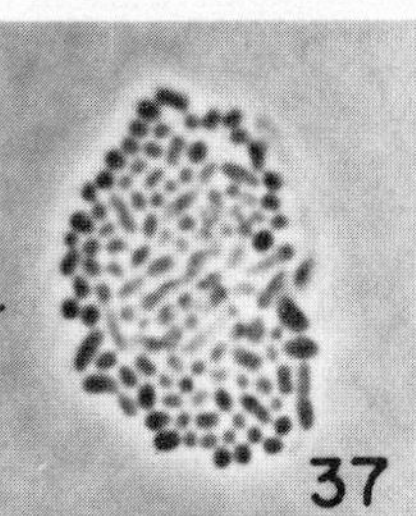

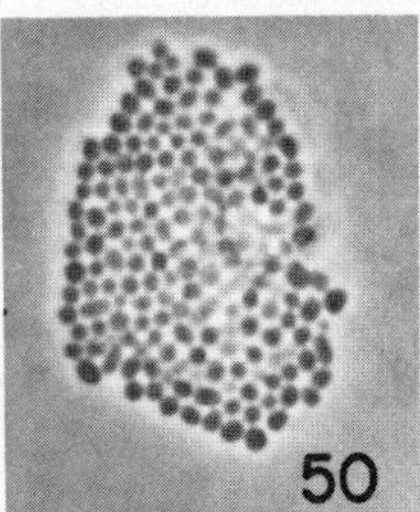

FIGURE 12-7
Phase contrast photomicrographs of *Arthrobacter crystallopoietes* colonial development from a single cell. Note the change in shape from coccus to rod to coccus. The numbers indicate time in hours; 10 μ = 10 μm. (J. C. Ensign and R. S. Wolfe, *J. Bacteriol.* 87:924, 1964.)

Endospore-forming Bacteria

Endospore formation, page 232

Only a few genera of bacteria are known to produce endospores, but they are very widespread in the environment as well as in the intestines of animals. Endospore-formers include two genera of Gram-positive, rod-shaped bacteria, *Bacillus* and *Clostridium. Bacillus* species are aerobic or facultatively anaerobic, whereas clostridia are obligate anaerobes, in some cases killed by even the faintest traces of oxygen. Endospores represent the most resistant form of life known; they tolerate extremes of heat and dryness, the presence of disinfectants, and radiation. Some members of *Bacillus* and *Clostridium* play a role in fixing atmospheric nitrogen, and others cause serious infectious diseases. Thermophilic strains of *Bacillus* can grow at temperatures above 70°C (158°F).

Branching Filamentous Forms

The actinomycetes are responsible for decomposition of much of the complex organic material in soil. They are characterized by their branching multicellular filaments. Members of the genera *Nocardia* and *Streptomyces* are mainly responsible for the characteristic odor of soil. A number of *Streptomyces* species are important antibiotic producers (Chapter 33). A distinctive feature of the *Streptomyces* is the presence of strings of specialized reproductive spores, called *conidia,* at the ends of their filaments (Figure 12-8). The filaments themselves (called *hyphae*) are coenocytic, meaning that the cytoplasm of the cell is continuous with that of adjacent cells. By contrast, members of the genus *Nocardia* (Figure 12-9) lack conidia, and their hyphae readily fragment into individual rod-shaped cells. *Nocardia* species often show acid-fast staining and are closely related to two groups of unicellular bacteria, mycobacteria and corynebacteria.

Nonbranching Filamentous Forms

Some bacteria that utilize organic energy sources occur as multicelled unbranched filaments. Members of one genus, *Simonsiella,* are harmless residents of the mouths of humans and numerous other warm-blooded animals. Species of *Simonsiella* occur as short chains of flattened cells in a ribbon-like arrangement (Figure 12-10). When in contact with a solid surface, these microorganisms show *gliding motility,* a slow progressive movement across the surface. Extrusion of a mucus-like material from the surface of the bacterium is thought to be responsible for this motion, but its exact mechanism is unknown.

Sheathed bacteria comprise another group of multicelled filamentous organisms. They differ from other nonbranching filamentous forms in that their chain of bacterial cells is enclosed in a sheath composed of a lipoprotein-polysaccharide complex that is chemically distinct from the bacterial cell walls. *Sphaerotilus* (Figure 12-11) is a very widespread genus in this group. The filaments of members of this genus often produce masses of brownish scum that may be seen beneath the surface of polluted streams; they interfere with sewage treatment proc-

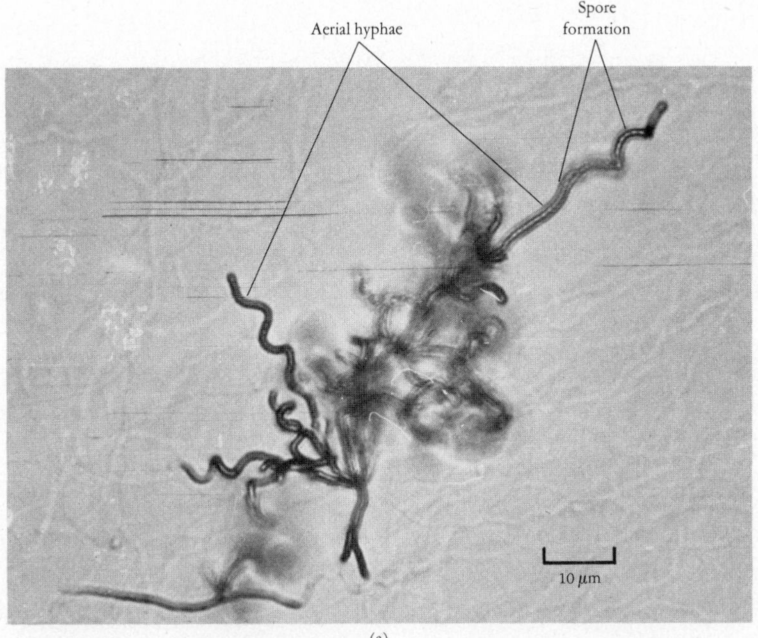

(a)

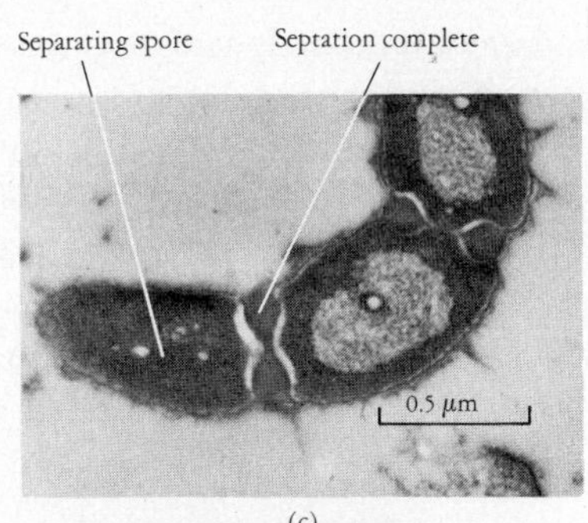

(b) (c)

FIGURE 12-8
Reproduction of *Streptomyces* species. (a) *Streptomyces fradiae*. (Courtesy of M. P. Lechevalier and H. A. Lechevalier.) (b, c) Electron micrographs showing steps in formation of spores called conidia (From M. W. Rancourt and H. A. Lechevalier. Reproduced by permission of the National Research Council of Canada. From *Canad. J. Microbiol.* 10:311–316, 1964.)

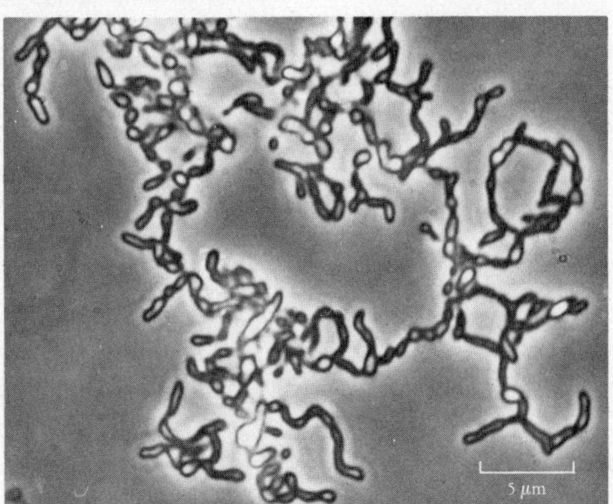

FIGURE 12-9
Phase contrast photomicrograph of *Nocardia asteroides*. Note the presence of branching. (Courtesy of J. T. Staley and J. P. Dalmasso.)

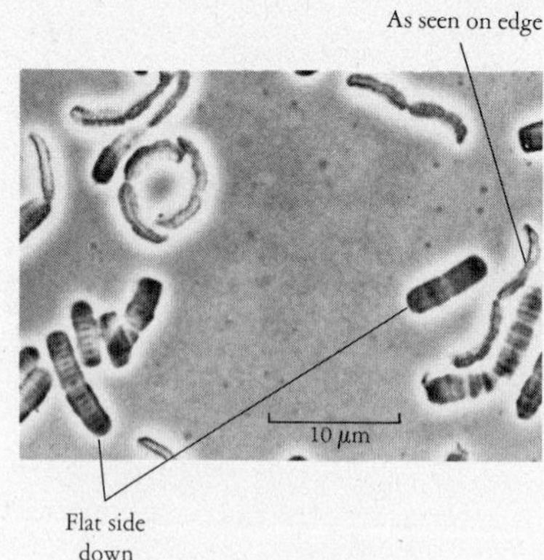

FIGURE 12-10
Simonsiella species, a nonbranching multicellular organism that uses organic energy sources. (Courtesy of J. T. Staley and J. P. Dalmasso.)

esses by plugging pipes. Reproduction of *Sphaerotilus* occurs by binary fission, polarly flagellated bacilli being released from the ends of the sheath. These bacilli then reproduce, secreting a new sheath. In freshwater habitats containing iron or manganese, ferric hydroxide or manganese oxide may be deposited on the sheaths of members of *Leptothrix,* the other common genus of sheathed bacteria.

Myxobacteria

Myxobacteria are strictly aerobic Gram-negative rods which employ gliding-type motility. They have a developmental cycle unique among procaryotes (Figure 12-12). At a certain time in the life of a culture of myxobacteria, presumably dictated by exhaustion of nutrients, some of the cells begin to congregate. The chemical signal for this congregation is thought to be cyclic AMP. The congregated cells pile up, finally producing a mass of cells which is supported above the growth medium by a stalk. This structure, called the *fruiting body* (Figure 12-13), may be brightly colored. The mass of cells on the stalk then differentiates into

Cyclic AMP, page 231

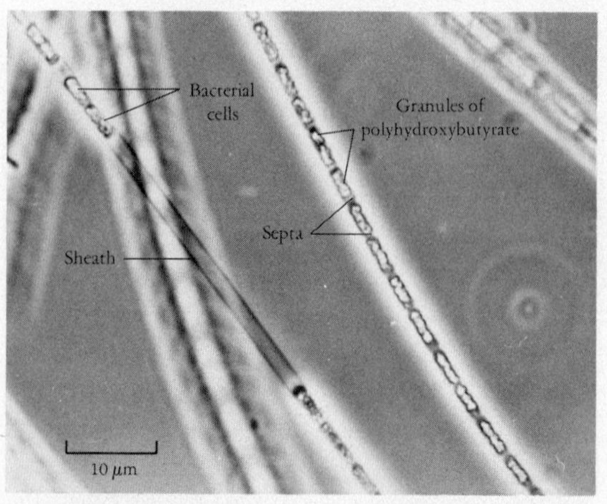

FIGURE 12-11
Sphaerotilus species, a sheathed bacterium. (Phase contrast photomicrograph courtesy of J. T. Staley and J. P. Dalmasso.)

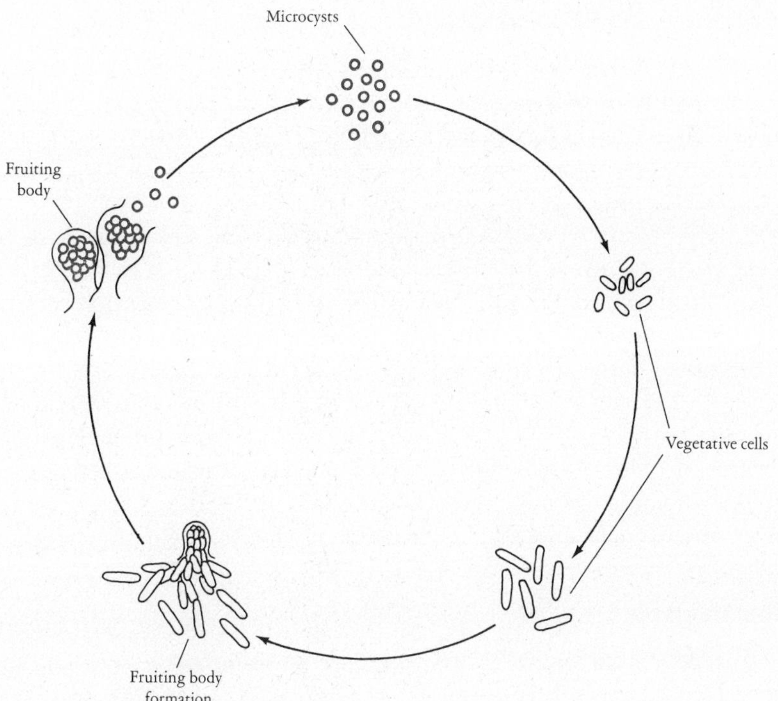

FIGURE 12-12
Life cycle of a fruiting myxobacterium.

spherical and refractory cells called *microcysts,* which are very similar to the cysts of *Azotobacter* species. The microcysts are considerably more resistant to heat, drying, and radiation than are the vegetative cells of myxobacteria but not as resistant as bacterial endospores. Myxobacteria are important in nature as degraders of complex organic substances; they can attack and digest other living or dead bacteria and even certain algae and fungi.

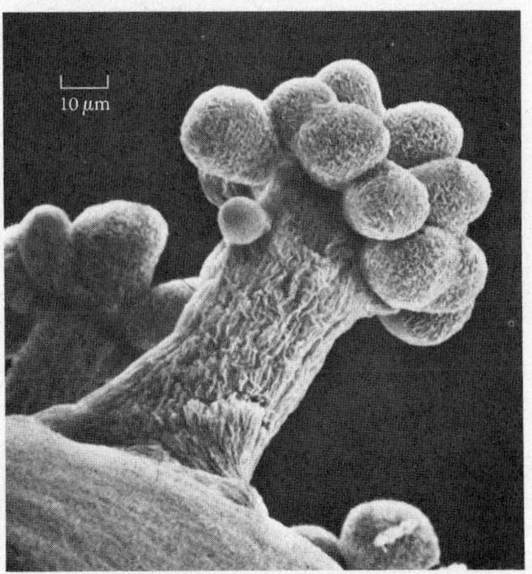

FIGURE 12-13
Chondromyces crocatus fruiting body. (Courtesy of J. Pangborn; P. L. Grilione and J. Pangborn, *J. Bacteriol.* 124:1558, 1975.)

Prosthecate and Budding Bacteria

Prosthecate and budding bacteria are characterized by projections (prosthecae) from the main body of their cells. These projections contain cytoplasm and have a cell wall. In the case of *Caulobacter* species, the prostheca serves to attach the organism to a solid substrate (Figure 12-14). Reproduction is asymmetrical (Figure 12-15); the daughter cell is flagellated and, upon separation from the parent, swims away to attach and repeat the reproductive cycle. Other species of prosthecate bacteria, such as *Ancalomicrobium* species, have two or more prosthecae (Figure 12-14) whose function is not known. During reproduction of *Hyphomicrobium* species (Figure 12-16), the developing prostheca carries the reduplicated genome outward. An enlargement (bud) is then formed, which subsequently reaches the size and shape of the parent cell. Formation of a septum in the remaining prostheca allows separation of the cells to occur, although extensive networks of cells can develop if attachment is maintained. These bacteria are commonly found in natural waters and can thrive on very low concentrations of single carbon organic substances. Some budding bacteria do not form prosthecae.

Spiral Bacteria

Spiral organisms fall into three distinct groups according to their cell wall structure and type of motility. For example, organisms of the genus *Saprospira* (Figure 12-17) occur as multicellular helical filaments of large size, measuring up to 500 μm in length. They show gliding motility and are found in natural waters, both salt and fresh. Except for their spiral shape, they resemble some nonbranching filamentous bacteria.

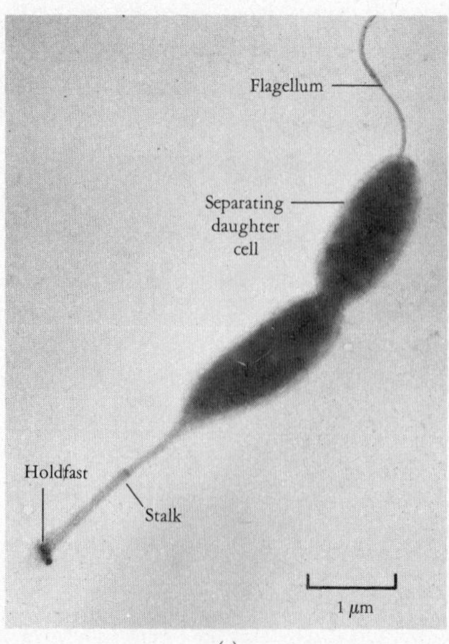

(a)

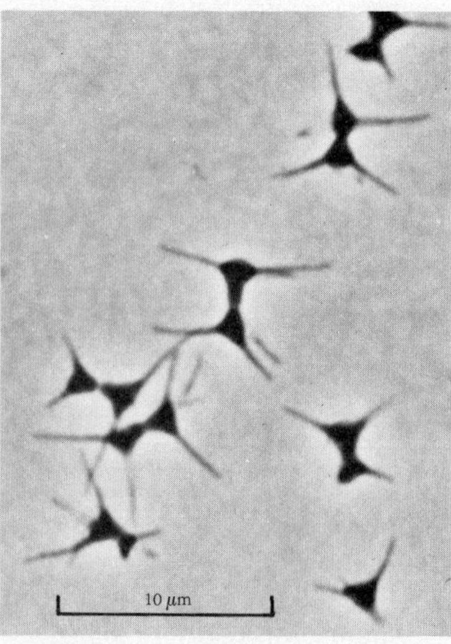

(b)

FIGURE 12-14
Prosthecate bacteria. (a) *Caulobacter* species and (b) *Ancalomicrobium* species. (Courtesy of J. T. Staley and J. P. Dalmasso.)

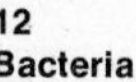

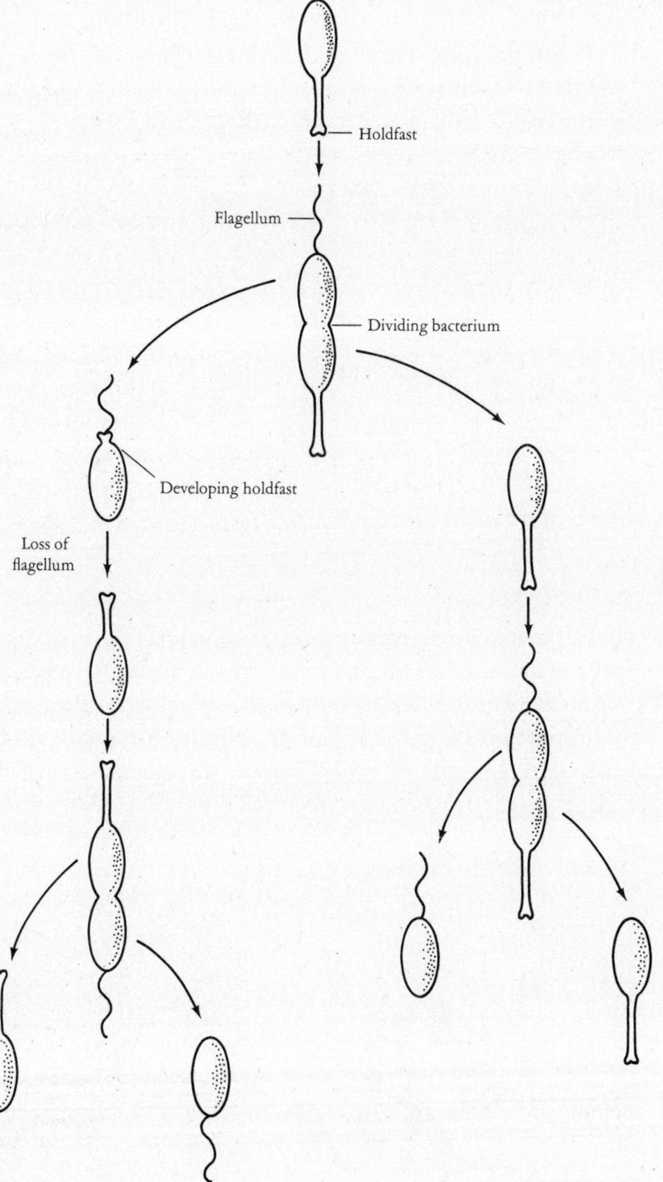

FIGURE 12-15
Life cycle of *Caulobacter* species.

A second group of spiral organisms, the spirochetes, is composed of helical and wave-shaped unicellular bacteria which have flexible cell walls and move by means of unique structures called *axial filaments* (Figure 12-18). Electron microscope study of some of these organisms has revealed that each axial filament is actually composed of fibrils identical in structure to flagella, originating near each end of the organism. The fibrils extend toward each other between two layers composing the cell wall and apparently overlap in the midregion of the cell. The

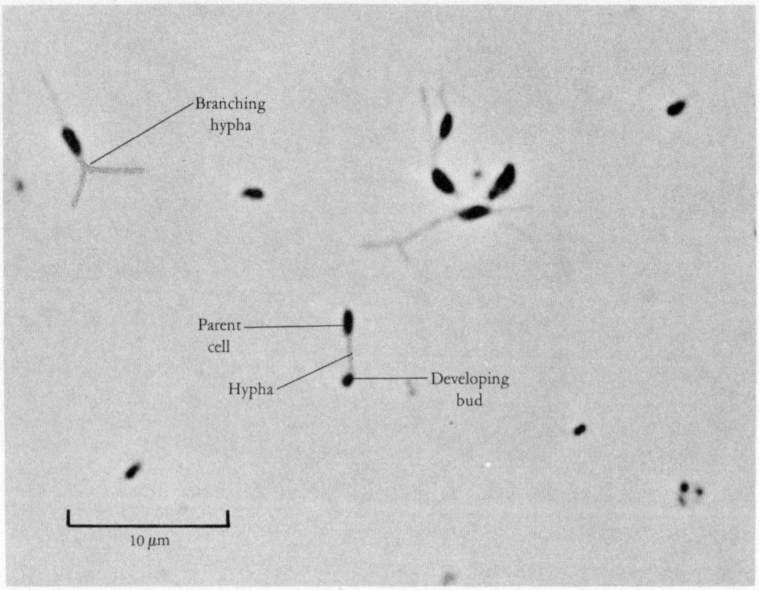

FIGURE 12-16
Hyphomicrobium species, a budding bacterium. (Courtesy of J. T. Staley, and J. P. Dalmasso.)

layer of cell wall material covering the axial filament tends to separate easily in laboratory studies and is often referred to as a sheath. As with flagella, the molecular basis for movement with axial filaments is not yet clear. Progression occurs with a wave propagated along the organism; spinning around the long axis and flexion are also seen. The smaller spirochetes are very slender and therefore difficult to see by the usual microscopic methods (Figure 12-17). Many are also difficult or impossible to cultivate, and their classification is based largely on their morphology and ability to cause disease. Difficulty in their cultivation may relate

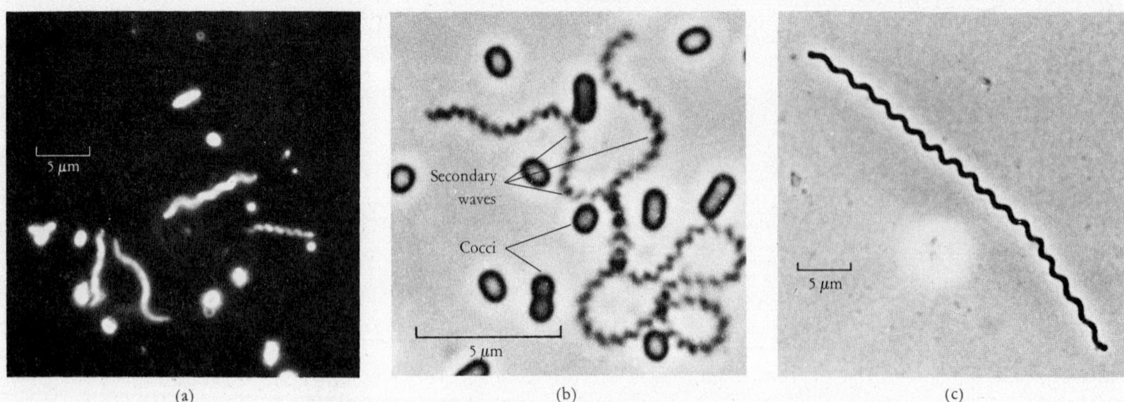

FIGURE 12-17 Spirochetes. (a) Darkfield photomicrograph of spirochetes from a monkey's mouth. (Courtesy of B. L. Williams.) (b) Phase contrast photomicrograph of a very long free-living spirochete showing secondary waves. Note the size relative to the cocci also present in the photograph. (Courtesy of J. T. Staley and J. P. Dalmasso.) (c) Although helical, this organism is not a spirochete; it is a member of a genus of multicellular gliding bacteria and lacks an axial filament. (Phase contrast photomicrograph courtesy of J. T. Staley and J. P. Dalmasso.)

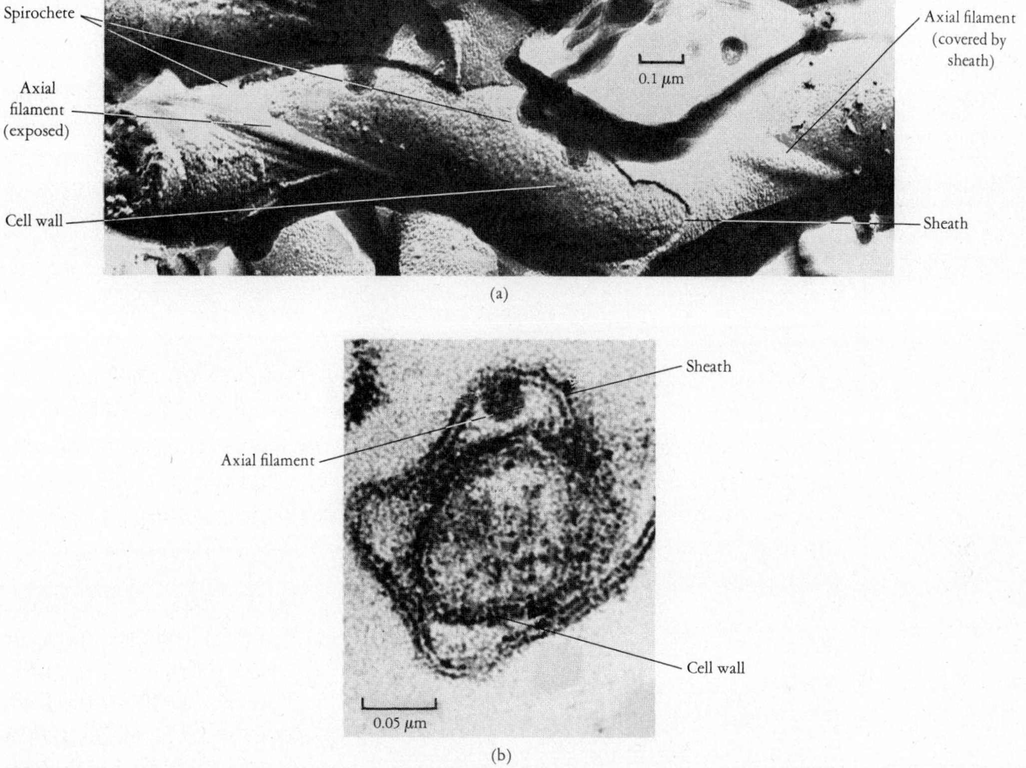

FIGURE 12-18 Electron micrographs showing the axial filament of a spirochete of the genus *Leptospira*. (a) Longitudinal view. (b) Cross section. (R. K. Nauman, S. C. Holt, and C. D. Cox, *J. Bacteriol.* 98:264, 1969.)

to the stringent anaerobic growth requirements of many spirochetes. Some species are widespread in aquatic habitats, and others parasitize warmblooded animals. Some *Spirochaeta* species are extraordinarily long (Figure 12-17), being up to 500 μm in length. Spirochetes were among the organisms observed by van Leeuwenhoek in 1683 in saliva.

Anaerobic culture methods, page 102

The third group of spiral organisms is exemplified by the spirilla, unicellular bacteria which differ from the first two groups in possessing a rigid cell wall. Movement is by one or more flagella located at one or both poles. Like the other spiral organisms, most spirilla are saprophytes residing in natural waters; at least one species, however, is pathogenic for humans.

Mycoplasmas

Mycoplasmas, because they *lack cell walls,* are plastic, easily deformed organisms that can pass through filters retaining most other bacteria. Nevertheless, different species have characteristic shapes (Figure 12-19). Mycoplasmas also are unusual among procaryotic cells in that their cytoplasmic membranes contain sterols, which stabilize the membranes and protect the cell against osmotic lysis. Growth

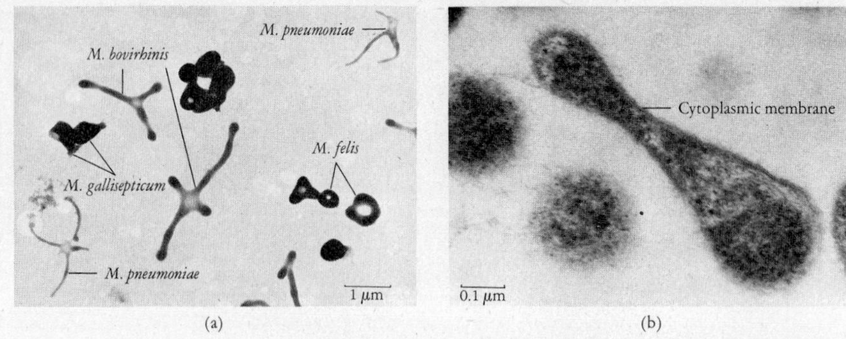

FIGURE 12-19
(a) Electron micrograph showing the morphology of different species of mycoplasma. (b) A thin section of *M. bovirhinis* showing absence of a cell wall. (Courtesy of E. S. Boatman.)

of mycoplasmas on solid media is quite distinctive (Figure 12-20). Their colonies are often so small that a lens or dissecting microscope is required for good visualization. The organisms grow down into the agar, often producing a dense central zone surrounded by a flat translucent area ("fried-egg appearance"). Most species require a medium containing serum, a source of sterols which must generally be supplied because the cells usually cannot synthesize them. Most strains of mycoplasmas also require vitamins, amino acids, purines, and pyrimidines. Carbohydrates can generally be utilized as sources of energy. Both aerobic and anaerobic species are recognized, and mycoplasmas appear to be a very diverse group in terms of their hereditary relationships. They all contain less DNA than any other free-living bacteria. Indeed, some scientists feel that the mycoplasmas are sufficiently distinctive to represent a group separate from the bacteria. Some species are found in soil and sewage, while others are responsible for animal and plant diseases. Formerly, mycoplasmas were called "pleuropneumonia-like organisms" (PPLO) because the earliest isolates were obtained from cattle with a disease called pleuropneumonia.

Organisms of the genus *Spiroplasma* (Figure 12-21), an interesting, newly discovered group of mycoplasma-like organisms, first came to the attention of scientists in the late 1960s. Superficially they resemble spirochetes in size and helical shape, and they also spin and flex in the manner of some spirochetes. However, other features link *Spiroplasma* more closely with species of *Mycoplasma:* they

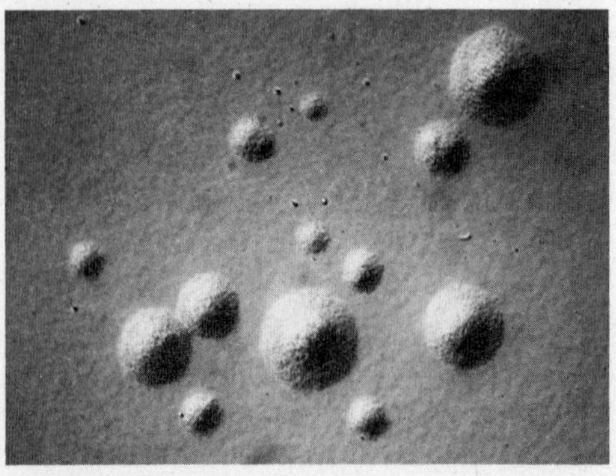

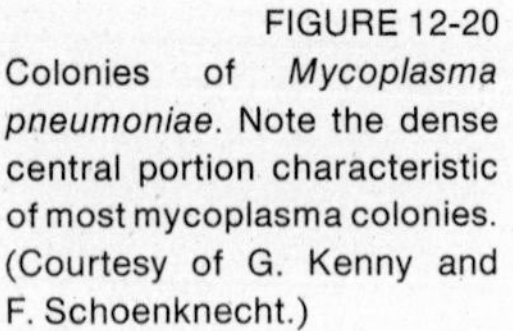

FIGURE 12-20
Colonies of *Mycoplasma pneumoniae*. Note the dense central portion characteristic of most mycoplasma colonies. (Courtesy of G. Kenny and F. Schoenknecht.)

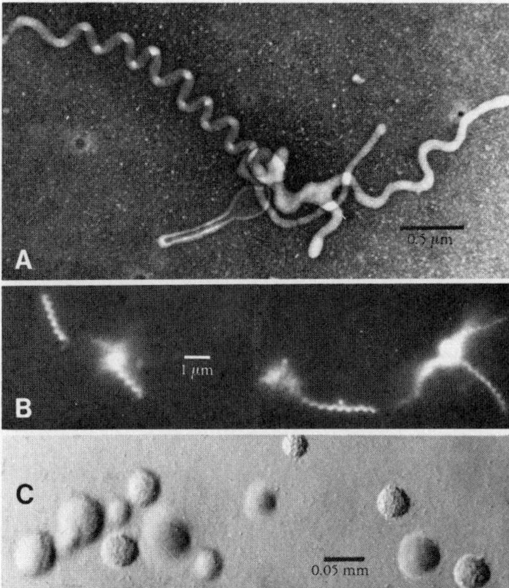

FIGURE 12-21
Spiroplasma species isolated from infected corn plants. (a) Electron micrograph showing helical shape. (b) Spiroplasmas from liquid culture, observed by darkfield microscopy. (c) Colonies of spiroplasmas on agar medium after 14 days incubation. (Courtesy of Tseh An Chen; T. A. Chen and C. H. Liao, *Science* 188:1015. Copyright © 1975 by The American Association for the Advancement of Science.)

lack cell walls and axial filaments, form mycoplasma-like colonies, and pass through filters having very small pores. They are known to cause certain plant diseases and also infect the insects that transmit these diseases. In 1976 a spiroplasma was shown to cause cataracts in baby mice, evoking speculation that these organisms might also prove to cause some perplexing diseases of other animals.

BACTERIA USING INORGANIC ENERGY SOURCES

Many bacteria utilizing an inorganic energy source have the unique ability to grow in the dark on a medium containing only inorganic substances, using carbon dioxide as their carbon source. As with other autotrophs, members of this group often have well-defined intracytoplasmic structures called *carboxysomes,* which contain the CO_2-fixing enzymes. Some of these bacteria require one or more organic compounds in the form of vitamins or other growth factors. Although mentioned only briefly here, bacteria using inorganic energy sources are extremely important in the natural environment and will be covered in greater depth in later chapters.

Nitrifying and Sulfur Bacteria

Thiobacillus and *Nitrosomonas* are two genera of Gram-negative, rod-shaped, aerobic soil organisms that obtain energy by oxidizing inorganic chemicals. *Thiobacillus* species (Figure 12-22) oxidize sulfur or sulfur compounds, while *Nitrosomonas* species oxidize ammonium ions. These bacteria play important roles in the sulfur and nitrogen cycles, as described in Chapter 34. In addition, oxidation

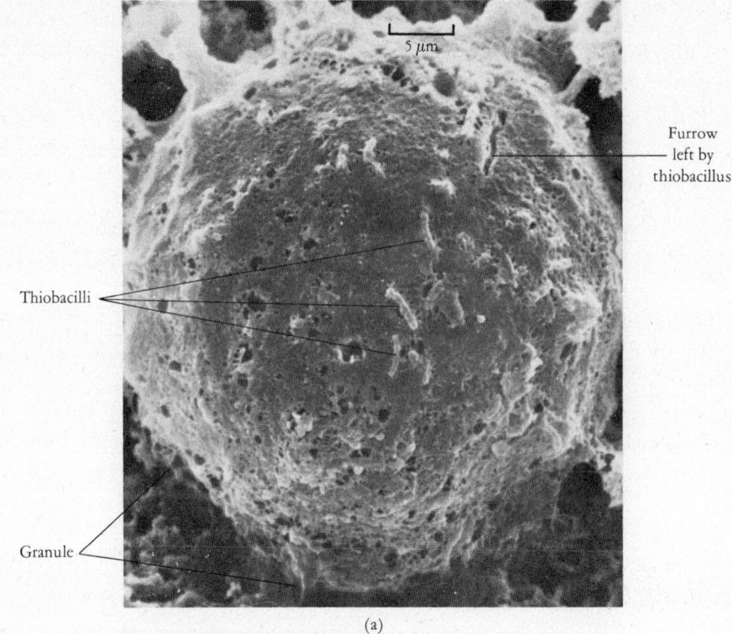

(a)

FIGURE 12-22
(a) *Thiobacillus denitrificans* grown on a granule of sulfur. (Scanning electron micrograph courtesy of J. LeGall.) (b) Electron micrograph of the internal structure of *Thiobacillus* species. Note the presence of multiple spherical bodies called *carboxysomes*. (c) Higher power of the carboxysomes; they are the site of CO_2 fixation. (Courtesy of J. M. Shively.)

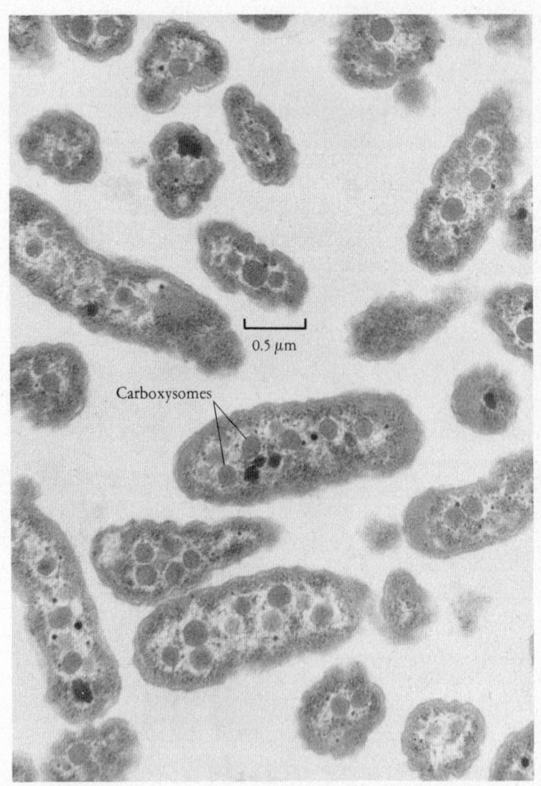

(b)

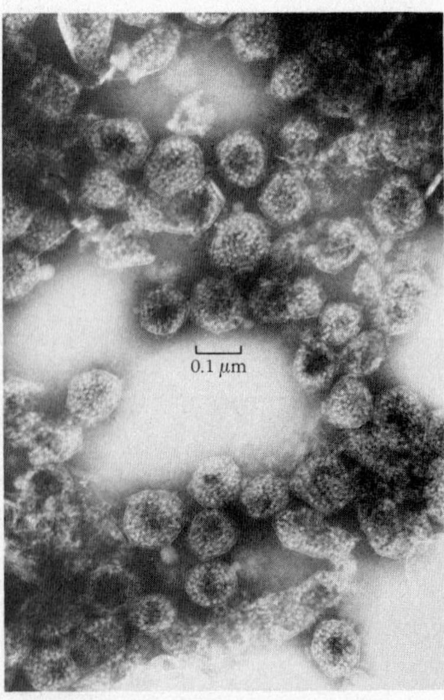

(c)

of sulfur by *Thiobacillus* species produces sulfuric acid in mine waters, and this is of major importance in dissolving uranium, copper, and other metals from deposits deep in the ground.

Members of the genera *Beggiatoa* and *Thiothrix* (Figure 12-23) are Gram-negative, multicellular, organisms occurring as unbranched filaments and living in sulfur springs and in sewage-polluted water. These microorganisms obtain energy by oxidizing H_2S to granules of elemental sulfur that are deposited intracellularly. Besides H_2S, some beggiatoas can use certain organic chemicals for energy. The filaments of beggiatoas are flexible and show gliding motility. The filaments of *Thiothrix* species are immobile and are attached to rocks or other solid surfaces. However, *Thiothrix* cells possessing gliding motility detach from the ends of the filaments and disperse to new locations. Commonly, several such cells will then attach to a solid object in a rosette arrangement. As they multiply, these cells then develop into new filaments.

Bacteria Utilizing Iron and Manganese

Acidic freshwaters may contain dissolved iron and manganese in a reduced state (Fe^{2+}, Mn^{2+}). These reduced ions can be oxidized by some bacteria, and the energy gained can be used to fix carbon dioxide into cellular components. For example, some members of the genus *Thiobacillus* oxidize ferrous ions as an alternative to utilizing sulfur compounds. Other genera are also involved in the oxidation of these chemicals.

Methanogenic Bacteria

Methane-producing bacteria grow in the digestive tracts of human beings and other animals and in sewage, swamps, and other places where there is organic material decomposing under strictly anaerobic conditions. They are of various shapes, some being Gram positive and some Gram negative, but all can obtain energy by reducing carbon dioxide with hydrogen to give methane. Their potential importance in the production of fuel is discussed in Chapter 39.

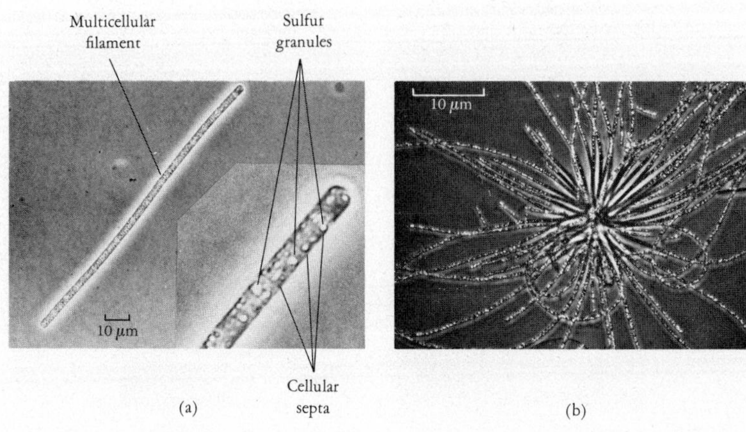

FIGURE 12-23
Phase contrast photomicrographs of gliding bacteria. (a) *Beggiatoa* species. (Courtesy of J. T. Staley and J. P. Dalmasso.) (b) *Thiothrix* species. Note rosette arrangement. (Courtesy of F. Palmer and E. Ordal.)

PROCARYOTES UTILIZING LIGHT FOR ENERGY (PHOTOBACTERIA)

Anaerobic Photobacteria

Photosynthesis, page 160

Anerobic photobacteria comprise the only group of photosynthetic organisms that grow anaerobically, and they do not produce oxygen as a by-product of photosynthesis. The chlorophyll pigments of anaerobic photobacteria differ from those of cyanobacteria, algae, and plants. Anaerobic photobacteria inhabit a restricted ecological niche characterized by oxygen lack and good light penetration. They also require reduced inorganic or organic compounds, such as hydrogen sulfide or fatty acids, as hydrogen donors. Thus they may be found in certain wet soils and aquatic or marine habitats. To achieve the location that provides optimum conditions for growth, some anaerobic photobacteria can change their position in a fluid environment by using their polar flagella or by altering the number of gas flotation vesicles in their cytoplasm. Some can fix atmospheric nitrogen and therefore require only light, water, carbon dioxide, molecular nitrogen, and some minerals for growth. Their ecological role as the primary producers of organic from inorganic material is insignificant in comparison with algae, except for a few special habitats. The anaerobic photobacteria of today are genetically diverse and include rods, cocci, helical, and other forms. *Rhodomicrobium vannielii* (Figure 12-24) is a budding anaerobic photobacterium which is capable of aerobic growth under certain conditions.

Aerobic Photobacteria (Cyanobacteria)

A second group of photobacteria are the cyanobacteria. In contrast to the anaerobic photosynthetic bacteria, cyanobacteria are aerobic, and oxygen is evolved as a

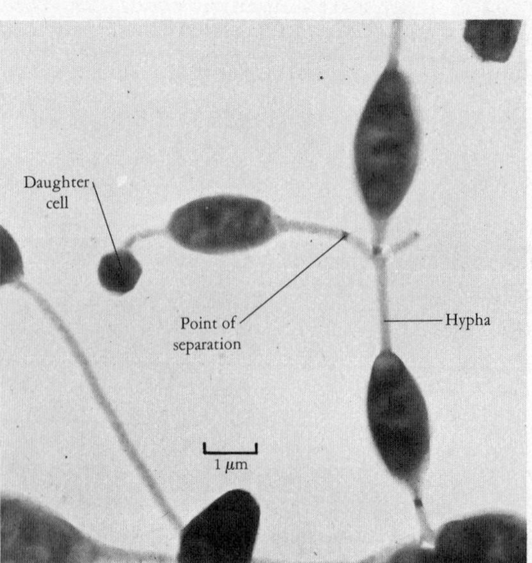

FIGURE 12-24
Rhodomicrobium vannielli, a budding photosynthetic bacterium. (Courtesy of H. Douglas; E. Duchow and H. Douglas, *J. Bacteriol.* 58:409, 1949.)

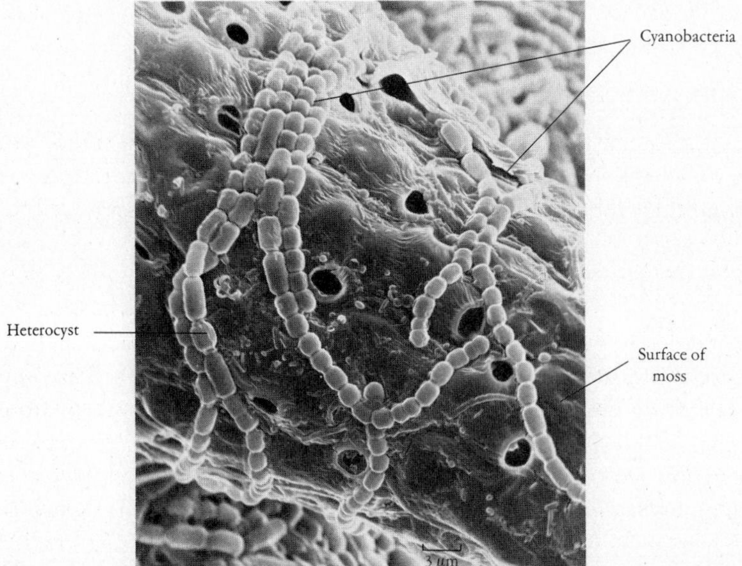

FIGURE 12-25
Scanning electron micrograph of cyanobacteria (probably *Anabaena cylindrica*) on the surface of a moss. (Courtesy of D. C. Jordan. Reproduced by permission of the National Research Council of Canada. From *Canad. J. Microbiol.* 22:897–907, 1976.)

by-product of their photosynthesis. Their chlorophyll pigments and photosynthetic process are of the same type as those of algae and plants. However, many cyanobacteria have the capacity to fix nitrogen, resembling the anaerobic photobacteria in this respect but differing from all algae and plants. As might be expected from their simple growth needs, cyanobacteria are extremely widespread in nature and are found where more fastidious microbes are unable to grow. In natural waters polluted with nitrogenous compounds and phosphates, they often grow to enormous numbers, producing a colored scum on the surface. The color of such cyanobacterial masses, ranging from blue-green to red, violet, and black, depends on the types of light-trapping pigments which a particular species possesses. Morphology of cyanobacterial species includes all the procaryotic types, rods, cocci, and spiral, as well as filamentous forms. Motility when present is of the gliding type, although, as with many other aquatic bacteria, the cyanobacteria can alter their depth of flotation by changing the number of their intracytoplasmic *gas vesicles.* However, two other cyanobacterial structures, *thylacoids* and *heterocysts,* are unique among procaryotes. Thylacoids are platelike membranes containing chlorophyll, whereas heterocysts (Figure 12-25) are specialized cells whose primary function is probably nitrogen fixation. The genetically diverse cyanobacteria usually reproduce by binary fission. Ancestors of this group of organisms were probably responsible for the initial appearance of atmospheric oxygen on the earth.

Halobacteria

A third group of photobacteria, the halobacteria, lacks chlorophyll and the usual photosynthetic apparatus. Instead, the light energy these organisms use is absorbed by *bacteriorhodopsin,* a substance related chemically to visual purple, the light sensitive pigment in the human retina. The absorbed light energy is used to generate the ATP needed for growth. Halobacteria can live only in very strong salt solutions, such as salt ponds and brines used for curing fish, on which they may produce red patches of growth. They are able to grow utilizing chemical energy from organic material, but only in the presence of oxygen; under anaerobic conditions they rely on their light-trapping pigment for energy.

SUMMARY

All procaryotes are now considered to be bacteria. Collectively they comprise a remarkably diverse form of life; some obtain energy for biosynthesis from sunlight, while others make use of numerous chemical reactions involving organic or inorganic materials. A number of bacteria even have metabolic machinery for using more than one type of energy source. Some bacteria can synthesize all their needs from inorganic substances and can live in harsh environments unsuitable to other life forms. Many other bacteria make use of organic materials synthesized by other organisms. Endospore-forming bacteria are highly resistant to heat, radiation, drying, and toxic chemicals. The various strict anaerobes grow only in the absence of oxygen. Some bacteria can grow only within the living cells of other organisms; in many instances this harms the other organism, but in some it confers benefit. Besides having great diversity of biochemical activity, bacteria have great diversity of form and mode of reproduction. A few exhibit a type of life cycle.

QUESTIONS

REVIEW

1. What sources of energy do different bacteria employ to carry out biosynthesis?
2. Classify the bacteria on the basis of their microscopic appearance.
3. By what mechanism do bdellovibrios conserve the energy required for biosynthesis?
4. Give the principal characteristics of enterobacteria.
5. Describe the three principal groups of bacteria that use light as an energy source.

THOUGHT

1. You are a scientist assigned the responsibility for establishing life on another planet. What are the tolerable limits of temperature, atmosphere, soil composition, and availability of water that would permit the use of naturally occurring (earth) bacteria?
2. What mechanisms allow different bacteria to grow both in the presence and absence of oxygen?

Colonial morphology, an aid in identifying bacteria, may vary considerably. Some of the different colonial types are illustrated here. (Courtesy of S. Eng and M.F. Lampe.)

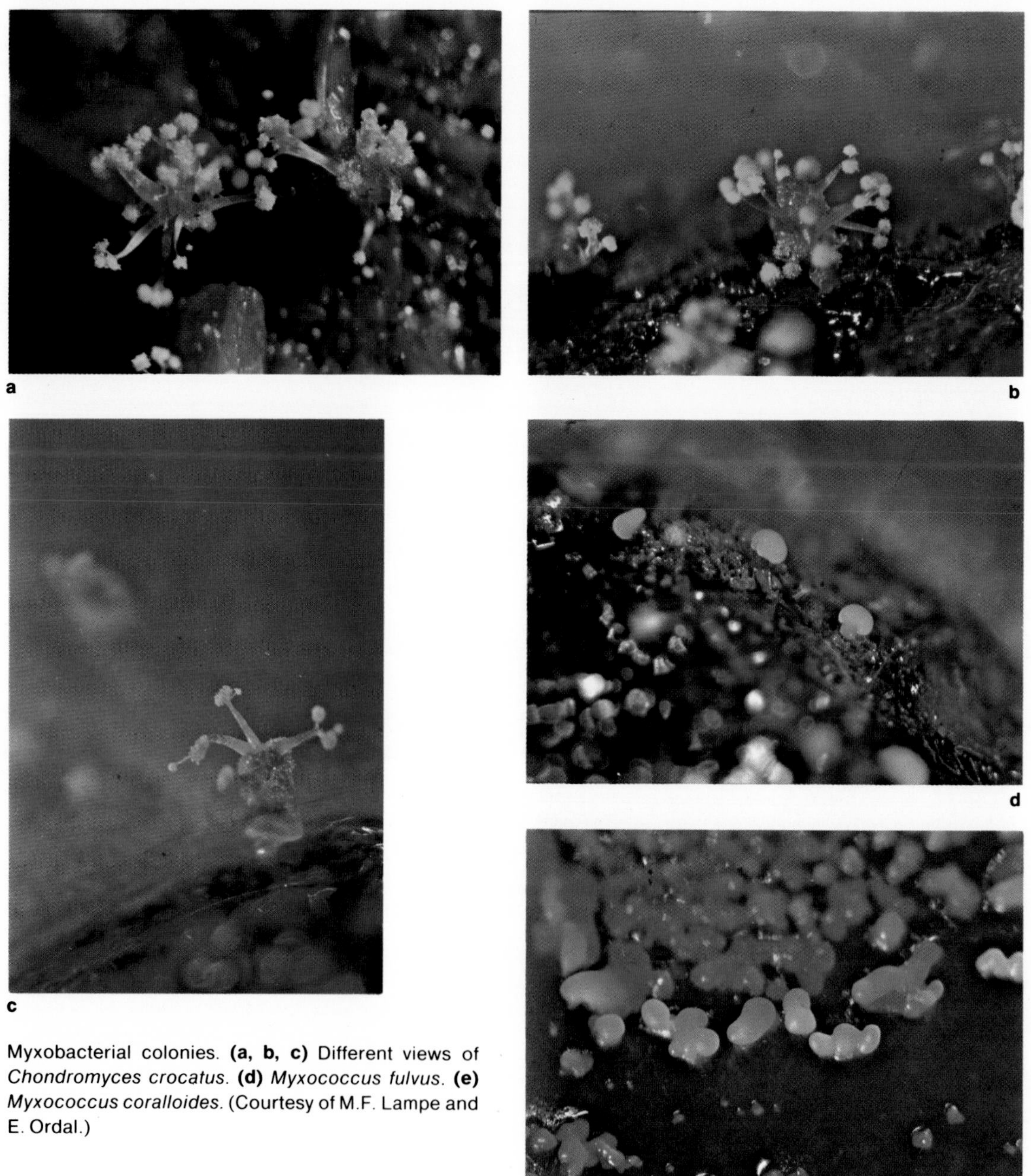

Myxobacterial colonies. **(a, b, c)** Different views of *Chondromyces crocatus.* **(d)** *Myxococcus fulvus.* **(e)** *Myxococcus coralloides.* (Courtesy of M.F. Lampe and E. Ordal.)

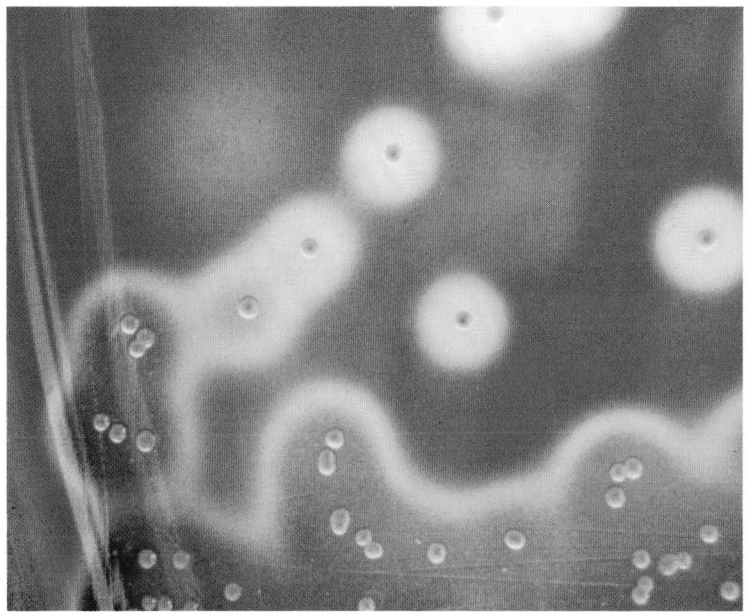
a

(a) *Streptococcus pyogenes* colonies on sheep blood agar medium incubated anaerobically. Note the zones of beta hemolysis surrounding the colonies. (Courtesy of C.E. Roberts.)

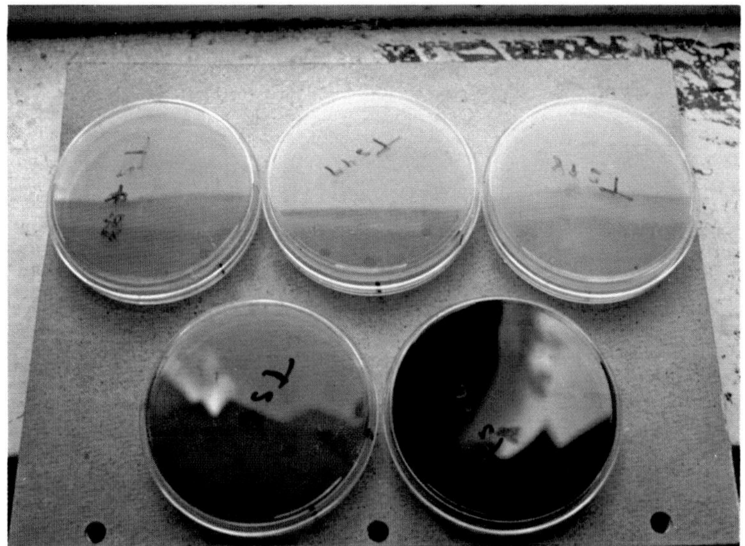
b

(b) Cultures of different strains of *Pseudomonas aeruginosa* from infected patients. Notice the differing colors produced by water-soluble pigments. (Courtesy of C.E. Roberts.)

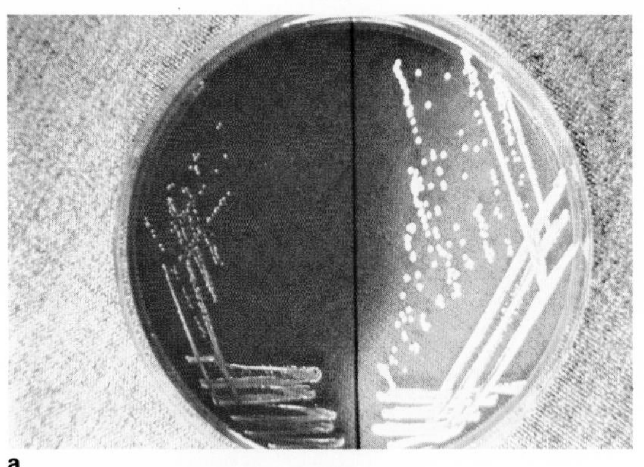

a

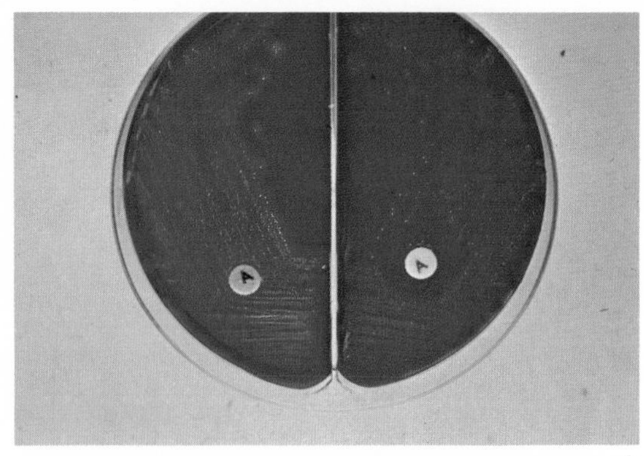

b

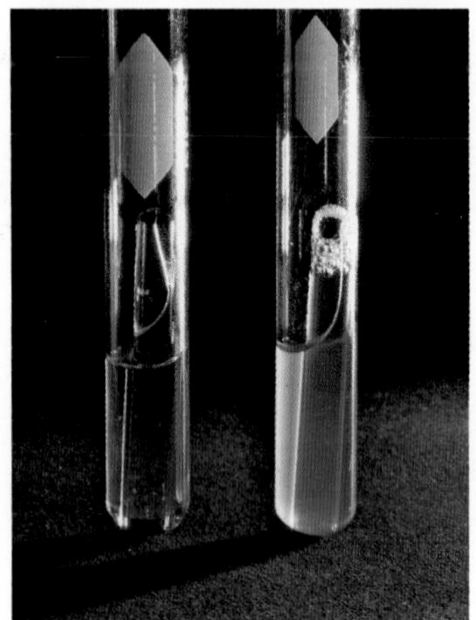

c

Biochemical tests. **(a)** Staphylococci on mannitol-salt agar. **(b)** Beta-hemolytic streptococci with a clearing around a bacitracin-containing disc, indicating a susceptibility to bacitracin. **(c)** Acid and gas production from sugars. **(d)** Breakdown of urea. **(e)** Complex changes of enterobacterial growth in triple-sugar-agar medium. (a, b courtesy of J. Portman; c, d, e courtesy of M.F. Lampe.)

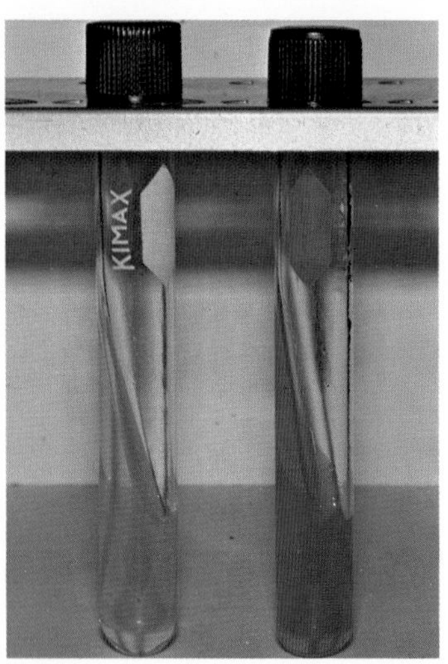

d

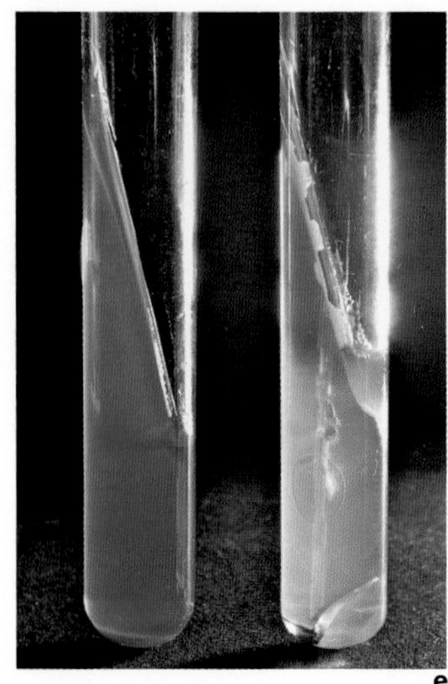

e

FURTHER READING

BUCHANAN, R. E., and N. E. GIBBONS, eds., *Bergey's Manual of Determinative Bacteriology.* 8th edition. Baltimore: Williams and Wilkins, 1974. A reference work describing the species of bacteria.

MCELROY, W. D., and H. H. SELIGER, "Biological Luminescence," *Scientific American* (December 1962). A general discussion of light generation by living creatures.

STANIER, R. Y., E. A. ADELBERG, and J. L. INGRAHAM, *The Microbial World.* Englewood Cliffs, N.J.: Prentice-Hall, 1976. A readable, comprehensive, and current text on microbiology.

STOECKENIUS, W., "The Purple Membrane of Salt-loving Bacteria," *Scientific American* (June 1976). An interesting paper on the role of bacteriorhodopsin in the energy metabolism of halobacteria, written by its discoverer.

CHAPTER 13

ALGAE

Algae are ubiquitous eucaryotic protists, living in many locations where sunlight is available. In general algae are characterized by the presence of a specific type of chlorophyll, chlorophyll-*a,* and other pigments, and they carry out plantlike photosynthesis.

OCCURRENCE

Algae are abundant in both fresh and salt water, and many inhabit soil, plants, and other terrestrial locations. Unicellular algae make up most of the phytoplankton of the ocean, in part because nutrients are not as concentrated in the sea as they are in soil, and single cells which have relatively large absorptive surfaces and which can move about may have an advantage over static multicellular organisms. Algae reproduce rapidly when conditions are favorable, and huge numbers in a "bloom" can color the water of ponds or areas of several square miles in the sea.

Because of their adaptability and their photosynthetic capacities, algae can often grow where other forms of life cannot thrive. In barren areas they can provide organic materials necessary for the establishment of higher forms of life, just as their predecessors must have pioneered in the first emergence of terrestrial life.

Some algae can be cultured in the laboratory in much the same way as bacteria. When this is done, media are prepared to simulate the environmental nutrients in which the algae normally live; for example, marine species are grown in an artificial seawater medium. Light of the proper wavelength is supplied, and temperatures, gases, and other factors of the environment are adjusted to reproduce the natural habitat of the algae as nearly as possible. Mixed cultures, in

LABORATORY CULTURES

which algae of more than one kind live together, are easily prepared, usually in the presence of bacteria or with many other microbial species. Of course, for metabolic studies or other experiments in which pure cultures of algae are essential, single cells or spores must be isolated and allowed to grow alone. The preparation of pure algal cultures is considerably more difficult than preparing mixed cultures.

MORPHOLOGY

Algae occur in a variety of forms, some of which are diagramed in Figure 13-1. The single-celled algae may be nonmotile or motile by means of ameba-like pseudopodia (rhizopodial algae; see Figure 13-1a), flagella (flagellate algae) or other means. The cells of algae are often associated in colonies, filaments, or other structures. Chains of algae may form by continuous division in one plane, producing filaments which are usually enclosed by a mucilaginous sheath (Figure 13-1b). Often the filaments branch by lateral division of the cells. Hollow-tube forms (Figure 13-1c) do not produce cross walls but instead become coenocytic, having many nuclei within a single, nonseptate organism.

The cell walls of algae vary considerably according to phylum, as indicated in Table 13-1. Different algal species secrete various kinds of polysaccharides which have several functions. They may aid in motility, and they can adhere to rocks or plants and hold algal colonies in positions advantageous for growth.

PHYSIOLOGY AND METABOLISM

Algae are aerobic photosynthetic organisms, classified into six major and several minor phyla. Table 13-1 lists the principal pigments of each major phylum of algae. The photosynthetic pigments absorb light energy, and both their color and the amount of energy they take up depend on the wavelength of light they can utilize. Three major kinds of pigments are found in algae: chlorophylls, carotenoids (carotenes and xanthophylls), and phycobilins (Figure 13-2). Slight differences in the molecular structure of the chlorophylls (*a, b,* and so on) determine the wavelengths of light these pigments can employ, and consequently the amount of light energy absorbed. Chlorophyll-*a,* present in all algae, absorbs red and blue light and transmits light of a green color, which may be masked or altered by other pigments. For example, some algae appear brown because they contain a relative abundance of xanthophylls and carotenes that transmit brown light and mask the green color transmitted by chlorophyll-*a.* Other algae appear reddish or purplish because of their phycobilins. The photosynthetic pigments of the algae are found in membrane-enclosed chloroplasts.

Chloroplasts, page 83

The nature of the substances stored by various phyla of algae may be related to their evolutionary development toward higher plants. For example, green algae, which have a number of properties in common with green plants, store a plantlike starch. Algae of other phyla polymerize the same simple sugars in differ-

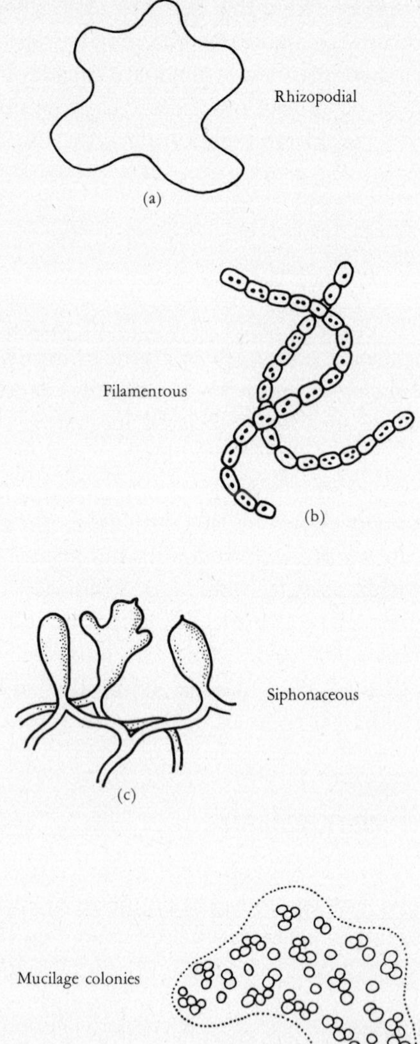

FIGURE 13-1
The gross morphology of various algae.

ent ways to produce other kinds of starches. Some species of algae store oils or fats, a property that is probably directly related to the survival of these species. Those making up much of the marine phytoplankton often store fat globules that make them buoyant and help keep them floating in the sea at levels where light for photosynthesis is available.

REPRODUCTION

Most algae can undergo vegetative reproduction by cell division or fragmentation without the production of any specialized structures. Other forms of asexual

TABLE 13-1
Characteristics of Six Major Phyla of Algae

Phylum	*Usual Habitat*	*Principal Pigments*[a]	*Storage Products*	*Cell Walls*	*Mode of Motility (If Present)*	*Modes of Reproduction*
Chlorophyta	Fresh and salt water; soil; tree bark; lichens	Chlorophyll-*b*; carotenes; xanthophylls	Starch (α-1, 4-glucan)	Cellulose and pectin	Mostly nonmotile (except one order), but some reproductive elements may be flagellated	Asexual, by multiple fission; spores; or sexual
Euglenophyta	Fresh water	Chlorophyll-*b*; carotenes; xanthophylls	Fats; starchlike carbohydrates	Lacking, but have elastic pellicle	One to three anterior flagella	Asexual only, by binary fission
Chrysophyta	Fresh and salt water; soil; higher plants	Carotenes	Starchlike carbohydrates (β-1, 3-glucan); oils	Pectin, often impregnated with silica or calcium	Unique diatom motility; one, two, or more unequal flagella	Asexual or sexual
Pyrrhophyta	Mostly salt water but common in fresh water	Carotenes; xanthophylls	Starch; oils	Cellulose and pectin	Two unequal lateral flagella in different planes	Asexual; rarely sexual
Phaeophyta	Salt water	Xanthophylls, especially fucoxanthin	Starchlike carbohydrates; mannitol; fats	Cellulose and pectin; alginic acid	Two unequal lateral flagella	Asexual, motile zoospores; sexual, motile gametes
Rhodophyta	Mostly salt water but several genera in fresh water	Phycoerythrin and other phycobilins; carotenes; xanthophylls	Starchlike carbohydrates	Cellulose and pectin; agar; carrageenan	Nonmotile	Sexual, gametes; asexual spores

[a]In addition to chlorophyll-*a*, which is present in algae of all phyla.

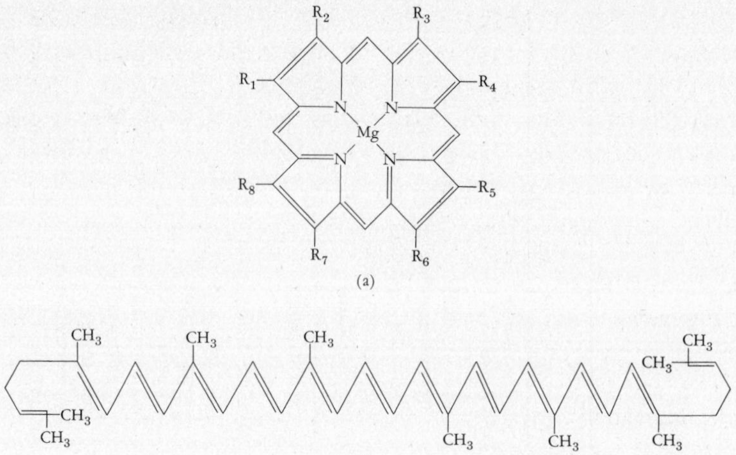

FIGURE 13-2
The molecular structure of some photosynthetic pigments. (a) Chlorophyll (R indicates an organic group). (b) A carotenoid pigment.

reproduction common among the algae involve the formation of specialized cells called *spores,* which germinate without uniting with other cells. These asexual spores may either be motile cells without cell walls (*zoospores*) or nonmotile cells enclosed by a wall.

The modes of sexual reproduction in algae vary so greatly, even within groups, that it is difficult to make generalizations about them. A few specific examples are included in later sections.

PROPERTIES OF THE PHYLA OF ALGAE

Some of the general characteristics of the six major phyla of algae are summarized in Table 13-1. This classification is based on a number of properties, including the principal pigments of each phylum, the storage products of its members, their mechanism of motility, and their modes of reproduction.

CHLOROPHYTA (GREEN ALGAE)

Both fresh and salt water, soil, plants, and tree bark serve as natural habitats for green algae of the phylum Chlorophyta. All members of this phylum contain chlorophyll-*a* and -*b,* carotenes, and xanthophylls, and they form starch as a storage product. The cells can divide vegetatively, or they may form asexual flagellated zoospores. Most of the vegetative forms of green algae are haploid.

A common green alga often found in ponds is the single-celled, motile *Chlamydomonas.* The single cells of this alga are oval in shape (Figure 13-3) and typically contain one chloroplast. Two identical flagella protrude from the anterior portion of the cellulose cell wall. The cells have two excretory contractile vacuoles, a red eyespot, and a pigment-containing light receptor that permits the organism to move toward light. During asexual reproduction the haploid cells lose motility and divide to give 4 (or, in some species, 8 or 16) zoospores which escape from the wall of the original cell to become independent and motile. Species of *Chlamydomonas* have proved valuable in studying the molecular biology and genetics of simple eucaryotic cells.

Another genus of the phylum Chlorophyta that has also been of great value in the study of molecular biology is the huge, single-celled alga *Acetabularia* (Figure 13-4), which may grow as tall as 10 cm. The single nucleus near the base of its stemlike stalk can be removed by cutting the stalk. Sections of the alga that contain a nucleus can regenerate a new cap (or leaflike portion) readily, but sections that lack a nucleus have limited powers of regeneration. *Acetabularia* cells of different species, each with a distinctive cap, can be grafted together, as indicated in Figure 13-4. When two nucleated stalks of different species are grafted, the resulting hybrid will form a cap with some of the characteristics of each species. However, if a stalk without a nucleus is grafted to the nucleated stalk of a different species, the hybrid will form a cap characteristic of the nucleated species. In other words, the nucleus of the organism clearly determines the ability to regenerate a cap and the nature of the cap that is made. Experiments such as these have been valuable in studying the relative functions of nuclear and cytoplasmic constituents of the cell.

EUGLENOPHYTA

A small phylum of unicellular flagellated algae, the Euglenophyta, includes a few marine members but is comprised predominantly of freshwater forms. Members of the prototype genus *Euglena* have a single obvious anterior flagellum, but close examination may reveal a smaller vestigial second flagellum. A rigid cell wall is lacking, and an elastic pellicle covers the cell. Chlorophyll and carotenes are contained in chloroplasts, and the cells of *Euglena* store fats and a

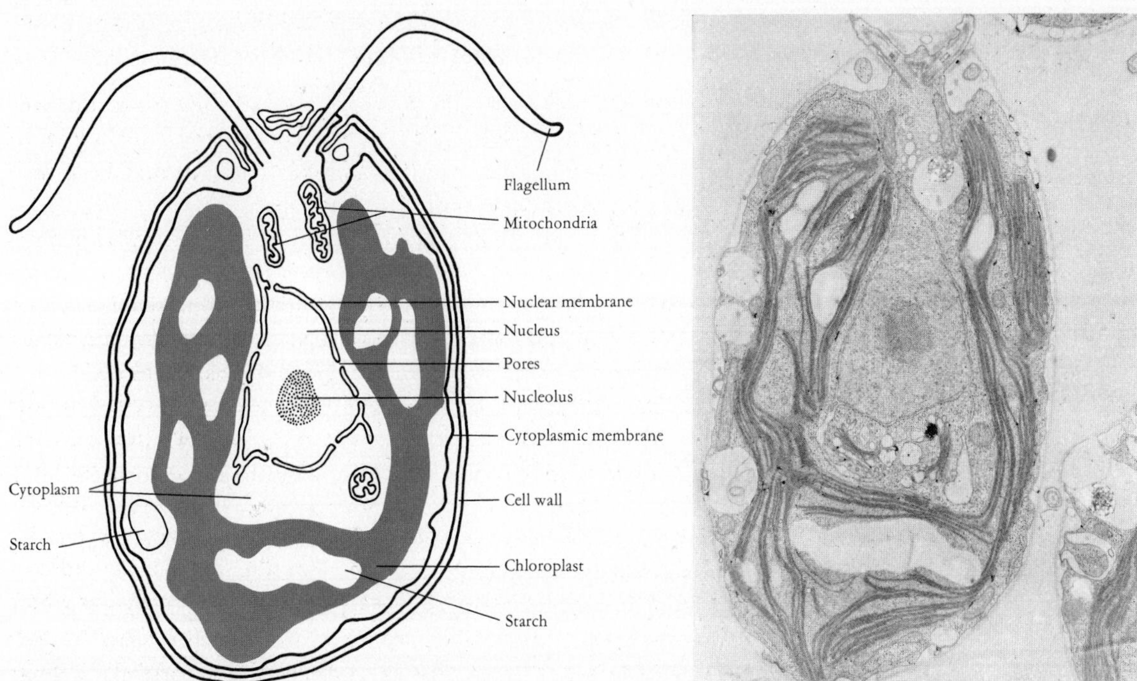

FIGURE 13-3 *Chlamydomonas,* a representative green alga. (Photo courtesy of V. Goodenough; V. Johnson and K. Porter, *J. Cell Biol.* 38:403, 1968.)

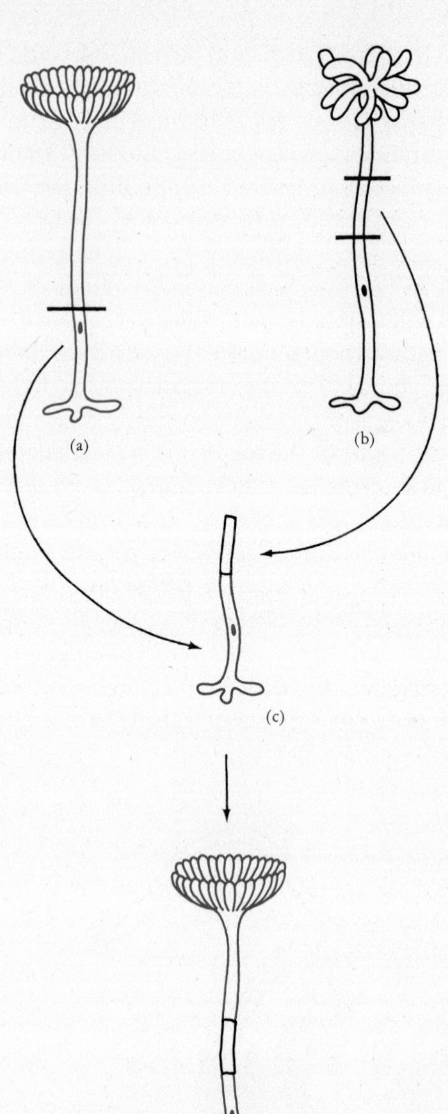

FIGURE 13-4
Cells of different species of *Acetabularia* (a,b) can be grafted together (c). The cap formed by the grafted alga is characteristic of the species from which the nucleus was derived (d). Experiments such as these have shown that the nucleus determines both the ability to regenerate cellular constituents and the nature of the regenerated cell components.

starchlike carbohydrate. The cells divide by binary fission, and sexual reproduction very rarely, if ever, occurs.

Many biologists regard members of the phylum Euglenophyta as protozoans for reasons discussed in Chapter 15.

CHRYSOPHYTA

The phylum Chrysophyta includes several classes. In one class some genera have calcium-containing hard walls; members of another class (Bacillariophyceae) contain silica in their walls. This latter class is of particular interest because it includes the diatoms that are abundant in plankton or in films on

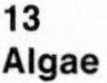

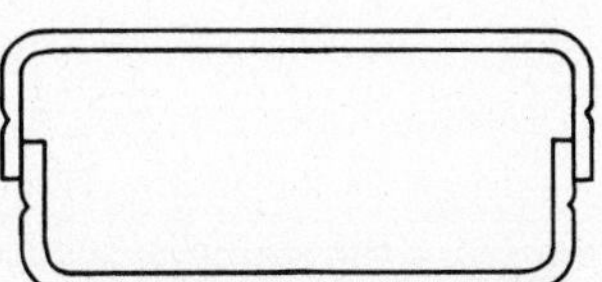

FIGURE 13-5 The construction of a diatom frustule. The hard coverings of diatoms (frustules) resemble covered boxes or dishes. When the cell divides, each daughter cell retains half of the parent frustule and subsequently forms another corresponding half.

rocks, plants, and wood within both salt and fresh water. The diatoms have a unique morphology, being constructed like a petri dish (centric diatoms) or a covered box (pennate diatoms), as illustrated in Figure 13-5. Their pectin-containing cell walls are impregnated with siliceous compounds to make an extremely hard shell-like covering called a *frustule.* Pores in the frustule are covered by a membrane in a number of species, but in many pennate diatoms the pores are open in a slit, or *raphe,* associated with an unusual form of motility (Figure 13-6). Contact with a solid surface is required for motility.

A third class of Chrysophyta includes rhizopodial ameba-like algae. Some of these can engulf food with pseudopodia in the same manner as the protozoan amebae.

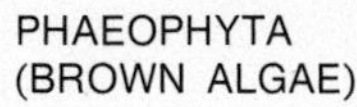

PHAEOPHYTA (BROWN ALGAE)

Xanthophyll and carotene pigments give algae of the phylum Phaeophyta their brown color. These brown algae are multicellular, and they are usually macroscopic. They are the sessile seaweeds of the cold ocean waters, but some are found in warmer areas such as the Sargasso Sea, named for *Sargassum,* a brown seaweed floating in it.

Members of the genus *Nereocystis* are common brown algae of the Pacific Coast which are attached at depths of from 6 to 24 meters (18–72 ft) in the cold waters between Alaska and southern California. They look and feel like long,

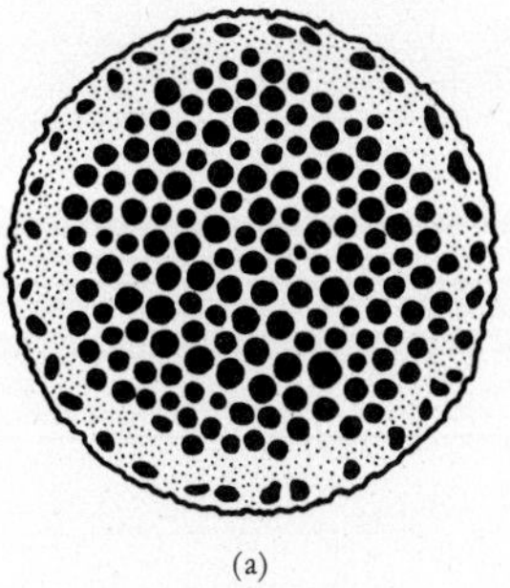

(a)

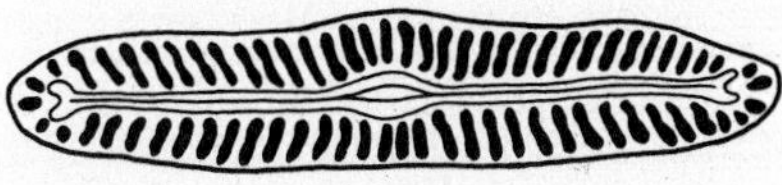

(b)

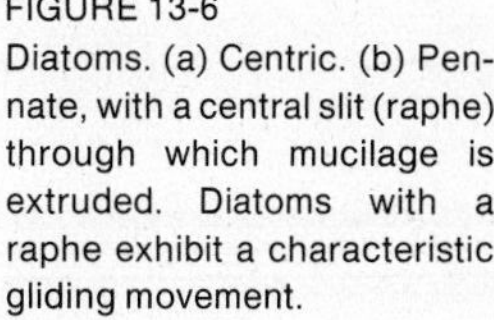

FIGURE 13-6
Diatoms. (a) Centric. (b) Pennate, with a central slit (raphe) through which mucilage is extruded. Diatoms with a raphe exhibit a characteristic gliding movement.

tough, brown rubber tubes, with an enlarged spherical bulb, or *bladder,* at the anterior end (Figure 13-7); leaflike ribbons extend from the bulb. In the ocean the bladder is filled with gas and holds the leaflike ribbons near the surface where they are exposed advantageously to the sunlight needed for photosynthesis.

RHODOPHYTA (RED ALGAE)

Algae of the phylum Rhodophyta owe their various colors to the presence of several pigments, especially the red phycoerythrin. Algae of this phylum store a unique starchlike carbohydrate. They never form flagella. Members of Rhodophyta are unicellular or multicellular, and they have complicated modes of sexual reproduction which usually involve separate male and female organisms. Nonmotile cells from the male are carried by the water to the germ cell of the female, and a zygote is formed.

Most members of Rhodophyta are seaweeds which grow abundantly, in warm, deep salt waters; some species may be found growing as deep as 260 meters (780 ft). One class (Corallinaceae) has hard, calcareous cell walls. Some of the earliest algal fossils are of this coralline class, and its members help make up the present-day coral reefs of tropical waters. Photosynthesis is essential for the life of the reef because the coralline algae, along with some species of green algae, are the primary producers of nutrients for the coral organisms of the reef. This explains why coral reefs develop only in shallow waters where sunlight can penetrate. The algae also help to precipitate calcium carbonate in the reef by removing carbon dioxide, thereby shifting the equilibrium of the following equation to the right:

$$\underset{\text{calcium bicarbonate}}{Ca(HCO_3)_2} \rightleftharpoons \underset{\text{calcium carbonate}}{CaCO_3\downarrow} + \underset{\text{water}}{H_2O} + \underset{\text{carbon dioxide}}{CO_2\uparrow}$$

PYRRHOPHYTA (DINOFLAGELLATES)

The dinoflagellates of the phylum Pyrrhophyta are primarily unicellular marine algae. They contain carotenes and xanthophylls and store oils and starch. Asexual reproduction is the general rule in this phylum.

One genus of dinoflagellates, *Gonyaulax,* may be lethal to man. Some spe-

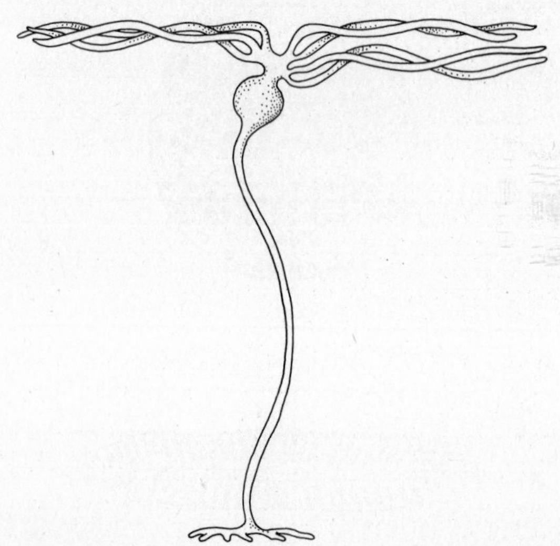

FIGURE 13-7
The kelp *Nereocystis* has a bulblike float that keeps the alga near the surface of the water, where the leaflike projections are exposed to light and can carry out photosynthesis.

cies of this genus produce a neurotoxin that causes a paralysis known as "paralytic shellfish poisoning" associated with "*red tides,*" ocean waters so abundant in red algae that fish are killed. This problem may occur wherever conditions allow abundant proliferation of *Gonyaulax* or a number of other toxin-producing algae. *Gonyaulax* and a few other species of marine algae are luminescent by means of a biochemical reaction similar to the one responsible for light emission by fireflies.

Not all of the eucaryotic algae can be classified within the major phyla. A small group of organisms which have not been positively classified are considered by some to constitute a separate phylum, Cryptophyta. Some other algae that are difficult to classify have been placed tentatively in another phylum of minor importance, Chloromonadophyta.

LICHENS AND OTHER SYMBIOTIC ALGAE

In addition to the coralline algae, other algae sometimes grow in symbiosis with certain organisms, including fungi. Lichen is the name given to a combination of an alga and a fungus growing together in a state of mutualism, in which each organism benefits. In a few lichens, however, the fungi appear to benefit at the expense of their algal partners. Green algae, such as *Trebouxia* species, are often found in lichens. However, the photosynthetic component of lichens is not necessarily algal; blue-green bacteria (formerly called blue-green algae) also grow together with fungi to form lichens.

Although the fungus and the alga of a lichen can be cultured separately they usually cannot be made to grow in symbiosis in the laboratory. The algal partners may be filamentous, intertwined with fungal filaments (hyphae), or they may be single-celled organisms supported by the hyphae of the fungus. The supporting fungal hyphae provide a safe haven for the alga, offering it protection from the environment, as well as nutrients extracted from the rocks, soils, or other surfaces on which the hyphae grow (Figure 13-8). The contributions of the alga are equally important. A green alga is the primary producer of nutrients for the pair, being able to utilize carbon dioxide and sunlight to produce carbohydrates. The fungus is able to absorb the nutrients from its partner by means of tiny rootlike fungal projections (haustoria) which appear to penetrate the algal cell.

Although they are widely distributed in nature, lichens are not often found near large cities because they are readily killed by the smokes and gases produced by civilization. However, they frequently grow in areas where many other kinds of organisms or vegetation cannot survive. Consequently, lichens are an important source of food, especially in extreme northern climates. The reindeer moss, which supports herds of reindeer, are lichens of the genus *Cladonia* (Figure 13-9). The stalked, cuplike structures of these lichens grow as high as 15 centimeters and are commonly found on ground which may be frozen during much of the year. *Cladonia* species and other lichens serve as food for man in various parts of the world.

Some of the lichens also make a significant contribution to soil formation

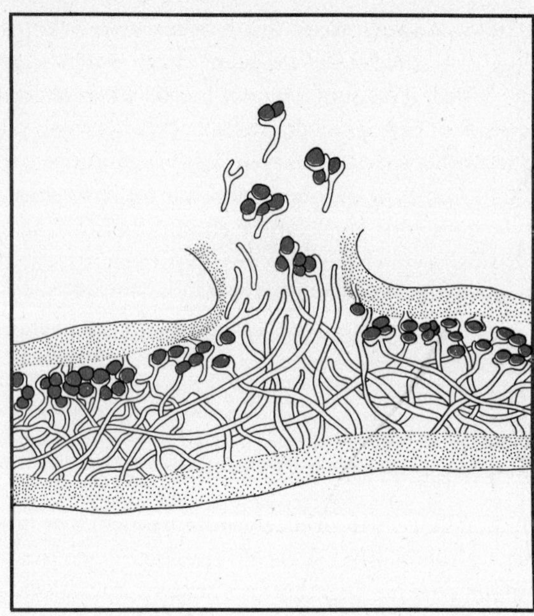

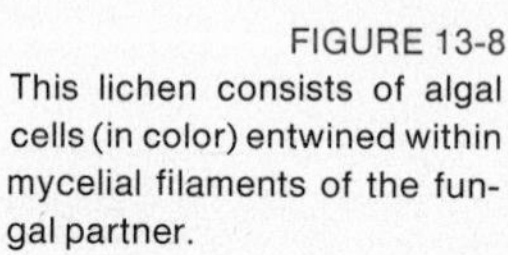

FIGURE 13-8
This lichen consists of algal cells (in color) entwined within mycelial filaments of the fungal partner.

by gradually eroding rocks or other strata on which they live. Unique acids made by the lichens aid in this breakdown of materials and also function to protect the lichens from destruction by insects and other small animals.

In addition to coral reefs and lichens there exist other examples of algal symbiosis. One species of green algae has been reported which lacks photosynthetic pigments but contains cyanobacteria within its cytoplasm. In this symbiotic arrangement the cyanobacteria supply the photosynthetic apparatus for both organisms.

ROLE OF ALGAE IN THE FOOD CHAIN

In an aquatic environment algae are the primary producers in the food chain, serving as the principal food source for the small animals that participate in the

FIGURE 13-9
A species of *Cladonia*, the lichen known as reindeer moss. (Courtesy of Turtox/Cambosco; Macmillan Science Co., Inc.)

chain. Thus, because fish thrive on the rich food supplies derived from algae, it is possible to show a direct correlation between the numbers of algae and fish present in a given area. In fresh waters and on land a similar chain of events takes place—the algae synthesizing carbohydrates, proteins, and other foodstuffs while at the same time producing oxygen, which supports animal respiration.

OTHER USES OF ALGAE

In addition to their vital activities in the food chain and in the production of oxygen, there are a number of other ways in which algae are of use to man.

Diatomaceous earth (diatomite, kieselguhr) consists of the frustules of dead diatoms. This silica-containing material is a mild abrasive, and it is the polishing agent in many metal polishes. Its resistance to temperatures of 600°C and higher makes it a good heat-insulating material, it is also valuable in making certain kinds of filters.

The red algae yield two polysaccharide products of major economic importance, agar and carrageenan, useful as jelling, emulsifying, and thickening agents. Alginates are another useful group of polysaccharides, derived from some of the brown algae such as the genus *Macrocystis.* Huge beds of *Macrocystis* off the coast of California, estimated to cover about 100 square miles, are "farmed" for the production of alginates. The use of algae as food is discussed in Chapter 39.

Algae as food, page 692

Vitamins A and D, produced by diatoms, are of importance both medically and commercially. Fish eat diatoms and concentrate the ingested vitamins in their livers. Man then extracts the vitamins from fish livers (for example, cod-liver oil) for human consumption.

HARMFUL ASPECTS OF ALGAE

Algal toxins, referred to previously, probably represent the most significant danger afforded by the algae. Although birds and mammals are occasionally affected by such algal toxins, fish are by far the most frequent victims. In Israel, for example, extensive use of fertilizers that drain from fields into fish ponds has enriched the ponds to the point where algal blooms are common. As a result thousands of fish have been killed each year. Consequently, extensive studies on the nature of the algal toxins have been undertaken, and several of the toxins have been well characterized. Perhaps some day man will be able to utilize these toxins for medicinal purposes, as snake venoms and other toxins are today being utilized. If that day comes, the benefits of algae to man will be even greater than they are today.

SUMMARY

Algae are ubiquitous photosynthetic organisms. Many are aquatic, but others occur in soil and on vegetation, trees, and rocks. Algae may be multicellular or unicellular and are often arranged in colonies, filaments, or other forms.

The characteristics of algae vary considerably among the six major phyla. These phyla include Chlorophyta (the green algae), Euglenophyta (unicellular flagellates), Chrysophyta (several classes, including the diatoms), Phaeophyta (the brown algae), Rhodophyta (the red algae), and Pyrrhophyta (the dinoflagellates).

Algae may exist in a number of symbiotic relationships, as, for example, with fungi in lichens or with corals in coral reefs. The food chains of the ocean depend on algae as the primary food producers. In addition to this contribution, algae or their products are put to many uses; thus algae are economically important. Some algae produce toxins that are lethal for man or other animals.

QUESTIONS

REVIEW

1. What major characteristics distinguish algae from bacteria?
2. Compare and contrast photosynthesis as carried out by algae and by the anaerobic photobacteria.
3. How are the six major phyla of algae distinguished?

THOUGHT

What might the world be like if all algae were killed?

FURTHER READING

CARSON, R. L., *The Sea Around Us.* New York: New American Library, 1950. A marine biologist presents a fascinating story of the seas and their inhabitants.

GIBOR, A., "Acetabularia: A Useful Giant Cell," *Scientific American* (November 1966).

ISAACS, J. D., "The Nature of Oceanic Life," *Scientific American* (March 1969).

PRESCOTT, G. W., *The Algae: A Review.* Boston: Houghton Mifflin, 1968.

ROUND, F. E., *The Biology of the Algae.* London: Edward Arnold, 1965. A scholarly and general account of the biology of algae and cyanobacteria.

SHILO, M., "Formation and Mode of Action of Algal Toxins," *Bacteriological Reviews* 31:180–193 (1967). A review in depth of algal toxins.

TAYLOR, D. L., "Algal Symbionts of Invertebrates," *Annual Review of Microbiology* 27:171–187 (1973). Emphasizes the biology of the algal partner in many symbiotic relationships.

FUNGI

Yeasts, molds, and mushrooms are part of the major group of eucaryotic protists called the *fungi.* These relatively simple organisms lack chlorophyll and are non-photosynthetic. Most of them live as *saprophytes* on dead organic materials, often in soil, but some are widespread parasites of living plants or animals. The fungi include many hundreds of important plant pathogens but relatively few species (less than 100) that are pathogenic for animals and man.

OCCURRENCE

The fungi are found in all parts of the world, wherever organic materials are available as nutrients. Some species are very specialized; occurring naturally, for example, only on a particular strain of one genus of plants. Others are extremely versatile metabolically and are able to subsist on (and destroy) many kinds of unlikely media such as leather, cork, hair, wax, ink, or even synthetic plastics, especially polyvinyls. Other species are able to grow in concentrations of salts, sugars, or acids high enough to kill most bacteria; these fungi are responsible for spoiling pickles, preserves, and certain other foods. Some fungi are resistant to pasteurization temperatures, and others can actually grow below the freezing point and rot bulbs and destroy grass in frozen ground.

Fungal reproductive cells, or spores, are found virtually all over the earth and are also abundant in the air. These minute forms occur in tremendous numbers in air near the earth's surface and have also been recovered from air currents at altitudes of more than seven miles.

CLASSIFICATION

The principal characteristics of the fungi are mycelial growth (described in the next section), spore formation, and the lack of chlorophyll. The fungi are divided into four major classes (Table 14-1), largely on the basis of their method of sexual reproduction or lack thereof.

MORPHOLOGY

Fungal cells and filaments are enclosed by *rigid cell walls* that usually contain the polysaccharide chitin. A fungal reproductive spore is typically a single cell about 3 to 30 μm in diameter, depending on the species, although some fungal spores are as small as 1 μm and others may be 300 μm long. When a mold spore reaches a favorable environment, it germinates and sends out a projection called a *germ tube* (Figure 14-1). This tube continues to grow into a long filament called a *hypha* (about 2 to 10 μm in diameter) which branches and intertwines with other hyphae. The complex tangled mass of hyphae which results is known as a *mycelium.* Mycelial formation is a principal distinguishing characteristic of many, but not all, fungi. The cottony growths of molds on bread and fruits are familiar examples of mycelia. The visible portion of the mycelium usually represents just a small part of the total mold growth, with submerged hyphae growing into the substrate making up the bulk of the growth.

Three of the classes of fungi (the Ascomycetes, Basidiomycetes, and Deu-

TABLE 14-1
Some Characteristics of the Major Classes of Fungi

Group	*Usual Habitat (Representative Groups)*	*Mycelia*	*Asexual Spores*	*Sexual Spores*
Phycomycetes	Water (watermolds); soil, bread, (black bread mold), etc.	Nonseptate	Sporangiospores, occasionally conidia	Various forms; may have flagellate gametes; zygospores; oospores
Basidiomycetes	Soil (mushrooms); grains (rusts, smuts)	Septate	Commonly not formed	Basidiospores (borne on clublike basidium)
Ascomycetes	Fruits, other organic media (true yeasts, *Neurospora*)	Septate	Conidia	Ascospores (borne in saclike ascus)
Deuteromycetes (Fungi Imperfecti)	Soil (most of the pathogens for man)	Septate	Conidia Arthrospores	Absent

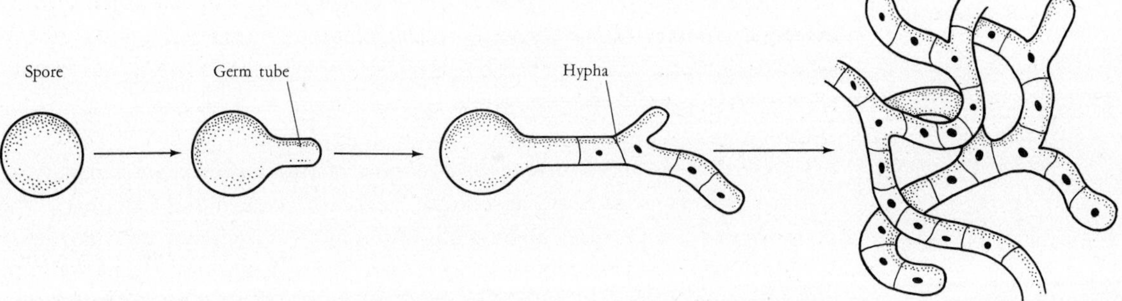

FIGURE 14-1 Spores of fungi germinate by forming a projection from the side of the cell (a germ tube) which elongates to form hyphae. As the hyphae continue to grow, they form a tangled mass called a mycelium.

teromycetes) form mycelia that are divided into cell-like compartments (septate mycelia) having one or two nuclei each. However, the dividing cross wall, or *septum,* is not completely closed, and cytoplasm can circulate from one compartment to another. Members of the fourth class of fungi, the Phycomycetes, have mycelia without cross walls (nonseptate) so that their entire mycelium may actually consist of a single multinucleated cell.

Yeast is the name given to single-celled ascomycetes or deuteromycetes that lack a mycelium, while mold refers to fungi with a mycelium. The true yeasts never grow in mycelial form, but other species of fungi, known as *dimorphic fungi,* are capable of growing in either the yeast or the mold form, depending on environmental conditions. A rich medium, 37°C temperature, and increased carbon dioxide in the atmosphere favor growth of the yeast form; the mold form of dimorphic fungi tends to grow under less favorable conditions. The morphology of a typical yeast cell is shown in Figure 14-2.

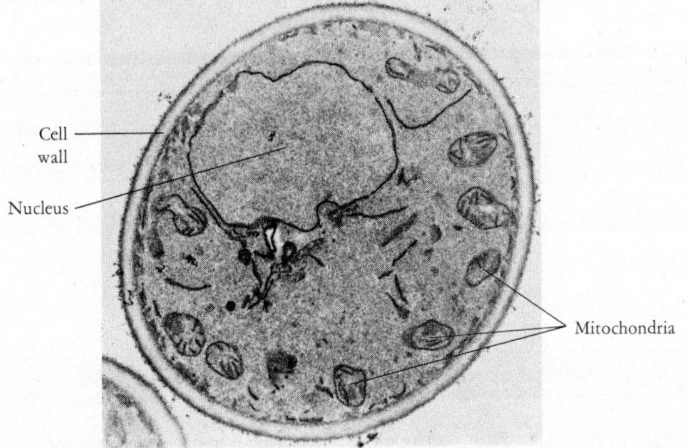

FIGURE 14-2 Morphology of a yeast cell as seen with the electron microscope. (×48,000) (Courtesy of E. S. Boatman and H. C. Douglas.)

PHYSIOLOGY AND METABOLISM

The fungi secrete a wide variety of enzymes which degrade organic materials, especially complex carbohydrates, into small molecules that are readily absorbed as nutrients. Most fungi are aerobic, but some of the yeasts are facultatively anaerobic and carry out important fermentation reactions. There are no obligately anaerobic fungi. Light is not necessary for fungal growth; in fact sunlight will often kill the vegetative cells of fungi. Fungal spores, however, are generally resistant to the ultraviolet rays of sunlight. Although some molds are aquatic, the majority of fungal species prefer a slightly moist environment with a relative humidity of 70 percent or more. Most fungi thrive at temperatures ranging from 20°C, or approximately room temperature, to 35°C, just below body temperature.

The pH at which fungi can grow varies widely, ranging from as low as 2.2 to as high as 9.6, but they usually grow best at an acid pH of about 5.0.

REPRODUCTION

Both the asexual and sexual reproductive spores of fungi occur in a number of forms, as indicated in Table 14-2 and diagramed in Figure 14-3. These fungal

TABLE 14-2
Some Characteristics of Fungal Spores

Type of Spore	*Characteristics*	*Examples*
Asexual Spores		
Conidiospores	Single-celled (microconidia) or multicellular (macroconidia): formed by extrusion of single cells from specialized areas of hyphae	*Penicillium* *Alternaria*
Arthrospores	Single-celled; formed by disjointing of hyphal cells	*Coccidioides*
Sporangiospores	Single-celled; formed within sacs at end of hyphae	*Rhizopus*
Blastospores	Buds; formed on yeast cells	*Candida*
Chlamydospores	Thick-walled, single cells; highly resistant to adverse conditions	*Candida*
Zoospores	Single-celled; motile with flagella	*Saprolegnia*
Sexual Spores		
Ascospores	Single cells in ascus (often eight per ascus)	*Neurospora* (class Ascomycetes)
Basidiospores	Single cells borne on basidium (often four per basidium)	*Agaricus* (class Basidiomycetes)
Zygospores	Large spore encased in a thick wall	*Rhizopus* (class Phycomycetes)
Oospores	Develop within oogonium (1 to 20 or more per oogonium)	*Saprolegnia* (class Phycomycetes)

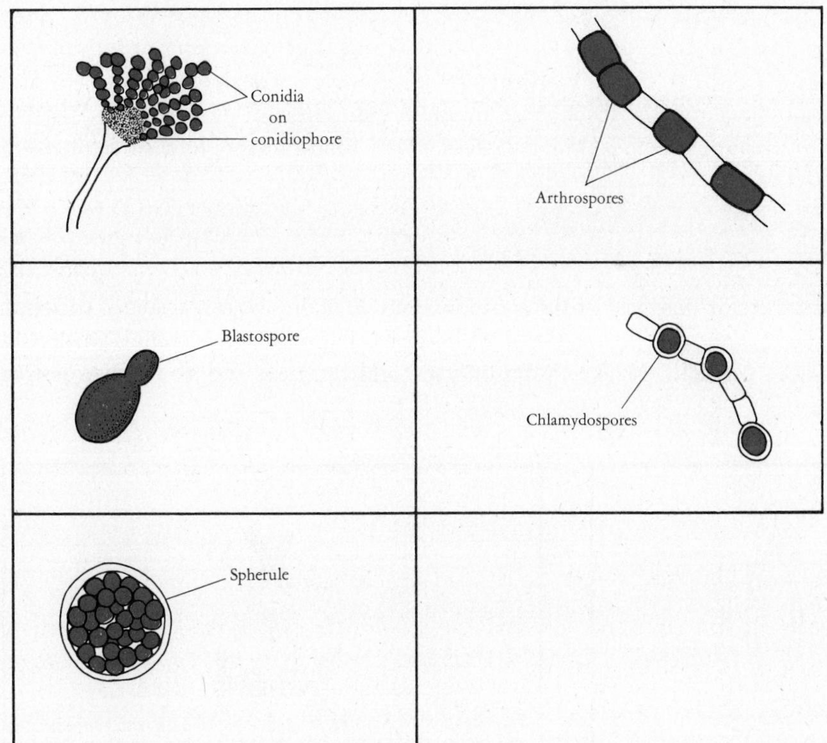

FIGURE 14-3
The morphology of various kinds of fungal spores.

spores should not be confused with the bacterial endospores, which are highly resistant to environmental influences. The fungal spores function as reproductive cells and are usually not significantly more resistant than other fungal cells. The sexual spores of most fungi are borne on specialized structures called *fruiting bodies.* Sexual reproduction is common to most fungi but often occurs only after many generations of asexual division. The Basidiomycetes, however, usually reproduce sexually and do not often form asexual spores. Examples of modes of sexual reproduction are given in the discussions of the various classes of fungi.

SOME CHARACTERISTICS OF THE CLASSES OF FUNGI

Phycomycetes

Members of this class are often called the *lower fungi.* Some of them, appropriately called the *water molds,* are aquatic and may be simple or complex; other groups are terrestrial. The simpler water molds are unicellular and do not produce a mycelium; however, other species do develop mycelial growth. The water molds reproduce asexually by means of motile zoospores which have one or two flagella and closely resemble certain protozoa. More highly developed genera of phycomycetes reproduce both sexually and asexually.

A familiar example of a terrestrial phycomycete is the genus *Rhizopus,* which includes the common black bread mold. It reproduces by means of asexual sporangiospores, as shown in Figure 14-4a; however, sexual reproduction is also common (Figure 14-4b). Organisms of the genus *Rhizopus* do not cause disease in healthy human beings, but can cause a fatal illness in people with debilitating diseases.

Basidiomycetes

Members of this class of fungi produce sexual spores on a specialized structure called a *basidium* (Figure 14-5). Well-known Basidiomycetes include the mushrooms, puffballs, bracket fungi of trees, and the rusts and smuts that destroy

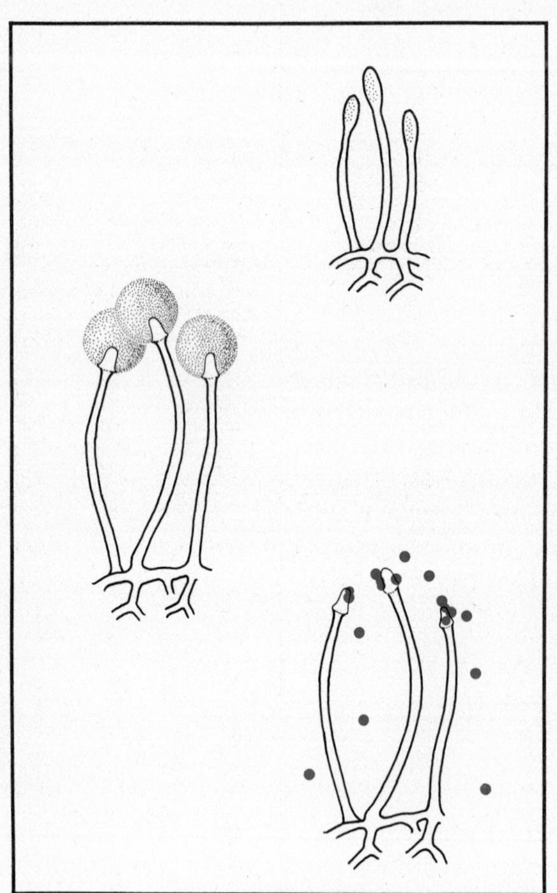

(a)

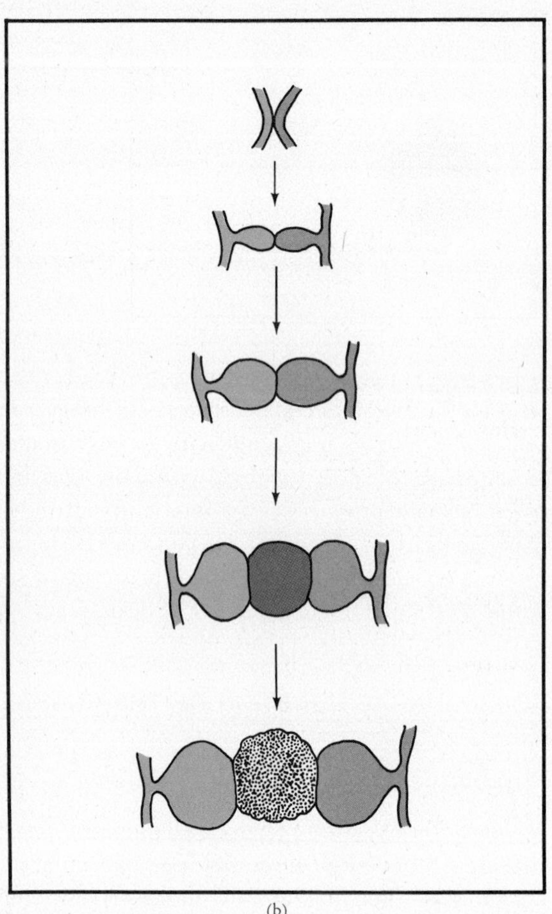

(b)

FIGURE 14-4 (a) Asexual sporangiospores of *Rhizopus*. The spores develop within a sporangium, which ruptures to release the mature spores. (b) Sexual reproduction in *Rhizopus*. Two mycelia of opposite sex (+ and −) come together and each forms a side branch. Following division of the two branches, the cells in contact fuse to form a zygospore, which becomes surrounded by a thick, black wall.

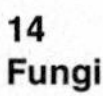

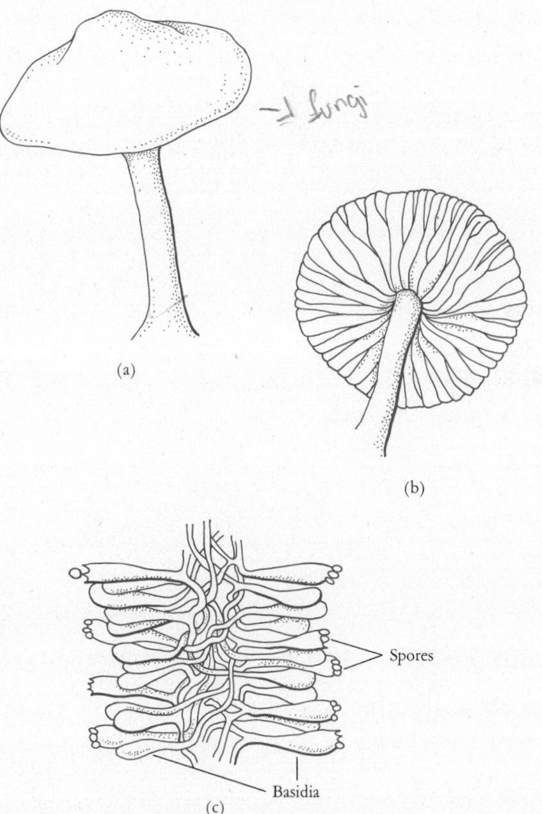

FIGURE14-5
(a) A typical Basidiomycete fruiting body. (b) The underside of the cap is composed of radiating gills. (c) A magnified view of the surface of a gill, showing a mass of basidia, bearing spores.

grains. The familiar mushrooms are merely the fruiting bodies of large subsurface mycelia, and they bear huge numbers of sexual spores. Figure 14-5 also illustrates the gilled structure of these fruiting bodies, with spores borne on the sides of the gills.

The life stages of the meadow mushroom, *Agaricus campestris,* are shown in Figure 14-6. Spores that land in favorable soil germinate to form a haploid mycelium which then spreads through the soil. Two such haploid mycelia can fuse to give rise to a dikaryon mycelium in which two nuclei are associated but not fused.

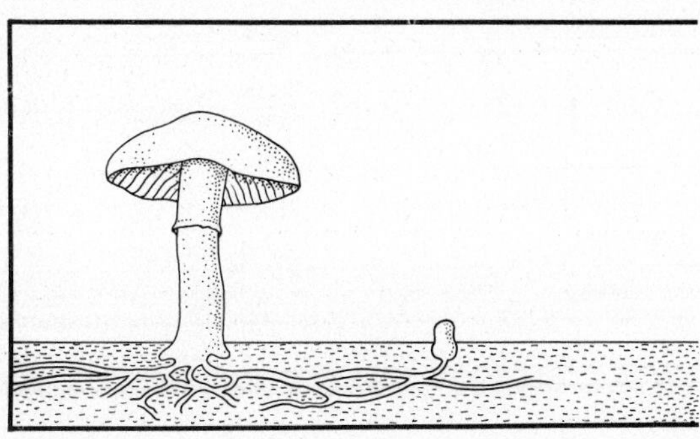

FIGURE 14-6
The meadow mushroom, *Agaricus campestris,* has an extensive underground mycelium. The fruiting body emerges as a small button mushroom (right), which rapidly grows into a typical mushroom with a gilled cap (left).

The mycelium secretes enzymes that break down complex organic compounds into simple nutrients it can absorb. After a period of growth the fruiting bodies (formed from dikaryon mycelia) emerge above the ground. Fusion of the nuclei within the fruiting bodies is followed by meiosis to produce haploid spores. The tremendous numbers of haploid spores produced along the gills of the fruiting bodies eventually become airborne, and each can give rise to a new organism.

Ascomycetes

This very large class includes most of the yeasts and certain other well known fungi such as morels and truffles. The sexual spores of Ascomycetes, formed within a structure called an *ascus,* often resemble peas in a pod (Figure 14-7). A simplified version of the life cycle of a species of *Neurospora,* a member of the Ascomycetes, is shown in the figure.

Practical uses of yeast, pages 672–676, 681, 683, 684, 692, 693

The life cycle of the yeast most often used for making bread, wine, and beer, *Saccharomyces cerevisiae,* is outlined in Figure 14-8. Actually, the cycle is more complex than indicated. This yeast divides asexually by budding, but it is also capable of conjugation and sexual reproduction.

Deuteromycetes (Fungi Imperfecti)

This group has been called the taxonomic dumpheap of the fungi, because it comprises those species which lack an observed perfect, or sexual, stage of repro-

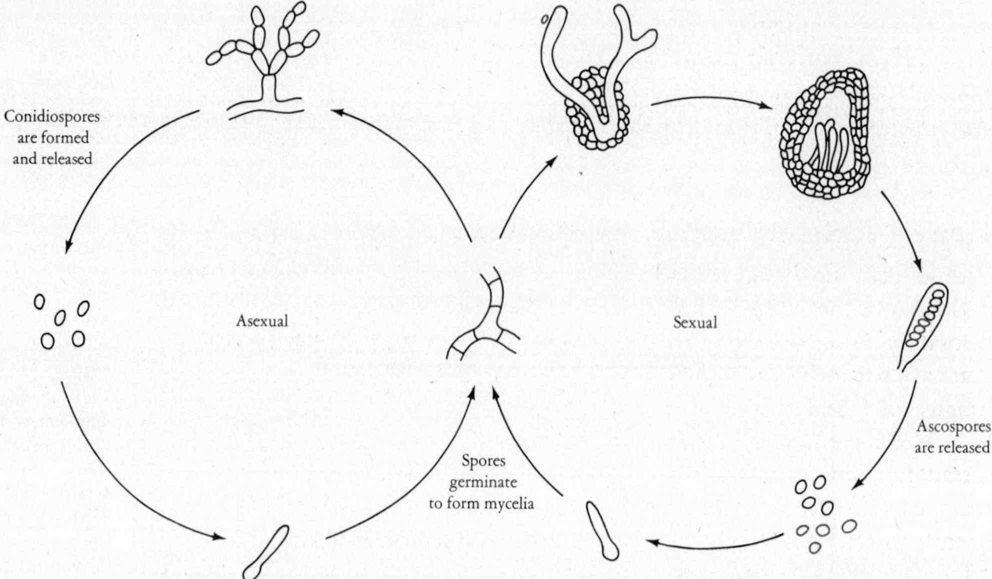

FIGURE 14-7 The life cycle of an ascomycete (a species of *Neurospora*). During the sexual cycle cells of opposite mating types (indicated by colors) fuse and give rise to ascospores enclosed within asci (singular: ascus). Study of this fungus has contributed significantly to understanding genetic recombination in eucaryotic cells.

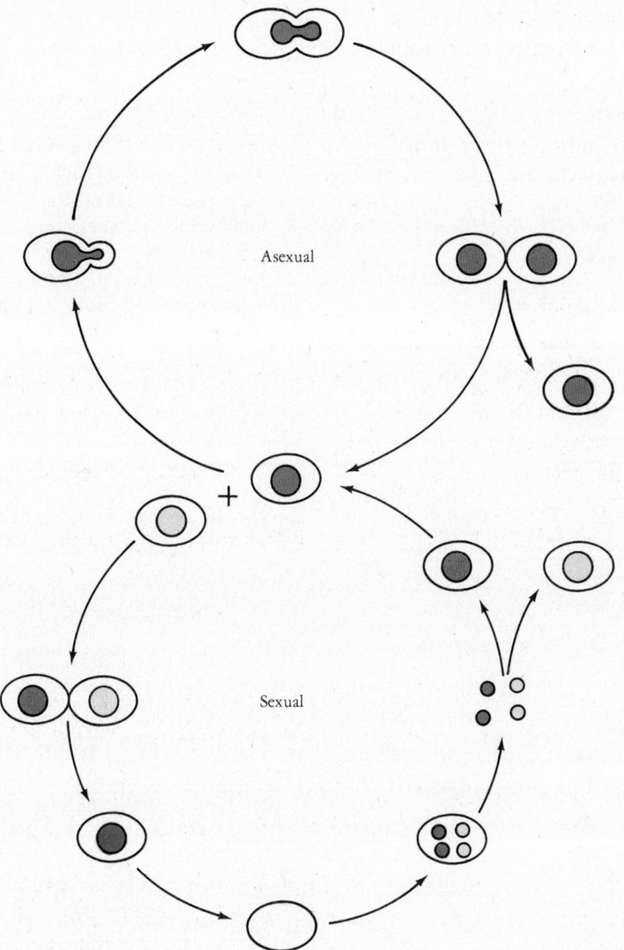

FIGURE 14-8
The life cycle of the yeast *Saccharomyces cerevisiae*. During asexual reproduction the yeast cells divide by budding. During sexual reproduction two cells of opposite mating types (indicated by colors) fuse, and an ascus is formed. By the process of meiosis four haploid ascospores are formed, two of each mating type.

duction and which therefore cannot be placed in any other fungal class. The Fungi Imperfecti apparently have become so adept at maintaining their species without utilizing sexual processes that they have either lost the capacity for sexual reproduction or else use it so seldom that it has not been observed in the laboratory. From time to time sexual sporulation has been discovered in fungi that were previously considered to be Fungi Imperfecti. In most of these cases the fungus was found to be an ascomycete.

Molds of the genera *Penicillium* and *Aspergillus* are deuteromycetes of widespread distribution and commercial importance. In addition, most of the fungi that are pathogenic for man are deuteromycetes. They form asexual spores—frequently several varieties within the same species—and these spores are often helpful in identifying the pathogenic fungi.

An overall view of the fungi of major medical importance, all of them members of the group Deuteromycetes, is presented in Table 14-3; some of these organisms are also illustrated in other chapters.

Medically important fungi, pages 470, 493–497, 534, 535, 557, 558

TABLE 14-3
Some Deuteromycetes (Fungi Imperfecti) of Major Medical Importance

Genus or Genus and Species	*Disease Produced*	*Important Characteristics*
Epidermophyton *Microsporum* *Trichophyton*	Superficial fungal infections	Infect superficial skin, hair, or nails; cause tinea (infections such as athlete's foot, ringworm, and others)
Histoplasma capsulatum	Histoplasmosis	Dimorphic yeast; infections described in Chapter 26
Coccidioides immitis	Coccidioidomycosis	Large spherules (15–75 μm) filled with endospores; infections described in Chapter 26
Blastomyces	Blastomycosis	Blastospores
Sporothrix	Sporotrichosis	Rosette-arrangement of conidia (Chapter 30)
Cryptococcus neoformans	Cryptococcosis (frequently meningitis)	Usually nonpathogenic but can cause severe or fatal infections
Candida albicans	Candidiasis (thrush, moniliasis)	Usually nonpathogenic but can invade (Chapters 25 and 28)

SYMBIOTIC RELATIONSHIPS OF FUNGI

Lichens, page 285
Mycorrhizas, page 613

In addition to their association with algae in lichens, fungi have developed many other symbiotic relationships. *Mycorrhizas,* fungal symbioses of particular importance, are formed by the intimate association between fungi and the roots of certain plants. The fungi involved commonly occur only in association with the roots, and they often allow their plant partners to grow where these plants could not otherwise survive. With the world facing major shortages of food and forest products, better understanding of these symbiotic relationships is urgently needed. Many trees have mycorrhizal roots, including the conifers that are able to live in sandy soil by means of their extensive mycelial mycorrhizas. Truffles form mycorrhizas with either oak or beech trees. Indian pipe and some other mycorrhizal plants have lost the ability to form chlorophyll and thus have become completely dependent on their fungal partners for nutrition.

Certain insects also depend on symbiotic relationships with fungi. For example, some ants and termites grow fungi for food, nourishing the fungi with leaves which they carry into their nests. Also, woodboring beetles introduce fungal spores into wood where the fungi grow and subsequently serve as food for the beetles.

UTILIZATION AND PRACTICAL IMPORTANCE

Fungal plant pathogens, pages 630, 631

Many of the fungi are important commercially. They participate in the production of many alcoholic beverages, of foods and chemicals, and of some antibiotics and drugs. Fungal plant pathogens have always been of great economic and social importance. The Irish potato famine in the 1800s was caused by fungi. This major disaster resulted in starvation and mass migration to other countries and had a profound effect on the history of the United States. A variety of fungal plant pathogens are currently responsible for large economic losses each year. In addition, fungi are essential as degraders of organic wastes.

Fungi in soil, pages 612–614

THE SLIME MOLDS (MYXOMYCETES)

The unattractive name of slime molds has been given to a group of unique eucaryotic protists that bears some resemblance to both fungi and protozoa. Four diverse orders comprise this group: two of them are amebae during their vegetative phase, and the other two are acellular, consisting of masses of protoplasm. Members of all four orders often produce funguslike fruiting bodies, and all can form slimy vegetative bodies which are not differentiated into tissues—thus conferring the name slime molds.

The *vegetative form* of the acellular slime molds (called the *plasmodium*) *is a multinucleate mass of protoplasm* which oozes like slime over the surface of the substrate, usually decaying wood and leaves. Furthermore, a slime mold, unlike other fungi, can ingest microorganisms, fungal spores, and small particulate matter encountered in its movement. Although it looks like slime, the plasmodium of the acellular slime molds is actually very similar to a fungal mycelium. It is, in fact, one very large cell, up to 7 cm in diameter, containing many diploid nuclei. The large cell is enclosed by a cytoplasmic membrane within which the cytoplasm can flow. However, in contrast to a fungal mycelium, the cytoplasm of slime molds is not surrounded by a rigid cell wall and so is free to ooze over the substrate. As long as environmental conditions remain favorable for vegetative development, the plasmodium continues to enlarge. But under conditions of starvation and in the presence of light, the plasmodium develops into dozens of small, intricately constructed and often brilliantly colored fruiting bodies, each one consisting of many *sporangia* situated on top of a stalk (Figure 14-9). Each sporangium is a saclike structure containing many uninucleate spores, formed by the walling off of sections of the plasmodium. These resting spores are generally blown by wind, and upon arrival in a moist environment they germinate. Each spore gives rise to one or more cells which can divide vegetatively for a time. Some haploid sexual spores are also formed, and under appropriate conditions two such sexual cells fuse to form a zygote, which in turn develops into a new plasmodium.

The cellular slime molds resemble the acellular forms in that they also have a life cycle which includes a vegetative as well as a fruiting body stage (Figure 14-10). However, they differ from the acellular slime molds in that the vegetative stage of the cellular slime molds consists of uninucleate amebae which are almost indistinguishable from true amebae. When the food supply becomes exhausted, some of these amebae begin to excrete a hormonelike substance, now known to be cyclic AMP, which attracts other amebae, causing them to stream toward a common point. When a large number of the amebae have come together, a "pseudoplasmodium," or slug, is formed. This community of amebae, each of which retains its individuality, acts as a single unit. The slug lifts itself, its posterior end flattened on the surface and its anterior end forming a nipplelike structure. The amebae at the top begin to migrate downward through the center of the slug, secreting cellulose as they migrate, thereby forming a cylinder from the top to the base of the slug. When the cellulose cylinder reaches the base of the slug, the amebae at the base begin to migrate up the stalk and finally form a mass at the tip. Each of these amebae becomes enveloped in a thin cellulose wall to form the mature spores of the organism, which are held together by a droplet of viscous liquid. When released, the spores germinate and develop once again into amebae. In certain respects, this life cycle resembles that of the myxobacteria.

Myxobacteria, pages 262–263

The various slime molds are important links in the food chain in soil. They ingest bacteria, algae, and other organisms as nutrients and in turn serve as food for larger predators. Slime molds have been valuable also as unique models for studying cellular differentiation during their aggregation and especially during the formation of their fruiting bodies.

(a) (b) (c) (d)

FIGURE 14-9 Fruiting bodies of slime molds. (a) Species of *Stemonitis* growing on a log. (Courtesy of J. P. Dalmasso). (b) *Dictydium cancellatum*. (c) *Metatrichia vesparium*. (d) *Lepidoderma tigrinum*. (b, c, and d courtesy of E. F. Haskins.)

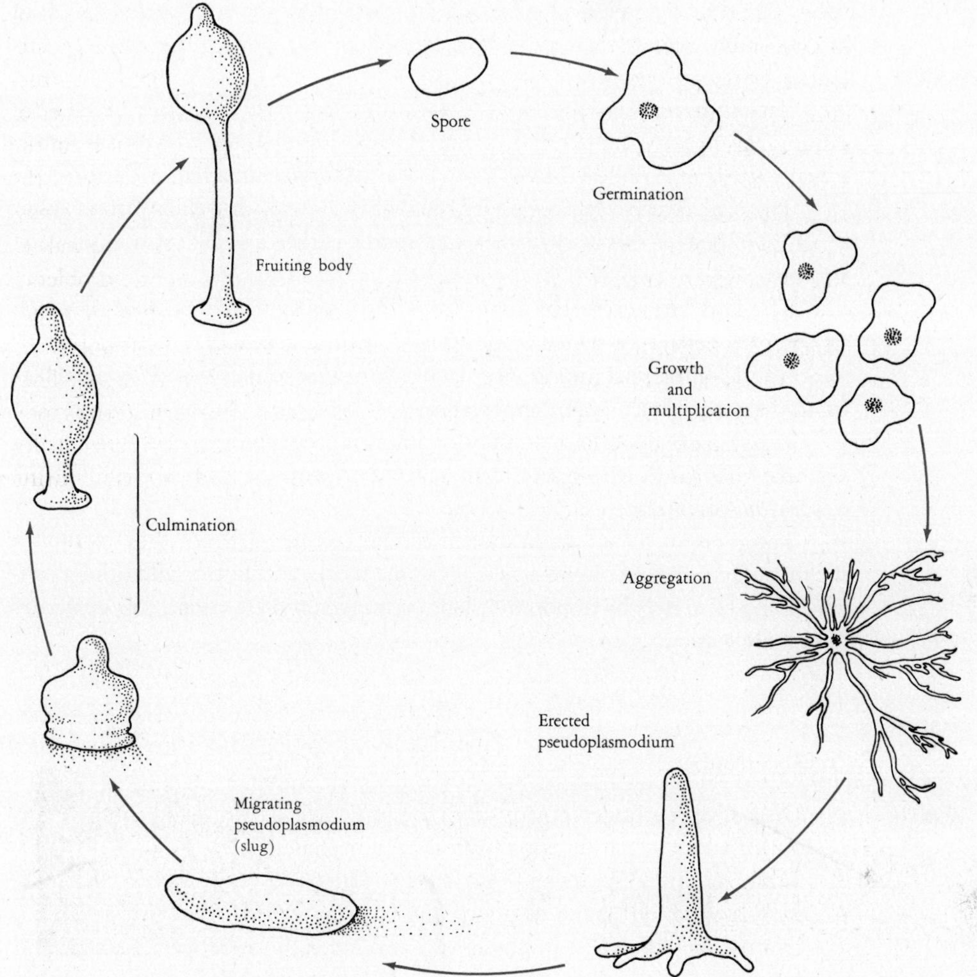

FIGURE 14-10 The life cycle of a cellular slime mold.

SUMMARY

The fungi are nonphotosynthetic, eucaryotic protists which include yeasts, molds, and mushrooms. A majority of them are saprophytes, living on dead organic materials, but some are parasitic on living plants or animals. Mycelial growth is characteristic of many fungi. Mycelial filaments and other fungal forms are enclosed by rigid cell walls, which usually contain chitin. Fungi reproduce by means of asexual spores, sexual spores, or both. Fungi are classified, largely on the basis of their method of sexual reproduction or lack thereof, into four classes: Phycomycetes, Basidiomycetes, Ascomycetes, and Deuteromycetes.

Phycomycetes, often called the lower fungi, may be aquatic or terrestrial. A group of phycomycetes, the aquatic water molds, reproduce asexually by means of motile zoospores. More highly developed genera reproduce asexually and sexually. The terrestrial phycomycetes are represented by the common black bread mold *Rhizopus,* which reproduces by means of asexual sporangiospores or by formation of sexual zygospores.

Basidiomycetes produce sexual spores on a specialized structure called a basidium. Mushrooms are familiar examples of Basidiomycetes. The mushroom is actually the fruiting body formed from large subsurface mycelia.

The class Ascomycetes includes most of the yeasts, as well as certain other fungi. Members of this class form sexual spores within a saclike structure called an ascus.

The Deuteromycetes, or Fungi Imperfecti, are so named because they lack an observed perfect, or sexual, stage. They comprise genera of widespread distribution and commercial importance, such as *Penicillium* and *Aspergillus,* as well as most of the fungi of medical importance.

Fungi participate in a number of important symbiotic relationships; for example, they join with algae to form lichens and with the roots of certain plants to form mycorrhizas.

Fungi are of great significance in certain commercial processes, including the manufacture of alcoholic beverages, some foods, antibiotics, and other products. Fungi also may be important plant pathogens and are essential as degraders of organic wastes.

QUESTIONS

REVIEW

1. Discuss the pathogenicity of fungi for animals and plants, and compare this with the disease-producing capabilities of bacteria.
2. What are fungal spores and how do they function?
3. How are the four major classes of fungi distinguished?
4. Discuss the differences in physiology and metabolism between bacteria and fungi.
5. Discuss several symbiotic relationships of fungi with other organisms.

THOUGHT

Weigh the values of fungi (economically, industrially, in the food chain, and so forth) against their deleterious effects for human beings (deterioration, disease production, and so on). Would the world be a better place without fungi?

FURTHER READING

Ainsworth, G. C., and A. S. Sussman, *The Fungi. An Advanced Treatise.* Vols. I and II. New York: Academic Press, 1965.

Bonner, J. T., "Hormones in Social Amoebae and Mammals," *Scientific American* (June 1969).

Christensen, C. M., *The Molds and Man. An Introduction to the Fungi.* Minneapolis: University of Minnesota Press, 1951.

EMMONS, C. W., C. H. BINFORD, and J. P. UTZ, *Medical Mycology.* 3d ed. Philadelphia: Lea and Febiger, 1977. The third edition of a classic text in mycology.

FULLER, J. G., *The Day of St. Anthony's Fire.* New York: Macmillan, 1968. A story of an unusual event thought to have resulted from pharmacologically active products of fungi.

Human Mycoses. Scope monograph. Kalamazoo, Mich.: The Upjohn Company, 1968. A well-illustrated manual of the major fungi responsible for human diseases.

KAVALER, L., *Mushrooms, Molds, and Miracles.* New York: Signet Books (division of John Day), 1965. A fascinating book about fungi, written in nonscientific language but containing much sound information.

MCKENNY, M., and D. E. STUNTZ, *The Savory Wild Mushroom.* Seattle: University of Washington Press, 1971. Scientifically accurate and esthetically pleasing, this well-illustrated book was written to enable the mushroom hunter to answer the question "Is it good to eat?"

Microbiology-1975. Ed. D. Schlessinger, Washington, D.C.: American Society for Microbiology, 1975, pp. 343–396. The report of a symposium on mycotoxins, such as aflatoxins, patulin, and others.

MYRVIK, Q. N., N. N. PEARSALL, and R. S. WEISER, *Fundamentals of Medical Bacteriology and Mycology.* Philadelphia: Lea & Febiger, 1974. The last four chapters present an abbreviated overview of medical mycology.

NEWALL, P. C., "Cellular Communication during Aggregation of the Slime Mold *Dictyostelium,*" *Microbiology-1975.* Ed. D. Schlessinger. Washington, D.C.: American Society for Microbiology, 1975, pp. 426–433. A review of studies on the biochemical and genetic aspects of the aggregation of slime molds.

WILSON, J. W., and O. A. PLUNKETT, *The Fungous Diseases of Man.* Berkeley and Los Angeles, Calif.: University of California Press, 1965. An excellent work concerned with human diseases caused by fungi.

CHAPTER 15

PROTOZOA

In addition to algae and fungi, the eucaryotic protists include another major category, the protozoa. There are not always clear dividing lines among the algae, fungi, and protozoa, and many species of these eucaryotic protists have the properties of more than one group. The slime molds, for example, have some characteristics of all three groups. Another example occurs with the Euglenophyta, a group of unicellular flagellates that is often considered to be protozoa. Most euglenae have chlorophyll and carry out photosynthesis, clearly properties of the algae; on the other hand, some species lack photosynthetic pigments, and all of the species in this group are capable of existing on complex nutrients when in the dark, both characteristics of the protozoa. Mutant strains of the algal genus *Chlamydomonas* which have lost chlorophyll might be classified in the protozoan genus *Polytoma,* but some protozoologists consider all *Chlamydomonas* to be protozoa.

In general, however, protozoa differ from algae and fungi in various ways, including their lack of photosynthetic capability, their motility, their means of reproduction, and other characteristics. Most of the protozoa are microscopic unicellular organisms, with some as small as 2 μm in diameter, while other species are easily visible to the unaided eye.

The life cycles of protozoa are sometimes complex, involving more than one habitat or host. It is usual to find morphologically distinct forms of a single protozoan species at different stages of the life cycle; such organisms are said to be *polymorphic* (Figure 15-1). This polymorphism can be compared to differentiation of various cell types to form the specialized tissues of plants and animals.

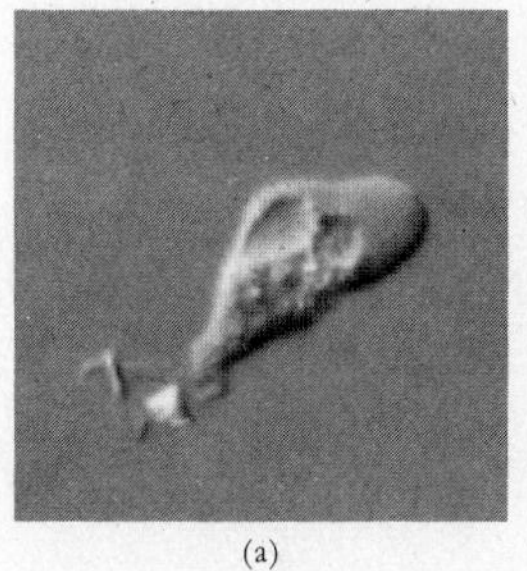
(a)

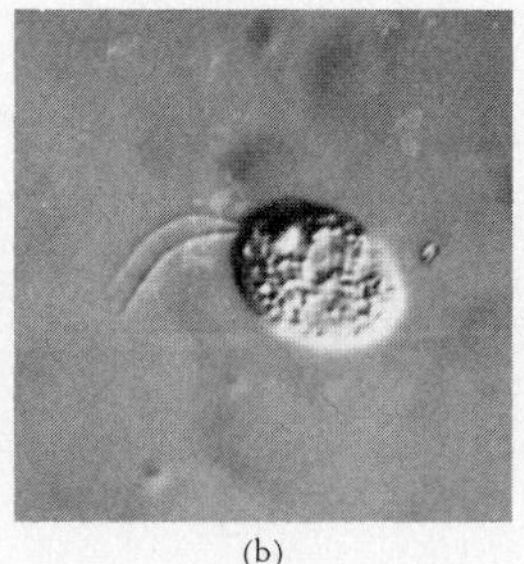
(b)

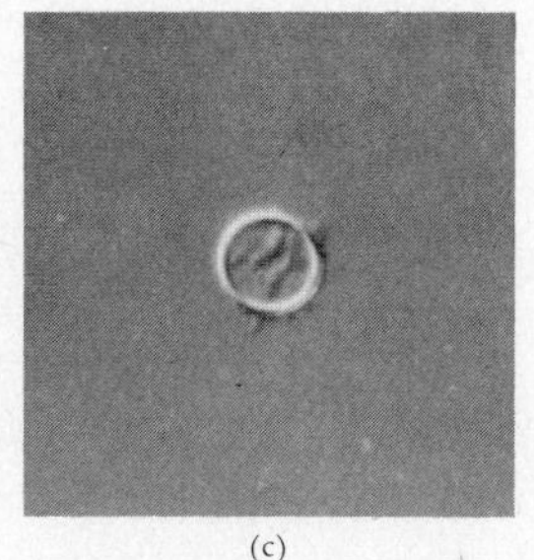
(c)

FIGURE 15-1
Polymorphism in a protozoan. This species of *Naegleria* may infect man. In human tissues the organism exists in the form of (a) an ameba (10 to 11 μm in its widest diameter). After a few minutes in water the flagellate form (b) appears. Under adverse conditions a cyst (c) is formed. (Courtesy of M. Roth and F. Schoenknecht.)

OCCURRENCE

All protozoa require large amounts of moisture, no matter what their habitat may be. Various species are found in marine, freshwater, or terrestrial environments, and a number are parasitic, living on or in other host organisms. The hosts for protozoan parasites range from simple organisms, such as algae, to complex vertebrates, including human beings.

Some marine protozoa make up part of the *zooplankton,* where they feed on algae of the *phytoplankton.* Other marine protozoa live attached to objects at the bottom of the water or along the shores. Freshwater species occupy similar habitats in lakes, streams, ponds, and even puddles of stagnant water.

On land, protozoa are abundant in soil as well as in and on plants and animals. Some specialized protozoan habitats include the gut of termites, roaches, and ruminants such as cattle.

MORPHOLOGY

Although polymorphic life cycles are common among the protozoa, the mature stage of each species is characteristic of the class to which it belongs. The four classes are listed in Table 15-1.

Organelles of locomotion in protozoa are either flagella, cilia, or pseudo-

TABLE 15-1
Some Properties of the Major Classes of Protozoa

Class	*Mode of Motility*	*Mode of Reproduction*
Mastigophora	Flagella (one or more)	Longitudinal fission
Sarcodina	Pseudopods (some have flagella also)	Binary fission
Sporozoa	Often nonmotile; may have flagella at some stages; creeping motion in some	Multiple fission; asexual reproduction in one host; some species have sexual reproduction in a second host
Ciliata	Cilia (multiple)	Transverse fission, asexual; also sexual reproduction involving micronucleus

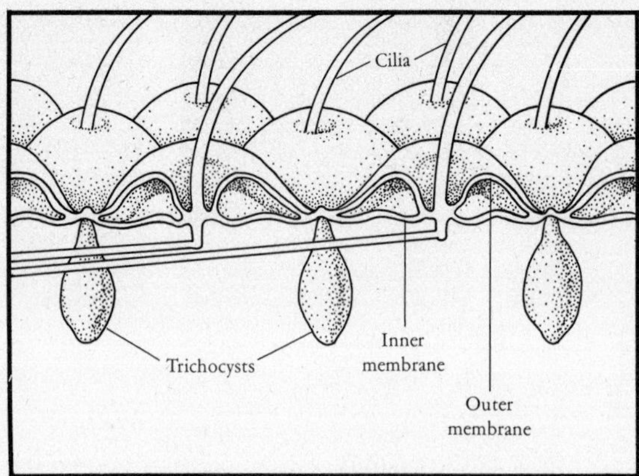

FIGURE 15-2
A cross section of the pellicle of *Paramecium* as seen by electron microscopy. The many cilia emerge through the inner and outer membranes of the pellicle. Undischarged trichocysts extend below the pellicle.

pods, depending on the species. The cilia differ from the flagella in being shorter and having their basal bodies connected by fibers which some believe may coordinate the ciliary motion. In ciliates the multiple cilia also help to propel food into the cell. Pseudopod formation in amebae is useful in these organisms' collection and ingestion of food, as well as in their locomotion.

Many protozoa form fibrillar or skeletal structures that give shape or rigidity to their cells. Slender fibrils in the cytoplasm of some species, such as *Amoeba proteus,* are associated with pseudopod-mediated motion. The ciliates commonly possess a thickened nonrigid *pellicle* consisting of two outer layers of membranes with a less dense middle layer (Figure 15-2).

Foraminifera of the class Sarcodina often have shells (sometimes called *tests*) impregnated with calcium or silicon compounds. Hard protozoan coverings known as agglutinated shells are made of an organic material in which granules of hard, inorganic materials are embedded. Frequently these coverings are com-

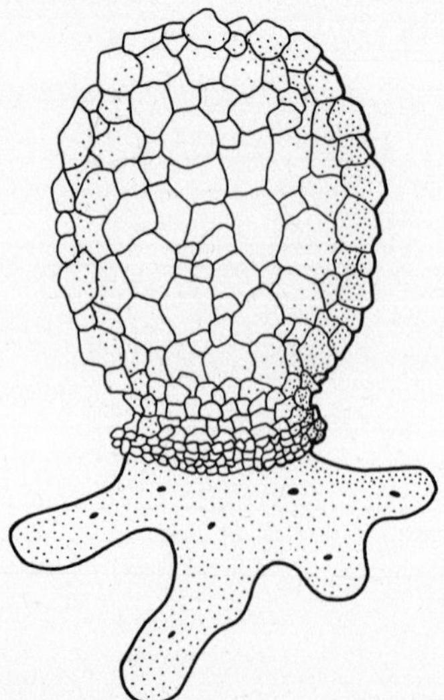

FIGURE 15-3
Difflugia, an ameba with an agglutinated shell (test) composed principally of small sand grains.

posed of polysaccharide material, similar to the chitin of fungal cell walls, encrusted with granules of inorganic silicon-containing quartz or calcium-containing limestone, as seen in Figure 15-3. One species of protozoa ingests sponge *spicules* and later deposits them in its agglutinated shell.

Periods of shell growth, alternating with rest periods when no shell is formed, give rise to many-chambered shells enclosing some of the foraminifera (Figure 15-4a). Other exquisitely beautiful geometric forms of rigid structure are seen in the marine radiolarians and freshwater heliozoans (Figure 15-4b). In contrast to the shelled forms just mentioned, these two types of organisms produce radial skeletons that are within the cell and extend to the outside. They are usually composed of silica, but some species have skeletons made of strontium compounds. The spicules produced by protruding skeletal segments lower the specific gravity of these organisms, allowing them to float. Flotation is also aided by the formation of globules of fat or fat storage products in their cytoplasm.

Certain protozoa have specialized intracellular organelles that function in defensive and offensive roles. For example, in some of the ciliates miniature dart-like structures (Figure 15-5) are ejected from the cell as a defense mechanism. Other species contain similar organelles which can paralyze or poison their targets.

NUTRITION

Protozoa are generally nonphotosynthetic and must assimilate preformed nutrients. Members of the classes Mastigophora and Sporozoa usually obtain dissolved organic materials through their cytoplasmic membranes. Other protozoa engulf

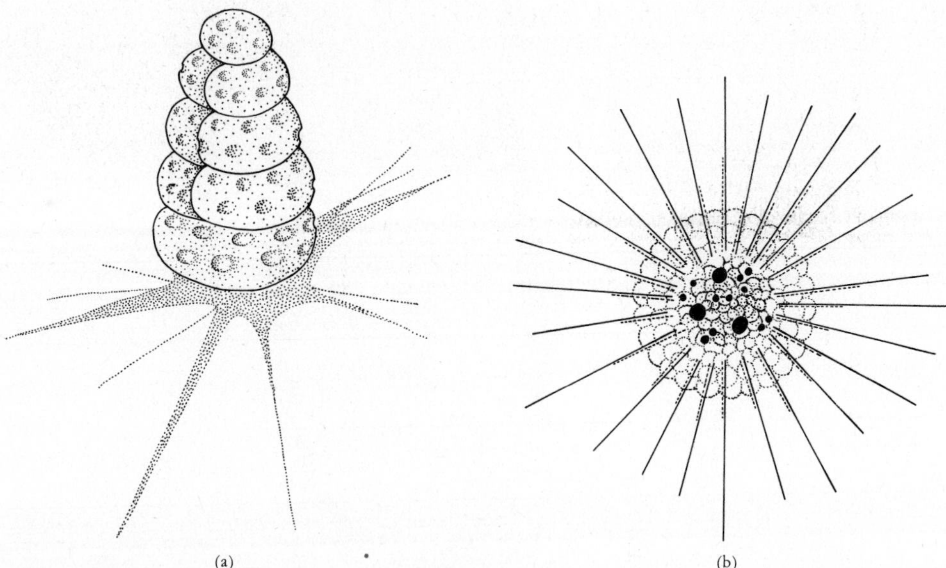

FIGURE 15-4 (a) Protozoan form with a multichambered shell, characteristic of some foraminifera. (b) A freshwater heliozoan. The marine radiolarians have forms similar to the heliozoan.

and ingest (phagocytize) particulate matter (Figure 15-6). Ameboid organisms of the class Sarcodina phagocytize solid food and enclose it in membrane-bound vesicles containing a variety of enzymes. Thus both the food and the enzymes capable of digesting it become enclosed in a common vacuole where digestion takes place. The products of digestion can then be used by the cell in various metabolic processes. It is of interest to note that virtually the same sequence of events occurs in some human cells that can phagocytize particulate matter. Both phagocytosis and another process, pinocytosis, are used by many protozoa as a means of ingesting nutrients. Pinocytosis is a process analogous to phagocytosis, except that soluble materials, rather than particles, are engulfed in tiny vacuoles (Figure 15-6e).

Following digestion, vacuoles containing indigestible particles or undigested residue are ejected through the cytoplasmic membrane in a reverse phagocytic process. In some species they are egested through a fixed area of the cell. Freshwater forms of protozoa often contain contractile vacuoles which appear to function in regulating osmosis by discarding excess water or wastes from the cell.

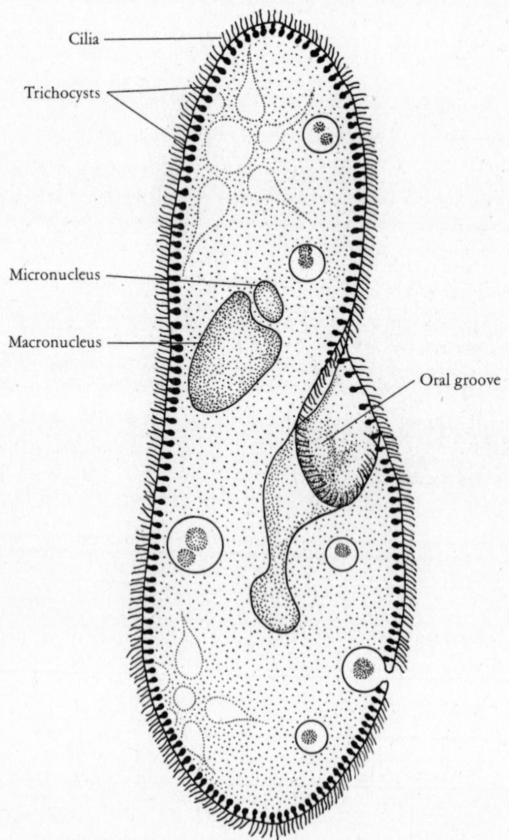

FIGURE 15-5
The morphology of *Paramecium*. The trichocysts are dartlike structures that can be ejected as a defense mechanism.

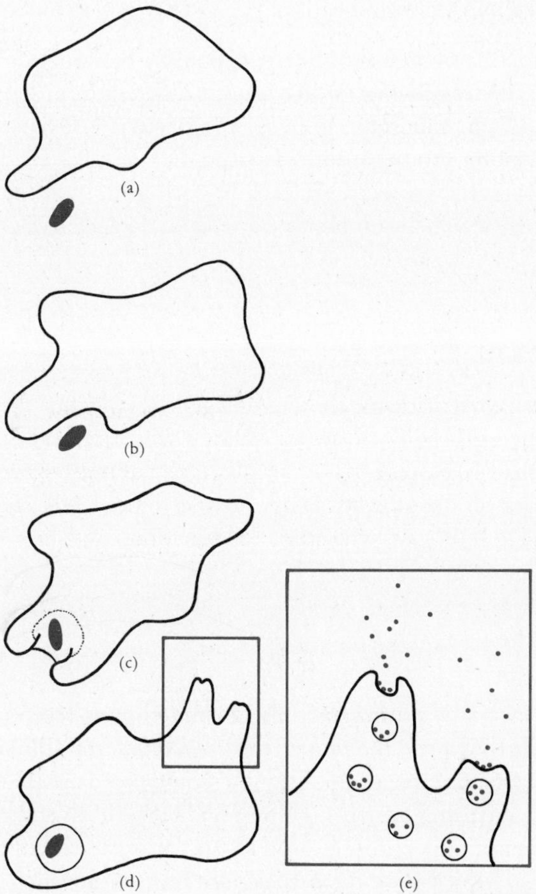

FIGURE 15-6
Phagocytosis and pinocytosis by an ameba. Phagocytosis (a through d) entails the engulfment and ingestion of particulate material. Pinocytosis (e), an enlargement of part of the cell shown in (d), is a similar ingestion of soluble molecules.

Most of the protozoa are strict aerobes or facultative anaerobes. Only a few strictly anaerobic species have been identified. Thus aerobic metabolic pathways are frequently followed in utilizing absorbed or ingested nutrients. Excess nutrients are converted into food reserves of various kinds, such as carbohydrates, polyphosphates, and lipids.

ENCYSTMENT

Many protozoa have the capacity to form *protective cysts.* When conditions are unfavorable for continued growth, the *vegetative forms* (trophozoites) of these protozoa become dehydrated and enclosed within a thickened, warty covering, forming the cyst. Some protozoan cysts have remained viable in dried materials for 40 or 50 years. This form is important in dissemination of protozoa, especially of a number of parasitic species, because it allows the organisms to survive until they reach a new, suitable host.

REPRODUCTION

Protozoa may divide by means of binary fission, by budding, or by a process of *multiple fission* called *sporulation* (which is quite different from spore formation in most fungi and from endospore formation in bacteria). Sexual reproduction is also common among protozoa, and a few specific examples are discussed in following sections. Some protozoa are diploid during most of their existence, in contrast to the algae which commonly have haploid vegetative cells.

REGENERATION

Many protozoa have remarkable abilities of regeneration when they are injured or mutilated. Some of the ciliates can regenerate a whole cell capable of reproduction from less than 10 percent of their original cell volume, provided the nucleus is intact. Besides its obvious advantage to the injured cell, this property has proven of value to biologists who study cell function.

BEHAVIOR

Certain components of protozoan cells can function as receptors for stimuli, causing the cells to express the property of *irritability,* the ability to respond to stimuli from the environment. For example, some protozoa (like certain algae) have pigment-containing eyespots that respond to light. In addition, protozoan cilia and flagella not only propel but can also act as organelles of touch, prompting the protozoa which have them to retract from or advance toward various objects that they encounter.

SOME CHARACTERISTICS OF THE CLASSES OF PROTOZOA

Mastigophora

The Mastigophora typically have cells that are ovoid or elongated and characterized by the presence of one or more flagella. They occupy diverse habitats, either as free-living organisms or as parasites.

A well-known example of this class of protozoa is the species *Trypanosoma gambiense,* which, along with a closely related species, is the causative agent of African sleeping sickness. These parasites infect tens of thousands of persons in the world every year. Their morphology, as they appear in one stage of development in the human bloodstream, is illustrated in Figure 15-7. As is the case with many other protozoa, *T. gambiense* has a complex life cycle (Figure 15-8) and

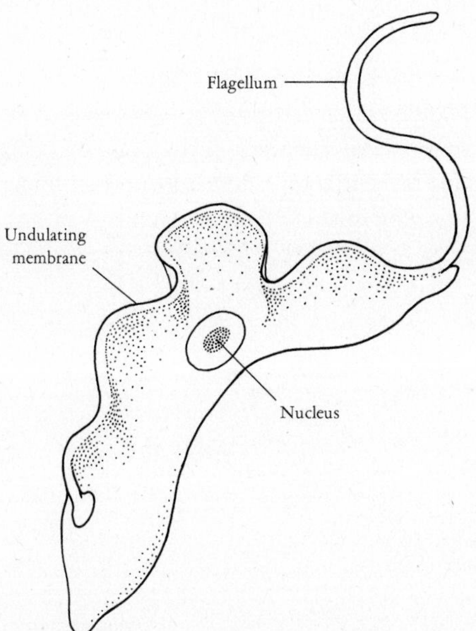

FIGURE 15-7
A trypanosome as it appears in the human bloodstream. (Courtesy of A. Frazier.)

requires two hosts. The different stages of *T. gambiense* may differ drastically in their antigenic composition, host range, and susceptibility to medications.

Species of *Giardia* and *Trichomonas* are among other members of the class Mastigophora that often cause human disease.

Sarcodina

Members of this class of protozoa possess one or more pseudopods and are characteristically ameboid. Some have skeletons or shells. The Sarcodina reproduce by binary fission. The amebae, members of this class, do not reproduce sexually; however, other Sarcodina do. Some species of *Amoeba* can produce protective cysts when exposed to unfavorable conditions.

Amoeba proteus, a freshwater species of Sarcodina (Figure 15-9), is a relatively large protozoan, over 300 μm in length when extended (as compared with *T. gambiense,* which is about 20 μm long). Extensive studies of phagocytosis and pinocytosis have been carried out using *A. proteus.* This has also been a useful organism for studying the relationship between nucleus and cytoplasm and the effects of each on cell function, because *A. proteus* has a very large nucleus which can be removed and replaced with the nucleus taken from another ameba. *Amoeba proteus* does not cause human disease, but a few species of smaller amebae are pathogenic for man. One such disease-producing species is *Entamoeba histolytica.*

Also included in the class Sarcodina are the foraminifera with their multichambered shells (Figure 15-4a), protozoan species with aggregation shells (Figure 15-3), and the radiolarians and heliozoans (Figure 15-4b), described previously.

Sporozoa

Some of the Sporozoa form pseudopods. Flagella are found only in *gametes,* if they occur at all. Many species can develop cysts, which protect them during transfer from host to host. All members of this class are parasitic, some for a single host and others for more than one host during their life cycle. An example of those which live in a single host is *Monocystis,* which parasitizes earthworms. Examples of the group of Sporozoa requiring more than one host are the malarial

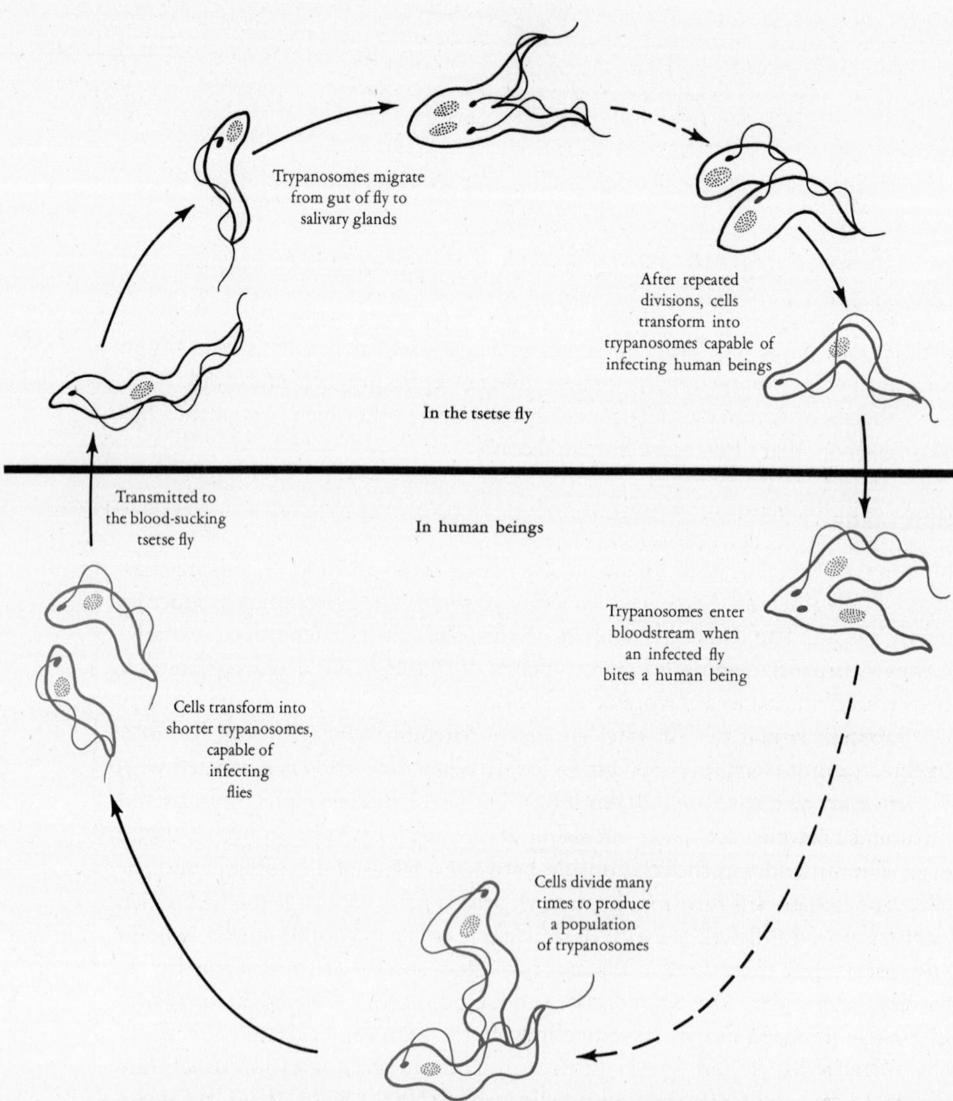
Trypanosomes migrate
from gut of fly to
salivary glands
After repeated
divisions, cells
transform into
trypanosomes capable of
infecting human beings
In the tsetse fly
Transmitted to
the blood-sucking
tsetse fly
In human beings
Trypanosomes enter
bloodstream when
an infected fly
bites a human being
Cells transform into
shorter trypanosomes,
capable of
infecting
flies
Cells divide many
times to produce
a population
of trypanosomes

FIGURE 15-8 Diagram of the life cycle of a species of *Trypanosoma* that causes African sleeping sickness. Development within the blood-sucking tsetse fly is required in order to produce forms that can infect human beings. The disease is transmitted only by the bite of an infected fly.

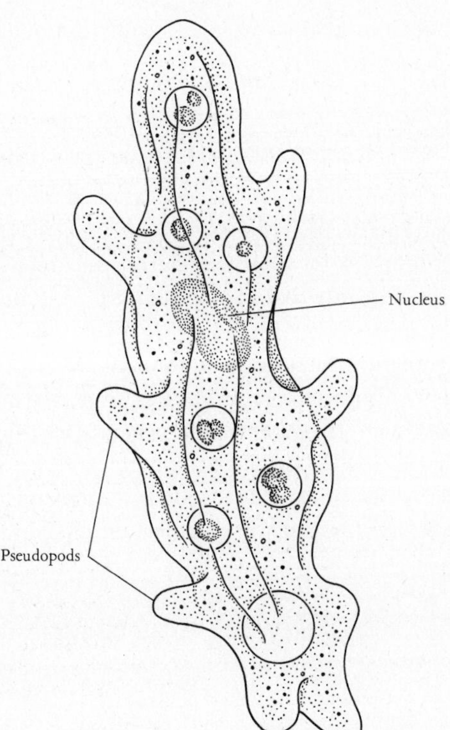

FIGURE 15-9
Diagram of *Amoeba proteus*, a large free-living ameba found in fresh water.

parasites of the genus *Plasmodium* and *Toxoplasma gondii*, a species that can parasitize man and many other mammals.

Ciliata

Ciliates represent the most highly developed forms of unicellular organisms; in this class of protozoa each single-celled organism is remarkably complex. About 6000 species are known. The class can be divided into two groups: the members of one group generally have cilia over only part of their cells while members of the other group have cilia distributed evenly over their entire cells. *Paramecium* is a familiar example of the latter type.

The various species of *Paramecium* are generally found in fresh water. In fact, puddles and transient ponds abound in paramecia. The paramecia are large protozoa, about 150 μm in length (Figure 15-5). Parallel rows of cilia cover the exterior of the cell, and their rhythmic beating is responsible for the motility of the organism. When the *Paramecium* is not moving, cilia around the oral groove create a whirlpool effect which helps bring food into the mouth.

One important characteristic of the class Ciliata, readily seen in paramecia, is the presence of a large macronucleus and one or more small micronuclei. (There are also multiple macronuclei in some ciliates.) Both kinds of nuclei are necessary because they have different functions. The macronucleus is concerned with cell growth and most cellular functions, whereas the micronucleus is involved in sexual reproduction. The macronucleus probably has many copies of all of the cell's genetic information because in some species it contains up to 500 times as much DNA as the micronucleus. And because it contains many copies of each chromosome, the macronucleus cannot undergo meiosis and produce haploid cells; the micronucleus is therefore necessary for the replication of genetic information and its distribution to daughter cells. The complex method of reproduction in *Paramecium* is shown in Figure 15-10. Note that during asexual reproduction the cell divides by transverse fission and that conjugation of two cells occurs during sexual stages of reproduction.

Interesting cytoplasmic inclusions are found in some paramecia. For example, *P. bursaria* is host to the green alga *Chlorella.* In this symbiotic relationship the alga has a safe haven and the paramecium benefits from algal photosynthesis. Also, self-replicating particles of nucleic acid, probably remnants of bacteria, are found in the cytoplasm of some paramecia. It is thought that originally certain bacteria lived in symbiosis within the paramecium and that during evolution

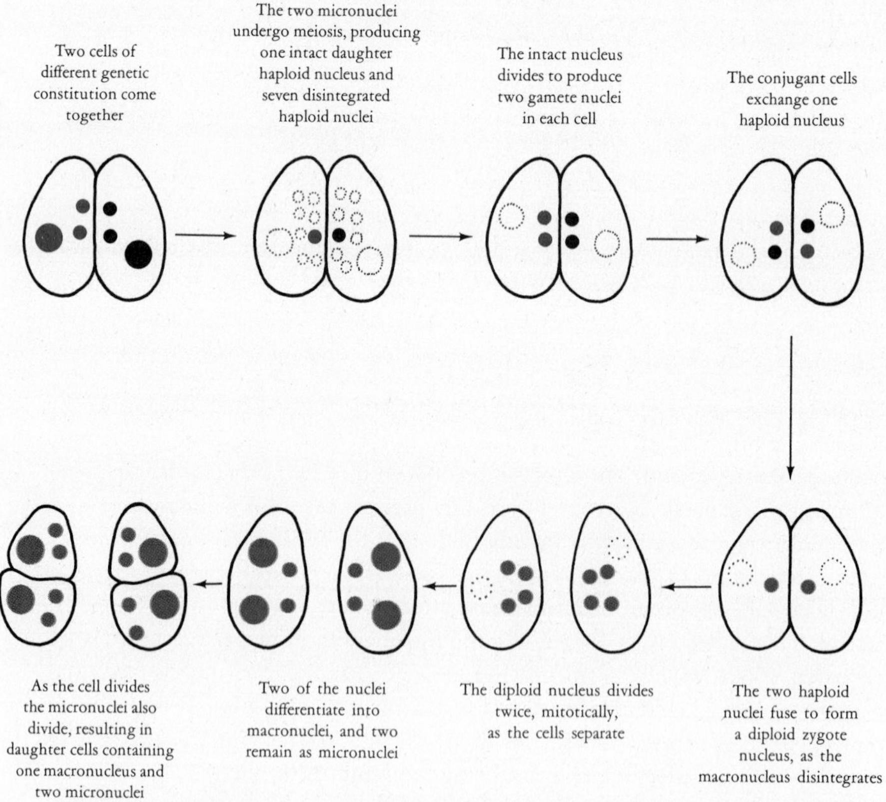

FIGURE 15-10 Reproduction in *Paramecium* species can occur by the process of conjugation. Cells of different genetic constitution are indicated by the color of the micronuclei. The daughter cells produced after conjugation contain a mixture of genes from each of the original conjugating cells.

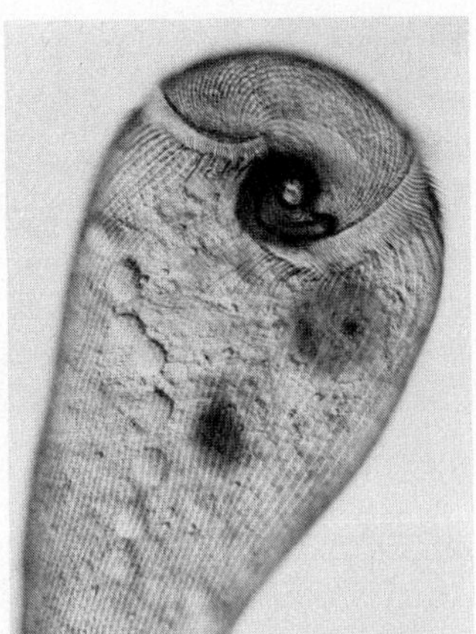

FIGURE 15-11
Stentor, a common large ciliate. (Courtesy of A. Frazier.)

these bacteria gradually lost portions of their cells until now only part of their nucleic acid remains, replicating along with the host cell. Paramecia possess a variety of *endosymbionts,* and evidence indicates that these are bacteria. One example of these is the kappa particle of *P. aurelia.* Paramecia containing kappa particles are called *killers,* because they produce a toxin that kills sensitive paramecia which lack the particles. Another example of these endosymbiotic particles, the lambda particle, has been grown in vitro and has been shown to be a Gram-negative, rod-shaped bacterium.

Another ciliate of interest is *Stentor.* These large (about 1 mm) trumpet-shaped protozoa are barely visible to the unaided eye, but a view through a low-power microscope reveals them to have a beautiful symmetry and gracefulness of movement (Figure 15-11). They are large enough to be easily manipulated in the laboratory and consequently have been studied extensively.

The only ciliate that is known to be pathogenic for man is *Balantidium coli* (Figure 15-12). It lives in the gastrointestinal tract, and both vegetative forms (trophozoites) and cysts are passed in the feces. The cysts can survive in soil or water, often long enough to be transmitted to another human host.

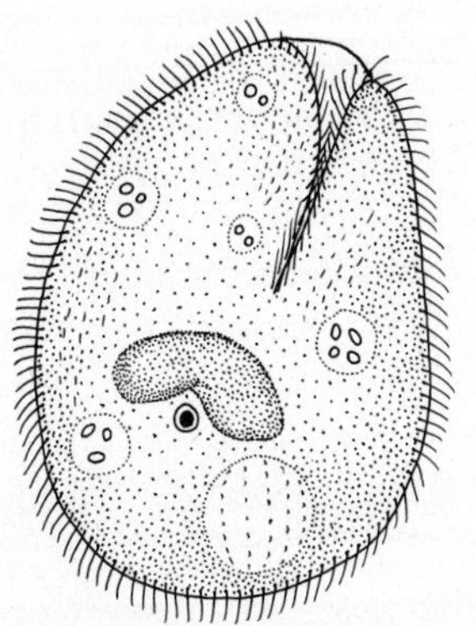

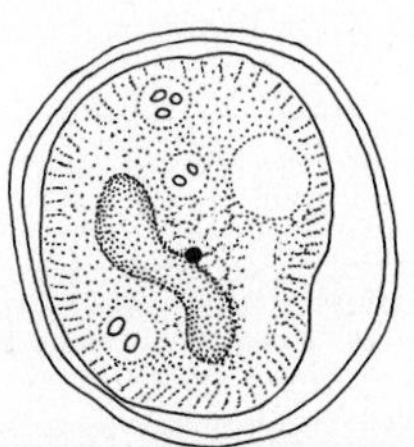

FIGURE 15-12
Balantidium coli, the only ciliate known to infect man. Trophozoite (left) and cyst (right) forms.

PRACTICAL IMPORTANCE

The protozoa are important to man in many ways. Many participate directly in the food chain, largely by feeding on bacteria and algae and in turn serving as food for larger species. The protozoa also play a major role in maintaining an ecological balance in the soil by devouring tremendous numbers of bacteria and other microorganisms. For example, a single paramecium can easily eat as many as 5 million bacteria in one day. This function of protozoa is also important in sewage disposal because most of the food protozoa consume is used for energy; the solids are largely converted to carbon dioxide and water, resulting in a large decrease in total sewage solids.

The major threat posed by protozoa derives from their ability to parasitize, and often to kill, a wide variety of animal hosts. Human infections with the sporozoan *Toxoplasma gondii* and the flagellate *Trichomonas vaginalis* are common the world over. In many parts of the world malaria and trypanosome infections are prevalent. Malaria alone has been one of the greatest killers of man through the ages. At present at least 150 million people in the world contract malaria each year, and 1.5 million die of it. These are conservative estimates; in 1970 one authority stated over 300 million of the world's people had some kind of malarial infection. Protozoan infections of animals are so common that it is impossible even to estimate their extent or underestimate their economic importance.

SUMMARY

The protozoa are eucaryotic protists which lack chlorophyll and cell walls, although some of them have a thickened pellicle or a shell-like covering. They are widespread in aquatic and terrestrial habitats, and many species are parasitic, their various hosts ranging from simple algae to man.

Most of the protozoa are aerobic or facultatively anaerobic, and only a few strictly anaerobic species have been identified. Nutrients are frequently ingested by phagocytosis or pinocytosis and wastes excreted by the reverse process. Many protozoa have the capacity to form protective cysts which enable them to survive for long periods in an unfavorable environment. Reproduction among the protozoa may be either asexual, sexual, or both, depending on the species and environmental conditions.

Some protozoa have cellular components that permit them to respond to stimuli, therefore expressing the property of irritability. Cilia or flagella propel the protozoan cells which possess them and also act as tactile organelles.

The four classes of protozoa are Mastigophora, Sarcodina, Sporozoa, and Ciliata. Organisms of the class Mastigophora are characteristically ovoid or elongated and possess flagella. An example is *Trypanosoma gambiense,* which causes African sleeping sickness. Members of the class Sarcodina include the amebae, foraminifera, radiolarians, and heliozoans. All members of the class Sporozoa are parasitic; the malarial parasite is an example of these parasites. Protozoa of the class Ciliata possess cilia and are examplified by the familiar *Paramecium.*

Protozoa are of practical importance in many ways. They participate in the food chain and play a major role in maintaining an ecological balance in the soil. Protozoa have been valuable as subjects for biological research. Their major threat is the ability of some of them to parasitize human and animal hosts, causing disease or death.

QUESTIONS

REVIEW

1. How do protozoa differ from algae and fungi?
2. Discuss several modes of locomotion that occur among the protozoa.
3. Describe several protective mechanisms that various protozoa have developed.
4. Compare and contrast a paramecium and a typical bacterium.

THOUGHT

Consider the kinds of information that can be gained from experiments using *Amoeba proteus.*

FURTHER READING

CHEN, T. T., ed., *Research in Protozoology.* Vols. I–IV. New York: Pergamon Press, 1967–1971.

CORLISS, J. O., "Systematics of the Phylum Protozoa," *Chemical Zoology.* Ed. M. Florkin and B. T. Scheer. Vol. I. New York: Academic Press, 1967, pp. 1–20.

CURTIS, H., *The Marvelous Animals—An Introduction to the Protozoa.* New York: Natural History Press, 1968.

DOGIEL, V. A., *General Protozoology.* Rev. ed., Trans. G. I. Poljansky and E. M. Chejsin. London: Oxford University Press, 1965.

GARNHAM, P. C. C., *Malaria Parasites and Other Haemosporidia.* Oxford: Blackwell Scientific Publishers, 1966.

HALL, R. P., *Protozoa—The Simplest of All Animals.* New York: Holt, Rinehart and Winston, 1964. A small paperback, containing clearly written basic information.

JAHN, T. L., and F. F. JAHN, *How to Know the Protozoa.* Dubuque, Ia: William C. Brown, 1949.

KIDDER, G. W., ed., *Protozoa.* Vol. I. *Chemical Zoology.* Ed. M. Florkin, and B. T. Scheer. New York: Academic Press, 1967.

KUDO, R. R., *Protozoology.* 5th ed. Springfield, Ill.: Charles C. Thomas, 1966.

LEMKE, P. A., "Viruses of Eucaryotic Microorganisms," *Annual Review of Microbiology* 30:105–145 (1976). An excellent discussion of the viruses that infect protozoa, fungi, and algae. Especially noteworthy are viruses that are found in the endosymbiotic bacteria of paramecia.

PITELKA, D. R., *Electron-Microscopic Structure of Protozoa.* New York: Pergamon Press, 1963.

PREER, J. R., Jr., L. B. PREER, and A. JURAND, "Kappa and Other Endosymbionts in *Paramecium aurelia,*" *Bacteriological Reviews* 38:113–163 (1974).

TARTAR, V., *The Biology of Stentor.* New York: Pergamon Press, 1961.

VICKERMAN, K., and F. E. G. COX, *The Protozoa.* Boston: Houghton Mifflin, 1967.

WEISMAN, R. A., "Differentiation in *Acanthamoeba castellanii,*" *Annual Review of Microbiology* 30:189–219 (1976). A detailed study of the many events that occur during the formation of cysts and the transition from the cyst to trophozoite stage in common amebae.

THE NATURE AND STUDY OF VIRUSES

During the late nineteenth century many infectious agents were identified as being bacteria, fungi, or protozoa. Most of these organisms were readily seen with the microscopes available at that time, and they could frequently be cultured in vitro. It was observed, however, that some infectious agents, unlike bacteria, were so small that they could not be seen with the light microscope; furthermore, they passed through filters that retained almost all known bacteria. These same agents could also not be grown in cell-free media, despite many attempts to do so, but would grow only in conjunction with intact, living host cells. The agents were called *filterable viruses,* viruses being a general term (meaning poison) that had once been applied to all infectious agents. Gradually the adjective was dropped and the word virus was used only in connection with the filterable agents. It was later discovered that these agents were obligate-intracellular parasites with a unique mode of reproduction.

As experimental data accumulated, it became apparent that viruses had certain features that were more characteristic of complex chemical substances than of cells. For example, tobacco mosaic virus could be precipitated from a suspension by using ethyl alcohol and would still remain infective, while this same treatment destroyed the infectivity of bacteria and other cells. In 1935 tobacco mosaic virus was crystallized by Dr. Wendell Stanley of the University of California, which emphasized that the physicochemical properties of viruses and cells are quite different. Crystals are formed only by purified chemical compounds and not by cells or by mixtures of dissimilar molecules. Nevertheless, crystallized tobacco mosaic retained the capacity to infect cells and promote its own replication, producing still more virus—clearly properties associated with life.

Efforts to learn more about the nature and characteristics of these curious agents, some of which infected animals and others plants, were greatly aided by the discovery of other viruses that infect only bacteria. Early in the twentieth century F. W. Twort in England and F. d'Hérelle in France, working independently, demonstrated the existence of filterable agents which infected and destroyed bacteria. These *bacterial viruses* became known as *bacteriophages,* or simply *phages.* Because bacterial cells can be grown with relative ease, phages can be cultivated much more readily than the viruses that infect animals or plants. For this reason bacteriophages have been studied extensively, and the knowledge gained has contributed enormously both to an understanding of viruses in general and to the molecular biology of all organisms.

Finally, the contributions of the electron microscope to virology cannot be overemphasized, because only a few of the largest viruses are visible in the light microscope. The development of techniques for staining and visualizing viruses by electron microscopy has been crucial for elucidation of viral morphology and its relation to the infective process.

THE NATURE OF VIRUSES

All viruses are chemical particles that differ from cells in many ways. Each virus particle, called a *virion,* consists of a single type of nucleic acid (DNA or RNA) surrounded by a protein coat (*capsid*) and, in some cases, also by an outer layer or envelope which contains carbohydrates and lipids (Figure 16-1). The presence of *either DNA or RNA, but not both,* is an important property that distinguishes viruses from cells. Thus there are RNA viruses and DNA viruses. Furthermore, *the nucleic acid of a virus may occur as either double-stranded DNA, single-stranded DNA, double-stranded RNA* or *single-stranded RNA.* Another important feature that distinguishes viruses from cells is the viruses' lack of components necessary for energy generation and protein synthesis (for example, ribosomes). Certain viruses do contain enzymes involved in nucleic acid synthesis. Generally, however, the enzymatic capabilities of viruses are extremely limited and confined to enzymes involved in the viruses' entry into cells and the replication of their own nucleic acid. Because of their metabolic limitations viruses are unable to replicate independently, and they must invade a living cell and utilize the cellular ribosomes, energy sources, and certain other components of this host cell in order to produce new virions. The process, *viral replication,* frequently causes profound changes in the host cell, often death of the cell.

The replication of viruses is unique. At an early stage of replication the viral nucleic acid separates from the protective protein capsid and envelope. Only this naked nucleic acid is required for subsequent steps in viral replication. In contrast, a cell maintains its morphological integrity during cell division. Furthermore, whereas a cell customarily divides to give rise to two identical cells, infection by a single virion can result in the simultaneous synthesis of dozens or even thousands of similar progeny virions within a single host cell.

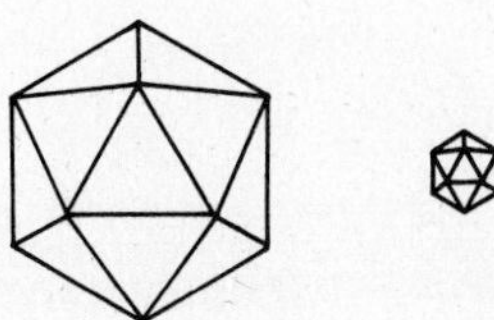

(a) Polyhedral, naked

(b) Helical, naked

(c) Enveloped

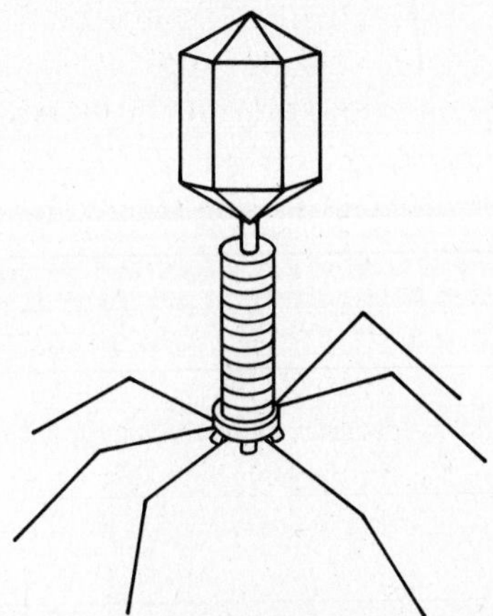

(d) Combination of polyhedral and helical, naked

FIGURE 16-1
Virus morphology.

FIGURE 16-2
Diagram of an icosahedral virion, showing the symmetrical arrangement of the capsomeres.

VIRUS ARCHITECTURE

Virus particles are generally either polyhedral or helical in structure, or they are sometimes rather complex combinations of these two shapes (Figure 16-1). Polyhedral viruses often appear to be almost spherical, but closer examination shows that the capsids of these viruses are actually composed of identical subunits arranged in patterns of icosahedral symmetry (that is, 20-sided polyhedrons in which each side is an equilateral triangle).

The symmetry of viruses is a property of the protein capsids that enclose and protect the viral nucleic acid genome. Each capsid is composed of subunits called *capsomeres.* Each capsomere in turn is made up of a number of protein molecules. Although a capsid may be composed of hundreds of capsomeres, the simplest possible icosahedral virion contains only 60 identical protein molecules arranged in five identical capsomeres (Figure 16-2). Helical viruses, such as the tobacco mosaic virus (Figure 16-3), consist of nucleic acid within a cylindrical capsid composed of many identical capsomeres in a spiral arrangement. Many virions have a much more complicated morphology (Figure 16-1c). The nucleic acid of some animal viruses, the *enveloped viruses,* is contained within a helical or polyhedral protein capsid, which in turn is surrounded by a membranous outer structure called an *envelope.* This membranous structure can be complex and consist of several layers of lipid and protein.

Some of the bacterial viruses are also structurally complex. For example,

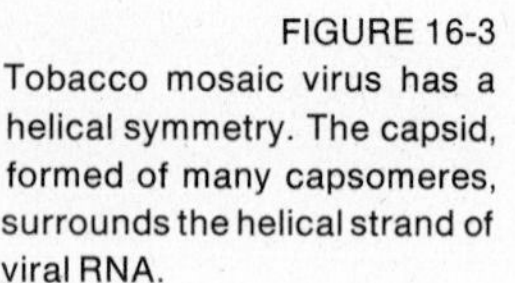

FIGURE 16-3
Tobacco mosaic virus has a helical symmetry. The capsid, formed of many capsomeres, surrounds the helical strand of viral RNA.

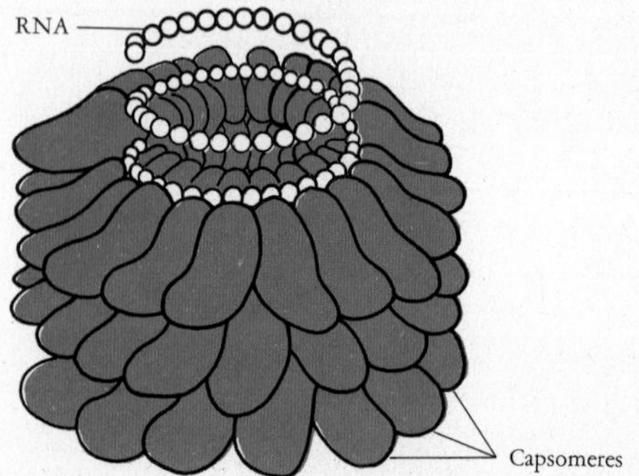

T-even bacteriophage, which infects certain strains of *Escherichia coli,* is composed of a polyhedral head attached to a helical hollow tail (Figure 16-1d). The nucleic acid of this phage is a single molecule of double-stranded DNA packed tightly within the head. Both the efficiency of packing and the incredible amount of DNA contained within the tiny virion are indicated in Figure 16-4, which shows the DNA released following gentle disruption of the phage head.

Viruses differ considerably in size. The smallest viruses are similar in size to large protein molecules or ribosomes, and their nucleic acid codes for only a few genes. The more complex virions may be larger than some of the most minute bacteria.

VIRUS REPLICATION

The basic steps in virus replication are similar for all viruses that have been examined, whether they infect bacterial, plant, or animal cells. First the viral nucleic acid must enter the host cell. In the case of bacterial and animal viruses the virion adsorbs specifically to receptors on the host cell and it is after this adsorption that the viral nucleic acid penetrates into the cell. Either free nucleic acid enters the cell (in the case of most bacteriophages) or whole virions (for all other viruses) enter and then release their nucleic acid. Replication of viral nucleic acid and synthesis of other viral constituents follows. These viral nucleic acid and protein constituents are made separately within the host cell and are then assembled into complete virions during the stage of virus maturation. This process is similar to

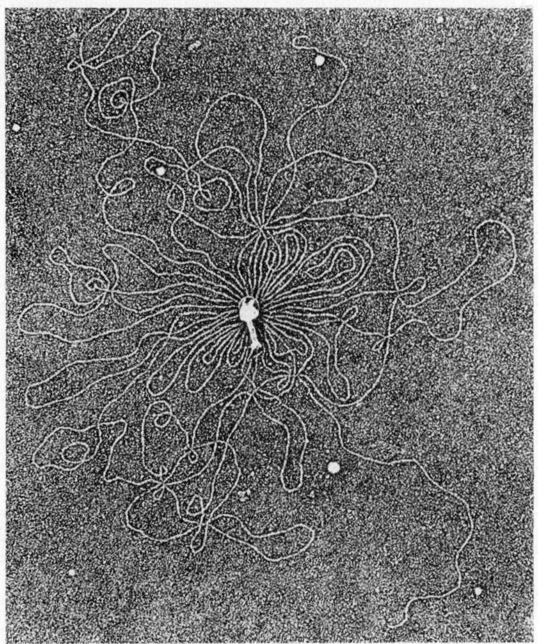

FIGURE 16-4
Osmotically shocked T-even bacteriophage. The single molecule of DNA containing the genetic information of the phage has been released from the phage coat (center). (Courtesy of A. K. Kleinschmidt; A. K. Kleinschmidt, D. Lang, D. Jacherts, and R. K. Zahn, *Biochim. Biophys. Acta* 61:857, 1962.)

assembly-line production in a factory where automobiles, for example, are manufactured from parts made separately; it is quite efficient. Finally, the newly formed mature virions are released from the host cell. The method of release varies depending on the virus in question: cells may be lysed, releasing many mature virions, or virus particles may be gradually extruded from the living host cell.

The details of replication in T-even phage have been studied extensively, both morphologically and on a molecular basis, and they offer an excellent model for the general process of infection by a phage which destroys the host cell by lysis. The term T-even phage designates any of a group of phages including T2, T4, and T6. Data collected from studies of any of these phages are generally applicable to all three. The following steps in replication are diagramed in Figure 16-5.

Adsorption

When a suspension of T-even phage particles is mixed with a susceptible strain of *E. coli,* the phage collide by chance with the bacteria. Fibers at the end of the phage tail are the adsorption sites of the phage that bind to specific receptors on the bacterial cell wall (Figure 16-5).

Penetration

Lysozyme, page 63

Following adsorption, an enzyme (phage lysozyme) located in the phage tail degrades a small portion of the bacterial cell wall. The tail sheath of the phage then contracts, and the tail core mechanically penetrates the cell wall. The tip of the phage tail then opens, and the viral DNA in the head of the phage is free to pass through the channel of the phage tail. The DNA is literally injected through the cell wall and, by some unknown mechanism, penetrates the cytoplasmic membrane, entering the interior of the cell, while the protein coat of the phage remains outside the cell. Thus the phage head protein protects the DNA while it is outside a host cell, and the phage tail proteins serve to attach and penetrate the host cell wall, allowing viral nucleic acid to enter the host cell.

Replication of Phage DNA

Transcription, page 174

Within minutes after entry of phage DNA into a host cell, all transcription of RNA from the host chromosome ceases. In fact the host DNA is actually degraded within minutes after phage infection. All RNA subsequently synthesized is mRNA transcribed from the phage DNA. By this mechanism the phage subverts all metabolism of the bacterial cell to its own purpose—the synthesis of more phage. Host enzymes continue to function, however. They supply energy for phage replication through the breakdown of glucose, synthesize the subunits of protein and nucleic acid of the replicating phage, and even participate in the synthesis of phage nucleic acid and phage coat protein. The production of phage proteins also requires the participation of bacterial ribosomes. In addition to the

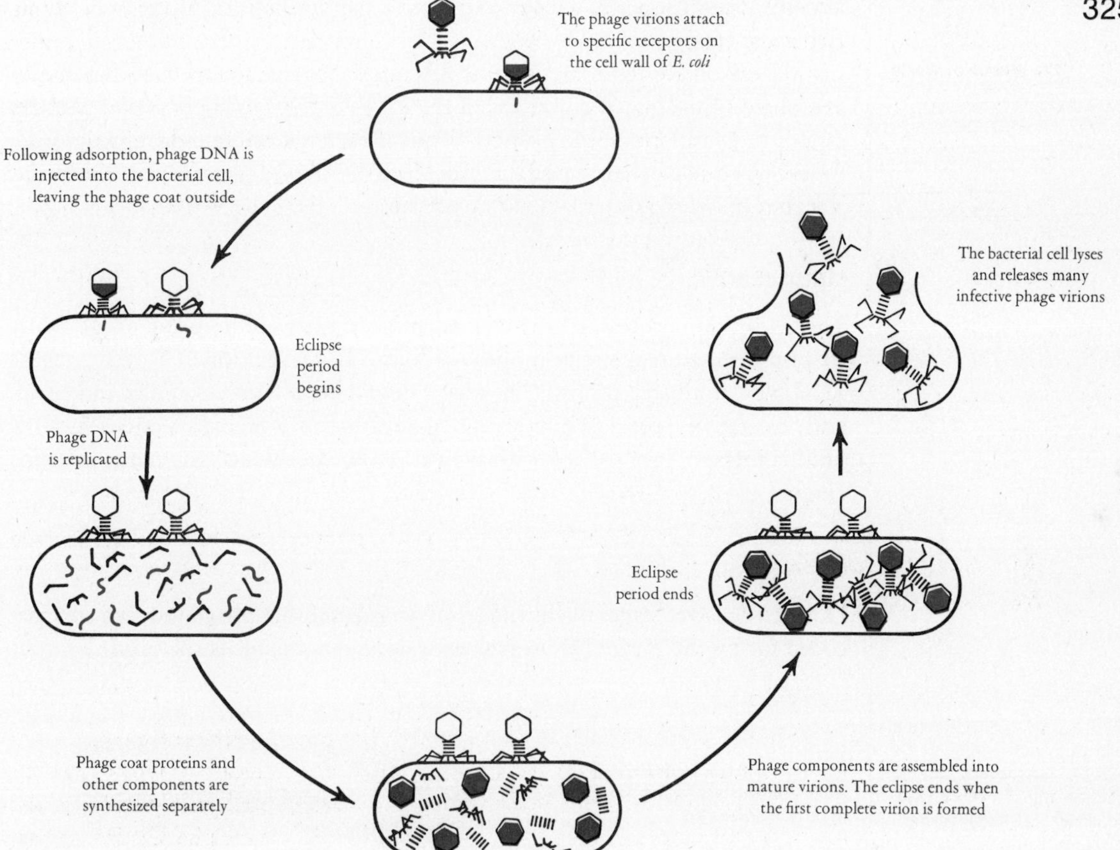
The phage virions attach
to specific receptors on
the cell wall of E. coli
Following adsorption, phage DNA is
injected into the bacterial cell,
leaving the phage coat outside
Eclipse
period
begins
Phage DNA
is replicated
The bacterial cell lyses
and releases many
infective phage virions
Eclipse
period ends
Phage coat proteins and
other components are
synthesized separately
Phage components are assembled into
mature virions. The eclipse ends when
the first complete virion is formed

FIGURE 16-5 Steps in the replication of T-even phage during the infection of *Escherichia coli*.

usual enzymes of the host, many enzymes appearing in the infected host cell are *phage-specific,* that is, *they are encoded in the phage DNA and are not present in the uninfected host cell.* Without these phage-specific enzymes, the synthesis of phage DNA and phage coat protein could not proceed. One such phage-specific enzyme degrades the host DNA. Another phage-induced enzyme catalyzes the synthesis of a pyrimidine base (hydroxymethyl cytosine) which is incorporated in place of cytosine into the phage DNA. At least 10 other phage-induced enzymes are concerned with phage DNA synthesis. In addition, the phage DNA codes for enzymes concerned with the assembly of the phage protein capsid.

During the replication of any virus the viral protein and nucleic acid components separate completely from each other. In fact the separation of nucleic acid from its protective protein coat is an obligatory and characteristic feature of virus reproduction. In the the case of T-even phages these components separate at the time the DNA is injected into the cell, leaving the protein coat outside. Thus the DNA of the T-even phage has two independent functions: it provides the template for replication of more phage DNA, and it serves as the template for transcription of mRNA which functions in the synthesis of phage-induced enzymes and capsid protein.

The mode of reproduction just described accounts for the fact that during one phase of phage infection, called the *eclipse period,* complete phage particles cannot be found inside the host cell. Thus if bacterial cells are disrupted during the early period of phage reproduction, phage DNA and protein coats are detectable, but no whole phage virions are present.

Maturation

The maturation process involves the assembly of phage protein and phage DNA to form the mature infective progeny virions. The maturation of T-even phage is a complex, multistep process. The phage heads and tails are assembled independently by stepwise processes involving aggregation of protein subunits. Once the head is formed, it is packed with phage DNA, after which the tail is attached (Figure 16-6).

Release

During the latter stages of the infection period, another phage-induced enzyme, coded for by the phage DNA, makes its appearance. This is the phage *lysozyme*

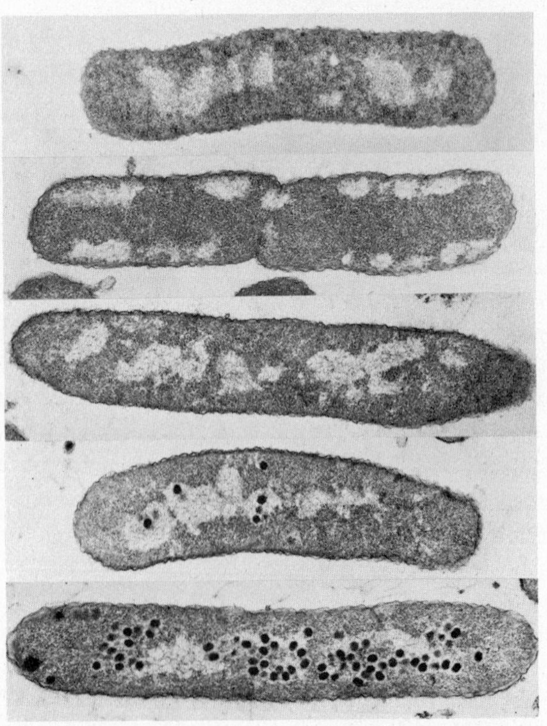

Intact *E. coli* cell at the time of infection. The light areas within the cell represent bacterial DNA.

By 2 min after infection changes in the bacterial DNA are apparent. Empty phage coats remain attached to the bacterial cell walls.

At 8 min after infection, bacterial DNA has been destroyed and phage DNA is being made (light areas).

By 12 min, condensed phage DNA is apparent (dark areas).

At 30 min the bacterial cell is nearly full of condensed DNA and completed phage heads. After assembly of many mature virions, the bacterial cell lyses and releases the phage particles.

FIGURE 16-6
Cells of *E. coli* sectioned for electron microscopy during infection with T-even phage. (Courtesy of E. Kellenberger.)

that digests the host cell wall from within, resulting in cell lysis and release of the many complete phage virions. The virions which are released infect available susceptible cells, and the process of virus replication is repeated. It is significant that lysozyme is never synthesized early in the infection process, presumably because it would lyse the host cell before any mature phage could be formed. There are controls regulating the regions of the viral nucleic acid that will be transcribed into mRNA at any time during the infection process. Thus there are "early" messages which are translated into "early" phage proteins (the enzymes of phage DNA synthesis), and "late" messages for "late" phage proteins (coat proteins and lysozyme). This control of transcription is mediated by changes in the enzyme RNA polymerase, which in some manner determine the site at which the RNA polymerase initiates transcription.

Translation,
page 135

RNA polymerase,
page 182

HOST RANGE OF VIRUSES

It is probable that any kind of cell can be infected by some viruses. To date, viruses have been shown to infect cells of at least some species of algae, fungi, and protozoa and of most species of bacteria, plants, and animals. However, any given virus has a limited host range and generally will infect cells of only one or a few species, and often just a few strains of a species. Thus the T-even phage will infect only certain strains of *E. coli,* and polioviruses naturally infect only cells of man and a few related higher primates. This limited host range reflects the fact that animal and bacterial viruses must interact with specific receptor sites on the host cell surface in order to invade the cell. The receptor site varies in chemical properties and location, depending on the nature of the infectious virus. In the case of phages the receptor site is usually a chemical constituent of the bacterial cell wall. Some phages, however, attach to receptors on pili and others to sites on flagella. Animal viruses attach to specific receptor sites on the cytoplasmic membranes of appropriate host cells. Mutations in a host cell that result in altering or deleting receptor sites render the cell resistant to attack by the particular virus, because the virus can no longer attach to and infect the cell.

The attachment of virus to host cell receptors is a chemical interaction. Its specificity depends on a close complementary fit between the molecules of the virus and those of the receptor; the two are held together by many weak chemical bonds (such as hydrogen bonds) which form when they associate. The complementary fit of virus and receptor molecules must be close in order to allow formation of enough weak bonds to hold the virus and host cell together. In the laboratory it is possible to bypass the initial steps of adsorption and penetration and to infect cells with viral nucleic acid alone. In some cases, however, viral DNA or RNA, once inside the cell, will not be replicated. Thus the host range of a virus depends on several factors.

Hydrogen bonds,
page 30

The general principles governing the host range of viruses have been studied on the molecular level with the T-even bacteriophages. The chemical nature of both host cell receptors and phage attachment sites has been elucidated. Additionally, the dependence of the host range of these phages on the chemical structure of their DNA has also become evident. The T-even DNA contains glucose bonded to specific areas of the DNA molecule (glucosylated DNA). Several enzymes are required for the incorporation of glucose into the DNA; some are bacterial and others are phage-induced. Normally, host *E. coli* cells contain these necessary bacterial enzymes, and glucosylated phage DNA is replicated upon infection. However, if the phage invades *E. coli* cells which lack these glucosylating

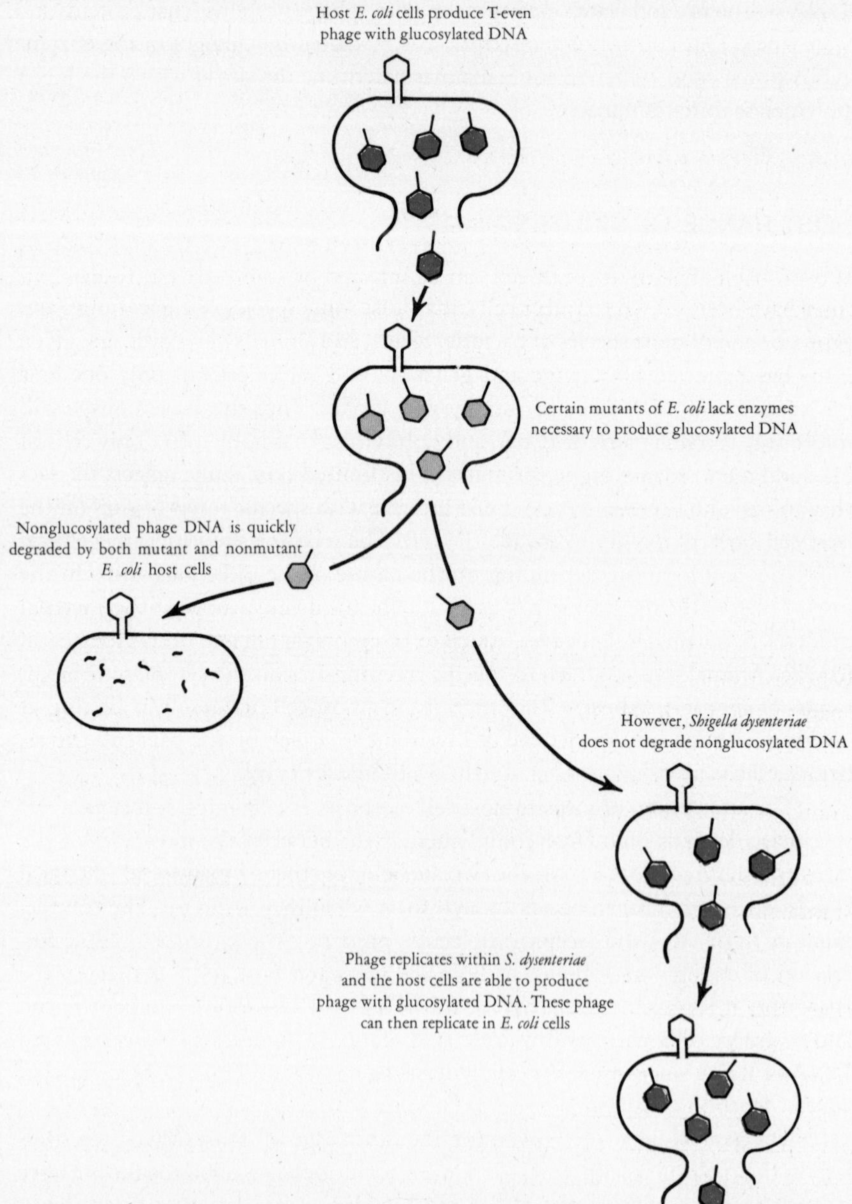
Host *E. coli* cells produce T-even
phage with glucosylated DNA
Certain mutants of *E. coli* lack enzymes
necessary to produce glucosylated DNA
Nonglucosylated phage DNA is quickly
degraded by both mutant and nonmutant
E. coli host cells
However, *Shigella dysenteriae*
does not degrade nonglucosylated DNA
Phage replicates within *S. dysenteriae*
and the host cells are able to produce
phage with glucosylated DNA. These phage
can then replicate in *E. coli* cells

FIGURE 16-7
Host-range modification of T-even phage.

enzymes because of a mutation in the bacteria, the DNA of the progeny phage is not glucosylated. The progeny phage then cannot replicate in new host cells because *E. coli* has a deoxyribonuclease in its membrane which degrades any DNA that is not glucosylated. However, the same phage can replicate in certain other Gram-negative rods that lack the deoxyribonuclease enzyme. These observations indicate that *the bacterial host is capable of modifying the structure of phage DNA, recognizing unmodified DNA and consequently controlling the host range of the progeny phage* (Figure 16-7). Other examples of this phenomenon of *host-controlled modification* of viruses have been studied in bacteriophage systems. However, analogous examples in animal and plant virus systems have not yet been described.

VIRUS-HOST RELATIONSHIPS

The relationships between even the most extensively studied mammalian cells and viruses are poorly understood. However, for bacteriophages three major types of relationships are recognized: the lytic, the lysogenic, and a state where viruses are released without cell lysis. Other, less well-defined relationships undoubtedly exist.

Lytic

The lytic relationship is exemplified by the T-even infection previously described. Here the virus is referred to as a *virulent bacteriophage* because infection results in lysis and death of the infected cell.

Lysogenic

Whereas virulent bacteriophages can only replicate and cause cell lysis, other phages exist which are capable of *either replicating and causing cell lysis or having their DNA integrated into the host cell DNA.* These are known as *temperate phages.* When the phage DNA integrates with the host cell DNA, it is called *prophage,* and the bacterial cell carrying a prophage is a *lysogenic cell.*

The most thoroughly studied temperate bacteriophage is the phage lambda (λ) of *E. coli.* This phage is similar in appearance to T-even, and the initial stages of infection with it are the same (Figure 16-8). The phage adsorbs and its DNA penetrates into the interior of the host cell, leaving the protein coat on the outside. One of two events then occurs: either the phage DNA replicates and phage-specific proteins are synthesized with a lytic cycle ensuing, or the phage DNA may become integrated into a specific region of the bacterial chromosome (Figure 16-9). Therefore in any culture infected with temperate phage, some bacteria will be lysogenized and others will be lysed. As a prophage, the phage DNA is replicated as part of the bacterial chromosome and reveals few signs of its existence; consequently it is difficult to know when a bacterium is lysogenic. Occasionally, however, the prophage leaves the host cell DNA and begins to replicate and release infectious phage. The presence of phage in a cell culture is the best indicator that the cells are lysogenic.

For some time this ability of temperate phages to become integrated and their capacity to exist as either integrated prophage or as ordinary lytic phage seemed very mysterious. During the last two decades the availability of sophisti-

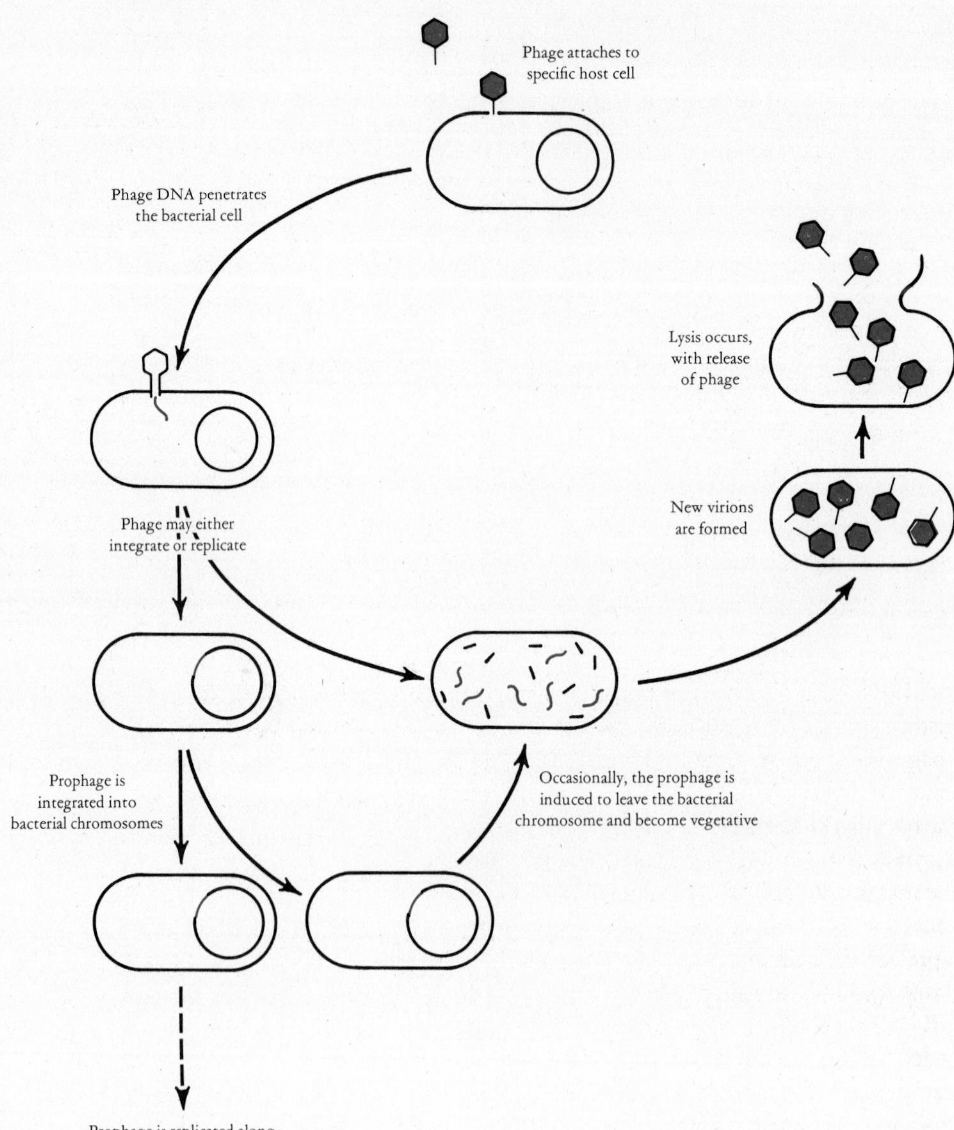

FIGURE 16-8 Temperate phage cycle. The lysogenic cycle is diagramed on the lower left and the lytic cycle on the right.

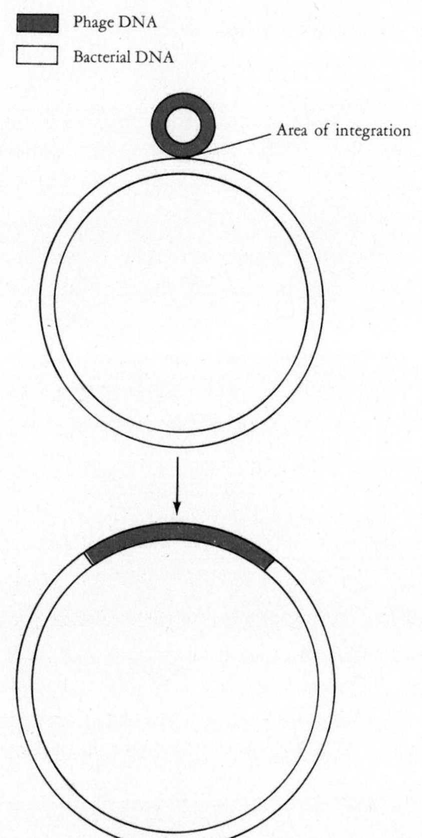

FIGURE 16-9
The integration of temperate phages. Nucleotide sequences of both phage and host are recognized by an enzyme system that allows phage DNA to be integrated with the host bacterial chromosome. When bacterial DNA is replicated, the phage DNA is also replicated.

cated techniques has made it possible to gain some understanding of the mechanisms involved. It was learned that short nucleotide sequences in the DNA of phage and host are capable of recognizing each other, allowing pairing of the phage and bacterial DNA, followed by the integration of the phage DNA into the bacterial chromosomes (Figure 16-9). To establish and maintain lysogeny the enzymes devoted to vegetative phage replication must be repressed. One gene in the integrated viral DNA codes for the synthesis of a protein repressor that prevents the synthesis of the proteins required for the lytic cycle. As long as this repressor is synthesized, the integrated phage chromosome is maintained as a prophage because the repressor prevents transcription of phage DNA. However, if the synthesis of the repressor stops or if the repressor is inactivated, an enzyme coded for by the virus is synthesized, and this enzyme excises the viral DNA from the bacterial chromosome. This excised viral DNA then initiates the vegetative state, replicates, and induces synthesis of phage proteins. Mature phage is then formed, and the cell lyses. Under ordinary conditions of bacterial growth the release of the phage DNA from the bacterial chromosome is a rare event, occurring perhaps once in 10,000 divisions of a lysogenic bacterium. However, if a

Complementary
structures,
page 31

lysogenic culture is treated with any agent that interferes with DNA replication (such as ultraviolet light), all of the prophage will enter the lytic cycle. This process, termed *induction,* can result in complete lysis of the culture.

The repressor produced by a prophage also prevents infection of a lysogenic cell by the same phage whose DNA is already carried by the lysogenic cell. Repressor binds to the DNA of the attacking phage and inhibits its replication. The cell is thus "immune" to infection by this phage but not by other phages.

Except for the property of being immune to reinfection with the same phage, lysogenic and nonlysogenic cells usually appear identical. However, in a few well-documented cases prophages have had profound effects on the properties of the host cell, a phenomenon called *lysogenic conversion.* For example, the ability of *Corynebacterium diphtheriae,* the causative agent of diphtheria, to synthesize the toxin responsible for this disease can depend upon lysogenization with a particular phage. Similarly, the ability of certain lysogenic Group A streptococci to produce the toxin that causes scarlet fever, and of lysogenic strains of *Cl. botulinum* to produce certain types of botulinum toxin, can also be governed by specific phages. In all of these cases loss of the prophage results in loss of the toxin-producing ability.

Episomes, page 209

F^+, Hfr, pages 203–207

Temperate bacteriophages are episomes, and their relations with host cells bear many similarities to the F^+-Hfr situation. As the Hfr can be excised from the bacterial chromosome to become an F^+ particle, the prophage can be excised from the bacterial DNA to become a *vegetative phage.* In a few cells a piece of the bacterial DNA remains attached to the piece of phage DNA which is excised. If this happens, the genes in the bacterial DNA fragment replicate as part of the phage DNA, become incorporated into mature phage, and are released following lysis of the bacterial cell. When the phage infects another bacterial cell, both phage and bacterial DNA may be integrated into the new host chromosome. Thus is produced a lysogenic cell which has bacterial genetic information from the previously lysogenized cell. This mechanism of genetic transfer is called *transduction* (Figure 16-10).

Lambda is a restricted or specialized transducing phage because it normally becomes integrated at a particular region on the host DNA and can transfer only the limited group of genes that are physically linked to the region in which the prophage is integrated.

Generalized transduction, Figure 9-21, page 212

Other phages, the generalized transducing phages, can transfer any region of the bacterial chromosome. This transfer mechanism has been considered in Chapter 9. In generalized transduction the phage DNA is not necessarily integrated into the host cell chromosome as it is in specialized transduction.

Transducing phages are usually *defective phages* because the DNA of these phages has an incomplete information content. In specialized transduction by phage, such as lambda, the bacterial genes which are integrated into the phage DNA replace a few phage genes. In generalized transduction, a segment of the bacterial DNA is packaged in the phage head in place of part or all of the phage DNA. The resulting defective phage cannot proceed through its entire replicative cycle because it lacks some of the genetic information required for various

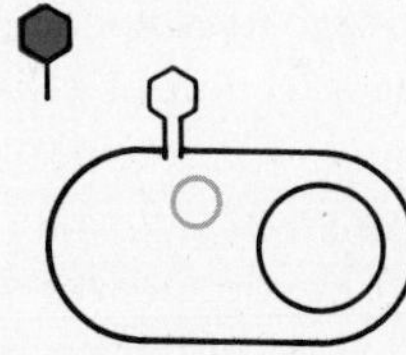

The DNA of a temperate phage penetrates into the bacterial host cell

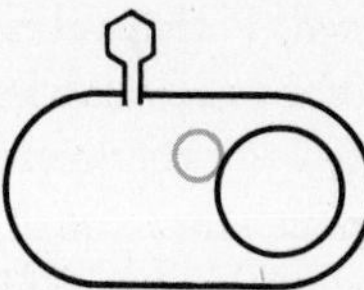

The phage DNA may become integrated with host cell DNA as a prophage

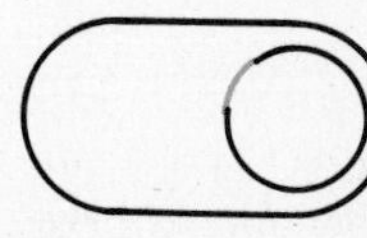

When the prophage is induced to leave the bacterial chromosome, it may exchange a bit of DNA, thereby carrying with it a few bacterial genes

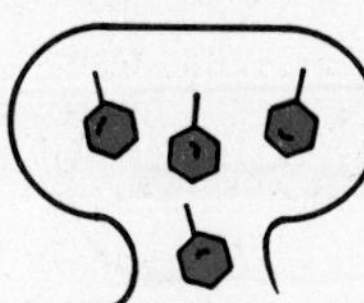

The phages that are replicated are defective because they lack vital genes that have been replaced by bacterial DNA

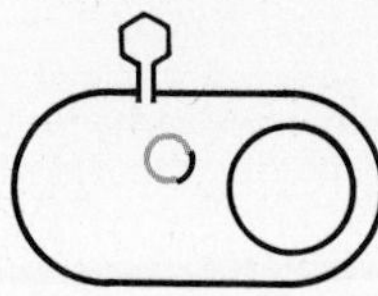

The defective phage DNA penetrates new host cells but cannot cause the production of new phage particles

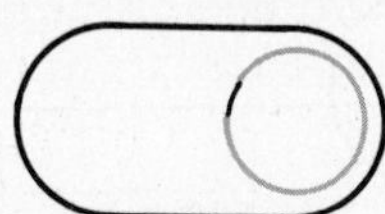

Bacterial genes introduced into the new host cell are integrated into the DNA, become a part of the bacterial chromosome, and are replicated along with the rest of the bacterial DNA

FIGURE 16-10
Specialized transduction by phage.

parts of the cycle. For example, if the gene that specifies the phage head protein is not present, the phage will not be formed. In order for a defective virion to complete its infective cycle, it is necessary that the missing genes be supplied. This can be done by infecting the same cell in which the defective virus is replicating with another virus particle which can supply the missing functions. The second phage in this example can supply the functional gene for the head protein, permitting phage maturation and allowing the defective phage to escape from the infected cell (Figure 16-11). It is highly probable that infection of cells in nature by phages and other viruses may be far more common than it appears to be because special efforts are necessary to recognize defective viruses.

The lysogenic relationship is of profound importance for several reasons. Lysogenic conversion is significant in the production of the few human diseases mentioned previously (diphtheria, scarlet fever, and botulism), and it is probable that other similar situations remain undiscovered. Perhaps the most important aspect of lysogeny comes from the fact that a number of viruses which cause cancer in animals may be defective viruses. For this and other reasons it is strongly suspected, although not proved, that some human cancers may be caused by viruses which can establish a state in human cells analogous to lysogeny in bacteria (see Chapter 18).

Virus Release without Cell Lysis

Both the T-even and lambda phages lyse and therefore kill their host cell when they are released. However, some bacterial viruses are known which are released

FIGURE 16-11
Complementation. Defective phages can cooperate to produce complete virions.

Defective viruses (those that lack a functional gene) cannot replicate all viral components. Consequently, no complete virions are formed in cells infected with a defective virus.

In cells infected with different defective viruses, complementation can occur. Each virus supplies the gene missing from the other virus, and complete virions are formed.

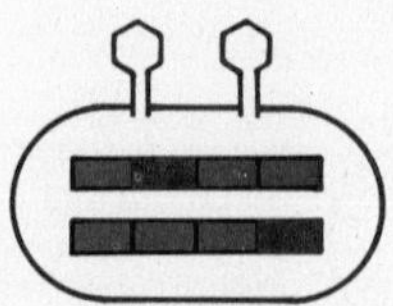

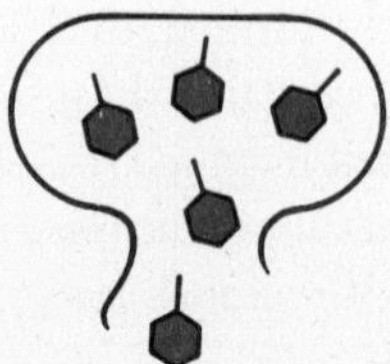

without harming the host cell. These filamentous phages have a long protein coat enclosing one molecule of single-stranded DNA. Their hosts are Gram-negative bacteria which possess sex pili, to which these phages adsorb. Once inside the host, the viral DNA replicates, and viral protein is synthesized. Several aspects distinguish this infection from infection by T-even phages. First, the filamentous phage does not completely take over its hosts' metabolism, because the infected bacteria continue to multiply. Second, neither an intracellular pool of filamentous phage nor the phage coat can ever be detected in the host cells' cytoplasm. This suggests that the phage coat is stored in the cytoplasmic membrane of the host cell, and the virus is assembled as it is extruded from the cell. Third, virus is continuously extruded rather than being released in a burst as is T-even phage when its host cell lyses. Fourth, infected bacteria are *carrier cells;* they can be subcultured and stored in the same way as noninfected strains, and they will continuously release virions into the medium. Infection of many animal cells by certain viruses also leads to virus release without concomitant cell death.

METHODS USED TO STUDY VIRUSES

Virus Purification

It is necessary to have quantities of purified viruses for many purposes. These include chemical and physical studies as well as the production of vaccines. With bacterial or plant viruses such purification is usually relatively easy because it is possible to obtain large quantities of infected cells. The purification process consists of separating the viral particles from the constituents of the host cell. One method that is often used involves breakage of the host cells and sedimentation of their contents in an ultracentrifuge. Centrifugation at low speed removes particles of cell debris, whereas higher speeds of centrifugation are necessary to sediment the smaller viral particles.

In animal viruses the primary difficulty is not so much in purifying the virions but in obtaining enough infected cells. Some viruses still defy efforts to cultivate them in other ways than in animal hosts, while others are more conveniently grown in cell cultures or embryonated chicken eggs.

Culture of Human and Other Host Cells

Human and animal cells have been grown under laboratory conditions since the early part of the century. However, the methods used for culturing them have become routine only during the last 25 years—since the advent of antibiotics (used to suppress growth of bacteria and fungi in cell cultures) and more sophisticated aseptic techniques. Currently almost any type of animal cell that normally divides in the body can be grown or maintained in culture. Such cell cultures provide the most convenient means for studying the growth of animal viruses or for producing them in large quantities.

Before the development of methods for cell culture, the most useful host for the cultivation of animal viruses was the embryonated chicken egg. Embryonated eggs are infected through an opening made in the egg shell about a week or two after fertilization. Depending upon the virus in question, growth occurs in one or more of the embryonic membranes, in the yolk sac, or in the embryo itself.

Quantification of Viruses

For any detailed study of viruses it is necessary to develop methods for accurately determining their number. The methods used vary greatly, ranging from the counting of virions to studying the consequences of their interaction with living cells. A summary of the most commonly used methods follows.

Plaque method. The detection of plaques is at once the most flexible, the most general, and the most informative method for quantification of virions because it is a simple and reproducible assay for infectivity. The method was first perfected for bacterial viruses and consists of the following operations. A sample of bacteriophage is mixed with a concentrated suspension of host bacteria and a few milliliters of melted agar at about 44°C. The agar is then poured into a petri dish containing a hardened layer of nutrient agar, on which it solidifies into a thin sheet containing a random distribution of viruses and bacteria. Each viral particle infects a bacterium, multiplies, and releases several hundred new virions. These infect other bacteria in the immediate vicinity, which again release virions. After a few multiplication cycles, all the bacteria in an area surrounding the initial viral particle are destroyed. During the same period of time uninfected bacteria in regions of the plate without viruses multiply rapidly, giving a dense, opaque background. Thus the net result of the plaque assay consists of clear plaque areas contrasting with areas of bacterial growth (Figure 16-12). Because each of the dilute virus plaques corresponds to a single viral particle in the initial sample suspen-

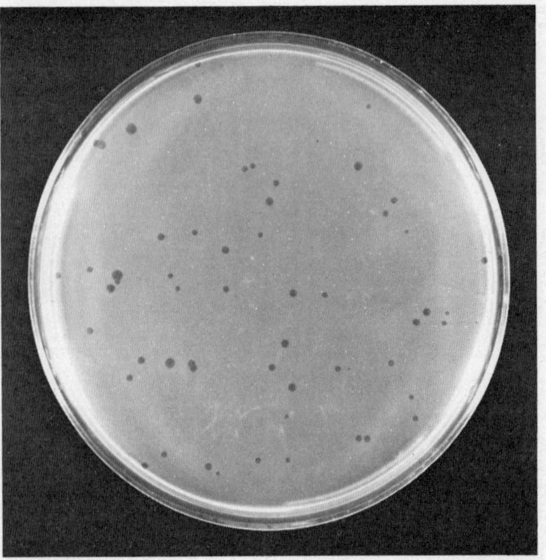

FIGURE 16-12
Phage plaques. The opaque background represents a lawn of bacteria. Each clear area is a plaque formed as a result of lysis of the bacteria by the progeny of a single phage particle.

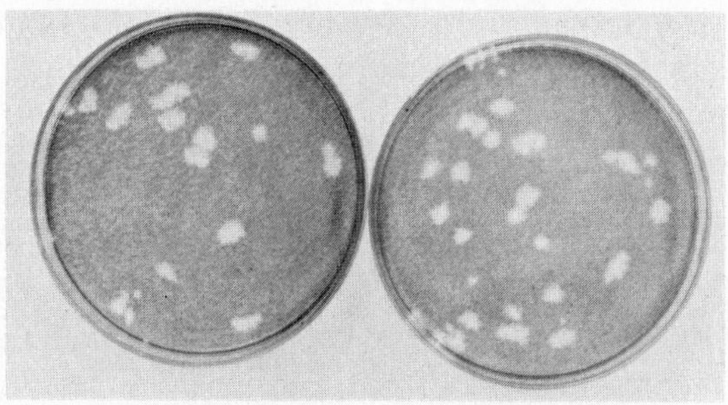

FIGURE 16-13
Plaques caused by infection of human embryonic tonsil cells with herpesvirus, type 1. (Courtesy of B. B. Wentworth; B. B. Wentworth and L. French, *Proc. Soc. Exp. Biol. Med.* 131:590, 1969.)

sion, the number of plaques is a direct assay for virus concentration. The plaque count is analogous to the bacterial colony count.

Plaque assays are equally useful for quantifying animal viruses that are lytic (Figure 16-13). In these assays the viruses added to animal host cells produce clear plaques in the monolayer of cells.

Electron microscopic counting. If reasonably pure preparations of virions are obtainable, their concentration may be readily determined by counting the viral particles in a specimen prepared for the electron microscope. A great inherent disadvantage in this method is that it cannot distinguish between infective and noninfective viral particles.

Quantal assays. Combined with dilution, *quantal assays* can yield an approximation of viral concentration. Several dilutions of the virus-containing preparation are made and administered to a number of animals, cell cultures, or chick embryos, depending on the host specificity of the virus. The titer of virus, or endpoint, is taken to be that dilution at which 50 percent of the inoculated hosts are infected. This is referred to either as the ID_{50}, or infective dose, or the LD_{50}, or lethal dose.

Hemagglutination. A large variety of viruses are able to agglutinate red blood cells by chemical reactions between virions and surface components of the cells (*hemagglutination*). In this process a virion attaches to two red cells simultaneously and causes them to clump together. Sufficiently high concentrations of virus cause large aggregates of red blood cells to form. Hemagglutination can be measured by mixing serial dilutions of the viral suspension with a standard red blood cell suspension. The highest dilution showing complete agglutination is taken as the endpoint or titer. One group of animal viruses able to agglutinate red blood cells is the myxoviruses, as exemplified by the influenza virus.

ARE VIRUSES ALIVE?

The discussion so far clearly shows that viruses are not cellular organisms because they have no cellular organization, but rather are obligate parasites at the genetic level. Whether or not they can be regarded as living depends on the definition of life. If life is considered to be a complex set of processes initiated by the transcription of nucleic acid, then a virus has both living and nonliving stages. It is alive when it is replicating inside an infected cell and dead as a virion outside a cell. Using this consideration only, a transforming fragment of DNA must also be alive once it becomes integrated into the DNA of a recipient cell. There is, however, at least one important difference between the transferability of bacterial-transforming DNA and that of viral nucleic acid. The transfer of the bacterial-transforming DNA occurs at a low frequency and is not an essential part of the existence of the species; however, the transfer of the viral nucleic acid is an essential part of viral existence and occurs with a very high frequency, at least among bacterial viruses. Furthermore, viruses have evolved to entities highly specialized for transfer of their genetic material. Thus they are "more living" than fragments of DNA. The fact that they are entities highly specialized for DNA or RNA transfer also makes them more like organisms than other cellular organelles, such as mitochondria and chloroplasts, that also contain protein and nucleic acid. The question of whether or not viruses are living, then, is more philosophical than scientific.

SUMMARY

Viruses are obligate intracellular parasites that require a host cell in order to replicate. Each virus particle, or virion, consists of either RNA or DNA surrounded by a protein coat. Virus particles are generally either helical or polyhedral in structure, and are sometimes combinations of these two forms of symmetry. In addition to their coat, some viruses have an exterior envelope. A virion contains few enzymes, usually concerned with viral replication. Viruses lack enzymes concerned with energy production, and they contain no ribosomes or other organelles essential for replication. Consequently they depend on host cells to supply these needs.

Virus replication is unique. Most viruses attach to specific receptors on host cells—the process of adsorption. Penetration of viral nucleic acid into the cell follows. Replication of the various viral components is the next step, followed by their assembly into mature, complete virions. The period while the viral nucleic acid directs the replication of viral constituents, during which no infective virions are present within the host cell, is called the eclipse phase. The eclipse phase ends when the first mature infective virion is formed. Eventually, mature virions are released from host cells. This may occur in a burst when the host cell lyses and releases many mature virions, or it may occur with mature virions being released gradually over a period of time without lysis of the host cell.

Viruses are specific with regard to their host cells. The host range of a virus depends in part on the presence of specific receptors on the host cell surface. However, other factors which determine host range include the ability of the viral nucleic acid to survive within the host cell and its ability to subvert cellular metabolism.

Three kinds of virus-host relationships are recognized among bacterial viruses, and these probably occur with other kinds of viruses as well. They are the lytic relationship, the lysogenic relationship, and the case where viruses are released while host cell viability continues. The lysogenic relationship involves a temperate virus that is an episome. Depending on certain conditions, a temperate virus can either replicate within the cell, resulting in the formation of many new virions and lysis of the cell, or alternatively the temperate viral nucleic acid may become integrated within the host cell chromosome. This integrated viral genome is referred to as a prophage, and the host bacterium is described as a lysogenic bacterium. When the prophage leaves the host chromosome, it can once again replicate and become lytic. The lysogenic relationship is responsible for several important biological phenomena, including lysogenic conversion and transduction.

Viruses can be studied by cultivating them in specific host cells, either bacterial, plant, or animal cells, depending on the virus. Quantification of viruses may be achieved by various methods including electron microscopic counting, plaque assay, quantal assay, and hemagglutination.

QUESTIONS

REVIEW

1. Summarize the features that distinguish viruses from cells.
2. What methods are used to quantify viruses?
3. People are susceptible to the influenza virus, the cause of the "flu." Why is it that some other animals, such as swine, are vulnerable to the same virus while still other animals or plants are not?
4. Explain the difference between temperate and virulent bacteriophage.

THOUGHT

Some bacterial viruses have RNA as their genome. Can such viruses be lysogenic?

FURTHER READING

HORNE, R. W., "The Structure of Viruses," *Scientific American* (January 1963). Clear discussion, with excellent electron micrographs and diagrams.

KELLENBERGER, E., "The Genetic Control of the Shape of a Virus," *Scientific American* (December 1966). A sequel to the Horne article, with data on the genes responsible.

LURIA, S. E., "The Recognition of DNA in Bacteria," *Scientific American* (January 1970). Explanation of how cells can react to foreign DNA.

WATSON, J. D., *Molecular Biology of the Gene.* 3d ed. Menlo Park, Calif.: W. A. Benjamin, 1976. The most authoritative and current text in molecular biology. Easy to read, with good illustrations and a glossary.

WOOD, W. B., and R. S. EDGAR, "Building a Bacterial Virus," *Scientific American* (July 1967). Another excellent article on viral assembly, focused on bacteriophage T-4.

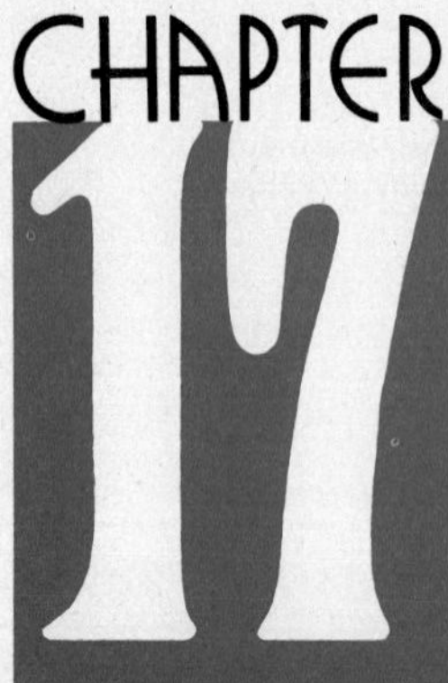

CHAPTER 17

ANIMAL AND PLANT VIRUSES

Much of the basic biology of viruses as elucidated in the study of bacteriophage systems also applies to animal and plant viruses; however, there are also certain properties unique to each group. In this chapter a general approach to the classification of viruses is presented, followed by a discussion of the animal viruses, their variety, mode of replication, and effects on host cells. In addition, a brief introduction to the plant viruses is included.

CLASSIFICATION OF VIRUSES

The taxonomy of viruses has been subject to change over the years. For one thing, as more is learned about the properties of different viruses, their classifications change. For this reason only the principles of viral taxonomy are considered here. A current classification scheme is included in Table 17-1 for reference, while a survey of presently recognized groups of animal viruses can be found in Appendix III. Figures 17-1 through 17-3 show the morphology of several well-known viruses that infect man.

The most widely used taxonomic criteria for viruses depend upon the structure of a virus itself. Four major criteria are used: (1) the nature of the nucleic acid—DNA or RNA, single-stranded or double-stranded; (2) particle structure—helical, icosahedral, or complex; (3) presence or absence of a viral envelope; and (4) dimensions of the viral particle. Beyond these physical characteristics, other criteria (immunologic, cytopathologic, or epidemiologic) are used to subdivide the groups. Such a classification provides great convenience and utility, although it is not necessarily based upon the evolutionary origin of individual viruses.

TABLE 17-1
Classification of Viruses

Nucleic Acid	*Capsid Shape*	*Envelope*	*Dimensions of Capsid (nm)*	*Bacterial*	*Plant*	*Animal*	*Special Features*
RNA	Polyhedral	Naked	18–38	Coliphage[a] f2	Bushy stunt virus	Picornaviruses	
			70–77		Wound tumor virus	Reoviruses	Double-stranded RNA
	Helical	Naked	17 × 300		Tobacco mosaic virus		
		Enveloped	30–75			Some of the Arboviruses	
			90–300			Myxoviruses	Multiple-piece genome
			37–220			Paramyxoviruses	
DNA	Polyhedral	Naked	22	Coliphage ØX174			Single-stranded DNA
			18–26			Parvoviruses (Picodna-viruses)	Some have single-stranded DNA
			40–57			Papovaviruses	
			70–80			Adenoviruses	
			140			Tipula insect viruses	
		Enveloped	120–250			Herpes viruses	
	Helical	Naked	5 × 800	Coliphage fd			Single-stranded
	Complex (polyhedral and helical)	Naked	Head 60 × 90 Tail 17 × 120	Coliphages T2, T4, T5, T6			Polyhedral heads, helical tails
	Complex	Enveloped	200 × 350			Poxviruses	Brick-shaped or ellipsoid

[a] A coliphage is a bacteriophage that infects *E. coli.*

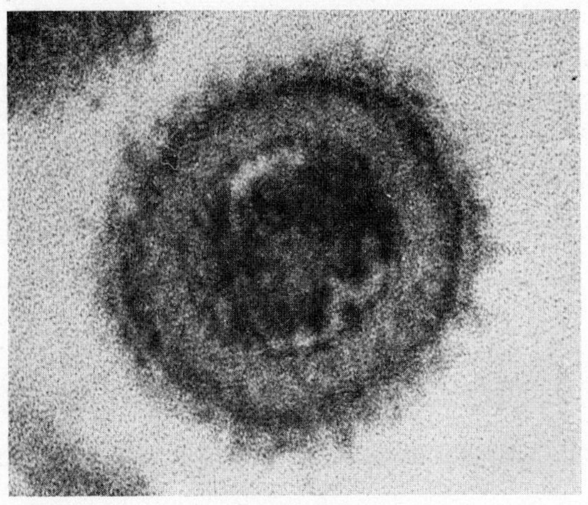

FIGURE 17-1
The morphology of herpes virus. The nucleic acid and protein show icosahedral symmetry and are surrounded by an envelope. (Courtesy of B. Roizman; *Hospital Practice,* April 1972.)

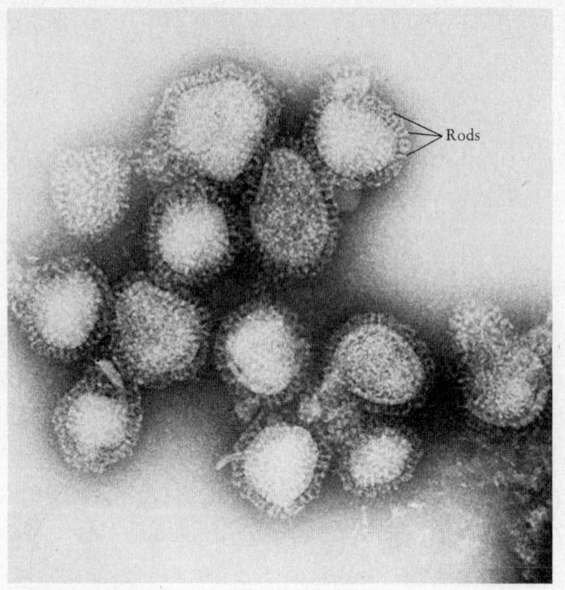

(a)

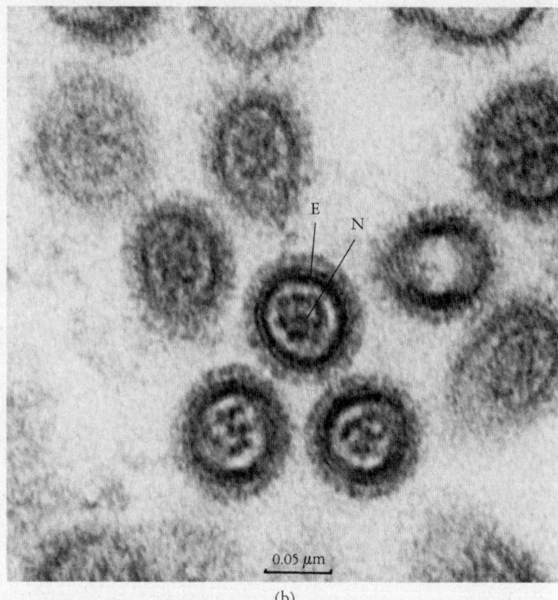

(b)

FIGURE 17-2 Morphology of influenza virus. (a) Electron micrograph of influenza A virions of the Hong Kong strain. Note the spikes or rods inserted into the envelope. (Courtesy of F. Murphy; from the cover of *Science* 163, 1969.) (b) Thin section of influenza virions. The exterior envelope (E) and internal nucleic acid and protein (N) can be distinguished. (Courtesy of E. S. Boatman.)

INTERACTION OF VIRUSES WITH ANIMAL CELLS AND TISSUES

In many cases the ultimate result of infection of an animal cell by a virus is death of the cell. The various kinds of cell damage which may lead to death or abnormal appearance of cells in culture are referred to as *cytopathic effects.* Damage to the cell is not only caused by the formation of vast numbers of viral particles; of even greater importance are the effects of virus-specified proteins on normal cellular processes. For example, viral proteins commonly induce changes in the permea-

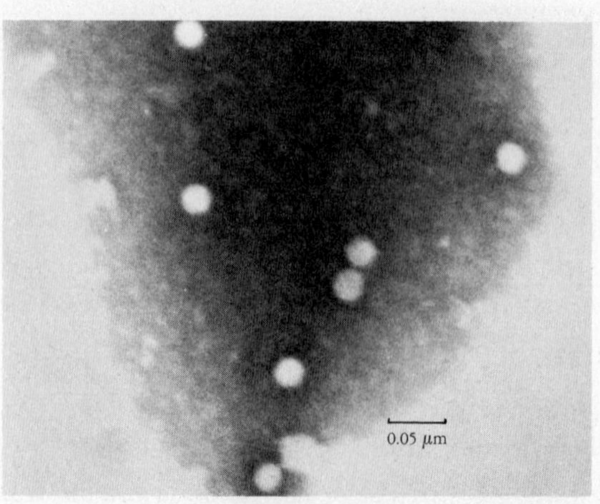

FIGURE 17-3 Morphology of rhinovirus. These are typical viruses of the picornavirus group. The small naked virions consist of RNA enclosed in an icosahedral capsid. (Courtesy of E. S. Boatman.)

bility of one or more of the cellular membranes. Furthermore, some viral proteins can specifically inhibit host DNA, RNA, or protein synthesis. Infected cells may also develop abnormalities in their chromosome structure.

INTERFERON

A very important response of some virus-infected cells is the synthesis of *interferons,* proteins that interfere with the replication of viruses. This protein is induced by viral infection but is coded for by cellular DNA. Interferon is released from infected cells and protects neighboring uninfected cells. It acts intracellularly to inhibit viral replication, in contrast to antibody, which acts extracellularly to inactivate virus particles and prevent their entry into cells. Normal cell functions are unaffected by interferon activity. Most viruses will initiate the interferon response from cells; in fact, the response initiated by one virus will act upon other, concurrent or subsequent, viral infections. The stimulus for a cell to produce interferon appears to be the presence of the viral replicative form of double-stranded RNA or even added synthetic RNA. Because this natural defense mechanism may be artificially induced by administering synthetic RNA, the protection of laboratory animals from some viral diseases has been accomplished in this manner. This presents an exciting future possibility for antiviral chemotherapy.

Antibody, page 380

Adsorption of Animal Viruses to Cells

Like bacterial viruses, animal viruses attach to host cells by means of a complementary association between attachment sites on the virion and receptor sites on the host cell surface. When such receptors do not exist, as in cells of types or species other than those of the host, or when the receptors are destroyed by an experimenter by means of enzymatic treatment, no attachment occurs. Animal viruses usually do not contain specific appendages corresponding to bacteriophage tail fibers; instead, attachment sites appear to be distributed over the surface of the virion. The chemical composition of host cell receptors varies according to the virus in question; the receptors for poliovirus, for example, are lipoproteins. Within a given species of animal, receptors occur in some cell types but may be absent or few in number in other cell types.

Uncoating of the Virus

In the case of most viruses the uncoating process, whereby the viral nucleic acid is released from the protein coat, is poorly understood and apparently quite variable. Complete virions are engulfed into animal cells by *phagocytosis* and are subsequently uncoated. This is true for virions of all sizes, shapes, and taxonomic assignment. The viral particles then exist temporarily within membrane-bound vacuoles of the cell and later may be released within the cytoplasm or nucleus. Uncoating appears to be the result of degradation of viral capsid proteins by enzymes of the host cell.

With viruses containing more than one coat or layer, uncoating may occur immediately after adsorption and fusion of the viral envelope with the cell membrane. However, in the case of enveloped poxviruses, uncoating is a more complex, stepwise process. The pox virus is first phagocytized, and after its outer viral membrane is removed, the phagocytic vacuole in which it is contained dissolves. Later, after the partially uncoated poxvirus particles have penetrated the cytoplasm of the host cell, a specific uncoating enzyme, coded by the viral DNA and present in the virion, appears and removes the inner core protein.

Replication of DNA Viruses

Eclipse, page 326

The period of eclipse between infection with a DNA virus and the appearance of progeny virus may be divided into several stages. The first of these stages involves the synthesis and translation of early mRNA. Just as for bacteriophage, parts of the genetic material of DNA-containing animal viruses are transcribed early and others later. The early messengers code for enzymes, particularly those involved in DNA synthesis, such as thymidine kinase and DNA polymerase. In the next stage of infection, viral DNA is replicated by a process which depends on the prior synthesis of the early enzymes. In all cases, except that of poxviruses, viral DNA is synthesized in the cell nucleus. In the case of the poxviruses, the largest and most complex viruses known, the site of DNA replication is the cytoplasm of the cell. Some DNA viruses possess genetic capability to cause the shutdown of host RNA and protein synthesis during viral replication.

At later stages of infection late messengers are produced; these code for viral capsid proteins and other proteins involved in virus regulation or assembly. Virus maturation occurs over a considerable period of time. Some viruses are released from the cell, without cell death, by egestion, whereas others are released when the cell dies and disintegrates. Again, for the larger viruses, the maturation process is more complex. In particular, the maturation of poxviruses appears to involve changes that take place after the viral components are assembled within the viral membrane. One of these changes is that the DNA is enclosed by many layers of viral membrane which later sort themselves out into inner and outer membranes.

Replication of RNA Viruses

With single-stranded RNA viruses the predominant route of replication involves the production of a *replicative form* consisting of double-stranded RNA; in this process a new RNA strand complementary in base sequence to the infecting strand is produced (Figure 17-4). This intermediate form then acts as a template for new viral RNA strands. Other RNA viruses have adopted an even more novel strategem, in which a DNA copy of the infecting RNA strand is synthesized. This is necessary because mechanisms for replicating RNA are lacking in noninfected cells. Therefore an infecting RNA molecule from one of these viruses must first specify the synthesis of the required DNA-synthesizing enzyme before it can replicate; alternatively, the viral particle itself must carry such an enzyme. As examples of ways in which these problems are mastered, the replication of a single-stranded RNA virus, poliovirus, will be considered first, and other viruses, in which the solution to reproduction differs, will be discussed later.

In poliovirus RNA the infecting strand can act both as mRNA and as a template for the replication of the viral RNA. The incoming strand rapidly attaches to host cell ribosomes, where it serves as a messenger for protein synthesis. When this happens, host RNA and protein synthesis are severely inhibited, apparently as a result of the synthesis of a new viral-specified protein. Among the other viral proteins produced in this process is RNA synthetase, an enzyme that catalyzes the replication of the viral RNA in a process commencing within half an hour after infection. The viral RNA synthesis then proceeds exponentially for about 3 hours. One novel feature of the translation of poliovirus messenger RNA is that the entire mRNA molecule codes for one very large polypeptide chain which is later cleaved into several proteins, including the viral RNA synthetase, the viral capsid proteins, and the proteins responsible for the curtailment of host RNA and protein synthesis. When a large pool of mature virions has accumulated as the result of association of the newly synthesized viral capsid proteins with the viral RNA, the host cell lyses and releases them.

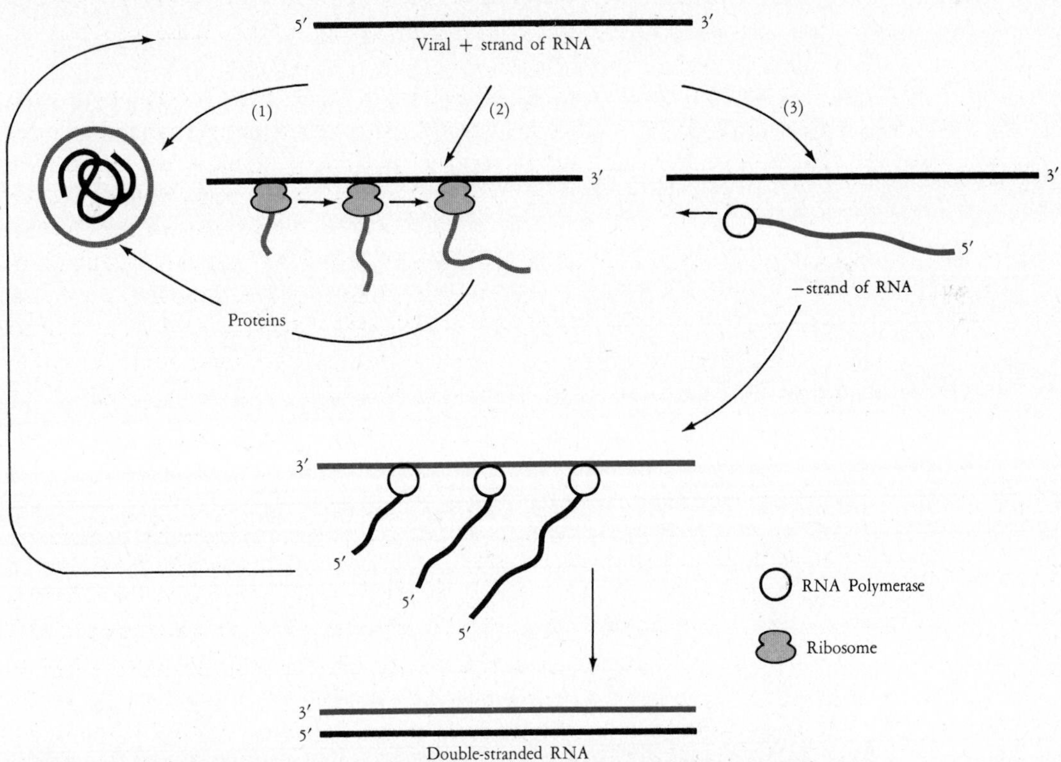

FIGURE 17-4 The polio virus + strand of RNA can be (1) incorporated into capsid protein to form a new virion or (2) serve as a template for translation of capsid protein or (3) serve as a template for transcription of a − strand of viral RNA. In the third case, the − strand can either be used as a template for transcription, resulting in the synthesis of more + strands of viral RNA, or become part of double-stranded RNA.

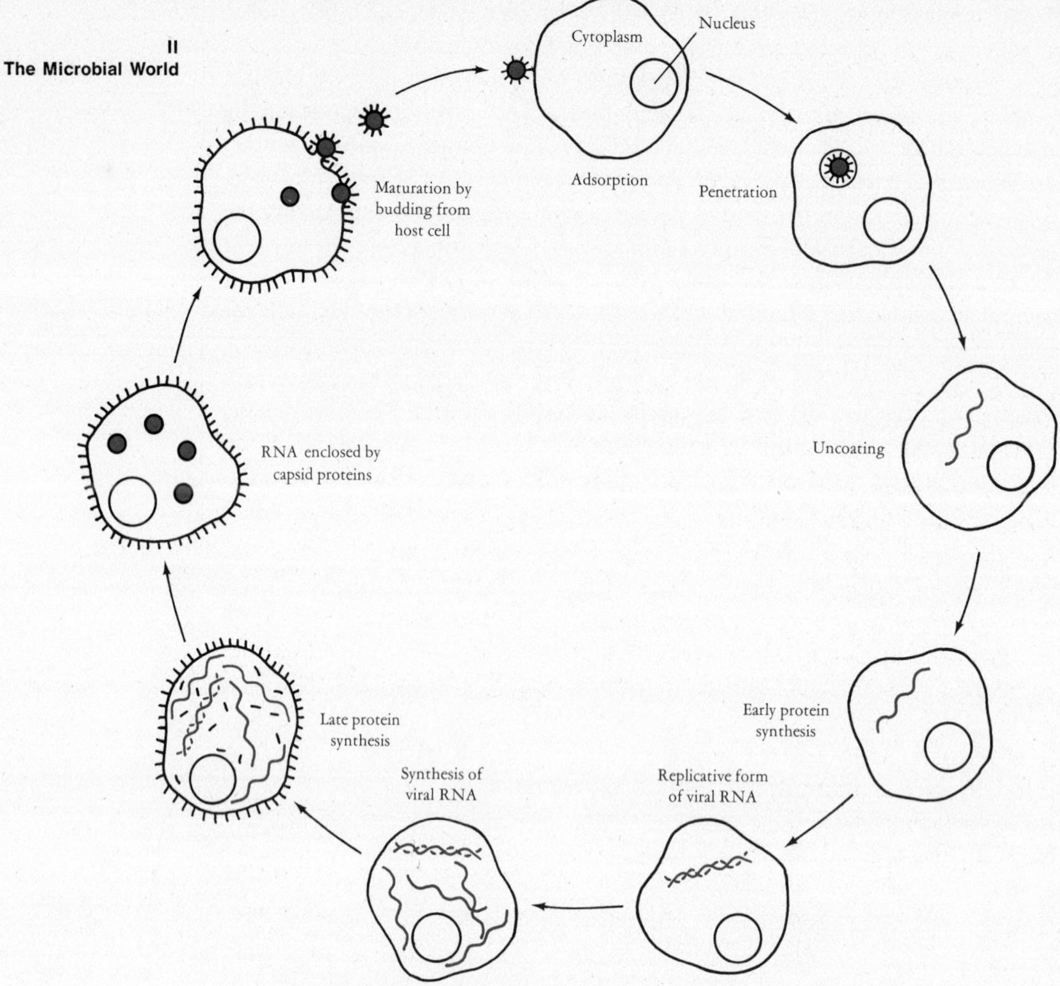

FIGURE 17-5 The replication of an enveloped virus, the influenza virus.

The double-stranded RNA of certain viruses, such as reovirus, consists of seven to ten pieces. In reovirus infection each of these pieces is apparently transcribed into a messenger for protein synthesis and replicated by an enzyme or enzymes carried within the virion itself. Thus reovirus-infected cells contain several species of single-stranded fragments of the viral RNA.

In the extracellular existence, many other RNA-containing animal viruses are enclosed by a lipoprotein envelope. This is acquired by the virus at the time of its release from the cell membrane through a process called *budding* (Figure 17-5). In this process the nucleocapsid of the virus is expelled through the plasma membrane, which has been altered by the incorporation of viral protein. Thus only extracellular RNA-containing viruses contain the complete envelope.

PLANT VIRUSES

A great number of plant diseases come from viral infections. Many of these are of considerable economic importance, particularly those occurring in crop plants. Virus infections are particularly prevalent among perennial crop plants or those propagated vegetatively, such as potatoes. Other crops in which viruses cause considerable loss of productivity include wheat, soybeans, and sugar beets. A serious virus infection may reduce productivity of these crops by more than 50 percent.

Infection of plants by viruses may be recognized through various signs (Figure 17-6). There may occur localized abnormalities in which there is a loss of green pigments, and entire leaves of the plant may turn yellow; in many cases rings or irregular lines appear on the leaves and fruit on the plant. Individual cells or specialized organs of the plant may become necrotic, and calluses or tumorous growths may appear. Usually infected plants become stunted in growth, although in a few instances growth is stimulated, leading to deformed structures. In the vast majority of cases plants do not recover from viral infections, for unlike animals, plants are not capable of developing specific immunity to rid themselves of invading viruses. On occasion, however, infected plants develop new growth in which visible signs of infection are absent, even though the infecting virus is still present. The reasons for this are not understood. In severely infected plants virus particles may accumulate in enormous quantities. For example, as much as 10 percent of the dry weight of an infected tobacco plant may consist of tobacco mosaic virus.

In a few instances plants have been purposely maintained in a virus-infected state. The best known example of this occurs in the case of tulips, where a virus transmitted through the bulbs can cause a desirable color variegation of the flow-

FIGURE 17-6 Lesions caused by plant viruses. *Nicotiana glutinosa* plants 11 days after inoculation with lettuce necrotic yellows virus. On the right is a healthy plant for comparison; left, a plant given a large dose; center, a plant given a small dose of the virus. Visible effects of the viral infection include loss of pigment in some areas, necrotic lesions, and distortion and stunting of infected leaves. (Courtesy of J. W. Randles; J. W. Randles and D. F. Coleman, *Virology* 41:459–464, 1970.)

ers. The infecting virus was transmitted in this way for a long time before the cause of the variegation was even suspected. The multiplication of plant viruses is analogous to that of bacterial and animal viruses in many respects.

Routes of Infection by Plant Viruses

Many plant viruses are extraordinarily stable; tobacco mosaic virus apparently retains its infectivity for up to 50 years. This stability is an important feature in maintaining the virus because processes of infection are generally very inefficient. The same feature is important in considering the difficulty of eradicating a viral infection. Viruses may spread from plant to plant by a variety of means. Some are transmitted through soil contaminated by prior growth of infected plants. For perhaps some 10 percent of the known plant viruses, transmission is possible through plant seeds or tubers or by pollination of flowers on healthy plants with pollen from diseased individuals. Seed transmission may occur in some hosts but not in others. Virus infections may also spread through grafting of healthy plant tissue onto diseased plants. Another, more exotic transmission mechanism is effected by plants such as the parasitic vine, dodder, which establishes connections with the vascular tissues of host plants and transmits virus from plant to plant.

Other important infectious mechanisms involve vectors of various types. These include insects, worms, man, and fungi. For example, tobacco mosaic is a serious disease of the tobacco crop, yet it has no known insect vectors; instead, man himself is the major vector of this disease. In this instance virus is transmitted to healthy seedlings on the hands of workers who have previously picked up the virus from infected plants or tobacco products. However, the most important plant virus vectors are probably insects, and insect control is a potent tool for controlling the spread of viruses.

Arthropod Transmission of Plant Viruses

Several distinct types of arthropod vector transmission of plant viruses are recognized. First, *external or temporary transmission* involves the association of a virus with the external mouth parts of the vector. In this case the ability to transmit virus lasts only a few days. Second, in *circulative transmission* the virus circulates in the body of the insect and may be infective for the lifetime of the insect. Third, the *transmission may involve actual multiplication of the virus within the insect.* In this case the viral agent is truly an insect virus as well as a plant virus. In many instances of an insect infection, such as with leaf hoppers, viruses persist through a large number of insect generations and may be transmitted to plants at any time. The existence of insect-transmitted plant viruses raises several interesting questions about viral evolution. In particular, it seems that plant and animal viruses may not be so different as they appear at first sight. Several of these plant- and insect-infecting viruses show similarities to animal viruses. Potato yellow dwarf virus is very similar to animal myxoviruses, and wound tumor virus of plants is morphologically very similar to the reoviruses of animals.

Control of Plant Virus Infection

Normally no attempts are made to cure virus infections of plants, because treatments are both ineffective and uneconomic. Instead, prevention of the spread of virus infection is the preferred tactic. This is accomplished by burning or otherwise destroying infected plants and sterilizing the soil. An exception to this general rule exists for long-lived plants, such as fruit trees, where heat treatment of young plants is sometimes used to eliminate viruses. For diseases transmitted by insect vectors, destruction of the vector itself, by insecticides, may be effective. Alternatively, new genetic strains of the plant in question, which are resistant to the virus or unappealing to the insect vector, may be developed.

Viroids

The term viroid has recently been introduced to categorize a group of subviral pathogens. Only a few viroids have yet been identified, all of them infecting plants. Those which have been characterized consist solely of a small single-stranded RNA molecule with a molecular weight of about 100,000. This RNA molecule seems to be capable of infection and replication without any incorporation into a virion. In some hosts disease results from infection. The agent causing *potato spindle tuber* disease can be isolated as a free single-stranded RNA molecule. The viroids are the smallest known agents of infectious disease.

The question of the mechanism of viroid replication has not been answered. However, it seems unlikely that viroids are satellites requiring a helper virus for replication, as in the case of *tobacco necrosis satellite virus.* At the present time it appears that viroids are capable of autonomous replication in susceptible plants.

The existence of viroids in animals remains speculative. However, it has been suggested that such agents are responsible for certain infectious animal diseases in which conventional viruses have not been identified as causative agents.

INSECT VIRUSES

For a long time plant viruses carried by and multiplying within insect vectors were thought to be without serious effect on their insect hosts. However, several cases are now known in which the insect vectors are themselves damaged or even killed by the viruses they carry. In addition to these kinds of viral infections, insects themselves are subject to their own viral diseases. In nature viral diseases are often important as a means of control of insects that are for one reason or another considered pests by man. Insects are also infected by a variety of viruses pathogenic for human beings. Some of these viruses are discussed in Chapter 32.

SLOW VIRUS INFECTIONS

The typical acute virus infection leads to the death of the infected cell within a period of from 12 to 48 hours. However, rapid death is not the inevitable result of virus infection. In some cases viruses can persist in host cells without any of the usual signs of infection. This possibility has led to the concept of slow virus infections. These insidious disorders are characterized by a progressive degeneration to a usually fatal outcome. It should be stressed that the adjective "slow" refers to the course of the disease, not to the virus. In fact the same virus may be capable of both slow and acute infection.

In slow virus infections the infecting virus can replicate and persist without inhibiting host RNA or protein synthesis. A well-known case is that of herpes simplex virus, which causes cold sores. Such infections can remain dormant for years until activated by fever or sunburn. During intervening periods the cell and virus live in apparent harmony without cell death.

In recent years it has become clear that certain neurological diseases of both humans and animals result from virus infection. Among these are kuru and Creutzfeldt-Jakob disease in human beings and scrapie in sheep. These are now known as transmissible diseases of viral origin. Many other degenerative diseases, such as multiple sclerosis and diabetes, may also be slow virus diseases, although only scant evidence supports this possibility.

Kuru and Creutzfeldt-Jakob disease are similar degenerative diseases of the human central nervous system. Fortunately, both are rare and kuru is restricted to the Foré tribe in New Guinea. In fact, as a result of the elimination of ritual cannibalism by this tribe, the incidence of kuru in the tribe has dropped. Before 1957 the disease was transmitted primarily to women and children as they prepared the brains of dead relatives for cooking. Infected brain tissue is also highly infectious when introduced through cutaneous inoculation. The infectious nature of kuru introduced in this way was directly demonstrated by Dr. Carleton Gajdusek and his colleagues at the National Institute of Neurological Diseases and Stroke. They inoculated chimpanzees with the homogenized brain tissue of patients who had died of kuru. Within from one to five years after inoculation, the chimpanzees developed a neurological condition very similar to that in human victims. Dr. Gajdusek was awarded the 1976 Nobel Prize in Medicine for this pioneering work.

Scrapie is a chronic infection of the central nervous system of sheep. As in the case of kuru, this disease is transmissible and is assumed to be of viral origin. However, the transmissible agents of scrapie have not yet been characterized. In the case of scrapie the infectious agent is very resistant to the chemical and physical treatments which destroy most viruses; accordingly, it may be a viroid rather than a conventional virus.

The agents of slow virus diseases also include viruses of the conventional type. In addition to the herpes virus already mentioned, it appears that the measles virus may be responsible for a rare form of encephalitis: subacute sclerosing panencephalitis (SSPE), which appears to result from chronic infection by measles virus. In fact this virus can be isolated from the brains of infected patients. The pathogenesis of subacute sclerosing panencephalitis is not understood since it is not clear why measles virus, which usually kills infected cells, can also act as a persistent nonlethal virus. It has been suggested that defective virus particles, perhaps lacking some of their normal genetic information, are responsible for the persistent virus infections. These defective particles, arising by chance mutation, may compete with the normal virus replication processes, thus slowing the course of infection. It is also surmised that tissue damage in such slow virus diseases derives from an immune response to antigens representing the defective particles.

SUMMARY

Animal viruses are generally classified on the basis of their nucleic acid type, capsid structure, presence or absence of an envelope, and the dimensions of the virion.

Virus replication depends on properties of both the virus and the host cell. Infected cells may produce interferon, which interferes with subsequent viral replication. In general animal viruses adsorb to specific receptors on host cells and are phagocytized by these cells. Within the cell the virus is uncoated, and viral components are replicated separately. The components are then assembled into virions. Enveloped viruses gain their envelope from either the nuclear or the cytoplasmic membrane of the host cell. Mature virions are released by budding from the cell, or by lysis of the host cell.

Insects act not only as vectors for plant and animal viruses, which may multiply in the insect host, but they themselves may be infected by specific insect viruses.

Some virus infections of animals persist over a long period of time, resulting in "slow virus" diseases such as kuru in human beings and scrapie in sheep.

QUESTIONS

REVIEW

1. What criteria are used for classifying animal viruses?
2. Explain the term "viroid," and indicate the features which distinguish these agents from viruses.
3. Give some examples of insect viruses which can be transmitted to plants or to vertebrates.
4. Compare the replication of the naked virion of poliovirus with that of the enveloped virion of influenza virus.

THOUGHT

How does the mechanism of virus transmission differ between animals and plants?

FURTHER READING

FENNER, F., et al., *The Biology of Animal Viruses.* 2d ed. New York: Academic Press, 1974. Authoritative two-volume reference on viruses of warm-blooded animals, including their molecular biology, classification, ecology, and the pathogenesis of the diseases they cause.

FENNER, F., and D. O. WHITE, *Medical Virology.* 2d ed. New York: Academic Press, 1976. Concise, up-to-date text covering most aspects of medically important viruses and viral diseases, including persistent infections and viroids.

HOLLAND, J. J., "Slow, Inapparent and Recurrent Viruses," *Scientific American* (February 1974). Well-illustrated explanations of the mechanisms and possible role of persistent viral infections in a number of troublesome diseases.

RAFFERTY, K. A., Jr., "Herpes Viruses and Cancer," *Scientific American* (October 1973). Gives details of the relationship between herpes viruses and cancers in various animals. Good illustrations of the hybridization technique for detecting viral DNA in infected cells.

SPECTOR, D. H., and D. BALTIMORE, "The Molecular Biology of Poliovirus," *Scientific American* (May 1975). Clearly written and well-illustrated. Gives details of the replicative process of one of the best-studied viruses.

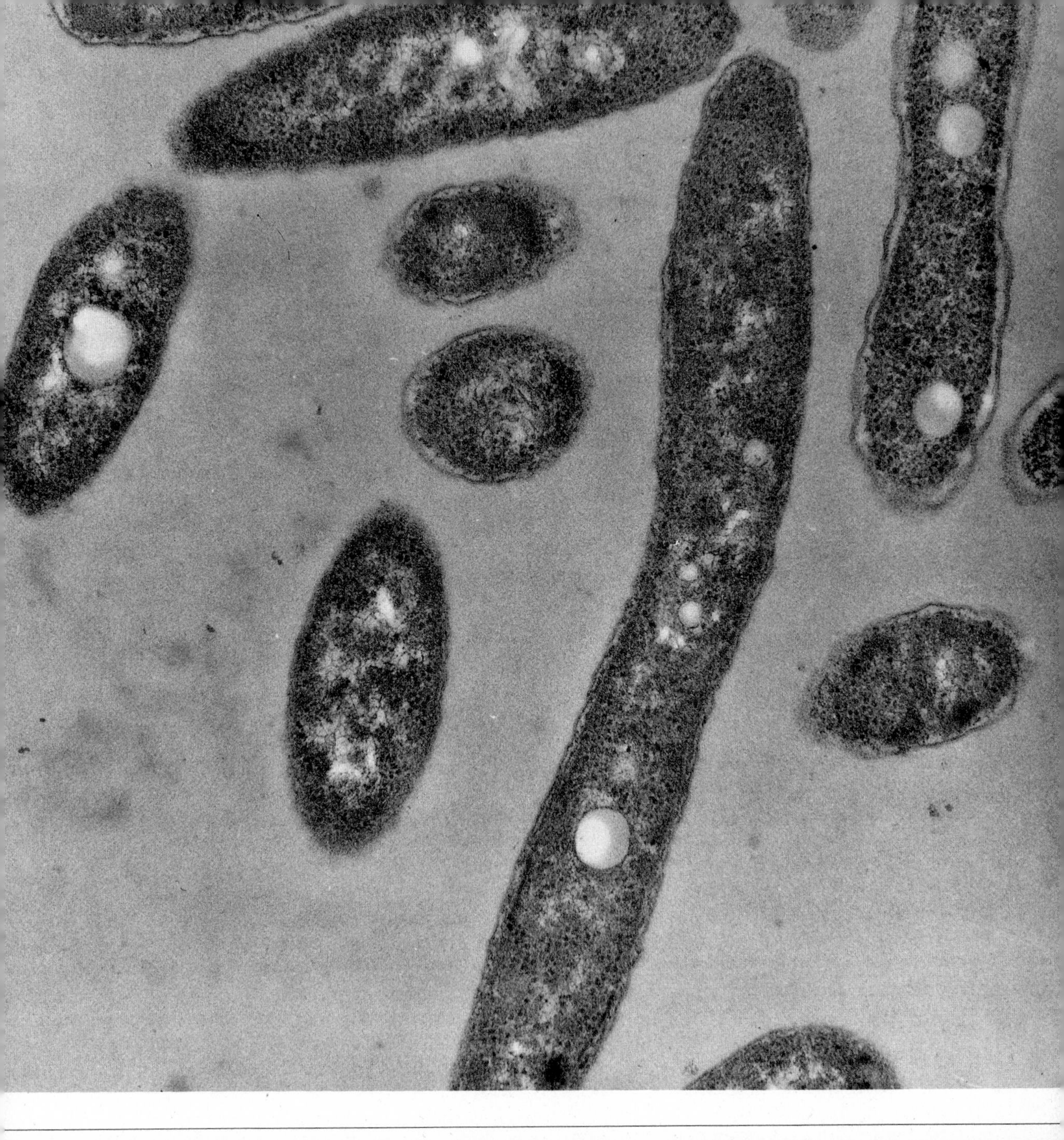

III MICROBES AND HUMAN BEINGS

VIRUSES AND CANCER

The most common result of infection of an animal cell by a virus is a cytopathic effect which disturbs the metabolism of the cell and may eventually destroy it. In such cases virus multiplication occurs and the progeny virus may also infect neighboring cells. It is this cytopathic effect which is responsible for the symptoms of most viral diseases.

Besides the cytopathic effect of some viruses, other types of viruses can produce quite different changes in animal cells, leading to the eventual appearance of a tumor. The process by which normal cells are altered to become tumor cells as a result of virus infection is called *transformation.* This designation should not be confused with the unrelated process of bacterial transformation (Chapter 9). These transformed cells often undergo a dramatic change in morphology and metabolic properties which precedes their rapid multiplication.

Bacterial transformation, page 200

At the present time no viruses have been shown unequivocally to be the cause of human cancer. Nevertheless, the large number of well-documented examples of virus-induced tumors in other animal species has suggested that at least some forms of human cancer are likely to be of viral origin. In this chapter we consider the role of viruses in the induction of tumors and the mechanisms by which this may occur.

TUMORS

Tumors vary considerably in the effect which they have on the organism. At one extreme, *benign tumors* are those whose growth is localized; they are not invasive into neighboring tissues or organs. Moreover, benign tumors have no tendency

BENIGN VERSUS MALIGNANT TUMORS

to give rise to secondary tumors in other parts of the body, although they may interfere with the functioning of an organism by reasons of their sheer size. Usually, benign tumors can be completely removed by surgery. At the other extreme are *malignant tumors,* or cancers, which can invade normal tissues and organs and destroy them. This process may be accompanied by *metastasis,* which involves the separation of cancerous cells from the main body of a malignant tumor and their passage to other parts of the body to initiate the growth of secondary tumors.

Several quite different kinds of agents can play a role in the induction of tumors. A well-known example is provided by carcinogenic (cancer-causing) chemicals which occur in tobacco and other smoke. The evidence which points to these substances as causal agents of human lung cancer is now overwhelming. In fact these carcinogenic agents may be used to produce many different kinds of cancer in laboratory animals.

Another class of chemical carcinogens are those introduced into the body by consumption of processed or adulterated food. In many cases the chemicals in such foods are themselves not carcinogenic, but become chemically or enzymatically transformed into carcinogens after their entry into cells. For example, nitrate, a preservative commonly added to meat, can be converted within cells to carcinogenic nitrosamines. Similarly, hydrocarbons introduced into food by charcoal broiling are converted in the liver to carcinogenic compounds. In addition to chemicals, sources of radiation, such as ultraviolet light and X-rays, are effective inducers of tumors. It is well known that humans who are exposed to excessive doses of such radiation—X-ray technicians or sailors, for example,—are more prone to develop certain kinds of cancer. In the case of both the chemical carcinogens and radiation, it is likely that malignancy is brought about by direct effect of these agents upon the genetic material, the nucleic acids of the cell, leading to a heritable change. Thus it is the mutagenic activity of these agents that is probably the key to their cancer-inducing ability. The observation that over 90 percent of carcinogens are mutagens gives credence to the Ames test in which substances are screened for carcinogenicity by testing their mutagenicity for bacteria.

Ames test, page 197

Finally, there are the tumor viruses, the subject of this chapter. Partly because of the many cases of virus-induced tumors in animals and partly because of the experimental advantages of working with viruses in animals, research on tumor viruses is currently among the most active of all areas of research encompassed by tumor biology.

EARLY HISTORY OF TUMOR VIROLOGY

There are many highlights in the history of tumor virology. In fact over the past 50 years the field has displayed several waves of activity and popularity. The idea that viruses may be involved in cancer dates from the earliest discoveries of viruses themselves, although the proposition that human cancer may be of viral origin was distinctly unpopular until the last 15 years or so. The first successful transfer of the capacity to induce a tumor was reported in 1908 by V. Ellerman and O. Bang. They showed that leukemias of chickens could be transferred by means of cell extracts. However, at the time that Ellerman and Bang did their work, there was no apparent connection between this blood disease and tumors. Soon after, however, in 1911 P. Rous demonstrated a similar transfer of a chicken sarcoma (connective tissue cancer). Even then, this important finding was generally regarded as a peculiarity of the domestic fowl.

The viral induction of a mammalian tumor was first reported by Richard Shope in 1932, who demonstrated the transfer of tumor-inducing capacity by cell-free filtrates of a rabbit papilloma (wartlike tumor of epidermal tissue). These tumors were generally benign in the wild cottontail rabbit population in which they occurred naturally, but when the virus was transferred to the domestic rabbit, malignant skin tumors called *carcinomas* appeared. A few years later John J. Bittner discovered a virus in mice which could induce murine mammary gland carcinoma and which was transmitted from a mother mouse to its progeny through the milk. This discovery was very important for the development of tumor virology because of the great advantages of the mouse as a laboratory animal. The discovery of many other mouse tumor viruses followed thereafter, most of which are responsible for leukemias as well as for solid tumors called *lymphomas,* and more recently for the connective tissue tumors called *sarcomas.*

CHARACTERISTICS OF VIRUS-ASSOCIATED ONCOGENESIS

Even before the viruses themselves were studied or any biochemical studies were initiated, several important characteristics of virus-associated tumor induction (or *oncogenesis*) became apparent. One of these was that in many instances a considerable period of time may elapse between introduction of the virus and the appearance of tumors. For example, mice infected as sucklings with the Bittner virus may not develop a mammary tumor for some months. Furthermore, although the disease is contagious, the route of infection may be obscure, as in this case. In addition, the efficiency of viral transformation is often extremely low, so that the vast majority of viruses infect cells without transforming them. Likewise, many animals infected with tumor viruses do not develop tumors. The reasons for such resistance are not at all well defined, although it is clear that immune mechanisms and genetics play a protective role.

As more methods became available, the tumor viruses were purified and their composition studied. Both the chicken and the mouse viruses were found to contain RNA. In the late 1950s several new tumor viruses, discovered in rodents and primates, were found to contain DNA. One example of such a DNA virus is simian virus 40 (SV_{40}), which was discovered in the cultures of Rhesus monkey kidney cells used for growing large quantities of poliovirus. Perhaps most surprising was the finding that human adenoviruses, common in respiratory infections, could induce tumors in various rodents.

The following discussion presents a brief summary of the special features of some of the known tumor viruses.

ROUS SARCOMA VIRUS

Rous sarcoma virus (RSV) is one of a large group of viruses prevalent in domestic fowl. Some of these viruses are apparently nonpathogenic, while others cause leukemia. The Rous virus produces sarcomas. Some strains of the Rous sarcoma virus are defective. These are unable to specify the synthesis of a component of the viral particle. In such cases complete RSV can only be produced under conditions of coinfection with another virus which supplies the missing viral component. If, for example, the helper supplies the virus coat (protein), the virions produced will resemble the helper virus antigenically. This feature of RSV growth is referred to as *defectiveness,* and the virus that supplies the missing component is known as the *helper.* However, defective Rous viruses are capable of transforming cells even though they produce no infectious progeny virus. There also exist strains of RSV which are not defective for replication or transformation.

See defective bacteriophage, page 332

Members of this group of avian viruses all have the same structural features. They contain RNA, which represents about 2 percent of the viral particle weight. The viruses are relatively large (Table 18-1) and contain a large lipid envelope similar to the myxoviruses. All of the viruses may be grown in chick embryos or in tissue cultures of embryonic chick cells. RNA tumor viruses that cause cancer in mammals, such as the Rauscher leukemia virus, have properties very similar to those of the Rous sarcoma virus. All RNA tumor virus particles, with the exception of mouse mammary tumor virus, have a similar structure. They consist of a centrally located, approximately spherical nucleoid surrounded by a membrane envelope and are known as C-type particles.

Myxoviruses, Table 17-1, page 341

C-PARTICLES

Cell Transformation

A very interesting and important feature of the Rous sarcoma virus is the high efficiency with which it can promote cell transformation. The addition of the virus to a culture of chicken fibroblasts results in the appearance of many foci of transformed cells. These can easily be recognized and distinguished from normal cells.

Normal fibroblasts tend to grow in parallel arrays and cease multiplication when they have produced a confluent monolayer. This property is characteristic of all normal cells and is referred to as *contact inhibition,* a term which implies that cell-contact regulates cell proliferation. On the other hand, transformed cells do not obey these restraints, growing irregularly to form colonies of cells containing more than one layer. The foci of transformed cells may easily be recognized by microscopic examination, and they provide a means of quantifying the number of RSV particles. In fact it has been shown that *all* cells which are infected with RSV are transformed whether or not RSV is produced in the transformed cells. Thus each focus of transformed cells results from the multiplication of a single transformed cell that may be produced by infection with a single RSV particle.

When cells are infected with a large number of RSV particles, they are

TABLE 18-1
Physical Properties of Some Tumor Viruses

Virus	*Nucleic Acid*	*Molecular Weight*	*Particle Diameter (nm)*	*Other Properties*
Rous sarcoma	RNA (single-stranded)	10×10^6	80	Contains 35% lipid in envelope
Mouse leukemia	RNA (single-stranded)	13×10^6	100	Has envelope similar to myxoviruses
Polyoma	DNA (circular double-stranded)	3.5×10^6	45	Icosahedrons; no envelope
SV_{40}	DNA (circular double-stranded)	3.5×10^6	45	Icosahedrons; no envelope
Adenoviruses	DNA	20×10^6	70	Icosahedrons; contains proteins other than capsomere

transformed, and they also produce more infectious RSV particles. However, the process of multiplication of RNA tumor viruses does not seem to severely affect normal cellular processes. Thus the Rous virus–infected cells are not necessarily killed and may release thousands of progeny virus as they continue to live and multiply.

SOME INFECTED HOSTS FAIL TO PRODUCE INFECTIOUS VIRIONS

The Rous virus can even transform mammalian cells, either in the animal (producing tumors) or in tissue cultures. These transformed cells are nonproducers. They evidently contain the viral genome, however, as evidenced by the fact that if hamster cells transformed by RSV are transferred to chicks, they produce the same characteristic tumors. Moreover, these tumors can be induced by various means to produce mature virus.

Persistence of the Viral Genome

See lysogenic phage, pages 330–332

Such observations about the biology of the transformed cell show that the viral genome is present in all transformed cells and that it must multiply along with the cell as the colony of cells, or the tumor grows. It has been suggested that the viral genetic material is integrated with that of the host cell in some way. Until recently this was rather difficult to visualize since the viral genome consists of RNA rather than DNA. However, it is now clear that after infection the RNA genome is copied into a DNA form, and this is physically integrated with the host DNA (Figure 18-1).

Two American scientists, Howard Temin from the University of Wisconsin and David Baltimore from the Massachusetts Institute of Technology, have shown independently that the process of copying the viral RNA genome into DNA is accomplished by an enzyme which can be isolated from intact virions of Rous sarcoma virus. For this work they were awarded the Nobel Prize in Medicine in 1975. The enzyme, an RNA-dependent DNA polymerase called "reverse

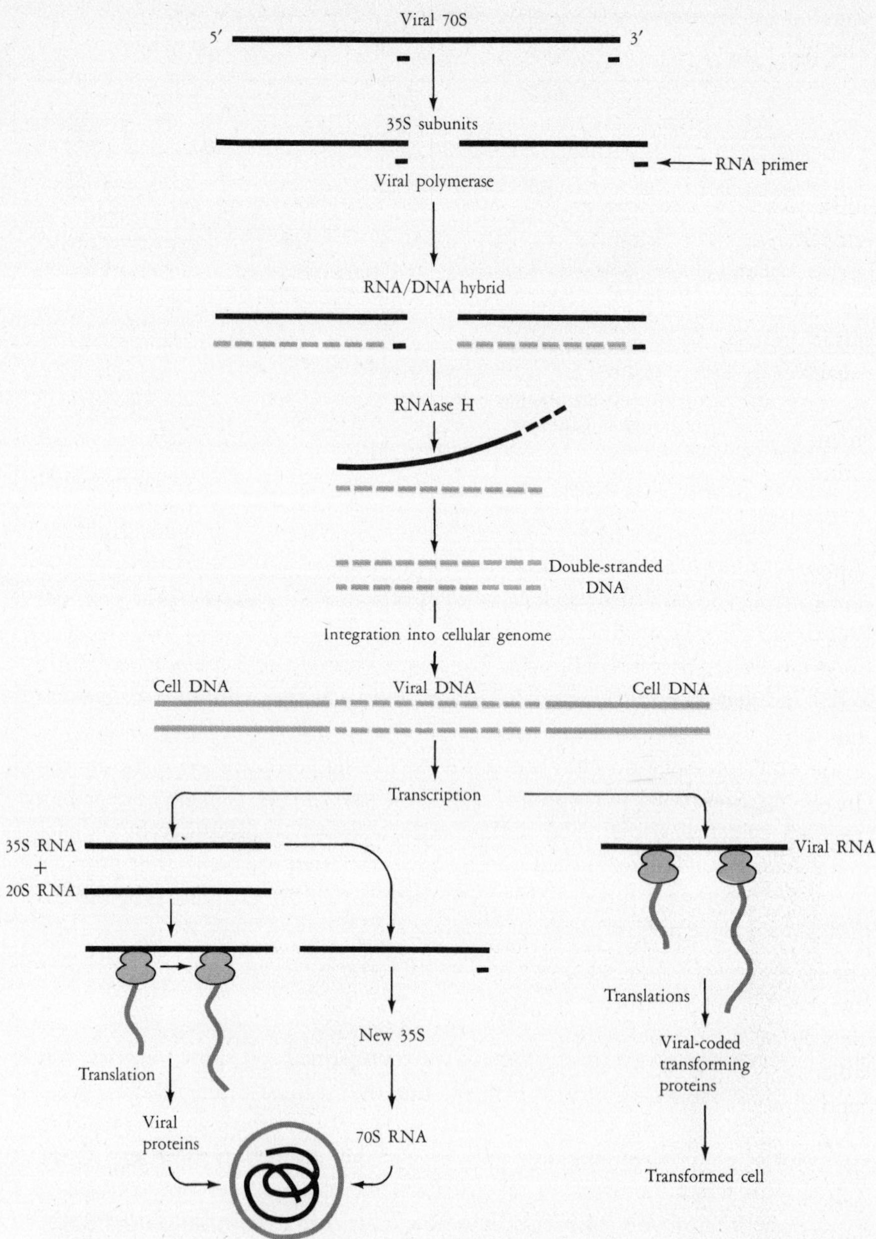

FIGURE 18-1 Two or three 35S RNA subunits are joined to form a viral 70S genome. The genome contains primer molecules, small pieces of RNA which can serve as starting points for DNA synthesis. When the virus infects a cell, the 35S subunits serve as templates for the viral reverse transcriptase, or polymerase. Using the primer molecules as starting points, DNA is synthesized with resulting DNA/RNA hybrid molecules. Ribonuclease H, which attacks only the RNA portion of these hybrids, leaves the DNA free to serve as a template for the synthesis of its complementary strand. The double-stranded DNA can then be integrated into the cellular genome and act as a template for synthesis of viral RNA. The viral RNA thus synthesized can, depending upon which portion is transcribed, act as new 35S viral genome subunits or as messenger RNA for either the viral-coded structural proteins or the viral-coded transforming proteins.

transcriptase," exhibits considerable specificity for the RNA genome of the virus. The initial reaction involves the formation of an RNA-DNA hybrid:

$$\text{viral RNA} + \text{deoxynucleotides} \xrightarrow[\text{enzyme}]{\text{Polymerase}} \text{RNA-DNA hybrid}$$

Subsequent to its formation the single-stranded DNA of the hybrid acts as a template for the synthesis of a complementary DNA strand. This DNA then circularizes and accumulates in the cytoplasm. These circular *proviruses* then move into the nucleus and integrate into the chromosomes of the host cell. Following this integration the viral genes behave as host genes and may be transcribed in the usual way.

It is clear that the viral genome persists in these transformed cells and that it enters its cryptic state with 100 percent efficiency. All infected cells are transformed and transmit the viral genome to their progeny. This efficiency contrasts with the very low rate of transformation of other tumor viruses.

REVERSE TRANSCRIPTASE

The discovery of the novel enzymatic mechanism involving reverse transcriptase has several important implications. Reverse transcriptase has been demonstrated in other RNA tumor viruses, such as those responsible for mouse leukemia. No such enzyme activity exists in nononcogenic viruses; however, an interesting exception to this general rule is the existence of reverse transcriptase in the so-called slow RNA viruses responsible for some neurological disorders. It thus seems that the existence of enzymes with such activity will be a useful diagnostic tool for detecting oncogenic virus infections. Furthermore, the very novelty of the RNA-dependent DNA polymerase reaction engenders some hope that specific viral enzyme inhibitors may be developed as therapeutic agents.

Slow virus infections, page 349

Genetics of RNA Tumor Viruses

The small size of the RNA tumor virus genome implies that it contains only a few genes, probably four. Three of these are known to code for components of the virus: reverse transcriptase, the envelope glycoprotein, and a series of lower molecular weight structural proteins known to be synthesized as part of a polypeptide which is a precursor of the coat protein. In addition, those viruses in the RNA tumor virus group which can transform cells have another gene, called the *sarc* gene. This finding strongly suggests that the product of the sarc gene is directly involved in the cell transformation process. Although the nature of this function has not been established, it may well be related to the control of cellular DNA synthesis. The elucidation of the role of the sarc gene will be a crucial step in understanding the mechanism of cell transformation.

Endogenous RNA Viruses

It appears that the genomes of all species of mammals and birds contain the genetic information for specifying virus construction. This can be demonstrated by several procedures. For example, chemical carcinogens induce the production of virus particles in mice or cultured cells. Infection of cells with exogenous viruses can also cause liberation of latent viruses. Presumably, then, virus-encoding genomes are integrated into the host chromosome in a latent form; consequently, the viruses formed are said to be endogenous RNA viruses. Such endogenous viruses are very similar to Rous sarcoma virus in structure but, except for some cases in inbred strains of mice, are not themselves oncogenic.

POLYOMA AND SV_{40} VIRUSES

Polyoma virus is so-called since it can produce a wide variety of different kinds of tumors. SV_{40} was discovered as a passenger virus in cultures of Rhesus monkey kidney cells used to grow poliovirus, and it was later shown to produce sarcomas in baby hamsters. The polyoma and SV_{40} viruses are basically similar, both structurally and in biological properties, and will therefore be considered together as examples of DNA tumor viruses. In both polyoma and SV_{40} viruses the genome is circular double-stranded DNA of about 3.5×10^6 daltons. The two viruses, however, do not appear to be closely related antigenically, and their DNA molecules differ in overall base composition.

Infection of Cells

Both the polyoma and SV_{40} viruses infect some types of cells productively with the appearance of mature infectious virus, whereas other cells become transformed. For example, SV_{40} infection of green monkey kidney cells leads to virus production, whereas infected hamster or human embryonic cells are transformed. There are, however, some cases where both virus production and transformation occur simultaneously.

Antigen, page 380

T-ANTIGENS

Early proteins, page 327

With polyoma or SV_{40} virus infection, virus particles are produced in the cell nucleus (several millions per cell). These are eventually released as the cells disintegrate. An antigen (distinguishable from the capsid protein) appears in the cell nucleus very early in infection. This same antigen is also present in transformed cells or tumors and is therefore useful as a means of identifying the prior infection of these cells by the virus. The antigen, referred to as the *T-antigen,* is virus- but not cell-specific. Thus the T-antigens occurring when the same cell is infected by either SV_{40} or polyoma are different, whereas different cells infected by the same virus contain the same T-antigen. Moreover, cells transformed successively by both SV_{40} and polyoma viruses contain the two different viral T-antigens. Consequently the T-antigen is coded by the viral genome and may correspond to one of the early proteins synthesized in certain bacteriophage infections. Its role appears to be to activate host DNA synthesis.

Transformation

Many kinds of cell cultures may be transformed by both the polyoma and SV_{40} viruses. Transformation produces foci of rapidly growing cells in random arrays. Although a single virus particle may transform a cell, the efficiency of this process is extremely low. The efficiency with which infection resulting in virus production occurs is 10,000 times that with which transformation occurs. This suggests that only a small fraction of the infective particles are capable of transformation, a situation which contrasts markedly with that for the Rous virus, in which most virus particles are capable of transforming their hosts.

The morphological and biochemical characteristics of polyoma- or SV_{40}-transformed cells are determined both by the cell type and the infecting virus. Because cells infected with both viruses display both sets of new properties, it is evident that the virus itself is responsible for the altered properties of transformed cells. However, transformed cells do not contain detectable quantities of virus or of capsid protein, although they must contain at least part of the viral DNA.

Integration of the Viral Genome

Recent experiments have demonstrated directly that transformed cells contain viral DNA and that this viral DNA is transcribed into viral messenger RNA. These experiments are of several types, each relying on recent advances in molecular and cell biology.

Molecular hybridization, pages 242–244

1. RNA molecules transcribed from the viral genome can be detected both in polyoma- and SV_{40}-transformed cells. By the use of molecular hybridization techniques, it can be shown that a small proportion of the RNA molecules produced in such cells have the base sequence characteristic of the virus genome in question. Moreover, in cells transformed successively with both SV_{40} and polyoma, RNA molecules are apparently transcribed from both viral genomes. These experiments demonstrate the presence of at least part of the viral genome in the transformed cell.
2. Similar experiments using radioactive DNA show more directly that the viral DNA itself is present. Again, using molecular hybridization techniques, it can be demonstrated that polyoma DNA reacts with cellular DNA extracted from transformed mouse cells much more efficiently than it does with normal mouse DNA. This reaction therefore provides a means for quantifying the amount of polyoma DNA actually present in the genome of the transformed cell. Current estimates vary, but the amount is placed between 2 and 60 copies of the viral genome per cell, assuming that the complete viral DNA molecule is present.
3. The fact that the complete viral genome is present in transformed cells has been most convincingly demonstrated by the liberation of mature infectious polyoma virus after transformed cells have been fused with other cells. This procedure releases the virus genes from their cryptic state in the transformed cells and allows them to specify the biochemical steps necessary for virus multiplication.

The actual state of the viral DNA molecule in the host genome is not yet clearly understood although the viral DNA is integrated and covalently bound to the host DNA. The integration of the viral DNA therefore appears to be analogous to the process of lysogenization by bacterial viruses. Insertion presumably occurs after an initial pairing event, and the viral DNA becomes covalently linked at each end to the host DNA. As in the case for lysogenic bacteriophages, release of the viral genome to initiate viral multiplication would be rare and might be triggered by radiation or physiological change. However, the molecular events involved in transformation are undoubtedly more complex than outlined in these few sentences. For example, infection with tumor viruses also induces host DNA synthesis, which appears to be crucial for subsequent cell transformation.

Induction, page 332

VIRUSES AND HUMAN CANCER

The widespread occurrence of virus-induced tumors in animals suggests the likelihood that at least some types of human cancer are of viral origin. Proof of this may be forthcoming within the next few years, and it is hoped that this will permit the development of methods for the prevention and treatment of some types of human cancer.

One current candidate for the role of human cancer virus is the Epstein-Barr (EB) virus. This herpes virus may be isolated from cell cultures of an unusual lymphoma which is common in patients in parts of Africa. Thus it is possible, although unproved, that the virus is a causative agent of the African lymphoma known as Burkitt's lymphoma. The lymphoma cells contain multiple copies of the EB viral genome, some as free circular DNA molecules and others integrated within the cell genome. The association of the virus with this and other distinct types of tumors suggests a causative role. However, as in other cases of viruses or viral genomes associated with tumors, a promotional or accidental relationship cannot be excluded. A close relative of EB virus causes the benign condition, infectious mononucleosis (Chapter 31).

Herpes viruses, Table 17-1, page 341

At the present time it appears that viruses may be causally implicated in human breast cancer. Virus particles virtually identical in structure to the mouse mammary tumor (Bittner) virus have been observed in samples of human breast milk. The occurrence of such particles is closely correlated with the incidence of breast cancer in the families of the women affected. Even more significant is the fact that these particles contain the RNA-dependent DNA polymerase characteristic of animal tumor viruses. It is, however, impossible at present to estimate what fraction of human cancer conditions is attributable to virus infection because there are many other obvious ways in which the genetic material of the cell may be permanently altered.

SUMMARY

Although viruses have not been unequivocally shown to be the cause of human cancer, it seems highly probable that at least some forms of human cancer will be shown to be of viral origin. This supposition rests on several observations. First, there is widespread occurrence of virus-induced tumors in animals. Second, virus particles which are virtually identical in structure to a virus that is responsible for mouse mammary tumor have been observed in human milk.

The study of tumor-causing viruses of animals has revealed some characteristics that will prove useful in the search for viruses causing tumors in man. Thus in the Rous and polyoma virus situations, only the viral genome is present in the tumor. The nucleic acid of the virus is integrated into the host DNA. In the case of RNA viruses, a unique enzyme that synthesizes DNA from the viral RNA is apparently responsible for synthesizing the viral DNA, which is then integrated. Also, a virus may be oncogenic in one host and cause no harmful effects in a closely related host. This creates the possibility that oncogenesis may be manifested by viruses now recognized only as causing infectious diseases. Indeed, the human adenoviruses, which are common in respiratory infections, can induce tumors in various rodents, but no evidence exists relating them to human tumors.

Although viruses undoubtedly are the cause of certain human tumors, it seems unlikely that they are the cause of all tumors. A wide variety of chemicals as well as X-ray and ultraviolet irradiation increases the frequency of cancer. It is likely that these agents bring about malignancy by directly affecting the DNA of the cell, leading to a heritable change.

QUESTIONS

REVIEW

1. Compare a lysogenic bacterial cell with an animal cell transformed by an oncogenic virus.
2. What is the significance of the discovery of the enzyme reverse transcriptase?
3. How have techniques of nucleic acid hybridization been useful in understanding the role of viruses in cancer?
4. Are there features in common between the known classes of tumor viruses?

THOUGHT

If a virus is recovered from a tumor, how can one establish that its presence is causally related to production of the tumor?

FURTHER READING

CAIRNS, J., "The Cancer Problem," *Scientific American* (November 1975). Clear, readable discussion of the cancer problem, including the possible contribution of oncogenic viruses.

GREEN, M., "Viral Cell Transformation in Human Oncogenesis," *Hospital Practice,* 10:91–104 (September 1975). Clear, comprehensive, up-to-date and well-illustrated description of the evidence of the possible role of viruses in human cancer.

PAGANO, J. S., C. HUANG, and Y. HUANG, "Epstein-Barr Virus Genome in Infectious Mononucleosis," *Nature* 263:787–789 (October 1976). This article discusses the heterogeneity of EB viruses.

TEMIN, H. M., "RNA-Directed DNA synthesis," *Scientific American* (January 1972). Well-illustrated and well-written explanation of the experiments demonstrating reverse transcriptase by a Nobel Prize-winning discoverer of the enzyme.

VIOLA, MICHAEL V., et al., "Reverse Transcriptase in Leukocytes of Leukemia Patients in Remission," *New England Journal of Medicine* 294:75–79 (January 1976). This report from scientists at Columbia University gives further evidence that human leukemia is caused by a virus. A thought-provoking article on the possible role of viruses in causing human cancer.

WATSON, J. D., *Molecular Biology of the Gene.* 3d ed. Menlo Park, Calif.: W. A. Benjamin, 1976. Easily read molecular biology text with a 48-page chapter on the viral origins of cancer.

INTERACTIONS BETWEEN MICROBES AND HUMAN BEINGS

THE NORMAL FLORA

Because microorganisms live in such a wide range of natural habitats, it is not surprising that some of them grow abundantly on surfaces of the human body, including the skin and gastrointestinal tract. Microorganisms that are well established on the external and internal surfaces of the body without producing overt disease are often referred to as the *normal flora;* those that inhabit the body only sporadically are the *transient flora.* The number of different species of microorganisms living in close association with man is very large and even includes microorganisms that have not yet been adequately characterized. Examples of the principal resident species are *Staphylococcus epidermidis, Propionibacterium acnes,* and *Pityrosporum ovale,* found on the skin; the viridans group of streptococci, *Branhamella catarrhalis*[1] and *Corynebacterium pseudodiphthericum*[1], found in the throat; and *Bacteroides* species and enterobacteria, found in the intestine.

The composition of this complex community at any one time represents a dynamic balance of many opposing forces that are constantly changing, both in response to external influences and as a result of the activities of the human host. The microbial flora and the host comprise an ecosystem, each member of which is influenced in some way by the others. For example, activity which increases perspiration, and thus skin moisture and nutrients, encourages proliferation of skin organisms, sometimes resulting in skin disease. Intestinal microbial populations change with variations in diet, gastric acidity, and the degree of intestinal activity. Through their combined metabolic activities intestinal microbes can also alter drugs and produce toxic compounds from foodstuffs.

[1] Also known as *Neisseria catarrhalis* and *Corynebacterium hoffmani,* respectively.

TERMINOLOGY

The terminology used to define associations among the large and dynamic community of microorganisms living with the human host can be confusing and tends to be inconsistent from one biologist to another. To ensure that the meaning is clear, several definitions are given in the following paragraphs.

A great variety of host and microbial factors interact to determine the effect of organisms on the human body, whether these organisms are established members of the normal flora or new arrivals from the environment or from other hosts. *Colonization* simply implies establishment of microbes on a body surface. If the microbe breaches the surface, enters body tissues, and multiplies, *infection* is said to have occurred. An infection which causes noticeable impairment of body function is called an infectious *disease.*

PATHOGENICITY

A *pathogen* is any disease-producing microorganism or virus. Unfortunately, the term "pathogen" is also loosely applied to those species of microbes that have the ability to cause disease under conditions that *usually* exist between the host and microorganism, and this results in confusion when infections by other organisms are encountered. (Table 19-1). *Pathogenic* means disease-causing, and *pathogenicity* is the ability to cause disease. An *opportunist* is a pathogen that is able to cause disease only in hosts with impaired defense mechanisms, which might result, for example, from wounds or alcoholism. These terms should not be used in the absolute sense because *pathogenicity always depends on both host and microorganismal factors.*

VIRULENCE

Virulence refers to properties of microbes that enhance their pathogenicity. These properties are responsible for different degrees of pathogenicity of different strains within a species. For example, pneumococci that have a capsule are *virulent* and are often pathogenic; those lacking a capsule are *avirulent* and have little pathogenic potential. The word "virulent" is also used in a quantitative sense, indicating that an organism has more disease-fostering attributes than other virulent strains of the same species. The implication is that such organisms are more likely to cause disease, or more likely to cause severe disease, than the other strains. In some instances the attribute responsible for virulence is known (such as the capsule of pneumococci or toxin production by *C. dipththeriae*), whereas in other instances it is not. However, the relative virulence of different strains of a species can be compared by determining the minimum doses which produce disease in laboratory animals of matched age and genetic background. Virulence measured in this way does not necessarily apply to human beings. For example, pneumococci with one type of capsule fail to produce disease in laboratory mice but are nevertheless pathogenic for human beings.

BALANCED VERSUS UNRESTRAINED GROWTH

Infection appears to occur more readily when competing microbes are eliminated or suppressed. Thus enhanced penetration of microorganisms may be related to their relatively high concentration, resulting from the unrestricted growth which

TABLE 19-1
Examples of Pathogenic Bacteria

Pathogenic in Normal People[a]	*Commonly Opportunistic*[b]	*Occasionally Opportunistic*[c]
Bacillus anthracis	*Actinomyces israelii*	*Acinetobacterium calcoaceticus*
Bordetella pertussis	*Bacteroides fragilis*	*Bacillus cereus*
Borrelia recurrentis	*Clostridium perfringens*	*Eikenella corrodens*
Brucella abortus	*Clostridium tetani*	*Enterobacter aerogenes*
Chlamydia psittaci	*Escherichia coli*	*Flavobacterium meningosepticum*
Clostridium botulinum	*Haemophilus influenzae*	*Propionibacterium acnes*
Corynebacterium diphtheriae	*Klebsiella pneumoniae*	*Pseudomonas cepacia*
Coxiella burnetii	*Neisseria meningitidis*	*Staphylococcus epidermidis*
Escherichia coli[d]	*Pasteurella multocida*	*Streptococcus agalactiae*
Francisella tularensis	*Proteus mirabilis*	*Streptococcus fecaelis*
Leptospira interrogans	*Pseudomonas aeruginosa*	
Mycobacterium tuberculosis	*Serratia marcescens*	
Mycoplasma pneumoniae	*Staphylococcus aureus*	
Neisseria gonorrhoeae	*Streptobacillus moniliformis*	
Salmonella typhi	*Streptococcus pneumoniae*	
Shigella dysenteriae		
Streptococcus pyogenes		
Treponema pallidum		
Vibrio cholerae		
Vibrio parahaemolyticus		
Yersinia pestis		

[a]Strains of these species cause disease in normal, nonimmune people.
[b]Commonly pathogenic only in cases of burns, wounds, viral infections, alcoholism, or similar conditions of impaired resistance.
[c]Commonly pathogenic only in cases of immaturity, birth trauma, defective immunity, large infecting dose, or when other factors overwhelmingly favor the microbe.
[d]Certain strains.

can occur in the absence of microbial competition. For example, the yeast *Candida albicans* is commonly found in relatively small numbers in the intestinal tract. However, if an antibiotic suppresses the competing normal bacterial flora of the intestine, the concentration of the yeast may increase manyfold, and the organism is then prone to invade the host. Another example of the ability of large numbers of organisms to overcome host defenses and invade is seen in certain eye infections. The eyes are frequently bombarded with small numbers of saprophytic organisms of the genera *Bacillus* and *Pseudomonas,* with no evidence of any deleterious effects. However, if contaminated eye medication, in which one of these organisms has grown to large numbers, is introduced into the eyes, serious invasion and destruction of the eye may result.

Mutual inhibition often occurs among the resident flora. It is likely that the inhibition of one microorganism by another results simply from competition for essential nutrients, such as vitamins, amino acids, and iron. However, other mechanisms of inhibition are known; for example, in the absence of blood or another source of catalase the pathogen *Corynebacterium diphtheriae* is inhibited by hydrogen peroxide produced by the viridans group of streptococci. In addition, fatty acids produced by the normal flora of the skin and intestine are inhibitory for many pathogens. Other microbes produce antibiotics or antimicrobial proteins (bacteriocins) of more restricted activity than antibiotics. In other instances the oxygen consumption and acid metabolic byproducts of one organism may inhibit the growth of others. Although most of these mechanisms of inhibition can be easily demonstrated in vitro, their importance in the natural setting is a matter of conjecture. It is well established, however, that the suppression of one component of the normal flora can allow its other components to produce disease and can also facilitate invasion by exogenous pathogens.

HOST FACTORS INFLUENCING RESISTANCE TO INFECTION

A number of mechanical and physiological functions of the host tend to restrict the growth of microorganisms on its body surfaces and also to isolate or destroy them should they enter its tissues. Complex mechanisms also operate to minimize the entrance of microorganisms into body cavities such as the lung and bladder. Examples of these mechanisms, which generally involve a single organ system, are discussed in subsequent chapters dealing with infectious diseases. Host factors which operate throughout the body will be discussed here.

Inflammation

Microbes or other foreign materials penetrating host tissue initiate a complex process called *inflammation.* The principal events of the inflammatory process (shown in Table 19-2) are marked by the appearance of redness, heat, and swelling of the affected part. These changes result from alterations in tiny blood vessels too small to be seen by the unaided eye. In response to tissue cell injury caused by the invader, substances are released which act on these blood vessels, making them enlarge and become more permeable to plasma components. Initially this effect is probably due to histamine, although the effect is perpetuated by powerful substances called *kinins* which are activated from plasma components that leak into the inflamed area. Certain derivatives of fatty acids, called *prostaglandins,* also play a role in maintaining the inflammatory effect and are responsible for pain as well as swelling.

Further changes occur as some of the vessels in an inflamed area become clogged because of activation of the blood clotting mechanism. Moreover, vessels affected by the inflammatory process become "sticky," causing the adherence of circulating white blood cells with multilobed nuclei, called *polymorphonuclear leukocytes* ("polys" or PMNs). These leukocytes crawl through the walls of the now permeable vessels and engulf the invading microbes (a process called *phagocytosis*). Phagocytosis is accompanied by intense respiratory activity in the PMNs

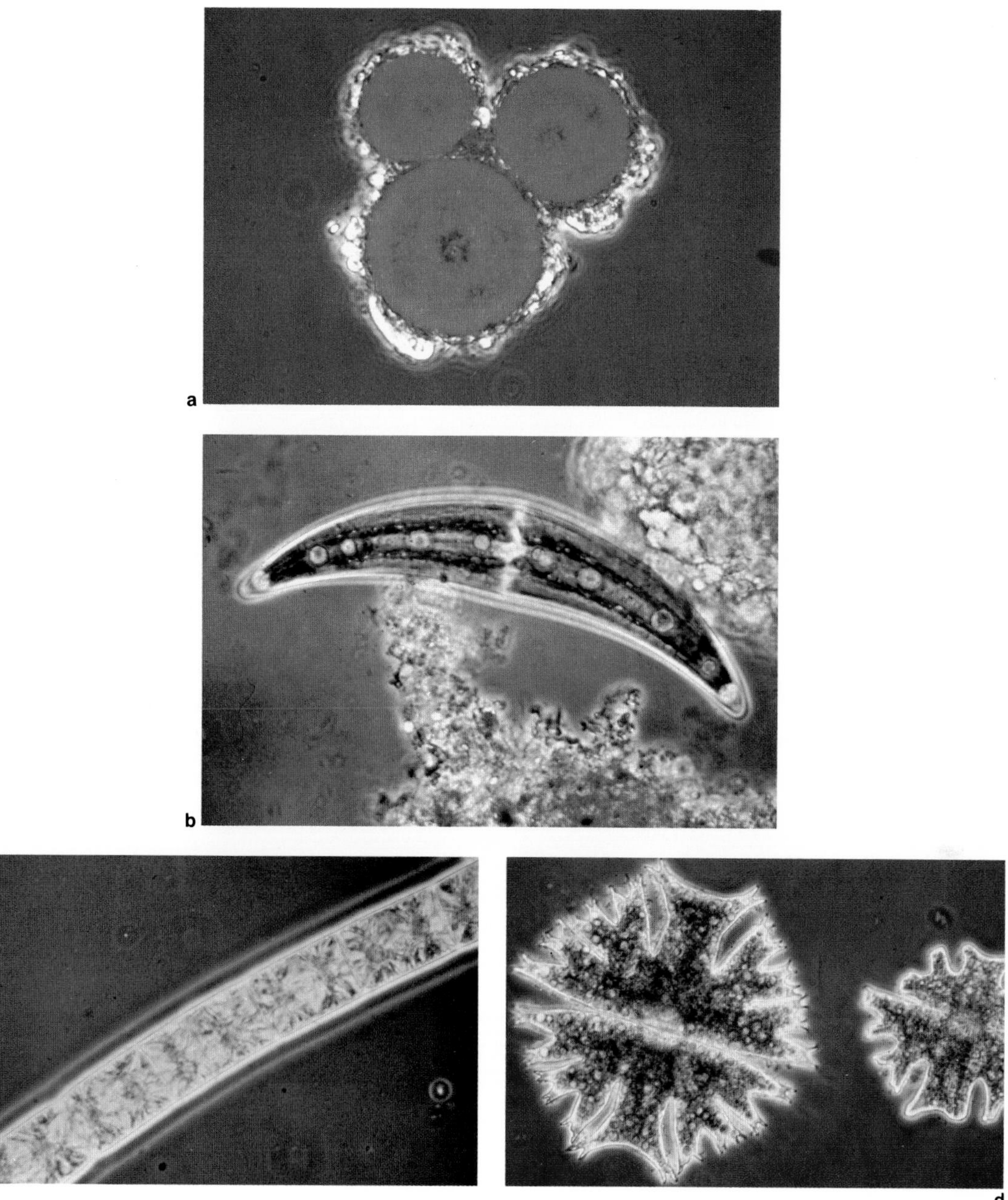

Representatives of four genera of algae. **(a)** *Mycanthococcus.* **(b)** *Closterium.* **(c)** *Zygnema.* **(d)** *Micrasterias.* (Courtesy of J.T. Staley.)

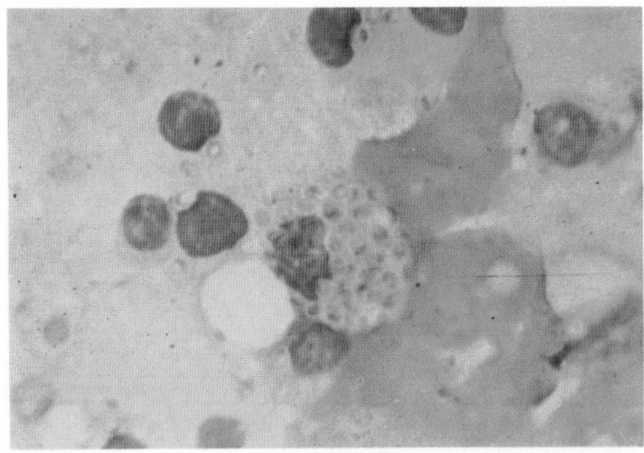

a

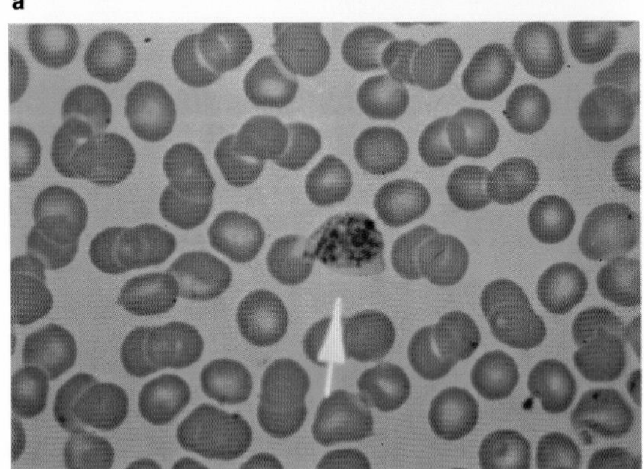

b

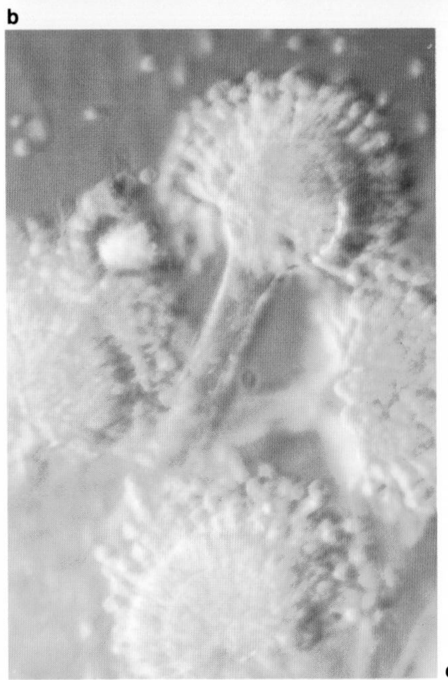

c

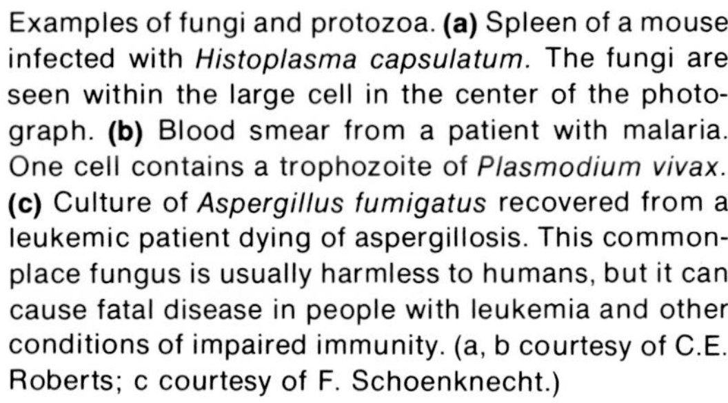

Examples of fungi and protozoa. **(a)** Spleen of a mouse infected with *Histoplasma capsulatum*. The fungi are seen within the large cell in the center of the photograph. **(b)** Blood smear from a patient with malaria. One cell contains a trophozoite of *Plasmodium vivax*. **(c)** Culture of *Aspergillus fumigatus* recovered from a leukemic patient dying of aspergillosis. This commonplace fungus is usually harmless to humans, but it can cause fatal disease in people with leukemia and other conditions of impaired immunity. (a, b courtesy of C.E. Roberts; c courtesy of F. Schoenknecht.)

(a) Smear of brain tissue from a rabid dog (Seller's stain). The arrow points to one of several Negri bodies within a motor nerve cell. **(b)** Natural virus-induced tumor. A wild cottontail rabbit showing a naturally acquired Shope papilloma on the lip. **(c)** Tissue cell culture of malignant human epithelial cells used for cultivating certain viruses. (a, c courtesy of C.E. Roberts; b courtesy of C.A. Evans.)

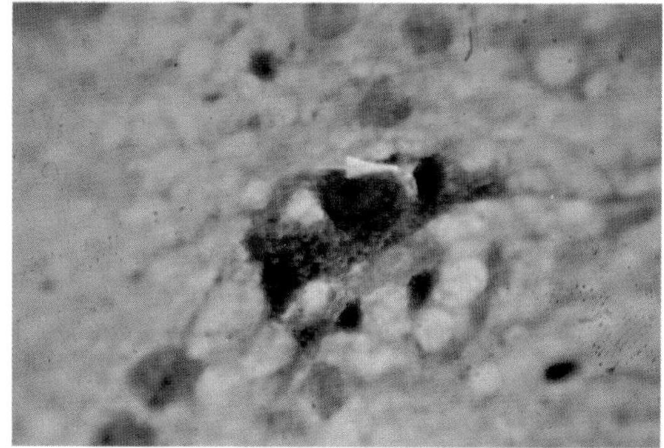

a

b

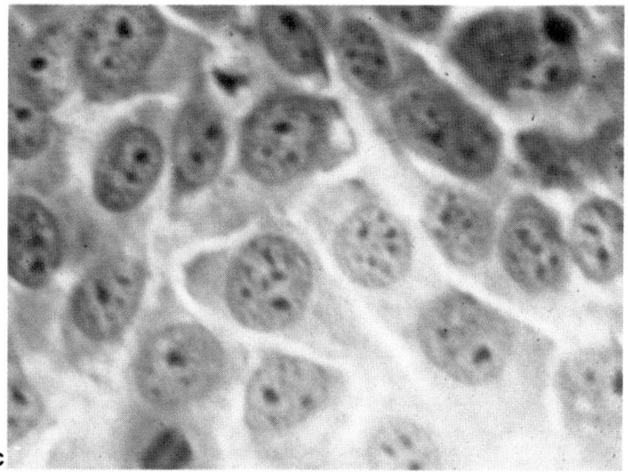

c

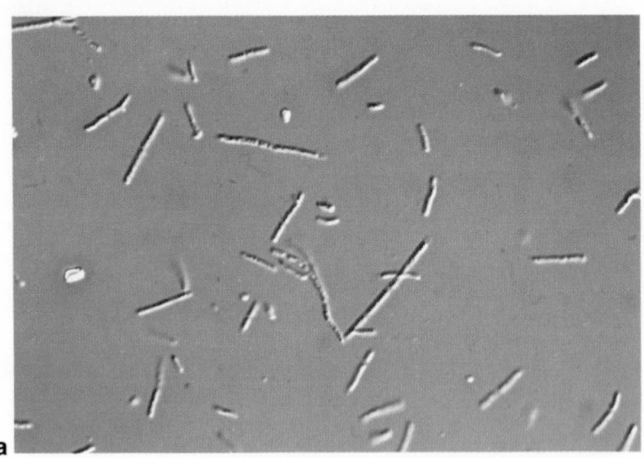

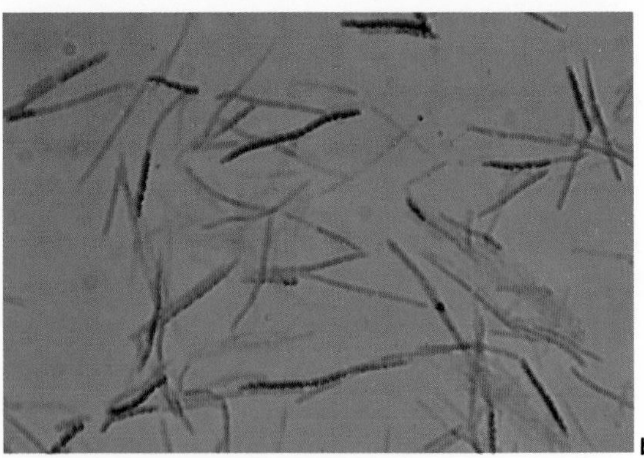

Two methods of viewing *Leptotrichia* species isolated from the mouth: **(a)** wet mount and **(b)** dry smear stained by the Gram method. Two methods of staining sputum from a patient dying of *Nocardia asteroides* infection: **(c)** Gram stain and **(d)** acid-fast stain. (a, b courtesy of M.F. Lampe and F. Schoenknecht; c, d courtesy of C.E. Roberts.)

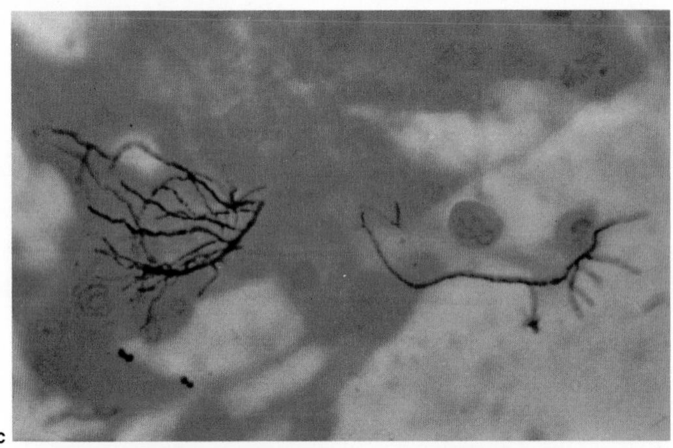

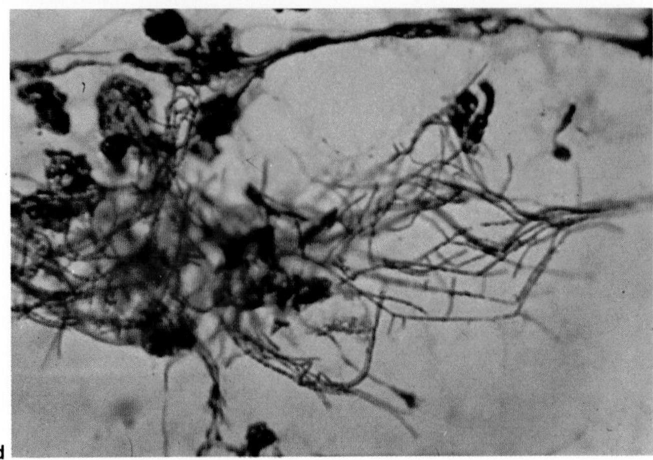

TABLE 19-2
Principal Events of the Inflammatory Process

Event	*Effects*
Tissue injury	Release of kinins, prostaglandins, histamines, and other mediators to act on adjacent blood vessels
Blood vessels dilate and show increased permeability to plasma which may clot	*Swelling* of the tissues results from the leakage of plasma; *elevated temperature* of the region may occur as a result of increased blood flow through the dilated vessels; *redness* may appear for the same reason; *pain* may result from increased fluid in the tissues and from direct effect of mediators on sensory nerve endings
Circulating white blood cells adhere to the walls of the altered blood vessels	The white blood cells migrate chemotactically through the vessel walls and to the area of injury; they are responsible for phagocytosis of foreign material and tissue debris, and for initiating antibody production

resulting in production of superoxide radicals and hydrogen peroxide. These and other intracellular substances cause the death of the engulfed susceptible microbes within minutes. Within vacuoles in the cytoplasm of the phagocytes, digestive enzymes, including mucopeptidases, lipases, and proteases, attack and partially or completely digest the engulfed organism. Many PMNs are killed in the process, and their death is accompanied by release of inflammatory substances that chemotactically attract other phagocytes called *macrophages.* Within a short time a third type of leukocyte, the *lymphocyte,* appears. The macrophages and lymphocytes together initiate an immune response to the invader; this will be discussed in detail in the next chapter.

The Mononuclear Phagocyte System

Inflammation serves to trap infecting microbes at the site of their invasion. However, if the invading microbes escape the localizing and destructive action of the inflammatory response and enter the blood or lymph vessels, the microbes may still be removed and destroyed by collections of macrophages distributed along the circulatory system, especially in the lymph nodes, spleen, liver, and bone marrow. These phagocytic cells are collectively known as the *mononuclear phagocyte (reticuloendothelial) system.*

Nonspecific Factors in Body Fluids

In addition to the inflammatory response and the action of phagocytes of the mononuclear phagocyte system, other nonspecific mechanisms contribute to defending the host against infection. For example, tissues and body fluids, such as saliva and blood plasma, contain compounds able to kill many species of bacteria. Two of these substances, *lysozyme* and *beta-lysin,* are also found in high concentrations in white blood cells or platelets and are released by a variety of conditions. These include inflammatory reactions and entrance of bacteria into the bloodstream. Lysozyme degrades the peptidoglycan composing the bacterial cell wall, whereas beta-lysin attacks the cytoplasmic membrane of the bacterial cell. These and other substances may work along with immune responses to lead to the destruction of invading bacteria.

Peptidoglycan, page 60

Another important group of substances are the *interferons.* These proteins help limit viral infections and perhaps those due to other parasites as well. Their induction and mode of action were discussed previously.

Interferons, page 343

Impairment of Defense Mechanisms

The factors just discussed are important in maintaining the ecological balance in favor of the host. A significant impairment in any of these factors may result in the extension of microorganisms into the host's tissues and the establishment of disease. For example, some people are born without the capacity to produce functional phagocytes or antibodies. Others lose these capabilities later in life. Still others, such as those receiving heart or kidney transplants from another person, may be given medicines which impair the functioning of their inflammatory response or immune response or both to prevent rejection of the transplant. Some patients are given antiinflammatory medicines, such as corticosteroids, to control the symptoms of one disease, only to contract infection because of the interference of these medicines with the inflammatory response. Then too, certain conditions, including cancer, sugar diabetes, and malnutrition may cause defects in protective mechanisms of the host. In all of these instances members of the normal flora assume an overtly parasitic role more frequently than they would with hosts having normal defense mechanisms. Patients with depressed or defective immunity can sometimes be protected from microbial invasion for long periods by being placed in a sterile plastic-walled chamber, with their normal flora suppressed by antimicrobial medicines. Of course, all of their food and even the air they breathe must be freed of microbes to prevent introducing into their bodies pathogenic organisms resistant to the action of the antimicrobial agents.

Racial Factors

Marked differences in susceptibility to certain infections occur among different strains of laboratory animals. Human beings also show racial differences in susceptibility to certain infections, and recently some of these differences have been explained scientifically. For example, 70 percent of American black people are naturally resistant to *Plasmodium vivax,* one of the causes of malaria, whereas most white people are susceptible to this parasite. In order for *P. vivax* to cause malaria it must first attach to a certain component of the surface of the host's red blood cells. This component is a specific receptor for a complementary surface component of *P. vivax.* Most white people have the receptor on their red blood cells and are susceptible to *P. vivax* infection; 70 percent of black people lack the receptor and are resistant.

MICROBIAL PROPERTIES WHICH INFLUENCE PATHOGENICITY

The pathogenicity of microorganisms stems from their ability to alter the structure and function of the tissue cells of the host. Such alterations are accomplished through actual penetration of the tissue by the parasitic organisms or by the penetration of their toxic by-products. However, these harmful effects of microorganisms depend on a number of factors besides penetration of tissue. Undoubtedly of general importance is the ability of a microorganism to survive environmental stress and transfer to susceptible hosts. The ability to adhere to specific host tissues is also a factor of primary importance in many instances, such as dental caries, gonorrhea, and certain diarrheas. Specific adherence serves at least two important functions: (1) it allows the pathogen to resist the cleaning and flushing mechanisms that protect the respiratory, gastrointestinal and genitourinary tracts of the host; and (2) adherence determines the site of microbial invasion and therefore many features of the disease process. Colonization by some pathogens is also aided by their ability to suppress or kill competing flora by elaborating antimicrobial substances.

Following penetration, some organisms cause remarkably little damage to host tissue, but because of their ability to stimulate a marked inflammatory response, normal function of the host is seriously impaired. This is true in most cases of pneumonia due to *Streptococcus pneumoniae,* in which the tremendous outpouring of fluid and inflammatory cells into the air sacs of the lung produces impairment of respiratory function. The ability of many organisms to grow within host organs or tissues is related to their ability to circumvent host defense mechanisms. For example, in some instances a bacterial capsule or other surface component will inhibit engulfment by phagocytes. Some microorganisms produce substances which kill or injure the inflammatory cells, thus interfering with their phagocytic function. At least one organism, *Borrelia recurrentis* (a cause of relapsing fever), is for a time able to bypass the antibody defense mechanism of the host by readily undergoing mutation to give different surface chemical structures which will not be attacked by antibodies. These mutants account for the recurrence of relapsing fever after the host antibody has suppressed the originally infecting strain. In diseases such as tuberculosis, typhoid fever, brucellosis, and tularemia, the infecting pathogens resist the killing action of phagocytosis and the digestive action of intracellular enzymes. They are therefore able to grow within the phagocytes and be safely transported, intracellularly, to other parts of the body. Finally, a number of pathogens, such as the viruses of influenza and infectious mononucleosis, can suppress host immune mechanisms.

Toxins

Pathogenic microorganisms are often armed with poisonous substances (*toxins*) that produce harmful effects on the host. The toxins produced by such bacteria fall into two groups, *exotoxins* and *endotoxins,* whose properties are given in Table 19-3.

EXOTOXINS

Important exotoxin producers include the Gram-positive *Corynebacterium diphtheriae,* the cause of diphtheria; *Clostridium tetani,* the cause of tetanus (lockjaw); and *Clostridium botulinum,* the cause of botulism, a type of paralysis. The causative bacteria of diphtheria and tetanus actually have little invasive tendency, but their exotoxins, absorbed from a localized area of infection, are responsible for the symptoms of diphtheria and tetanus. Therefore either disease can be prevented by immunizing against the causative exotoxin. In botulism the responsible bacterium grows in food and releases its toxin there; in most cases the organism does not infect the host. A few important exotoxins, such as the enterotoxin of *Staphylococcus aureus,* resist boiling temperatures for 30 minutes or more.

Highly potent exotoxins are also found in certain Gram-negative pathogens. In general they differ somewhat from the examples just given in that they are not as readily released into culture media, and immunization with their toxoids usually does not protect against the disease they promote. Some important Gram-negative exotoxin producers are *Vibrio cholerae,* the cause of cholera, *Bordetella pertussis,* the cause of whooping cough, *Shigella dysenteriae,* a cause of dysentery, and *Yersinia pestis,* the cause of plague. In some of these diseases the role of the exotoxin is in doubt; the exotoxin effect has been demonstrated only in selected species of laboratory animals.

Besides the exotoxins, species of bacteria may release other products that could possible contribute to their virulence. Some of these products have traditionally been called *toxins,* and others are known only by the kinds of effects they produce in vitro. In contrast to exotoxins, they are generally not responsible for any major effects in diseases, although it seems likely that they contribute in a minor way. A number of nonpathogens release similar products, emphasizing the

TABLE 19-3
Important Properties of Bacterial Toxins

Property	*Exotoxins*	*Endotoxins*
Bacterial source	Gram-positive and -negative species	Gram-negative species only
Location in bacterium	Synthesized in cytoplasm and released from cell	Component of the cell wall
Chemical nature	Protein	Lipopolysaccharide containing lipid A
Ability to form toxoid	Present	Absent
Stability	Generally heat labile, 60–100°C for 30 min	Heat stable
Action	Each has distinctive effect	All have the same effect; fever and circulatory system damage

TABLE 19-4
Some Bacterial Extracellular Products That May Contribute to Virulence

Description	*Function*	*Example of Producing Bacterium*
Phospholipase	Breaks down lecithin, a lipid component of mammalian cell membranes	*Clostridium perfringens*
Hemolysin	Destroys red blood cells	*Clostridium perfringens*
Coagulase	Clots plasma	*Staphylococcus aureus*
Collagenase	Breaks down collagen, a tissue fiber	*Clostridium perfringens*
Hyaluronidase	Breaks down hyaluronic acid, a tissue component	*Staphylococcus aureus*
Lipase	Breaks down fat	*Staphylococcus aureus*
Deoxyribonuclease	Breaks down DNA	*Staphylococcus aureus*
Leukocidin	Kills white blood cells (leukocytes)	*Staphylococcus aureus*

auxiliary role such substances probably play in disease. Many of these substances are enzymes. Several examples are given in Table 19-4 and in the chapters on specific diseases.

ENDOTOXINS

As shown in Table 19-3, the second major group of bacterial toxins, the endotoxins, differs greatly from the exotoxins. Most Gram-negative organisms, both pathogens and nonpathogens, possess endotoxins, in contrast to the few bacterial species that produce exotoxins; endotoxins generally play a contributing rather than a primary role in pathogenesis.

Microbes other than bacteria may also produce toxins. For example, a substance similar in action to endotoxin has been demonstrated in some fungi. Other fungal toxins, such as aflatoxins, are well studied. Algae and viruses may also show toxic properties. Some of these microbial toxins are discussed in other chapters.

INFECTIOUS DISEASES

Historically, the ubiquity of microorganisms made it difficult to prove that certain species could cause disease. Proof of the existence of infectious diseases was finally accomplished by using the following criteria for a suspected pathogen:

1. The microorganism must be present in every case of the disease.
2. The microorganism must be grown in pure culture from the diseased host.
3. The same disease must be reproduced when a pure culture of the microorganism is inoculated into a healthy susceptible host.
4. The microorganism must then be recovered from the experimentally infected host.

These criteria were skillfully employed by Robert Koch and his associates in the late 1800s and thus became known as "Koch's postulates."

When microbial numbers and virulence outweigh host defenses, infection occurs and disease is a likely consequence. In subsequent chapters numerous infectious diseases are discussed along with the organisms causing them. These organisms are the traditional pathogens which frequently are responsible for diseases in normal persons who have not had previous exposure to them. It will be well to keep in mind, however, that many other microorganisms have the capacity to cause disease in an abnormal host or in a normal host under special conditions favoring the microbe. Moreover, some viral infections of great importance produce very insidious disease, as with the slow virus and oncogenic virus infections (Chapters 17, 18, and 39). Furthermore, some microorganisms and viruses can cause latent infections.

Latent Infections

In some instances microorganism may live in the host's tissues without manifesting its presence. Some such agents may remain latent for many years, only revealing their pathogenic capacity when the balance of the host-parasite ecosystem shifts in their favor. Such is the usual case with the virus of herpes simplex ("cold sores"), and it also occurs with chicken pox virus and the intracellular bacterium *Rickettsia prowazekii,* the cause of typhus. Another example is *Mycobacterium tuberculosis,* the cause of human tuberculosis. The mycobacteria are often confined within a small area of tissue by host defense mechanisms, but they may eventually begin growing and destroying tissue when the host experiences malnutrition, the effects of antiinflammatory medicines, or other poorly understood factors that influence the host's ability to combat the pathogen.

SUMMARY

Large numbers of microorganisms of many varieties inhabit the human body surfaces. Normally these microorganisms keep each other in check by poorly understood mechanisms, but when this balance is disturbed, disease of the host may result. This may occur following administration of an antibiotic active in killing or suppressing some components of the host's normal flora but not others, or by alterations in host physiology.

Multiplication in the tissues of the host—infection—occurs when resident or invading microorganisms circumvent or overcome the host's defenses. In addition to the specific defenses of each organ system of the body, the defensive actions of inflammation and antibody response work to stop the spread of microorganisms and eliminate them from the tissues. Virulent microbes more readily overcome body defenses than others, often because they possess toxic extracellular products or have other attributes that interfere with phagocytic killing. Some possess powerful toxins which can cause serious disease, even though they have no apparent role in assisting microbial invasion.

Organisms which cause human disease originate from the normal flora of an individual or are transmitted from other persons, animals, or the environment. Because disease results from an imbalance between body defenses and the disease-producing capabilities of the organism, even saprophytes can be pathogenic under some circumstances.

QUESTIONS

REVIEW

1. In what ways is the normal human bacterial flora beneficial to human beings?
2. Explain why the normal human bacterial flora is a potential threat to human health.
3. List the ways in which the body defends itself against microbial invasion.
4. List the microbial properties that foster pathogenicity.
5. Compare and contrast exotoxins and endotoxins.

THOUGHT

1. What is the difference between "infection" and "infectious disease"?
2. Explain why virulent microorganisms commonly fail to produce disease, and why avirulent strains may cause severe disease.

FURTHER READING

ALLISON, A., "Lysosomes and Disease," *Scientific American* (November 1967).

BURNET, MACF., and D. B. WHITE, *Natural History of Infectious Disease.* 4th ed. London: Cambridge University Press, 1972. A popular, readable book with many historical examples.

FINKELSTEIN, R. A., et al., "Pathogenic Mechanisms in Bacterial Diseases," *Microbiology-1975,* Ed. D. Schlessinger. Washington, D.C.: American Society for Microbiology, 1975. Pages 105–340 cover up-to-date information on pathogenesis of infectious diseases.

PIKE, J. E., "Prostaglandins," *Scientific American* (November 1971).

SKINNER, F. A., and J. G. CARR, eds., *The Normal Microbial Flora of Man.* London: Academic Press, 1974. Description and ecological relationships of skin, mouth, vaginal, and other flora.

CHAPTER 20 IMMUNOLOGY: ANTIGENS, ANTIBODIES, AND CELL-MEDIATED IMMUNITY

Long before microorganisms were discovered it was known that one attack of certain illnesses, such as smallpox, left the survivor specifically immune to the same disease. Before 1800 Edward Jenner capitalized on the common knowledge that people who had been exposed to cowpox were often not susceptible to the similar but much more serious smallpox. He introduced a method, similar to that in use today, of vaccinating people with cowpox materials to induce in them immunity against smallpox. During the nineteenth century, Louis Pasteur, Robert Koch, and others were quick to investigate the possibilities of immunizing humans with the microorganisms that cause anthrax, rabies, and other diseases. They found that immunization was possible if the organisms used were either modified so that they were no longer capable of causing disease (attenuated) or were killed. Thus it was that immunology developed within the discipline of microbiology.

THE SCOPE OF IMMUNOLOGY

Since the time of Koch and Pasteur, tremendous progress has been made in understanding the theoretical basis of immunological processes. Although certain nonspecific defenses are found in both higher and very primitive animals, the capacity for highly specific immunological interactions arose during the evolution of the vertebrates and is lacking in animals below the level of the vertebrates in the phylogenetic scale. The specificity of these immunological reactions has opened exciting areas of investigation into the means by which the molecules involved in such reactions recognize one another. It has become apparent that the same kinds of specific mechanisms may be responsible for a wide variety of reactions that are not related to antimicrobial defense. Thus under certain circumstances immunological interactions can cause tissue damage or malfunction (allergic reactions) in the host; such reactions also commonly account for the failure of organ transplants. Specific immunological reactions are used in typing blood, as an aid in diagnosing many diseases, in classifying bacteria, and even in identifying human beings. This last application is used in criminal investigation.

The scope of immunology has broadened to encompass many subdisciplines, including immunochemistry, immunogenetics, transplantation immunology, and tumor immunology, as well as the immunology of aging and fetal-maternal relationships. The discussions in the following sections indicate how the same basic mechanisms may account for the diversity of immunological phenomena.

INNATE (NATURAL) IMMUNITY AND ACQUIRED IMMUNITY

Some characteristics of innate and acquired immunity are presented in Table 20-1. Innate immunity is inborn, independent of previous experience, and often depends on the activities of phagocytes and of the nonspecific defense factors dis-

TABLE 20-1
Examples of Innate and Acquired Immunity

Innate	Inborn as a result of the genetic constitution of a species or an individual; independent of previous experience		For example, immunity of man to distemper viruses of cats and dogs; protection provided by phagocytes and protective substances such as lysozyme
Acquired	Naturally acquired	Active	Antibodies acquired following natural exposure to a foreign agent, for example, after a case of poliomyelitis; long lasting, specific
			Interferon produced following exposure to viruses; temporary; nonspecific
		Passive	For example, placental transfer of immunity to poliomyelitis from mother to fetus; temporary; specific
	Artificially acquired	Active	Acquired following immunization, for example, with poliovirus vaccine, long lasting; specific
		Passive	Acquired by administration of protective antibodies; for example, the transfer of preformed antibodies against poliovirus; temporary; specific

cussed in Chapter 19. Acquired immunity, on the other hand, is gained as the result of experience with a foreign agent. As a general rule, *acquired immunity depends on the production of or acquisition of antibodies or specialized immune cells and is therefore highly specific* (*the specific immune response*). For instance, immunity is acquired following infection or immunization with one type of poliovirus; specific antibodies are formed which protect against reinfection with the same type of poliovirus but offer no protection against infection with mumps or measles virus, streptococci, or even another type of poliovirus.

ANTIBODIES AND ANTIGENS

Antibodies and antigens must be defined in terms of one another. *An antibody* (*Ab*) *is a protein produced in the body in response to the presence of an antigen and able to combine specifically with that antigen. An antigen* (*Ag*) *is a substance that can incite the production of specific antibodies and can combine with those specific antibodies.* The simple diagram in Figure 20-1 illustrates some of the properties of antigen and antibody molecules. The antibody-reaction sites on antibody molecules represent the areas of these molecules that combine with antigen. The antigenic-determinant sites on the surfaces of antigen molecules are the specific chemical groups that combine with antibodies. The *specificity* of an antigen or antibody molecule is determined by the size and shape of the antigenic determinant and the corresponding antibody-reaction site, because there must be a *close physical complementary fit* between the two. The *valency* of an antibody molecule refers to the number of specific antibody-reaction sites on its surface.

The diagram in Figure 20-1 indicates that a molecule of antigen may have

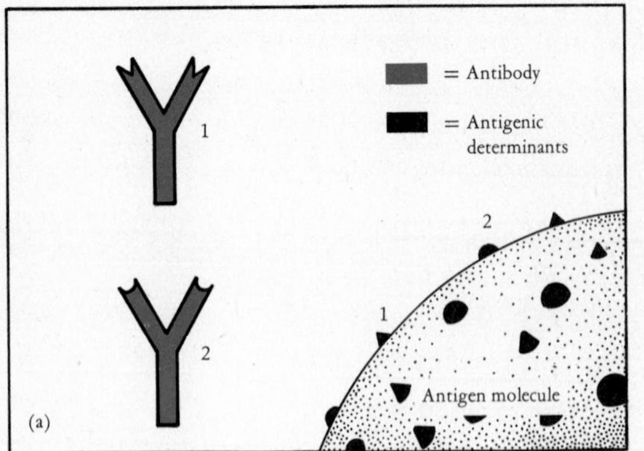

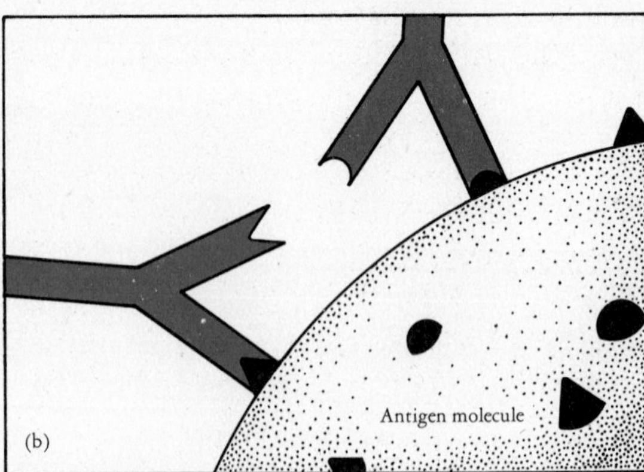

FIGURE 20-1 Schematic representation of antibody and antigen molecules. The antibody molecules represented each have two antibody reaction sites that are complementary to the antigenic determinant for which they are specific. Note that a single antibody molecule has reaction sites for only one specificity of antigenic determinant. The antigen molecule has more than one kind of antigenic determinant, so it is multispecific, and more than one of each kind of determinant, so it is also multivalent.

TABLE 20-2
Important Properties of Antibodies and Natural Antigens

Property	*Antibodies*	*Natural Antigens*
Specificity	Always monospecific	Usually multispecific
Valency	Bivalent (or multivalent, depending on class of immunoglobulin)	Multivalent
Chemical nature	Immunoglobulin (protein, usually gamma globulin)	Protein, Polysaccharide } Often Lipid, Nucleic acid } Occasionally
Molecular weight (approximate)	160,000–900,000	Greater than 10,000

more than one kind of antigenic-determinant site; in other words, the antigen may be multispecific. Each kind of antigenic-determinant site contributes a single specificity. As shown in the diagram, antigen molecules are often multivalent, that is, have more than one antigenic determinant.

In contrast to antigens, antibody molecules are diagramed as monospecific, because all known *antibodies have the property of combining with only one kind of antigenic determinant, that is, with one specificity of antigenic determinant.* Figure 20-1 shows antibody molecules that are bivalent, or capable of reacting with two antigenic determinants of the same specificity. In human beings more than 80 percent of all antibodies are bivalent. The antibody-reaction site is drawn as an inverted region, in agreement with physicochemical evidence indicating that this site is actually recessed.

Figure 20-1 shows that there is a close configurational "lock-and-key" fit between antigenic-determinant 1 and antibody 1, and between antigenic-determinant 2 and antibody 2. Thus antibody 1 would not be expected to combine with antigenic determinant 2, which lacks a close complementary conformation to it. The close fit accounts for the specificity of an antigenic determinant for particular antibody molecules or, conversely, for the specificity that antibody molecules exhibit for complementary antigenic determinants.

The need for a close fit between the antigenic-determinant and the antibody-reaction site is understandable when the nature of the binding between the two molecules is examined. The forces that result in the antigen-antibody interaction are weak and short-range, such as hydrogen bonding; thus a number of weak bonds must participate to hold the two molecules together. As would be expected with noncovalent binding forces, antigen-antibody interaction is *reversible* under the proper circumstances.

Some additional properties of antibodies and natural antigens are included in Table 20-2. Antibodies are large protein molecules, immunoglobulins (Ig), usually found in the gamma globulin portion of serum when blood proteins are separated. Five major classes of human immunoglobulins have been distinguished, as outlined in Table 20-3.

TABLE 20-3
Properties of the Various Classes of Human Immunoglobulins

Class	*IgG*	*IgM*	*IgA*	*IgE*	*IgD*[a]
Ability to combine with specific antigen	+	+	+	+	
Functions					
Protection within the body	+	+			
Protection of mucous membranes			+		
Responsible for certain allergies				+	
Biological properties					
Complement-fixation	+	+			
Placental transfer	+				
Secretion into saliva, mucus, and other external secretions			+		
Specific attachment to phagocytes	+	+			
Specific attachment to mast cells and basophils				+	

[a]The functions of IgD are unknown, but there is some evidence that it functions during the development and maturation of the immune response.

The Structure of Antibodies

Antibodies, like other proteins, are made up of polypeptide chains, each kind of chain coded for by a particular gene or genes. Each of the five classes of immunoglobulins differs from the others in the kinds of chains it has in its molecule.

Immunoglobulins of the most abundant class, IgG, have been studied extensively. A diagrammatic representation of the structure of an IgG protein molecule is given in Figure 20-2. Each IgG molecule comprises two identical halves, each half consisting of a heavy (H) and a light (L) polypeptide chain held to-

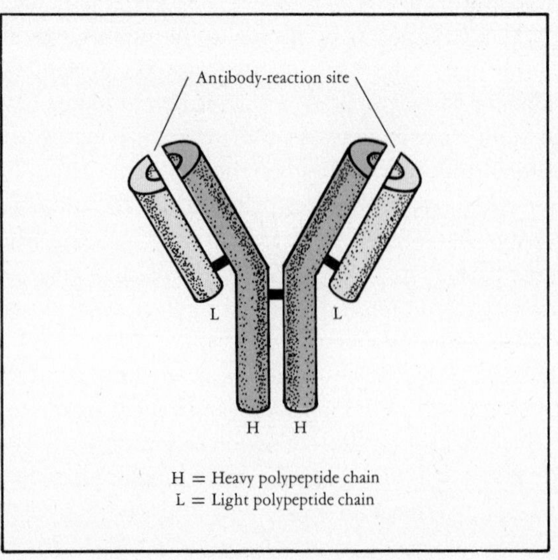

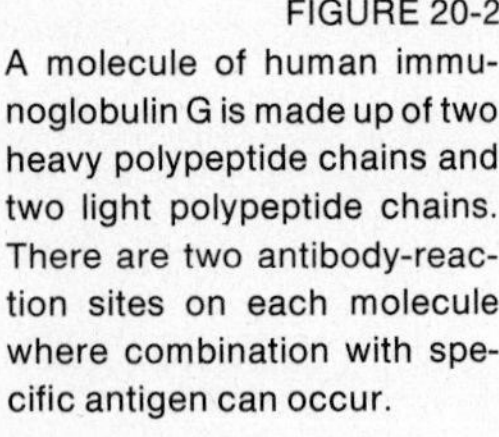
FIGURE 20-2
A molecule of human immunoglobulin G is made up of two heavy polypeptide chains and two light polypeptide chains. There are two antibody-reaction sites on each molecule where combination with specific antigen can occur.

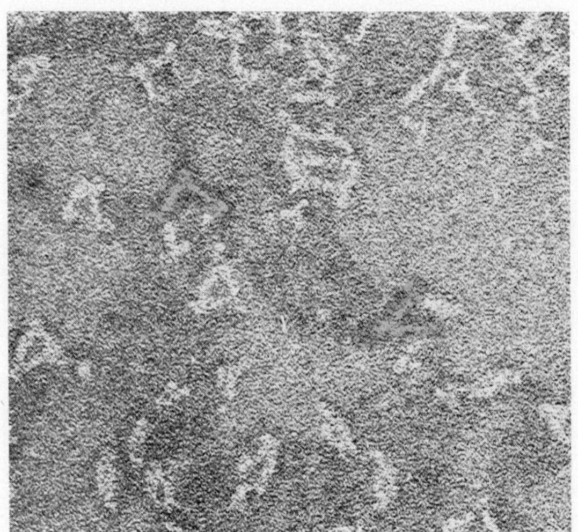

FIGURE 20-3
Electron micrograph showing rabbit antibodies (IgG) that have combined with a specific hapten. The Y shape of the antibodies is readily apparent. (×500,000) (Courtesy of N. M. Green.)

gether with both covalent and noncovalent bonds; in turn the two halves are similarly bonded together. Thus there are 2H and 2L chains per molecule of IgG. *The H chain is characteristic and different for each immunoglobulin class.* The Y shape diagramed in Figure 20-2 is seen in actual electron micrographs of antibody molecules (Figure 20-3). Similar electron micrographs have revealed the flexibility of the molecule around a "hinge area" at the center of the Y; antibody molecules combined with large antigen molecules form different angles around this central hinge area. In fact the arms of the Y may form a 180° angle, giving the appearance of a T rather than a Y.

The H and L polypeptide chains of the immunoglobulin molecule have both variable and constant sequences of amino acids. The constant amino acid sequences are virtually the same in all human antibodies of the same class. This constant portion is responsible for many of the biological properties of the molecules (Table 20-3). However, the amino acid sequences in one part of the immunoglobulin molecule varies in antibodies of different specificities. Thus the L chains of a particular type will each have virtually identical amino acids linked together to form half of the L chain (the constant region), and the amino acid sequence of the other half (the variable region) will vary from one L polypeptide chain to another. Figure 20-4 shows that the two identical antibody-reaction sites of an antibody molecule are formed by the variable regions of both the H and L chains.

The structures of the larger and more complex immunoglobulin molecules, belonging to certain classes, have not yet been fully determined. Immunoglobulins of the class IgM, for example, are known to consist of five subunits (Figure 20-5), each of which resembles IgG.

The study of immunoglobulin structure has contributed greatly to an understanding of the ways in which antibody molecules function. It has also given insight into the genetics and evolution of immunological responses.

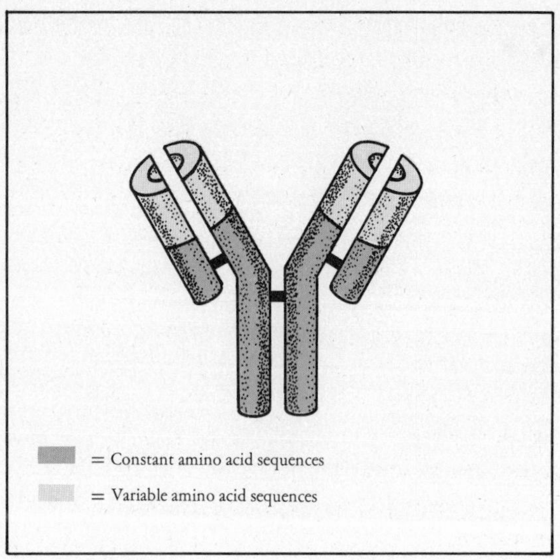

FIGURE 20-4
Variable amino acid sequences at the ends of both heavy and light polypeptide chains cooperate to make up the antibody-reaction sites. The variability accounts for the wide range of different antibody specificities.

The Structure of Antigens

Most antigens, like antibodies, are macromolecules. Substances with low molecular weights are usually not antigenic. Antigens may be proteins, polysaccharides, complex molecules combining various substances (for example, lipopolysaccharides), or occasionally lipids or nucleic acids. However, proteins and polysaccharides are usually much better antigens than lipids and nucleic acids. Another im-

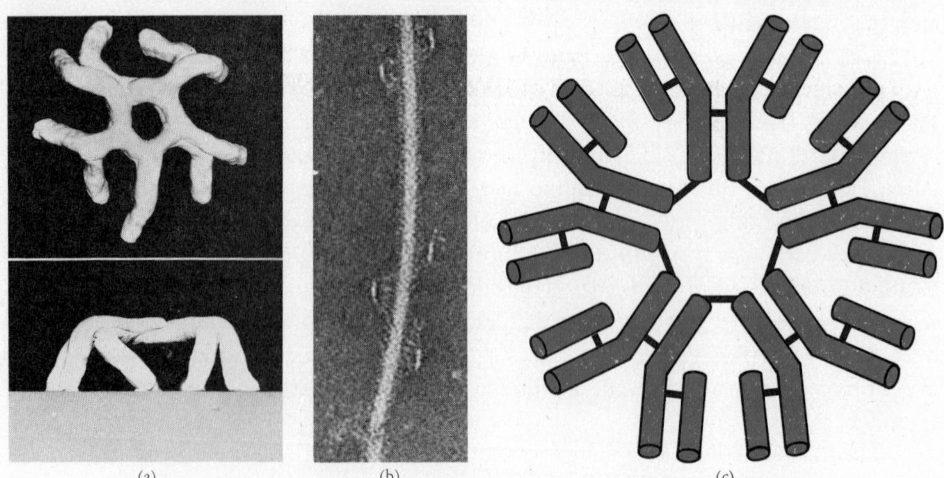

FIGURE 20-5 The structure of immunoglobulin M. (a) Model of the proposed structure of IgM. When viewed from the top, it is apparent that the molecule consists of five Y-shaped units similar to IgG. The side view of IgM attached to an antigen gives the impression of a "staple." (b) Actual electron micrograph of IgM attached to specific antigen determinants on a flagellum of *Salmonella paratyphi*. Note the "staple" appearance. (c) A schematic representation of IgM. Note that each Y-shaped subunit consists of two heavy and two light polypeptide chains. (Photos courtesy of E. A. Munn and M. J. Hobart; a from A. Feinstein and E. A. Munn, *Nature* 224:1307, 1969.)

portant characteristic of antigens is that they are usually foreign to the host that forms the specific antibodies to them. If this were not so, an individual would respond immunologically against his own body constituents, with the potential of producing tissue damage. Under unusual circumstances this can occur, resulting in *autoimmune* diseases. It is important, however, to note that the body components of one host may be antigens in other hosts in which they are foreign.

HAPTENS

Although low molecular weight substances are by themselves not often antigenic, they can serve as *haptens. A hapten is a substance that can react with specific antibodies but cannot incite the production of antibodies unless it is chemically combined with a large carrier molecule.* Thus a hapten may be simply an antigenic determinant or part of one. For example, penicillin is a small molecular weight substance that is not antigenic alone; in the body, however, its breakdown product acts as a hapten and can combine with large protein molecules of the host to form an antigenic hapten-carrier complex. Antibodies are produced in response to this complex, and some can react with the penicillin hapten alone. Obviously, the ability of many small molecules to act as haptens greatly increases the possibility of immunological reactions developing against foreign substances.

IMMUNOLOGIC TOLERANCE

Many factors, including genetic constitution, are involved in determining whether or not an individual host responds to a particular antigen. The failure to develop an immune response to a potential antigen, known as *immunologic tolerance,* can be induced in various ways. Recognition of one's own tissues (self) and consequent failure to respond immunologically to substances of self is a form of tolerance. Substances that are encountered during embryonic development and that persist after birth usually do not stimulate an active immune response. The developing individual instead becomes immunologically tolerant to such substances and remains tolerant as long as the substances are accessible to potential responding cells.

A number of intricate control mechanisms operate during states of tolerance. Only a few of these are adequately understood at present; however, immunologic tolerance is currently under intensive study. Some possible explanations for this phenomenon will be discussed in later sections.

ANTIBODY PRODUCTION

Although not all of the details of antibody production are fully understood, it is known that antibody molecules are produced in lymphoid tissues by specialized cells called *lymphocytes* and *plasma cells.* Because animals below the vertebrates in the phylogenetic scale lack these cells, they also lack the capacity to form antibodies.

The lymphocytes responsible for immune responses generally fall into one or the other of two populations called T and B lymphocytes. Both populations develop from precursor cells in the bone marrow, but precursor lymphocytes differentiate into T lymphocytes under the influence of the *thymus* gland (hence T cells), and B lymphocytes mature under the influence of other, undefined tissues known as the *bursa-equivalent* (hence B cells). Cells of the B lymphocyte line produce and secrete antibodies and further differentiate into plasma cells that are highly specialized for secreting large amounts of antibody. Whereas some T lymphocytes are concerned with cell-mediated immunity (discussed in a later section), others cooperate with B cells during antibody production. However, the T cells do not secrete antibodies. Lymphocytes of the T cell series are important in the regulation of many immune responses; one subset, called *suppressor cells,* serves to check or suppress certain responses.

In humans, lymphoid tissues are widespread throughout the body and include the lymph nodes, spleen, tonsils, and others. The principal cells of these tissues, the lymphocytes, also circulate in the blood and lymph (Table 20-4 and Figure 20-6). In addition to lymphocytes, the lymphoid organs contain many phagocytic *macrophages,* which can engulf foreign materials and remove them from circulation. Macrophages cooperate with lymphocytes during the immune response in several ways. The lymphoid tissues are strategically located to protect the body against invasion by foreign agents via virtually any route.

When a foreign antigen enters the body, it soon reaches the circulation, is

TABLE 20-4
Cells Involved in Body Defense

Cell Type	*Morphology*	*Location in Body*	*Functions*
Polymorphonuclear neutrophils (PMNs, polys)	Lobed nucleus; granules in cytoplasm; ameboid appearance	Account for part of the leukocytes in the circulation; few in tissues except during inflammation	Phagocytosis and digestion of engulfed materials
Monocytes; macrophages	Single nucleus; abundant cytoplasm	Macrophages present in all tissues and in lining of vessels; monocytes are less mature circulating forms	Phagocytosis and digestion of engulfed materials; can participate in killing foreign cells that are not engulfed
Lymphocytes	Single nucleus; little cytoplasm	In lymphoid tissues (such as lymph nodes, spleen, thymus, appendix, tonsils); also in the circulation	Participate in immunological responses
Plasma cells	Single nucleus pushed to one side of ovoid cell; cytoplasm packed with ribosomes	In lymphoid tissues	Antibody synthesis
Basophils; mast cells	Lobed nucleus; large granules in cytoplasm contain histamine	Basophils in circulation; mast cells present in most tissues	Release histamine and other mediators of inflammation

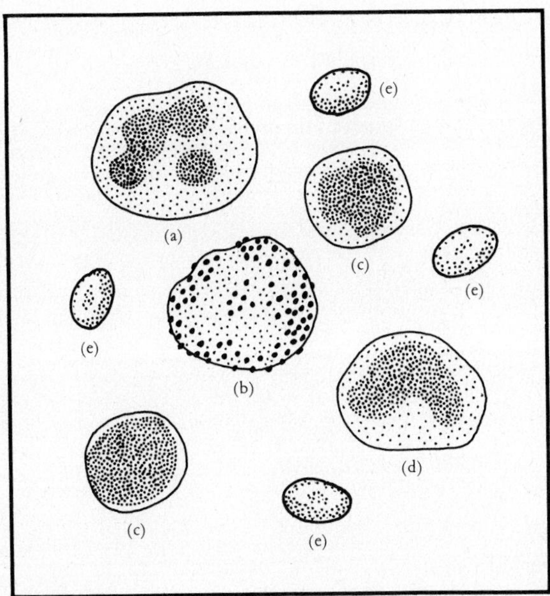

FIGURE 20-6
Cells of human blood. (a, b, c, d) Leukocytes. (a) Polymorphonuclear neutrophil (also referred to as PMN, poly, or granulocyte). (b) Basophil. (c) Lymphocyte. (d) Monocyte. (e) Erythrocytes (red blood cells).

channeled through various depots of lymphoid tissue, and is eventually removed from the circulation, usually by macrophages. Inside the macrophage, many kinds of large molecules and particles are digested by enzymes into smaller antigenic molecules, which can reach potential antibody-forming cells. The process of digestion of ingested materials within macrophages is analogous to that described in amebae (Chapter 15). Macrophages themselves lack the capacity to synthesize antibodies but often are important in the antibody response because they prepare antigens and present them to populations of responsive lymphocytes called *immunocompetent cells.*

Phagocytosis and digestion by amebae, pages 308–309

Among the many millions of lymphocytes in the body, only a small number are capable of responding to any given antigenic determinant; however, a multitude of antigens can be recognized because the total population of lymphocytes comprises thousands of subpopulations of immunocompetent cells, each with different recognition sites for antigens. Any antigenic determinant can select and stimulate only those lymphoid cells that recognize and respond to it by virtue of the specific antigen-binding receptors fixed to their surface membranes. On a B cell, the antigen-binding receptors are antibodies synthesized by the B cell that carries them; each B cell has many such receptors, all with the same specificity. Following the interaction of an antigen with its specific immunoglobulin surface receptors on a group of B cells, and the subsequent cellular proliferation of these B cells induced by this interaction, the B lymphocytes differentiate into highly efficient antibody-producing cells called *plasma cells* (Figure 20-7a). It is these plasma cells (Figure 20-8) that *synthesize and secrete* the bulk of the antibodies present in blood and other body fluids, the *humoral antibodies.*

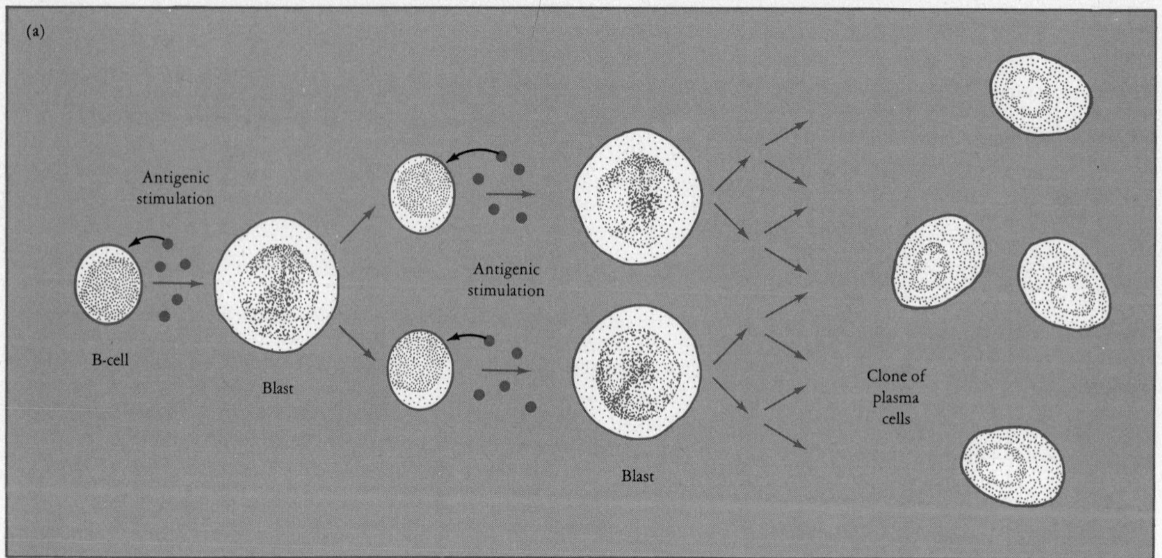

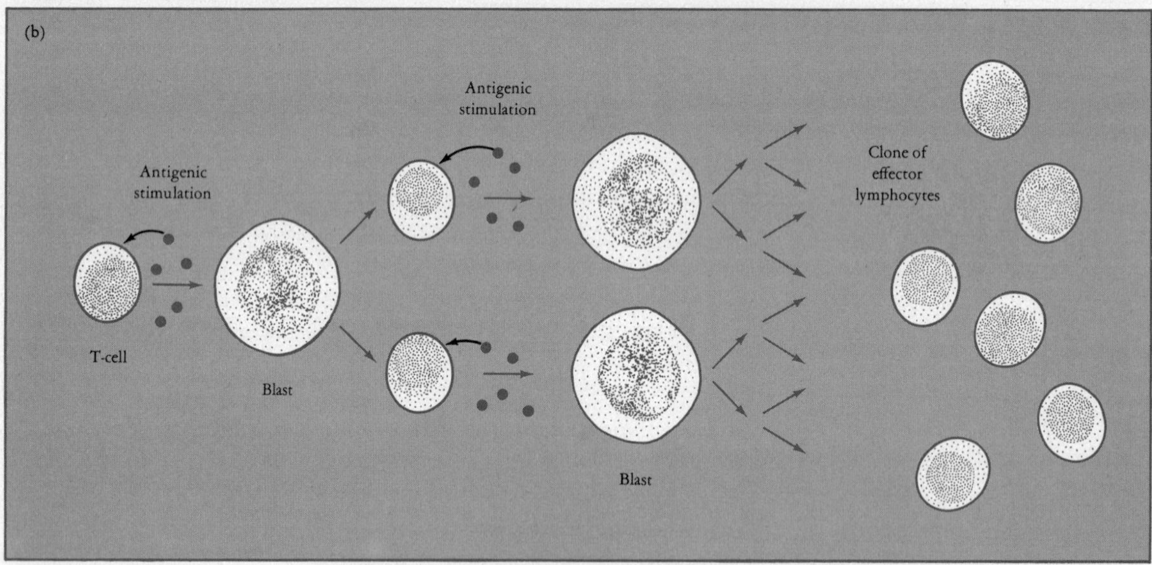

FIGURE 20-7 The postulated development of the immune response. (a) Under the influence of the lymphoid bursa equivalent, stem cells from the *bone marrow* become *B-cells*. Upon stimulation with specific antigen, B-cells enlarge to become blast cells, which divide to form two small lymphocytes. The process is repeated if antigen is available. During proliferation the B-cells differentiate, finally yielding a clone of plasma cells synthesizing antibody molecules. (b) Under the influence of the *thymus*, stem cells become *T-cells*. Following antigen-stimulated proliferation, T-cells yield clones of effector lymphocytes that are active in cellular immunity. Memory cells are also produced during both T- and B-cell responses.

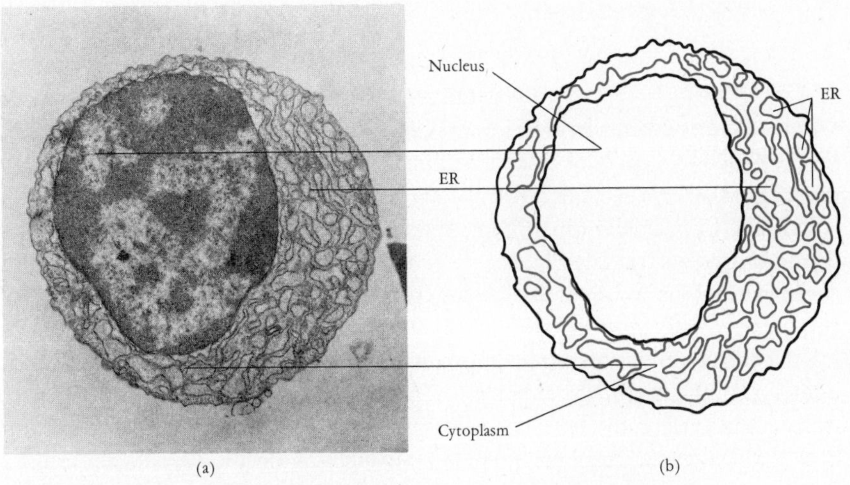

FIGURE 20-8
(a) Electron micrograph of a plasma cell. (b) Diagram of the same cell, indicating the channels of rough endoplasmic reticulum (ER) in the cytoplasm, where antibody molecules are synthesized. Plasma cells are responsible for the synthesis of most antibody molecules. (Photo courtesy of U. Storb.)

THE INTERACTION OF ANTIGENS AND HUMORAL ANTIBODIES

Humoral immunity refers to the *in vivo protection provided by the activities of humoral antibodies.* Humoral antibodies aid body defenses in many ways; alternatively, they sometimes cause more harm than benefit.

Interactions between antigens and antibodies occur in vitro, as well as in vivo, thus permitting the development of valuable laboratory tests. For example, the concentration of antibody in a serum or other fluid can be estimated by making serial dilutions of the fluid and testing it with antigen; the greatest dilution that gives a positive test for antibody activity represents the *titer* of the fluid. Therefore the titer of a fluid is a semiquantitative measure of the antibodies it contains. Serum, which is the liquid portion that remains after blood has clotted, is easy to obtain and a good source of antibodies, so it is often tested; the term *serology* has arisen to designate the study of humoral antibody reactions in vitro.

An increase in the amount of specific antibodies (a *rising titer*) in the serum is often useful for diagnosing infectious disease. Early during infection the titer of specific antibodies is usually very low or nonexistent; however, by 10 days or 2 weeks later the titer commonly increases as a result of antigenic stimulation by the infecting organisms. For instance, in the case of suspected typhoid fever, serum samples are taken as soon as possible during the illness and again approximately 2 weeks later. An appreciable increase or "rise in titer" of antibodies specific for the typhoid organism indicates that the illness is indeed typhoid fever.

Various tests have been developed for measuring quantities of antibodies present, or for determining the location of antibodies or antigens in tissues, by use of tagged antigen or antibody reagents. Radioactive substances offer a tag that can be readily measured by radioimmunoassay, the most sensitive of all immunological tests. Substances that fluoresce when exposed to ultraviolet light are frequently used as tags that permit either antigens or antibodies to be visualized, as discussed in Chapter 28.

Fluorescent antibody test
pages 531, 532

The Precipitin Reaction

Antibodies that combine with soluble antigens in vitro to form a visible precipitate are called *precipitins.* This is a functional term which describes an observable activity of antibody molecules; under different circumstances the same molecules may engage in a variety of other reactions. The precipitin reaction occurs in two stages. First, within seconds after they are mixed, antigen and antibody molecules collide and form small primary complexes. The second stage, which occurs over a period of from minutes to hours, involves the formation of large, precipitating complexes as the result of latticelike cross-linking between the small primary complexes (*lattice formation,* Figure 20-9). Whereas most chemical interactions involve the combination of fixed proportions of each reactant, antigens and antibodies combine in *multiple proportions.* This occurs because antigen molecules are usually multivalent and have many antigenic-determinant sites available to combine with bivalent antibodies; one large antigen molecule may therefore bind from 1 to as many as 40 antibody molecules, depending on the antibody concentration.

This unique ability of antigens and antibodies to combine in multiple proportions is well illustrated by the quantitative precipitin test. In this test constant amounts of antiserum are placed in a series of tubes, and antigen is added in increasing amounts. Thus the first tube contains relatively little antigen and an excess of antibody; the second tube contains the same amount of antibody, with more antigen, and so on until the last tubes in the series contain an excess of antigen. At some point between the zones of antibody excess and antigen excess, a *zone of optimal proportion* is established, in which all of the antigen and antibody molecules are present in the precipitate (Figure 20-10). If all of the tubes are then centrifuged, supernatant fluid from the first tubes can be shown to contain free

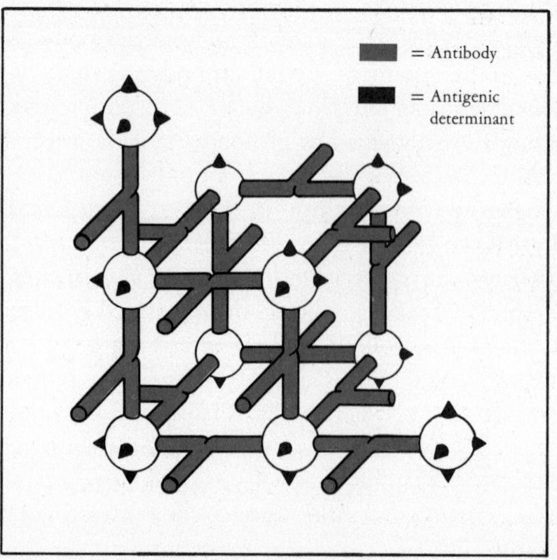

FIGURE 20-9
Lattice formation by antigen and antibodies leads to precipitation. Bivalent antibodies combine with multivalent antigens, linking the two in an insoluble complex lattice formation that precipitates.

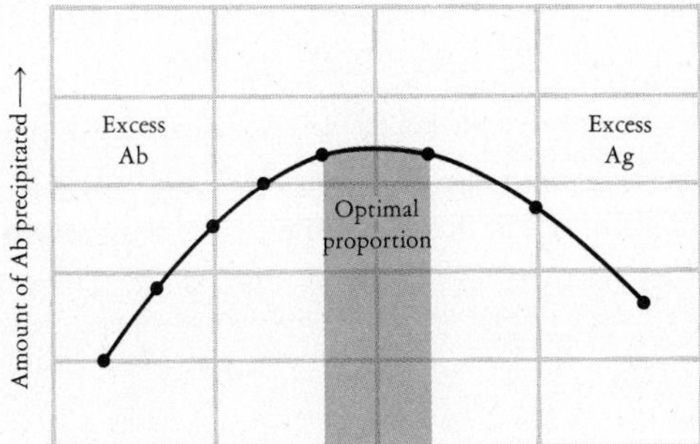

FIGURE 20-10
A precipitin curve. The maximum amount of precipitate forms near the zone of optimal proportion where no free antigen (Ag) or antibody (Ab) exist. Soluble Ag-Ab complexes form in the region of Ag excess.

antibodies, whereas supernatant fluid from the last tubes contains free antigen. Because antibody molecules have a known average nitrogen content, then, if the antigen lacks nitrogen, the amount of antibody in an antiserum can be determined indirectly by measuring the total nitrogen of the precipitate near the point of optimal proportion where the quantity of precipitate is greatest.

The quantitative precipitin reaction illustrates a number of important properties of antigen-antibody interactions, but it is too cumbersome and time-consuming to be used routinely. Consequently, a variety of simplified techniques utilizing the principle of the precipitin reaction have been developed. Many of these utilize precipitation in agar or other gels. One method, involving double diffusion in agar, is called the Ouchterlony technique. Antigen and antibody are placed in separate wells cut in the agar and are allowed to diffuse toward each other. When a specific antigen and its antibody meet at optimal proportions between the wells, a line(s) of precipitate forms. This method is useful for identifying unknown substances, as indicated in Figure 20-11a.

Immunoelectrophoresis is a variation of the precipitation-in-gel technique; it combines precipitation with electrophoresis, a method for separating mixtures of proteins (Figure 20-11b). A number of sophisticated versions of this technique have been developed.

The Agglutination Reaction

Agglutination reactions are similar in principle to precipitation reactions; however, in agglutination reactions *the antigen is particulate* rather than soluble, so much larger aggregates of antigen and antibody are formed. Figure 20-12, which illustrates an agglutination reaction between red blood cells and specific antibodies (agglutinins), indicates how readily the agglutinates can be seen. For comparison the figure also shows a preparation without any agglutination. A semiquantitative agglutination reaction is illustrated in Figure 20-13.

Both agglutination and precipitation aid in the removal of antigens from the circulation in vivo, because agglutinates or large antigen-antibody complexes are much more quickly and efficiently removed and disposed of by phagocytic cells than are single antigen particles or molecules.

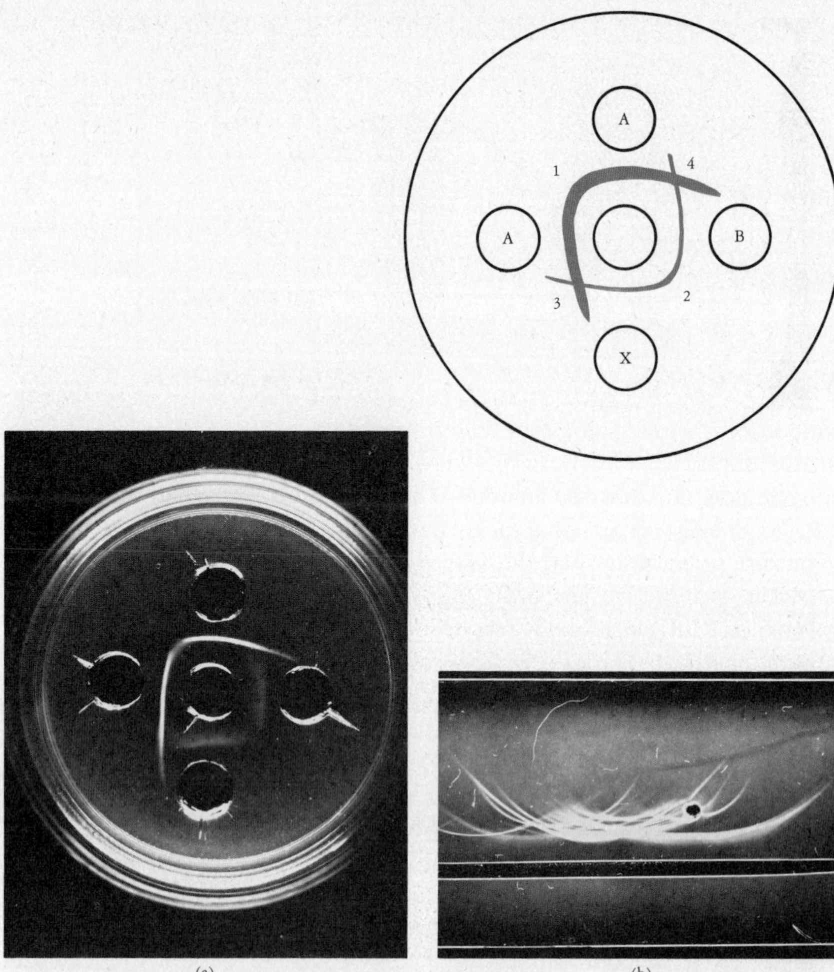

FIGURE 20-11 Antigen-antibody precipitation in agar gels. (a) Using the Ouchterlony double diffusion in agar method, antigen (Ag) is placed in one well cut in the agar and antibody (Ab) in another well. Each diffuses through the agar and a line of precipitate is formed where the two meet in proper proportions. This method can be used to determine whether an unknown substance is identical with or shares antigenic determinants with a known substance. If the known and unknown are identical in terms of antigenic determinants, the lines of precipitate fuse (a reaction of identity); if they are not the same, the lines do not fuse but cross each other (a reaction of nonidentity). In the center well are anti-A Ab and anti-B Ab; Ag A and Ag B are in outer wells as shown. X indicates an unknown Ag. Fusion of the lines at 1 and 2 shows reactions of identity (Ag A = Ag A; Ag B = Ag X). Reactions of nonidentity at 3 and 4 indicate that Ag A and Ag B are not the same and that Ag A and Ag X are not the same. (Photo courtesy of M. Tam.) (b) The immunoelectrophoresis technique combines double diffusion in gel with electrophoresis. A mixture of many protein antigens, in this case human serum, is partially separated by electrophoresis before diffusion occurs, thereby permitting many more lines of precipitate to be distinguished. Human serum was subjected to an electric current; the various serum proteins present migrated to different areas, depending on their electric charges and other properties. Following electrophoresis, rabbit antiserum against human serum was placed in the trough below and the two diffused. More than 30 different proteins in serum, each represented by a distinct line of precipitate, can be distinguished by this method. (Photo courtesy of B. C. Gilliland.)

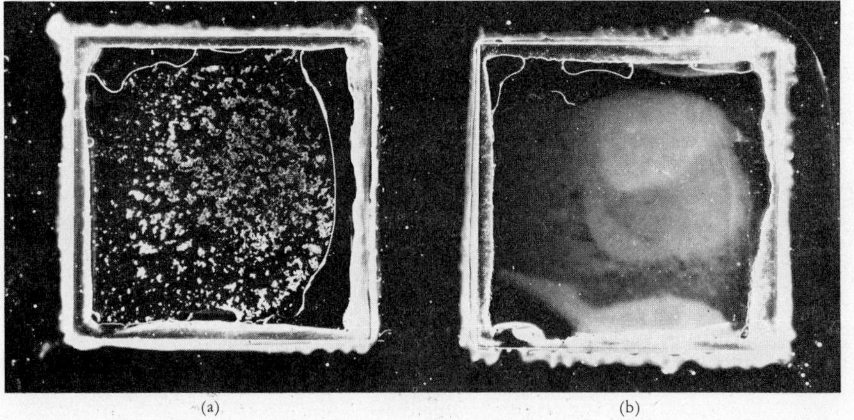

FIGURE 20-12
Agglutination of erythrocytes. (a) Agglutinated red blood cells and antibody. (b) Non-agglutinated red blood cells (control).

Complement-Fixation Reactions

Bacteria or other cells may sometimes undergo *lysis* as the result of an interaction between antigenic components of the cell, specific humoral antibodies, and *complement* (C). Actually C is not a single substance, but a complex system of at least 11 serum proteins that participate nonspecifically in a temporally defined se-

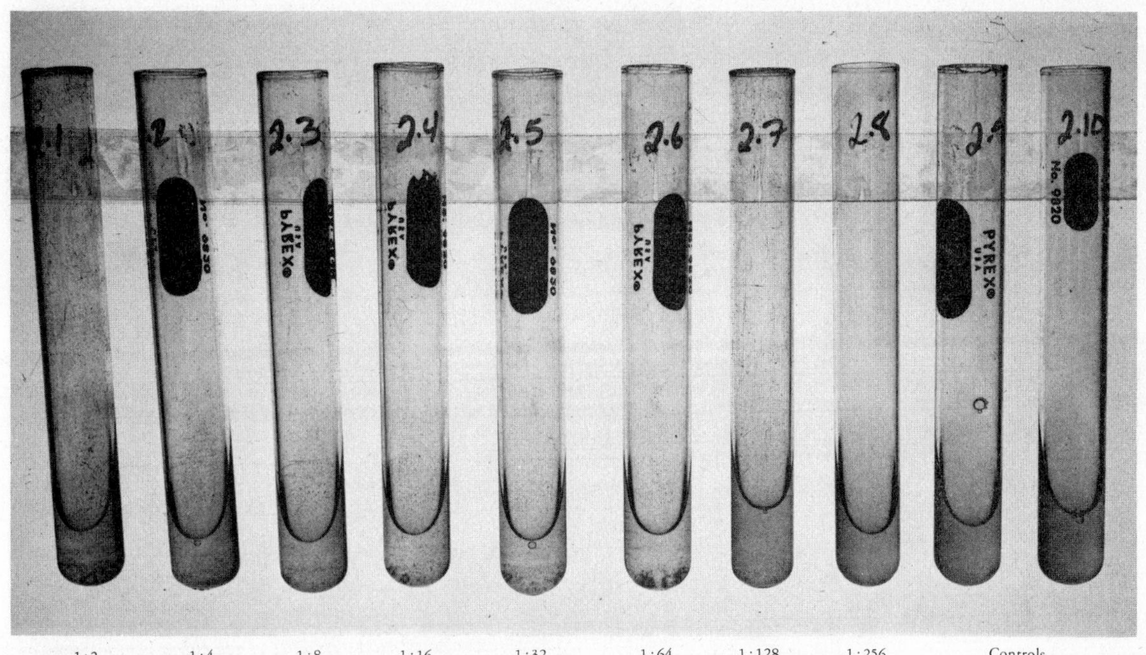

FIGURE 20-13 Tube agglutination test. Dilutions of serum are mixed with equal volumes of antigen-containing particles. The control tubes contain serum alone or particles alone. The test shown has visible agglutination through 1:128 and none in dilution 1:256; therefore the titer is 128. (Courtesy of B. C. Gilliland.)

quence in various immunological reactions initiated by antibodies. Cell lysis thus depends on specific antibody acting together with nonspecific components of the complement system. As a rule, the C sequence is activated by antibody only after an antigen-antibody interaction has occurred. Complement can also become activated by antigen-antibody complexes in a number of ways to cause results other than cell lysis. Opsonization is one such important consequence of the activation of certain components of the complement system.

Many tests utilize complement-fixation as an indirect measure of antibody concentration in a given material. The advantage of using a complement-fixation test is that some antibodies do not give a visible reaction after combining with antigen; however, if they react with, or *fix,* C after reacting with an antigen, these antibodies can be assayed by determining the amount of C that has been fixed. Figure 20-14 diagrams the complement-fixation test procedure.

In performing the complement-fixation test it is first necessary to heat the patient's (test) serum (56°C, 30 min) to *inactivate* the unknown amount of endogenous C present. Serial dilutions of the serum being tested are then mixed with the antigen preparation and an appropriate amount of C. After the specific antibodies have combined with the antigen and have fixed C, the indicator system is added. Usually this consists of sheep red blood cells plus specific antibodies against the red cells. When the red cells, antibodies against the red cells, and C combine, the various components of C act sequentially, resulting in damage to small areas of the red cell membranes. Minute holes formed in the membranes allow water to enter the cell, resulting in red-blood-cell lysis (hemolysis) which causes a visible change in color of the indicator solution. If C has been fixed in the first (test) reaction, it is not available to cause hemolysis in the second (indicator) reaction. However, if no antigen-antibody-C interaction has occurred in the test system, C remains available to cause hemolysis in the indicator reaction. The amount of this hemolysis can be quantitated to give a measure of the amount of C fixed in the test reaction and, indirectly, the amount of specific antibody present in the test serum.

The Coombs Antiglobulin Test

The *Coombs test* is another indirect way of demonstrating antibodies that do not give a visible reaction with antigen. For this test, antiserum (Coombs antiglobulin serum) is made in goats, rabbits, or other foreign species against human serum globulin. Because both human antibodies and complement are serum globulins, the Coombs serum will react with them. Suppose a few human antibody molecules specific for certain red blood cells have combined with the red cells but have not been able to agglutinate them under the conditions employed. When antiglobulin serum is added to such red blood cells, it will react with the antibodies on the red cell surfaces and cause cross-linking and, consequently, visible agglutination (Figure 20-15).

Positive complement-fixation reaction

Test system:
Antigen +
heated test serum
containing specific
antibodies +
unheated serum as a
source of complement } Ag-Ab-C complex

Indicator system added:
Red blood cells +
antibodies specific
for the red cells } Red cell-Ab complex, but no C left to lyse red cells because C was fixed by test system

Result: No hemolysis

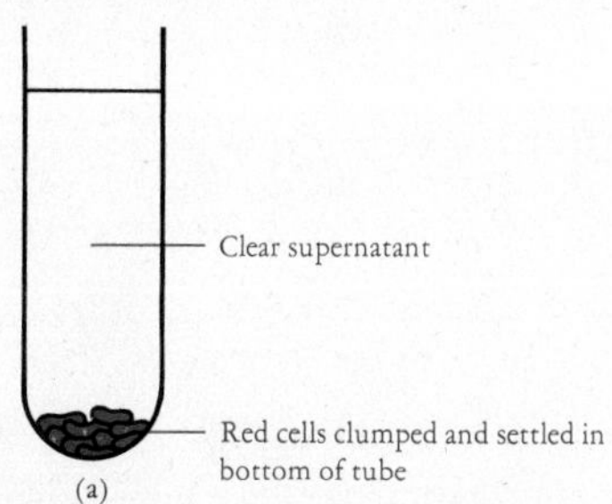

Negative complement-fixation reaction

Test system:
Antigen +
heated test serum
lacking specific
antibodies +
unheated serum as a
source of complement } Ag + C

Indicator system added:
Red blood cells +
antibodies specific for
the red cells } Red cell-Ab-C complex with C still available from test system

Result: Hemolysis

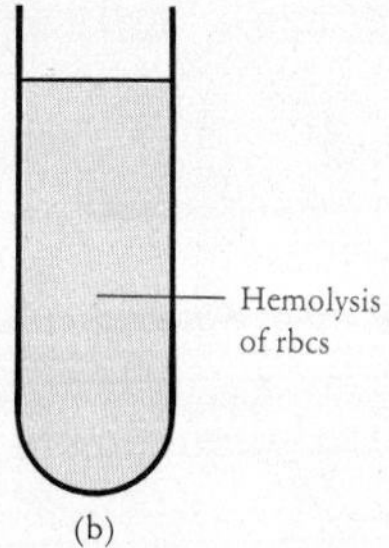

FIGURE 20-14
Complement-fixation test procedure.

Other Methods of Antibody Detection

Antibodies against viruses can be detected by various methods. One of these is *virus neutralization.* The interaction of neutralizing antibodies with virus particles results in the loss of ability of the virus to infect host cells, either in intact animals or in cell cultures (Figure 20-16).

Virus plaque formation, pages 336–337

Virus hemagglutination, page 337

Another test used for measuring antiviral antibodies is the hemagglutina-

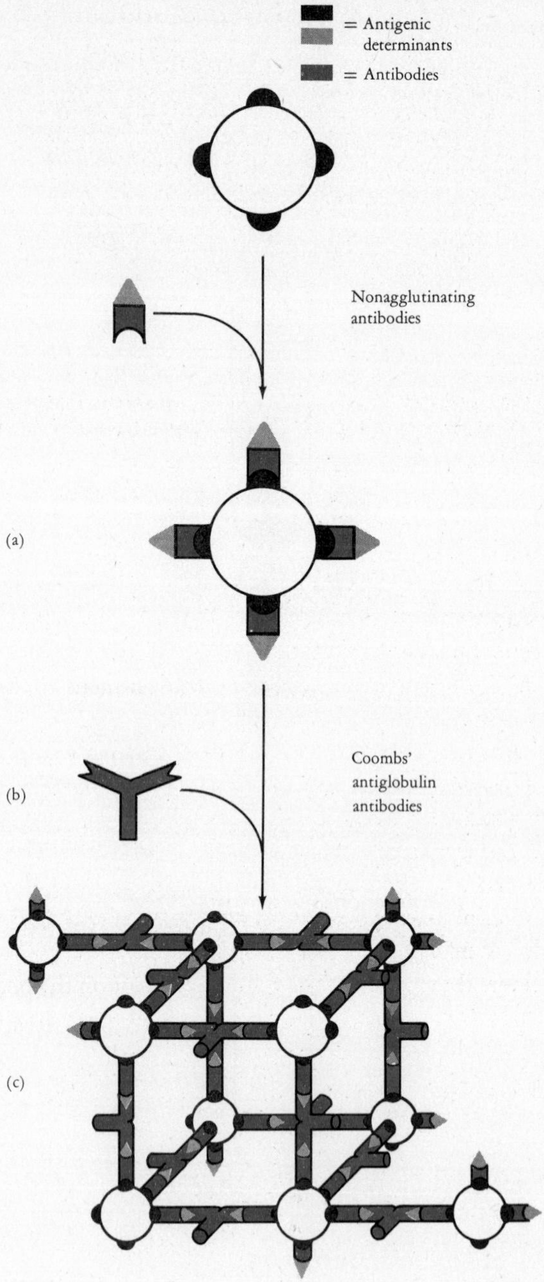

FIGURE 20-15
The Coombs antiglobulin test. (a) Antibodies sometimes combine with antigen in such a way that agglutination or precipitation cannot occur to give a visible reaction. (b) Specific antiglobulin antibodies can be used to link the antigen-antibody complexes together. (c) The result is lattice formation and a visible reaction.

tion-inhibition reaction. Some viruses contain substances that agglutinate normal red blood cells (hemagglutination). Specific antiviral antibodies can react with these substances and prevent the hemagglutination reactions; these antibodies can therefore be measured by the hemagglutination-inhibition test.

Antibodies specific for soluble microbial toxins are called *antitoxins.* These

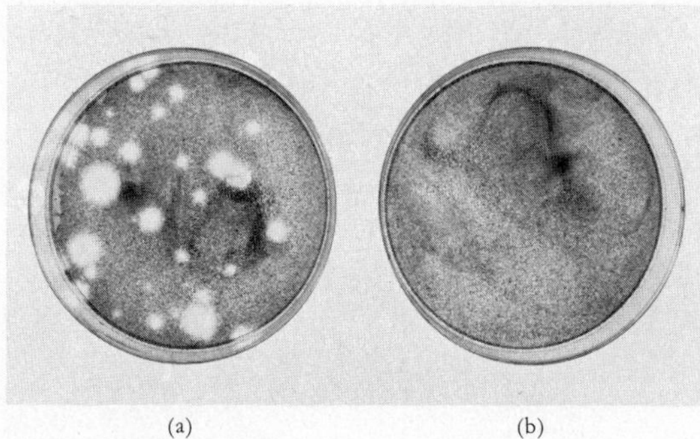

(a) (b)

FIGURE 20-16 Virus neutralization test. (a) When a susceptible cell culture is exposed to poliovirus particles, cells become infected and are destroyed, producing visible clear areas (plaques) in a lawn of cells. (b) If specific neutralizing antibodies are mixed with polioviruses before exposure to the cells, the viruses cannot combine with host cell receptors. As a result, the cells are not infected and no plaques appear in the lawn of cells.

often cause *toxin neutralization* by combining with toxins in such a way as to block the reaction of the toxic site on the toxin molecule with its target cell (Figure 20-17). Antitoxin molecules may also act by forming toxin-antitoxin complexes or precipitates, which in vivo are readily removed from the circulation by phagocytic cells, thus preventing toxin from reaching sensitive target cells.

Humoral antibodies often react with surface antigens to coat bacteria or other particles, causing them to be engulfed, or phagocytized, more efficiently. This phenomenon is called *opsonization,* and the antibodies that promote phagocytosis are called *opsonins.* Sometimes opsonins act by attaching to the phagocyte itself (a *cytophilic* attachment) via the constant amino-acid region of the antibody in the tail of the Y-shaped antibody molecule (Figure 20-18). The attached antibody, upon combining with antigen, is then readily engulfed. Opsonizing antibody may also function by reacting with antigen and activating components of the complement system to form antigen-antibody-complement complexes that adhere to phagocytes through receptors for complement on the phagocyte membrane. Opsonization is highly important in the defense against many pathogens.

CELL-MEDIATED IMMUNITY

The immunity mediated by the activities of specifically immune cells (rather than by humoral antibodies) is termed *cell-mediated immunity* (CMI). Whereas humoral immunity can be passively transferred to a nonimmunized recipient by transferring antibody-containing serum or antibody-producing cells, CMI can be passively transferred only with sensitized cells. Both T lymphocytes and macrophages participate in CMI. Immunocompetent T cells carry cell-surface molecules (receptors) that recognize the specific antigen to which they can respond (much as immunoglobulins on B cell surfaces recognize and interact with specific antigen to trigger antibody production); however, the nature of the surface receptors on T cells is not known and is highly controversial. Following

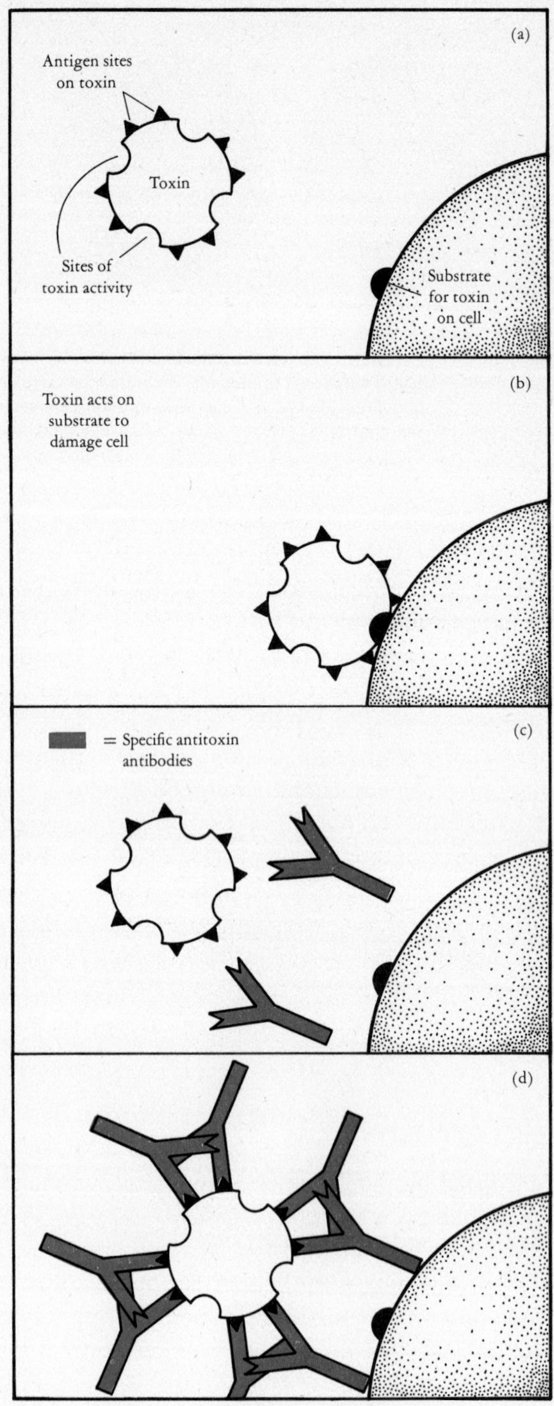

FIGURE 20-17
Steric hindrance of toxin activity by antitoxin action. (a) Sites of toxin activity, indicated by inverted areas on the toxin molecule, are specific for substrate on the cell surface. (b) Interaction of toxin with substrate leads to cell damage or death of the cell. (c) Antitoxin antibody molecules are specific for everted antigenic determinant sites on the toxin molecule. (d) Reaction of toxin and antitoxin leads to steric hindrance of the sites of toxin activity, thereby blocking the toxic action of the molecules.

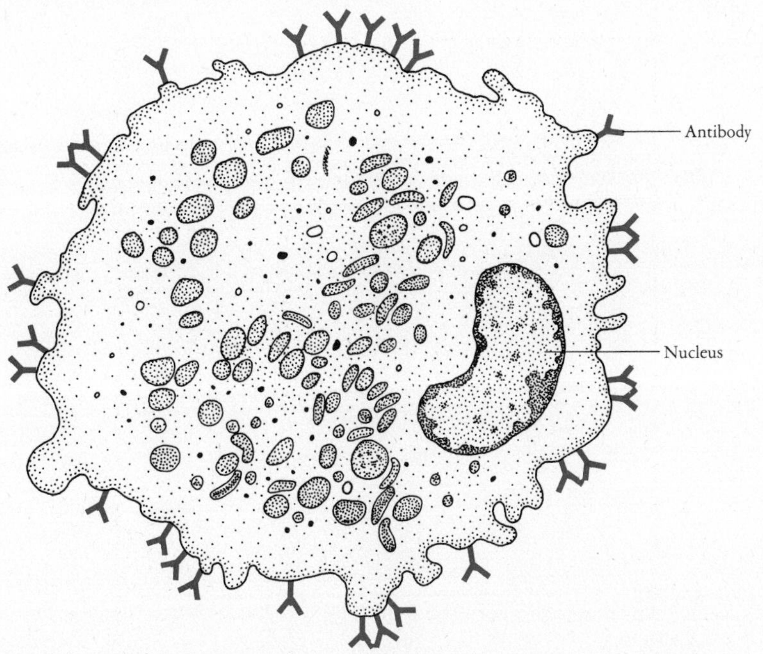

FIGURE 20-18
Cytophilic antibodies attach to the surface of macrophages or other cells via the constant amino acid portions of the antibody molecules. The antibody molecules are drawn much larger than they actually are, with respect to the macrophage, to emphasize their attachment to the cell surface.

the interaction of antigen with T cell receptors, the T cells are stimulated to proliferate and form expanded clones (Figure 20-7b). In addition, antigen-stimulated T cells are the principal producers of *lymphokines,* substances which participate in CMI in a nonspecific manner (Table 20-5). Some of the lymphokines act on macrophages to make the macrophages highly active cells that can inhibit or kill intracellular microorganisms (Figure 20-19).

TABLE 20-5
Lymphokines (Products of Sensitized Lymphocytes):[a] Probable Mediators of Cell-mediated Immunity and Delayed Hypersensitivity

Lymphokine	*Activity in Vitro*	*Probable Action in Vivo*
Chemotactic factor	Attracts macrophages	Causes influx of macrophages into area where antigen is present
Migration-inhibitory factor	Inhibits migration of macrophages	Keeps macrophages immobilized in area near antigen
Lymphotoxin	Kills many different kinds of cells nonspecifically	Kills foreign cells; also kills cells of host, resulting in tissue damage
Macrophage-activating factor	Causes macrophages to become metabolically active and to synthesize many degradative enzymes	Results in macrophage activation; activated macrophages are able to kill foreign cells and to degrade ingested materials efficiently

[a]Lymphokines are synthesized and released by sensitized lymphocytes following stimulation by specific antigen. Only a few of the lymphokines are listed in this table.

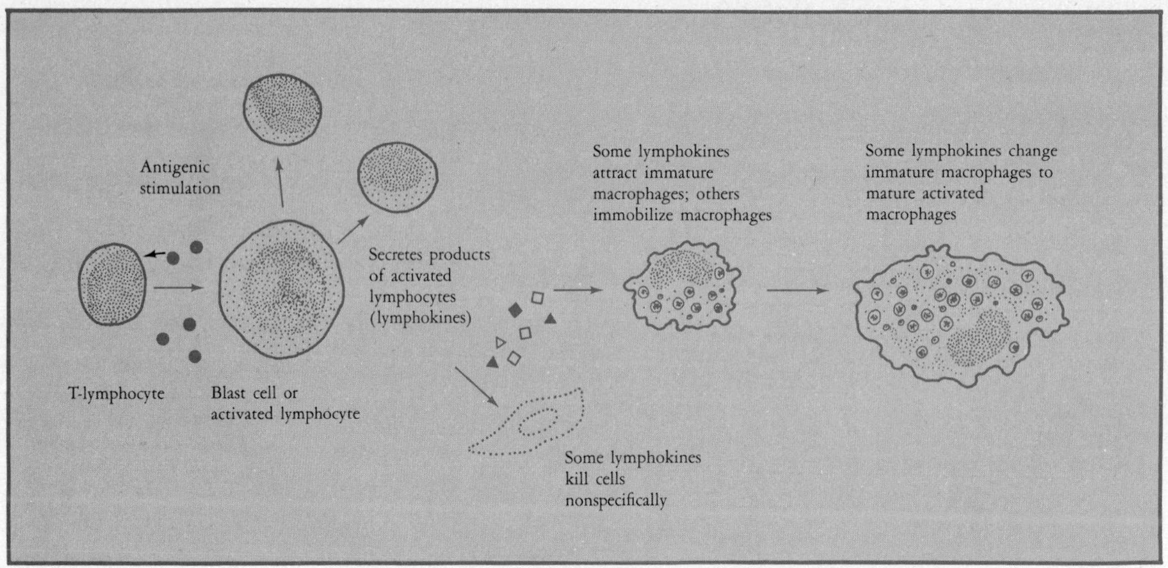
Antigenic
stimulation
T-lymphocyte
Blast cell or
activated lymphocyte
Secretes products
of activated
lymphocytes
(lymphokines)
Some lymphokines
attract immature
macrophages; others
immobilize macrophages
Some lymphokines change
immature macrophages to
mature activated
macrophages
Some lymphokines
kill cells
nonspecifically
FIGURE 20-19 The production of lymphokines by antigen-stimulated sensitized T-lymphocytes.

Cell-mediated immunity is involved in tuberculosis, a disease caused by the tubercle bacillus, *Mycobacterium tuberculosis.* This organism is engulfed by and multiplies within macrophages of the nonimmune host. However, although humoral antibodies against *M. tuberculosis* are produced by the host, they do not control the infection. Resistance to tuberculosis depends on effective CMI, which rests on the increased antimicrobial activity of activated macrophages that engulf and destroy the tubercle bacilli (Figure 20-20).

During tuberculosis the following sequence of events probably occurs: specifically sensitized T lymphocytes are stimulated by antigens of the infecting tubercle bacilli to produce lymphokines, which cause nonactivated macrophages to migrate to the infected area and to become immobilized and highly activated. These activated macrophages then kill or inhibit the growth of the intracellular tubercle bacilli they ingest, thereby limiting the infection. Furthermore, in tuberculosis and some other diseases antigen-stimulated T lymphocytes and macrophages can participate in *granuloma* formation, an operation in which the immune cells surround and wall off foreign materials that cannot be disposed of in other ways.

A similar sequence of events is postulated to occur in other infectious diseases that are limited by antimicrobial CMI. These include leprosy, brucellosis, some of the diseases caused by fungi, and many others.

Cell-mediated immunity is also of major importance in the response to most tumors and to foreign cells in tissue transplants, such as kidney or skin grafts. In this antitissue CMI, however, the mechanisms differ somewhat from those in antimicrobial CMI: recognition of the foreign antigens of tumors or organ grafts occurs via T lymphocytes; the antigen-stimulated T cells proliferate

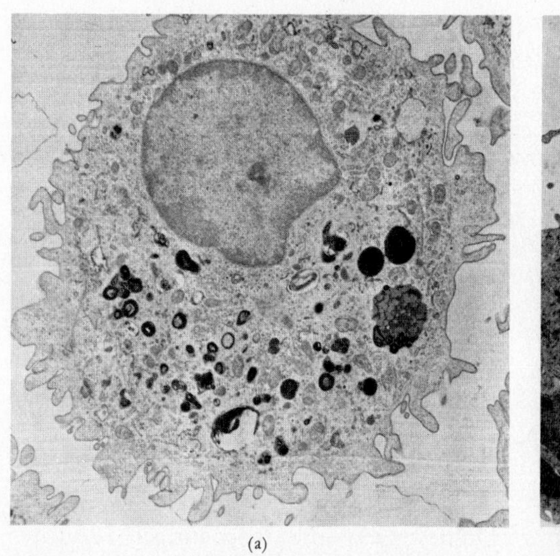

(a)

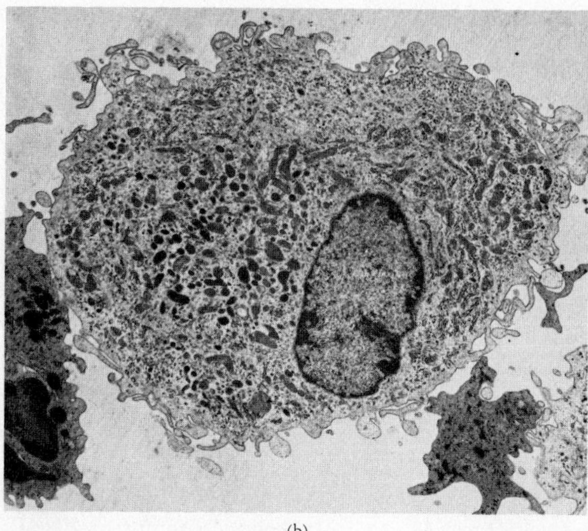

(b)

FIGURE 20-20 Normal and activated macrophages. (a) A macrophage from the lung of a normal nonimmunized rabbit. Few phagocytic vacuoles (dark areas) are present. (b) A macrophage from the lung of a rabbit 5 days after immunization with mycobacteria. The cell is highly activated, as evidenced by the abundant phagocytic vacuoles and lysosomes (dark areas) present in the cytoplasm. (Courtesy of Q. N. Myrvik and E. S. Leake.)

and produce lymphokines, as they do during antimicrobial cell-mediated immunity; and lymphokines affect additional lymphocytes, macrophages, and other cells. However, the principal mode of destruction of foreign cells in antitissue CMI is contact killing of these foreign target cells by T lymphocytes. Macrophages and antibodies may contribute, but to a lesser degree than T cells. In order for a foreign target cell to be destroyed by an immune T cell, a close contact must exist between the immune cell and the antigen-containing target cell. The mechanisms of target cell destruction in cell-mediated immunity have not been determined in spite of intensive research; however, it is fairly certain that more than one mechanism is involved, because it is known that immune cells may be members of distinguishable populations, including macrophages as well as lymphocytes. It is thus likely that the killing mechanisms of these diverse cells vary from one cell to another.

IMMUNOLOGIC ENHANCEMENT

Foreign cells, like most other antigens, induce both humoral and cell-mediated immunologic responses. Under some circumstances specific antibodies and complement can kill foreign cells; however, it is paradoxical that under many circumstances specific antibodies not only fail to kill such cells but actually inhibit effective killing of target cells by immune cells. Evidently, then, humoral antibodies can react with their target cells without causing harm, perhaps because they interact with these cells in such a way that complement lysis of the target cells cannot occur. Whatever the reason, the target cells are not harmed but instead are protected from destruction by immune lymphoid cells. This phenomenon has been called *immunologic enhancement,* because the survival of the protected target cells is enhanced. Some possible mechanisms for immunologic enhancement will be further considered in Chapter 21.

THE ANAMNESTIC RESPONSE

The immune response, whether it is predominantly cell-mediated or humoral, occurs much faster and more efficiently after a secondary or subsequent exposure to an antigen than after the first exposure. After the primary exposure to an antigen, the host is spoken of as being *primed* or *sensitized* to that antigen. Following the first contact of the host with the antigen, that is, during the primary response, immunocompetent lymphocytes proliferate extensively. The *secondary response* (memory or anamnestic response) to an antigen designates the events which occur when the antigen is again contacted, whether this is for the second or the hundredth time; this response is accelerated and more intense than the primary response. One factor that contributes to this phenomenon is that some of the immunologically competent cells that proliferate during the primary response survive as memory cells and add to the total population of cells capable of responding to the specific antigen. Also, memory cells appear to recognize and respond to reintroduced antigen in an accelerated manner. Figure 20-21 illustrates the differences in extent and in timing of humoral antibody production during the primary and anamnestic responses.

The anamnestic response is the basis of immunization methods to prevent disease. After being immunized, the host is prepared to give a prompt and accelerated response when the antigen is again encountered, either in the form of infecting organisms or as a "booster" dose of antigen. A table of commercially prepared immunizing agents and additional information about immunization is presented in Appendix IV.

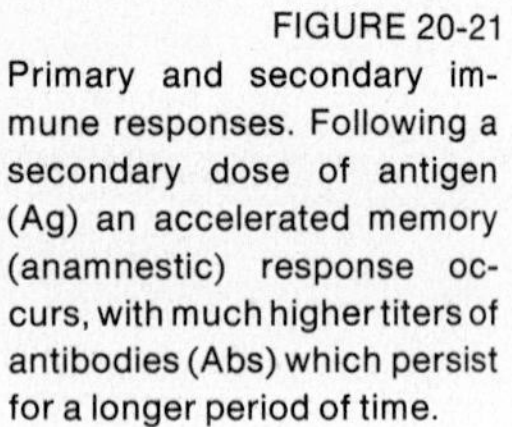
FIGURE 20-21
Primary and secondary immune responses. Following a secondary dose of antigen (Ag) an accelerated memory (anamnestic) response occurs, with much higher titers of antibodies (Abs) which persist for a longer period of time.

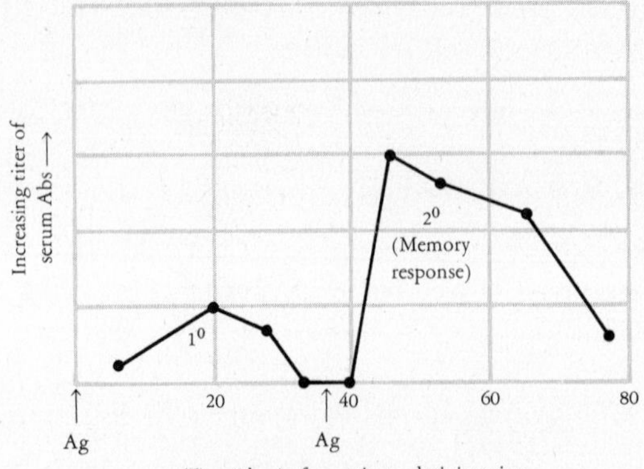

The anamnestic response results in the rapid synthesis of antibodies in response to antigenic stimuli, such as pathogenic microorganisms, and thereby obviates the necessity of producing and maintaining large quantities of each specific antibody at all times. Because antibodies are usually produced only following specific antigenic stimulation of immunocompetent lymphoid cells, persistence of the antigen is required for continued proliferation of antibody-forming cells and high levels of antibody in serum. When the antigen is no longer available, the corresponding antibodies are formed for a limited time and then gradually decrease in quantity as they are degraded.

Almost all antibodies possessed by the newborn (except in the case of infections of the fetus) are of maternal origin; they are essentially all IgG molecules because only antibodies of the class IgG are able to cross the placenta from mother to fetus. Following birth, maternal antibodies are soon replaced by antibodies of the various classes, synthesized by the infant itself. The amounts of immunoglobulins in the blood increase during childhood to reach the adult level at about 12 years of age or earlier, depending on the immunoglobulin class.

Immunoglobulins in the blood of adults normally remain within a remarkably constant range of approximately 1500 to 2000 mg per 100 ml of blood plasma, but following prolonged or intensive antigenic stimulation, the blood level of immunoglobulins may become elevated. However, this elevation is limited because the specific antibodies formed usually suppress the synthesis of excessive amounts of immunoglobulin antibodies (feedback suppression). One means by which antibodies restrict their own synthesis is by combining with the antigen to form antigen-antibody complexes that are readily phagocytized and destroyed. T suppressor cells also help to limit the response.

Antibody molecules have a finite lifespan. They may be lost from the body in secretions or excretions, and those which are not lost are eventually degraded by proteolytic enzymes. The *half-life* of antibody molecules varies considerably, depending on their class and location (that is, whether they are free, cell-fixed, or transported into secretions). The average half-life of most antibody molecules of the class IgG in human circulation is about 20 to 25 days. A unique property of IgG is that this immunoglobulin exerts a regulatory influence on the rate of its own catabolism; the higher the serum concentration of IgG the more rapidly the molecules of IgG are degraded.

Besides feedback and other immune-regulating mechanisms, control of immune responses is also exerted at the genetic level. As with other proteins, the ability to synthesize any given antibody molecule depends ultimately on a cell having the genetic information which codes for that protein. In addition to the genes that code for antibodies, a number of other genes exist that exert regulatory effects on immune responses. Because immune responses are complex, often requiring cooperation between several cell types, genetic and other immune-controlling influences operate at a number of levels, providing an extremely complex regulatory system that is only beginning to be understood.

SUMMARY

Immunity may be either innate or acquired. Innate immunity is inborn and often depends on the activities of phagocytes and other nonspecific defense factors. Acquired immunity is gained in response to a foreign agent and usually involves specific immune cells or antibodies or both. An antibody is a protein that is produced in response to an antigen and which in turn can combine specifically with that antigen. The interaction between determinant sites on antigen molecules and the corresponding reaction sites on antibody molecules requires close complementary fit between the two and the formation of a number of weak bonds. The interaction is highly specific, and it is reversible.

Certain portions on antibody molecules of any given class are structurally variable from molecule to molecule because the sequences of amino acids in such portions vary. This variation encompasses the antibody-reaction sites and is responsible for the tremendous variety of specificities found among antibody molecules. Antibodies also have constant regions in which the sequences of amino acids do not vary from molecule to molecule; these constant regions are responsible for certain biological properties of the molecules, such as the ability to pass the placenta and to fix complement. The five classes of antibody molecules differ in structure and, to some degree, in function.

Antigens are generally large molecules, foreign to the host who responds to them immunologically. Haptens are low molecular weight substances that can combine with specific antibodies but cannot stimulate immune responses unless the hapten becomes coupled to a high molecular weight carrier molecule.

Antibodies are synthesized by lymphoid cells, principally plasma cells. It is postulated that antigen stimulates a few specifically immunocompetent lymphocytes which respond by proliferating and differentiating into clones of antibody-producing lymphocytes and plasma cells. Antibodies synthesized and secreted into body fluids are known as humoral antibodies.

Humoral antibodies aid in body defense by interacting in vivo with foreign antigens in various ways, leading to the destruction or removal of these antigens. Depending on the circumstances, the in vivo interaction of antigen and antibody may lead to precipitation, agglutination, opsonization, complement-fixation, virus neutralization, or toxin neutralization. These reactions can also be observed in vitro and serve as means of demonstrating the presence of antibodies in body fluids, such as blood, or in tissues.

The protection provided by humoral antibodies is called humoral immunity. In contrast, the immunity carried out by the activities of specifically immune lymphoid cells is known as cell-mediated immunity. The development of cell-mediated immunity depends on the activities of specific T lymphocytes. Both lymphocytes and macrophages cooperate in effecting cell-mediated immunity.

The immune response occurs faster and is more efficient following a secondary stimulus than after a primary stimulus with any given antigen. The accelerated secondary response is called the memory, or anamnestic, response.

Antibody synthesis and degradation are regulated by a variety of controls which are not fully understood at present.

QUESTIONS

REVIEW

1. Make a diagram illustrating the important characteristics of antigens and antibodies.
2. The constant amino-acid sequences of antibody molecules are responsible for many biological functions. List some of these functions. What is the role of the variable amino-acid sequences of antibody molecules?
3. What is the cause of the precipitin reaction? In a quantitative precipitin test why might some tubes that contain small amounts of antigen and large amounts of antibody fail to show any precipitate? Draw an explanatory diagram.
4. Describe a complement-fixation test. What are its advantages over the simpler agglutination test?

THOUGHT

1. Are the bonds responsible for antigen-antibody interactions covalent or noncovalent? How might these interactions differ if both covalent and noncovalent bonds participated? Would this be likely to benefit or harm the host?
2. Speculate about the development of the immune system and ways in which the diversity of antibodies might have arisen. (The Jerne article listed in the bibliography for this chapter will lead to interesting speculation.)

FURTHER READING

COOPER, M. D. and A. R. LAWTON, "The Development of the Immune System," *Scientific American* (November 1974).

EDELMAN, G. M., "The Structure and Function of Antibodies," *Scientific American* (August 1970).

EISEN, H., *Immunology*. Hagerstown, Md.: Harper & Row, 1974. An excellent introduction to the molecular and cellular principles of immune responses.

FEINSTEIN, A., N. E. RICHARDSON, and E. A. MUNN, "Structure and Function of Immunoglobulins." *The Proceedings of the Third John Innes Symposium*. Amsterdam: North-Holland, pp. 111–127.

JERNE, N. K., "The Immune System," *Scientific American* (July 1973).

KABAT, E. A., *Structural Concepts in Immunology and Immunochemistry*. 2d ed. New York: Holt, Rinehart and Winston, 1976. A sophisticated immunochemical approach toward understanding basic immunology.

MAYER, M. M., "The Complement System," *Scientific American* (November 1973).

PORTER, R. R., "The Structure of Antibodies," *Scientific American* (October 1967).

ROITT, I., *Essentials of Immunology*. 2d ed. Oxford: Blackwell Scientific Press, 1974. A simplified but fairly comprehensive overview of the essentials of immunology.

ROSE, N. R., and FRIEDMAN, H., *Manual of Clinical Immunology*. Washington, D.C.: American Society for Microbiology, 1976. A comprehensive manual dealing with applications of immunology for the detection and analysis of a wide variety of diseases.

SELL, S., *Immunology, Immunopathology*, and *Immunity*. 2d ed. Hagerstown, Md.: Harper & Row, 1975. A fundamental approach to immunopathology, along with basic immunology, written primarily for medical students and upper-level biology students.

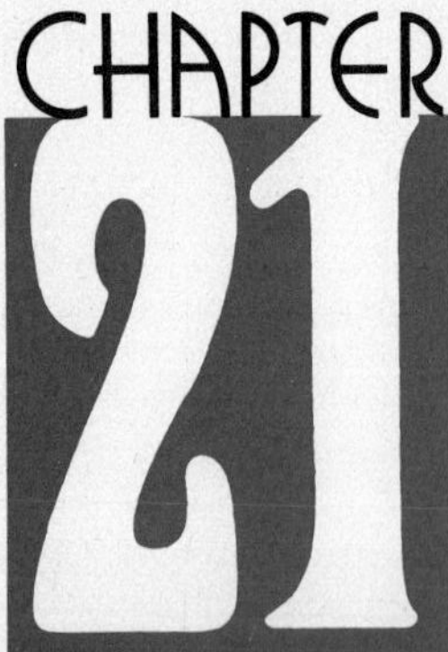

CHAPTER 21

IMMUNOLOGY: HYPERSENSITIVITIES

The ability to make specific immunological responses has developed during thousands of centuries of evolution, and most of these responses have obvious survival value. Nevertheless we find that they are not always protective but may sometimes harm or even kill the individual making the response. *The term hypersensitivity has been applied to immunological reactions that cause tissue damage or malfunction in the host. A synonym for hypersensitivity is allergy.* It is difficult to imagine why man should possess the capacity to develop uncontrolled allergic reactions that become life-threatening, even though local allergic damage to tissues frequently offers definite advantages for survival.

IMMEDIATE AND DELAYED ALLERGIES

Hypersensitivities can be classified into two broad categories: immediate and delayed. Immediate reactions occur rapidly after exposure of a sensitized individual to a specific antigen, and they reach their maximal effects from within minutes to a few hours, depending on the type of reaction. Delayed-type reactions, however, are not apparent until from six to twelve hours after exposure to the antigen and reach their peak from two to three days later. Immediate hypersensitivity reactions depend on the activities of humoral antibodies, whereas those of delayed hypersensitivity result from the activities of specifically sensitized lymphoid cells rather than free humoral antibodies. As a rule, blood serum that is collected from a person with an immediate-type allergy to a specific antigen and then transferred to a normal person will make the recipient temporarily allergic to that antigen. This *passive transfer* of allergy results from the transfer of specific antibodies in the serum. Delayed hypersensitivities, on the other hand, can be transferred with living lymphoid cells from an allergic individual but not with serum; therefore some activities of these lymphoid cells (*specifically sensitized lymphoid cells*) must be responsible for this type of allergy.

Some of the characteristics of the major categories of hypersensitivities are outlined in Table 21-1. Immediate hypersensitivities are classified into three categories on the basis of the mechanisms responsible for the allergic reaction in each. An understanding of these mechanisms is important in order to know how to treat each kind of condition. This chapter is chiefly concerned with the IgE-mediated and cell-mediated hypersensitivities because they account for a large majority of all allergic reactions. Immediate cytotoxic hypersensitivities, in which cells are destroyed directly, are exemplified by blood transfusion reactions (discussed later in this chapter).

The immunological response to any antigenic stimulus is commonly multifaceted, involving protective and hypersensitivity reactions of various kinds. However, most hypersensitivity reactions are predominantly one or another of the four types listed in Table 21-1, with varying lesser degrees of other responses occurring simultaneously. Thus "serum sickness" is a disease that usually results chiefly from an immediate-type, *immune-complex reaction,* although some components of the other types of immune responses may also be demonstrable in this condition. Serum sickness is an allergic reaction to foreign blood serum, such as horse antitoxin given to counteract the toxin of tetanus. In this reaction humoral

TABLE 21-1
Some Characteristics of the Major Types of Hypersensitivities

Type	*Time of Reaction, Following Challenge of Sensitized Individual with Antigen*	*Mediated by Activities of:*	*Examples*
Immediate			
IgE-mediated	Immediate, within seconds or minutes; fades by an hour or several hours	IgE antibodies, fixed to mast cells in tissues and to basophils	Hayfever, hives, asthma, anaphylactic shock
Cytotoxic	Immediate, within seconds or minutes	Humoral antibodies and complement,[a] reacting with antigen that is a part of a cell	Blood transfusion reaction with mismatched blood
Immune complex	Immediate, within minutes to few hours	Humoral antibodies and complement,[a] reacting with soluble antigens that are not part of a cell	Serum sickness, Arthus-type reactions, farmer's lung
Delayed			
Cell-mediated	Delayed; first visible at about 6 hr, peaks at 24 to 48 hr; gradually declines over a period of days	Specifically sensitized lymphoid cells	Positive tuberculin skin test; poison ivy rash

[a] Complement may not always be necessary.

antibodies of the class IgG are formed and combine with the soluble horse serum antigens and complement; the soluble immune complexes formed in this way then interact with polymorphonuclear white blood cells, leading to release of leukocyte enzymes that are directly responsible for the tissue damage seen in the condition. On the other hand, the administration of foreign serum may also cause a reaction that is predominantly IgE-mediated.

IgE-mediated, Immediate-Type Hypersensitivity

Anyone who has ever seen a person with hayfever, hives or allergic asthma is familiar with the manifestations of IgE-mediated allergy; these conditions are so common that almost everyone recognizes them.

Hayfever occurs when a suitably sensitized person inhales a specific antigen; the itching, tearing eyes, sneezing, and runny nose that result are all evidences of extensive insult to the upper respiratory tract. *Hives* is a skin disease characterized by redness, itching, and the formation of *wheals,* skin lesions that in some ways resemble mosquito bites but are much larger and more troublesome. Hives often occur after ingestion of the offending antigen; for example, a person allergic to lobster may experience intense itching and "break out with hives" soon after eating even small amounts of lobster. Hives can occasionally be a severe disease. The localized swelling in reaction to an antigenic challenge may produce giant hives; when the swelling impinges on the larynx, it may even obstruct the air flow; fortunately, however, this occurs rarely. *Eczema* is another skin condition that is often caused by IgE-mediated allergic reactions.

Asthma, with its symptoms (wheezing, coughing, and difficulty in breathing, particularly in exhaling) is perhaps not as familiar to many people as are hayfever and hives; asthma, however, is a fairly common and sometimes serious affliction. An IgE-mediated reaction is only one of a number of possible causes of asthma.

Generalized (*systemic*) *anaphylaxis* is another important manifestation of IgE-mediated allergy. It may be fatal, but fortunately it is extremely rare. Generalized anaphylaxis is similar to other IgE-mediated allergic reactions, such as hives and asthma, but instead of being localized, it is manifested systemically throughout the body. Swelling occurs not just in the skin but also in the bronchi (resulting in difficulty in breathing) and in many other internal tissues. Death often ensues from a combination of causes, including generalized loss of fluid from the blood vessels because of increased vascular permeability, producing *shock.* The symptoms and signs of generalized anaphylaxis become apparent within seconds or minutes after exposure to the causative antigen and may include itching and flushing of the skin, difficulty in breathing, and subsequent collapse as a result of shock.

Haptens, page 385

In the United States today penicillin is the substance most often responsible for generalized anaphylactic reactions. As mentioned previously, this small molecular weight antibiotic is by itself nonantigenic; however, a metabolic product of penicillin can act as a hapten. This combines within the body with certain large molecular weight proteins, and the resulting hapten-carrier complex is capable of inciting production of antibodies, some of which are specific for penicillin alone. Penicillin is often injected, rather than taken by mouth, so that a large quantity of this antibiotic gets into the bloodstream very quickly and is immediately circulated throughout the body. If the recipient has previously become allergic to penicillin, an immediate reaction can occur systemically and may be fatal within minutes. Although systemic anaphylaxis is extremely rare in patients receiving penicillin, less severe allergic reactions to penicillin are not uncommon and are estimated to occur in as many as 10 percent of recipients of this antibiotic. Because of the danger of systemic anaphylaxis, any person who has had an IgE-mediated reaction to penicillin—or to any other foreign material—should avoid subsequent treatment with the reaction-inducing agent.

For many years the mechanisms common to the various IgE-mediated allergic diseases were obscure. Many investigators, who observed that these allergic manifestations were often intensified by emotional upsets, were convinced that these diseases were caused by emotional and psychological factors; however, evidence that antibodies are responsible has accumulated over the years. It was not until 1965 that the class of immunoglobulins responsible for these allergies was discovered and named IgE. This immunoglobulin class was extremely difficult to isolate and identify because it is present in the blood serum in almost infinitesimal quantities; in fact, the amount of IgE present in normal blood plasma represents only about one fifty-thousandth of the amount of IgG present!

The property of IgE antibodies responsible for their unique activity is their capacity to attach to certain types of cells, called *mast cells* and *basophils,* found in most parts of the body. These cells are particularly abundant in the nose, lungs, and skin, which helps to explain the frequent localization of IgE-mediated reactions to these areas.

Mast cells and basophils (Figure 20-6) contain many large granules packed with histamine and other substances. The attachment of IgE to mast cells represents a cytophilic association in which a specific site on the constant amino-acid portion of the IgE molecule attaches to a receptor on the mast cell, as shown in Figure 21-1. Once attached, the IgE molecule can survive for many weeks with its antibody-reaction sites readily available to interact with a specific antigen. The person with an IgE-mediated hypersensitivity, such as an allergy to ragweed pollen, will have many IgE antibodies specific for this pollen fixed to mast cells throughout his or her body, especially in the upper respiratory tract. When not exposed to the antigen, these antibodies are harmless, but when ragweed pollen is inhaled, it combines rapidly with the cell-fixed IgE antibodies. Within seconds the antigen-IgE reaction on the mast cell surfaces leads to the release, from the mast cell granules, of histamine and other chemicals which mediate the tissue damage of the hypersensitivity. The result is an attack of hayfever. If instead of being inhaled the ragweed antigen has been accidentally injected into the bloodstream of a highly allergic individual, systemic anaphylaxis and death may occur within minutes.

Figure 20-6, page 387

Because histamine is the principal mediator in hayfever and hives, antihista-

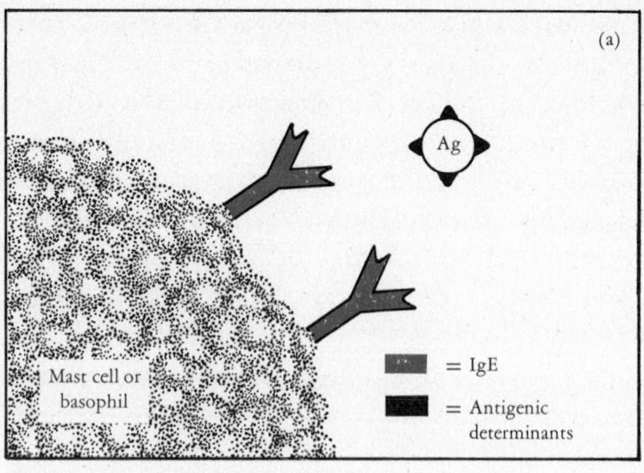

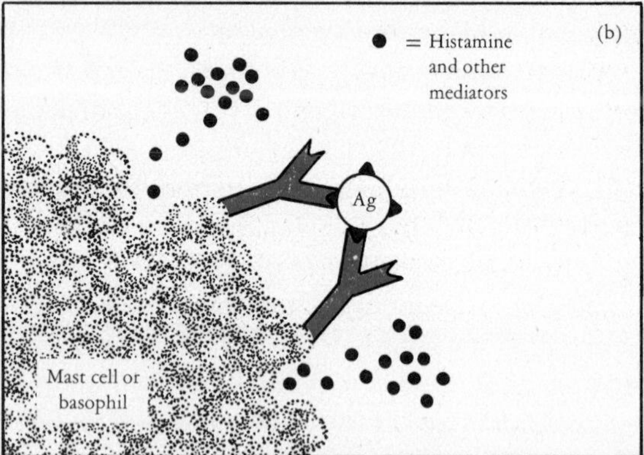

FIGURE 21-1
Schematic representation of an immunoglobulin E-mediated hypersensitivity state. (a) IgE molecules form a cytophilic attachment to mast cells or basophils. (b) A molecule of antigen bridges two molecules of IgE, causing release of mediators from granules of the cell.

mine drugs are very effective in treating these allergic diseases. Other mediators are more important in asthma, and antihistamines are consequently of limited value in counteracting asthmatic attacks. It is easy to show that the symptoms of hives are caused by the action of histamine simply by injecting a small quantity of histamine into the skin: A reaction identical to hives occurs immediately. Other factors that cause histamine release may also lead to the development of hives or hayfever; however, antigen-IgE interactions are the most frequent cause of these diseases.

Skin tests are often done to determine the substances to which a person is sensitive, or allergic. In these tests, extremely small quantities of antigen are introduced directly into the skin. If the subject is sensitive, an immediate reaction resembling hives occurs: A fluid-containing wheal forms, surrounded by a reddened area. The reaction reaches its peak within 20 to 30 min and fades within a few hours. Great care must be taken to avoid inducing systemic anaphylaxis while skin testing highly allergic persons.

In view of the danger of generalized anaphylaxis following exposure to antigen, it seems paradoxical that one of the common ways of treating severe allergies involves multiple injections of a specific antigen. This procedure is often referred to as desensitization; however, the preferred term for it is *immunotherapy.*

Its purpose is to make the patient less hypersensitive (or allergic) to the offending antigen. The procedure starts with extremely small doses of antigen, with the amount of injected antigen being very gradually increased over a period of months. Usually, as treatment continues, the patient becomes less sensitive or nonreactive to the particular antigen. One rationale for immunotherapy is that repeated doses of antigen cause the formation of IgG and other classes of antibody. The IgG antibodies then combine with the antigen before it can react with IgE, thereby preventing the patient's allergic reaction. There is no solid proof that this phenomenon actually accounts for desensitization; it is known, however, that IgG antibodies are formed during immunotherapy. Other evidence suggests that immunotherapy also results in an actual suppression of specific IgE production.

It has been observed that only a small percentage of persons exposed to potential antigens—such as ragweed pollen, for example—become allergic to them. Furthermore there is a familial tendency to develop IgE-mediated allergic diseases. Although the reasons for these phenomena are not known, it is thought that "allergic individuals" are genetically predisposed to produce abnormally large amounts of IgE antibodies. In addition, some people have a tendency to release histamine and other mediators more readily than normal individuals and consequently are more prone to allergic disorders.

Delayed Hypersensitivity (Cell-mediated)

Common examples of delayed hypersensitivity (DH) are reactions to poison ivy or poison oak, and positive reactions to the tuberculin skin test. Because most people in the United States have been skin-tested with tuberculin at one time or another, the tuberculin skin test is probably familiar; in this test, very small quantities of proteins from tubercle bacilli (the causative agents of tuberculosis) are introduced into the skin in the same manner as described for the IgE-mediated skin tests. Instead of an immediate reaction, however, individuals who are allergic to these bacterial proteins usually develop a delayed-type reaction that first becomes apparent, as a reddened area, about 12 hr after the tuberculin antigen is injected. This reddened area then gradually becomes *thickened, or indurated,* reaching a *peak at from 24 to 48 hr* after the inoculation. There is no formation of fluid-containing wheals, such as is seen in the IgE-mediated skin reaction.

A positive tuberculin skin test does not indicate that the tested individual has tuberculosis; it simply means that he has had enough exposure to tubercle bacilli or related organisms to become allergic to their bacterial protein. The test is useful as an aid in diagnosing doubtful cases of tuberculosis, in epidemiological studies, and in several other ways. Similar tests are used for studying other diseases.

Reactions that operate by the same mechanisms as the prototype tuberculin reaction are responsible for cell-mediated rejection of transplanted organs and immune reactions to cancer, for autoimmune diseases, and for antimicrobial cell-mediated immunity (CMI); in fact, DH usually accompanies CMI, because

Cell-mediated immunity, pages 397–401

the two are manifestations of the same process. In cell-mediated, as in humoral immune reactions, the term *immunity* describes protective reactions, whereas the term *hypersensitivity* refers to reactions that involve tissue injury which may or may not contribute to protection.

Delayed hypersensitivity reactions occur only in sensitized hosts; that is, the host must experience a primary response and produce expanded clones of sensitized lymphocytes before a reaction can be demonstrated. When lymphocytes from sensitized hosts (sensitive lymphocytes), are cultured *in vitro* with a specific antigen, they produce lymphokines (see Table 20-5); such lymphokines presumably function, as previously described, in mediating the observable DH reaction in vivo.

Table 20-5, page 399

Lymphocytes and their products are not the only factors involved in DH reactions; macrophages and other cells also participate. When an antigen, such as tuberculin, is injected into the skin of a sensitized person, it is soon encountered by a few sensitive lymphocytes, for these cells circulate constantly. It is thought that upon recognizing and interacting with an antigen the sensitive lymphocyte is stimulated to transform and it is during this process that lymphokines are secreted. These substances act on many other cells in the vicinity; some of the secreted lymphokines, for example, attract macrophages and immobilize them at the reaction site. The resulting accumulation of cells is in large part responsible for the hard swelling (induration) of the skin seen between 24 and 48 hr of a delayed hypersensitivity reaction. If this hypothetical sequence of events is truly what happens during delayed reactions, as now seems likely, then the delay of 12 hr or so before the reaction becomes visible represents the time required for the transformation of lymphocytes and the subsequent steps involved.

The question of how DH is related to CMI has not been fully answered. It is known that a state of DH is usually present when effective CMI also exists and it seems likely that the products of stimulated lymphocytes may lead to both protection against microorganisms and to tissue damage in the host.

One other kind of cell-mediated hypersensitivity reaction that is of particular interest is *contact hypersensitivity.* The prototype of this reaction is the familiar poison ivy rash. When a person is first exposed to poison ivy, no rash develops, but sensitivity to a particular hapten in the poison ivy plant may be established. Upon re-exposure to poison ivy, the allergic reaction is triggered. The delayed reaction begins hours after contacting poison ivy, reaches a peak after two or three days, and gradually subsides if the hapten does not remain present. Contact allergies can be caused by many other kinds of substances, some of them simple chemicals, such as the chromium salts used in tanning leather or the nickel in watch bands. The allergic individual may remain sensitive for many years or may gradually lose sensitivity if the antigen is not reencountered.

A distinction should be made here between substances that are toxic and those which are allergenic. Certain soaps are good examples; some contain toxic materials that damage the skin and cause a rash on virtually all persons who are exposed to them; others may be very mild and harm only the skin of the few individuals who are allergic to some of their constituents. In the latter case the rash may be extensive and severe, even though the soap is nontoxic for normal people.

MEDICAL PROBLEMS INVOLVING HYPERSENSITIVITY STATES

Having examined some of the basic mechanisms of immediate-type and delayed-type hypersensitivities, we can now discuss some allergy problems in terms of the kind of allergy involved, and the balance between the protective and harmful aspects of immunological responses.

Transplantation

It is common knowledge that successful organ transplantation depends largely on overcoming or bypassing the body's very efficient mechanisms for rejecting or destroying foreign materials. Although there are differences in the ways in which various kinds of grafts are rejected, this discussion is concerned only with transplants of the human tissues commonly employed. Most tissue grafts in man are transplanted from a normal donor to a genetically and hence antigenically different human recipient who has not been previously exposed to the donor's antigens (primary grafts). As a rule primary graft rejection does not result from the activities of humoral antibodies. Instead, it results principally from the action of sensitized lymphocytes and accessory cells acting in much the same sequence described in DH reactions. Killing of the grafted cells results from their direct contact with specifically sensitized T lymphocytes.

Several methods are used to partially avoid or overcome the rejection reaction. Avoidance of as much of the reaction as possible is preferable, and tissue matching is therefore done to minimize rejection. Tissue matching is possible because human tissue can be classified into major groups, called histocompatibility groups. These are analogous to the major blood groups discussed later. In the laboratory the donor tissue is typed for both its major histocompatibility antigens and for its major blood group antigens, using known antisera. If these antigens are compatible with (match) those of the recipient, the rejection response will be limited to only minor antigens and will be much weaker than if the major antigens are incompatible. Before grafting an organ or tissue, other matching tests are also done, using mixed lymphocytes of the donor and recipient. Even so, *immunosuppressive agents* must be used to overcome the immune response to the numerous weaker antigens for which matching is not presently feasible.

Immunosuppression is a descriptive term used to designate the *nonspecific* suppression of an individual's immune responses. It differs from immunological tolerance, which involves lack of response to a *specific* antigen. Because immunosuppression results in an overall immunological depression, the transplant recipient's response to many if not all antigens is inhibited. In tissue transplantation specific tolerance to all antigens of the transplant would be ideal, but unfortunately, since no one knows how to achieve this, nonspecific immunosuppression with agents such as drugs must be used. However, such generalized immunosuppression also results in the loss of effective protective responses to many harmful agents; consequently, infections are a major cause of death in transplantation patients. In situations such as kidney transplantation, where it is possible to choose donor organs whose antigens match those of the recipient fairly well, mild immunosuppression may be sufficient to overcome the rejection process. Consequently, a majority of kidney transplants are highly successful. Furthermore, because kidneys are paired organs, it is possible for a related donor, whose tissues are antigenically similar to the recipient's, to donate one kidney and continue in good health with his or her one remaining kidney. However, it is almost impossible to obtain a vital single organ (such as a heart) from a related donor; therefore, such organs cannot always be carefully matched. Large doses of immunosuppressive agents must be used to prevent rejection of poorly matched grafts, and this contributes to the poor survival rates of such patients.

There are many ways to bring about immunosuppression; most of them depend on preventing the proliferation of lymphoid cells. This may be done with certain kinds of irradiation, drugs, or by means of antilymphocyte globulin, prepared from antiserum (made in animals) against human lymphocytes. Usually transplant recipients are given a combination of more than one of these immunosuppressive agents.

Mention has been made of enhancing antibodies which paradoxically seem to enhance the survival of foreign cells rather than to oppose it. In the following section on immunity to tumors, this mechanism is discussed in more detail. It appears that enhancing antibodies and other enhancing factors may sometimes function during primary organ transplantation, resulting in prolonged survival of the graft; at present, however, no procedure has been developed that will consistently encourage immunologic enhancement in the transplant recipient and at the same time suppress or overcome those of his or her immune responses that destroy the graft. When and if this is ever achieved the problem of transplantation rejection may essentially be solved.

Immunity to Tumors

Nature of tumors, pages 353–354

Because tumor cells arise from host tissue, it might be expected that they would be recognized as self and would not incite an immune response. However, this is not the case, for it is well established that *tumor cells have unique antigens* (*tumor antigens*) *that cause them to be recognized as foreign.* A cell-mediated immune response analogous to transplantation rejection takes place, with tumor cells being killed following direct contact with immune lymphocytes or macrophages. It is possible that this protective response eliminates many cells that would otherwise grow into detectable tumors. There are several possible explanations for the occurrence of tumors in spite of an active immune response against them. First, tumor cells proliferate actively and may simply outgrow and overwhelm the capacity of immune cells to destroy them. Furthermore, tumor cells commonly have a coating substance on their surfaces (somewhat analogous to the capsules of bacteria) that inhibits their effective contact with immune cells. Alternatively, *blocking factors* comprising enhancing antibodies specific for the tumor cells, specific antibody-antigen complexes, or other enhancing factors may block effective CMI.

Recently it has been found that in most of the tumor systems studied, both in animals and in man, the following results generally hold true: Subjects with tumors have immune lymphocytes (in their circulation and in their lymphoid tissues) that are capable of killing or inhibiting specific tumor cells in vitro. For example, when blood lymphocytes are collected from an experimental animal with a tumor and mixed with its own tumor cells in culture, the tumor cells are usually killed or inhibited, as compared with control tumor cells mixed with lymphocytes from a normal animal. Figure 21-2 illustrates this experimental pro-

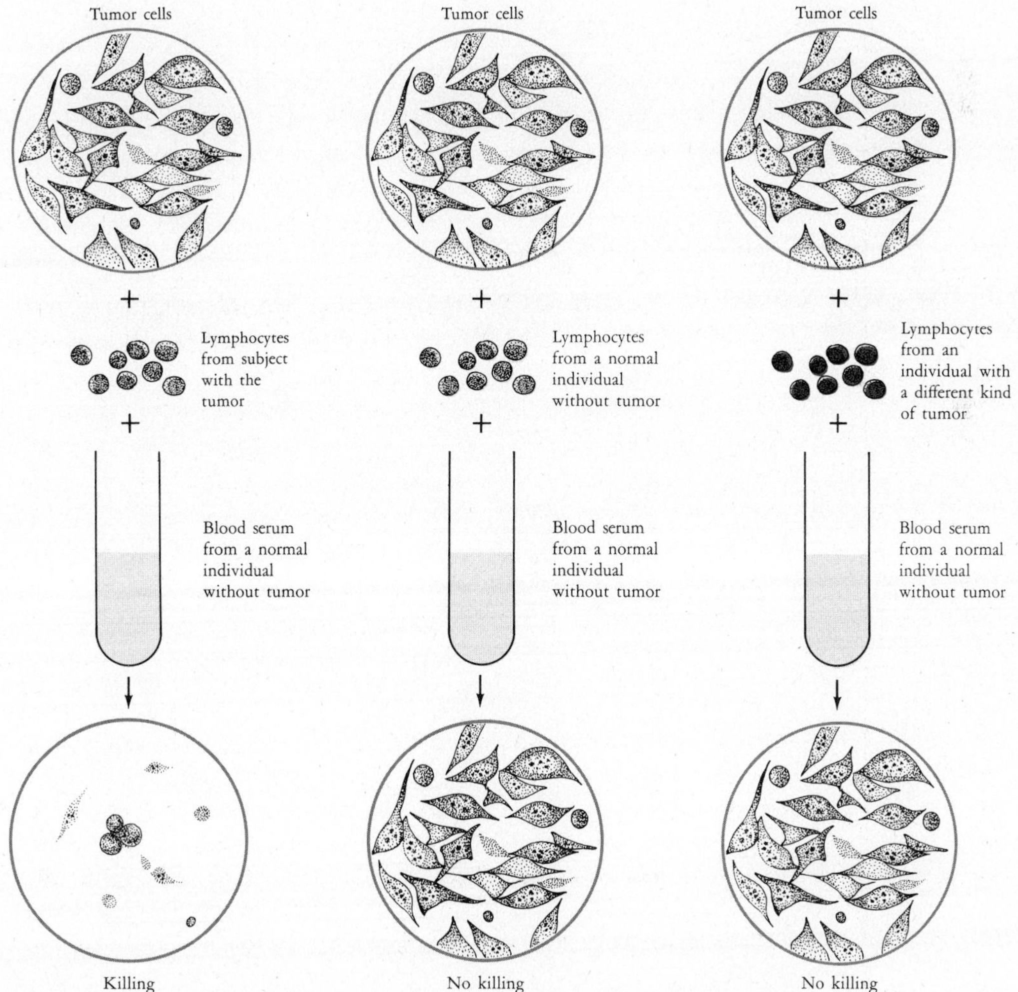

FIGURE 21-2 Lymphocytes from a tumor-bearing subject can kill specific tumor cells in culture.

cedure. If, however, blood serum containing specific humoral antibodies against the tumor is added to the mixture, the ability of lymphocytes to kill tumor cells may be abrogated (Figure 21-3). Specifically, when the tumor in an animal is regressing, indicating an effective immunological response, addition of the animal's serum generally does not decrease the killing of its tumor cells by sensitized lymphocytes. On the other hand, when the tumor is progressing, addition of the animal's serum tends to counteract or block the killing of its tumor cells by lymphocytes, leading to survival of the tumor cells. This has been shown to result from the presence of specific humoral blocking factors, possibly enhancing antibodies or antigen-antibody complexes. It is thus possible that fatal progressive tumors (cancers) evade cellular immune mechanisms in large part because of the interference provided by such blocking factors. Here again, as was the case with

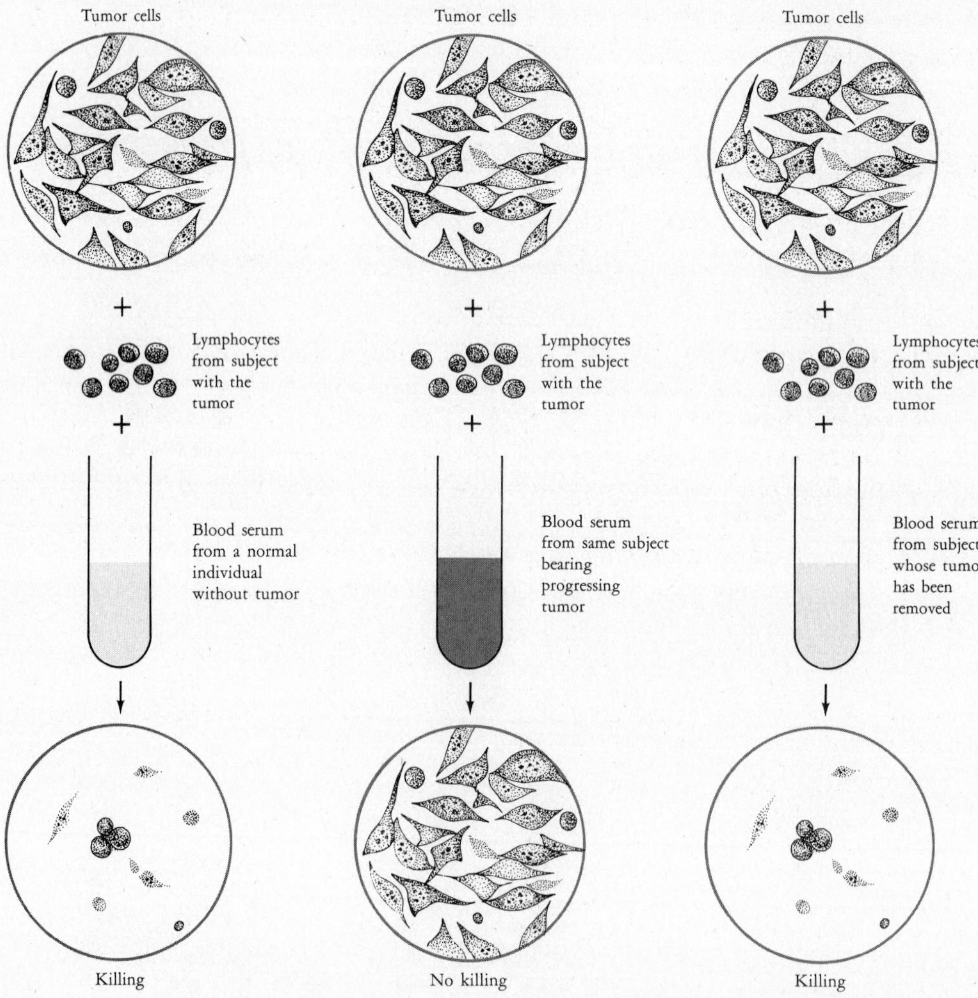

FIGURE 21-3 Serum from a tumor-bearing subject may interfere with tumor-cell killing by that subject's lymphocytes.

tissue transplantation, an understanding of how to control immunological responses, both cellular and humoral, would provide a rational approach toward controlling a major medical problem.

Although direct-contact killing of tumor target cells by *lymphocytes* is of primary importance in immunity to most tumors, it is of interest that *macrophages* can also inhibit or kill tumor cells, either *specifically* as the result of bearing cytophilic antibodies or other specific antigen-recognition factors, or *nonspecifically* as the result of becoming activated; activated macrophages have a selective toxicity for tumor cells over normal cells.

BLOOD GROUPING AND BLOOD TRANSFUSION

Blood transfusion is a form of tissue transplantation, with foreign cells being transferred to a recipient of the same species; however, the cells in blood are free and fully exposed to both the antibody and complement in plasma. For this and other reasons, reactions to blood transfusions usually result from the action of specific humoral antibodies and complement, and are humoral rather than cell-mediated reactions.

ABO BLOOD GROUP SYSTEM

The reason for the feasibility of transfusing human blood, first discovered by Landsteiner and his coworkers early in the twentieth century, is that the strongest and by far the most common transfusion reactions are caused by incompatibilities between the few antigenic types of a single, relatively simple blood group system known as the ABO system. It is important to remember that there are hundreds of different antigens in blood plasma or serum, and that still other antigens are present on both erythrocytes and leukocytes; in fact there are dozens of minor blood group systems of red cell antigens in addition to the major ABO system. Fortunately, most of these minor antigens do not incite a strong antibody response, otherwise blood transfusions would not be practical.

Because of the complexities of blood group antigens, this discussion is limited to simplified explanations of two of the most important systems, the ABO and the Rh (Rhesus) systems.

In the ABO system the presence or absence of only two antigens, called A and B, accounts for the four major blood groups, as shown in Table 21-2. Anti-

TABLE 21-2
Some Characteristics of the ABO Blood Group System

Blood Type (Phenotype)	*Antigen(s) Present on the Erythrocytes*	*Antibodies Normally Present in the Plasma*	*Percent in a Mixed Caucasian Population*[a]
O	Neither A nor B	Both anti-A and anti-B	45
A	A	Anti-B	41
B	B	Anti-A	10
AB	Both A and B	Neither anti-A nor anti-B	4

[a] The incidence of ABO blood groups may vary greatly between genetically different populations.

gens A and B are polysaccharides found in abundance on the surfaces of erythrocytes of the appropriate blood groups. Group O red cells lack both A and B antigens; group A red cells have A (but not B) antigen; group B erythrocytes possess B (but not A) antigen; and group AB red cells have both A and B antigens on their surfaces. It is a general immunological rule, mentioned previously, that a normal individual will not produce antibodies against his own antigens; therefore, if the individual has either A or B antigen, or both, he or she will not produce antibodies against these antigens. A unique feature of this system is that (contrary to the usual situation) antibodies are normally present against the A and B antigens that are lacking on the red cells. These are called natural antibodies, meaning antibodies that are formed without *deliberate* antigenic stimulation.

These basic facts about the ABO system are sometimes stated as *Landsteiner's principle:* whenever antigen A or antigen B is missing from the red cells of a normal individual, the corresponding specific antibody for the missing antigen(s) is present in the individual's plasma.

Natural antibodies, such as those in the ABO system, account for only a very small part of the total immunoglobulins. It is thought that the natural anti-A and anti-B antibodies are probably formed as the result of repeated stimulation with common substances in the environment that contain the A and B antigenic determinants. These determinants are known to be simple sugars, chemical groupings that are widespread in many foods, dusts, bacteria, and other substances to which man is constantly exposed.

Table 21-2 includes a summary of the plasma antibodies normally present in human beings of each of the four major blood groups. There are some subgroups and exceptions to the simplified information in Table 21-2, but for the most part this is a less complex blood-group system than many others, such as the rhesus system discussed later.

In transfusing blood, the antigens on the donor's red cells and the antibodies in the recipient's plasma are of greatest importance because transfusion reactions are usually the result of destruction of the donor's erythrocytes by the recipient's antibodies. Certainly the reverse situation can occur, and antibodies in the donor's plasma can react with erythrocytes of the recipient, but this is much less likely to happen simply because of the relative quantities of each. When a pint of blood is transfused, the donor plasma is immediately diluted by the 10 to 12 pints of blood in the adult recipient; therefore the transferred donor blood must contain an uncommonly large amount of antibodies in order to cause a detectable reaction as the result of destruction of the recipient's red cells. Donor red cells, however, are constantly bathed in the recipient's plasma, and therefore have a high risk of being promptly destroyed by the recipient's antibodies. Thus, before transfusions are given, compatibility tests (other than blood typing) between the bloods of a donor and a recipient must be conducted to minimize the possibility of reactions in vivo that might result if these persons' ABO groups only were determined.

The genetics of the various blood group systems has been extensively studied by immunogeneticists and immunohematologists. Many practical applications of these studies have been helpful in law, anthropology, and other diverse areas. For example, knowledge of the genetics of the ABO system allows ABO blood groups to be accepted as evidence in court under certain conditions. The ABO blood group locus is found on one pair of the 23 pairs of human chromosomes; the ABO genes on each chromosome of this pair are codominant, that is, the A and B genes are both expressed if both are present. Because one chromosome of each pair is inherited from the mother (maternal) and one from the father (paternal), an individual's blood group is determined by the ABO genes obtained from both parents (Figure 21-4). Table 21-3 shows the possible genotypes of individuals of the major ABO blood groups and the genotypes of their parents. The simplest case is seen with group O individuals. Their genotype can only be OO, and they must obviously have received one O gene from each parent, so that the only possible parental genotypes for these people are those with at least one O gene. When both parents are O, there is no possibility that the children will be anything except O (Figure 21-5). The situation becomes slightly more complex with the other blood groups (Figure 21-5), but it is still a simple matter to calculate the probability of children of a given blood group being born of any ABO genetic mating.

The ABO blood groups are accepted as evidence in cases of disputed parentage only when they exclude an individual as being the parent. Thus if a group O woman bears a group O child, a man with AB blood can be excluded as being the father of the child, since he could have transmitted only an A gene or a B gene and the child has neither. If the male in question were A, B, or O, he could not be excluded as a possible father on the basis of the child's phenotype alone.

RHESUS BLOOD GROUP SYSTEM

The Rh (Rhesus) system is the most important of the minor blood group systems. Although this system is complex and many antigens are involved, only the strongest one, called the D antigen or simply the Rh antigen, will be discussed. If the Rh antigen is present on his or her red cells, a person is Rh-positive; if it is lacking, the person is Rh-negative.

The Rh-positive adult is not at risk with respect to most important Rh

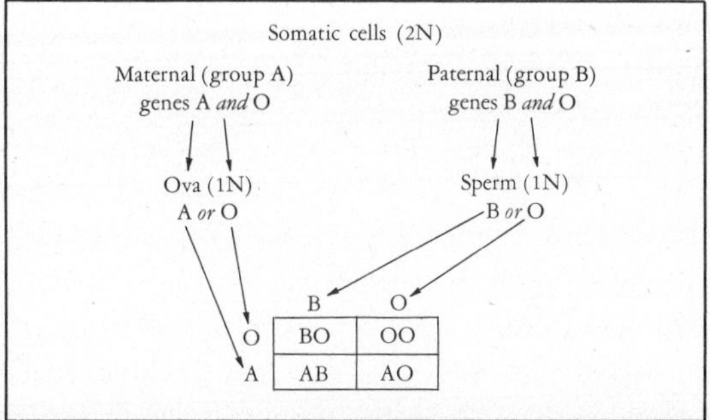

FIGURE 21-4 Codominant inheritance of ABO blood groups. One pair of chromosomes in each diploid somatic (body) cell carries the genes determining the ABO group. Ova and sperm cells are haploid (IN), containing only one chromosome of each pair and therefore only one gene for the ABO group. At mating the ova and sperm combine to give a random distribution of all possible combinations. The ABO genes are codominant and both are expressed. Consequently, the overall frequency of blood groups in offspring from this combination will be 25 percent each of groups O, B, A, and AB.

TABLE 21-3
Possible Parental Genotypes[a] for Individuals of the Major Blood Groups

Blood Type (Phenotype)[b] of Individual	*Possible Genotype(s) of Individual*	*Possible Parental Genotypes*
O	OO	OO
		AO
		BO
AB	AB	AB
		AA
		BB
		AO
		BO
A	AA	AA
	AO	AO
		BO
		OO
		AB
B	BB	BB
	BO	BO
		AO
		OO
		AB

[a] Genotype: genetic constitution, determined by both of a pair of genes.
[b] Phenotype: expressed characteristics, determined by the dominant gene or genes of a pair.

blood group problems. Such a person can receive either Rh-positive or Rh-negative blood transfusions with impunity, because Rh-positive blood is compatible with other Rh-positive blood and Rh-negative red cells simply lack the Rh antigen altogether. It is the Rh-negative adult who may experience an Rh blood transfusion reaction as a result of being immunized against the Rh antigen, either from transfusions, organ grafts or, in the case of females, from pregnancies which introduce this antigen.

Rh DISEASE (HEMOLYTIC DISEASE OF THE NEWBORN)

The antibodies to the Rh antigen that are formed by a pregnant woman may damage her offspring. The disease that results is called hemolytic disease of the newborn, or simply Rh disease (Figure 21-6). While an Rh-negative mother is carrying an Rh-positive fetus, a few fetal cells may pass across the placenta to the mother but usually not enough to stimulate a *primary* antibody response. At birth, however, the baby's blood cells often enter the mother's circulation in sufficient numbers to incite a vigorous immune response. Also, abortion procedures are very likely to cause large numbers of blood cells to pass from the fetus into the mother. The anti-Rh antibodies formed by the Rh-negative mother cause her no harm because they cannot damage her cells (which lack Rh antigens). However, problems may arise for any subsequent Rh-positive fetus that the mother may carry. With such a fetus, the few Rh-positive cells that cross the placenta from the fetus's to the mother's blood may be enough to initiate an *anamnestic response.* Large quantities of anti-Rh antibodies of the IgG class are produced, which cross the placenta to the fetus and cause extensive disease due to

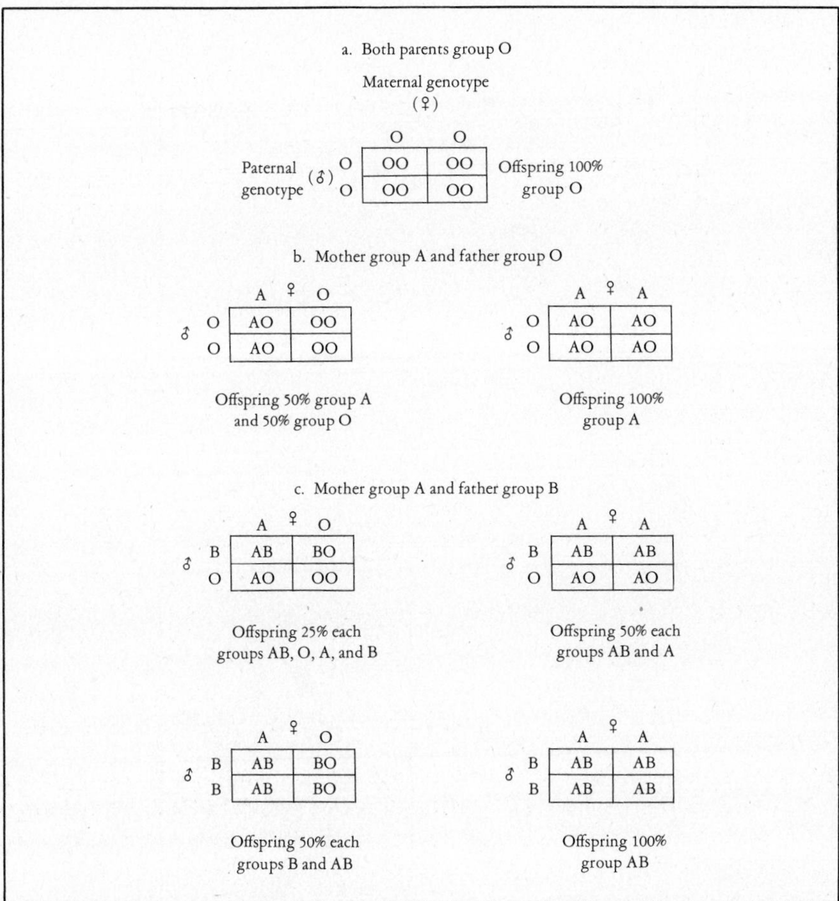

FIGURE 21-5
Probabilities of offspring of various ABO blood groups from different matings.

fetal red cell damage. Although miscarriage and loss of the fetus can result, the damage from this phenomenon is often not apparent until very soon after birth, as indicated by its name, *hemolytic disease of the newborn.*

The child whose Rh antigens are attacked by its mother's antibodies usually survives *in utero* because harmful products of red cell destruction can be eliminated from its system by certain enzymes, present in large amounts in the mother but only in very small amounts in the fetus and newborn. Soon after the exchange of materials between the circulations of mother and fetus is interrupted at birth, however, the child becomes acutely ill. Not only are the products of erythrocyte destruction toxic, but the baby is also seriously anemic and may not have enough red blood cells to survive. In this critical situation, replacement transfusions are life-saving. The baby's blood is gradually withdrawn while fresh *Rh-negative blood is transfused.* The benefits of this are twofold: First, toxic products, maternal antibodies, and the infant's Rh-positive red cells are all removed; and second, the infant's red cells are replaced by Rh-negative erythrocytes which can-

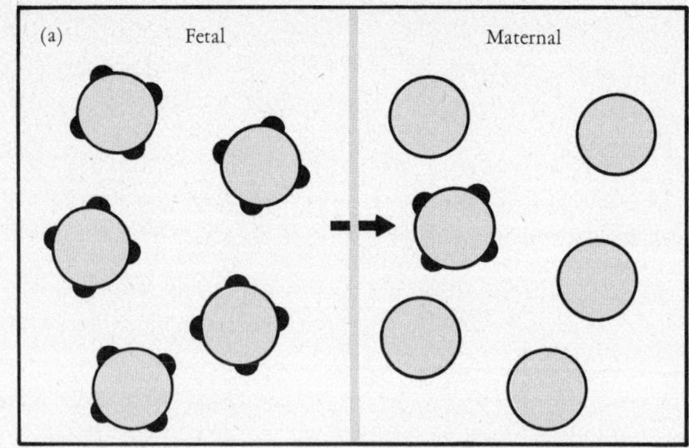

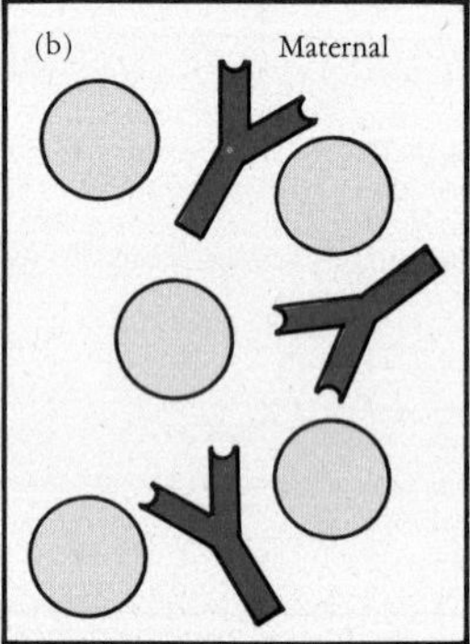

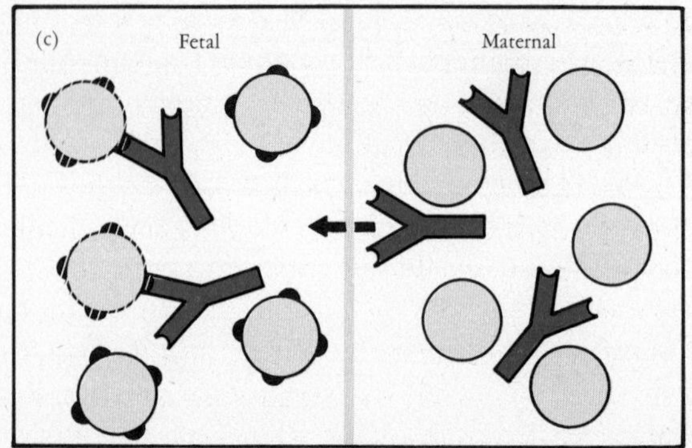

FIGURE 21-6
Events leading to Rh disease (hemolytic disease of the newborn). (a) Fetal red blood cells bearing the Rh antigen determinant (designated by black half circle) enter the maternal circulation, and the Rh antigenic determinants become available to stimulate an immune response. (b) Anti-Rh antibodies are formed. (c) During subsequent pregnancy with an Rh-positive fetus, the anti-Rh antibodies can cross the placenta, leading to destruction of fetal red blood cells and Rh disease.

not react with any remaining anti-Rh antibodies. The child does not produce large numbers of Rh-positive red cells until the transfused cells age and are removed from the circulation, a matter of weeks. By this time the maternally derived anti-Rh antibodies have also decreased to a harmless level.

Anti-Rh antibodies are routinely given to susceptible Rh-negative mothers at or soon after the birth of an Rh-positive child. This procedure can prevent the mother from becoming immunized by her infant's Rh-positive cells at the time of birth and will thereby prevent Rh disease in her subsequent children.

Although it might be expected on the basis of the overall incidence of Rh-positive and Rh-negative people in the general population that nearly 15 percent of all newborns would develop Rh disease, in fact less than 1 percent are so affected, for many reasons including the following: First, many Rh-positive males are heterozygous and transmit an Rh-negative gene, so that children of an Rh-positive father and an Rh-negative mother are often Rh-negative and consequently not subject to Rh disease. Second, the mother may not become immunized, even when incompatibilities exist.

Autoimmune (Autoallergic) Diseases

It was stated earlier that the immunologically competent cells of an individual may, occasionally and for various reasons, fail to recognize certain of the individual's own components as self and may respond immunologically to them. The result is commonly known as autoimmune disease. A better designation is *autoallergic* disease.

Antibodies and immune or sensitized cells that react with components of self can be formed during conditions that break down and degrade tissue. Minor changes in the resulting damaged tissue components can then produce substances that are not recognized as self and that cause immunologic responses. The antibodies or sensitized cells that are formed in this process can cross-react with the very similar normal components of self. Such reactions may be inconsequential and disappear as soon as the original predisposing conditions are corrected, but autoallergic disease occurs when an individual's tissues are damaged by their interaction with his or her own antibodies or sensitized cells. It is commonly difficult to prove that a disease is caused by an autoimmune reaction, however, since the evidence for such a reaction is often circumstantial.

Among the proven examples of autoimmune diseases are some forms of hemolytic anemia and of thyroiditis. Also of probable autoimmune causation is rheumatoid arthritis, a common form of inflammatory joint disease.

Autoimmunity has also been postulated to play an important role in aging; however, definitive proof of this hypothesis is lacking. It is known that T lymphocytes decrease in numbers and functional capacity with age, and that immune capabilities generally wane with old age. Because T lymphocytes are often important in controlling or suppressing immune responses, it is possible that the lack of T cell control that accompanies aging permits the occurrence of immune responses against components of self.

RELATIONSHIPS BETWEEN PROTECTIVE IMMUNE REACTIONS AND ALLERGIC REACTIONS

The extent of protection or harm that results from an immunological response depends on a combination of factors: The nature and dose of the antigen, its route of entry and fate in the body, the numbers and state of the lymphoid cells that respond to it, the classes of antibodies formed against it, the sites of interaction between the antibodies and the antigen, and numerous others. In virtually no case is the immunologic response restricted only to antibody production, cell-mediated hypersensitivity, or any other single state. As more is learned about immunological mechanisms, it may become apparent that hypersensitivity reactions, even though they often damage tissues, still offer essential protection to the body as a whole, in conjunction with conventional forces of immunity.

The subject of immunity against particular microorganisms is discussed in subsequent chapters. It is well to remember that the basic concepts discussed in this chapter must be applied when choosing immunization procedures. Immunizing antigens must be administered so that they maximize protective benefits and limit the chances of inciting hypersensitivity reactions.

SUMMARY

The term hypersensitivity has been applied to immunological reactions that cause tissue damage or malfunction in the host. A synonym for hypersensitivity is allergy. Allergic reactions may be classified into two broad categories: immediate and delayed.

Immediate hypersensitivities include allergies such as hayfever, hives, and allergic asthma. These allergies are mediated by antibodies of a particular class of immunoglobulins, IgE, which are cytophilic for mast cells. Interaction of an antigen with cell-fixed IgE causes the release of chemical mediators such as histamine from mast cells, resulting in the observed allergic reactions. Other kinds of immediate hypersensitivities result from the action of antigen with humoral antibodies belonging to immunoglobulin classes other than IgE.

Delayed hypersensitivity is mediated by specifically sensitized cells, principally lymphocytes. The mechanism of lymphocyte activity in delayed hypersensitivity reactions is similar, or perhaps identical, to the action of lymphocytes during protective cell-mediated immune reactions. The tuberculin skin test is the prototype of delayed allergic reactions. Contact hypersensitivity is another form of delayed hypersensitivity, exemplified by the familiar poison ivy rash.

Hypersensitivity reactions are involved in transplantation rejection, immunity to tumors, immunity to some infectious agents, blood transfusion reactions, and a variety of other medical problems. Autoimmune diseases also result from hypersensitivity reactions.

The immunological response to any foreign antigen usually involves a combination of humoral and cellular responses, including both protective and hypersensitivity reactions.

QUESTIONS

REVIEW

1. List some differences between immediate and delayed-type hypersensitivities. How do these relate to humoral and cellular immunity?
2. Describe IgE-mediated allergies and the role of IgE in causing these kinds of allergic reactions.
3. What kinds of cells participate in delayed-type hypersensitivity reactions? What is their probable mode of action?
4. What is the chance that a man with group A blood is the father of a group B baby whose mother is AB? Could this blood-group evidence be used in court in a child-support lawsuit? Explain.

THOUGHT

1. Consider some of the practical reasons for studying hypersensitivities. List at least five important applications of such studies.
2. The human fetus is actually a foreign transplant and yet it is not rejected immunologically. How might immunologic enhancement contribute to this lack of rejection? (The Beer and Billingham article listed in the references gives an intriguing glimpse of the immunology of maternal-fetal relationships.)

FURTHER READING

ALLISON, A. C., and J. FERLUGA, "How Lymphocytes Kill Tumor Cells," *New England Journal of Medicine* 295:165–167 (1976). A brief review of evidence pertaining to the mechanisms of killing of tumor cells by lymphocytes.

BEER, A. E., and R. E. BILLINGHAM, "The Embryo as a Transplant," *Scientific American* (April 1974).

CLARK, C. A., "The Prevention of Rhesus Babies," *Scientific American* (November 1968).

GELL, P. G. H., R. R. A. COOMBS, and P. J. LACHMANN, eds., *Clinical Aspects of Immunology.* 3d ed. Philadelphia: F. A. Davis, 1975. An excellent, comprehensive book presenting a multiauthored view of the clinical aspects of immunology.

GOOD, R. A., and D. W. FISHER, eds., *Immunobiology.* Stamford, Conn.: Sinauer Associates, 1971.

MCCLUSKEY, R. T., and S. COHEN, eds., *Mechanisms of Cell-mediated Immunity.* New York: John Wiley & Sons, 1974. Contains good chapters on tumor immunity and transplantation immunology.

OLD, L. J., "Cancer Immunology," *Scientific American* (May 1977).

PATTERSON, R., ed., *Allergic Diseases.* Philadelphia: J. B. Lippincott, 1972. Some basic descriptions of immunologic mechanisms are included, along with detailed clinical descriptions of allergic diseases.

SELL, S., *Immunology, Immunopathology, and Immunity.* 2d ed. Hagerstown, Md.: Harper & Row, 1975. Excellent simplified discussions of the immunopathology of the various types of hypersensitivities.

CHAPTER 22 BACTERIA OF MEDICAL IMPORTANCE

This chapter describes some of the numerous species of bacteria important in causing human disease. It is intended mainly for reference and review, although some may wish to read it through in preparation for the subsequent chapters on various diseases.

GRAM-POSITIVE COCCI

Staphylococci

Staphylococcal skin infections, pages 462–463

Staphylococcus aureus is easily the most important cause of wound infections, boils, and other skin infections as well as of abscesses in bones and other tissues of the human body. Moreover, certain strains of this organism are prominent causes of food poisoning. As seen in infected material and cultures, staphylococci are typically arranged in clusters, although single or paired staphylococci, as well as short chains of these organisms, are seen under some conditions. *Staphylococcus aureus* grows readily on the usual laboratory media, and well-developed colonies are present in 18 hr. These colonies are the color of thick cream and most strains of the organism cause a small zone of red blood cell destruction around their colonies on blood agar. In identifying the species of *Staphylococcus,* the most important item is a positive *coagulase test.* Virtually all strains of *S. aureus* produce an enzyme, coagulase, that causes plasma to coagulate (clot), whereas none of the other medically important cocci can accomplish this under the conditions of the test. The ability of *S. aureus* to ferment mannitol is also an important characteristic in its identification. Besides coagulase, this organism produces numerous other cellular products; several of these are enzymes involved in destroying components of animal tissues. Others are toxins, one of which causes damage to the skin, while another results in food poisoning. Of great importance is the enzyme *penicillinase,* which destroys penicillin and thus protects the bacterium from treatment with this medication. Presently penicillinase is found in about 50 percent of the strains of *S. aureus.* Indeed, strains of *S. aureus* have appeared that are resistant to each of essentially all of the antimicrobial medications in current use.

Staphylococci are more resistant to elevated salt (NaCl) concentrations than are many other bacteria, and special media making use of this property can be used to recover them from their mixtures with these other species. Also, resistance to drying fosters the transmission of staphylococci from one host to another. Normally, staphylococci live in the nose, and sometimes on moist areas of the skin of human beings and other animals. Also helpful in identifying different strains of staphylococci is the susceptibility of these strains to different bacteriophages. *Staphlococcus aureus* must be differentiated from *Staphylococcus epidermidis,* a coagulase-negative species almost universally present on the human skin, and from *Micrococcus* species, which are also common normal inhabitants of the human body. Of help in identifying micrococci is that they are nonfermentative (they do not produce acid byproducts from sugars anaerobically). *Micrococcus* species and *S. epidermidis* are occasional causes of infection, usually of the heart valves.

Streptococci

"Strep throat," pages 477–479

Streptococcus pyogenes is the cause of both "strep throat" and wound infections as well as infections of the skin, ear, lung, and other tissues. It may also be responsible for delayed effects on the body, one of which is a kidney disease (glomerulonephritis) and another, rheumatic fever (a heart disease). In infected material *S. pyogenes* most commonly occurs as diplococci or in short chains, while in culture, chains of this organism are characteristic. *S. pyogenes* is easily cultivated on the usual laboratory media, but prefers a medium enriched with blood and an atmosphere with increased carbon dioxide and water vapor. Its colonies on blood agar typically exhibit the phenomenon of "shoving"; that is, when touched with a bacteriological wire loop, an entire colony slides across the medium. The vast majority of strains of *S. pyogenes* also show a zone of *beta hemolysis* around their colonies, meaning that complete red cell destruction and clearing of the medium occurs. Two kinds of hemolysins, streptolysin O and streptolysin S, are usually released from *S. pyogenes* cells. Streptolysin O is reversibly inactivated by oxygen, and this explains why beta hemolysis is more pronounced under anaerobic conditions. Cultivation of *S. pyogenes* mixed with other species (as from a culture of the throat) may result in inhibition of growth of *S. pyogenes;* anaerobic culture conditions and selective media help minimize this effect. Another feature of *Streptococcus pyogenes* is that it is more likely to be inhibited by low concentrations of the antibiotic, bacitracin, than are other beta hemolytic streptococci, and this property aids in its identification. All members of the species *S. pyogenes* have a specific cell wall carbohydrate antigen called the *group A polysaccharide* (Figure 22-1).

Lactic acid bacteria, page 258

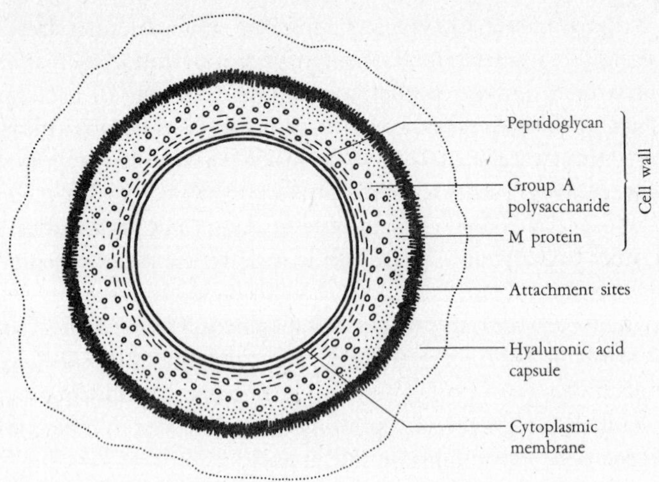

FIGURE 22-1
Stylized drawing of some important components of *Streptococcus pyogenes*.

Streptococcal impetigo, pages 463–465

Other cell wall antigens of importance in the species are the M and T proteins, which differ from strain to strain. Invasiveness of *S. pyogenes* is fostered by the M antigen, which is antiphagocytic, as well as by a capsule in some strains. Immunity to *S. pyogenes* is strain-specific and depends on antibody to the M protein.

Streptococcus agalactiae differs from *S. pyogenes* in possessing the group B polysaccharide antigen in its cell wall and in being less frequently beta hemolytic. Colonies of this organism almost always have a yellowish-brown pigment on

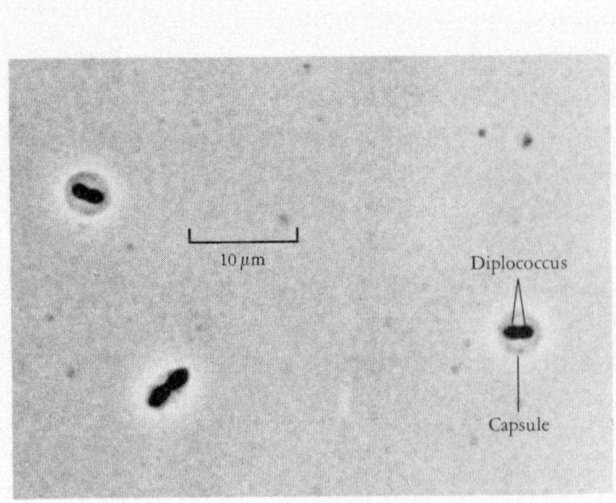

(a)

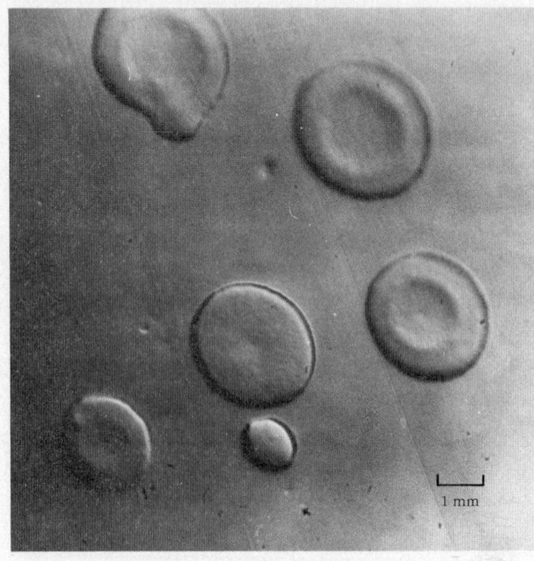

(b)

FIGURE 22-2 (a) *Streptococcus pneumoniae*. Note that the organisms are in pairs, accounting for the generic name *Diplococcus* formerly given to this species. The large capsule surrounding the organism is made more distinct because of specific antiserum added to the surrounding fluid. (Courtesy of J. T. Staley and J. P. Dalmasso.) (b) *Streptococcus pneumoniae*, umbilicated colonies. Note varying stages of collapse of the central portion of the colonies. (Courtesy of J. P. Dalmasso.)

appropriate media. Virtually all strains hydrolyse hippurate, whereas few other beta hemolytic streptococci do so. *Streptococcus agalactiae* is a prominent cause of meningitis in newborn babies and is also responsible for mastitis in cows.

Streptococcus pneumoniae is a common cause of bacterial pneumonia, ear infections, and meningitis. Its most striking characteristic is a large capsule, which is generally necessary for disease production. Also called the "pneumococcus," *S. pneumoniae* generally occurs as oval diplococci, with their adjacent sides flattened and with the long axis of each coccus the same as the long axis of the pair. Colonies of pneumococci grown aerobically on blood agar are surrounded by a zone of *alpha hemolysis,* meaning partial destruction of the red blood cells in the medium and green discoloration of the agar. Unlike other alpha hemolytic streptococci, pneumococci are strongly inhibited by optochin (a chemical relative of quinine), and are lysed by bile. Organisms in colonies of this species also tend to lyse spontaneously, causing collapse of the central portion of the colony (Figure 22-2). There are about 80 different types of *S. pneumoniae,* based on differing capsular antigens.

Pneumococcal pneumonia, pages 485–486

GRAM-POSITIVE RODS

Corynebacteria

Corynebacterium diphtheriae, the cause of diphtheria, is a club-shaped bacillus of variable size with a tendency to stain irregularly (Figure 22-3); its cytoplasm contains granules of volutin which stain metachromatically. The characteristics of shape and staining of *C. diphtheriae* develop best on Loeffler's medium, which is composed largely of solidified serum. Other useful media for identifying this organism contain potassium tellurite, a chemical that inhibits many other bacterial species and which is reduced intracellularly by *C. diphtheriae* to a black compound, giving the colonies of the organism a characteristic grey or black appearance. *Corynebacterium diphtheriae* grows aerobically, on these as well as on many of the usual microbiological media. Growth is generally rapid, although some strains do not develop visible colonies for several days. Three different types of *C. diphtheriae: gravis, mitis,* and *intermedius,* are seen, based on differences in colo-

Diphtheria, pages 479–480

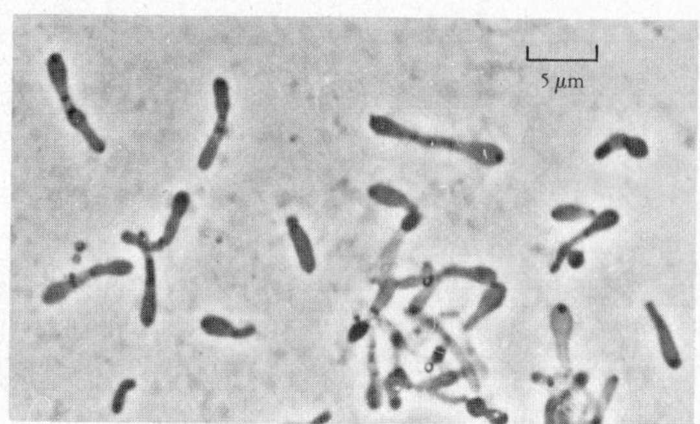

FIGURE 22-3
Phase contrast photomicrograph of *Corynebacterium diphtheriae* after 18 hr incubation on Loeffler's medium. *C. diphtheriae* causes diphtheria. (Courtesy of J. T. Staley and J. P. Dalmasso.)

nial appearance, staining, and biochemical activity. The identification of these types is often useful in tracing epidemic spread of *C. diphtheriae.* Most (but not all) strains of *C. diphtheriae* elaborate a powerful toxin that diffuses from the organism and is responsible for the seriousness of infection with it. It can be demonstrated in many strains of *C. diphtheriae* that production of this toxin depends on the presence of a bacteriophage, an example of *lysogenic conversion. C. diphtheriae* must be differentiated from bacteria of similar appearance among the normal flora, such as *Corynebacterium pseudodiphthericum* (which is biochemically inactive) in the throat, and *Propionibacterium acnes* (which is anaerobic) on the skin.

Bacillus Species

ANTHRAX

Bacillus anthracis is the cause of *anthrax,* a serious disease of sheep, cattle, and sometimes human beings. In this disease a malignant ulcer often forms at the site of infection or severe pneumonia develops if the organisms are inhaled. The serious nature of anthrax is due to toxins produced by the bacillus. *Bacillus anthracis* is a large rod that, as seen in infected material, is characteristically square-ended. Like many other *Bacillus* species, its cells remain loosely adherent following division, producing short chains. *Bacillus anthracis* grows well either aerobically or facultatively anaerobically on many of the usual laboratory media enriched with blood. Stained organisms from cultures on such media reveal long chains, the individual bacteria containing a single spore near the center of their rod-shaped cells. The cells synthesize a unique polypeptide capsule when incubated in carbon dioxide. Unlike most *Bacillus* species, *B. anthracis* is nonmotile. It can also be distinguished from other species of its genus by its susceptibility to a bacteriophage. The spores of *B. anthracis* are resistant to environmental conditions and remain fully infectious for years in soil, wool, hides, and other habitats.

Other *Bacillus* species are commonly present in the natural environment. One, *Bacillus cereus,* has caused serious infections of the eye when introduced by way of contaminated medications. Some strains of *B. cereus* produce an exotoxin responsible for food poisoning.

Clostridia

Gas gangrene, page 555

Clostridium perfringens is the cause of most cases of gas gangrene, some being caused by other species of clostridia. Certain strains of *C. perfringens,* which frequently possess remarkably heat-resistant spores, cause food poisoning, and others produce a severe gangrenous infection of the intestine. *Clostridium perfringens* is an encapsulated nonmotile rod that tolerates low concentrations of oxygen, in contrast to other medically important clostridia, which are strict anaerobes and are motile by peritrichous flagella. *Clostridium perfringens* ferments glucose with the production of a large amount of hydrogen; it also produces an impressive array of extracellular toxins (see Table 19-3). The effects of two of these toxins can be seen in the areas surrounding colonies of *C. perfringens* on blood agar, where a striking double zone of red blood cell destruction is apparent. The first of these zones, a wide zone of partial destruction, is the result of alpha toxin, which is most active at low temperatures. Closer to the colony a zone of complete red cell destruction occurs as the result of theta toxin. Alpha toxin is actually an enzyme, lecithinase, which degrades the lipid lecithin, a component of mammalian cell membranes. Because egg yolk contains abundant lecithin, the effect of alpha toxin can be demonstrated dramatically by inoculating *C. perfringens* into a medium containing egg yolk (Figure 22-4). The organism also actively degrades carbohydrates, releasing abundant amounts of CO_2 and H_2. The normal habitat of *C. perfringens* is the intestine of human beings and other animals, and soil is commonly contaminated with its spores. A number of different strains of *C. perfringens* can be distinguished by differences in their toxin production.

Clostridium botulinum, another organism of the genus *Clostridium,* is the cause of botulism, an often fatal disease characterized by severe paralysis. This organism is strictly anaerobic and is motile. Endospores are formed near the center of its cells. A powerful extracellular toxin produced by *C. botulinum* causes the symptoms of botulism, and identification of *C. botulinum* is based both on biochemical tests and the presence of this toxin. Indeed, six types of *C. botulinum* occur, based on differences in the toxin. In some types it has been demonstrated that lysogenic conversion is involved in toxin production. *C. botulinum* normally inhabits the soil.

Botulism, pages 518–520

Clostridium tetani, a third species of *Clostridium,* causes tetanus, a very severe spastic paralysis caused by a toxin produced by the organism. *Clostridium tetani* shows two striking features not seen in combination in other medically important clostridia. The first of these is the spherical spore that forms at the end of the bacillus, rather than the usual oval spore positioned near the center of the rod, as in the other species of *Clostridium.* The second feature is a swarming growth that may cover the surface of solid media. Final identification of *C. tetani* depends on demonstrating the presences of its toxin, which acts in a unique manner quite distinct from the toxin of *C. botulinum.*

Tetanus, pages 553–555

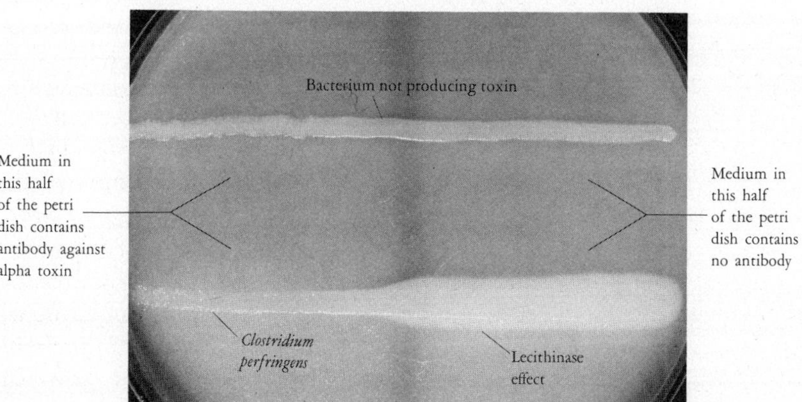
Bacterium not producing toxin
Medium in this half of the petri dish contains antibody against alpha toxin
Medium in this half of the petri dish contains no antibody
Clostridium perfringens
Lecithinase effect

FIGURE 22-4
Effect of clostridial alpha toxin (a lecithinase) on egg yolk incorporated into the growth medium. (Courtesy of M. F. Lampe.)

ACTINOMYCETES

Aerobic Actinomycetes

Branching bacteria, page 260

Nocardia asteroides grows in soil and is capable of producing serious lung or generalized infections, especially in people with debilitating diseases. It grows aerobically on a number of rich laboratory media, producing short filaments that project upward into the air from the colony surfaces. After about 4 days' incubation, the filaments, which are often acid-fast, fragment into rod-shaped bacteria, each of which can then develop into another colony of the organism.

Anaerobic Actinomycetes

Actinomyces infections, pages 555–556

Actinomyces israelii is the cause of actinomycosis, a chronic deep-tissue infection of the face or other parts of the body. This organism is similar to the nocardiae in that fragmentation of its filaments occurs and it forms no spores. However, *A. israelii* is generally nonacid-fast and is anaerobic. The organism normally lives in the mouth, upper respiratory passages, and intestine of human beings. A number of similar species of *Actinomyces,* of lesser disease-causing ability, can be differentiated from *A. israelii* on the basis of their antigens and their products of metabolism.

MYCOBACTERIA

Tuberculosis, pages 489–491

Mycobacterium tuberculosis is the cause of human tuberculosis. It is a slender, beaded-appearing, acid-fast rod. Its resistance to decolorization after staining is greater than that of the nocardiae and many other species of mycobacteria, in that both acid and alcohol fail to remove the stain. The cell walls of *M. tuberculosis* contain an abundance of waxy materials that are responsible for the organism's staining characteristics. *Mycobacterium tuberculosis* is much more resistant than most bacteria to killing by strong acid and alkali, as well as by certain antibiotics and disinfectants. This permits its recovery from mixed populations that include more rapidly growing bacteria. Although the organism has simple growth requirements, it often fails to grow on many ordinary laboratory media because of the presence of inhibitory substances. Even on the best media growth is very slow, the generation time being about 12 hr. Unless the medium on which they are growing contains a wetting agent, the organisms tend to adhere strongly to each other and under moist conditions of growth will form long tangled ropes composed of cells sticking side-to-side with their long axes parallel to that of the rope. Other characteristics that help identify this species include their production of excess niacin (a B-vitamin readily detected in the laboratory by a simple chemical test), their ability to reduce nitrate, and their susceptibility to certain antimicrobial medications. In addition to *M. tuberculosis,* several other species of mycobacteria can occasionally cause human infections.

Mycobacterium leprae, the cause of leprosy, resembles *M. tuberculosis* in being a slender rod, strongly acid-fast, with a tendency toward irregular staining. The acid-fast property of *M. leprae* can be removed by extraction of these organisms with pyridine. *Mycobacterium leprae* apparently grows only intracellularly; it has not been grown on artificial media but can be cultured in nine-banded armadillos and in the footpads of specially prepared mice. Growth is exceedingly slow, with a generation time of 12 days or more in mice.

GRAM-NEGATIVE COCCI

Neisseria gonorrhoeae (the "gonococcus") is the cause of gonorrhea. Like other neisseriae it is a Gram-negative diplococcus with adjacent sides flattened, and the long axes of the cells perpendicular to the long axis of the pair. In material from infected patients, *N. gonorrhoeae* characteristically occur within polymorphonuclear (PMN) body defense cells. They grow aerobically on specialized media such as chocolate agar, and growth requires, or is enhanced by, carbon dioxide. The organisms are oxidase-positive, meaning that their colonies give a positive test for cytochrome C oxidase, and metabolize glucose with acid byproducts. Disease-producing strains of *N. gonorrhoeae* generally possess pili, apparently because these bacteria must attach to specific types of body cells in order to establish infection. Other surface properties of the species may differ from one strain to another.

Neisseria meningitidis (the "meningococcus") is a prominent cause of meningitis and serious bloodstream infections. It is closely related to *N. gonorrhoeae* but is less demanding in its growth requirements. Meningococci are distinguished from gonococci on the basis of their ability to produce acid from maltose as well as glucose, and on antigenic differences. Seven antigenic types of *N. meningitidis* are recognized, based on differing surface polysaccharides, but infections are principally due to types A, B, and C. Strain differences occur within these groups.

GRAM-NEGATIVE RODS THAT GROW UNDER AEROBIC CONDITIONS

Pseudomonads

Pseudomonads, page 252

Pseudomonas aeruginosa is a common cause of hospital-acquired infections. It has a single polar flagellum. Growth occurs readily on a wide variety of laboratory media at 37°C, producing large oxidase-positive colonies within 18 hr. Various strains of this organism can also grow in a variety of aqueous solutions, including laboratory-distilled water and some disinfectants. The metabolism of *P. aeruginosa* is respiratory, but it is able to grow well anaerobically on many of the usual laboratory media because they contain nitrate, which can substitute for oxygen as a final electron acceptor. Most strains of *P. aeruginosa* produce a greenish discoloration of growth media due to production of two pigments, pyocyanin and fluorescein. Several toxins, including one especially potent exotoxin, are produced by some strains of the organism. Surprisingly, the exotoxin of *P. aeruginosa* has the same mode of action as diphtheria toxin. *Pseudomonas aeruginosa* is found in many environmental sources and in the human intestine; it is not highly invasive, primarily infecting persons with impaired defenses. However, treatment is difficult because of the resistance of *P. aeruginosa* to many antimicrobial medications.

Acinetobacters

Acinetobacter calcoaceticus,[1] like *Pseudomonas,* is another organism with respiratory metabolism that causes infections in hospitals. Like the previous species, it is commonly found in the environment and is of low invasiveness. Otherwise it differs markedly from *P. aeruginosa,* occurring as short rods or cocci in pairs, lacking flagella, being oxidase-negative, and failing to produce pyocyanin or reduce nitrate.

Other Gram-negative Rods with Respiratory Metabolism

Tularemia, page 565

Francisella tularensis[2] is the cause of *tularemia.* This organism is a tiny, nonmotile, nonencapsulated rod that often exhibits bipolar staining. It requires special media containing added glucose and cysteine or cystine for growth. It produces acid without gas from several sugars and can be identified by reaction with specific antibody. *Francisella tularensis* infects numerous species of wild animals and is transmitted to man by arthropods or upon handling or eating of infected animals.

Brucellosis, page 566

Brucella abortus causes the disease *brucellosis* ("undulant fever" or "Bang's disease"), a chronic, fever-producing disease. *Brucella abortus* is a tiny, often very short, nonmotile rod that may require medium enriched with serum and glucose as well as an increased carbon dioxide concentration (approximately 5 percent) for growth. Strains being recovered from infected material may take two or three weeks to develop colonies, and even when adapted to growth on laboratory media, appearance of well-developed colonies may take two to four days. *Brucella abortus* is generally acquired by drinking unpasteurized milk from infected cows, in which it causes abortions. Five other species of *Brucella* are recognized and are named according to their principal host animal, including the goat, pig, dog, and others. Differentiation of these species depends on biochemical tests such as oxidase and urease, on the pattern of their susceptibility to two dyes, basic fuchsin and thionin, on their susceptibility to bacteriophages, and on antibody tests for them.

Whooping cough, page 488

Bordetella pertussis, the cause of whooping cough (pertussis), resembles *B. abortus* in size, shape, and lack of motility, but it characteristically has a large capsule. A special medium, called *Bordet-Gengou Medium,* containing potato extract, glycerol, and 20 percent rabbit blood, is generally used for cultivation of the

[1] Also widely known as *Herellea vaginicola.*

[2] Formerly *Pasteurella tularensis.*

species. *Bordetella pertussis* fails to produce acid from sugars and most other biochemical tests for it are negative. Different antigenic types of *B. pertussis* occur and vaccines against pertussis must include the prevalent type of organism if they are to be effective. *Bordetella pertussis* inhabits the human respiratory tract, and most, but not all, of those persons that are infected with it develop pertussis.

Enterobacteria

Enterobacteria, pages 255–257

Escherichia coli is the best known of the enterobacteria and probably the most thoroughly studied of all living organisms. It causes urinary tract infections, diarrhea, and meningitis. This organism grows readily at 37°C on most laboratory media, producing large colonies overnight. Most *E. coli* strains ferment lactose and degrade tryptophan to indole but fail to grow on media containing citrate as the sole carbon source. Generally, when metabolizing sugars such as glucose and lactose, *E. coli* produces not only acid but gas. Gas production depends on an enzyme (hydrogen lyase) which degrades formic acid to carbon dioxide and hydrogen. Within the species *E. coli* there are hundreds of different strains which vary in their O, H, and K antigens, metabolic capabilities, antibiotic susceptibilities, and disease-producing potential. In strains that cause diarrhea, at least three factors may play a role in the condition: toxin production, the ability to attach to the intestine, and the ability to invade intestinal-lining cells. The ability of some strains of *E. coli* to cause meningitis is related to a certain type of K antigen. *Escherichia coli* lives in the intestine of human beings and other animals; those of its strains that are pathogenic for animals are generally harmless to people.

Shigella dysenteriae causes bacillary dysentery. This organism is a close relative of *E. coli,* but differs in being consistently nonmotile and usually failing to ferment lactose. With rare exceptions it does not produce gas from glucose. *Shigella dysenteriae* produces an exotoxin which acts on the intestine and the central nervous system of laboratory animals (and probably human beings as well). Three other species of *Shigella* are recognized: *S. flexneri, S. boydii,* and *S. sonnei.* However, although all four species are capable of invading intestinal-lining cells, only *S. dysenteriae* is known to produce an exotoxin. The normal habitat of the shigellae is the intestine of human beings.

Salmonella enteritidis is another relative of *E. coli* that does not ferment lactose. It is representative of hundreds of species of *Salmonella* that cause *gastroenteritis* (diarrhea, vomiting, and often fever). Most species of the genus are able to utilize citrate as their sole carbon source, and they characteristically produce hydrogen sulfide from sulfur-containing amino acids, a property they share with only a few genera of enterobacteria. Many salmonellae demonstrate the interesting property of *phase variation,* in which each clone of microorganisms contains some organisms with one antigenic type of flagellum, while others in the clone possess a second antigenic type. By means of suitable laboratory techniques the two types of organisms can be separated, but if each is allowed to multiply, some of the progeny of each will always show the flagellar type of the other. Salmonellae inhabit the intestine of an enormous number of warm- and cold-blooded vertebrae species, including chickens and pet turtles. They are classified on the basis of their O and H (and sometimes K) antigens, in conjunction with their patterns of metabolizing different sugars and other substrates. Traditionally, salmonellae of each set of antigenic characteristics (referred to as a "serotype") have been given a species name, and this has resulted in approximately 1000 officially recognized species of this genus.

Typhoid fever, pages 513–514

Salmonella typhi requires special mention, because it represents those salmonella that cause *enteric fever* (see typhoid fever in Chapter 27). *Salmonella typhi* differs biochemically from most other salmonellae in that it lacks hydrogen lyase and therefore does not produce CO_2 and H_2. It also produces little hydrogen sulfide. Also unlike most others in its genus, it has only one phase antigen, and it possesses a capsular (K) antigen designated Vi. Different strains of *S. typhi* can be distinguished by their susceptibility to different bacteriophages.

Salmonella cholerae-suis represents a third group of salmonella that cause *septicemias* (severe blood stream infections). These organisms are differentiated on the basis of their serotype and biochemical tests.

Proteus mirabilis is a cause of urinary and blood stream infections. Like most *Salmonella* species, it produces hydrogen sulfide from sulfur-containing amino acids and does not ferment lactose. Its distinctive characteristics include the ability to degrade phenylalanine to phenyl-pyruvic acid, to hydrolyse urea rapidly, and to break down proteins. It is characterized by a distinctive, objectionable odor and by *swarming.* A swarming proteus may cover the entire surface of the laboratory media onto which it is inoculated; spreading from the site of inoculation occurs in a discontinuous fashion, however, presumably because the genes responsible for swarming are activated by conditions at the site of heavy growth and become inactive when organisms reach fresh medium. *Proteus mirabilis* often lives harmlessly in the intestine of human beings and other animals.

Yersinia pestis[3] causes the dreaded disease, plague. This organism resembles *Shigella* species in many respects, being a nonmotile, lactose nonfermenting organism that ferments glucose without gas production and that fails to hydrolyze urea or utilize citrate as its sole carbon source. It differs from shigella antigenically and in its slow growth and optimum growth temperature of 28°C. Staining of *Y. pestis* is more intense at both poles than in the center of the cell (bipolar staining). The organism produces a powerful toxin active against tissue cells. Other species of *Yersinia* cause abdominal infections.

Klebsiella pneumonia, page 486

Klebsiella, Enterobacter, and *Serratia* are other genera of medically important enterobacteria, causing urinary tract, blood, and lung infections. Most species in this group are able to use citrate as their sole carbon source, and produce 2,3-butanediol from fermentation of glucose. *Klebsiella* species are nonmotile and encapsulated, and *Serratia* species often produce a red pigment. All have been responsible for serious infections acquired in hospitals from contaminated water and aqueous solutions. All three organisms may inhabit both normal human beings and the natural environment.

[3] Formerly *Pasteurella pestis.*

TABLE 22-1
Biotypes of *Vibrio cholerae*

	Classical	*El Tor*
Hemolysis	−	Usually +
Polymyxin susceptibility	+	−
Voges-Proskauer	−	+

\+ = positive test.
− = negative test.

Vibrios

Cholera, pages 574–575

Vibrios, page 257

Vibrio cholerae is the cause of human cholera. Like other vibrios, it is generally a curved rod with polar flagella. It grows poorly or not at all on most selective media used for cultivating pathogenic enterobacteria, but grows readily on media designed for vibrios, producing large oxidase-positive colonies after overnight incubation. Two important characteristics of *V. cholerae* are its preferences for a high salt concentration and its tolerance of severely alkaline conditions (growth can occur at pH 10). A number of different strains of *V. cholerae* can be distinguished on the basis of their metabolic and antigenic features. Most strains causing disease of human beings have the O group I antigen and either the "classical" or El Tor biotype (Table 22-1). Another species, *V. parahaemolyticus,*[4] is characterized by its requirement for sodium chloride. It is widespread in seawater and is a prominent cause of gastrointestinal disease of human beings. Recently *V. parahaemolyticus* has been shown to produce an exotoxin active against heart muscle.

Other Facultative Anaerobes

Haemophilus influenzae is a prominent cause of ear and respiratory infections, meningitis, and epiglottitis. The *Haemophilus* species, so named because they require components of blood, either hemin (X-factor) or NAD (V-factor) or both (Table 22-2), are small nonmotile rods. Hemin is a component of the cytochromes and of the enzyme catalase, while NAD is an oxidizing agent for various metabolic reactions. *Haemophilus influenzae* requires both NAD and hemin. These growth requirements of *Haemophilus* account for the phenomenon of *satellitism,* wherein *Haemophilus* colonies develop in the area immediately sur-

TABLE 22-2
Differentiation of *Haemophilus* species

Species	*X-factor*	*V-factor*
H. influenzae	+	+
H. parainfluenzae	−	+
H. aphrophilus	+	−

\+ = required for growth
− = not required.

[4] Also known as *Beneckea parahaemolyticus.*

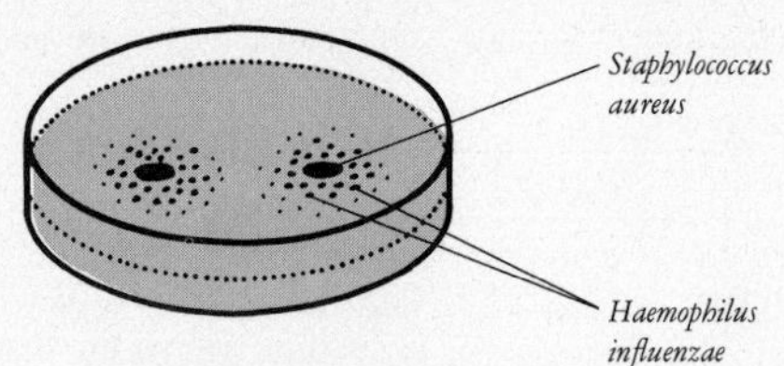

FIGURE 22-5 Colonies of *Haemophilus influenzae* growing as satellites around *Staphylococcus aureus.* The entire surface of the medium has been inoculated with *H. influenzae,* but growth only occurs adjacent to the staphylococcal colonies, which supply both X and V growth factors.

rounding a colony of another bacterium which is producing the required substance (Figure 22-5). *Haemophilus influenzae* is subdivided into several types on the basis of a surface antigen, type b being responsible for most severe illnesses. Some strains of *E. coli* have the same antigen.

Pasteurella multocida, a common infecting agent in animal bites, (Chapter 30), is a small, short nonmotile rod which often exhibits bipolar staining and encapsulation. It is oxidase-positive and ferments several sugars without producing gas. Growth is enhanced by the addition of blood to the medium. The normal habitat of *P. multocida* is the mouth and upper respiratory tract of a wide variety of animals.

RAT BITE FEVER

Streptobacillus moniliformis, a cause of *rat bite fever,* is highly pleomorphic, varying from coccoid forms to very long, irregularly staining filaments. This organism is nonmotile, nonencapsulated, catalase-and oxidase-negative, and ferments glucose. Growth generally requires media supplemented with high concentrations of blood or serum and incubation in an atmosphere of increased carbon dioxide and water vapor. The most remarkable feature of *S. moniliformis* is its capacity to spontaneously develop variants (called *L-phase variants* or *L-forms*) which lack a normal cell wall. These variants are plastic, easily deformed cells that produce characteristic colonies when grown on solid media. These colonies of wall-less *Streptobacillus moniliformis* are smaller than their parent colonies and have a "fried egg" appearance, similar to colonies of mycoplasma (Figure 12-20) with a darker central portion where the bacterial cells penetrate below the agar surface. The wall-less variants of the species are susceptible to osmotic lysis and are also resistant to penicillin, whereas the normal parent cells are susceptible. Continued cultivation of a variant is marked by the appearance of parent-type organisms in the cultures. *Streptobacillus moniliformis* lives in the mouths of rats. L-phase variants of many other species of bacteria, can be produced in the laboratory by treatment of the bacteria with substances active against the cell wall, or by injecting the organisms into a laboratory animal.

ANAEROBIC GRAM-NEGATIVE RODS

Bacteroides fragilis, an organism that causes abscesses and blood stream infections, is nonmotile, and like other bacteroids tends to be variable in shape. It is a strictly anaerobic Gram-negative rod that produces succinic acid as a major end product of metabolism. Growth is rapid at 37°C on a medium containing blood or serum and colonies are well developed in 24 hr. Incorporation of bile in the medium stimulates the growth of *B. fragilis.* Like the majority of medically important Gram-negative rods, this organism is resistant to penicillin. The normal habitat of *B. fragilis* is the large intestine, where it occurs in high concentrations. Several distinct biotypes of *B. fragilis* are known, distinguished by differing patterns of biochemical reactions.

Fusobacterium nucleatum is another nonmotile strict anaerobe, with a characteristically long slender shape and pointed ends. Like other fusobacteria, its major metabolic product is butyric acid. It normally inhabits the mouth and upper respiratory tract. Fusobacteria are usually penicillin-susceptible. Both *B. melaninogenicus* and *F. nucleatum* cause abscesses and blood stream infections, and there is evidence that both may contribute to periodontal disease.

OBLIGATELY INTRACELLULAR BACTERIA

Rickettsiae

The rickettsiae (Figure 22-6) cause a variety of generalized infections or pneumonia (Table 22-3 and Chapter 32). All of these agents are short nonmotile rods, transmitted by arthropod vectors (insects, arachnids). In general, rickettsiae do not survive well in the environment, the exception being *Coxiella burnetii.* The relative resistance of *C. burnetii* to drying and heat allows it to be transmitted by air and food products. Rickettsiae can be cultivated in embryonated eggs or tissue cell cultures. Interestingly enough, some species stimulate an antibody in the host that reacts with certain cell wall antigens of *Proteus* species. The appearance of such antibodies can be used diagnostically.

Chlamydiae

Chlamydiae (Figure 22-7) differ from rickettsiae in being coccoid rather than rod-shaped, and in being independent of arthropod vectors for transmission. Two species are recognized.

Chlamydia trachomatis strains are responsible for *trachoma* (the world's lead-

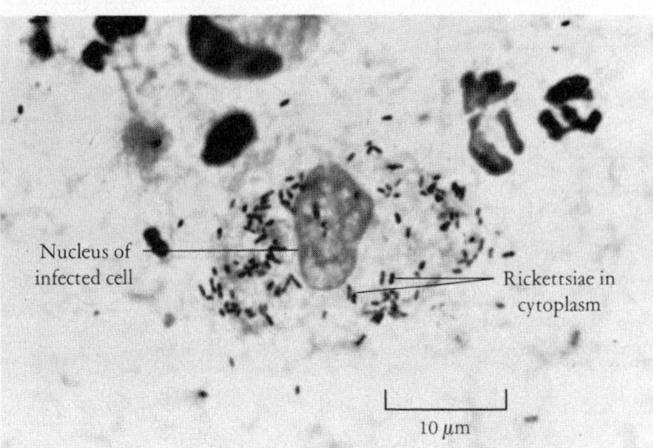

FIGURE 22-6
Rickettsia rickettsii growing within a cell of an infected rodent. (Courtesy of W. Burgdorfer.)

TABLE 22-3
Rickettsial Diseases

Species	*Disease*	*Principal Hosts*	*Vector*	Proteus *Antibodies* *OX-2*	*OX-19*	*OX-K*
R. rickettsii	Rocky Mountain spotted fever	Rodents, dogs.	Tick	+	+	−
R. akari	Rickettsial pox	Mice	Mite	−	−	−
R. prowazekii	Typhus	People	Louse	+ (weak)	+	−
R. tsutsugamushi	Scrub typhus	Rodents	Mite	−	−	+
C. burnetii	Q fever	Rodents, cattle, sheep, goats	Tick (Air)	−	−	−

+ = antibody elicited.
− = antibody not elicited.

ing cause of blindness), *inclusion conjunctivitis* and *urethritis,* pneumonia in infants, and the venereal disease *lymphogranuloma venereum.* This species produces a compact area of growth ("inclusion body") within the host cell. This inclusion contains a carbohydrate matrix and can be recognized after staining with iodine or certain dyes.

Chlamydia psittaci is the cause of a lung infection called *ornithosis* or *psittacosis.* It differs from *C. trachomatis* in failing to develop inclusion bodies, and the genetic relationship of the two species is distant. The individual organisms of *C. psittaci* can be seen by using light microscopy and appropriate staining, and they resemble those of *C. trachomatis. Chlamydia psittaci* infects a variety of mammals and birds, different strains producing a variety of diseases.

SPIRAL-SHAPED BACTERIA

Spiral Bacteria with Polar Flagella

Rat bite fever, page 438

Spirillum minor, another cause of *rat bite fever,* is 5 μm long or less and is a spiral with only two or three coils. It is motile by tufts of flagella at each pole. This organism has not been grown on laboratory media. It inhabits the mouths of rats.

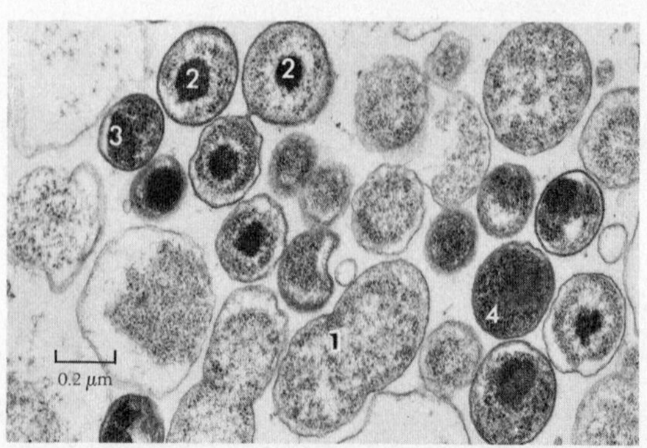

FIGURE 22-7
Chlamydiae growing in a tissue cell culture. The numbers indicate development from dividing form (1) to mature infectious bacterium (4). (Courtesy of R. R. Friis; R. R. Friis, *J. Bacteriol.* 110:706, 1972.)

Campylobacter fetus infects the genital tract of a variety of animals and can cause abortions. Its infections in cattle are venereally transmitted and offer a model for studying chronic genital tract infections. Blood stream infections with *C. fetus* occur rarely in human beings. Many strains of this organism are truly microaerophilic, growing only in atmospheres containing from 3 to 15 percent oxygen.

Spirochetes

Syphilis, pages 530–534

Spiral bacteria, pages 264–267

Treponema pallidum, the cause of syphilis, is a slender spirochete with tightly wound coils and ranges up to 20 μm in length (Figure 22-8). Because this organism is difficult to visualize by staining, darkfield microscopy is generally employed to see it. Typically, the coils of *T. pallidum* are somewhat flattened from side to side, giving the treponema organisms a wavelike appearance. The action of axial filaments is thought to produce the stately rotational and flexing motions of the organism. Although *T. pallidum* has not been cultivated in cell-free media, it can be grown in laboratory rabbits. Normally it is a parasite only of human beings. Species of similar appearance to *T. pallidum* live in the human mouth without causing disease.

Borrelia recurrentis, a cause of *relapsing fever,* is a Gram-negative organism with loosely wound spirals that are thicker than those of *T. pallidum.* Some strains of this organism have been grown anaerobically in complex media enriched with serum or other animal protein. *B. recurrentis* is a parasite of human beings and the louse, *Pediculus humanus.* Relapsing fever in the United States is also caused by other *Borrelia* species that are tick-borne; these have different species names, depending on the species of tick that carries them.

LEPTOSPIROSIS

Leptospira interrogans (so named because its shape resembles a question mark) causes *leptospirosis.* This organism is a slender spirochete with tightly wound coils. Like *T. pallidum,* it is best visualized by darkfield microscopy. However, *L. interrogans* grows well aerobically in liquid media containing serum, and frequently the organism shows bending near its ends, to give the appearance of hooks. A large number of different antigenic types of *L. interrogans* exist, and normally live in the urinary systems of dogs, rats, cattle, and other animals. The

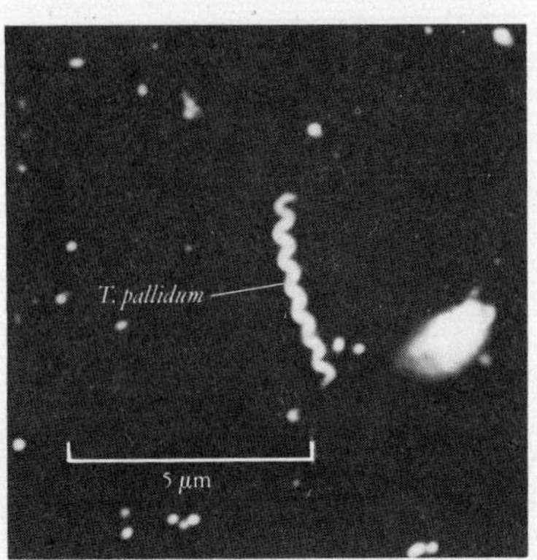

FIGURE 22-8
Darkfield photomicrograph of *Treponema pallidum,* the cause of syphilis. (Courtesy of F. Schoenknecht and P. Perine.)

triple vaccine given routinely to dogs is designed to protect them against leptospirosis (as well as distemper and canine hepatitis). Human leptospirosis is usually a mild self-limiting fever, but severe forms, involving the eye, nervous system, and liver sometimes occur.

MYCOPLASMAS

Mycoplasmas, pages 267–269

Mycoplasma pneumonia, pages 487–488

Mycoplasma pneumoniae is a common (perhaps the most common) cause of bacterial pneumonia. Its plasticity and lack of distinctive staining make direct microscopy of little value in its identification. However, these organisms are easily grown on media specially prepared for this species, producing tiny colonies in about 10 days (see Figure 12-20). The colonies of *M. pneumoniae* are typical of bacteria lacking a cell wall, having a dense central portion that grows into the medium, and often a "fried egg" appearance. The best growth occurs under aerobic conditions, and abundant hydrogen peroxide is produced. Hemolysis of red blood cells occurs. The surface of the organism has an area that reacts with neuraminic acid-containing receptors on cells lining the respiratory tract, causing its firm attachment to these cells (Figure 26-2). Although several other species of mycoplasmas inhabit the human mucous membranes, they appear to be unrelated to *M. pneumoniae,* most requiring microaerophilic or anaerobic growth conditions, or showing other biochemical differences. They have relatively little disease potential. Mycoplasmas are not related to other bacteria, including L-phase variants.

SUMMARY

The bacteria of medical importance include representatives of many different groups, including the Gram-positive cocci, Gram-positive rods, actinomycetes, mycobacteria, Gram-negative cocci, Gram-negative rods, spiral bacteria, and mycoplasmas. Some are normally free-living, whereas others require an animal or human host. At least several species contain tens or hundreds of different strains; sometimes only a few strains within a species are virulent. Some pathogenic bacteria cannot be cultivated on cell-free media, either because they are obligate intracellular parasites or for other reasons that are not understood. Others can be cultivated but require media specially designed to meet their growth requirements.

QUESTIONS

REVIEW

1. What properties of *S. aureus* aid us in isolating it from cultures that contain other bacterial species?
2. What properties of *S. pyogenes* enhance its ability to cause disease?

3. Why is it that some strains of *C. diphtheriae* are capable of causing disease while others are not?
4. What microbiological properties help differentiate the various medically important enterobacteria?
5. Why does *H. influenzae* require a medium containing blood products?

THOUGHT

1. How many different kinds of microbiological media would a hospital laboratory need in order to cultivate all possible disease-causing species of bacteria from an undiagnosed infection?
2. Many different strains are present within each bacterial species; some strains cause disease and others do not. Would it be better for a diagnostic laboratory to look for disease-causing properties (such as capsules or toxins) of isolated organisms or to identify their species?

FURTHER READING

DAVIS, B. D., et al., *Microbiology*. 2d ed. New York: Harper & Row, 1973. Comprehensive medical microbiology text.

LENNETTE, E. H., E. H. SPAULDING, and J. P. TRUANT, eds., *Manual of Clinical Microbiology*. 2d ed. Washington, D.C.: American Society for Microbiology, 1974. Authoritative work on medically important microbes and their identification.

ROUECHE, B., "A Man Named Hoffman." *Annals of Epidemiology*. Boston: Little, Brown, 1967. A popular story that concerns anthrax.

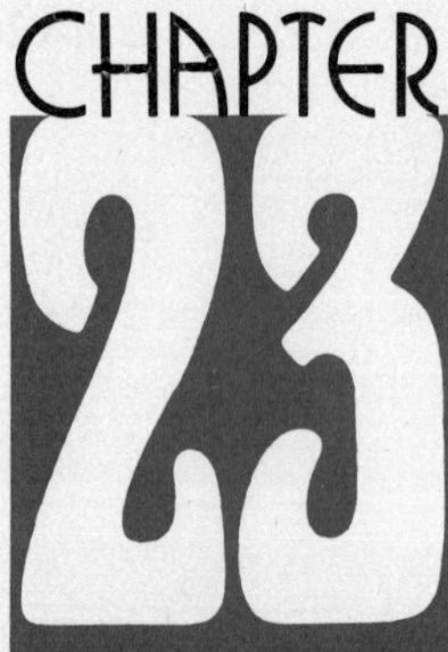

CHAPTER 23

EPIDEMIOLOGY OF INFECTIOUS DISEASES

The field of science that deals with factors determining the frequency and distribution of disease is called *epidemiology*. As the population of the world increases, the importance of the epidemiology of microbial infections is heightened. This arises not only as a result of the increasing density of human populations, which facilitates the spread of infectious diseases, but also because of mass production and distribution practices. A single food supplier, for example, may send his products not only to a community but to several states or nations. If the food is contaminated with a pathogenic microbe, the resulting illness may thus be widespread. Furthermore, efforts employed in infectious disease control, such as vaccination or chlorination of a water supply, often involve millions of people in a single program. Our species tends to rely increasingly on such measures, as well as on highly organized surveillance of public health, to control infectious diseases. It is well to note, however, that such reliance on advanced technology and social organization is vulnerable to war, natural disasters, and other catastrophies, and that a single irrational act may affect thousands of people.

EPIDEMICS: DETECTION AND GENESIS

The occurrence within a short period of time, in a given area, of an unusually large number of infections caused by a single agent is called an *epidemic*. The following sections elucidate a few factors involved in the origin, spread, and control of epidemic diseases of microbial origin.

Recognition of Epidemics

One may suspect an epidemic if an unusually large number of cases of a given disease appear within a short time of one another. However, many different agents may sometimes produce the same disease and, conversely, a single agent may sometimes produce a wide variety of diseases. For example, more than 80 different infectious agents can produce the common cold, whereas a single agent, poliovirus, can produce illness ranging from the symptoms of a cold to fatal paralysis. Thus, the factors involved in any given epidemic can usually be elucidated only when a sizable number of cases, coming from the same agent, have been segregated. In most instances it is necessary to identify the causative agent with great precision, and even to identify individual strains within a species of the agent. Some agents, however, cause diseases so characteristic that they immediately arouse suspicion of the causative organism. An example is *Vibrio cholerae,* the cause of cholera. Indeed, many years ago, a serious epidemic of cholera in England was solved by noting that almost all persons with the disease reported drinking from a single water source, while almost all those escaping the disease drank water from other sources.

Chlorea,
pages 514–515

Precise Identification of Infectious Agents

In some instances careful identification of a species may be essential to identify the existence of its epidemic spread. In one large hospital for example, it had been the practice to name a group of lactose nonfermenting enterobacteria "paracolons." A number of enterobacterial species fall into this group and occur in cultures from sick people, and it did not seem particularly remarkable that patients occasionally died with such organisms in their bloodstreams. With the introduction of new techniques in the hospital's microbiology laboratory, however, it became noted that paracolons from blood cultures almost always fell within a single species, *Serratia marcescens.* This prompted special investigations that led to the discovery that hospital equipment contaminated with *Serratia* was infecting the patients.

Serratia,
page 436

The value to epidemiology of identification of types within a species of organism is shown by another episode. An unusually large number of *Streptococcus pyogenes* infections had appeared over the course of several months, but because they had been reported from different hospitals, among patients attended by many different medical and other personnel, it was not clear whether an epidemic existed. When the surface antigens of the streptococci were identified by using antisera, it was found that about three-quarters of the patients had been infected with the same type of *S. pyogenes,* whereas the remainder had been infected by miscellaneous types of the organism. Furthermore, all of the patients infected with the first type of streptococcus had been attended by the same physician, who proved to be the disease carrier.

Streptococcal antigens,
page 428

Epidemics due to *Staphylococcus aureus* or *Pseudomonas aeruginosa* are even more difficult to trace than streptococcal epidemics because these two species are so common among the normal body flora and in the environment. Individual strains of staphylococci can be identified by bacteriophage typing (Figure 23-1), and *Pseudomonas* strains can be distinguished using bacteriocin typing (Figure 23-2). Even greater precision can be obtained by using both antibiotic suscepti-

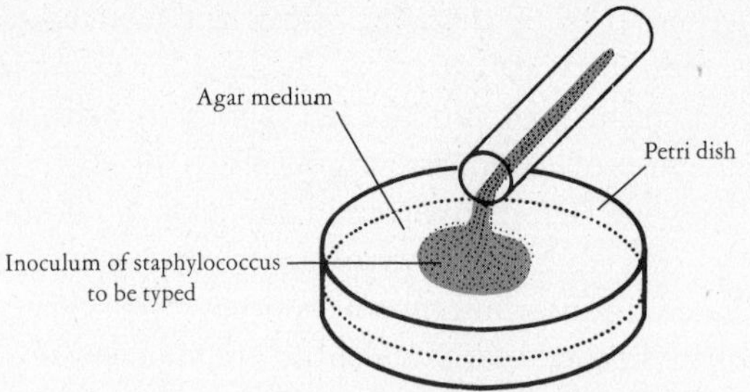

(a) Spreading an inoculum of *S. aureus* over the surface of agar medium.

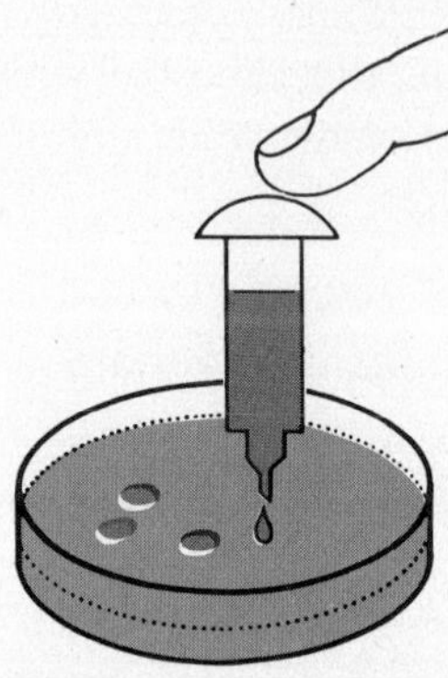

(b) Dropping bacteriophage suspensions on the inoculated surface. Twenty-three different suspensions are thus deposited in a fixed pattern. After incubation, different patterns of lysis are seen with different strains of *S. aureus.*

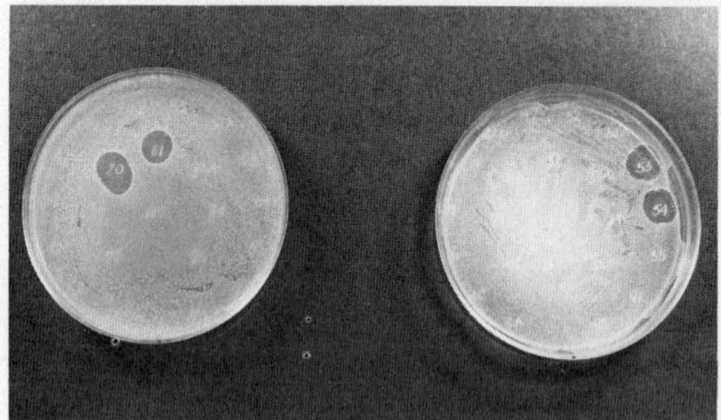

(c) Photograph of two different strains of *S. aureus*, 53/54 and 80/81.

FIGURE 23-1
Bacteriophage typing of *Staphylococcus aureus*.

Testing antibiotic susceptibility, page 604

bility patterns (antibiograms) and bacteriophage typing, as shown in Table 23-1 for *S. aureus.* In this table, the first eight patients shown were very likely to have had the same epidemic strain of staphylococcus, because all the isolates of the organism had the same antibiogram and the same phage type. Patients 9 and 10 and patients 11 and 12 were very likely not to have been part of the epidemic since their isolates differed either in phage type or antibiogram.

Biochemical tests, page 247

Strains of bacteria may also differ in their ability to metabolize various chemicals (see Chapter 11). Identification of strains of an organism by patterns of their biochemical activity against a variety of substrates is called *biotyping*.

The reservoir of any infectious agent is the sum of all of the potential sources of the agent. For example, the reservoirs for diphtheria, syphilis, and typhoid fever consist of other human beings; for botulism and coccidioidomycosis, the soil is the reservoir; and for more than 100 other diseases, animals are the reservoirs. The spread of a disease from its reservoirs may involve vectors (arthropods or other animals), vehicles (food or other nonliving materials), and human beings (through direct person-to-person contact). When agents spread from sources relatively far from humans (either in geographic distance or genetic relatedness), it is sometimes possible to detect their spread before they significantly involve human beings. Encephalitis viruses of wild birds and mosquitoes, for example, may be detected by placing a "sentinel" (such as a chicken) in a cage, and exposing it to the mosquito vectors. The sentinel animals can then be tested periodic-

Coccidioidomycosis, pages 493–495

Encephalitis, pages 575–576

FIGURE 23-2
Bacteriocin typing of *Pseudomonas aeruginosa.*

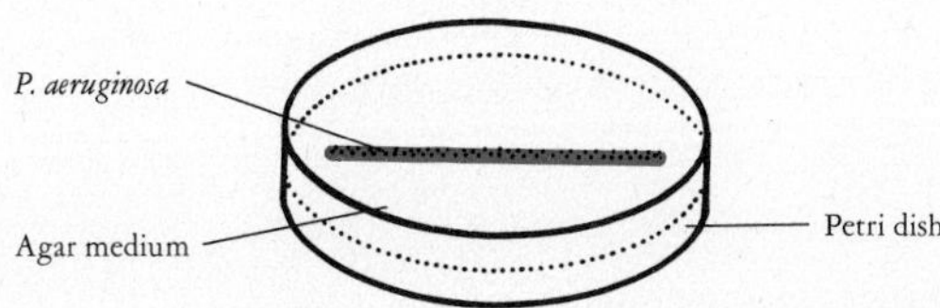

(a) Strain of *P. aeruginosa* inoculated in a straight line and allowed to grow.

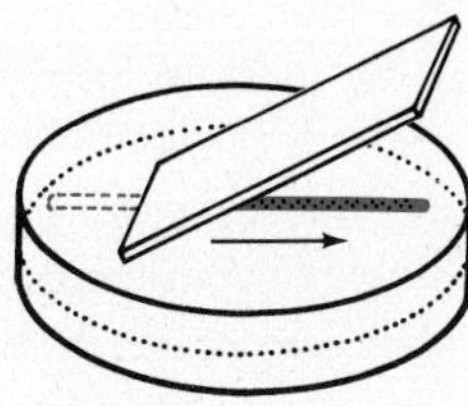

(b) The growth of *P. aeruginosa* is being scraped off the medium. Any remaining bacteria are subsequently killed with chloroform vapor. Known strains of *P. aeruginosa* are then streaked at right angles to the area from which growth has been removed. The known strains are allowed to grow.

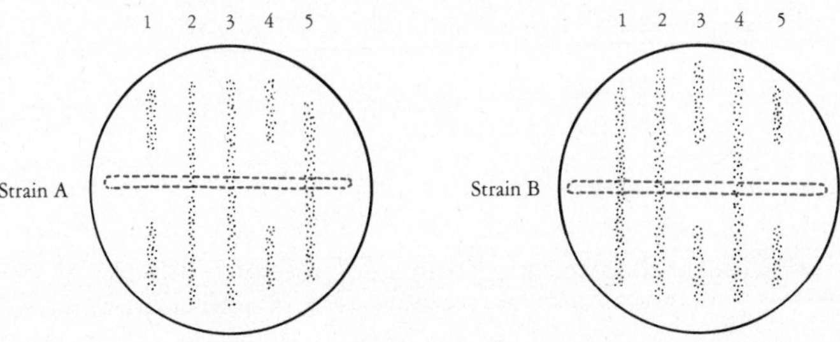

(c) Two different strains of *P. aeruginosa* distinguished by their differing pattern of killing against the same set of five known strains. *P. aeruginosa* strain A kills known cultures 1 and 4; strain B, 3 and 5.

TABLE 23-1
Use of Antibiogram and Bacteriophage Type in Distinguishing Epidemic Strains of Staphylococci

Patient	*Staphylococcus* Antibiogram[a]	Bacteriophage Type
1	RRSRR	80/81
2	RRSRR	80/81
3	RRSRR	80/81
4	RRSRR	80/81
5	RRSRR	80/81
6	RRSRR	80/81
7	RRSRR	80/81
8	RRSRR	80/81
9	RRSRR	54/83
10	RRSRR	54/83
11	RSSSR	80/81
12	RSSSS	80/81

[a] R = resistant; S = sensitive to the antibiotics penicillin, tetracycline, chloramphenicol, erythromycin, and streptomycin, respectively.

ally to see whether the encephalitis arborvirus has infected them. The use of sentinels is especially helpful in such epidemic diseases as encephalitis, because the extensive spread of the disease within the human community can occur without its detection, owing to the high ratio of infected cases that are asymptomatic to those with overt disease. The time it takes to detect the spread of an infectious agent is also influenced by the incubation period of the disease. Diseases with relatively long incubation periods can spread extensively before the first cases of the disease appear. The *dose* of infecting agent received by susceptible individuals is also important in epidemiology because smaller doses generally result in longer incubation periods and higher percentages of asymptomatic infection.

Typhoid fever, pages 513–514

The extent and speed of its epidemic spread are influenced by the *mode of transmission* of a disease. A dramatic example of the importance of this factor was the spread in 1963 of typhoid fever from a ski resort in Switzerland. As many as 10,000 people had been exposed to drinking water containing small numbers of typhoid bacilli. The long incubation period of the disease allowed widespread dissemination of its agent by skiers flying home to various parts of the world. As a result, more than 430 cases of typhoid developed in at least six countries.

Air as a Vehicle of Infectious Agents

Microorganisms are discharged into the air by sneezing, coughing, talking, and singing. Many are enclosed within large droplets of saliva or mucus, most of which fall quickly to the ground within a distance of 3 or 4 ft from the mouth or nose. Large-sized droplets of spray are particularly effective sources of contamination in crowded conditions such as schools and military barracks. Thus, desks or beds in such locations should ideally, be spaced 8 or 10 ft apart to minimize the transfer of infectious agents in this manner. Smaller droplets of mucus or

DROPLET NUCLEI

saliva dry quickly, leaving one or two organisms attached to a thin coat of the dried material. These "droplet nuclei" can remain suspended in the air by minor air currents for indefinite periods, and because of their small size may escape the trapping mechanisms of the upper airways, entering the lungs; but under usual conditions, such droplet nuclei are rapidly diluted by the movements of the air and only highly infectious agents, such as the viruses of measles and smallpox, are readily disseminated in this manner. The unreliability of the dilution factor of air under conditions of inadequate ventilation, however, is demonstrated by a recent hospital epidemic of tuberculosis. In this epidemic an unsuspected case of tuberculosis had been cared for in a large ward for 2½ days; during this time 13 of 44 persons working in the ward became infected with the tubercle bacillus. As is often the case, the air-conditioning unit in the ward had merely recirculated most of the room's air, without filtering it.

Tuberculosis, pages 489–491

MICROBES IN THE AIR OF BUILDINGS

Large buildings present special problems in the airborne transmission of disease agents. Here, respiratory pathogens are not only disseminated as a result of sneezing, coughing, and talking, but persons with skin infections, such as boils, may discharge the pathogens from these lesions as body movement shakes clothing and rubs off skin cells; passage of flatus disseminates intestinal species. The number of organisms in a body of air can be estimated by using machines which propel measured volumes of this air against the surface of a medium contained in petri dishes. This technique has shown that the number of colonies that develop per cubic foot of air sampled rises in proportion to the number of people in a room. Partly for this reason, most modern public buildings have elaborate ventilation systems that constantly change the air. In some hospitals the air flow can be regulated so that the operating room is under a relative positive pressure, preventing contaminated air from flowing into the room; microbiology laboratories may be placed under negative pressure so that the escape of microbes to other parts of the building cannot occur. Special contamination problems may arise from the pumping action of elevators and the chimneylike effect of laundry chutes extending from the warm air (and dirty laundry) areas in the basement of a building. Air-conditioning systems themselves may be sources of infectious agents, since the locations of their air intakes may be such that they draw in contaminated dust. In addition, molds, actinomycetes, and other organisms may grow in humidifying devices within the ductwork of an air conditioning system and be distributed throughout a building. Some people have expressed concern that during wartime, saboteurs could infect large numbers of people in this way. Consequently, military authorities have suggested that, in the event of threatened bacteriological warfare, vaccines be sprayed into the air intake systems of public buildings.

The survival of organisms in air varies greatly with the species of organism and with air conditions such as relative humidity, temperature, and exposure to light. Subdued light promotes survival of microbes, but the effects of temperature and humidity vary with the species. Organisms in still air and those in large particles eventually fall to the floors of buildings but are readily resuspended in the air on dust particles. Thus, sweeping is not permitted in most hospitals; mopping with a dampened mop is used instead. Although the use of ultraviolet light or filtration can markedly reduce the number of viable organisms in air, these methods are generally expensive and unnecessary. Under usual conditions satisfactory control of airborne infection is achieved by good ventilation and dust control. Filtration and ultraviolet irradiation do, however, have important uses in laboratory work and in some special types of hospital units where the risk of microbial contamination is great.

Water and Food as Vehicles

Water and food have been responsible for numerous epidemics, frequently involving thousands of people. Although agents often infect very large numbers of people, the percentage of those exposed who actually do develop disease is generally low. Moreover, the incubation period of waterborne epidemic diseases is apt to be long. These features of waterborne epidemics generally result from the enormous dilution by oceans, lakes and streams of the infecting agent. Attack rates (the number of cases developing per hundred people exposed) of foodborne infections tend to be higher than those from water because of the lack of dilution, and food may also be nutrient for the infectious agent, allowing its multiplication. A sizable increase in numbers of the agent may therefore occur before the food is eaten by the recipient. Feces, directly or indirectly, is the most common origin of water contamination. Foods can be infected at their source (pork with *Trichinella,* eggs with *Salmonella,* milk with *Brucella*), or during their preparation, as by the ameba *Entamoeba histolytica,* or by bacteria such as *Salmonella* species and *Staphylococcus aureus.*

Trichinella, page 581

Entamoeba, pages 516–517

Person-to-Person Transmission

IMPORTANCE OF HANDS IN TRANSMISSION

Hands are important agents in the transfer of disease-producing organisms, and in modern hospitals and elsewhere they generally exceed the importance of air. Hands readily become contaminated from contact with the nose, with areas of skin infection, or with fecal discharge. Dysentery bacilli readily contaminate the hands through toilet tissue. Hands are thus of prime importance in the transfer of organisms from one person to another, both directly and through such intermediates as foods and fomites (inanimate objects such as doorknobs and towels). For such reasons, scrubbing and washing are of major importance in reducing the numbers of potentially pathogenic microbes on the hands, thus reducing the possibility of transferring an infectious dose of an organism.

Carriers

The human reservoir for a particular agent is often much larger than is suggested by the size of the population that is sick from the agent. In fact the most important transmitters of infectious agents are individuals who themselves have only mild or inapparent infections. In contrast to those who are ill, these people, the unsuspected disseminators of epidemic agents, move freely about their daily activities. In addition to individuals with mild or atypical disease, the human reservoir of an infection includes persons still in the incubation period who have not yet become ill, persons who have recovered from the illness, and transient and chronic carriers colonized without infection—all of whom may disseminate disease agents. With many diseases the infectious agent continues to be present in body secretions during the convalescent period, gradually decreasing in concentration over periods ranging from days to months. Transient carriers of infectious disease are those who are contaminated with an agent for only a brief time, whereas chronic carriers continue to excrete an infectious agent constantly or intermittently for months, years, or even a lifetime. After a disease such as typhoid fever has been controlled in a population, for example, the presence of chronic carriers poses a threat of its recurrence which continues for many years. Under conditions of good sanitation, where disease spread is difficult, deaths from the diseases of advancing age may gradually eliminate chronic carriers.

FACTORS INFLUENCING THE CHARACTER OF EPIDEMICS

Acquired Immunity

Influenza, pages 491–493

The attack rates of an epidemic disease agent are influenced by previous exposure of the population to that agent or to agents antigenically related to it. Influenza, for example, is unlikely to spread in a population in which 90 percent of the people have previously been infected by the same agent and have acquired antibody to it; a strictly human disease of this type requires a continuous source of susceptible hosts. Resistance to epidemic disease may also be acquired by exposure to related microbes, as exposure to cowpox protected the mildmaids of Jenner's time against smallpox. Partly because of this factor, the age distribution of a population influences the attack rate of an infectious disease. A high proportion of old people in a population may mean a better chance of low disease rates with many infectious agents, because older people with active immunity may act as barriers to their spread. A high proportion of young children may provide a better chance of epidemic spread, although some of the maternal antibodies cross the placenta, giving protection to an infant for the first three to six months of its life. This protection of an infant depends on the immune status of its mother and the nature of the infectious agent; antibodies protective against *Bordetella pertussis* (the cause of whooping cough), for example, do not cross the placenta. The persistance of acquired immunity to an infectious disease also varies greatly with the nature of its causative agent, being perhaps lifelong for poliovirus or only lasting two years, as with rhinovirus.

IMPORTANCE OF AGE DISTRIBUTION

Whooping cough, pages 488–489

Poliomyelitis, pages 545–547

Genetic Background

Natural immunity in laboratory animals varies greatly with their genetic background and the same is true of human beings. Differences in the incidence of epidemic disease among different racial groups are often pronounced, but it is frequently difficult to determine the relative importance of genetic, cultural, and environmental factors to these differences.

Environmental and Cultural Factors

Other things being equal, infectious diseases tend to hit economically disadvantaged groups of human beings hardest. Malnutrition, overcrowding, and fatigue may contribute to greater human susceptibility to some diseases because this has also been observed with laboratory animals. Paradoxically, economically deprived people may escape epidemics of some diseases (of poliomyelitis for example), because overcrowding, inadequate sanitary practices, and lack of medical care sometimes promote the constant spread of the infectious agent to the very young, leading to the early development of immunity in the population. Furthermore, in the young, maternal antibody (through placental transfer) and other factors often favor the development of mild or unnoticed infections and lasting immunity.

A wide variety of cultural traits also markedly influence the rates of many diseases. For example, people in some European countries drink relatively little water, thus largely escaping waterborne typhoid fever. Some cultural or ethnic groups savor dishes prepared from raw meat or fish, becoming subject to high attack rates of *Trichinella* and other parasitic worms.

Trichinosis, page 581

Virulence of the Epidemic Strain

Marked variations in the incidence of infectious disease often appear in different epidemics caused by the same microbial species. For instance, the spread of *N. meningitidis* in a population will produce low disease rates at one time, whereas at another it will produce high rates of meningitis and significant mortality. Similarly, one bacteriophage type of *S. aureus* may spread among hospital personnel and patients causing little disease, whereas another bacteriophage type of the organism causes boils, pneumonia, and many infections of surgical wounds. It is likely that such differences in the rate of an infection are sometimes due to differences in the virulence of strains within the species or subspecies of its causative organism. Unfortunately good laboratory tests of their virulence for human beings are available for only a few microbial pathogens, and so the importance of this factor cannot readily be quantified.

Meningitus, pages 540–543

Intensity of Exposure

The probability of developing infection and disease may be low or absent if the environmental concentration of microbial pathogens is low. On the other hand, there is probably no human infection for which immunity is absolute. Heavy exposure to a microbial pathogen may produce serious or even fatal disease in a person who has specific immunity to ordinary doses of the pathogen. *Even immunized persons should therefore take precautions to minimize exposure to epidemic agents.* This principle is especially important to laboratory workers, who may be exposed to unnaturally high inocula as a result of accidents or faulty technique, and to medical workers who attend patients with infectious diseases.

Increasing Host Resistance

Methods which promote good general health appear to result in increased resistance to infection by some agents (such as those that cause tuberculosis) and to less development of severe disease. Host resistance can also be increased by inducing specific immunity against pathogenic agents by using vaccines and toxoids (Appendix IV). In the United States, among children entering kindergarten in 1974, 73 percent had been immunized against poliomyelitis, 85 percent against diphtheria, pertussis and tetanus, and 72 percent against measles. The chief problem with such immunization is to maintain immunity once the possibility of recurrent natural exposure to a pathogen becomes low. Fortunately, where the risk of diseases such as tetanus and diphtheria exists, "booster" injections of toxoid usually produce a prompt anamnestic antibody response, protecting even those who were last immunized a decade earlier. Finally, in some instances, host resistance may be increased temporarily by passive immunization (administration of gamma globulin to prevent infectious hepatitis, for example), or by administration of certain medications. As an example of the latter, persons exposed to influenza can be protected by giving a medicine that interferes with viral penetration of host cells.

Anamnestic response, page 402

Hepatitis, pages 508–510

Reduction of Reservoirs and Vectors

Control of infectious agents can sometimes be achieved most readily by an attack on their reservoir. In the case of *Plasmodium vivax* (a cause of malaria) or the encephalitis viruses, reduction of the mosquito host population reduces both the reservoir and the vector. Searching out and treating patients with malaria also decreases the reservoir. In the case of *Schistosoma* (a parasitic worm), treating patients and poisoning the intermediate host snail with copper sulfate helps decrease the reservoir. Carriers of *Salmonella typhi* can be detected by screening populations for antibodies to the Vi capsular antigen of this organism and by instituting treatment for carriers of the disease. Control of tuberculosis can similarly be achieved through identification of cases by the combined use of X-rays and skin tests, followed by medical treatment. Modern processing of sewage eliminates much of its pathogenic potential, thereby rendering it unlikely as a source of epidemic disease. Purification of water sources by filtration and chlorination, and destruction of foods or other materials containing epidemic agents, is also highly effective.

Malaria, pages 573–575

Schistosomiasis, pages 581–583

Segregation, Isolation, and Quarantine

Separating persons with infectious disease from those susceptible to it interferes with transmission and is helpful in preventing the spread of certain diseases. A singing teacher with a tuberculous infection of his larynx, for example, might soon infect his entire class unless he were segregated from them. Quarantine of susceptible unexposed persons is also helpful in some cases where these persons can be reliably identified, such as, perhaps, a village near an area experiencing plague. For many epidemic situations, however, segregation, isolation, and quarantine are by themselves inadequate for gaining control of the contagion. The failure of these measures comes from the usual difficulty in identifying more than a fraction of the number of infected persons: those with mild or atypical illnesses that go undetected will continue to disseminate the epidemic agent to susceptible people.

Control of Infectious Diseases in Hospitals

As indicated earlier, hospitals pose special problems in infection control. In fact, a hospital can be regarded as a high-population-density community made up largely of uniquely susceptible people, and into which the most virulent microbial pathogens of a region are continually introduced. Furthermore, extensive use of antimicrobial medications in hospitals fosters the selection of those microbial strains that resist such therapy, and members of the hospital staff are likely to become carriers of these strains. Beyond this, modern medical practices often require transgressions of normal host defenses; these transgressions may take the form of surgical wounds, interference with normal microbial flora, and medications which impair the inflammatory response and the immune mechanism. For these reasons many modern hospitals employ a hospital epidemiologist—a specially trained nurse or doctor—to investigate infections acquired in the hospital. The hospital epidemiologist also keeps abreast of regional infectious diseases (such as influenza) which might pose a special threat to hospitalized patients, supervises surveillance of hospital equipment and medications to ensure that they do not become contaminated with microbial pathogens, and helps prevent spread of resistant hospital pathogens (such as *Staphylococcus aureus*) to the families of patients and the community.

HOSPITAL EPIDEMIOLOGIST

The Public Health Network

Specific approaches to infectious disease control such as those mentioned above pale in significance compared to the modern public efforts which put them into practice. Massive programs involving international cooperation are promoted by the World Health Organization, and national institutions, such as the United States Public Health Service, exercise constant surveillance of health. Doctors, nurses, laboratory workers, and teachers participate in reporting infectious diseases to public health authorities. Expert teams are ready on short notice to investigate outbreaks and to assist in instituting control measures. The public health network extends far beyond such organizations, however. Education and maintenance of community awareness involve the communications media and many people. Politicians, teachers, engineers, industrialists, and home economists all participate. The dramatic declines in incidence of the terrifying scourges of former years could not have occurred without the collaboration of many of the units comprising our social structure.

SUMMARY

Epidemiology is the science dealing with factors determining the frequency and distribution of disease. The rapidly increasing density of human populations, the popularity of mass production and distribution methods, and the magnitude of artificial (as opposed to natural) immunization techniques are some of the modern factors influencing disease occurrence. Epidemic spread of infectious agents is not always easy to detect, and precise identification of the causative organism is usually of great importance. The nature of reservoirs of infection and routes of disease spread from them are important to understand. Vectors, air, water, food, and people are the principal modes of infectious disease transmission. Control involves reduction of the reservoir, interfering with transmission and protecting susceptible hosts by increasing their resistance or by separating them from exposure. The extent and distribution of diseases caused by microbes are influenced by many characteristics of the exposed population, including age distribution, previous exposure to the same or related agents, and cultural practices. Effective control of epidemics caused by infectious agents requires participation of many sectors of our social structure.

QUESTIONS

REVIEW

1. What methods can be used to identify individual strains within a species of organism?
2. List the principal reservoirs of microbial pathogens.
3. In what ways does the size of the infecting dose of a microbial pathogen affect the early detection and scope of an epidemic?
4. What are the principal methods for the prevention and control of epidemics?
5. Are infected persons who lack symptoms important in epidemics? If so, how?

THOUGHT

1. Discuss the importance of precise identification of microbial strains in defining an epidemic and tracing its source.
2. Discuss how the characteristics of the host population (density, mobility, age distribution, cultural traits, and other factors) influence the scope of an epidemic.

FURTHER READING

BENENSON, A. S., ed., *Control of Communicable Diseases in Man.* 12th ed. Washington, D.C.: The American Public Health Association, 1975. A small handbook giving each disease, including its reservoir, mode of transmission, incubation period, period of communicability, and methods of control.

GREENBERG, B., "Flies and Disease," *Scientific American* (January 1965).

Infection Control in the Hospital. 3d ed. Chicago: American Hospital Association, 1975.

Includes types of infections encountered, factors involved in their appearance, and details of control measures.

ROUECHE, B. *Annals of Epidemiology.* Boston: Little, Brown 1967. Entertaining and informative popular accounts of the tracing of specific epidemics to their sources.

THOMPSON, M. *The Cry and the Covenant.* New York: New American Library, 1955. Historical novel about Ignaz Semmelweis and his discovery that a streptococcal disease was spread by physicians.

CHAPTER 24

SKIN INFECTIONS

The major part of the body's confrontation with the outside world occurs at the surface of the skin. As long as it is physically and functionally intact, this tough, flexible, and by no means inert outer covering is remarkably resistant to infection. On the other hand, because of its exposed state, it is frequently subject to mechanical, thermal, and chemical injury. As indicated in Chapter 21 substances in the environment can stimulate hypersensitivity reactions (allergies), many of which involve the skin. Infections of the skin and subcutaneous tissues most frequently arise because of such alterations, although a few organisms are capable of directly penetrating apparently unbroken skin. Microbes carried by the bloodstream after entering the body via the respiratory or gastrointestinal system, frequently infect the skin from within.

ANATOMICAL AND PHYSIOLOGICAL CONSIDERATIONS

The surface layer (Figure 24-1) of the skin is composed of flat scalelike material made up of cells containing *keratin,* a tough protein which is also found in hair and nails. The more superficial cells of this layer are dead and continually peel off, to be replaced from a deeper layer consisting of cells more nearly polyhedral in shape. The latter become flattened and die as keratin is formed within them. Quite obviously, from the microbial point of view, the outer layer of the skin is far from having the smooth durable surface it presents to us (Figure 24-2). These two superficial layers of the skin are called the *epidermis.* Supporting the epidermis is a second, deeper, layer of skin cells forming a matrix through which a large number of tiny nerves, blood and lymphatic vessels penetrate. This layer, called

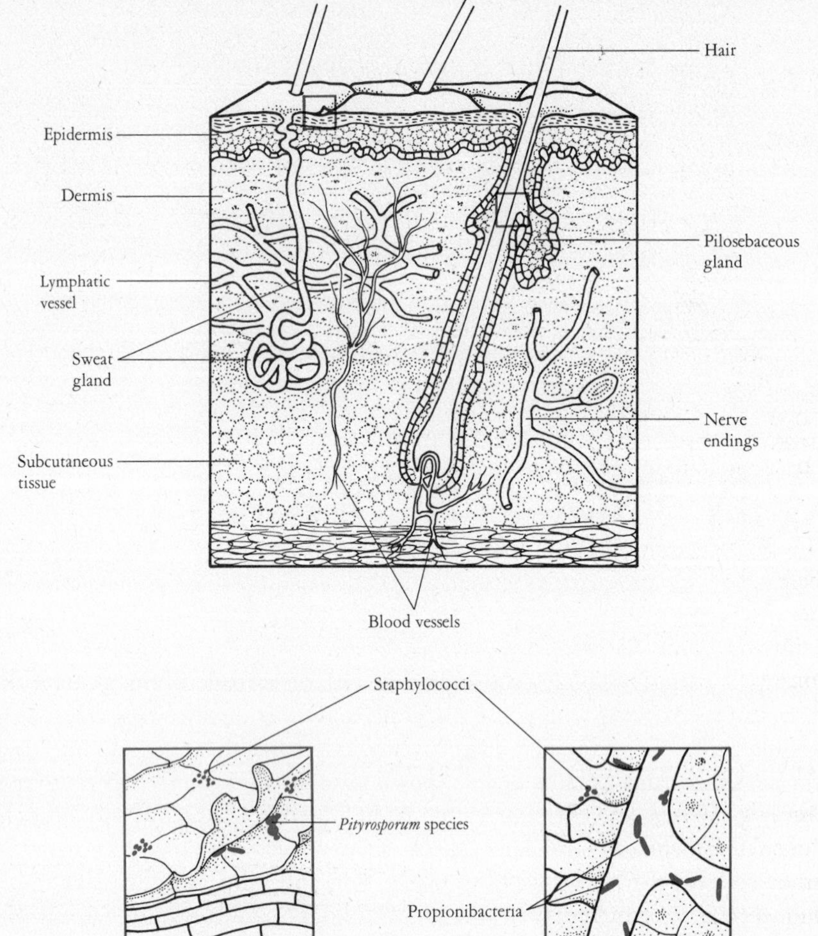

FIGURE 24-1
Microscopic anatomy of the skin.

the *dermis,* blends in a very irregular fashion with the fat and other cells which make up the subcutaneous tissue.

Almost completely traversing the dermis and epidermis are the fine tubules of sweat glands and hair follicles. Both kinds of structure are potential passageways through which microbes can pass below the skin to reach the deeper body tissues. Feeding into the sides of the hair follicles are tiny *pilosebaceous* glands that produce an oily secretion. This secretion normally flows up through the follicle and spreads out over the skin surface.

The secretions of the sweat and sebaceous glands are very important to the microbial population of the skin because they supply water, amino acids, and lipids which serve as nutrients for microbial growth. Breakdown of the lipids by the normal skin's microbial residents results in fatty acid by-products, and these inhibit many potential disease-producers. In fact, the normal skin surface is a rather unfriendly terrain for most pathogens, being too dry, too acid, and too toxic for their survival.

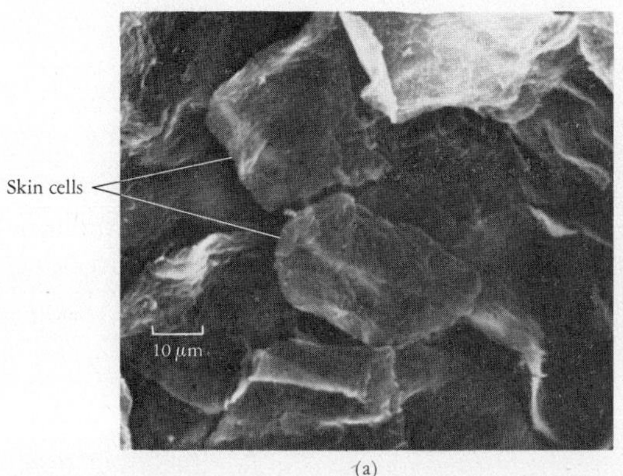

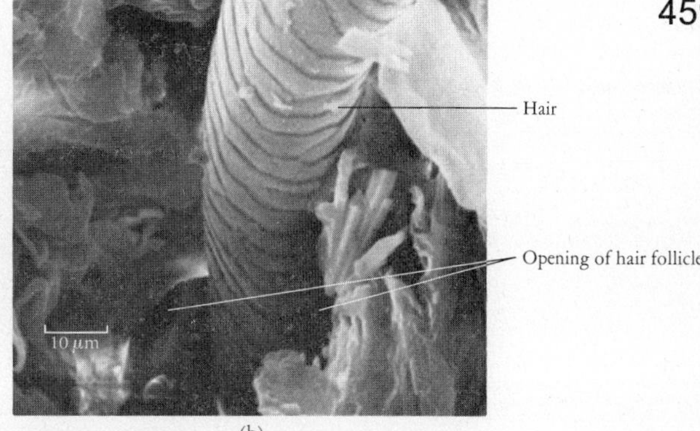

(a) (b)

FIGURE 24-2 Scanning electron micrographs of normal skin. (a) Surface of the epidermis. (Courtesy of C. M. Papa, M.D., Johnson and Johnson Co., Research; C. M. Papa and B. Farber, *Arch. Derm.* 104:262. Copyright © 1971, American Medical Association.) (b) Hair follicle. (Courtesy of C. M. Papa, M.D., Johnson and Johnson Co., Research. SEM Photo copyright © 1970 by C. M. Papa, M.D.)

Factors which derange the normal functioning of the skin tend to nullify these antibacterial properties. For example, an allergic condition that results in the weeping of plasma onto the surface of the skin neutralizes the antimicrobial action of the skin's fatty acids. For reasons not yet clear, several disorders of body hormones also seem to make the skin more prone to pathogenic invasion.

THE MICROBIAL INHABITANTS OF THE NORMAL SKIN

NUMBER OF MICROBES ON THE SKIN

The microorganisms living amongst the various components of the normal human skin (Table 24-1) are startlingly numerous. Indeed, it is very difficult to remove and count the skin's viable microbial inhabitants because they come off in clumps or clusters, each of which may produce only one colony when grown on laboratory medium. In one study a sharp knife was used to scrape off the superficial layer of the skin over the shoulder blade, and more than 200,000 viable organisms were found per square centimeter. Depending on the body location, amount of skin moisture, and other factors, gentle washing with a detergent solution has yielded counts of superficial bacteria ranging from only about a thousand organisms per square centimeter on the back to more than 10 million on the scalp and in the axilla (armpit).

TABLE 24-1
Principal Members of the Normal Skin Flora

Name	*Characteristics*
Diphtheroids	Pleomorphic, nonmotile, Gram-positive rods of the genera *Corynebacterium* and *Propionibacterium*
Micrococci and Staphylococci	Gram-positive cocci arranged in packets or clusters; micrococci are nonfermentative, staphylococci, fermentative
Pityrosporum species	Small yeasts which require oily substances for growth

In recent years there has been renewed interest in studying the skin's flora because of the possible derangements in it which might occur as a result of space flight. Thus, the United States National Aeronautics and Space Administration has supported extensive studies of the effects on the skin of a space diet, high atmospheric oxygen, and confinement in space suits with minimal opportunity for bathing. Under these conditions shifts have been found to occur in the various components of the skin's microbial flora, and these shifts can result in skin disease.

Diphtheroids

Corynebacterium diphtheriae, pages 429–430

The principal organisms inhabiting the normal human skin fall into three main groups. The first group can be conveniently labeled "diphtheroids," or organisms resembling, if only to a limited degree, the diphtheria bacillus *Corynebacterium diphtheriae.* Diphtheroids are small Gram-positive pleomorphic ("many shaped") rods of very low virulence. They are nonmotile, do not form spores, and are distinguished from the diphtheria bacillus by their different morphology, their carbohydrate fermentation patterns, and by their nonproduction of toxin. Rarely, diphtheroids cause abscesses and heart valve infections.

The diphtheroid species found on the skin in largest numbers is *Propionibacterium acnes,*[1] present on virtually all human beings that have been studied. Surprisingly enough *P. acnes* is anaerobic (or sometimes aerotolerant); the primary site of its growth in the skin is not on the skin surface but within the hair follicles. Growth of *P. acnes* is enhanced by the oily secretion of the sebaceous glands, and the organisms are usually present in large numbers only in areas of the skin where these glands are especially well developed—the face, upper chest, and back. These are also the areas of the skin where acne develops, and the frequent association of *P. acnes* with acne undoubtedly inspired their name, even though most people who carry the organisms do not have acne. However, the lipases of *P. acnes* and other organisms can release fatty acids from the accumulated oily secretion of the pilosebaceous glands. These fatty acids inflame the glands and probably contribute to the severity of acne. *Propionibacterium acnes* can be subdivided into several different biotypes.

The more superficial skin areas also contain large numbers of aerobic diphtheroids, representing an even more heterogeneous group than the anaerobic diphtheroids. These aerobic diphtheroids are especially numerous in moist skin areas, such as in the axillae and between the toes; they show a wide variety of shapes and biochemical actions, and a few have been assigned species names. Some, members of the species *Corynebacterium xerosis,* metabolize fatty acids and may contribute to maintaining the normal level of the skin's fatty acids.

The main significance of diphtheroids is that they are commonly present in cultures of clinical material and must be distinguished from pathogens.

[1] Formerly called *Corynebacterium acnes.*

Micrococci and Staphylococci

The second large group of microorganisms universally present on the normal skin are Gram-positive cocci which occur in groups or clusters and resemble the pathogen *Staphylococcus aureus.* They are distinguished from that organism by their differing action on sugars and by their lack of coagulase enzymes. Like the diphtheroids, these Gram-positive organisms are rarely pathogenic and represent a large, heterogeneous group, the taxonomy of which is a matter of continuing debate. On most parts of the body, staphylococci and micrococci comprise the most numerous population of those skin bacteria able to grow in the presence of air.

The chief importance of the skin's ordinary micrococci and staphylococci is probably in preventing colonization of the skin by pathogens and in maintaining the balance among the skin's various flora. At least in some instances the two groups have been shown to produce diffusible antimicrobial substances highly active against *P. acnes* and other Gram-positive bacteria, and this may be a factor in maintaining the ecological balance on the skin. Their relative resistance to salt and drying also gives them an advantage over potential pathogens in the competition for nutrients. Like its diphtheroids, the skin's micrococci and staphylococci commonly appear in cultures from patients in which they have no medical significance.

Pityrosporum

The third group of organisms inhabiting the normal skin are not bacteria but tiny yeasts. They are round, oval, or sometimes short fat rods. A constriction in the middle of these organisms has given rise to the name "bottle bacillus" for them. They grown on media containing fats such as olive oil, and belong to the genus *Pityrosporum.* Two species are generally recognized and are almost universally present on human skin. The first, *Pityrosporum ovale,* is an oval organism, about 4 μm in length, and divides by the formation of a large broad-based bud. A wall, or septum, is then formed between the bud and the parent organism and separation ensues. The second species, *Pityrosporum orbiculare,* is a spherical, thicker-walled organism about 3 μm in diameter, and divides (in a more conventionally yeastlike fashion) by pinching off polar buds. The two species also differ in the kinds of fats which they can use for growth. As with some of the diphtheroids large numbers of both organisms are found in certain skin conditions, and attempts have been underway for years to define what role they may play in the causation of dandruff and *tinea versicolor.* The latter is a common skin disease that causes no symptoms other than a patchy increase in pigment in light-skinned persons, or a decrease in pigment in those with dark-skin, as well as a slight scaliness. *Pityrosporum orbiculare* can be cultured in large numbers from such patches of involved skin. In contrast to those growing on normal skin, however, the *P. orbiculare* organisms growing in masses on diseased skin surfaces put down

TINEA VERSICOLOR

tubelike projections (pseudomycelia) between the skin scales, and these may even enter the layer of living skin cells. With some other yeasts, production of such pseudomycelia is taken to mean parasitic invasion rather than saprophytic growth, and the same may be true with *P. orbiculare.*

SKIN DISEASES CAUSED BY BACTERIA

Furuncles and Carbuncles

Staphylococci, pages 426–427

The familiar pimple is sometimes the result of an infection of the skin. Presumably such infection occurs when a virulent organism, such as *Staphylococcus aureus* (Figure 24-3), is rubbed into the opening of a pilosebaceous gland; the organisms then multiply and penetrate the wall of the neighboring hair follicle. An inflammatory response results, followed by the formation of pus and death of tissue. Pressure increases as the result of the breakdown of host materials, with a resulting increase in osmotic activity. Protrusion of the infection outward through the superficial skin layers follows, with eventual rupture, discharge of pus, and healing of the lesion. With pimples this process occurs relatively close to the skin surface; there is not much swelling or effect on blood vessels and nerves to give pain. If, however, the infection penetrates deeper than the dermis, a boil (furuncle) results, with more dramatic and disturbing results. Deeper skin infections may even spread to the area around adjacent follicles, with pus being pushed by the increasing pressure through the tissues to discharge at multiple points on the skin. This kind of infection is called a carbuncle, and is the most dangerous of all skin infections because of the possibility that it will be carried by the blood vessels to areas such as the heart, brain, or bones. Indeed, every location where infection is established, tissue destruction and pus formation occur, as in a boil; this is why it is more than flirting with danger to squeeze pimples, boils, or carbuncles. Although pimples may or may not be the result of infection, boils and carbuncles are almost always caused by the pyogenic (pus-causing) bacterium *Staphylococcus aureus* (Table 24-2).

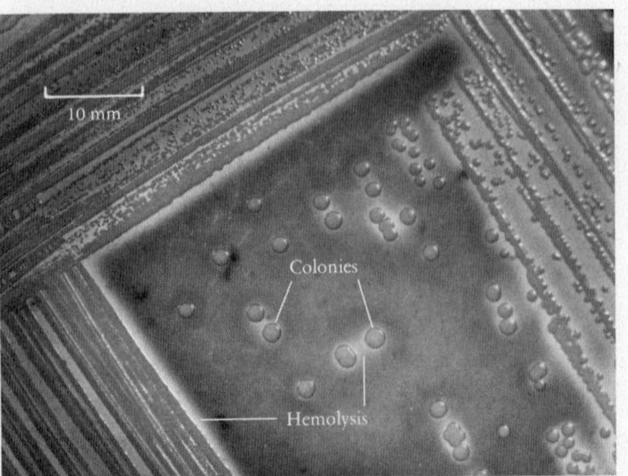

FIGURE 24-3
Staphylococcus aureus colonies growing on sheep blood agar medium. Note the zone of clearing around the colonies as a result of destruction of red blood cells in the medium. (Courtesy of F. Schoenknecht.)

TABLE 24-2
Characteristics of *Staphylococcus aureus*

Identification	*Extracellular Products*	*Pathogenic Potential*
Gram-positive cocci arranged in clusters; cream-colored colonies; produce catalase and coagulase and most ferment mannitol; strain identification using bacteriophages	Hemolysins, leukocidin, hyaluronidase, deoxyribonuclease, staphylokinase, proteinase, lipase, penicillinase, and others	A prominent cause of boils, wound infections, abscesses, impetigo

Impetigo

Impetigo, another bacterial skin disease, is not uncommon in school children, among whom it can spread epidemically. Unlike boils, the infection of impetigo is superficial, involving patches of epidermis just beneath the skin's dead scaly layer. The growth of the causative bacteria typically results in thin-walled blisters which break and are replaced by crusts and by weeping of plasma through the skin. Usually there is no fever or pain accompanying impetigo, although lymph nodes near the involved areas often enlarge, indicating that bacterial products enter the lymphatic system.

Staphylococcus aureus in large numbers is the organism most commonly isolated in pure culture from the blister fluid of persons with impetigo. Most such cases have, furthermore, been caused by staphylococci of bacteriophage type 71, rather than of the more common phage types. This indicates that the organisms that cause impetigo must have some peculiar properties which allow them to infect the intact superficial skin layers. So far, these traits have not been defined, perhaps because their expression occurs only in the human "medium" and not in laboratory media. It is known, however, that the staphylococcal strains causing impetigo produce larger amounts of hyaluronidase than other strains of staphylococci, and they also produce a diffusible bactericidal substance which acts against other Gram-positive organisms, including the impetigo-causing streptococci discussed later. These factors may aid their colonization and spread within the skin, although they are not solely responsible for these organisms' virulence.

In many persons with impetigo, the blister fluid contains the bacterium *Streptococcus pyogenes* rather than *S. aureus.* This organism (Table 24-3) can be recovered in pure culture on laboratory media or, later in the disease, can be recovered mixed with *S. aureus* strains of various bacteriophage types. Although material from streptococcal infections usually shows these organisms as small Gram-positive cocci occurring in pairs, the organisms in culture produce the chains of cocci typical of other streptococci. Furthermore, despite the fact that beta hemolysis is not unique to streptococci, it is very useful for detecting possible colonies of *S. pyogenes* in mixed cultures. In such cases, one need select only those colonies which are beta hemolytic for further identification procedures. The often-used term "beta hemolytic streptococcus" includes many different kinds of streptococci, some requiring anaerobic growth conditions and some with little or

Streptococci, pages 427–429

TABLE 24-3
Characteristics of *Streptococcus pyogenes*

Identification	*Extracellular Products*	*Pathogenic Potential*
Gram-positive cocci in chains or pairs; colonies beta-hemolytic and small: catalase negative; cell wall contains group A polysaccharide; strains distinguished by cell wall proteins (M and T antigens)	Hemolysins (streptolysins) O and S, streptokinase, DNAase, hyaluronidase, and others.	Causes impetigo, pharyngitis, wound infections, puerperal fever; late complications of infection include glomerulonephritis and rheumatic fever

no virulence, as well as the highly pathogenic *S. pyogenes.* The latter organism is distinguished from the others by using antiserum to identify its specific cell wall antigen, the group a polysaccharide.

Like *S. aureus, S. pyogenes* produces a wide variety of extracellular products. For example, in infections with *S. pyogenes,* streptolysin O is released by the organisms and is absorbed by the body. Because streptolysin O is antigenic, antistreptolysin O antibodies often appear in the serum during the course of a *S. pyogenes* infection and can be detected and measured in the laboratory. An increase in the amount of antistreptolysin O in a blood specimen taken 1 to 2 weeks after the infection begins, compared to the amount of antistreptolysin O at the beginning of the infection, is good evidence that the infection was caused by *S. pyogenes.* Antibodies to other extracellular products of *S. pyogenes* are also diagnostically useful.

A number of extracellular products may contribute to the virulence of *S. pyogenes.* These include enzymes which break down protein, nucleic acid, and the hyaluronic acid of host tissues. As with staphylococci it is doubtful whether any one of these factors plays an essential role in streptococcal pathogenicity. However, the pathogenic role of the surface antigens of *S. pyogenes* is quite well defined, at least in experimental infections in laboratory animals. The most important of these substances is the heat and acid resistant *M-protein,* which interferes with phagocytosis of the organism by white blood cells of the host, thus preventing its destruction and permitting it to continue to multiply. M-protein is thus very important to the virulence of *S. pyogenes,* and antibody to it plays a big role in immunity by opsonizing the organism. However, because there are many different strains of *S. pyogenes* and a number of different M-proteins, immunity to one strain does not necessarily provide immunity to another. In some strains, hyaluronic acid (the same substance normally present in body tissue) is produced by the streptococcus itself, in the form of a capsule. This substance also interferes with phagocytosis and results in increased virulence.

GLOMERULONEPHRITIS

Sometimes persons convalescing from infections caused by *S. pyogenes* suddenly develop fever and high blood pressure, and excrete blood and protein in their urine. This potentially fatal sequel to streptococcal infection, *acute glomerulonephritis,* is caused by inflammation of small tufts of tiny blood vessels (glomeruli) within the kidney (nephros). In this condition, soluble streptococcal antigens combine with specific antibodies to form antibody-antigen complexes which interact with complement. These immune complexes are deposited in the kidney, where they result in tissue damage. However, this complication occurs only after infection with certain types of *S. pyogenes* and not with others. Another late complication of streptococcal infection, rheumatic fever, occurs much more commonly as a result of throat rather than skin infections, whereas in recent years the reverse has generally been true of glomerulonephritis. In both of these complications it is unusual to find *S. pyogenes* or other bacteria in the diseased tissues since the pathological changes in the tissues represent a hypersensitivity reaction induced by the streptococci, rather than direct attack by these organisms.

SKIN DISEASES CAUSED BY VIRUSES

Smallpox (Variola)

Smallpox epidemics have been a scourge of mankind for many centuries, but nowhere have they been more severe than on the North American continent following the first arrivals of Europeans. A smallpox epidemic swept the Massachusetts coast during 1617–1619, killing about nine-tenths of the native American inhabitants. A visitor to the friendly Indian chief Massassoit in 1621 found many skeletons lying on the ground because not enough people had remained alive to bury the many dead.

The rash in a typical case of smallpox begins as small red spots, which quickly develop into small, firm pimples, with the pimples finally becoming small blisters as the infected cells break down. Initially these blisters contain clear fluid, but they later become filled with pus. In the earliest stages before pus forms in the blisters, scrapings of the involved skin can be smeared onto glass microscope slides, stained, and inspected with the light microscope. The causative virus particles of smallpox are then often visible in large numbers as small bodies near the limits of resolution of the microscope. In addition, infected skin cells also show commonly cytoplasmic inclusion bodies poorly defined. Electron microscopic examination for the smallpox virus can also be made with scrapings, or even with the pus and crusts of later stages of the disease. This approch has been used in the rapid diagnosis of skin diseases which might possibly be smallpox, although it can only either identify or rule out viruses of the pox group, and not distinguish among them. However, specific identification of smallpox virus antigen can be made by preparing smears of infected material on microscope slides and staining them with fluorescent smallpox antibody.

Early attempts at smallpox control made use of the interesting principle that introduction of an infectious agent into the host via an *abnormal route* may often result in a milder form of the disease. Thus ancient Chinese and Indian people rubbed material from smallpox cases into their nostrils or skin (variolation) in an effort to contract such mild disease. If successful, these people would recover from the resulting illness with a lasting immunity. Unfortunately, the smallpox produced was sometimes severe, or even worse, started an epidemic

VARIOLATION

among the members of a community, as happened as recently as 1970 in an area of Africa where variolation was still occasionally employed.

Edward Jenner's observation that the pox virus of cows could infect human beings, producing only a mild illness but substantial immunity to smallpox, was, of course, a landmark in the control of infectious diseases. This "experiment of nature" demonstrated clearly that an organism of little or no virulence could confer immunity against a related organism of dangerous virulent potential. Following Jenner's work artificial inoculation with cowpox became a widespread practice which was highly effective in controlling the spread of serious smallpox.

VACCINIA

The conquest of smallpox, however, was achieved utilizing vaccinia—a different virus—the origin of which is buried in the obscurity of time. Strangely enough, present-day laboratory studies show the vaccinia virus to be more closely related to the smallpox viruses than to those of cowpox. The use of vaccination eliminated endemic smallpox from Europe and North America during the first half of the twentieth century. However, hundreds of millions of doses of vaccine given in mass immunization programs failed to eliminate the disease elsewhere, and in 1967 thirty countries on three continents still had endemic foci of smallpox. In that year, however, the World Health Organization began an eradication program with the goal of eliminating smallpox from the earth by 1977. Health workers switched from mass immunization programs to a policy of surveillance and containment. Outbreaks were pinpointed by trained searchers from health departments and by local people who were rewarded for reporting cases. Infected and exposed persons were then restricted to their local areas, and all susceptible people were vaccinated. Success was dramatic. India, which had had 49,000 cases in May 1974, had 43 in May 1975 and none thereafter. By 1977 the last existing foci of smallpox in the world were being eliminated in Somalia.

What are the chances that smallpox will reappear? Could infectious smallpox virus, in the hands of a variolator in some remote village, or lyophilized and forgotten in some laboratory, still exist? Could related viruses in monkeys or some other animal produce a mutant with virulence for humans? Could latent infection exist in some human being somewhere? These questions are intriguing, but present knowledge indicates that smallpox will never again be a fearful scourge of our species.

Chicken Pox (Varicella)

A few years ago in a large suburb of Seattle, Washington, a 29-year-old woman complained to her physician of a painful rash which had appeared on one side of her chest. She had had chicken pox years ago while in grammar school, and it had cleared up in the usual manner. The doctor examined the rash on her chest and noticed that it consisted of small red bumps, blisters, and scabs, and that it seemed to follow the branches of one of the nerves of skin sensation. He informed her that she had shingles (herpes zoster), and that it represented a reactivation of the chicken pox infection she had had many years ago. This seemed somewhat unlikely to the patient, but her skepticism lessened appreciably when

HERPES ZOSTER

the first of her four children developed chicken pox two weeks later, followed by the other three in succession, and then their friends, until half of the younger children in the immediate community had become involved.

PATHOGENESIS OF VARIELLA

The varicella virus enters the body when air contaminated with virus particles by a person with chicken pox is inhaled by another person. The virus multiplies in cells lining the second person's respiratory tract, invades the bloodstream, and from there ultimately reaches the cells of the epidermis. The infected skin cells initially enlarge, then break down, leaving a fluid-filled blister beneath the skin. This later becomes filled with pus, ruptures, and forms a crust or scab before finally healing. Most cases of chicken pox are mild and recovery is uneventful. The varicella virus is, however, a threat to babies of mothers who contract chicken pox near the time of birth, and to persons whose immune mechanisms are impaired. In such instances the virus causes extensive damage to internal organs, which often results in death. In older children and adults the virus sometimes causes severe pneumonia.

So far, the development of herpes zoster is not fully understood. However, it is quite clear that in some persons who develop chicken pox, the varicella virus is not entirely eliminated from the body at the time of recovery. Surviving virus is thought to enter the cells of some nerves of sensation. These nerves, responsible for transmission of sensation from the skin, are very complex and interesting structures. The nuclei of the nerve cells in these nerves are located in small bodies called *ganglia,* lying near the spinal column. From its nuclear area the cell extends an extremely long process that runs all the way to the skin, as well as a shorter process that runs to the spinal cord. In the nerve, a number of these extended cell processes are enclosed together in a sheath of other cells, and each individual nerve fiber within the sheath is further enclosed by a layer of specialized cells.

The first change occurring when herpes zoster begins is an inflammatory process in the ganglion of a nerve; this is then followed by the appearance of infectious virus in the skin supplied by the nerve cells originating in the ganglion, as well as by virus in the cerebrospinal fluid in some cases. High levels of antibody appear within a few days, but despite this, skin blisters containing infectious virus are present in herpes zoster for much longer periods than they are in patients with chicken pox, where antibody is slower in appearing.

The latent varicella-zoster virus resides in or adjacent to the nerve cells of a sensory ganglion, but the mechanism by which replication of the virus is kept under control is unknown. Herpes zoster probably results when the varicella virus is actively replicated in the nerve cell nuclei in a ganglion, and is then conveyed to the skin through the nerve cell processes, where it is inaccessible to neutralizing antibody.

The mechanism of persistence of the varicella virus may represent an important adaptive solution to the problem of its survival in small isolated populations. It has been noticed, for example, that when a highly infectious virus such as that causing measles is introduced into such a population, it spreads quickly and infects most of the susceptible individuals, who then either become immune or succumb to the infection. The measles virus then disappears from the community. By contrast, when the chicken pox virus is introduced to a community, it persists indefinitely, creating recurrent epidemics whenever sufficient numbers of susceptible children have been born. Cases of shingles are probably the cause of these recurring epidemics.

Herpes virus, Figure 17-1, page 341

Significantly, the varicella virus is not a true pox virus but a member of the herpes group, which includes the virus of herpes simplex (fever blisters, cold sores). Like other herpes viruses, the virus responsible for chicken pox and herpes zoster has an icosahedral shape with a DNA core and a lipid-containing envelope. If material is scraped from the base of the blisters of chicken pox or shingles, smeared on microscope slides and stained, an important finding can often be made with a light microscope. This finding is that although the individual virus particles cannot themselves be clearly identified, they cause the host to produce large, multinucleated cells. Finding these cells is therefore very useful in identifying infections with herpes viruses. Also sometimes visible under the light microscope are herpes virus-infected host cells with inclusion bodies within their nuclei, the site of viral assembly. Specific antiserum can also be used to identify the viral antigen in scrapings from chicken pox or shingles, and this is most quickly and conveniently used as a fluorescent antibody.

Gamma globulin, page 381

Because chicken pox and shingles are rarely fatal for normal children, and most people have been infected by the varicella virus by the time they reach adulthood, there has been little effort at developing control measures for these diseases. Indeed, chicken pox is so much more severe in adults than in children that it may be wise to promote childhood infection. Very susceptible people (such as patients with leukemia) can be protected by injecting them with gamma globulin from persons who have recovered from shingles.

Measles (Rubeola)

BACTERIAL SUPERINFECTION

Measles (rubeola) develops in a manner similar to smallpox and chicken pox, but its major effects involve the respiratory tract as well as the skin. Serious consequences of measles are quite common and result from secondary infection by bacteria such as *Streptococcus pyogenes, Haemophilus influenzae,* and *Streptococcus pneumoniae.* Measles is now controlled effectively by using a living vaccine derived from measles virus considerably lessened in virulence (attenuated) by prolonged growth in the laboratory. Some cases of measles can be followed years after their occurrence by a very rare disease marked by progressive degeneration of the brain, and lasting months or years. Measles virus antigen can be demonstrated in the brains of such patients, and high levels of measles antibody are present in their blood.

German Measles (Rubella)

German measles (rubella) develops in a manner similar to rubeola, but its effects on both the skin and respiratory tract are mild, and it is less contagious than rubeola. For this reason, many people escape childhood without developing rubella and are therefore susceptible to it as adults; rubella epidemics may thus occur among college students and military recruits. Unfortunately, during the stage when the virus is circulating in an infected pregnant woman, her fetus contracts the infection too. In fact, if the mother contracts rubella during the first eight weeks of her pregnancy, there is at least a 90 percent chance that her fetus will become infected. During this stage of development the tissues of the fetus are easily damaged by the virus, which arrests tissue cell multiplication and produces chromosome abnormalities. Miscarriages often result from such infection, but more commonly the child lives and is born with deafness, a defective heart, poor vision, or mental retardation. These defects, however, do not always follow infection of the fetus. The risk of serious defects among live-borne children has ranged from only 30 to 50 percent when rubella was contracted during the first month of pregnancy, with the risk declining with infections occurring later in pregnancy. Some of the affected children are unable to develop neutralizing antibody to rubella virus and continue to excrete the virus for many months after birth (a possible example of immune tolerance). The virus can also persist for years in tissues inaccessible to antibody, such as the lens of the eye.

BIRTH DEFECTS

Immunologic tolerance, page 385

To reduce the possibility of rubella infection in pregnant women, control of rubella is now underway using a live vaccine to immunize children. This vaccine was derived from the German measles virus through many subcultures of the virus in tissue cell cultures. It produces milder disease than the "wild type" rubella virus. It should not be administered to women who might be pregnant since it is remotely possible that the vaccine virus, like the wild virus, might damage the fetus.

The principal features of the viral skin diseases discussed above are given in Table 24-4.

TABLE 24-4
Viral Diseases Involving the Skin

	Variola (*Smallpox*)	*Varicella* (*Chicken Pox*)	*Rubeola* (*Measles*)	*Rubella* (*German Measles*)
Causative agent	Variola virus, a pox virus	Varicella-zoster virus, a herpes virus	Rubeola virus, a paramyxovirus	Rubella virus, a togavirus
Importance	Very contagious; variola major has a high mortality	After recovery from chicken pox, the infection may become latent: recurs as herpes zoster	Damages the respiratory mucosa; thereby predisposes to serious bacterial infections of respiratory system	Infections that develop during pregnancy commonly cause fetal damage; virus excretion continues after birth for extended periods of time
Control	Vaccination with vaccinia virus; medications given shortly after exposure can prevent the disease	No vaccines available; passive immunization for highly susceptible persons such as leukemics	Vaccine: a live, attenuated rubeola virus	Vaccine: a live, attenuated rubella virus

FUNGOUS DISEASES OF THE SKIN

See Table 14-3, page 298

Earlier in this section we mentioned the possible role of one of the components of the normal skin flora, yeast of the genus *Pityrosporum,* in causing a mild skin disease, tinea versicolor. Other fungi are responsible for more clear-cut and often more serious infections of the skin, although even in these cases there are strong host factors which play a role. The yeast *Candida albicans* is commonly found among the normal flora of the skin, yet in some people it invades the nails, skin, and subcutaneous tissues. In many people with candidal skin infections, no precise cause for the invasion can be determined. Similarly, a variety of mold-type fungi may invade the skin, hair, and nails, producing conditions with such colorful names as athlete's foot, jock itch, and ringworm, but only in a minority of the persons they colonize. Other persons colonized by the same organisms shown no abnormality. These skin-invading molds belong to the genera *Trichophyton, Microsporum,* and *Epidermophyton,* and are collectively called *dermatophytes.* They are peculiar in that they are unable to grow in living tissues but instead multiply in inert structures containing keratin. Most of their unpleasant effects result from secondary bacterial invasion or allergic reactions to the presence of these fungi, but the thickening of nails and the loss of hair they promote give rise to complaints as well. Some of these agents have major reservoirs in soil or on the skins of animals, while others are known to attack only human beings.

DERMATOPHYTES

SUMMARY

The normal skin is scaly and uneven at the microscopic level. Oily secretions and sweat are produced by glandular elements in the skin and support the growth of a rich aerobic and anaerobic normal flora. The most important microbial groups on the skin are diphtheroids (including *Propionibacterium acnes*), micrococci and *Staphylococcus epidermidis,* and yeasts of the genus *Pityrosporum.*

Bacterial pathogens most frequently invade skin damaged by trauma or other factors. *Staphylococcus aureus* is one of the most important of these invaders, producing furuncles, carbuncles, and impetigo. It characteristically produces tissue destruction and abscesses if it escapes localization in the skin and gains access to other parts of the body. *Streptococcus pyogenes* (group A beta hemolytic streptococci) also causes skin infections such as impetigo. Superficial infections by certain strains of this bacterium sometimes lead to glomerulonephritis, an inflammatory condition of the kidneys, developing during healing of the skin disease.

A number of viral diseases involve the skin, although the mode of entry of the viruses into the body is usually respiratory or gastrointestinal. Smallpox (variola) has until recent years been the most dreaded of these diseases because of its severity and extremely infectious nature. Vaccination with the related virus, vaccinia, has now conquered this disease. Chicken pox (varicella) is caused by the varicella-zoster virus, a member of the herpes virus group, to which the herpes simplex (cold sore) virus also belongs. These two viruses can exist in tissues in an inapparent form, only to produce active disease years after their initial infection.

Measles (rubeola) characteristically produces its most severe damage to the respiratory system, resulting in superinfection by bacteria. An unusual degenerative brain disease is now thought to be a late complication of measles virus infection. German measles (rubella) is a mild disease, chiefly of importance because of its ability to infect the fetus and produce birth defects.

A number of fungi infect the skin. *Candida albicans* is one of the most invasive. Others, of the genera *Trichophyton, Microsporum,* and *Epidermophyton,* can attack only keratinized structures such as the outer layers of epidermis, the hair, and the nails. Inflammatory reactions of the skin may result from allergy to fungal products or superinfection by bacteria.

QUESTIONS

REVIEW

1. Give the attributes of skin that make it resistant to infection.
2. Describe the causative microorganism and pathogenesis of carbuncles.
3. What is the relationship between varicella and herpes zoster?
4. Compare and contrast the complications of rubella and rubeola.
5. Name three genera of dermatophytes, and describe how they cause skin disease.

THOUGHT

1. Taking a hot shower often causes a transitory *increase* in the numbers of bacteria on the skin surface. Why?
2. Are there any risks involved in eliminating a virus such as the variola virus from the earth?

FURTHER READING

HENDERSON, D. A., "The Eradication of Smallpox," *Scientific American* (October 1976).

LANGER, W. L., "Immunization against Smallpox before Jenner," *Scientific American* (January 1976).

MARPLES, M. J., "Life on the Human Skin," *Scientific American* (January 1969).

ROGERS, F. B., "'Pox Acres' on Old Cape Cod," *New England Journal of Medicine* 278:21–23 (1968).

ROTHMAN, S., and A. L. LORINCZ, "Defense Mechanisms of the Skin," *Annual Review of Medicine* 14:215–242 (1963).

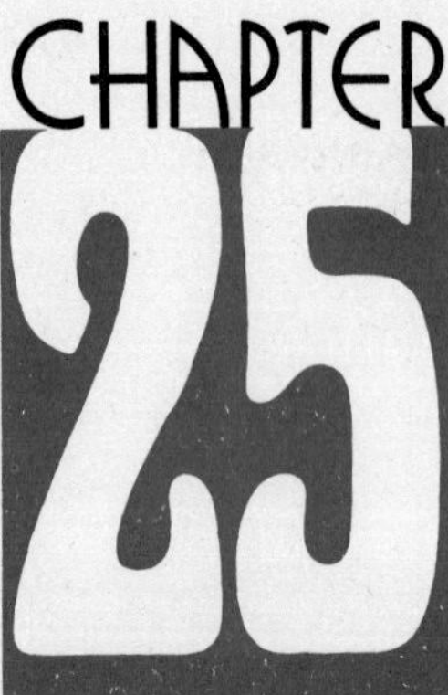

CHAPTER 25

INFECTIONS OF THE UPPER RESPIRATORY SYSTEM

Infectious diseases of the upper respiratory system are by far the most common afflictions of mankind, and far outweigh any other diseases in cumulative misery and loss of productivity. The following sections illustrate the features of microbial interactions with the upper respiratory tract and its appendages.

ANATOMICAL AND PHYSIOLOGICAL CONSIDERATIONS

A person normally breathes about 16 times per minute, taking 500 cc of air with each breath, or over 11,500 liters of air per day. This enormous volume of air, with its accompanying microbes, flows into the respiratory system (Figure 25-1) as a result of the vacuum produced in the lung by chest and abdominal muscles. The air enters the respiratory system at the nostrils (a common site of colonization by *Staphylococcus aureus*), flows into the nasal cavity, and curves downward through the throat. It enters the lower respiratory tract below the epiglottis, a muscular fold of tissue which closes off the lower respiratory tract during swallowing (and is the site of a serious *Haemophilus influenzae* infection of children). The nasal cavity is a chamber above the roof of the mouth, incompletely divided into right and left halves by a vertical wall extending almost back to the throat. Spongy masses of tissue bulge into each half of this chamber from its outside walls. These tissues are similar to the erectile tissues of the genitalia in that they expand and contract with alterations in blood flow controlled by nervous reflexes. With extreme enlargement these spongy tissues contribute to "nasal congestion" or obstruction of the airways. The stimuli for these changes are not entirely well defined, but probably include temperature, humidity, and emotional factors, as well as the irritating effect of infections.

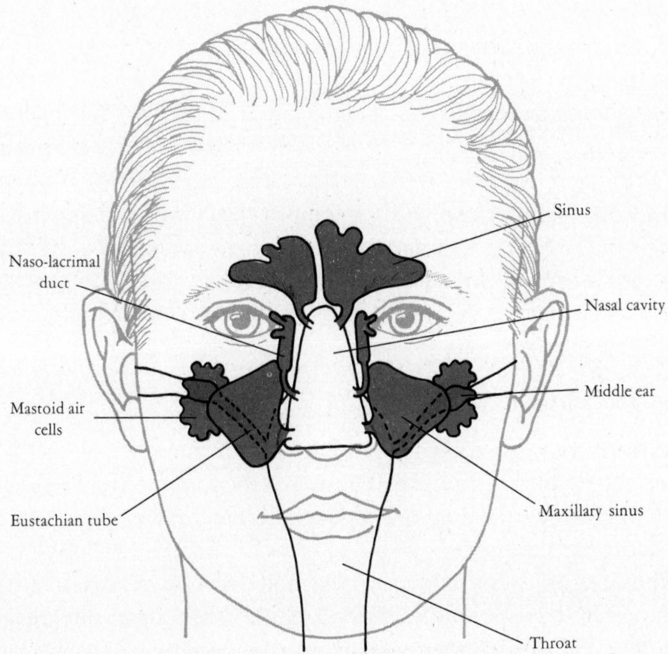

FIGURE 25-1
Upper respiratory system.

Openings of the tear ducts, sinuses, and eustachian tubes from the ears enter the nasal chamber on each side. Thus the conjuctiva of the eye, middle ear, and sinuses of the skull (including the air spaces in the mastoid bones) are connected to the nasal chamber. Collections of lymphoid tissue, the tonsils and adenoids, are located where the mouth and nasal chamber join the throat. Such tissue is important in the production of immunity to infectious agents, but paradoxically it can also be the site of certain infections, and with enlargement can contribute to infections of the ears by interfering with normal drainage through the eustachian tubes. The lining cells of the nasal chamber and its appendages (including the middle ear) have tiny hairlike projections (cilia) along their exposed, free border. These cilia beat synchronously and continually propel a film of mucus (secreted by other cells) into the throat and nose. The rich and varied normal microbial flora of the nasal passage and throat often includes such potential pathogens as *Streptococcus pneumoniae* and *Haemophilus influenzae.* The ciliary action of the lining cells, which continuously moves mucus outward through small passages into the nasal cavity, normally keeps the ears and sinuses free of these and other microorganisms. The pH of the nasal mucus is usually about 6.5; this slight acidity favors the growth of some respiratory pathogens.

MUCO-CILIARY DEFENSE MECHANISM

The primary functions of the upper respiratory tract are to regulate the temperature and humidity of inspired air and to remove from it or destroy undesirable microbes and other foreign material. When cold air enters the upper tract, nervous reflex mechanisms immediately increase the blood flow and decrease it with entry of warm air. Transfer of heat between the blood and the air usually adjusts the temperature to within two or three degrees of the normal body temperature by the time the air reaches the lung. Inspired air is also saturated with water vapor, the nasal passage having the ability to give up more than a quart of water per day to inspired air. Expired air is cooled on passing out of the respiratory system, giving back much of its water vapor to the system's mucous membranes. Trapping of inspired particles by the upper respiratory system is also very efficient. Turbulence produced by nasal hairs causes larger particles to impinge on the system's mucous film and become trapped. Even with small particles, 1 to 5 μm in size, only 50 percent escape trapping by the upper airways. It seems likely that infections of the upper respiratory tract interfere with these primary functions and thereby promote lower respiratory tract infection.

NORMAL FLORA OF THE UPPER RESPIRATORY SYSTEM

Even though the eye is exposed to a multitude of microbes, about half of all normal healthy people yield negative cultures of the conjunctiva. Presumably this sparsity of microbes results from the frequent mechanical washing of the eye with lysozyme-rich tears and from the blinking reflex of the eyelid which clears the eye surface. Thus, organisms impinging on the moist membrane covering the eye are almost immediately swept into the nasopharynx. When organisms are recovered in conjunctival cultures, they are usually scant in number and consist of species found among the ordinary flora of the nasal chamber of skin (Table 25-1). Micrococci, staphylococci, diphtheroids, *Branhamella catarrhalis*[1] and *Haemophilus* species are the groups most commonly encountered.

The secretions of the nasal entrance usually contain large numbers of diphtheroids, micrococci and staphylococci, and smaller numbers of environmental microbes such as *Bacillus* species. About a third of all normal human beings also carry *S. aureus* in this area. The nasal passages have a relatively sparse aerobic flora, the species being those found in the nasal entrance or in the nasopharynx. The nasopharynx, however, contains large numbers of microorganisms, mostly alpha streptococci of the viridans group, gamma streptococci, *Branhamella catarrhalis,* and diphtheroids. Anaerobic Gram-negative bacteria, including the genus *Bacteroides,* are also present in high concentrations in the nasopharynx. In addition, pathogens such as *S. pneumoniae, H. influenzae,* and *N. meningitidis* are commonly found in this area, especially during the cooler seasons of the year. Carriage of *Streptococcus pyogenes, Mycoplasma pneumoniae, Corynebacterium diphtheriae, Bordetella pertussis,* and other respiratory pathogens in the nasopharynx occurs less commonly.

VIRAL INFECTIONS OF THE UPPER RESPIRATORY SYSTEM

The Common Cold (Acute Afebrile Infectious Coryza)

The all too familiar symptoms of the common cold are the result of upper respiratory tract infections by any of approximately 120 known viruses and a few bac-

[1] Formerly *Neisseria catarrhalis.*

TABLE 25-1
Normal Flora of the Upper Respiratory System

Name	*Appearance*	*Comments*
Micrococci and staphylococci	Gram-positive cocci	Commonly includes the pathogen, *S. aureus*
Corynebacterium	Pleomorphic Gram-positive rods; nonmotile; nonsporeforming	Nonpathogens collectively referred to as "diphtheroids"
Branhamella	Gram-negative diplococci	Resembles *Neisseria* species
Haemophilus	Small pleomorphic Gram-negative rods	Commonly includes the pathogen, *H. influenzae*
Bacteroides	Small pleomorphic Gram-negative rods	Strict anaerobes
Streptococcus	Gram-positive cocci in chains	Alpha (especially viridans streptococci), beta, and gamma types; commonly includes the pathogen, *S. pneumoniae*

MULTIPLE CAUSES OF COLDS

teria. Myxoviruses (influenza, parainfluenza, and respiratory syncytial viruses), adenoviruses, and enteroviruses (Coxsackie and ECHO viruses) have all been recovered from persons with colds. However, the vast majority of common colds are caused by rhinoviruses—small acid-sensitive RNA viruses of the picorna group. These viruses were initially very difficult to cultivate in the laboratory since they failed to infect laboratory animals or tissue cell cultures. Workers in England, however, then discovered that if cell cultures of monkey or human origin were incubated at 33°C instead of at body temperature, and at a slightly acid pH instead of at the alkaline pH of body tissues, positive cultures were obtained in many cases. It is noteworthy that the lower temperature and pH of this method are conditions which normally exist in the upper respiratory tract.

PATHOGENESIS OF COLDS

Human beings are the only significant reservoir of those rhinoviruses that produce colds, and close contact with an infected person appears to be necessary to transmit these agents. Once the viruses lodge on the respiratory lining cells, infection of a few of these cells becomes established, virus is replicated intracellularly, is discharged, and infects adjacent cells. The resulting irritation stimulates nervous reflexes which cause an increase in nasal secretions, sneezing, and swelling of the mucosa and nasal erectile tissue causing the airway to become partially or completely obstructed. Later, an inflammatory reaction occurs, with dilation of blood vessels, oozing of plasma, and congregation of white blood cells in the infected area. Secretions from the area may then contain pus and blood. The infection may be limited, probably by interferon release, but it can extend into the ears, sinuses, or lower respiratory tract before it is halted by the development of humoral and cellular immunity. Locally secreted IgA antibody is thought to be important in this immunity; plasma antibodies alone do not give protection.

Interferon, page 343

There is as yet little evidence that the scientific proof of the 1960s has convinced the average person that colds are caused by viruses. The conviction that colds are caused by cold is deeply ingrained, and scientific studies have been repeatedly undertaken to define the relationship between exposure to cold and the common cold. Studies as long ago as the 1920s showed that the incidence of colds increases dramatically in all parts of a region with the onset of cold weather. Another important study, reported in 1933, showed that colds disappeared from the Arctic island of Spitzbergen during the long winter, when no ships came, and reappeared in Spitzbergen when ships arrived in the late spring. This indicated that new sources of infectious virus were necessary to produce colds, regardless of the temperature of a region. Indeed, studies have shown that the incidence of colds was the same when a rhinovirus was administered to nonimmune volunteers exposed to chilling as it was in the case of those not exposed to chilling. Finally, it has been shown that in semitropical areas a sharp increase in colds occurs with the onset of the rainy season, even though the mean temperature stays about the same. The influence of season on the incidence of colds and other respiratory diseases has been clearly demonstrated, but the mechanism for this phenomenon has not. It seems possible that the physiological state of the nasal chamber (including its temperature, pH, air velocity, and mucus flow) is influenced by meteorological conditions, and might be important in aiding the growth and dissemination of virus, in increasing the severity of the symptoms produced by infection, or in decreasing the number of viral particles necessary to produce illness. More important, however, is the fact that people tend to congregate indoors when the weather is bad, increasing the opportunities for infectious viruses to spread.

ROLE OF WEATHER

Specific control measures are not available for colds: the large number of different causative agents of the cold make the development of vaccines impractical, and rhinoviruses, like other viruses, are unresponsive to antibiotics and other medications that control bacterial infections. Generally speaking, a person is most infectious to others during the first day or two of a cold, when symptoms are the worst, and, at least in the case of adults, a person with severe symptoms is much more likely to transmit a cold than someone whose symptoms are mild. For the brief period of one to two days after onset, very high concentrations of virus are found in the nasal secretions and *on the hands* of infected persons, with substantial amounts also in the saliva. The nasal mucosa is exquisitely susceptible to infection, and the disease is contracted by another person when the virus is inhaled, or is unwittingly rubbed into the eyes or nose by contaminated hands. Colds, however, are not highly contagious; less than half of nonimmune persons exposed in a family or dormitory setting contract colds. Washing the hands, even in plain water, readily removes rhinoviruses.

Adenovirus Infections

Children and young adults may develop an illness resembling a cold, but with high fever, very sore throat and severe cough, and swelling of the lymph nodes of the neck. A whitish-gray exudate may spread over the tonsils and throat. Occasionally, pneumonia develops, or infection of the eye (conjunctivitis). Recovery occurs spontaneously in from 1 to 3 weeks.

The cause of this illness in most cases is an adenovirus, a moderate sized

(75-nm diameter) icosahedral virus with a DNA core. More than 30 strains of adenoviruses infect man, and all share a common antigen that can be demonstrated by complement fixation tests. The individual strains of the virus are differentiated by hemagglutination-inhibition or neutralization tests with specific antisera. The adenoviruses can be cultured from respiratory secretions and feces by using animal cell cultures.

Little is known about the pathogenesis of the disease caused by adenoviruses. The source of the infectious agent is human beings; related viruses infect animals, but not man. Once inside the cells of the host, the virus grows in the cell nuclei, producing inclusion bodies. In severe infections extensive cell destruction and inflammation occur. Different types of adenoviruses vary in their predilection for different tissues. The acute respiratory illness described above, for example, is likely to be caused by adenoviruses 4, 7, or 21, whereas type 8 is likely to cause extensive eye infection with few other symptoms. Adenoviruses 1 and 2 produce a mild throat infection in young children and then become latent in the tonsils and other lymphoid tissues where they remain for years. Adenoviral throat infections may resemble infectious mononucleosis and "strep throat." Some human adenoviruses cause tumors when injected into baby hamsters, but are not oncogenic in human beings.

TISSUE SPECIFICITY

Killed vaccines have been successfully used to prevent acute respiratory disease caused by adenoviruses 3, 4, and 7. Live vaccines, administered orally, infect the intestine and produce immunity in adults. Both types of vaccine have been useful in preventing adenovirus-caused epidemics among military recruits. The vaccines are not marketed for civilian use, however, because adenovirus infections are ordinarily not serious enough to require widespread vaccination. Antibiotic treatment is of no value for treating adenovirus infections and sometimes does harm by suppressing some of the body's normal bacterial flora and allowing growth of resistant opportunists from among the normal flora or elsewhere. In addition to cold and adenoviruses, various other agents (including the bacteria *S. pneumoniae* and *Haemophilus* species) may also cause conjunctivitis and infection of the nose and throat.

Vaccines, page 682

BACTERIAL INFECTIONS OF THE UPPER RESPIRATORY SYSTEM

"Strep Throat"

The most important bacterial infection of the throat is caused by *Streptococcus pyogenes,* the group A beta hemolytic streptococcus described previously as being a cause of impetigo. Streptococcal infections of the throat often resemble adenoviral infections very closely, but generally cause greater enlargement and tenderness of the lymph nodes in the neck. Conjunctivitis and pneumonia are not usually present. Although abscess formation and other local complications may prolong the illness, most patients with streptococcal sore throat recover spontaneously after about a week. In fact many infected persons have only the symp-

Streptococci, pages 427–429

toms of a mild cold. Some strains of *S. pyogenes* produce a toxin (erythrogenic toxin), which is absorbed and carried by the bloodstream to the skin, resulting in a red rash. When this happens, the disease is called *scarlet fever.* Production of erythrogenic toxin is under the genetic control of a temperate bacteriophage carried by the bacterium, an example of *lysogenic conversion.*

Streptococcal throat infections are spread both by the respiratory route and by contaminated food. The incidence of *S. pyogenes* as a cause of sore throat varies greatly with age, time of year, and geographic location. Among students with sore throats at a large West Coast university, *S. pyogenes* was isolated from less than five percent; among some groups of military recruits, however, the incidence has been 25 percent.

Glomerulonephritis, page 464

Streptococcal throat infections, like skin infections, may lead to glomerulonephritis as a result of antibody reacting with streptococcal products. In addition, persons with severe untreated *S. pyogenes* throat infections carry about a 2.5 percent risk of developing acute rheumatic fever. Persons with untreated *mild* infections may also develop rheumatic fever, although the risk is very much lower than it is for those with severe infections. Mild infections do result in many cases of rheumatic fever, however, because such infections are relatively common and are therefore much more likely to go untreated. For this reason, persons with fever and sore throat should get a throat culture to rule out *S. pyogenes* before deciding that antibiotic treatment of their condition is unnecessary.

RHEUMATIC FEVER

Rheumatic fever usually occurs about three weeks after the onset of a streptococcal sore throat, and is characterized by inflammatory changes in the joints, heart, skin, and other tissues. Heart failure and death may ensue, but the process usually subsides with the help of medications. Unfortunately, the heart valves may be left permanently damaged by rheumatic fever, and if the afflicted person develops subsequent *S. pyogenes* infections, rheumatic fever often recurs promptly, producing still more damage to the heart.

In contrast to acute glomerulonephritis, which is caused by only a few types of *S. pyogenes,* infection with any M-protein-bearing type of *S. pyogenes* may result in rheumatic fever. The disease is far more common following streptococcal throat infections than it is after infections of the skin or other body sites. Indeed, in many cases, by the time rheumatic fever develops, *S. pyogenes* can no longer be cultured from the throat, and cultures of the blood, heart, and joint tissue are also usually negative. These facts suggest that rheumatic fever, while a consequence of the initial infection, is due to an immunologic response which takes time to develop. The mechanism involved in this process is still a mystery, however; possibly, a reaction between antibody and some streptococcal product is responsible for the ensuing tissue damage. Alternatively, antibody, increasing in response to some streptococcal antigen, may cross-react with the infected individual's tissues. Indeed, some streptococci have been shown to have a cytoplasmic membrane antigen in common with human heart muscle.

Adequate ventilation and avoidance of crowding helps to control the spread of streptococcal infections. But today, persons suspected of having streptococcal sore throats have throat cultures performed, and if the diagnosis is confirmed, they are treated with penicillin. The organisms are eliminated by such treatment in about 90 percent of these cases and the risk of rheumatic fever then becomes infinitesimal. The remaining 10 percent of infected persons may become permanent carriers, but their systems gradually tend to select avirulent mutants lacking M-protein. The reasons for this are unknown, but presumably involve acquired immunity to the M-protein. Persons with rheumatic fever are usually advised to take penicillin for at least five years, and sometimes for life, to prevent reinfection and the high risk of recurrent heart disease.

The incidence of rheumatic fever in the United States has been dropping steadily for many years, this may perhaps be due in part to the widespread practice of administering penicillin for sore throats. Currently, only about 2500 cases of rheumatic fever are reported in the United States each year, as opposed to about 9000 cases per year in 1960.

Diphtheria

Diphtheria usually begins with a mild sore throat and slight fever, accompanied by a disproportionately great amount of fatigue and malaise. Swelling of the neck is often dramatic. A whitish-gray membrane forms on the tonsils, throat, or in the nasal cavity. Heart and kidney failure and paralysis may follow these symptoms.

The cause of diphtheria is *Corynebacterium diphtheriae* (Table 25-2), a nonmotile Gram-positive rod of variable shape, often showing metachromatic staining. Most strains of *C. diphtheriae* produce a very powerful exotoxin which can be identified using specific antiserum. To date, only one type of this exotoxin has been identified. Lytic or lysogenic infection of avirulent strains of *C. diphtheriae* by certain temperate bacteriophages induces toxin production, another example of lysogenic conversion.

Corynebacteria, pages 429–430

Lysogenic conversion, page 332

Human beings, either carriers or persons with active diphtheria, are the source of *C. diphtheriae.* The organisms are inhaled and establish infection in the upper respiratory system. They have very little invasive ability, but the powerful toxin which they elaborate is absorbed by the bloodstream. The gray-white membrane which forms in the throat is made up of host mucous membrane and inflammatory cells which have been killed by the growing diphtherial organisms. This membrane may come loose and obstruct the airways so that the patient may sometimes smother to death. Absorption of the toxin by body cells results in cessation of cellular protein synthesis, this effect occurring at the stage of activated amino acid transfer from tRNA to the growing peptide chain of a protein. Studies using radioactive tracer have shown that in the presence of diphtheria toxin, a fragment of NAD becomes tightly bound to one of the enzymes that carry out such amino acid transfer in protein synthesis. The enzyme is then unable to carry out its vital function in protein synthesis, and injury or death of the host's cells results. The principal pathologic effect of the toxin occurs in the heart, kidneys and nerves; fatalities are common.

PATHOGENESIS OF DIPHTHERIA

Toxoid, prepared by formalin treatment of diphtheria toxin, is used to immunize against diphtheria. Antibodies produced in response to toxoid adminis-

TABLE 25-2
Characteristics of *Corynebacterium diphtheriae*

Identification	*Reservoir*	*Pathogenesis*
Pleomorphic, nonmotile Gram-positive rods that show metachromatic staining; grow on media containing tellurite as grey or black colonies; diphtheria toxin produced, identified by specific antiserum	Human beings with nose, throat, or skin colonization	Growth in the upper respiratory tract results in formation of a membrane which can cause respiratory obstruction; absorbed toxin damages heart, nervous system and kidneys

tration specifically neutralize the diphtheria toxin. Because serious damage in diphtheria results from toxin absorption rather than microbial invasion, control of diphtheria can be effectively accomplished by immunization with toxoid. The well-known childhood "shots" (DPT) consists of diphtheria and tetanus toxoids and pertussis vaccine, all three generally given together. Unfortunately, these immunizations are often neglected, and particularly among socioeconomically disadvantaged groups, serious epidemics of the diseases periodically occur. As an example, 66 cases of diphtheria with three deaths occurred in San Antonio during the first eight months of 1970. Seventy-five percent of these cases occurred in persons who had had no previous immunization. A particularly refractory epidemic, occurring in the Pacific Northwest in 1972 and 1973, was associated with the presence of skin ulcerations in many of its cases. Persons with diphtheritic ulcers represented an important reservoir in this epidemic because their ulcers were chronic, caused few symptoms, and discharged large numbers of *C. diphtheriae.*

Earache (Otitis Media) and Sinus Infections (Sinusitis)

Infections of the middle ear (the space just behind the eardrum) result when infectious agents from the nasal passages and throat spread upward through the eustachian tube. This may occur simply by cell-to-cell spread of an infection, undoubtedly assisted at times by changes in airway pressure forcing infected secretions upward in the tube; most people have experienced such sensations of pressure change while driving down a steep hill or descending in an aircraft. Because of the infection, damage to the ciliated cells occurs, resulting in inflammation and buildup of pressure from fluid and pus collecting behind the eardrum. The throbbing ache of a middle ear infection is produced by pressure on nerves supplying the middle ear. Normal drainage through the eustachian tube is impaired by the poor ciliary action of its mucous membranes and by inflammation, which tends to decrease the effective diameter of the tube.

As one would expect, the organisms causing otitis media are those which infect the upper respiratory system. The majority of bacterial middle ear infections come from *S. pneumoniae,* and a large additional share are caused by

H. influenzae. Streptococcus pyogenes may also cause severe otitis media, but less commonly than the other two species. Occasional cases of otitis media are caused by species of *Mycoplasma.* Cultures for bacteria are negative in about half the cases of otitis media and such infections are presumed to be caused by respiratory viruses.

The bone of the skull behind the ear is honeycombed with small air cells called *mastoid cells;* these connect with the middle ear and can be infected through its connections with the nasal chamber. The respiratory viruses associated with the common cold frequently also involve the sinuses, and bacterial sinus infection with *S. pneumoniae, H. influenzae,* or anaerobes such as *Bacteroides* species (Table 25-3), are not uncommon. Bacterial infections of the ear may result in perforation of the eardrum, and rarely, those of the ear and mastoid spaces may extend into the bone to involve nerves passing through the skull or to produce infection of the brain and its covering membranes (meningitis). Occasionally, bacterial infections of the sinuses spread in a similar fashion.

Haemophilus influenzae, page 437

Scientific studies of the relationship between viral and bacterial upper respiratory diseases are as yet incomplete. However, upper respiratory viral infections often precede superinfections with *S. pneumoniae, H. influenzae, S. pyogenes, N. meningitidis,* and other bacterial respiratory pathogens. Indeed, as mentioned in Chapter 24, one of the greatest hazards of measles (rubeola) is bacterial superinfection of the respiratory system, because the measles virus infects the mucous membranes just as it does the skin. Damage to the membrane caused by the virus is conducive to bacterial invasion.

SUMMARY

The middle ears, mastoids, sinuses, and nasal corners of the eyes all connect with the nasal passages and throat to comprise the upper respiratory system. Inhaled organisms establish infections in the lining cells of the nasal airways, and these infections may spread along common membranes to other parts of the upper respiratory system and to the lung. Rhinoviruses, adenoviruses, and enteroviruses as well as the bacterial agents *Streptococcus pneumoniae, Haemophilus influenzae,* and *Streptococcus pyogenes* are frequent causes of infections in this area. The principal complications of such infections are structural damage to the eardrums or drainage channels, or extension to the bone of the skull and the nervous system. There

TABLE 25-3
Principal Bacterial Causes of Sinus Infections

S. pneumoniae	*H. influenzae*	*Bacteroides species*
Gram-positive diplococci often encapsulated; bile soluble; alpha hemolytic colonies	Gram-negative pleomorphic rods; facultative anaerobes; require X- and V-factors for growth	Gram-negative pleomorphic rods; strict anaerobes

is an increased hazard of bacterial infections after viral infections of the upper respiratory system. Diphtheria is a serious disease that can be completely controlled with adequate immunization. Two late complications of streptococcal throat infections are rheumatic fever, which can cause severe heart damage, and glomerulonephritis, a serious kidney disease.

QUESTIONS

REVIEW

1. What are the primary functions of the nasal passages and how might infection interfere with them?
2. What portions of the upper respiratory system are normally free of microorganisms?
3. What measures can be taken to minimize the chances of catching or transmitting a cold?
4. What is rheumatic fever and how does it relate to infections caused by *Streptococcus pyogenes* (group A beta hemolytic streptococci)?
5. Why is diphtheria such a dangerous disease when *Corynebacterium diphtheriae* rarely invades beyond the respiratory mucosa?

THOUGHT

1. The common cold probably causes a greater loss of human productivity than almost any other infectious disease. Why has so little progress been made in controlling colds?
2. By what mechanisms might a viral respiratory infection pave the way for a bacterial complication?

FURTHER READING

ANDREWES, C. H., "The Viruses of the Common Cold," *Scientific American* (December 1960).

D'ALESSIO, D. J., J. A. PETERSON, C. R. DICK, and E. C. DICK, "Transmission of Experimental Rhinovirus Colds in Volunteer Married Couples," *Journal of Infectious Diseases* 133:28–36 (January 1976).

FREIMER, E. H. and M. MCCARTY, "Rheumatic Fever," *Scientific American* (December 1965).

WOOD, W. B., JR., *From Miasmas to Molecules.* New York: Columbia University Press, 1961.

LOWER RESPIRATORY TRACT INFECTIONS

ANATOMICAL AND PHYSIOLOGICAL CONSIDERATIONS

The lower respiratory system (Figure 26-1) consists of the windpipe (*trachea*) and its various branching divisions and subdivisions (*bronchi* and *bronchioles*), ending in the tiny, thin-walled air sacs (*alveoli*) that make up the lungs. The lungs are surrounded by two membranes, one of which adheres to the lung and the other to the wall of the chest and diaphragm. These membranes (*pleura*) normally slide to and fro against each other as the lung expands and contracts. There is thus a potential space between the two membranes where products of infection can accumulate and compress the lungs.

DEFENSE MECHANISMS

In contrast to their presence in portions of the upper respiratory system, microorganisms are normally absent from the lower tract. As with the upper tract, however, much of the lower respiratory system is lined with ciliated cells and with a film of mucus. This film is constantly swept upward from the bronchioles and bronchi toward the throat at the rate of about an inch per minute under normal conditions, and its function is to trap and remove microorganisms and other foreign material. Extraneous factors such as tobacco smoke and chilling may decrease the ciliary action of the lower tract, and prolonged exposure to irritants leads to loss of cilia and flattening of the lining cells. The cough reflex, which also aids in expelling foreign materials, is activated by irritants and excessive secretions, and can be depressed by external factors, such as alcohol and narcotics. Finally, phagocytic macrophages are numerous in the lung tissues and move readily into the alveoli and airways in response to foreign substances, such as microbes. These protective mechanisms are very efficient, especially against bacterial pathogens, and it is unusual to see bacterial lung infections, except in those persons whose defenses are impaired.

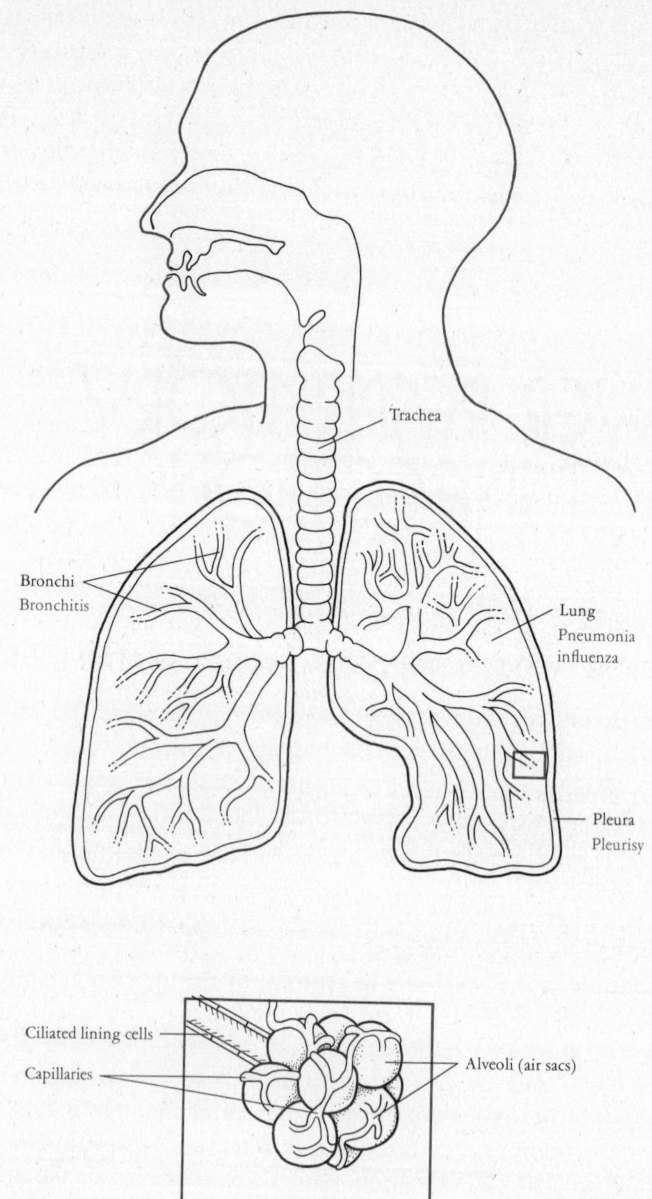

FIGURE 26-1
The lower respiratory system.

PNEUMONIA

Pneumonia results from inflammation of the lung and is usually manifested by fever, cough, chest pain, and production of sputum. Pneumonia coming from microbial infection may result when inhaled pathogenic organisms escape the trapping mechanisms of the upper and lower respiratory system. Perhaps more commonly, such microbes first establish themselves in the upper respiratory system and a large inoculum is then accidentally carried into the lung with a ball of mucus during a deep breath or when the cough reflex is depressed. Thus upper respiratory infection is often a prime factor in the development of lower tract infection. Serious pneumonias in babies and adults also occur when large numbers of organisms of little virulence are inhaled from air conditioning or inhalation equipment containing contaminated water. Most pneumonias are bacteria or viral, but other microbes, chemicals, and allergies may also cause pneumonias.

BACTERIAL PNEUMONIAS

Pneumococcal Pneumonia

One of the most common bacterial causes of lung infection is the Gram-positive, encapsulated diplococcus, *Streptococcus pneumoniae* (pneumococcus).[1] Pneumococci (Table 26-1) are obligate parasites of human beings, and both encapsulated or nonencapsulated variants of these organisms are commonly present among the upper respiratory flora in normal persons. When encapsulated (virulent) pneumococci enter the alveoli of a susceptible host, they multiply rapidly and elicit an inflammatory response. Serum and phagocytic cells pour into the air sacs of the lung, causing difficulty in breathing and production of sputum. This increase in fluid produces abnormal shadows on X-ray films of the chest in patients with pneumonia. The inflammation may involve nerve endings, causing pain, and when this pain comes from the pleura, the result is *pleurisy.* Pneumococci commonly enter the bloodstream from the inflamed lung and occasionally produce *septicemia* (infection of the bloodstream), *endocarditis* (infection of the heart lining), or meningitis. If infected persons fail to develop such complications, then within a week or two sufficient specific antibody usually arises to permit trapping and destruction of the infecting organisms by lung phagocytes, and complete recovery usually results. Most pneumococcal strains have little ability to destroy lung tissue.

Streptococci, pages 427–429

PATHOGENESIS

Treatment of most pneumococcal infections is very effective if given early

TABLE 26-1
Characteristics of *Streptococcus pneumoniae*

Identification	*Normal Habitat*	*Pathogenic Potential*
Gram positive cocci in pairs; colonies umbilicated, show alpha hemolysis; bile soluble	A common inhabitant of the upper respiratory tract of human beings	Encapsulated strains are opportunistic pathogens; cause eye, sinus and ear infections, and pneumonia; infections sometimes spread to heart or brain; immunity results from antibody to the capsule, but there are a large number of capsular types

[1] Formerly known as *Diplococcus pneumoniae*.

enough. Pneumococci are uniformly susceptible to penicillin; erythromycin or tetracycline is also usually effective against these organisms, but resistant strains of pneumococci are occasionally now encountered.

As with several other common bacterial infections renewed interest has been shown in developing preventive measures against pneumonia, especially for high risk patients, such as those suffering from chronic lung diseases or alcoholism. Vaccines which stimulate the production of anticapsular antibodies to the causal organisms can be prepared, giving immunity to pneumococcal disease. More than 80 capsular types of pneumococcus are known, but about three-fourths of all cases of pneumonia are due to one of only 12 types. To be fully effective the vaccines against pneumonia must include pneumococci of each of the capsular types prevalent in a community. In the United States alone an estimated 20,000 deaths each year from pneumococcal disease could be prevented by immunization, and the benefits of immunization would be even greater in countries with less well-developed health facilities.

Klebsiella pneumonia

Enterobacteria, pages 435–436

Among the less common but more serious causes of pneumonia are Gram-negative rods such as enterobacteria of the genus *Klebsiella* (Table 26-2). The pathogenesis of the pneumonia caused by these organisms is similar to that of pneumonia caused by pneumococci, except that permanent damage to the lung usually results and complications are more frequent. Most cases of *Klebsiella* pneumonia occur in persons who drink heavily, for alcohol is known to interfere with lung defense mechanisms at several levels, including ciliary action, cough reflex, and phatocytosis.

R-factors, pages 209–210

Klebsiella organisms persist well in the environment relative to other enterobacteria, but are killed by potent disinfectants. They are susceptible *in vitro* to some antibiotics, although the response of *Klebsiella* infections to therapy is often slow. *Klebsiella* are also prone to develop resistance to antimicrobial medications very quickly because of their high mutation frequencies. They also commonly contain R-factors, which are responsible for drug resistance. Indeed, in some hospitals, *Klebsiella* are now considered to be the principal source of R-factors transferrable to other pathogens. The virulence of *Klebsiella* is due partly to the antiphagocytic property of their capsules, and endotoxin may well play a role in the tissue destruction characteristic of the pneumonia they cause. Laboratory

TABLE 26-2
Characteristics of *Klebsiella pneumoniae*

Identification	*Normal Habitat*	*Pathogenic Potential*
Encapsulated, nonmotile enterobacteria; lactose fermenting; usually positive citrate and Voges-Proskauer tests	Intestinal tract of human beings and other animals; numerous environmental sources	Causes severe pneumonia in alcoholics and persons with debilitating illnesses; can cause urinary and other infections

TABLE 26-3
Characteristics of *Mycoplasma pneumoniae*

Identification	*Normal Habitat*	*Pathogenic Potential*
Grows best under aerobic conditions, with colonies visible in 5 to 10 days; growth inhibited by specific antiserum	Human beings; present in the respiratory tract many weeks following infection	Common cause of pneumonia which is usually mild and self-limiting

diagnosis of this disease is complicated by the fact that *Klebsiella* may be present in the oropharynx of normal people, and may thus contaminate sputum specimens.

Mycoplasma Pneumonia

Mycoplasma pneumoniae, page 442

Pneumonia caused by *Mycoplasma pneumoniae* (Table 26-3), frequently referred to as *primary atypical pneumonia,* may resemble pneumococcal pneumonia very closely. However, patients are usually not so acutely ill as in the latter and serious complications are not so common.

M. pneumoniae attach to specific receptors on cells lining the respiratory tract (Figure 26-2), but the exact mechanism by which they produce pneumonia is unknown. One interesting aspect of the disease is the appearance of serum IgM antibodies called "cold agglutinins" which cause the clumping of human red blood cells at low temperatures. The clumps disaggregate when the suspension is warmed. The reason for the appearance of these antibodies is that *M. pneumoniae* possesses an antigen similar to one of the human red cell antigens, prompting an immunological cross-reaction. Tests for cold agglutinins are useful diagnostically, because approximately 80 percent of hospitalized persons with mycoplasma pneumonia have them. A more specific diagnostic aid for the disease is a complement-fixation test using the patient's serum and mycoplasma antigen. The causative organisms can also be recovered in cultures, but because of slow growth this may take two weeks or more to give a diagnostic result.

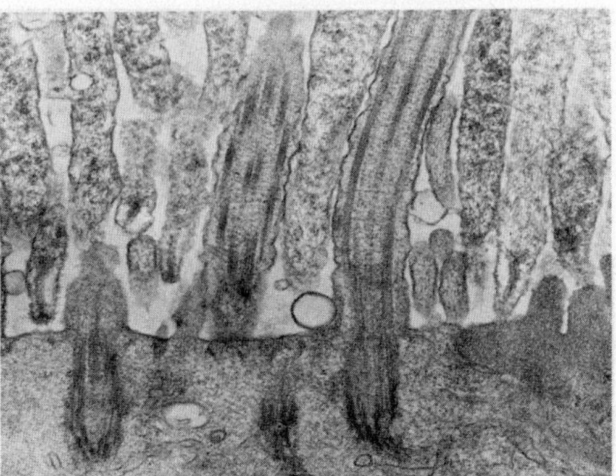

FIGURE 26-2
Electron micrograph showing attachment of *Mycoplasma pneumoniae* to respiratory membrane. Notice the distinctive appearance of the tips of the mycoplasmas adjacent to host membrane. The tips probably represent a site on the microorganism that is specialized for attachment. (Courtesy of J. B. Baseman; from P. C. Hu, A. M. Collier, and J. B. Baseman, *J. Exp. Med.* 145, 1328, 1977.)

TABLE 26-4
Characteristics of *Bordetella pertussis*

Identification	*Normal Habitat*	*Pathogenic Potential*
Small, encapsulated, nonmotile, Gram-negative rod; nonfermentative; requires special medium highly enriched with blood	Respiratory tract of human beings	Serious, long-lasting lung infections, especially in infants; causes violent intractable cough

Action of penicillin, page 599

Not much is known about the survival of *M. pneumoniae* outside the host. *Mycoplasma pneumoniae* is presumed to spread from person to person by droplet infections. The organisms are not susceptible to penicillin, because they lack a cell wall, but are susceptible to several other antibiotics, such as erythromycin and tetracyclines. Many viruses are capable of producing pneumonia resembling mycoplasma pneumonia, but, of course, antibiotic therapy is of no value in viral pneumonia.

OTHER BACTERIAL INFECTIONS OF THE LUNG

Whooping Cough

Whooping cough is primarily a childhood disease and is manifested at onset by nasal congestion and a mild cough. After a week or two, these symptoms are replaced by spasms of violent coughing followed by a loud gasping noise as the patient draws a breath. Vomiting and convulsions frequently occur.

Bordetella pertussis, pages 434–435

Fluorescent antibody, pages 389, 531–532

The causative agent of whooping cough is the bacterium *Bordetella pertussis* (Table 26-4), a small, encapsulated, strictly aerobic Gram-negative rod. Early in the illness it is present in enormous numbers in the respiratory secretions of the host. Specific identification of the organism is made using antiserum; indeed, fluorescent antibody can reliably identify *Bordetella pertussis* in smears of nasopharyngeal secretions.

Infection with *Bordetella pertussis* is confined entirely to the surface of the tracheobronchial system, in which the organisms grow in dense masses. The mucus becomes thick and tenacious and ciliary action is impeded. These factors result in the characteristically violent but relatively ineffective coughing of whooping cough. Some of the organisms disintegrate and release their toxin, causing death of some of the lining cells of the tract and stimulating a rise in the numbers of lymphocytes in the bloodstream of the host. About 10 percent of infants with pertussis die from the disease; the response of pertussis to antibiotic treatment is poor.

Bordetella pertussis do not tolerate drying or sunlight and die quickly outside the host. They are spread entirely by inhalation of droplets from the respiratory system of an infected person. Control is achieved by intensive vaccination of infants with a killed vaccine. The effectiveness of this vaccine is about 70 percent, and its widespread use early in life is probably responsible for the drop from

265,269 reported cases of pertussis in the United States in 1934 to only about 2000 in 1976. However, pertussis continues to be a major problem in poverty areas of the Southeast and the big city slums because of the low percentage of infants who receive vaccine in these areas.

Tuberculosis

Tuberculosis is characteristically a chronic infection of the lungs manifested by fever, weight loss, cough, and sputum production, lasting months or years. In some patients the infection spreads to the bones, meninges, kidneys, or other parts of the body. In rare instances tuberculosis begins as an infection of the skin or gastrointestinal tract, rather than the lung.

Early in the twentieth century the death rate from tuberculosis in the United States was about 200 per 100,000 population, and almost everyone in larger urban areas had contracted the infection by adulthood. Although the initial infection was usually arrested by body defenses and went unnoticed, reactivation during later life was very common. In the United States today new cases of tuberculosis occur at the rate of only about 20 per 100,000 population per year, with most permanently arrested by modern treatment. School children have a very low infection rate and fewer than five percent of young adults have been infected. The reasons for the dramatic decline in tuberculosis are incompletely understood, since the decline began before the availability of specific therapy and modern public health measures. Nevertheless tuberculosis is far from having been conquered, especially among the poor. In the United States there are an estimated 16 million people who have been infected in the past and are subject to possible reactivation of their disease. Overcrowding and malnutrition are important in the reactivation and in the spread of tuberculosis; and worldwide, tuberculosis is still high on the list of leading causes of death from infectious diseases.

Tuberculosis is caused principally by *Mycobacterium tuberculosis* (Table 26-5), the "tubercle bacillus," a slender acid-fast, rod-shaped bacterium. Unfortunately, the simple acid-fast staining procedure does not definitely identify *M. tuberculosis* because other pathogenic and nonpathogenic mycobacteria are common. Non-

Mycobacteria, pages 432–433

TABLE 26-5
Characteristics of *Mycobacterium tuberculosis*

Identification	*Source*	*Pathogenic Potential*	*Control*
Acid fast rods; aerobic; require special growth media; slow-growing; forms cords	Human beings with tuberculosis	Highly infectious; initial infection, often asymptomatic, is confined to small area by host defenses and becomes inactive; infection commonly reactivates later in life.	Relatively resistant to disinfectants; cases detected by tuberculin tests or chest X-rays and treated with antimicrobial drugs; prevention by BCG, a live vaccine

pathogenic mycobacteria are present in the secretions around the human urinary opening, in wax of the ear, and in grain products and other foodstuffs. They are frequently also present in tap water and air, and can thus contaminate stains and other reagents used in microbiological procedures. Moreover, acid-fast bacteria are difficult to remove from glassware, and washing and autoclaving often do not change their shape or acid-fast staining property.

Tubercle bacilli are inhaled in air which has been contaminated by a person with tuberculosis. The microbes lodge in the lung and produce an inflammatory response; they are then ingested by macrophages, where they survive destruction and may multiply, being carried to lymph nodes in the region and often to other parts of the body. As mentioned in Chapter 21, after about two weeks a delayed-type hypersensitivity to the tubercle bacilli develops. A brisk reaction then occurs at the sites where the bacilli have lodged. Macrophages collect around the bacilli and some coalesce to form large multinucleated cells (Langhans giant cells). Lymphocytes then collect around these multinucleated cells and tend to wall them off from surrounding tissue. This kind of inflammatory focus, called a *granuloma,* is the characteristic response of the body to microbes and other foreign substances that resist digestion and removal. The granulomas of tuberculosis are called *tubercles.* The mycobacteria cease to multiply in the tubercles, but a few may remain viable for many years. In areas where there are large collections of tubercles the blood supply may be so poor that the tissue cells die and autolyse. If this process involves a bronchus, the dead material may discharge into the airways, causing a cavity and spreading the organisms to other parts of the lung. Coughing and spitting transmits the organisms to other people as well. Indeed, a persistent lung cavity may remain with prolonged shedding of bacteria into the bronchus.

PATHOGENESIS

Delayed hypersensitivity, pages 411–413

Hypersensitivity to the tubercle bacilli is detected by injecting tuberculin, which is the supernatant fluid from a culture of *M. tuberculosis,* or by injecting a purified fraction of the supernatant (purified protein derivative, PPD) into the skin. Similar preparations are available to test for hypersensitivity to other species of pathogenic *Mycobacterium.* In persons who are sensitive, redness and swelling develop at the injection site, reaching a peak intensity in from 48 to 72 hr. This test is called the tuberculin test. A strongly positive reaction to the test is thought to indicate the presence of living tubercle bacilli somewhere in the body of the person tested. Such a positive test is associated with the possibility that impaired general health or certain treatments that suppress the immune response in the individual in question may result in renewed multiplication, enlargement of tubercles, and actively progressing tubercular disease. This may happen even years after the initial infection.

PREVENTION

In the United States today control of tuberculosis is carried out by identifying infected persons with chest X-rays and skin tests, and by giving antimicrobial medications to those who have active disease, since these persons are now the only important sources of tubercular infection in the population. Milk is no longer a significant source because of the widespread use of pasteurization and the surveillance of dairy cows to prevent spread of *Mycobacterium bovis,* the principal cause of bovine tuberculosis.

Drugs useful in the treatment of infections with *M. tuberculosis* are isonicotinic acid hydrazide (INH), para-amino salicylic acid (PASA), the aminoglycoside antibiotics (such as streptomycin), and rifampin. However, other species of mycobacteria are often highly resistant to these and other medications. Even among *M. tuberculosis,* drug-resistant mutants occur with high frequency. But since mutants simultaneously resistant to more than one drug have a much lower frequency of occurrence, two or more drugs are usually given together in treating tuberculosis. Because of the very slow generation times of most pathogenic mycobacteria and their resistance to destruction by body defenses, drug treatment of tuberculosis must generally be continued for one or more years to obtain permanent arrest of the disease. Even so, nonmultiplying bacilli enclosed within old tubercles may not be killed by treatment.

INH, PASA, page 589

Vaccination against tuberculosis has been widely used in Scandinavia and elsewhere and is of proven value. The vaccinating agent, a living attenuated mycobacterium known as Bacillus Calmette-Guerin (BCG), is probably derived from a bovine mycobacterium. Repeated subculture in the laboratory over many years resulted in selection of this strain of *M. bovis,* which has little virulence for human beings but does result in some immunity. The protective effect of BCG vaccination is of value to nurses and tuberculosis sanitorium personnel, has been about 80 percent effective (the incidence of tuberculosis being one-fifth that occurring in unvaccinated persons), and lasts for several years. Persons receiving the vaccine also develop delayed hypersensitivity to tuberculin. Public health authorities have discouraged routine use of the BCG vaccine in the United States because most people who now develop active tuberculosis do so as a result of activation of a tubercular infection acquired years ago, rather than new disease.

INFLUENZA: A VIRAL LOWER RESPIRATORY TRACT DISEASE

Respiratory infection by myxoviruses of the influenza group is called *influenza,* which differs from bacterial pneumonia in its tendency to involve the bronchi and bronchioles and their supporting tissue to a greater extent than the alveoli. Even though the vast majority of people recover completely from influenza without any treatment, the number of deaths from this disease is high because so many people become infected. Death may be caused by influenza itself, although it is more commonly the result of secondary invasion of the influenza-damaged lung by pathogenic bacteria. *Staphylococcus aureus, Haemophilus influenzae, Streptococcus pyogenes,* and *Streptococcus pneumoniae* are the chief bacterial offenders.

Influenza virus, Figure 17-2, page 342

The influenza virus enters the respiratory system by inhalation of air contaminated with virus by another infected person. Penetration of the mucous blanket is presumably aided by a viral enzyme, neuraminidase, attached to the viral envelope. Infection becomes established in cells lining the air passages and large amounts of virus are then released to infect other cells. Death of the ciliated cells, inflammation, and leakage of plasma are prominent features of influenza.

The influenza viruses can be isolated in a variety of tissue cell cultures in which they may produce cytopathic effects or hemadsorption (the adherence of red blood cells to tissue cells infected by the virus). The viruses hemagglutinate red blood cells from several animal species. Antisera specific for their viral nucleoproteins can be used to classify the influenza viruses into three major groups, A, B, and C; several subgroups and types are further identifiable using antibodies specific for their viral neuraminidases and hemagglutinins. For example, the group A influenza viruses have a common nucleoprotein antigen which differs from that of viruses in groups B and C; the Hong Kong group A virus of 1968 is further distinguished from other group A viruses by its specific hemagglutinin and neuraminidase.

PANDEMIC INFLUENZA

Pandemics (epidemics of worldwide scope) of influenza have swept the world at approximately 10 year intervals for centuries. The worst pandemic of modern times occurred in the fall of 1918 when the disease abruptly appeared in such diverse cities as Boston and Bombay and then spread over much of the world, involving an estimated 500 million people within a 6- to 8-week period. At its peak, hundreds of persons died each day in major cities. In the Seattle, Washington, area 1 out of every 175 persons between the ages of 20 and 30 years succumbed, and infants fared even worse. About one-fifth of the fatalities occurred within the first 4 days of illness. The virus responsible for this major influenza epidemic was never recovered because the science of virology had not yet developed. Antibody tests done years later on the survivors of the epidemics, however, showed that the virus had finally disappeared from the human population about 1929, and that meanwhile, a similar virus (swine influenza virus) had become established in pigs. The swine virus, perhaps a mutant of the epidemic human strain, has continued to be endemic in pigs to the present time. The swine virus is capable of infecting human beings, but does not cause illness or spread from person to person.

A more recent influenza epidemic, caused by the "Hong Kong" strain of influenza A virus, appeared in 1968 (Figure 26-3). Subsequent to 1968, related viruses, presumably mutants of the Hong Kong strain, continued to cause relatively minor influenza epidemics every 2 or 3 years. Then, in January 1976, there appeared at Fort Dix, New Jersey, an influenza epidemic that included a new virus. This new virus was antigenically related to the swine influenza virus, but it behaved quite differently, apparently spreading to involve about 500 soldiers and causing one death. Because it was known that very few persons born since the 1920s had antibody to the swine type virus, the Fort Dix influenza epidemic was an alarming development. A vaccine was prepared by crossing the new virus with one easily cultivated in the laboratory, and a massive campaign was initiated to immunize the population of the United States against the swine type virus. However, further epidemic spread of the swine virus failed to occur and the immunization campaign was later dropped.

GENETIC INSTABILITY

The genetic instability of influenza viruses, manifested by the repeated appearance of new strains of these organisms as the host population develops immunity to the old, is incompletely understood. Presumably, minor antigenic variations in the influenza viruses result from single mutations that cause the

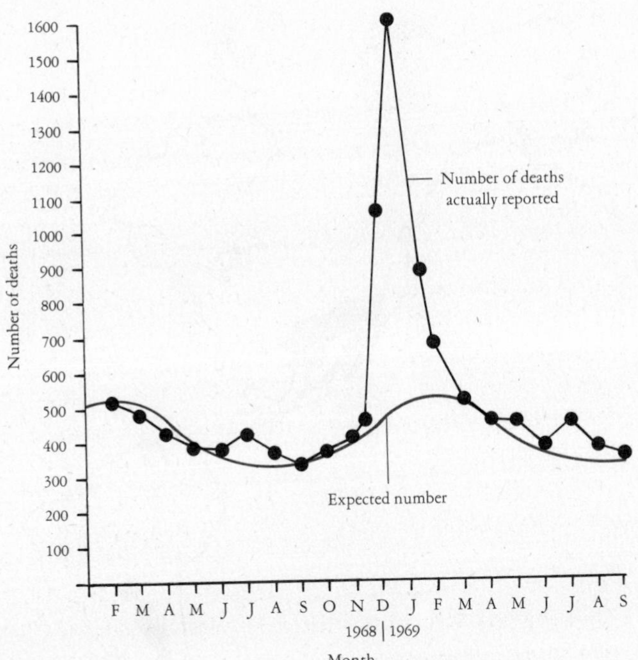

FIGURE 26-3
Increased number of deaths from respiratory disease during an influenza epidemic. (From Morbidity and Mortality Weekly Report, Annual Supplement, December 1969.)

appearance of new surface antigens without significantly altering other attributes of these organisms' virulence. However, another mechanism of genetic change can readily be demonstrated *in vitro* and is thought to occur on rare occasions in the natural setting. The genome of the influenza viruses consists of several distinct RNA fragments, and when infections are simultaneously established in a cell by two strains, exchange of these RNA fragments occurs. The resulting virus has some characteristics of each of the infecting strains. The known ability of both the Hong Kong and swine type influenza viruses to infect the same host suggests that the 1976 swine virus could have arisen by this mechanism.

FUNGOUS INFECTIONS OF THE LUNG

Coccidioidomycosis

Valley fever and desert rheumatism are the other names for coccidioidomycosis, which occurs commonly in certain hot, dry, dusty areas of the Americas. Fever, cough, chest pain, and loss of appetite and weight are common features of this disease. About 1 out of 10 persons suffering the illness develops hypersensitivity manifested by a rash on the shins or on other parts of the body and pain in the joints. The vast majority of persons afflicted with coccidioidomycosis recover spontaneously within a month and, in contrast to tuberculosis, have little risk of later reactivation of the disease. In a few persons, the disease closely resembles tuberculosis. Death of lung tissue may lead to the formation of cavities and, more

FIGURE 26-4
Geographical distribution of *Coccidioides immitis*.

rarely, the infection may spread throughout the body so that the skin, mucous membranes, brain, and internal organs become involved. About half of all persons with disseminated coccidioidomycosis die unless proper treatment, as described below, is given.

Dimorphic fungi, page 291

The causative agent of coccidioidomycosis, *Coccidioides immitis,* is a dimorphic fungus living in the soil. It grows only in limited areas of the Western Hemisphere (Figure 26-4). Infectious spores have unknowingly been transported to other areas (such as southeastern United States), but there is no evidence that the fungus can establish itself in other climates. In the areas where *C. immitis* is endemic, infections occur only during the hot dusty seasons when its spores are airborne, and disappear during periods of rain. Virtually all individuals who live in these areas show evidence of having been infected. Indeed, persons have become infected with *C. immitis* while traveling through contaminated areas.

The tissue phase of *Coccidioides immitis* is the form present in infected tissues and can be identified by microscopic examination of sputum or pus. It is a thick-walled sphere usually ranging from 20 to 80 μm in diameter; in contrast to most other fungi growing in tissue, it never buds. The larger spheres of the fungus contain several hundred small cells called "endospores," although they have little resemblance to the endospores of bacteria (Figure 26-5a). The mold form of the organism grows readily on most laboratory media at room temperatures and is the form which grows in soil. On Sabouraud's medium (a specialized medium for growing fungi), growth of the mold form of *C. immitis* usually occurs in from three to five days. Such cultures are extremely infectious since most of the hyphae of the organism develop numerous very light barrel-shaped arthrospores (Figure 26-5b), and these separate easily from the hyphae and become airborne. Because other molds may resemble *C. immitis,* cultures may need to be converted to the spherule form to complete the identification.

EXTREME INFECTIVITY OF MOLD FORM

Coccidioidomycosis is initiated by inhaling the arthrospores of *C. immitis* growing in the soil. Because these infectious arthrospores do not develop in man or animals, transmission of *C. immitis* from animal to animal does not occur. After lodging in the lung, the arthrospores develop into spherules which mature and discharge their endospores, each of which can then develop into another spherule. The pathogenesis of coccidioidomycosis is similar to that of tuberculosis. The immune response of an individual can be measured by skin testing with coccidioidin, a culture supernatant analogous to tuberculin. Most persons show a positive coccidioidin skin test within three weeks after inhaling the fungal spores of *C. immitis* and retain the capacity to give a positive skin test for life if they remain in the area where the fungus occurs. Those who move away for several years and those with disseminated disease often lose their skin reactivity to coccidioidin. Within the first month of illness, most persons with coccidioidomycosis will develop measurable precipitating antibodies, but these generally drop to low titers by the third month regardless of the progress of the disease. Complement-fixing antibodies to the organism arise later than precipitating antibodies and often remain detectable for years, although the titer of these antibodies falls when recovery occurs. However, titers of complement-fixing antibodies continue to rise in disseminated infections, and this laboratory finding means that treatment must be given to save the person's life.

Complement fixation, pages 393–394

The only effective control of *Coccidioides immitis* is attained by measures which decrease dust. Serious infections can often be arrested by intravenous administration of the antibiotic amphotericin B; this treatment must be continued for months, but reactivation of the disease can occur months or years after therapy is discontinued.

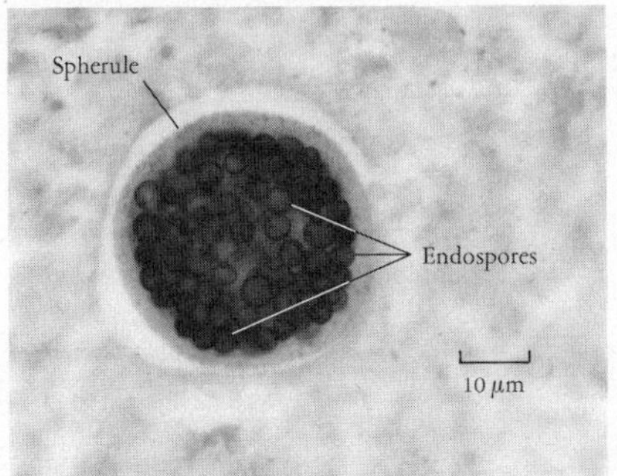

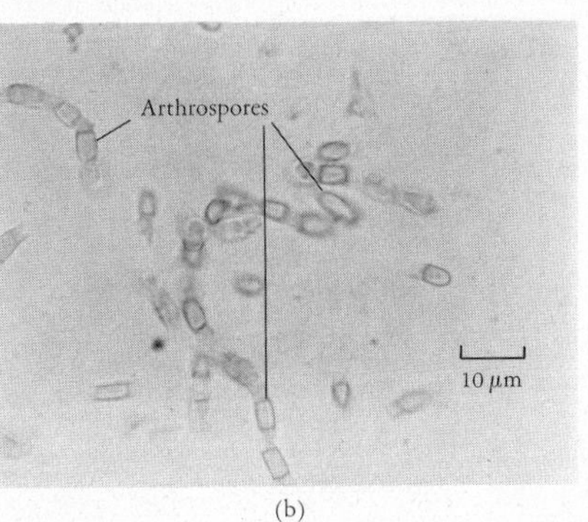

FIGURE 26-5
Coccidioides immitis. (a) Tissue phase. Spherule containing endospores (stained preparation). (b) Mold phase. The barrel-shaped arthrospores are characteristic of this species.

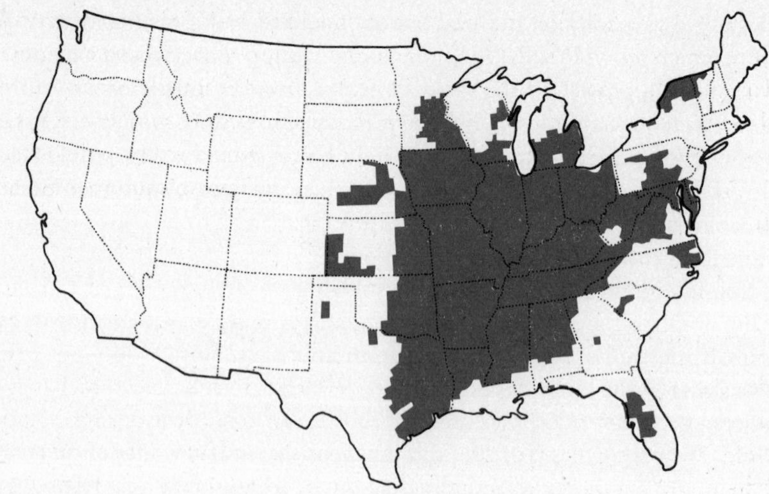

FIGURE 26-6
Geographical distribution of *Histoplasma capsulatum* in the United States, as revealed by high frequency of positive histoplasmin skin tests in human beings. Besides the United States, human histoplasmosis has been reported from more than 40 other countries, including Argentina, Italy, Thailand, and South Africa.

Histoplasmosis

This disease is very similar in many respects to coccidioidomycosis, except for its distribution. Histoplasmosis has a spotty occurrence in tropical and temperate zones around the world (Figure 26-6). The causative organism of the disease is the dimorphic fungus *Histoplasma capsulatum,* a fungus preferring soils contaminated by bat or bird droppings. Caves and chicken coops are notorious sources of the infection. In its tissue phase *Histoplasma capsulatum* is an oval yeast ranging from 2 to 5 μm in size, often growing within macrophage cells of the host (Figure 26-7a). The mycelial phase of the organism shows chlamydospores characterized by numerous projecting knobs (Figure 26-7b). Delayed hypersensitivity to the organism is detected by skin testing with histoplasmin, an antigen analogous to coccidioidin and tuberculin. Precipitating and complement-fixing antibodies develop in histoplasmosis as they do in coccidioidomycosis. Because cross-reactions of these antibodies often occur to antigens of *Coccidioides* and another pathogenic fungus, *Blastomyces dermatitidis,* skin tests and antibody determinations are helpful in establishing the cause of a fungal infection only if antigens from all three fungi are tested at the same time. The antigen corresponding

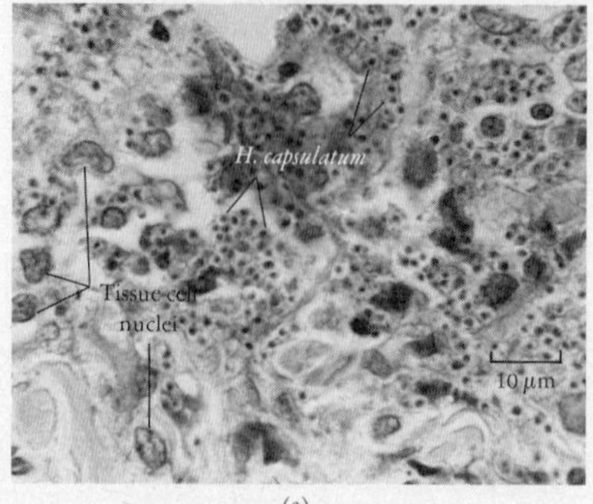

(a)

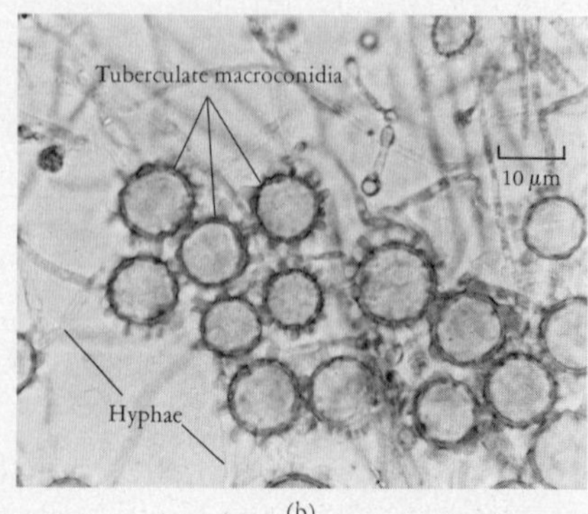

(b)

FIGURE 26-7
Histoplasma capsulatum. (a) Tissue phase. Numerous tiny yeast cells are present within the cytoplasm of macrophages. (b) Mold phase. The tuberculate macroconidia appear in cultures grown at 25°C.

to the infecting fungus usually shows the largest skin test and greatest antibody response. Probably half of the reported cases of histoplasmosis have occurred in the United States, primarily in the Mississippi River drainage area and in the South Atlantic states; skin tests reveal that millions of people living in these areas have been infected. As with *Coccidioides* only a tiny fraction of persons infected with *Histoplasma* develop serious illness, and the antibiotic amphotericin B is helpful in arresting such serious infections.

Amphotericin, page 597

OTHER LUNG DISEASES CAUSED BY MICROBES

Besides the organisms discussed above, numerous other bacteria, viruses, and fungi produce infections of the lung. Still other agents are unable to establish infection, but repeated exposure to them from inhaled dust may nevertheless result in serious damage from the inflammatory reactions they elicit. For example, repeated inhalation of hay dust containing certain thermophilic actinomycetes may produce an allergic reaction to the bacterial antigens. This is characterized by severe inflammation and granuloma formation in the airways and alveoli. Finally, spores of saprophytic fungi may be responsible for some cases of asthma. In this case there is little damage to lung tissue, but immediate hypersensitivity to the foreign antigen causes spasm of the airways and difficulty in breathing.

SUMMARY

Microorganisms enter the lung with inspired air, or in infected material from the upper respiratory tract. They may be removed by bodily defense mechanisms, or produce a variety of diseases depending on the nature of the infecting organism. *Streptococcus pneumoniae, Klebsiella* species, and other bacteria can produce lung infections, as can viruses, such as those causing influenza. Tuberculosis results from infection with *Mycobacterium tuberculosis,* and related organisms may sometimes cause a similar disease. Chronic illnesses resembling tuberculosis may also result from infection by certain fungi such as *Coccidioides immitis* and *Histoplasma capsulatum.* Severe lung disease can also result from inhaling saprophytic bacteria, as the result of hypersensitivity of lung tissues to these organisms' antigens. Vaccines can help in preventing some lower respiratory infections, including pneumococcal pneumonia, influenza, whooping cough, and tuberculosis.

QUESTIONS

REVIEW

1. What are the mechanisms by which the lower respiratory system is protected from infection?
2. On what factors does specific immunity to pneumococcal disease depend? Does immunity to one strain of *Streptococcus pneumoniae* give immunity to all strains of this species? Explain.

3. Give reasons why the total number of deaths in an influenza epidemic may be high despite the fact that influenza itself is usually a mild disease.
4. Describe *Mycoplasma pneumoniae.* What disease does it cause?
5. What bacterium causes whooping cough? How is the disease prevented?

THOUGHT

1. If some means were devised to completely prevent the transmission of *Mycobacterium tuberculosis,* how long would it take tuberculosis to disappear?
2. Which would you expect to be a more sensitive test for tuberculosis, a chest X-ray or a tuberculin test for delayed hypersensitivity? Why?

FURTHER READING

COMROE, J. H., "The Lung," *Scientific American* (February 1966).

DUBOS, R. and J., *The White Plague.* Boston: Little, Brown, 1952. The story of tuberculosis.

KILBOURNE, E. D., "National Immunization for Pandemic Influenza," *Hospital Practice* 11:15–17 (June 1976).

ORAL MICROBIOLOGY, GASTROINTESTINAL INFECTIONS, AND FOOD POISONING

The alimentary tract, like the skin, is one of the body's boundaries with the environment and is continually exposed to microbes of the endogenous flora as well as to extraneous organisms. It is, so to speak, our "inside outside," and it is one of the major routes of access for invading germs.

ANATOMIC AND PHYSIOLOGIC CONSIDERATIONS

The alimentary tract includes the mouth, esophagus, stomach, small and large intestines. The system also includes some very important appendages: the salivary glands, the liver and gall bladder, and the pancreas. These are attached to the alimentary tract by tubes through which pass the fluids the appendages produce. Figure 27-1 illustrates the relationships of the various parts of the alimentary system.

The Mouth

The teeth (Figure 27-2) are made up largely of calcium phosphate with some protein matrix. The outer portion, or enamel, is especially dense, and yet certain bacteria are able to penetrate it and produce cavities and tooth destruction. Both microscopic and large crevices in the surfaces of the teeth tend to collect food particles that are sites for microbial colonization. The gingival crevice is an ecological niche of great importance because both gingivitis and periodontal disease originate there. Saliva is secreted into the mouth from the various salivary glands at a rate of about 1500 ml per day. It serves to keep the mouth clean and lubri-

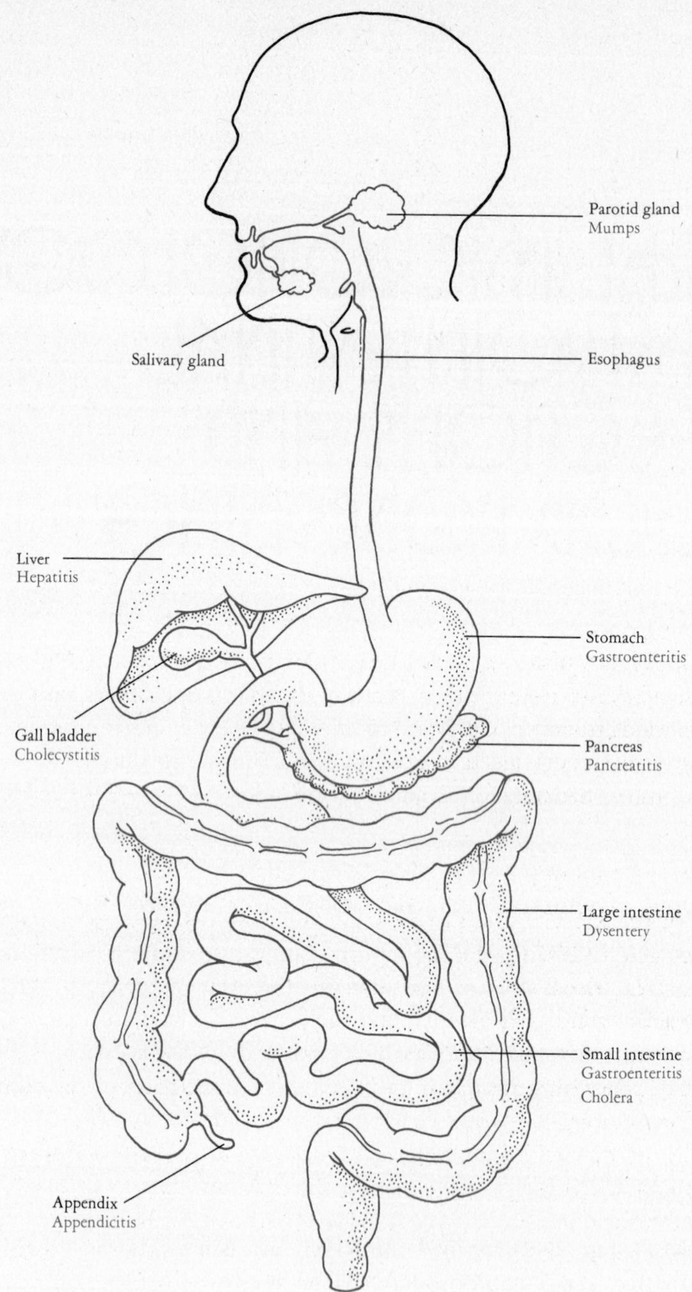

FIGURE 27-1
The alimentary system.

IMPORTANCE OF SALIVA

cated, helps maintain a neutral pH, and because it is saturated with calcium, tends to prevent the calcium phosphate of the teeth from dissolving. It has readily demonstrable inhibitory and killing powers against various groups of microorganisms, both members of the normal microbial flora and pathogens. Saliva also contains mucins and IgA antibodies, both of which can coat bacteria and inhibit their ability to attach to and colonize the teeth and mouth tissues.

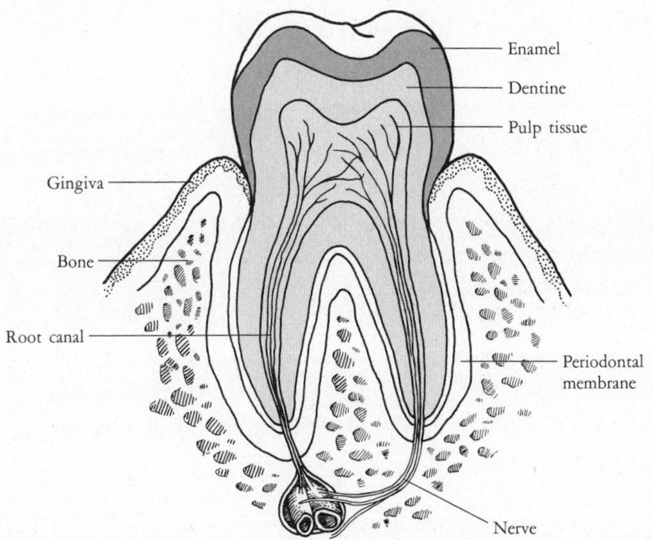

FIGURE 27-2
Stylized drawing of a tooth and adjacent structures.

The Stomach

The esophagus connects the mouth with the stomach, a distensible sac with a muscular wall and lining cells, some of which produce hydrochloric acid and pepsinogen, a precursor of the protein splitting enzyme, pepsin. In response to food the stomach begins a mixing action, bringing acid and enzymes into contact with the ingested material. Many types of microbes are destroyed by this action.

MICROBES ARE KILLED BY STOMACH JUICES

The Small Intestine

A valve at the end of the stomach controls the passage of the well-mixed gastric contents into the small intestine; the rate of passage slows down if there is considerable fat in the food or if it is markedly hypertonic or hypotonic. The small intestine in an adult is about 8 or 9 ft long and has an enormous absorptive and secretory area—about 30 sq ft. This large area results from the presence of many small fingerlike projections (villi)—20 to 40 per square millimeter, each 0.5 to 1 mm long—along the inside surface of the intestine. Microbial diseases may markedly interfere with functions of the small intestine, such as absorption and secretion, by affecting the cells lining the intestine and may therefore disrupt the normal transit of intestinal contents. Small patches of lymphoid tissue spot the walls of the small intestine and these may be selectively invaded by certain pathogens.

SITES OF ATTACK BY PATHOGENS

The liver has many functions, one of which is to remove the breakdown products of hemoglobin from the bloodstream; these yellowish-green products are excreted with the bile, the fluid produced by the liver. Interference with this excretory process, by infection or other factors, may produce jaundice (a yellow color of the skin and eyes). The gall bladder is simply a storage area for the bile, where this fluid is concentrated and its acidity neutralized. Infection of the gall bladder, *cholecystitis,* is often a life-threatening disease, but in other cases potentially pathogenic organisms can grow in the gall bladder for years without causing symptoms. The pathogens are discharged into the intestine with the bile and are passed with feces. Bile also helps emulsify fats in the intestine and has antibacterial activity against some microbes. The pancreas also has multiple functions, including the production of several digestive enzymes. About 500 ml of bile per day pour into the upper portion of the small intestine from the liver and gall bladder, and another 2000 ml of digestive juices are added by the pancreas. These fluids, as well as digestive juices of the small intestine itself, are alkaline and neutralize the stomach acid passed into the intestine. The pancreatic and intestinal fluids break down foodstuffs enzymatically into absorbable amino acids, fats, and simple sugars. These smaller molecules are then selectively taken up by the lining cells of the small intestine.

The Large Intestine

Because of the great absorptive capacity of the small intestine, about only 300 to 500 ml of gastrointestinal fluid normally reach the large intestine per day. The large intestine can also absorb water, electrolytes, vitamins and amino acids from foodstuffs. After this further absorption, the normally semisolid feces remain. The feces represents an important source of pathogens for other parts of the body, especially for the urinary tract and blood stream.

ORAL MICROBIOLOGY

It has been stated that if all the dentists in the United States were to work 24 hr a day filling cavities, as many new cavities would form each year as had been filled the previous year. Periodontal disease is no less significant, and the amount of money spent treating these two oral diseases clearly exceeds that spent for all other infectious diseases. Many other infectious processes involving the mouth are covered elsewhere in this book and will not be extensively discussed here; rather, certain hazards relating to oral microbiology are covered.

Bacteria of the Mouth

Lactic acid bacteria, pages 258–259

An enormous variety of bacteria inhabit the mouth, in concentrations ranging up to 10^{11} per gram in tooth deposits and to 10^9 per ml in the saliva. Two features of this oral bacterial population are especially noteworthy. First, of all the species of bacteria introduced into the mouth from the time of birth onward, only a relative few are able to colonize and persist in the oral cavity. Second, the distribution of these bacterial species is far from uniform; certain members colonize one area, and others another. Most prominent are species of the genus *Streptococcus,* fastidious parasites that metabolize carbohydrates with the production of lactic acid. *Streptococcus salivarius* preferentially colonizes the upper aspect of the tongue, *Streptococcus sanguis* prefers the teeth, while *Streptococcus mitis* likes the mucosa of the

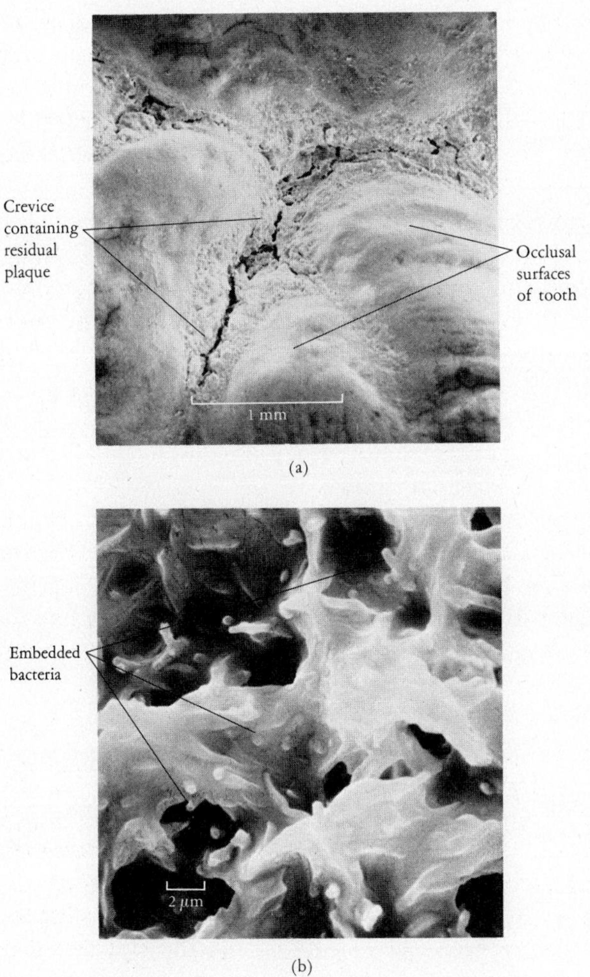

FIGURE 27-3
Scanning electron micrograph of tooth surface shortly after brushing. (a) Crevice showing residual plaque. (b) Higher magnification of plaque showing masses of bacteria embedded in an amorphous material (Courtesy of K. A. Galil; from K. A. Galil, *J. Canad. Dent. Ass.* 41:499, 1975.)

cheek. Tenacious collections of bacteria commonly occur on the smooth surfaces and in the crevices of the teeth. These sponge-like collections, called *dental plaque* (Figure 27-3), are composed chiefly of streptococci intermixed with filamentous Gram-positive branching and nonbranching bacteria (*Actinomyces, Rothia, Nocardia, Bacterionema, Leptotrichia* genera), and anaerobic Gram-negative cocci of the genus *Veillonella.* The gingival crevice characteristically is populated by strict anaerobes of the genera *Bacteroides* and *Fusobacterium,* and by smaller numbers of spirochetes.

Ecological Considerations

Colonization of the mouth is dependent upon the ability of bacteria to adhere to oral surfaces. Certain strains of streptococci and other bacteria have short surface filaments that attach specifically to certain host tissues, allowing these organisms to resist the scrubbing action of food and the tongue and the flushing action of salivary flow. However, the host limits the extent of bacterial colonization of its mucous membranes by constantly sloughing off the superficial layers of cells and replacing them with new ones; the rate of this sloughing correlates with the microbial burden. Furthermore, saliva and other oral secretions contain antibodies and certain glycoproteins that coat the bacterial surface and prevent or weaken its attachment to the oral tissues.

Colonization of the mouth by strictly anaerobic flora requires the presence of teeth because only these structures can provide niches of sufficiently low oxidation-reduction potential to allow these organisms' growth. The blind pockets created by the gingival crevices and fissures in the teeth are one such niche. But tooth surfaces also represent a nonsloughing tissue where enormous bacterial concentrations are able to build up. The metabolic activity of the various organisms growing on teeth consumes oxygen and thus provides another area where some anaerobic organisms can grow.

The basic pattern of bacterial colonization of the mouth as outlined above is relatively constant, although new strains of bacteria constantly replace the old. To what extent these organisms arise as mutants of existing oral strains or come directly from the environment is not known.

Dental Plaque and Caries

PATHOGENESIS OF DENTAL CARIES

The initiating event in the formation of human dental plaque is the adherence of a cluster of streptococci to the tooth pellicle, a thin film of adsorbed proteinaceous material from the saliva. The streptococci that adsorb to this film have a specific affinity for the tooth surface and are not the predominant streptococci in the saliva bathing the teeth. Their attachment is initially weak, but is considerably strengthened by the deposition of extracellular polysaccharide. Other organisms then adhere to the developing plaque. Initially these include aerobic organisms of the genera *Neisseria* and *Nocardia.* Growth of these organisms is accompanied by a reduction in the oxidation-reduction potential of the plaque, making it suitable for anaerobic species to colonize. Production of lactic and other acids by streptococci in the plaque provides a suitably acid medium for lactobacilli, and these organisms can thus also colonize dental plaque, but not other parts of the mouth. Plaque-colonizing anaerobes of the genus *Veillonella* also utilize the lactic acid produced from the metabolism of streptococci and lactobacilli. Filamentous organisms, too, colonize the plaque, and by the end of a week make up most of its bulk.

Scientists have placed tiny electrodes into dental plaque and studied what happens to its pH when sugar and other carbohydrates enter the mouth. The response is dramatic! The pH drops from its normal 7.0 to about 4.0 within minutes. This means a thousandfold increase in the acidity of the plaque, to a level where calcium phosphate dissolves. The duration of this acid state depends directly on the duration of exposure of the teeth to sugars, and to their concentration. Thus, people who eat frequent snacks are more prone to dental caries than are those who take large meals less frequently. After food leaves the mouth, the pH rises slowly to neutrality. This effect is often delayed by the ability of some plaque-inhabiting bacteria to store a portion of foodstuff as polysaccharide, which they then metabolize to acid. Plaques thus act as tiny acid-soaked sponges intimately applied to the teeth.

In other studies, using monkeys and germ-free rodents, various bacteria from human dental caries have been tested for their ability to cause this disease. These experiments have established that pure cultures of certain strains of streptococci are capable of producing dental plaque and caries. These strains are uniquely able to adhere firmly to the smooth surfaces of teeth, but only if the diet contains sucrose. Streptococcal enzymes convert this sucrose to a dextran-like polysaccharide which enables the bacteria to adhere firmly to the teeth, allowing sustained acid production. A number of these caries-producing streptococci have been isolated and have been designated "*Streptococcus mutans*" although they are a heterogeneous group, some members of which may represent strains of known streptococcal species. Because experimental and natural conditions differ considerably, there is controversy as to whether all human caries conforms to the experimental disease discussed here. However, there is little doubt that the *S. mutans* group of organisms is of fundamental importance in human caries. Note that *both* a cariogenic streptococcus *and* a suitable diet are required to produce dental caries. Plaques may develop on a sugar-free diet, but these are incapable of causing caries even when carbohydrates are introduced to the diet. Although sucrose is the most important caries-producing component of the human diet, other sugars can also support development of cariogenic plaques.

Dextran, page 43

Sucrose, page 41

Dental Plaque and Periodontal Disease

PATHOGENESIS OF PERIODONTAL DISEASE

Periodontal disease is a chronic inflammatory process involving the supporting tissues of the roots of the teeth. It is often an insidious process and the chief cause of loss of teeth from middle age onward. Formation of plaque is apparently a necessary condition for the development of periodontal disease, but considerably less is known about its mechanism. Plaque and calculus (calcified plaque) in the region of the gingival crevice are known to cause an inflammatory reaction (gingivitis) which at first involves the gum (gingiva), with swelling, redness, and cellular exudate. Severe acute cases of this condition are given the name "Vincents' angina" and are characterized by death of tissue (necrosis) at the edge of the teeth, pain, and fever. Associated with the presence of devitalized tissue is a marked overgrowth of fusiform and spirochaetal bacteria (*Fusobacterium* and *Treponema* species). When gingivitis is chronic, the gingival crevice becomes widened and deepened, allowing a marked overgrowth of bacteria. An abscess then develops and penetrates through the base of the gingival crevice. The large population of microorganisms in the abscess produces collagenase, protease, and hyaluronidase enzymes, as well as endotoxins, and these products penetrate into the tissues. An immune response of the host to the bacterial products undoubtedly contributes to the inflammatory process. The periodontal membrane be-

Spirochetes, pages 265–267

comes involved, and the bone surrounding the tooth is gradually absorbed, one of the effects of the inflammatory mediator prostaglandin E. Periodontal disease thus eventually leads to the loosening and loss of teeth, but it can be halted in its earlier stages by cleaning out the inflamed crevice and removing plaque and calculus.

Oral Bacteria and Endocarditis

Mononuclear phagocyte system, page 371

Endocarditis, pages 562–563

Small numbers of oral bacteria are forced into the bloodstream by a variety of dental procedures and even by brushing of the teeth or by chewing. Normally, these organisms are quickly removed by the body's mononuclear phagocyte system, but in individuals with abnormal heart valves they can cause bacterial endocarditis (Chapter 31). In an effort to prevent endocarditis dentists treat such patients with an antibiotic in order to lower the numbers of viable bacteria in the gingival crevices of infected areas of their mouths, and to kill any bacteria that might become implanted on a defective heart valve. When an antibiotic such as penicillin is given, there is a transitory drop in the numbers of oral bacteria, followed in a few days by a return of the oral flora to normal levels even if the antibiotic treatment is continued. The rise in numbers of oral bacteria is due to the ascendancy of strains resistant to penicillin, and this is why, in attempting to prevent endocarditis, antibiotic medication is started only an hour or two before a dental procedure, continued for two days afterward, and then promptly discontinued.

Herpes Simplex Virus and Hepatitis B Virus

Apparently healthy persons can excrete infectious viruses in their saliva over long periods of time and may thus represent a hazard to nurses, dentists and others who come in contact with these persons' saliva. One such excreted virus, the herpes simplex virus, causes cold sores (fever blisters) and ulcers of the mouth or eye. It can also cause a painful infection called paronychia, or herpetic whitlow, in the area of the fingernail; this condition must be differentiated from bacterial paronychia.

Another virus, the hepatitis B virus, is also excreted in the saliva of some persons and can contaminate the hands of dental workers. Indeed, dentists have a risk of infection with hepatitis B that is two to three times greater than that of the general population, and this may in part be owing to contamination from patients' saliva.

GASTROINTESTINAL MICROBIOLOGY

The rich microbial flora of the mouth is absent in the fasting stomach, since both hydrochloric acid and other antimicrobial secretions of the stomach play a role in microbial killing, depending on the species of microbe. The small intestine is also relatively free of microbes. Studies of healthy people usually show no more than 10^3 bacteria per milliliter in the upper small intestine and 10^5 or less per milliliter in the lower small intestine. The predominant organisms of the small intestine are usually bacteria able to grow in air, including facultatively anaerobic Gram-negative rods and various streptococci. Lactobacilli are fairly commonly found in small numbers, as are small numbers of yeasts such as *Candida albicans.* The main reason for such low concentrations of microorganisms is thought to be the flushing action of rapid passage of the various digestive juices through the small intestine.

In contrast to the scanty numbers of organisms in the stomach and small intestine, the colon contains very high microbial counts, usually in the neighborhood of 10^{11} per gram of feces. Numbers of anaerobic organisms of the genera *Lactobacillus* and *Bacteroides* generally exceed the numbers of all other organisms by about a hundredfold. Facultatively anaerobic Gram-negative rods, particularly *Escherichia coli,* and members of the genera *Enterobacter, Klebsiella,* and *Proteus,* predominate among fecal microbes able to grow aerobically. In contrast to the anaerobes these organisms are easy to cultivate and some have been studied in exhaustive detail.

Bacteroides, pages 438–439

Enterobacteria, pages 435–437

In the intestine, just as in the mouth, bacteria adhere to specific tissue receptors. This undoubtedly helps to explain the pattern of distribution of the normal intestinal flora as well as the sites of intestinal attack by pathogens.

Biochemical Activities of Intestinal Microbes

Because of the many different kinds of bacteria living together in the alimentary tract, some of which depend on each other for growth, it is very difficult to predict which of their metabolic effects observed in the test tube are important to their animal host. It is, however, clear that some intestinal bacteria can use up vitamin C (ascorbic acid) and several B vitamins (choline, folic acid, cyanocobalamine, and thiamine). Thus under certain conditions, intestinal bacteria can contribute to vitamin deficiency; they can also produce ammonia, organic acids from carbohydrates, and degrade bile and digestive enzymes.

"BLIND LOOP SYNDROME"

Some of these metabolic effects of the intestinal flora undoubtedly play a role in human disease. One example is an unusual condition called the "blind loop syndrome" in which, as the result of disease or surgery, a pocket of the small intestine becomes relatively isolated from the rest of the organ. In the pocket, the flow of intestinal juices is reduced and large populations of the normal intestinal flora build up. Individuals affected by this condition may become severely anemic and lose weight because the large numbers of bacteria that accumulate in the loop use up certain vitamins and degrade bile. The condition may be alleviated by correcting the structural abnormality or by giving antibiotics to reduce the numbers of organisms in the loop.

Although some intestinal microbes may perform potentially harmful metabolic actions, other common intestinal bacteria synthesize an excess of useful vitamins. These include niacin, thiamine, riboflavin, pyridoxine, cyanocobalamine (vitamin B12), folic acid, pantothenic acid, biotin, and vitamin K. These

TABLE 27-1
Hepatitis Viruses

Characteristics	*Pathogenic Potential*	*Epidemiology*
Small viruses of uncertain classification, relatively resistant to heat and disinfectants; HBV is a 43-nm particle containing double-stranded DNA; HAV is a 27-nm particle which probably contains RNA	Preferentially infect the liver, often producing jaundice and sometimes liver failure	Source: human cases and carriers; incubation period: usually 15 to 50 days for HAV; 43 to 180 days for HBV; transmission: usually fecal-oral for HAV; blood or blood products for HBV—as little as 0.0001 ml of blood may be infective

syntheses are of enormous importance for the nutrition of some animals. For human beings they are of doubtful importance except when there is an inadequate diet. Thus, when there is poor nutrition, oral antibiotics sometimes bring about vitamin deficiency, presumably by causing a decrease in vitamin production by intestinal microbes.

SOME VIRAL DISEASES OF THE ALIMENTARY SYSTEM

Mumps

Mumps is an illness characterized by fever and painful swelling beneath one or both ears. This condition comes from an infection of the parotid salivary glands, which are located between the angle of the jaw and the ear. The agent responsible for mumps is a paramyxovirus known as the mumps virus; it is transmitted by saliva, in which it may be found as much as a week before the onset of illness. In a few infected persons no illness occurs, and in others the major effects of mumps appear in the central nervous system, the testes, ovaries, pancreas, or thyroid gland, with or without involvement of the parotid glands.

Paramyxoviruses, Table 17-1, page 341

Because other infectious agents may sometimes involve the parotid glands, not every patient with fever and painful swelling of the glands can be assumed to have mumps. Identifying mumps virus infection is most commonly done by testing the patient's serum for development of complement-fixing antibodies against known mumps virus antigen prepared in the laboratory. Additionally, the mumps virus can be isolated by injecting saliva or urine from infected persons into chicken embryos. A vaccine can give partial protection against mumps, but no antimicrobial therapy is available for its prevention or treatment.

Complement fixation, pages 393–394

Hepatitis

Hepatitis, or inflammation of the liver, is caused by some allergic reactions, certain toxic chemicals and various microbial agents. Symptomatic cases of this disease typically show loss of appetitie and vigor, fever, and jaundice. Two viruses, the hepatitis A virus (HAV) and the hepatitis B virus (HBV) account for most cases of hepatitis. The principal characteristics of both these viruses appear in Table 27-1. Both are notable in their resistance to disinfectants and lack of satisfactory growth in vitro, but they are immunologically distinct and differ in their mode of spread and incubation period.

Hepatitis A virus disease (formerly called "infectious hepatitis") spreads in epidemic fashion, principally through the fecal contamination of hands, food or water. A prominent cause in recent years has been ingestion of raw shellfish since these animals concentrate the hepatitis A virus from polluted seawater. Some cases are severe, requiring many weeks of bed rest for recovery. However, most cases are mild and self-limited, and many are asymptomatic. The infection is sufficiently widespread in the population for many people to have hepatitis A virus antibody in their blood. For this reason, gamma globulin pooled from the bloods of many donors can be administered to susceptible exposed persons to provide passive immunity to the disease.

Gamma globulin, page 381

Hepatitis B virus disease (formerly called "serum hepatitis") is spread principally by blood and blood products. Minute amounts injected into the blood or rubbed into minor wounds can cause symptomatic infection, and ingestion of small amounts of plasma experimentally has also produced the disease. Hepatitis caused by HBV tends to be more severe than that caused by HAV, with mortality ranging from 1 to 10 percent with HBV, but many cases of HBV infection show few or no symptoms. Indeed, for every case with jaundice there are at least three without. However, the incidence of infection with HBV has been increasing progressively over the past decade and an estimated 150,000 cases per year occur in the United States. Many hepatitis B virus infections result from needles shared by drug abusers. Tatooing, ear-piercing, shared toothbrushes, razors and towels, and similar factors can also be responsible for the transmission of HBV infections. Blood transfusions have been a common source of HBV infection in the past, but with present precautionary practices, the risk is now less than 1 percent. Hepatitis virus B antigen can be demonstrated in from one to five of every 1000 prospective blood donors. Although present methods of blood testing fail to detect the virus in a small percentage of donors, tests now under development should be capable of identifying essentially all carriers.

After an individual has contracted HBV infection, virus is not detectable in the blood stream for a month or longer, and signs of liver injury take an additional week or more to develop. Because the virus does not cause a cytopathic effect, damage to the liver is probably due to an immune response against virus-induced cellular antigens. In most patients the hepatitis B virus disappears from the blood during the latter part of the illness, but in a few cases, both subclinical and symptomatic, the virus continues to circulate in the blood for months or years. Such cases are of major importance in the epidemiology of hepatitis B and there are an estimated 900,000 such carriers in the United States. The incidence of HBV carriers is even higher in parts of the tropics and in the Eastern hemisphere. Besides being present in the blood, the hepatitis B viral antigen is present in large amounts in saliva and breast milk, but these fluids are less likely to infect other persons than is blood. Five percent or more of pregnant carriers transmit the disease to their babies at the time of birth, and more than two-thirds of those with hepatitis late in pregnancy or soon after delivery do so. The infected babies may die of liver failure, but most survive, usually to become chronic carriers.

As many as 10^{13} to 10^{14} viral particles per milliliter of serum may be present in the blood of carriers, but only a small fraction of these particles are infectious. This small fraction, representing the mature hepatitis B virus, consists of 43 nm-diameter particles having an outer lipoprotein surface antigen and an inner core that contains DNA. The remaining HBV particles are mostly small spherical forms with the scattering of filaments which represent viral coat protein. This material, which is so abundant in the blood of some carriers, can be purified and used as a diagnostic antigen for detecting HBV antibody in patients. Moreover, it can be injected into laboratory animals to stimulate production of specific antiserum for use in detecting circulating HBV surface antigen in patients. Since this virus coat protein stimulates protective antibodies, it could be used for active immunization against hepatitis B.

Seriological tests, page 389

There is no specific treatment for hepatitis, but prospects for its better control have greatly improved as a result of the findings discussed above. Carriers of HBV can now be identified and instructed in measures that avoid transmission of the disease. Likewise, health workers and others at high risk for hepatitis B can undertake rational precautions in handling blood, wearing protective clothing, washing their hands and keeping them out of their mouths. Special precautions can also be taken during pregnancy. Finally, there are now good prospects for the use of active and passive immunization to assist in the control of hepatitis B.

ENTEROBACTERIAL INFECTIONS OF THE GASTROINTESTINAL TRACT

Diarrhea and Vomiting (Gastroenteritis) Caused by *E. coli*

Escherichia coli is an almost universal member of the normal intestinal flora of human beings (and a number of other animals) and was therefore long ignored as a possible cause of gastrointestinal disease. It was not until the early 1960s that *E. coli* became generally recognized as a cause of life-threatening epidemic gastroenteritis in infants. More recently, *E. coli* has been recognized as a cause of gastroenteritis in adults, notably as the agent responsible for many cases of travelers' diarrhea ("Aztec two-step," "Casablanca crud," "Delhi belly," "Hong Kong dog," "Montezuma's revenge," and the like). Understanding of the importance of *E. coli* to gastrointestinal disease has depended on the knowledge that there are in fact hundreds of distinct strains of the bacterium identifiable by serotyping, biotyping, and by other means, and that only those possessing certain virulence factors cause gastrointestinal disease. Table 27-2 gives some of the factors now known to be responsible for the virulence of *E. coli,* but there is strong evidence that yet other factors are to be identified. Some of the virulence factors, such as enterotoxin production and pilus formation, are controlled by plasmids transfer-

Enterobacteria, pages 435–437

TABLE 27-2
Virulence Factors of *E. coli* That Cause Gastrointestinal Disease[a]

Virulence Factor	*Characteristics*	*Laboratory Detection*
Labile toxin (LT)	Exotoxin, molecular weight 10^6; antigenic; inactivated at 60°C; acts by causing increase in cyclic AMP in intestinal cells	Visible change in certain tissue cultures; causes fluid accumulation in small intestine of rabbits
Stable toxin (ST)	Exotoxin, molecular weight 10^3; not antigenic; resists boiling 30 min; does not increase cyclic AMP	Fluid accumulation in small intestine of mice and rabbits
Invasive property	Permits invasion of intestinal cells	Invasion of eyes of guinea pigs or certain tissue cell cultures
Adherence property	Surface material complementary to intestinal cell receptors	Specific antiserum; electron microscopy

[a] *E. coli* strains that cause diarrhea may have none, or one or more of these factors.

able from strain to strain of *E. coli* by conjugation. *Escherichia coli* strains that cause gastroenteritis often have more than one such plasmid.

Dysentery Caused by *Shigella* Species

Another frequent source of intestinal infection are bacteria of the genus *Shigella* (Table 27-3). One of the most severe and extensive shigella epidemics of modern times began in Guatemala early in 1969 with antibiotic-resistant *Shigella dysenteriae* striking almost simultaneously at widely separated villages. Characteristically, the persons infected had fever, diarrhea, and vomiting, with pus and blood appearing in the feces from 12 to 72 hr after onset of the disease. The condition subsided spontaneously in from five to seven days in most instances. The causative organism of this epidemic was not immediately determined because the epidemic strain was unusually fastidious in its growth requirements and most cultural attempts failed to yield *Shigella.* State public health laboratories in the

TABLE 27-3
Enteric Pathogens of the Genera *Salmonella* and *Shigella*

	Characteristics	*Source*	*Pathogenic Potential*
Salmonella	Lactose nonfermenting enterobacteria; usually motile	Cold- and warm-blooded animals, including people	Gastroenteritis; sometimes invasion of the bloodstream; *S. typhi* causes typhoid fever
Shigella	Lactose nonfermenting enterobacteria; nonmotile	Human beings	Gastroenteritis; dysentery

United States began to isolate the agent from travelers returning from Guatemala during the same year, but spread of the agent from these people to those living in the United States occurred only rarely.

PATHOGENESIS OF DYSENTERY

Like *Salmonella* species, *Shigella* organisms enter the mouth on materials contaminated directly or indirectly by feces; their source, however, is essentially always human. The organisms invade and multiply in the cells of the intestinal lining. Death of these cells results in intense inflammation and formation of small intestinal abscesses and ulcerations. Severe dysentery is the result. Children infected with some *Shigella* strains often have headache, a stiff neck, and convulsions. Adults commonly have painful joints for weeks or months following recovery. Because shigellae are not commonly found in the general blood circulation during dysentery, the absorption of a bacterial toxin may explain the extra-intestinal symptoms of this disease. However, only in strains of *Shigella dysenteriae* has the presence of an exotoxin been established.

Toxins, pages 374–375

Control of the spread of shigellae is almost entirely accomplished by sanitary measures and surveillance of food handlers and water supplies. It is important to note in this regard that the Guatemalan epidemic mentioned above followed a severe drought and subsequent flooding during which sanitation suffered.

Gastrointestinal Diseases Caused by *Salmonella* species

Gram-negative rods of the genus *Salmonella* (Table 27-3) are a prominent cause of gastroenteritis, infecting over a million people in the United States every year. These organisms generally enter the gastrointestinal tract with food, although contaminated water, fingers, and other objects may also be sources. The food products most commonly contaminated with *Salmonella* are eggs or egg products, and poultry. Epidemics have also resulted from contaminated brewers' yeast, protein supplements, dry milk, and even a medication used to help diagnose intestinal disease! The reason for such a large variety of substances possibly being contaminated with *Salmonella* is not hard to explain: these organisms infect a wide variety of animals—from cows and chickens to pet turtles—and it is primarily the species of *Salmonella* that originate from animals that produce diarrhea in people.

Virulent salmonellae invade the lining cells of the colon and lower small intestine. An inflammatory response results and causes an increase in fluid secretion and a decrease in fluid absorption by the intestine. Exotoxins are not involved. A few species of *Salmonella* are prone to invade the blood stream and tissues, producing abscesses, fever, and shock with little or no diarrhea (see the following section on typhoid fever).

The causative organisms of salmonellosis are usually present in the feces in large numbers. Because many species of *Salmonella* are relatively resistant to toxic chemicals, selective media containing such substances as bismuth sulfite or brilliant green dye can be used in culturing, to suppress the normal fecal flora and allow the *Salmonella* to grow. In contrast to the majority of strains of *Escherichia,*

Enterobacter, and *Klebsiella,* most strains of *Salmonella* are lactose nonfermenters. Salmonellae are specifically identified by a series of biochemical tests and by reactions with specific antisera against their somatic and flagellar antigens. One of the anomalies of microbiological history is that a different species name has been assigned for each set of these antigens, and consequently about 1000 species of *Salmonella* have been recognized! Present feeling, however, is that only three are justified: *S. enteritidis* (containing the gastroenteritis-causing strains), and *S. typhi* and *S. cholerae-suis* (containing the invaders of the blood stream and tissues).

Control measures for these pathogenic bacteria have concentrated on establishing reporting systems for *Salmonella* epidemics, tracing of sources by careful identification of individual strains, and routine sampling of animal products for contamination. Unfortunately, the incidence of reported cases of salmonellosis has tended to rise, and the organisms have also shown increasing plasmid-mediated resistance to antimicrobial medicines. This is partly due to selection of resistant strains of *Salmonella* by the widespread use of such medicines in animal feeds.

R factors, pages 209–211

Typhoid Fever

People sick with typhoid typically develop a fever, which increases over a three-day period and is followed by severe headache and abdominal pain. In some cases, there is rupture of the intestine and shock from loss of blood. If treatment is not given, about one out of five persons dies of typhoid. The causative organism of typhoid, *Salmonella typhi,* infects only human beings. It usually enters the gastrointestinal tract in contaminated food or water, as do other salmonellae, and like them readily penetrates the intestinal lining. Phagocytic defense cells appear in the gut wall and ingest the invaders but the typhoid bacilli multiply inside these cells and are carried to other parts of the body by the bloodstream. *Salmonella typhi* also localize in the collection of lymphoid cells of the gut wall, and destruction of this tissue can lead to intestinal rupture and hemorrhage.

Salmonella typhi, page 436

Typhoid organisms can be recovered from the blood or bone marrow of an infected person, and later from the person's feces, by using selective and enrichment media. Over a 1- to 2-week period, a rise in antibody titer against the organisms can often be demonstrated in the blood of an infected person and may be helpful in identifying the cause of his illness.

This strictly human disease is maintained in nature by carriers, people who appear perfectly well but who may excrete as many as 10^{10} typhoid bacilli per gram of their feces. Because normally, far fewer of the organisms than this are required to infect, it is easy to see how dangerous typhoid carriers can be. One of the most notorious carriers was Typhoid Mary, a young Irish cook living in New York State in the early 1900s. She is known to have been responsible for at least 53 cases of typhoid fever transmitted during a 15-year period. At that time about 350,000 cases of typhoid occurred in the United States each year. Today, with improved sanitation and public health surveillance measures the reported incidence of typhoid fever in the United States is about 300 cases per year.

A vaccine against *S. typhi* and related organisms is widely used by the military and in countries of high typhoid prevalence and is effective in reducing the incidence and severity of this disease; it does not, however, offer complete protection. Searching out and treating typhoid carriers is another control measure. Carriers can be detected by isolating *S. typhi* from their stool cultures and by serological tests for antibodies against the Vi (capsular) antigen of *S. typhi.*

CHOLERA

Cholera is a unique form of severe diarrhea which was limited largely to India and Southeast Asia until only recently. From 1961 to 1972, however, extension of cholera into the Middle East, U.S.S.R., and Africa has occurred (Figure 27-4). Since then, the incidence has declined, but in 1975 29 countries were still involved, with 87,500 cases. Cholera is characterized by profuse outpouring of fluid from the intestine within a few hours, a loss often amounting to 15 percent of the body weight. So much of the body's water and electrolytes leave that the blood becomes thickened and reduced in volume, producing insufficient blood flow to keep vital organs, such as the kidney, working properly. Many people die unless this lost fluid can be replaced promptly.

Vibrio cholerae, page 437

Cyclic AMP, pages 231–232

Much research has been done to determine how *Vibrio cholerae* produces cholera. Unlike salmonella and shigella infections, there is no visible damage to the lining cells of the intestine in cholera. The pathogenesis of the disease depends on an exotoxin that acts by causing excessive secretion of water and electrolytes by the cells of the small intestine. As with the heat-labile toxin of *E. coli,* the toxin of *V. cholerae* induces increased levels of cyclic AMP in the intestinal

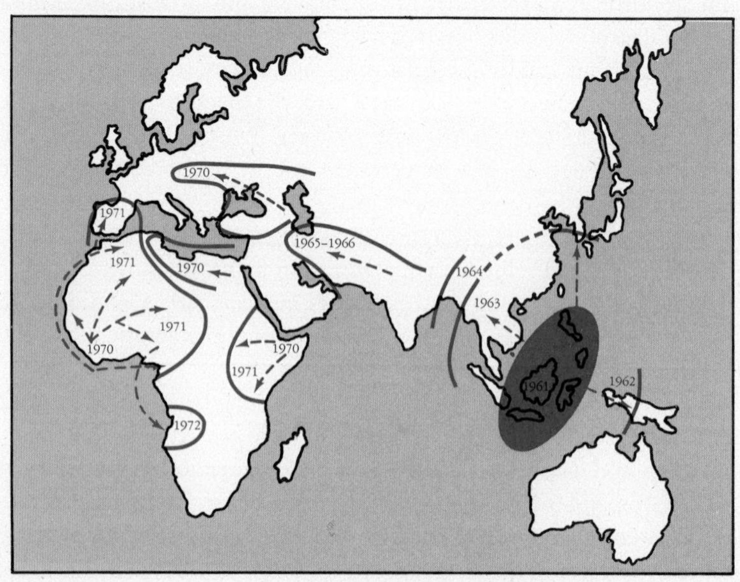

FIGURE 27-4
The spread of cholera from 1961 to 1972. (From Morbidity and Mortality Weekly Report, United States Public Health Service 21:171, 1972.)

cells, resulting in maximal secretory activity. The colon is not affected, but its absorptive capacity is exceeded by the huge volumes of fluid that rush through it.

Dogs may harbor *V. cholerae,* but human beings are the most important reservoir of this organism. Between epidemics of cholera the *V. cholerae* organisms persist in the intestines of their human carriers, who have been shown to excrete them for periods as long as six years. Control of cholera is aided by administering a vaccine of killed *V. cholerae;* this stimulates bactericidal and antitoxic antibodies in its recipients, but protection lasts only a few months. Well-nourished persons living under sanitary conditions appear to be quite resistant to cholera. Indeed, over the decade ending in 1970, only six cholera cases and no deaths were reported among the millions of American travelers in areas of the world where cholera existed. The requirement of cholera vaccination for travelers entering the United States has been dropped.

PROTOZOAN INFECTIONS OF THE GASTROINTESTINAL TRACT

Several species of protozoa produce intestinal infections and may be important in causing human disease. All of the major protozoan groups (Chapter 15) are represented, as, for example, *Giardia lamblia* (flagellate), *Balantidium coli* (ciliate), *Isopora belli* (sporozoan), and *Entamoeba histolytica* (ameba).

See Table 15-1, page 305

Giardia lamblia (Figure 27-5) is one of the most widespread of the intestinal protozoa, infecting both human beings and other animals. Numerous epidemics of giardiasis have occurred among travelers, especially to the U.S.S.R. and Southeast Asia, and in rural communities in the United States and Canada. The symptoms of this infection are usually chronic and mild and include diarrhea, nausea, indigestion, flatulence (gas), fatigue, and weight loss. The infecting organisms attach to the lining of the small intestine, and in heavy infestations may cause an inflammatory reaction and interference with the absorption of foodstuffs. Diagnosis of the infection is made by identifying the cysts or trophozoites of *Giardia lamblia* in the feces, although small intestinal biopsy is sometimes necessary. Surface water contaminated by wild animals or human beings is a common source of giardiasis, but person-to-person spread has occurred in nurseries.

Perhaps the most important of the protozoa producing intestinal infections of humans is *E. histolytica* (Figure 27-6). This ameba, one of the causes of dysentery, is found worldwide, being especially prevalent in areas of poor sanitation. Encysted *E. histolytica* organisms usually enter the alimentary tract with contaminated food or water and, once in the intestine, liberate trophozoites. Upon reaching the upper portion of the large bowel, these trophozoites begin feeding on mucus and cells lining the intestine. The amebae also produce several digestive enzymes which aid them in penetrating through the lining cells into the intestinal wall. In fact, they may burrow into blood vessels and be carried to the liver or other body organs. Continued multiplication and tissue destruction at intestinal and extraintestinal sites result in the formation of abscesses. The irritant effect of the amebae on the intestinal lining cells results in increased movement of the

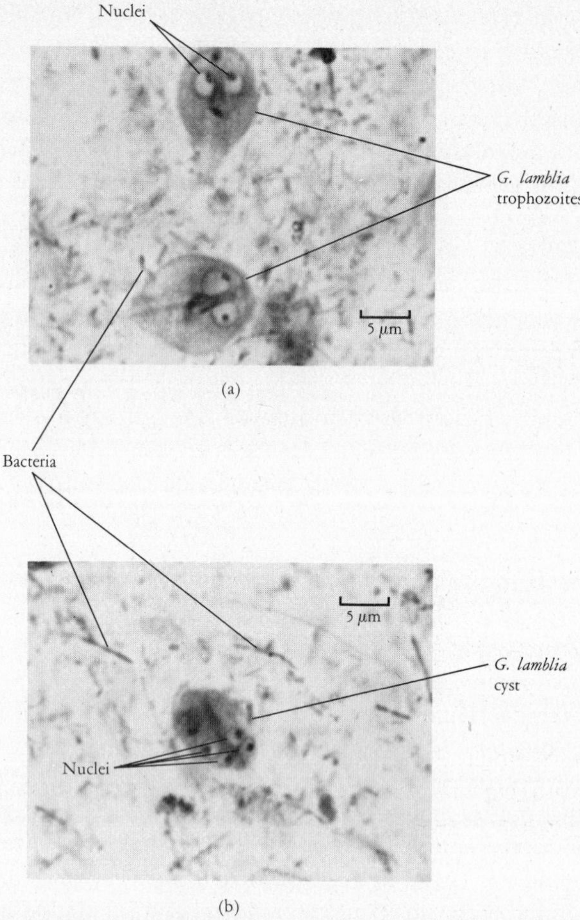

FIGURE 27-5
Giardia lamblia in feces. Photographs of stained smears. (a) Trophozoites. (b) Cyst. (Courtesy of M. F. Lampe.)

bowel and in production of fluid. Thus diarrhea—often tinged with blood—is an effect of amebic dysentery. Direct examination of diarrheal fluid on a warm microscope slide may reveal the motile trophozoites of *E. histolytica,* which generally range from about 20 to 40 μm in diameter. Many persons develop a chronic infection which may be asymptomatic and therefore go unnoticed. The trophozoites of *E. histolytica* are not usually seen in examination of the feces of such patients, but cysts of the organism are commonly present. On passage through the large bowel the cysts typically develop four nuclei, and in this form are infectious for the next host. The cysts of *E. histolytica* can be identified by direct microscopic examination of the feces of infected persons, and methods are also available for concentrating them from fecal material to aid in their detection.

Entamoeba histolytica can be grown anaerobically in pure cultures or in aerobic fluid cultures containing bacteria. The organisms grow better when certain bacteria are added to the medium, and a mixture of added bacteria gives better amebic growth than does a single bacterial species. Indeed, in studies with some germ-free animals, *E. histolytica* has caused little or no injury to the intestinal lining unless bacteria were also introduced. And in the treatment of persons with

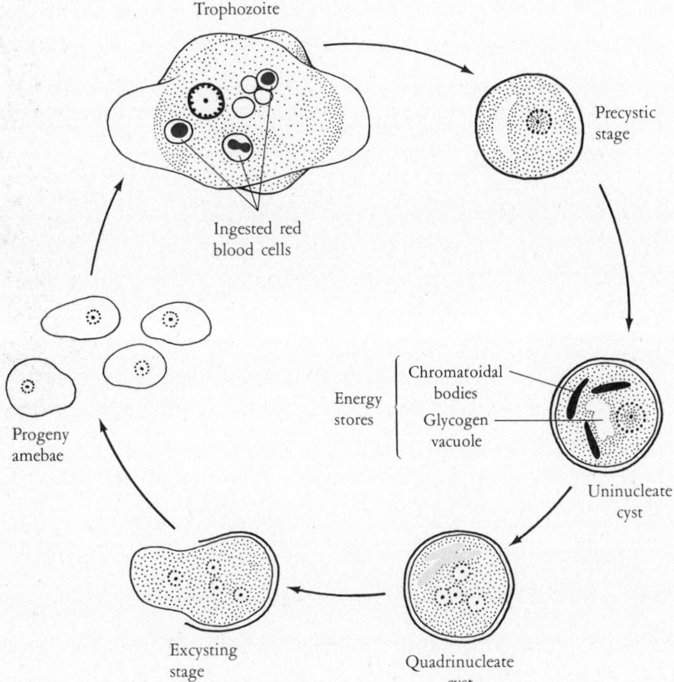

FIGURE 27-6
Life cycle of *Entamoeba histolytica*. The quadrinucleate cyst is the infectious stage.

ROLE OF INTESTINAL BACTERIA

amebic dysentery, antibiotics active against bacteria are often effective against the disease even though they have no direct activity against the amebae. This indicates that a synergistic action of amebae and bacteria is required to produce amebic dysentery.

FOOD POISONINGS

Illness resulting from microbial growth in food comes chiefly by two different mechanisms: (1) the contaminating microorganisms may infect the person who ingests the food, or (2) products of microbial growth in the food may be poisonous. Gastroenteritis, dysentery, and typhoid fever caused by bacteria of the genera *Salmonella* and *Shigella* may originate from the first mechanism, whereas a wide variety of common microbial species (including members of the bacterial genera *Clostridium, Bacillus,* and *Staphylococcus*) may be responsible for illnesses originating by the second mechanism. Although all illnesses coming from microbial growth in ingested food are popularly considered "food poisoning," the following discussion deals with those illnesses caused by microbial products rather than those coming from direct infection.

Staphylococcal Food Poisoning

One of the most common forms of food poisoning is due to *Staphylococcus aureus,* the same species of bacterium responsible for boils and other infections. The illness ensues a few hours after ingesting contaminated food, is manifested by nausea, vomiting, and diarrhea, and is followed by recovery within another few hours. This demoralizing but rarely fatal type of food poisoning results from an exotoxin produced when certain strains of *S. aureus* grow in a suitable foodstuff, usually one high in carbohydrates. The exotoxin is called an enterotoxin ("entero-" referring to intestine) and it is not to be confused with the endotoxin of the cell walls of Gram-negative rods. It may be produced quickly by staphylococci growing at room temperature, often within only 1 or 2 hr. Moreover the toxin is relatively heat stable, so that subsequent cooking may kill the staphylococci, but leave toxic activity. Several antigenically distinct varieties of enterotoxins are known to be produced by the various food-poisoning strains of *S. aureus*, and can be identified in extracts of contaminated foods by using specific antisera. *Clostridium perfringens*, which produces a heat-labile enterotoxin during its process of sporulation, is another common cause of food poisoning. The pathogenesis of the food poisoning caused by *Bacillus cereus* has also been ascribed to an exotoxin.

HEAT-STABLE ENTEROTOXIN

Poisonings Caused by Fungi

See Further Reading, page 203

In earlier ages, ergot poisoning ("St. Anthony's Fire") followed the ingestion of grain contaminated with certain fungi. Persons afflicted with this condition had agonizingly painful convulsions, and some developed gangrene of their hands and feet.

Proper standards of agricultural practice and public health surveillance of food grains now make it unlikely that ergot-producing fungi will contaminate food and give ergot toxicity. However, another kind of fungal toxin is of great current interest. Called *aflatoxin*, this fungal poison is produced by common molds of the genus *Aspergillus*. Aflatoxin can cause acute poisoning in many species of animal, affecting chiefly the young. A few milligrams, for example, may cause severe liver damage and death of a dog within 72 hr. Much interest in aflatoxin arose with the observation that tiny traces of this poison in the foodstuffs of certain animals caused tumors. In fact, for trout, aflatoxin is one of the most potent tumor inducing substances known. Aflatoxins have been demonstrated in many human foods, and high levels of an aflatoxin metabolite may appear in cow's milk. Aflatoxins produced by various strains of molds are not always the same and a family of related compounds is now also known. The possible role of these substances in human disease is still under study.

AFLATOXIN

Botulism

References were made in Chapter 13 to the paralysis of human beings and other animals that eat shellfish containing red-tide algae of the genus *Gonyaulax*. Botulism, another microbial poisoning that is characterized by paralysis, is one of the most feared diseases of mankind. Its cause is the Gram-positive, rod-shaped, strictly anaerobic bacterium, *Clostridium botulinum*, widely distributed in soils

Gonyaulax, page 285

around the world. The name *botulinum* comes from the Latin word for sausage, and was chosen because some of the earliest recognized cases of botulism occurred in people who had eaten contaminated sausage. However, many other foods have been sources of this food poisoning, including vegetables, fruits, meats, and cheese. Like other clostridia, *C. botulinum* produces endospores which may be unusually resistant to heat, and can thus persist in foodstuffs despite cooking and canning processes. These spores can later germinate and growth of the bacterium may result in the release of a powerful exotoxin into the food. When someone eats the contaminated food, the toxin is absorbed into the bloodstream where it may continue to circulate for as long as three weeks. This circulating toxin is carried to the various nerves of the body, where it acts by blocking the transmission of nerve signals to the muscles, thus producing paralysis. Twelve to 36 hr after a person has eaten toxin-containing food, blurred or double vision gives the first indication of this paralysis. All muscles may be affected, but respiratory paralysis is the most common cause of death, which ensues, despite treatment, in about one-fourth of the victims of botulism. Botulism has no effect on sensation or on most other functions of the brain, but the toxin causing it is one of the most powerful poisons known, a few milligrams being sufficient to kill the entire population of New York City. Indeed, botulism has resulted from eating a single contaminated string bean, or even from licking a finger contaminated with toxin. Fortunately, the toxin is highly heat-labile, and even high concentrations are completely inactivated by boiling of contaminated food for 15 min.

Clostridium botulinum, page 431

PREVENTION OF BOTULISM

In the five years preceding 1922 there were 83 outbreaks of botulism in the United States, many traceable to commercially canned foods. The work of the distinguished microbiologist Dr. Karl F. Meyer and his colleagues showed that canning methods then in use were inadequate to kill the heat-resistant spores of *C. botulinum.* Strict controls were then placed on commercial canners to ensure adequate sterilizing methods. Since then, outbreaks caused by commercially canned foods have been infrequent. The occasional lapses in proper canning technique point up the potential for disaster when defective processing occurs in large food-producing and distributing firms.

Most cases of botulism from inadequately processed canned food have been caused by organisms of the type A or B strains. Awareness of major outbreaks of type E botulism occurred in the early 1960s, primarily in Japan and the United States. Like other types of *C. botulinum,* type E strains are soil organisms, but they are more readily taken up by fish and sea mammals. Type E strains also differ in that they can produce toxin at lower temperatures and under greater exposure to aerobic conditions. For these reasons type E botulism is generally associated with ingestion of contaminated seafoods.

Botulism is treated by administering the appropriate antitoxin. This neutralizes toxin circulating in the bloodstream, but recovery of the nerves is slow, requiring weeks or months. Control of the disease depends on proper sterilization and sealing of food at the time of canning and heating it adequately preparatory to serving. One cannot rely on a "spoiled" taste or appearance to detect contamination, because such changes may not always be present.

Antitoxin, pages 396–397

INFANT BOTULISM

Proper food handling cannot completely eliminate the danger of botulism because *C. botulinum* can occasionally colonize wounds or the intestine and produce a mild form of the disease. In 1976, for example, four previously healthy infants were hospitalized in parts of California because of paralysis caused by botulism. *Clostridium botulinum* organisms and toxin were present in their feces, but the toxin was not detectable in their blood. Although one infant required respiratory support, and another tube feeding, all promptly recovered without receiving antitoxin treatment. Recovery was probably the result of their own antibody production. The source of these infants' infections was undoubtedly *C. botulinum* spores, which are ubiquitous in dust and foodstuffs, and *Clostridium botulinum* organisms and toxin persisted in the infants' feces for a time after their recovery, but were gradually replaced by competing normal flora.

SUMMARY

One of the major routes of microbial invasion of the human body is the alimentary tract; resulting diseases may involve any part of the body including the alimentary system itself. Members of the normal alimentary flora with little invasive ability can produce dental caries or the blind loop syndrome. Certain viral infections have specific tissue affinities, such as for the parotid glands in mumps and the liver in hepatitis. The genus *Salmonella* is responsible for a wide spectrum of human diseases, ranging from gastroenteritis (generally from species colonizing other animals) to the more serious typhoid fever from *Salmonella typhi,* a strain infecting only human beings. *Shigella,* another facultatively anaerobic Gram-negative rod, is a frequent cause of dysentery as is the ameba *Entamoeba histolytica.* The toxins of *Vibrio cholerae* and *E. coli* greatly increase intestinal fluid excretion, producing watery diarrhea and dehydration.

Microbial food poisonings result from eating foods contaminated with certain toxic microbes or with the by-products of their growth in the food. Bacteria of the genera *Bacillus, Clostridium,* and *Staphylococcus* are commonly responsible for poisoning manifested by nausea, vomiting, and diarrhea. In the case of *Staphylococcus aueus,* a heat-stable enterotoxin is responsible. Fungal toxins include the aflatoxins of *Aspergillus* species, which can produce liver damage and tumors in some animals. One of the most feared types of poisonings is botulism, an often fatal paralysis caused by the exotoxins of *Clostridium botulinum.*

QUESTIONS

REVIEW

1. Give the mechanisms by which microorganisms can interfere with the normal function of the gastrointestinal tract. Why is cholera such a devastating disease when *Vibrio cholerae* neither invades the body nor destroys the intestinal mucosa?
2. What is the relative importance of microbial and dietary factors in dental caries?

3. Should one consider blood agar prepared from human blood to be sterile? Explain. What danger is there in washing a test tube containing human serum?
4. Explain how a foodstuff responsible for staphylococcal food poisoning may fail to yield growth of *Staphylococcus aureus* when inoculated onto appropriate culture media.
5. What simple procedure, that can be carried out in every household kitchen, will assure that no one gets botulism from canned food?

THOUGHT

1. Discuss the role of selective adherence in the pathogenesis of infectious diseases of the alimentary tract.
2. Because enterotoxin production by *E. coli* is governed by plasmids, what is the likelihood that other enterobacteria will be found to produce enterotoxins?

FURTHER READING

BOWEN, W. H., "Prospects for the Prevention of Dental Caries," *Hospital Practice* 9:163–168 (May 1974).

GIBBONS, R. J., and J. VAN HOUTE, "Bacterial Adherence in Oral Microbial Ecology." *Annual Review of Microbiology.* Ed. M. P. Starr et al. Vol. 29. Palo Alto, Calif.: Annual Reviews, 1975.

GOLDMAN, P., "Therapeutic Implications of the Intestinal Microflora," *New England Journal of Medicine* 289:623–628 (September 1973).

HIRSCHHORN, N., and W. B. GREENOUGH, "Cholera," *Scientific American* (August 1971).

ROUECHE, B., *Annals of Epidemiology.* Boston: Little, Brown, 1967. Popular accounts of episodes of botulism, typhoid fever, and *Salmonella* gastroenteritis.

———, "The Simpsons and the Hepatitides," *The New Yorker Magazine* (August 21, 1971), pp. 72–86. Concerns hepatitis.

CHAPTER 28

INFECTIONS OF THE GENITOURINARY TRACT

The human body is efficiently constructed to prevent microbial invasion, a fact which is particularly evident in studying the anatomy of the genitourinary tract. Under certain circumstances, however, the genitourinary tract can be invaded by a variety of organisms from the normal flora, which act as opportunists, as well as by pathogenic species.

ANATOMICAL AND PHYSIOLOGICAL CONSIDERATIONS

The urinary tract consists of the kidneys (in the upper urinary tract), the bladder (in the lower urinary tract), and accessory structures. The kidneys act as a specialized filtering system to cleanse the blood of many waste materials, selectively reabsorbing other substances that can be reutilized. Figure 28-1 shows that each kidney is drained by a tube, the ureter, which connects it with the urinary bladder. The bladder empties through the urethra. Special groups of muscles near the urethra act as sphincters that keep the system closed most of the time and help prevent infection. The downward flow of urine also helps cleanse the system by flushing out microorganisms that may have invaded it, before they have a chance to multiply and infect.

Even when infection of the lower urinary tract occurs, the kidneys are still protected to some extent by the action of other sphincter muscles at the junctions of the bladder and ureters. It is fortunate that mechanisms exist to protect the kidney against infection, because many kidney infections can be very difficult to eradicate, are often chronic, and can lead to marked destruction of the kidneys. Death follows kidney failure promptly unless the patient is lucky enough either to be able to use one of the all too scarce artificial kidneys, or to receive a kidney transplant.

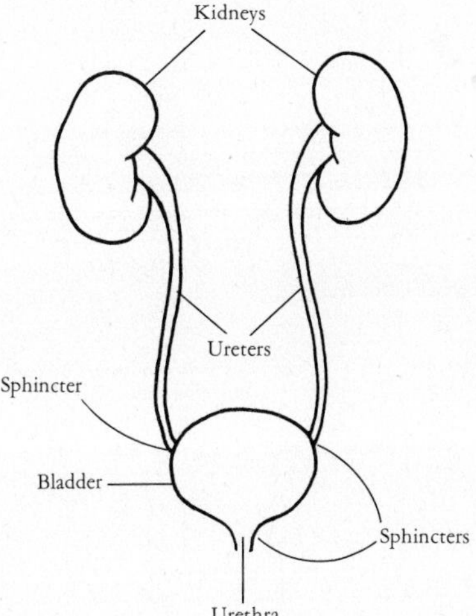

FIGURE 28-1
The anatomy of the urinary tract. Urine flows from the kidneys down the ureters into the bladder, which empties through the urethra. Sphincter muscles help to prevent any contamination from ascending.

The urinary tract is protected from infection by a number of mechanisms besides its anatomy. Normal urine contains organic acids and other antimicrobial substances, as well as small quantities of antibodies. During urinary tract infections larger quantities of specific antibodies can be found in the urine. There is also evidence that antibody-forming lymphoid cells infiltrate the infected kidneys or bladder and form protective antibodies locally at the site where they are needed. Also, during infection there is an inflammatory response in which phagocytes are of the utmost importance in engulfing and destroying the invading microorganisms.

The anatomy of the male and female genital tracts is shown in Figure 28-2. When an ovum is extruded from the ovary, it is swept into the adjacent fallopian tube, and then into the uterus, by the action of ciliated cells (Figure 28-2b). Normally the uterus is protected from infection by sphincter muscles at the cervical canal, which connects the cervix (lower third of the uterus) with the vagina. Because the vagina is the portion of the female genital tract connecting the female genital system with the environment, it is the most frequent site of infection of this system.

NORMAL FLORA OF THE GENITOURINARY TRACT

Normally the urine and the urinary tract above the entrance to the bladder are completely free of microorganisms; the urethra, however, has a normal resident flora. Species of *Streptococcus* (alpha hemolytic or enterococci), *Neisseria, Mycobacterium, Bacteroides,* and some of the enterobacteria are commonly isolated from the normal urethra.

The normal flora of the female genital tract is influenced by the action of estrogen hormones on the lining (epithelial) cells of the vaginal mucosa. When estrogens are present, glycogen is deposited in these epithelial cells. This glycogen

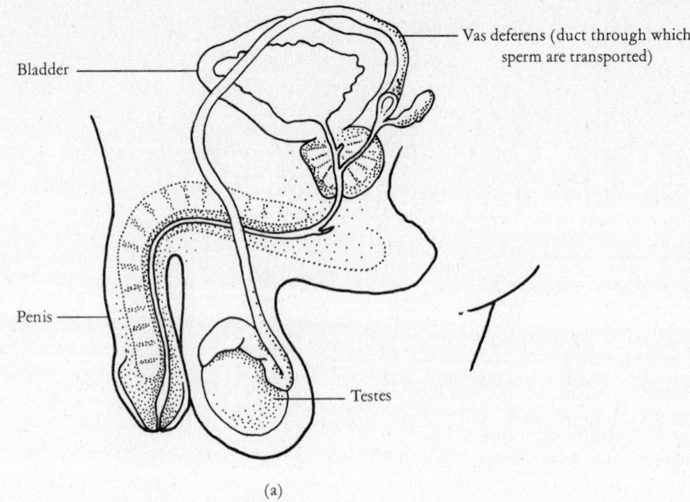

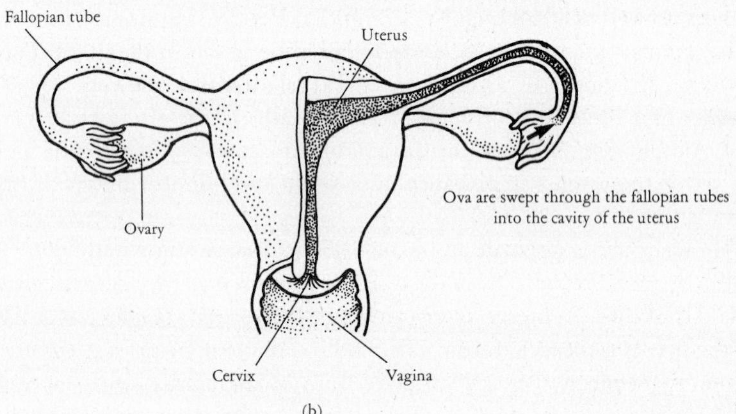

FIGURE 28-2
The anatomy of the genital tract. (a) Male. (b) Female.

is then converted to lactic acid by lactobacilli, resulting in an acid pH of the genital tract. Thus the normal flora and resistance to infection of the female genital tract vary considerably with the hormonal status of the female. For several weeks after birth the vagina of newborn females also remains under the influence of maternal hormones and has an acid pH. During this time lactobacilli are predominant in its normal flora. As the influence of the maternal hormones wanes, however, the pH of the infant vagina increases to neutrality and remains neutral until puberty. During this childhood period the normal flora of the tract consists of a variety of cocci and rod-shaped organisms, including species of *Streptococcus* and many other genera. At puberty lactobacilli again become predominant, although smaller numbers of yeasts and other bacterial species are also present. After the menopause, there is a return of the tract to a neutral pH and to the mixed flora typical of childhood.

URINARY TRACT INFECTIONS

A number of factors can predispose the urinary tract to infection. Any situation that leads to stasis of the urine increases the chances of such infection. After anesthesia and major surgery, for example, the reflex ability to void urine may be inhibited for a time. In this circumstance urine accumulates and distends the very elastic bladder. Even a few bacteria which have managed to evade the urinary tract defenses and enter the bladder can multiply during stasis, causing infection of the bladder. Obviously, urine is a good culture medium for many bacteria, providing abundant nutrients. Paraplegics and persons paralyzed in the lower part of their bodies also commonly suffer urinary tract difficulties. Because they lack nervous control, paraplegics are unable to void normally and almost invariably develop urinary tract infections.

Because the vagina is closely adjacent to the urethra in the female (Figure 28-2b), the pressure associated with coitus may permit the introduction of organisms into the urinary tract. The bladder infections which result are often referred to as "honeymoon cystitis," a not infrequent disease accompanying initial sexual experience.

Unless such infection of the lower urinary tract is overcome, either by successful treatment or by natural immune mechanisms, it can ascend through the ureters to involve the kidneys. Thus most kidney infections develop subsequent to bladder infection. Bacteria carried by the bloodstream can also infect the kidneys, but this occurs much less often than ascending urinary tract infections.

Infections of the urinary tract occur far more frequently in females than in males because the female urethra is short (about 1.5 inches, compared with 8 inches in the male), and closely adjacent to the genital and intestinal tracts. More than 90 percent of urinary tract infections are caused by bacterial species that are part of the normal intestinal flora and that are consequently readily available to invade the urinary tract when its natural defenses are interrupted (Figure 28-3). Serological studies of infecting organisms have shown that they are usually derived from the patient's own intestinal bacteria. In these studies, for example, *Escherichia coli* isolated from the urine during urinary tract infections was almost always of the same *serotype* as *E. coli* from the intestine of the patient.

Escherichia coli accounts for about 50 percent of urinary tract infections, but other Gram-negative rods of different genera, such as *Proteus* and *Enterobacter,* are also often the cause (Figure 28-3). *Pseudomonas aeruginosa,* an aerobic Gram-negative rod, is a particularly troublesome urinary tract pathogen because it is difficult to treat successfully. *Streptococcus faecalis* (enterococcus), found in the normal bowel flora, is the most frequently isolated Gram-positive organism responsible for urinary tract infections.

The causative organisms of urinary tract infections can usually be recovered easily and in large numbers from the urine of infected patients. In doing this, it is necessary to collect voided urine specimens carefully, in order to minimize external contamination from the patient's genitalia or intestinal tract. This is more difficult in the female than the male and is best accomplished by collecting a midvoiding sample after carefully cleansing the external genitalia. Normal urine

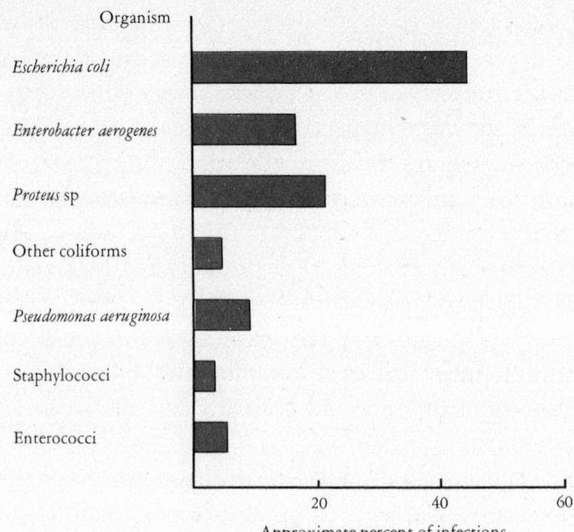

FIGURE 28-3
Common causes of urinary tract infections. The percentages vary considerably from one study to another. Often infections are caused by more than one kind of organism.

collected in this way usually contains less than 10,000 bacteria per milliliter, as determined by colony counts. Concentrations of bacteria greater than 100,000 per milliliter indicate infection. It is important to remember that bacteria multiply exponentially; and that cultures for colony counts must therefore be prepared promptly after the urine is collected in order to obtain accurate counts of urinary bacteria.

Urine specimens are cultured on media especially suitable for the growth of enterobacteria and on blood agar, which will also support the growth of enterococci and other Gram-positive pathogens. *Escherichia coli,* the organism most often responsible for urinary tract infection, is the prototype of the enterobacteria.

During some diseases in which bacteria primarily attack areas of the body other than the urinary tract, the causative organisms may also be found in the urine. Thus in typhoid fever, *Salmonella typhi* become disseminated throughout the body and are excreted in the urine beginning at about one to two weeks after infection, and sometimes establishing a chronic urinary tract infection. In like manner slender spirochetes of the genus *Leptospira* (the cause of leptospirosis) may become established in the urine. Contamination of water with urine from infected human beings or animals is the major means of transmission of leptospirosis. The *Leptospira* organisms may then continue to be excreted in the urine for many months after recovery from the illness. Leptospirosis is a common infection of animals and is one of three diseases against which dogs are commonly immunized.

BACTERIAL INFECTIONS OF THE GENITAL TRACT

Puerperal Fever

Group A beta hemolytic streptococci (*S. pyogenes*) are a major cause of puerperal fever (childbirth fever). They are transmitted by direct contact at childbirth with contaminated instruments or hands. In addition to *S. pyogenes,* the causative bacteria of puerperal fever may be some of the anaerobic species (clostridia, bacteroides, and anaerobic streptococci) that infect wounds and find a fertile area for growth in tissues injured by trauma.

The history of puerperal fever illustrates the general human tendency to resist change. For centuries it had been an accepted fact that many mothers developed fever and died following childbirth. About the middle of the nineteenth century in the hospitals of Vienna (the major medical center of the world at that time), about one of every eight women died of puerperal fever following childbirth. A Hungarian physician named Semmelweiss, who had come to Vienna to practice, studied this abhorrent situation and concluded that the disease was transmitted by the attending physicians or their instruments. He instituted a program of simple disinfection of all instruments used during delivery and was able to cut the maternal death rate at childbirth from about 12 percent to less than 1 percent. Instead of accepting these findings and the new technique of disinfection Semmelweiss used, however, his colleagues at Vienna refused to face the fact that they had been responsible, even unknowingly, for the death of so many patients. The work of Semmelweiss was attacked, and he was forced to leave Vienna and return to his native Hungary. There he was again able to use disinfection techniques and to achieve a remarkable reduction in the number of deaths from puerperal fever. The famous American physician, Dr. Oliver Wendell Holmes, independently made the same findings at about the same time.

Over the years, with increasing improvement (and acceptance) of disinfection and sterilization techniques, puerperal fever has become rare in countries with adequate medical facilities. Even today, however, outbreaks occasionally occur and it may be very difficult to identify the source. Not long ago a number of cases of puerperal fever occurred in the United States despite careful precautions and the use of proper antiseptic techniques. In one report it was found that a medical attendant was an anal carrier of *S. pyogenes,* which was disseminated by the movement of his clothing.

Veneral Diseases

Although there are a number of venereal diseases, by far the most prevalent is *gonorrhea.* At present in the United States, the incidence of gonorrhea is the highest of any reportable bacterial disease, and from 1963 to 1976 it increased at the alarming rate of 15 percent each year (Figure 28-4). The causative organism, *Neisseria gonorrhoeae* (gonococcus), is a Gram-negative diplococcus, identical in appearance to the closely related *N. meningitidis* (the cause of epidemic meningitis). The gonococci are flattened on adjacent sides, and are frequently found within certain white blood cells (PMNs) in pus, as indicated in Figure 28-5. Gonococci are parasites of human beings only, preferring to live on the mucous membranes of their host. They are susceptible to cold and drying and hence do not survive well outside their host. For this reason gonorrhea is transmitted primarily by direct contact, and as the bacteria live primarily in the genital tract, this contact is almost always venereal. There are certain exceptions to this general

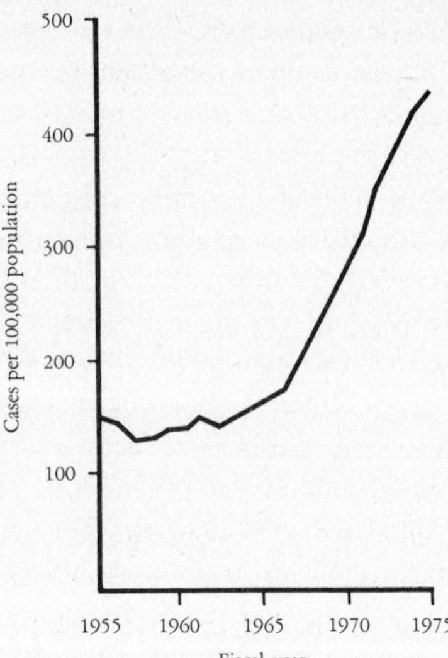

FIGURE 28-4
Reported civilian cases of gonorrhea in the United States from 1955 through 1974. (Morbidity and Mortality Report, United States Public Health Service 24:61, 1975.)

rule. One form of gonococcal disease, for instance, is an eye infection (ophthalmia neonatorum) that can be transmitted at birth from mother to infant during the infant's passage through an infected birth canal. This form of gonococcal infection has been controlled in the United States by laws requiring the use of silver nitrate or an antibiotic instilled directly into the eyes of all newborn infants.

Gonococci grow in the genitourinary tract (and sometimes in the pharynx) only in areas where a certain kind of epithelial cell—the columnar epithelial cell—is found. Pathogenic gonococci possess pili that attach selectively to columnar epithelial cells; thus there is a positive correlation between the presence of pili and the ability of gonococci to adhere to mucous membranes and establish infection. It has been postulated, although not proven, that the adhering bacteria may produce toxins that injure adjacent host cells. This is an interesting speculation, but in fact the mechanisms whereby gonococci cause disease are almost entirely unknown.

Gonorrhea differs in males and females, at least in part because the location

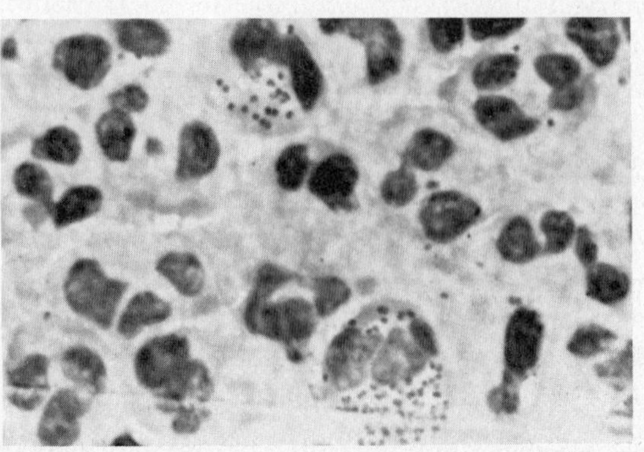

FIGURE 28-5
Neisseria gonorrhoeae in Gram-stained material from a patient with gonorrhea. The diplococci are abundant within polymorphonuclear leukocytes in the pus.

of columnar epithelial cells in the two sexes differs. In males the disease is characterized by inflammation of the urethra, with pain during urination and a thick, pus-containing discharge which becomes apparent a few days after infection. Usually the disease in males is self-limiting and clears of its own accord, but the infection may extend itself in the genital tract, producing complications. As an example, extensive inflammatory reaction to the infection can lead to the formation of fibrous tissue that obstructs the urethra, predisposing to future urinary tract infection. If fibrous tissue occludes the tubes that carry the sperm (vas deferens) or if tissue in the testes is destroyed by inflammation, sterility may result.

Gonorrhea follows a different course in women. Gonococci thrive in the cervix and fallopian tubes (Figure 28-2b) as well as in other areas of the female genital tract. The early stages of infection, involving the cervix, are commonly mild, and the victim is usually unaware of the disease. Females are thus especially prone to become carriers of gonorrhea. If the disease is not limited at this stage, infection frequently progresses upward into the uterus and the fallopian tubes. Scar and fibrous tissue formed in response to the infection may block normal passage of the ova through the tubes, and gonorrhea is thus a cause of sterility in females. Obstruction of a fallopian tube can also lead to a dangerous complication—ectopic pregnancy—in which the ovum is fertilized and develops in the fallopian tube or in the abdominal cavity outside the uterus.

Another common complication of gonorrhea, in both females and males, is gonococcal arthritis caused by the growth of gonococci within the joint spaces. Any joint may be affected, especially the larger ones. It has been shown that some strains of gonococci have a greater tendency than others to disseminate to the joints or other areas distant from the original site of their infection.

In both males and females, gonococci can be recovered from purulent discharge or from the secretions collected from infected areas. Gram-stained smears often reveal the characteristic Gram-negative diplococci within the PMN cells of pus, but in order to identify the organisms as *N. gonorrhoeae* it is necessary to culture them and to carry out certain biochemical and serologic tests (Chapter 22).

Gonococci,
page 433

Although gonococci are usually susceptible to penicillin, relatively resistant strains are being selected. Thus, 20 years ago 300,000 units of penicillin were usually sufficient to cure a case of gonorrhea, while today a dose of 4.8 million units is often necessary, and even this dose is not uniformly successful. In 1976 penicillinase-producing strains of gonococci were isolated from patients in several areas of the world. This penicillin resistance has been shown to result from the acquisition of a plasmid.

The widespread use of oral contraceptives ("the pill") has contributed to the increased incidence of gonorrhea and other venereal diseases in a variety of ways, both social and medical. Several studies have shown that, on the average, women taking oral contraceptives begin having intercourse at an earlier age and have more sex partners than was the case before the "pill." Then too, mechanical contraceptives, such as diaphragms and condoms, which impose a physical barrier to infection, are not as frequently employed today as they previously were. In addition, long-term use of some oral contraceptives leads to a migration of susceptible epithelial cells from upper areas of the cervix into more exposed areas of the outer cervix. Oral contraceptives also tend to increase both the pH and the moisture content of the vagina. All of these factors favor infection, not only with gonococci but with several other agents that cause venereal diseases as well. Women taking oral contraceptives are thus especially vulnerable to gonorrheal infection and are also more liable to develop serious complications if they contract the disease. Furthermore, such complications usually become apparent early in these women, often within a few days after infection.

Several other factors contribute to the increasing incidence of gonorrhea. Carriers of gonococci, both male and female, can unknowingly transmit these bacteria over months or even years. Identification of carriers is presently critical in controlling gonorrhea. Another factor contributing to the increased incidence of gonorrhea is the lack of solid immunity following recovery from the disease. The individual who has recovered from gonorrhea is susceptible to reinfection. Studies of the immunology of gonorrhea are incomplete, and it is uncertain whether repeated infection results from brief immunity or from a multiplicity of immunologically distinct strains of the gonococcus. A great effort, so far unsuccessful, is being made to develop a vaccine that will provide effective protection against gonococci.

Syphilis, another venereal disease of major importance, is caused by a spirochete, *Treponema pallidum.* The organisms (treponemes) of syphilis are motile, extremely thin, tightly coiled spirals. They cannot be seen on Gram-stained smears because they are so thin; thus it is necessary to use special methods, such as darkfield examination or impregnation with silver stains, to visualize them. These treponemes cannot be grown in culture, and under natural conditions are strictly parasites of man. They can, however, be grown in limited numbers in the testes of rabbits. Like the gonococcus, *T. pallidum* is exquisitely sensitive to destruction by drying and temperature change, and for this reason it is almost exclusively transmitted by sexual contact.

Syphilis, the disease caused by *T. pallidum,* occurs in three stages. As a rule, in *primary syphilis,* treponemes transmitted by sexual contact grow and multiply in local areas of the genitalia, spreading from there to the lymph nodes and bloodstream. About three weeks after infection an ulcer, called a hard chancre, appears at the site of infection. This represents a cellular response to the bacterial invasion, and examination of a drop of fluid expressed from the chancre reveals that it is teeming with infectious *T. pallidum.* Whether treatment is given or not, the chancre disappears within a period of from 4 to 6 weeks and the patient may mistakenly believe that the disease is cured. However, about 2 to 10 weeks later a rash and other manifestations of *secondary syphilis* may appear. By this time the spirochetes have spread throughout the body and infectious lesions occur on the skin and mucous membranes in various locations, especially in the mouth; therefore syphilis can be transmitted by kissing during this stage. The spirochetes also readily cross the placenta during this stage and can be passed from a pregnant mother to her fetus. The secondary stage lasts for weeks to months, sometimes as long as a year, and gradually subsides. About 50 percent of untreated cases never progress past the secondary stage; however, after a latent period of from 5 to 20 years, or even longer, some cases continue to the tertiary stage. Primary and secondary stages pass unnoticed in many cases.

Tertiary syphilis represents a hypersensitivity reaction to small numbers of T. pallidum that grow and persist in the tissues. The treponemes may be found in almost any part of the body, and the symptoms of tertiary syphilis depend on where the hypersensitivity reactions occur. If they occur in the skin, bones, or other areas not vital to existence, the disease is not life-threatening. However, if they occur within the walls of a major blood vessel, the vessel may rupture and instant death ensue; hypersensitivity reactions in the eyes cause blindness; in the central nervous system they can cause insanity, characteristic paralysis, and other manifestations. History and literature are replete with kings, queens, statesmen, and heroes who have degenerated into madness, mental incompetency or other serious debility as a result of tertiary syphilis; among these people have been Henry VIII, Ivan the Terrible, Catherine the Great, Arthur Schopenhauer, Heinrich Heine, Benito Mussolini, Emil von Behring, and Lord Randolph Churchill, to name just a few. It is difficult to comprehend the emotional, as well as the physical, anguish caused by this disease over so many centuries before the discovery of antibiotics. Syphilis remains a dread disease, but at least today adequate means exist for diagnosing and successfully treating it.

Primary or secondary syphilis can be diagnosed by finding the motile spirochetes in fresh material from lesions on the skin or the mucous membranes of the genitalia or the mouth. Darkfield illumination permits the thin spirals to be seen. Dried specimens can be examined by using fluorescent antibody techniques. In tertiary syphilis the patient is not infectious, and the internal lesions contain so few organisms that they cannot be detected by these means.

During all stages of syphilis serological tests are helpful in diagnosis. Although *T. pallidum* cannot be grown in culture and thus is not readily available as a source of antigen, it was accidentally found that antibodies formed during syphilis will react with certain substances extracted from beef heart. This fortuitous discovery allowed the development of tests for serum antibodies against *T. pallidum,* using the beef-heart substance as "antigen." At first, complement-fixation reactions, such as the Wassermann test, were widely used, but today most laboratories use a form of precipitation reaction called the *flocculation test;* examples of this are the Kahn test and the VDRL (Venereal Disease Research Laboratory) test. In these tests a fluffy "antigen"-antibody precipitate, or flocculate, is formed. Such flocculation tests are useful for preliminary testing of large numbers of people, but a positive test does not always indicate that an individual has syphilis, since false positive reactions occur fairly frequently. Other methods must be employed to confirm the suspected diagnosis, using specific antigens of *T. pallidum.* The fluorescent treponemal antibody test (FTA-ABS) is especially valuable in this respect (Figure 28-6). In this indirect immunofluorescence test, the patient's serum is allowed to interact under carefully controlled conditions with killed *T. pallidum* organisms that have been grown in the testes of rabbits. If specific antitreponemal

(a) Killed *Treponema pallidum* spirochetes, fixed to the slide, are incubated with the subject's serum. If specific antibodies are present they combine with the spirochetes.

(b) The slide is washed to remove excess serum, and fluorescent-labeled anti-human gamma globulin (anti-HGG) antiserum is added. The anti-HGG antibodies combine with any human antibodies already on the spirochetes.

(c) Following incubation, the slide is washed and examined microscopically using ultraviolet light. The *T. pallidum* organisms coated with antibodies fluoresce and can be visualized readily.

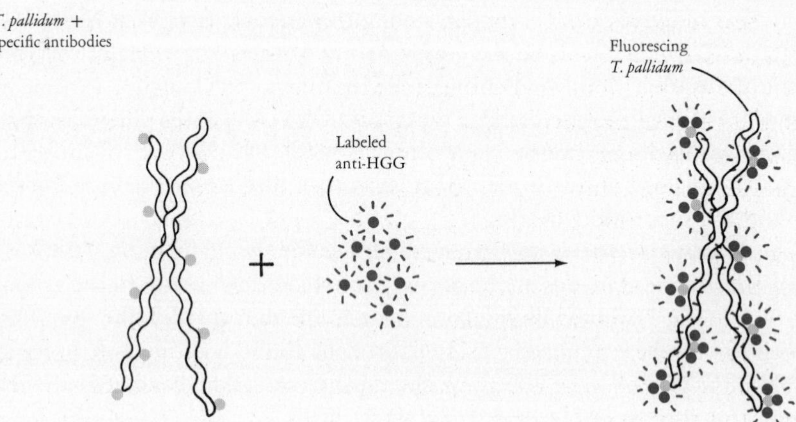

FIGURE 28-6
The fluorescent treponemal antibody (FTA-ABS) test. This test employs indirect immunofluorescence to demonstrate humoral antibodies against *Treponema pallidum*.

antibodies are present in the serum, they combine with the treponemes; excess serum proteins can then be washed from the slide. In order to detect the reaction, fluorescent dye-tagged antibodies against human gamma globulin are added. If specific antitreponemal antibodies have combined with them, the treponemes will fluoresce when examined with the fluorescence microscope under ultraviolet light.

Sometimes it is difficult to establish a diagnosis of syphilis. A chancre may fail to develop, so that the primary stage of the disease goes unnoticed, or the symptoms that are observed may be confused with symptoms of another disease. *Syphilis has been called "the great imitator" because it resembles or imitates many other diseases as it progresses.* Furthermore, congenital syphilis may go unrecognized if the disease is not apparent in the mother. It is possible for mothers who have passed through the infectious stages of secondary syphilis and have no symptoms of the disease to transmit treponemes across the placenta to their fetuses.

Treponema pallidum is always susceptible to low doses of penicillin, and penicillin is thus the treatment of choice for syphilis except for those persons who are allergic to penicillin. Treatment must be continued for a longer period during tertiary than during primary or secondary syphilis, probably because many of the organisms in tertiary syphilis are not actively multiplying. There is no lasting immunity to syphilis either after spontaneous recovery or treatment early in the disease, nor have effective methods of immunization been developed.

In general, syphilis can be controlled to some extent because of the use of effective antibiotics and an intensive program for tracing and treating the contacts of infected individuals. However, increases in the disease occur whenever precautions are relaxed. For example, the number of cases of syphilis in the United States each year dropped precipitously following the introduction of penicillin, as shown in Figure 28-7; there was, however, an increase in 1962. When more stringent precautionary measures were taken, the rate decreased again. Nevertheless, the number of cases almost tripled during the entire period from 1958 to 1975.

Several factors explain the differences in incidence of syphilis as compared with gonorrhea in the United States. Case workers carefully trace all reported cases of both syphilis and gonorrhea, attempting to reach and treat all sexual contacts of these patients. This approach usually allows contacts to be treated prophylactically during the long (1- to 3-week) incubation period before syphilis develops. However, the incubation period of gonorrhea is so brief (1 to 3 days) that this disease has usually developed in sexual contacts before they can be traced and treated. Frequently, a number of other contacts will also have been infected. A second important factor is that serological tests will identify subjects with unsuspected syphilis, but reliable serological tests for gonorrhea are not yet available for screening large numbers of people in order to detect both the male and female carriers of gonococci. These carriers, often women who have never had symptoms of gonorrhea, are the major reservoir of gonococci, and there is no easy way to identify them. It is necessary to culture secretions from the genital tract and to identify *N. gonorrhoeae* from the cultures—a laborious and

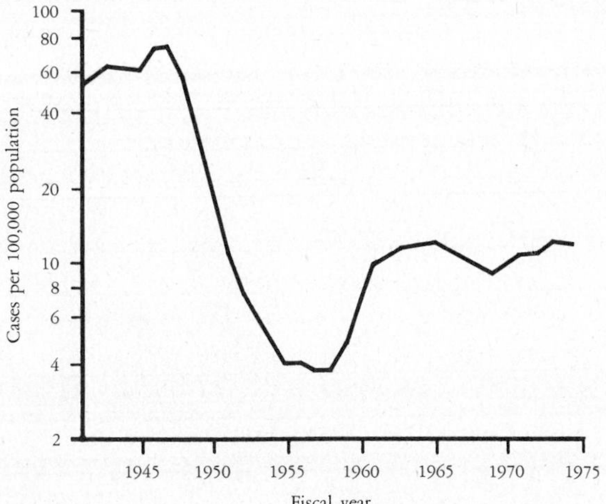

FIGURE 28-7
Reported cases of primary and secondary syphilis per 100,000 population in civilians in the United States from 1941 through 1974. (Morbidity and Mortality Report, United States Public Health Service 23:60, 1974.)

expensive procedure. Thus it appears that the incidence of gonorrhea will continue to increase even more than the incidence of syphilis.

Homosexual or bisexual men present a problem in the control of syphilis, because about a third of the male cases in the United States currently occur in this population. In fact, in urban areas such as Los Angeles, Seattle, and New York more than half of all reported syphilis cases involve homosexual men. Until the disease is controlled in this population, it is probable that its incidence will not decline significantly.

Chlamydia trachomatis, pages 439, 440

Although gonorrhea and syphilis account for a majority of venereal disease cases, other venereal diseases caused by bacteria occur frequently in the United States. A common cause of venereal disease is now recognized to be *Chlamydia trachomatis* (Chapter 22), an obligate intracellular bacterium that produces symptoms resembling gonorrhea. Nonvenereal transmission of this agent also occurs. Another bacterial cause of venereal disease, one which is not common, is the small Gram-negative rod *Haemophilus ducreyi,* the cause of chancroid.

NONBACTERIAL INFECTIONS OF THE GENITOURINARY TRACT

A common fungal cause of genitourinary disease is Candida albicans (Figure 28-8). This budding yeast is part of the normal flora of the mucous membranes in about 35 to 40 percent of the population, and it is usually nonpathogenic. The interaction between large numbers of bacteria and small numbers of these fungi results in an ecological balance between the groups. When this balance is upset, disease may occur. This happens most often after intensive antibacterial treatment, particularly with some of the more effective broad-spectrum antibiotics which suppress much of the normal bacterial flora of the mucous membranes and allow fungi to grow. The large numbers of *C. albicans* which are found in such a situation can cause extremely persistent and severe infections of the mucous membranes of the genitourinary tract and also of those of the alimentary tract. Other predisposing factors to *Candida* infection include pregnancy, hormone therapy with oral contraceptives, diabetes, and a variety of other conditions.

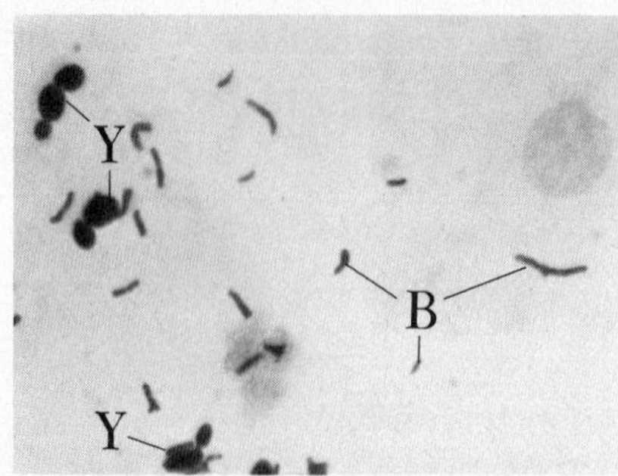

(a)

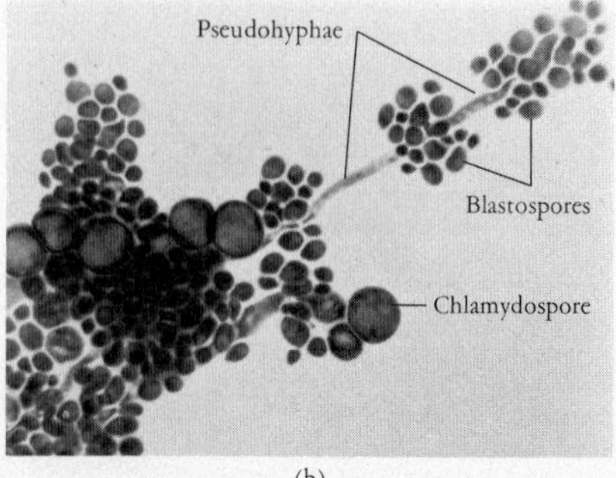

(b)

FIGURE 28-8
Candida albicans. (a) The morphology of *C. albicans* as it appears in the genitourinary tract. The budding yeasts (Y) are large (6 μm) and readily distinguished from bacteria (B). An epithelial cell from the genitourinary tract is seen in the upper right-hand corner (b) *C. albicans* grown in culture. (Courtesy of S. Eng.)

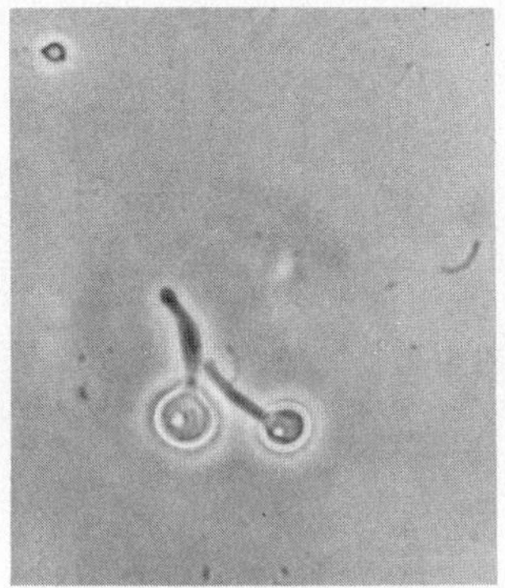

FIGURE 28-9
Candida albicans forms germ tubes when incubated in normal human serum or saliva. (Courtesy of L. H. Kimura.)

Candida albicans is an oval, budding yeast, about 6 μm in length (Figure 28-8a). When grown at 37°C on rich medium, it forms smooth, creamy colonies of budding yeasts. However, under less favorable conditions, such as a less nutritious medium at room temperature, pseudohyphae are formed along with the budding yeast cells. These pseudohyphae consist of very elongated budding cells in chains, resembling hyphae in a mycelium. Frequently round thickened spores (chlamydospores) form at the ends of the pseudohyphae of *C. albicans* (Figure 28-8b). These morphological characteristics help to distinguish *Candida* from other yeasts. Their carbohydrate fermentation patterns may also be used for identification of *C. albicans.* In addition, this organism produces germ tubes in human serum (Figure 28-9).

Certain protozoa of the genus Trichomonas, *often members of the normal flora of the genitalia, may invade the genitourinary tract to cause disease.* They frequently cause a mild infection of the vagina in females, and less often an infection of the prostate in males, or of the bladder in either sex. These flagellate, unicellular protozoa are easily identified by microscope examination of the urine or infected exudate (Figure 28-10).

A viral infection, genital herpes, is of particular interest because of its high incidence and its possible, or suspected, relation to cancer of the genital tract. Genital herpes is caused by herpes virus type 2, which is transmitted by venereal contact. (This

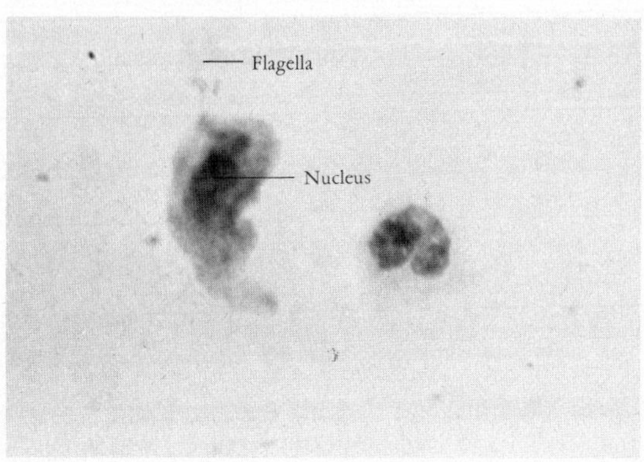

FIGURE 28-10
Flagellate of the genus *Trichomonas* in Gram-stained material from the human genitourinary tract. The protozoan is on the left and a leucocyte on the right. Identification of *Trichomonas* sp. is easier in a fresh specimen than in a stained smear because the living organisms have a characteristic motility. (Courtesy of S. Eng.)

herpes virus closely resembles herpes virus type 1, the cause of herpes simplex, commonly known as fever blisters or cold sores). Serological studies have shown that women with carcinoma of the cervix are much more likely than normal women to have serum antibodies against herpes virus type 2. This finding suggests that the virus could be involved as a predisposing factor or as a cause of carcinoma of the cervix. Alternatively, it is possible that some other factor predisposes to both carcinoma of the cervix and genital herpes. At present, extensive investigation is under way to attempt to clarify the relationship of herpes virus type 2 to human cancer.

SUMMARY

Urinary tract infections, much more common in females than in males, are usually caused by members of the normal intestinal flora, especially *Escherichia coli* and similar Gram-negative rods. The bacteria invade when the body's normal defenses are interrupted, and cause infection of the bladder or kidney or both. As a rule infections occur first in the bladder and if not limited there reach the kidneys by ascending through the ureters.

Historically important, but uncommon in the United States at present, is puerperal fever (childbirth fever), often caused by *Streptococcus pyogenes.* Among bacterial infections of the genital tract that are extremely common at present are some of the venereal diseases. Gonorrhea results from infection with *Neisseria gonorrhoeae,* and syphilis is caused by the spirochete, *Treponema pallidum.*

QUESTIONS

REVIEW

1. Describe the normal flora of the genitourinary tract.
2. Why are Gram-negative rods and enterococci the most common causes of urinary tract infection?
3. Gonorrhea can be treated effectively with antibiotics, and yet this disease is epidemic and increasing in incidence. Why?
4. Check the appropriate blanks below:
   Syphilis may be transmitted by the following:

	Primary Syphilis	*Secondary Syphilis*	*Tertiary Syphilis*
Sexual intercourse	_____	_____	_____
Kissing	_____	_____	_____
Blood transfusions	_____	_____	_____
Mother to fetus (via placenta)	_____	_____	_____

5. What are some possible reasons that there is no effective vaccine against syphilis?

1. What is a possible explanation for the development of recently reported penicillin-resistant strains of gonococci that produce penicillinase?
2. What are the epidemiologic implications of gonococcal strains that are totally resistant to penicillin?

FURTHER READING

BUCHANAN, T. M., "Antigenic Heterogeneity of Gonococcal Pili," *Journal of Experimental Medicine* 141:1470–1475 (1975). A report that the gonococcal pili that attach to host cells are antigenically different from strain to strain, suggesting that specific antibodies against multiple antigenic varieties of the organism may be necessary to prevent reinfection.

HANDSFIELD, H. N., T. O. LIPMAN, J. P. HARNISCH, E. TRONCA, and K. K. HOLMES, "Asymptomatic Gonorrhea in Men," *New England Journal of Medicine* 290:117–123 (1974). The authors report a startling incidence of asymptomatic gonococcal infections among males, showing that men, as well as women, are frequently carriers of gonococci.

MUDD, S., ed., *Infectious Agents and Host Reactions.* Philadelphia: W. B. Saunders, 1970. This book contains additional information about *Treponema pallidum* and syphilis.

ROSEBURY, T., *Microbes and Morals.* New York: Viking Press, 1971. A fascinating study of the relationship of society, culture, and literature to venereal diseases, written by a well-known microbiologist.

SWANSON, J., "Role of Pili Interactions between *Neisseria gonorrhoeae* and Eukaryotic Cells *in Vitro.*" *Microbiology—1975.* Ed. D. Schlessinger. Washington, D.C.: American Society for Microbiology, 1975, pp. 124–126. A brief review of studies showing that pili facilitate the adherence of gonococci to epithelial cells but not to phagocytes.

WARD, M. E., J. N. ROBERTSON, P. M. ENGLEFIELD, and P. J. WATT, "Gonococcal Infection: Invasion of the Mucosal Surfaces of the Genital Tract." *Microbiology—1975.* Ed. D. Schlessinger. Washington, D.C.: American Society for Microbiology, 1975, pp. 188–199. An excellent review of studies on invasion of the mucous membranes, and even entrance into the epithelial cells, by gonococci.

WILSON, J. W. and O. A. PLUNKETT, *The Fungous Diseases of Man.* Berkeley: University of California Press, 1967. In-depth discussions of *Candida albicans* and other fungi that cause infections.

CHAPTER 29

INFECTIONS OF THE NERVOUS SYSTEM

Infections of the nervous system are relatively uncommon but are apt to be serious. This is because they strike at our ability to move, to feel, or to think. Nervous system infections may be caused by bacteria, viruses, fungi, and even some protozoa.

ANATOMIC AND PHYSIOLOGIC CONSIDERATIONS

The brain and spinal cord make up the *central nervous system.* Both are enclosed in bony cavities, the brain in the skull, the spinal cord inside the spinal canal (Figure 29-1). The network of nerves throughout the body is connected with the central nervous system by large nerve trunks which penetrate the bony cavities that protect it. These larger nerves can therefore be damaged by infections of certain bones. Two kinds of nerves are involved in the body's nerve network: *motor nerves,* which activate different parts of the body, and *sensory nerves,* which transmit feelings, such as heat, pain, light, and sound. All of the body's nerves are made up of special cells with very long extensions that transmit electrical impulses. These long extensions can sometimes regenerate if damaged, but cannot be repaired if the nerve cell of which they are part is itself killed, as may occur, for example, in poliomyelitis.

CEREBROSPINAL FLUID

Deep inside the brain are several cavities filled with a clear fluid called *cerebrospinal fluid.* This fluid is continually produced in these cavities, flowing from them through small openings to spread over the surface of the brain and spinal cord and out along the nerve trunks which penetrate the bony cavities. The cerebrospinal fluid then flows into lymph channels and eventually enters the blood-

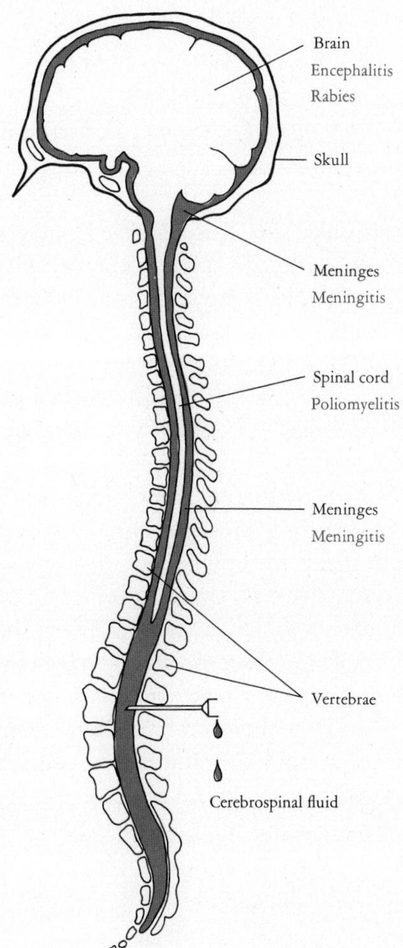

FIGURE 29-1
The central nervous system.

stream. In persons with suspected infections of the central nervous system, a needle can be safely inserted between the spinal vertebrae in the small of the back, and a sample of cerebrospinal fluid withdrawn from the spinal canal for examination (Figure 29-1). In this manner the causative agent of a central nervous system infection can often be found and identified, and appropriate treatment administered.

MENINGES

Covering the surface of the brain and spinal cord are two membranes called *meninges;* the cerebrospinal fluid flows between these. Inflammation of these membranes is called *meningitis.* Because both blood vessels and nerves must pass through both the meninges and through the film of cerebrospinal fluid between them, meningitis may involve these vessels and nerves and cause their malfunction.

The central nervous system lies in a well-protected environment, which prevents infectious agents from readily getting to it. When they do, it seems that their entry must almost have been accidental, because most microbes that infect

ROUTES OF MICROBIAL INVASION

the nervous system infect other parts of the body much more frequently. The routes by which microbes reach the brain and spinal cord are as follows:

1. *The bloodstream.* This is probably the chief source of central nervous system infections, although it is very difficult for infectious agents to cross from the bloodstream to the brain. This barrier to the passage of harmful agents is often called the "blood-brain barrier." Although its exact nature is unknown, it is effective in preventing microbes from entering nervous tissue in all but a few individuals afflicted with bloodstream infections. The reasons for these few failures of the blood-brain barrier are not known, but probably relate to the concentration of organisms in the bloodstream and to the length of time they circulate. Veins in the face may connect directly with those on the brain surface, and upper facial infections may therefore sometimes spread to the brain.
2. *Entrance by means of the nerves.* Some agents penetrate the central nervous system by traveling up the nerves. It is uncertain whether these organisms pass inside the processes of the nerve cells or move through passages between the cells that surround and protect the nerve cell processes. In some instances the agent grows in these surrounding cells.
3. *Direct extensions of infections elsewhere.* To reach the central nervous system, microorganisms must penetrate not only the bone that surrounds it but also the tough outer membranes that surround the meninges and bone. Infections in bone surrounding the central nervous system may rarely erode inward to reach the brain or spinal cord. Skull fractures may produce nonhealing injuries that predispose to infection of the central nervous system. Infections of the system may also extend to it from the respiratory sinuses, mastoids, or, more commonly, from the middle ear.

SOME BACTERIAL INFECTIONS OF THE NERVOUS SYSTEM

Meningococcal Meningitis

Meningococcal meningitis is often called epidemic meningitis although it typically appears at widely separated locations and throughout the year. The typical case first shows symptoms of a mild cold, followed by the sudden onset of a severe throbbing headache and fever, and marked pain and stiffness of the neck and back. Purplish spots may appear on the skin. The afflicted individual may develop shock and die within 24 hr of having felt and appeared completely healthy.

Neisseria meningitidis, page 433

The causative organism of meningococcal meningitis, *Neisseria meningitidis* (the meningococcus) (Table 29-1), can usually be found on smears made from cerebrospinal fluid or in scrapings of the purple skin spots of the disease. Meningococci are Gram-negative cocci occurring in pairs (diplococci), each member of which has a flattened side where it faces its neighbor. Most of the organisms seen

TABLE 29-1
Characteristics of *Neisseria meningitidis*

Identification	*Normal Habitat*	*Pathogenic Potential*
Gram-negative diplococci with characteristic pattern of sugar fermentations; types identified by specific antisera	Upper respiratory tract of human beings	Can cause serious infections marked by bloodstream invasion and meningitis

in cerebrospinal fluid occur within white blood cells which enter the fluid in response to the infection. *N. meningitidis* grow readily on media enriched with blood, especially if incubated in an atmosphere with an increased carbon dioxide concentration. Some meningococcal strains are killed quickly by moderate chilling, and this may account for some failures to recover the organism in cultures of material from patients.

N. meningitidis is a strictly human parasite, and infection is transmitted by exposure to a person carrying this species. The vast majority of meningococcal infections pass unnoticed and are only detected by the chance recovery, on culture, of a heavy growth of the causative organism from the throat of an infected person. If a selective medium is used to suppress the growth of other throat flora, meningococci can be recovered from about 15 percent of normal persons, and in closely crowded conditions, such as military camps, the organisms eventually spread to almost all persons present. Even so, the rates of cerebrospinal meningitis remain low. Figure 29-2 shows how susceptibility to the disease decreases dra-

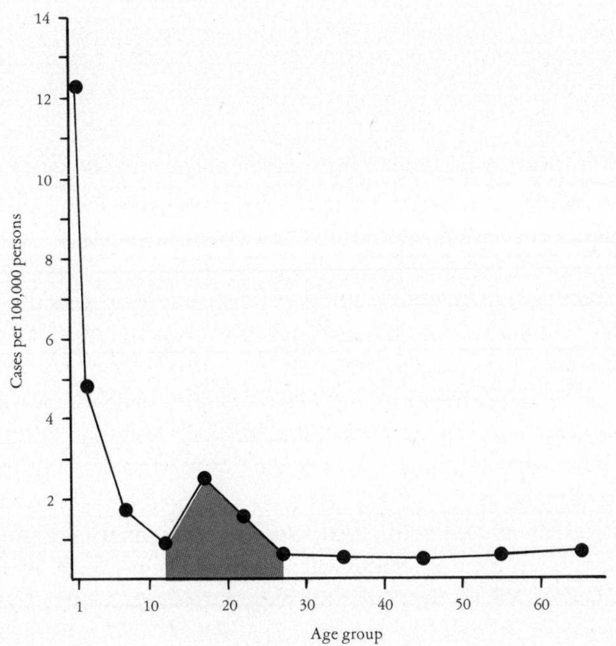

FIGURE 29-2
Meningococcal disease rates by age group in the United States in 1969. The rise in incidence indicated by the colored area is due to cases in the armed forces. (Morbidity and Mortality Weekly Report, United States Public Health Service 19:415, 1970.)

matically with age. The rise in incidence of meningococcal meningitis in the age range of from 15 to 25 years comes from cases in the armed forces. Factors such as fatigue and crowding, which are present during military training, apparently overcome the natural resistance developing with age.

Until the mid-1940s, *N. meningitidis* type A was responsible for most outbreaks of cerebrospinal meningitis, while types B and C generally caused the isolated cases that appeared from time to time. After a peak number of cases (5077) reported in 1953, there was a fall in the incidence of the disease until about 1962, when its rates began to climb again, and with this increase, type B strains of the organism began to appear in increasing numbers. By the early part of 1965, almost all reported cases of cerebrospinal meningitis were caused by type B and a few type C organisms. But a further shift then began to appear, so that by 1970 more than two-thirds of all reported cases of the disease were caused by type C organisms. After 1972, however, type B strains again became dominant, with several localized epidemics of type A appearing in 1975 and 1976. Furthermore, although virtually all meningococci had been exquisitely susceptible to sulfa drugs before 1953, a dramatic rise in resistance to these drugs appeared coincidentally with the increase in infections caused by types B and C organisms. This sulfa resistance peaked in 1970, when more than two-thirds of the strains of meningococcus were resistant to the drugs, and then declined to about 25 percent.

Sulfa drugs, page 588

Factors which may have played a role in these changing frequencies include (1) relative immunity of the population to type A strains, developed during previous epidemics, (2) the widespread use of sulfa drugs, which resulted in the selection of rare sulfa-resistant strains known to be present even before the introduction of sulfa drugs, (3) crowding of susceptible populations, and (4) genetic interchange among different strains of meningococcus.

Transformation, pages 200–203

The mechanisms by which meningococci produce cerebrospinal meningitis have been the subject of considerable study. The organisms probably infect when their victim inhales droplets from another person, almost always a carrier, or one with an unnoticed infection. They establish an infection in the upper respiratory tract and from this site are seeded into the bloodstream. The blood carries the organisms to the meninges and spinal fluid where they multiply faster than they can be engulfed and destroyed by the white blood cells. The inflammatory response, with its formation of pus and clots, may obstruct the normal outflow of cerebrospinal fluid, which squeezes the brain flat against the skull by its internal pressure. Infection of brain tissue, and later scar formation, may damage motor nerves and produce paralysis. Moreover, *N. meningitidis* circulating in the bloodstream injure the blood vessels so that they cannot maintain normal blood pressure, and shock results. The smaller blood vessels of the skin are also affected, with small hemorrhages producing the purple spots of the disease. Vital organs are also often damaged.

Blood vessel damage by *N. meningitidis* probably results from endotoxin. Not only does the organism's endotoxin cause shock but it also produces fever and heightened sensitivity to any further exposure to endotoxin. This increased sensitivity, called the *Shwartzman phenomenon,* can be demonstrated in animals

Endotoxin, Table 19-3, page 374

following injections of meningococci, and it may well be responsible for the damage to the small blood vessels seen in humans. The tendency of meningococci to autolyse (spontaneously rupture) may explain the release of their endotoxin.

Type-specific opsonins protect against meningococcal disease. In fact, vaccines composed of purified capsular polysaccharide have been prepared against *N. meningitidis* types A and C. These vaccines are effective in controlling meningitis and are currently being used to immunize certain high-risk human populations. Since epidemic meningococcal strains probably have high virulence, persons exposed to cases of meningococcal disease are given prophylactic treatment with the antibiotic rifampin.

Other Bacterial Causes of Meningitis

In addition to *N. meningitidis,* many other bacteria may cause infection of the meninges, and in most cases, just as with *N. meningitidis,* these organisms are commonly carried by healthy people, are transmitted by inhalation, and only rarely produce meningitis. Organisms in this category include the respiratory pathogens *Streptococcus pneumoniae* and *Haemophilus influenzae.* Although common inhabitants of the respiratory tract these organisms seldom cause meningitis in newborn babies, presumably because antibody against them is transferred across the placenta from the mother. Instead, Gram-negative rods such as *Escherichia coli* from the mother's intestinal tract, or even *Flavobacterium meningosepticum* from the nursery water faucet, are more likely to cause meningitis in the newborn. The current view is that the newborn infant lacks protection against Gram-negative rods such as these because the antibodies effective against these organisms are of the IgM type and are unable to pass across the placenta. Other causes of meningitis in newborn infants are beta hemolytic streptococci of serological group B, and a Gram-positive rod, *Listeria monocytogenes.* These two species of bacteria are carried among the vaginal flora of some women, and can act as opportunists in immunologically immature infants.

Haemophilus influenzae, pages 437–438

Streptococcus agalactiae, pages 428–429

LEPROSY, A BACTERIAL DISEASE OF THE PERIPHERAL NERVES

Leprosy was once common in Europe and America, and in 1868 G. A. Hansen of Norway demonstrated the causative bacteria (*Mycobacterium leprae*) in the tissues of leprosy patients. This is said to have been the first time a bacterium was causally linked to a human disease, and yet *M. leprae* have to date still not been grown in culture. Like tuberculosis, leprosy began to recede in Europe and America for unknown reasons, and today it is chiefly a disease of warmer and more economically underdeveloped countries, with an estimated worldwide incidence of between 10 to 20 million cases.

Mycobacterium leprae, pages 432–433

Because of the frustration involved in trying to grow *M. leprae* in vitro, there have been many attempts to infect laboratory animals with it. A tremendous advance was made in 1960 when it was demonstrated that the organisms would multiply in the footpads of immunosuppressed mice. More recently, evidence has accumulated that the nine-banded armadillo prevalent in Texas and Louisiana is susceptible. Wild armadillos infected with a bacterium indistinguishable from the leprosy bacillus have been found in Louisiana, but there is no evidence that these animals are the source of human leprosy.

From studies of infected mice it has been possible to acquire some interesting information about *Mycobacterium leprae.* One of the more striking findings has been that these bacteria have a very long generation time, estimated, in fact, to be about twelve days! This finding helps account for the very long incubation period of human leprosy, probably a minimum of two years and often ten years or longer. In addition, growth of *M. leprae* in mice has permitted studies aimed at finding new medicines for the treatment of leprosy in human beings.

The earliest detectable finding in human infection with *M. leprae* is the invasion of the small nerves of the skin. Indeed, *M. leprae* is the only known human pathogen which preferentially attacks the peripheral nerves. Thereafter the course of the infection probably depends on the immune response of the host. In most cases cell-mediated immunity and delayed hypersensitivity develop to the invading bacteria. Macrophages limit the growth of *M. leprae* and the bacteria remain very scanty, but nerve damage may progress and lead to disabling deformity. In most instances, however, the disease arrests itself after a few years and the nerve damage, although permanent, does not progress further. This limited type of leprosy is called the *tuberculoid* type. Persons with tuberculoid leprosy rarely, if ever, transmit the disease to others.

Cell-mediated immunity, pages 397–401

Delayed hypersensitivity, pages 411–413

In some infected persons, however, cellular immunity and delayed hypersensitivity to *M. leprae* either fail to develop or are lost, and unrestricted growth (rather than destruction) of *M. leprae* occurs in macrophages. This relatively uncommon form of leprosy is referred to as the *lepromatous* type. In addition to the nerves, it involves most tissues of the body. The tissues and mucous membranes then swarm with the leprosy bacteria, which can be transmitted to other persons by intimate direct contact. A similar disease can be produced in mice by removing their thymus glands and giving them X-ray treatment or antibodies against their lymphocytes to destroy their capability for cellular immunity. With lepromatous leprosy, there may be a general impairment of delayed hypersensitivity responses, not only to leprosy bacilli, but to many other antigens as well. On the other hand, a return toward normal immune function occurs when the disease is controlled by medication.

It is now widely known that leprosy is one of the least contagious of all infectious diseases, and the morbid days when lepers were forced to carry a bell or horn to warn others of their presence are fortunately past. Today persons with leprosy can expect a normal lifespan, provided ill-grounded fears and social pressures do not cause them to avoid early medical consultation. The risk of leprosy to persons in the armed services abroad appears to be small, perhaps 30 cases having been detected between 1940 and 1970. However, one of these, living in a northern state, is known to have transmitted the disease to his wife and two of his children. In Vietnam and neighboring countries the incidence of leprosy among native people has been estimated at about 5 per 1000. It will take a number of years before the impact of the Southeast Asian war on the incidence of leprosy in the United States and other countries is certain. In the past the average time between exposure and discovery of the disease in armed forces veterans has been about ten years. The disease is detected by microscopic inspection of scrapings or biopsies of infected tissues. Arrest of the disease can usually be achieved by long-term treatment with sulfone drugs.

Sulfones,
Figure 33-1, pages 589, 598

SOME VIRAL DISEASES OF THE NERVOUS SYSTEM

Poliomyelitis (Infantile Paralysis)

A person developing poliomyelitis usually first suffers the symptoms of meningitis: headache, fever, stiff neck, and nausea. In addition, pain and spasm of some muscles generally occur, later followed by paralysis, and finally by a relative shrinking of muscle and failure of normal bone development in the affected area. In the more severe cases the muscles controlling respiration are involved, and the victim requires a machine to pump air in and out of his lungs. Some recovery of function is the rule if life can be sustained during the acute stage of the illness. Sensation is not affected.

Historically, poliomyelitis has had a greater impact in more economically advanced countries. The disease does occur in poorer, more crowded nations, but epidemics of paralytic poliomyelitis generally appear to affect only small groups of geographically or culturally isolated people in these countries; among the remaining population the virus is widespread, and very few persons escape childhood without infection. The mothers in these countries therefore have antibody in their bloodstream, and this crosses the placenta and enters the fetal circulation. Newborn infants in these nations are thus partially protected against bloodstream infection and nervous invasion by poliomyelitis virus for as long as their mother's antibodies persist in their bodies—usually for about 2 or 3 months. During this time, because of widespread exposure to the poliovirus through crowding and unsanitary conditions, the infant is likely to develop a mild infection of his throat and intestine, and thereby to achieve life-long immunity. Even if exposure is delayed beyond two or three months, it is likely to occur in the earlier years when the disease is less likely to be severe.

In contrast, in such areas as suburban American communities in the days before vaccine was developed, the poliomyelitis virus sometimes could not spread fast enough to sustain itself because of efficient sanitation. If its reintroduction from another community were delayed sufficiently, some people of all ages, including older children and adults, would then lack antibody and would be susceptible. When the virus was finally reintroduced, a high incidence of paralysis resulted.

The causative agent of poliomyelitis, the poliovirus, can be recovered from

Cytopathic effect, page 342

Viral neutralization, page 395

See Appendix III, pages 711–713

the throat and feces early in the development of the illness, later in the blood, and finally, in some patients, in the cerebrospinal fluid. Excretion of the virus in the feces continues for weeks or months, but is transitory in the other sites. Introduced into appropriate tissue cell cultures (usually consisting of monkey kidney cells) the virus produces a cytopathic effect (plaques) which can readily be seen under the low power of the light microscope. Poliomyelitis is generally caused by one of three types of poliovirus, but occasionally may be caused by other enteroviruses. The three major polioviruses are designated by roman numerals I, II, and III, and have only a slight antigenic relationship to one another. The antiserum which prevents a cytopathic effect in tissue culture identifies the virus type.

Like other enteroviruses, polioviruses are small, RNA-containing viruses which are quite stable under natural conditions, but are inactivated by pasteurization and proper chlorination of drinking water. The viruses can infect a cell under natural conditions only if that cell has certain receptors, to which the virus can attach, on its surface. This helps explain the remarkable selectivity of the polioviruses for motor nerve cells of the brain and spinal cord, whereas they spare most other kinds of nerve cells. Destruction of the infected cell occurs upon release of the mature virus.

Until the late 1950s poliomyelitis was a terrifying threat, especially in economically advanced nations. Fear of the disease was widespread, and hundreds of thousands of people in the United States contributed dimes to the National Foundation for Infantile Paralysis to help victims and to support research to solve this problem. Franklin D. Roosevelt, himself a polio victim, helped dramatize the need for this support.

One of the most important developments toward overcoming polio was the demonstration, in 1949, by Dr. John Enders and his colleagues at Harvard University, that polioviruses grow readily in tissue cell cultures, producing clear-cut cytopathic effects. Details of the pathogenesis of polio and of the ecological relationships of the viruses were then worked out. By 1954 Dr. Jonas Salk had perfected a formalin-inactivated virus vaccine which was widely employed and soon dramatically lowered the incidence of paralytic poliomyelitis. Meanwhile, cell culture techniques had allowed the selection of mutant poliovirus strains, that lacked the ability to attack nervous tissue, but could nevertheless stimulate antibody formation against the wild-type, virulent virus. Living vaccines consisting of such attenuated viruses were then quickly developed and proved more effective than the killed-virus vaccines. The living vaccines prepared by Dr. Albert Sabin were chosen for use. As a result of these historic research developments, the United States now has generally less than a dozen cases a year of paralytic poliomyelitis as opposed to the many thousands it had in earlier years. Ironically, some of the more recent cases of polio have been caused by the live polio vaccine, which can produce paralysis in a rare recipient.

The production and delivery of adequate polio vaccine require the most highly sophisticated technology, and a susceptible adult population could become extremely dependent on this technology for protection from poliomyelitis and other diseases. On the other hand, two incidents in the development of inactivated poliomyelitis vaccine point out the dangers that can exist despite the extreme caution with which mass administration of biological agents is treated. In the first incident, formalin inactivation of the poliovirus was inadequate and vaccine recipients received some living wild virulent virus which caused cases of paralysis. In the second incident, the poliovirus used for the vaccine was unknowingly contaminated by another virus, called simian virus 40(SV_{40}), growing in the monkey tissue poliovirus cultures. This virus withstood the formalin exposures that killed the poliovirus, and many people received the living monkey virus. Although SV_{40} has been shown to induce tumors in laboratory animals (Chapter 18), many years have now gone by with no ill effects appearing in the vaccine recipients. This episode, however, demonstrated that the risks of large-scale tampering with our species cannot always be adequately foreseen.

Formalin, pages 134–135

Rabies

Rabies is one of the most feared of all human diseases because it is agonizing in its manifestations and essentially always fatal. It is an acute infectious disease involving the central nervous system and is usually transmitted to warm-blooded animals by saliva from animals infected with the rabies virus. In human beings the disease usually develops after an incubation period of from 30 to 60 days, although the extreme range may be from ten days to more than a year. A prominent feature of the disease is spasms of the muscles of the mouth and throat at the sight (or even the thought) of water, and thus the popular name "hydrophobia."

Identification of the causative virus of rabies is usually made most quickly by preparing smears from the brain of a person or animal that has died from the disease. These smears are stained with a combination of dyes called *Seller's stain,* or with fluorescent antirabies antibody, and inspected for nerve cells containing viral inclusions (*Negri bodies*). The rabies virus can be cultivated in animals such as laboratory mice or in tissue culture. The virus is large, "bullet" shaped and enveloped, and contains single-stranded RNA. It buds from the surface of infected cells. Differences in virulence and ability to produce Negri bodies exist from isolate to isolate.

Although rabies virus infection can occur by both the respiratory and oral routes under research conditions, the principal mode of transmission of rabies to human beings is the passage of infected saliva through the skin barrier by way of a bite or other break in the skin. Once through the skin, the virus passes slowly, by an unknown mechanism, along the course of nerves to the central nervous system, where it multiplies in some of the motor nerve cells without causing cell destruction. The slow travel of the virus via the nerve is thought to account for the very long incubation periods often observed in rabies. As a matter of fact, with laboratory animals, rabies deaths have been prevented by cutting the nerve coming from the site of injection of the virus. The mode of spread of the rabies virus through the nerves and central nervous system is still poorly understood, and the explanation of how the virus gets to saliva-producing glands and other tissues (such as the conjunctiva and fat) is obscure.

Rabies is widespread in wild animals (Figure 29-3), representing an enor-

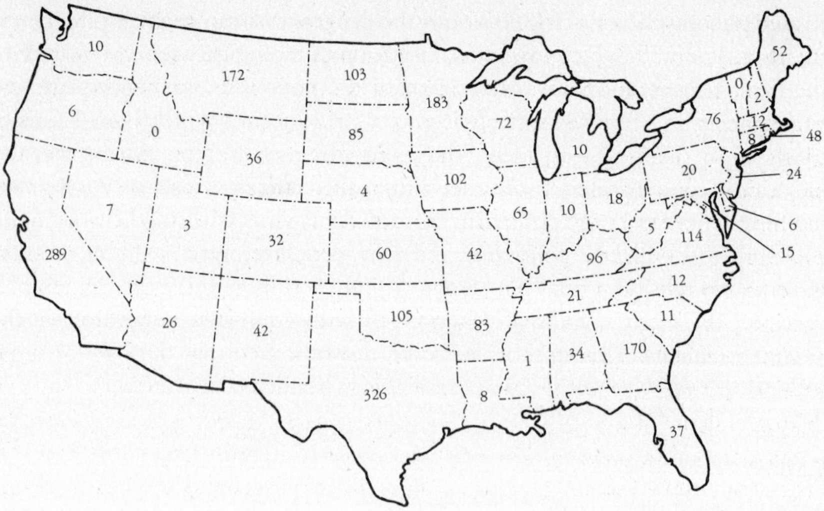

FIGURE 29-3
Number of cases of animal rabies reported by state in 1975. (Morbidity and Mortality Weekly Report Annual Supplement, August 1976.)

mous reservoir from which infection can be transmitted to human beings. In South America the vampire bat is often infected, and 700 deaths of cattle from this source were reported in 1969 from Brazil alone. Foxes and squirrels are slowly spreading the virus across Europe, and in the United States skunks constitute the chief reservoir host, with bats, foxes, and raccoons also being important. The majority of human cases of rabies in the United States have been caused by dog bites, and this is true in most areas of the world. The dog population in the United States is about 25 million, and an estimated 1 million people get bitten each year. Fortunately, since World War II the incidence of dog rabies (and human rabies) has dropped dramatically and now almost 85 percent of reported rabies infections are in wild animals. Unfortunately, the data on the natural history of rabies in wild animals are incomplete. In dogs anywhere from 10 to 100 percent will die of rabies following injection with wild rabies viruses from different sources. About three-quarters of those that develop rabies excrete rabies virus in their saliva and about one-third of these begin excreting it from one to three days before they get sick. Some dogs become irritable and hyperactive with the onset of rabies, salivate copiously, and attack people, animals, and inanimate objects. Perhaps more common is the "dumb" form of rabies, in which an infected dog simply stops eating, becomes inactive, and suffers paralysis of throat and leg muscles.

IMPORTANCE OF DOGS

A most dramatic decline in rabies incidence in people has resulted from immunizing dogs and other pets against rabies infection. In effect this places a partial barrier to the spread of the rabies virus from wild animal reservoirs to human beings. Because pets vary in their tolerance of rabies vaccines, depending on their age and species, several kinds of vaccines are available. In recent years these have consisted of both attenuated and inactivated rabies viruses.

RABIES VACCINE

The risk of rabies developing in people bitten by dogs having rabies virus in their saliva may be as high as 30 percent, and with bites by other wild animals, the risk may be higher still. Louis Pasteur discovered that this risk can be lowered considerably by administering rabies vaccine as soon as possible after exposure to the virus. Presumably, the effectiveness of this measure is based on the accessibility of the rabies virus and its susceptibility to inactivation by immune mechanisms at some time during its long sojourn to the motor nerve cells of the brain. By giving large doses of rabies vaccine immediately after a bite, sufficient immunity can be developed in most cases in time to prevent infectious virus from reaching the central nervous system. In fact, the risk of rabies can be reduced by this technique by about 85 to 90 percent. Unfortunately, however, there is a risk of central nervous system disease from the vaccine itself. This arises because the vaccine virus is cultivated in animal brain tissue containing an antigen held in common by both the animal and the human brain. The person receiving rabies vaccine may thus develop an immune response to antigens of his own brain, producing "allergic encephalitis" in about one of every 8000 instances. This risk drops to only about one in 25,000 with rabies vaccine cultivated in duck embryos instead of animal brains. In fact, the duck embryo vaccine is safe enough to be used routinely to immunize veterinarians and people living in areas of high rabies incidence before they are bitten by a rabid animal, or by laboratory workers who work with the rabies virus. In the United States about 30,000 people get post-exposure vaccine each year.

The risk of rabies can be lowered still further by injecting rabies antibody into an exposed person. Formerly this antibody was produced by injecting rabies vaccine into a horse, but horse antiserum is a foreign protein and many human recipients soon developed antibodies against the antiserum. This often resulted in serum sickness, or even in anaphylactic shock. Today, antiserum obtained from human volunteers who have been immunized against rabies is available, and the dangers of using horse serum are avoided. Preventive measures are not always successful, however. In 1976 a woman with a superficial bat bite on her finger died of rabies despite prompt treatment with both vaccine and antiserum.

Serum sickness, page 407

Anaphylaxis, page 408

SUMMARY

In comparison with other tissues the brain and spinal cord are relatively protected from infections, and central nervous system involvement generally occurs in only a fraction of the total infections produced by a microbial agent. In many cases examination of cerebrospinal fluid is helpful in recovery and identification of the infecting microbe. The respiratory pathogens *Neisseria meningitidis, Haemophilus influenzae,* and *Streptococcus pneumoniae* are the most frequent causes of bacterial meningitis, although in newborn babies the cause is more likly to be *Escherichia coli* or even a Gram-negative rod from the environment. *Mycobacterium leprae,* the cause of leprosy, is the only pathogen that preferentially attacks peripheral nerves. The disease exists in two main forms, tuberculoid and lepromatous, the latter associated with defective delayed hypersensitivity. Poliomyelitis and rabies are viral diseases of the nervous system. The former is generally caused by three enteroviruses, designated polioviruses I, II, and III, which preferentially attack the cells of motor nerves. Dramatic control of the disease has been achieved by development of vaccines. Rabies is not a common disease in human beings, but a huge reservoir of rabies virus exists in wild animals. The main source of human infections is the domestic dog. Human rabies control rests largely on vaccination of dogs against rabies virus. Routine vaccination of human beings is avoided because of a small danger of developing allergic encephalitis. Human vaccination is justified for persons with possible exposure to the virus.

QUESTIONS

REVIEW

1. Give the principal reservoirs for the bacteria that cause meningitis. Name three species.
2. Why are most newborn babies resistant to some agents of meningitis but not others?
3. What defect in host defenses is associated with the development of lepromatous (as opposed to tuberculoid) leprosy?
4. Explain the apparent paradox of a higher incidence of paralytic poliomyelitis in countries with good standards of sanitation.
5. Give reasons why only a few cases of rabies occur in human beings in the United States despite the widespread distribution of rabies in animals.

THOUGHT

1. The continuing interaction between *Neisseria meningitidis* and human beings has been accompanied by shifting patterns in the serological types of this organism responsible for meningococcal disease. Discuss possible explanations for these changes.
2. Explain why patients sometimes die from meningococcal infections despite intensive treatment with penicillin, a medication which is highly effective in killing meningococci.

FURTHER READING

BRUBAKER, M. L., C. H. BINFORD, and J. R. TRAUTMAN, "Occurrence of Leprosy in U.S. Veterans after Service in Endemic Areas Abroad," *Public Health Report* 84:1051–1058 (December 1969).

KIMBERLIN, R. H., "Recent Developments in Scrapie," *Nature* 257:641–642 (October 1975).

MARTIN, B., *Miracle at Carville.* New York: Doubleday, 1950. A famous novel concerning leprosy.

PAUL, J. R., *A History of Poliomyelitis.* New Haven, Conn.: Yale University Press, 1971. A fascinating account of the conquest of this ancient, terrible disease.

ROUECHE, B., "The Incurable Wound." *Annals of Epidemiology.* Boston: Little, Brown, 1967. A story about rabies.

WOUND INFECTIONS

Most infections can be compared to the invasion of Troy by the Greeks inside their wooden horse in the sense that microorganisms also use devious methods to overcome host defenses and invade. Infections of wounds, however, are perhaps better compared to the capture of a city where the fortifications have been broken down, giving the invaders relatively free access.

Whether a wound is caused accidentally, by trauma, or intentionally, such as by surgery, the open area is extremely susceptible to microbial invasion. Once infected, pus often forms in the injured area resulting in a wound abscess. An abscess is a localized collection of pus (including white blood cells, components of tissue breakdown, and any infecting organisms that may be present). A surrounding area of inflammation tends to isolate the abscess from normal tissue. Consequently, abscess formation helps to localize an infection and prevent its spread. However, microorganisms in abscesses are partially protected from antimicrobial medications, and they are a potential source of infection of the bloodstream and other parts of the body.

ORGANISMS OFTEN RESPONSIBLE FOR WOUND INFECTIONS

Virtually any kind of pathogenic or opportunistic organism can cause wound infections under appropriate conditions, but certain types of wounds are subject to infection by particular organisms. Staphylococci are by far the most frequent cause of wound infections. Aerobic or facultative bacteria also common in wounds include streptococci, enterobacteria, and pseudomonads. Other microorganisms that infect wounds less frequently include a variety of fungi that invade wounds in which conditions favor their growth and some animal pathogens that are transmitted by animal bites or scratches.

Anaerobic microorganisms are responsible for about 10 percent of all wound infections. Therefore it is necessary to use special culture media and methods for obtaining anaerobic conditions, along with the usual aerobic techniques, when examining material from wounds. One reason for the relatively frequent occurrence of anaerobe-induced wound infections is that anaerobic conditions often develop in tissues following wound injury, providing anaerobic microorganisms with an excellent environment for growth. Puncture wounds caused by contaminated nails, thorns, splinters, and other sharp objects introduce foreign material and microorganisms deep into the body. Bullets and other projectiles that enter at high velocity can carry contaminated fragments of skin or cloth into the tissues. Because of the force with which they enter, projectiles cause relatively small breaks in the skin, and these may close quickly, masking areas of extensive tissue damage. Another major reason for anaerobes frequently infecting wounds is that these organisms are included among the normal human flora and may be the dominant organisms numerically; they are therefore readily available to invade wounded areas.

AEROBIC INFECTIONS OF SURGICAL WOUNDS

Even when the most careful precautions are taken, a surgical wound may become infected. *Staphylococcus aureus* is the most common cause of infections of such wounds as well as of those resulting from accidental trauma. The presence of sutures in surgical wounds is an important predisposing factor to *S. aureus* infection. Experiments have shown that hundreds of thousands of *S. aureus* can be injected beneath the skin without causing serious damage in a normal individual, but that less than a hundred of these same organisms can cause abscess formation when introduced on a suture. Furthermore, tissue reactions to sutures and to other foreign bodies strongly favor the establishment of staphylococcal infection.

Many strains of *S. aureus* produce penicillinase, an enzyme that degrades penicillin; therefore, penicllins that are resistant to this enzyme (such as methicillin) are generally used to treat staphylococcal infections. Some staphylococcal strains resistant to methicillin have emerged, but infections with these strains can be successfully treated with other antibiotics.

Streptococcus pyogenes and enterobacteria are other common causes of surgical wound infections (Table 30-1).

INFECTIONS OF BURNS

Burned areas, with their damaged skin, offer ideal sites for infections by bacteria of the environment or of the normal flora. Almost any potential pathogen can infect burns, but one of the most prevalent and hardest to treat is the Gram-negative rod *Pseudomonas aeruginosa*. It is especially dreaded because of its resistance to

TABLE 30-1
Important Causes of Wound Infections

Type of Wound	*Causative Agent*
Wound resulting from accidental trauma	*Staphylococcus aureus* *Streptococcus pyogenes* *Clostridium* sp. *Sporothrix schenckii*
Surgical	*S. aureus* *S. pyogenes* *Pseudomonas aeruginosa* Enterobacteria
Burn	*P. aeruginosa* *S. aureus* *S. pyogenes* Enterobacteria
Animal-inflicted	Rabies virus *Pasteurella multocida*

antibiotics. In fact, infections with pseudomonas are a frequent cause of death in burn patients; however, a new and promising method has recently been introduced for combating these infections. Although more evidence is needed to demonstrate its efficiency, it offers hope that the death rate from infected burns may soon be lowered. The method involves immunization of badly burned patients with killed pseudomonas vaccines within the first few days after injury and at intervals during the first two weeks thereafter. Antibodies against the bacteria are produced, usually before the wounds have time to become infected, and the antibodies give some protection against pseudomonas invasion.

Pseudomonas aeruginosa is also of great importance as a cause of other kinds of infections, including those of surgical wounds, and treatment with antibiotics to control other invading bacteria often predisposes a patient to infection with antibiotic-resistant pseudomonas.

ANAEROBIC WOUND INFECTIONS CAUSED BY CLOSTRIDIA

Tetanus (Lockjaw)

One of the most dangerous diseases resulting from anaerobic infection of wounds is tetanus, caused solely by the action of a powerful exotoxin produced by *Clostridium tetani.* Even though this disease is easily prevented by immunization with tetanus toxoid, hundreds of people in the United States contract the disease and many die from it each year. During the period 1966–1974, 1455 cases of tetanus were reported to the Public Health Service, of which 762 were fatal. In 1974 alone, 101 cases and 44 deaths from tetanus were reported. It has been shown that 97 percent of the people who developed tetanus in two recent years *had never been immunized* with toxoid. *Adequate immunization,* a very simple procedure, *would have prevented their disease.*

Tetanus occurs much more frequently on a global basis than it does in the

United States. In many parts of the world it is a common cause of death in newborn babies, as a result of cutting the umbilical cord with instruments contaminated with *Cl. tetani.*

Clostridium tetani organisms are widely distributed in soil. They are also commonly found in the gastrointestinal tract of humans and other animals; hence, manured soil is apt to be especially rich in the spore-forming *Cl. tetani* (Figure 30-1). The organisms rarely invade past the wound area, and the widespread symptoms of tetanus result from the action of an *exotoxin* which is absorbed from the infected wound. The isolation of *Cl. tetani* from wounds is not definite proof that a person has tetanus; conversely, failure to find the organism does not eliminate the possibility of tetanus. The reasons for these observations are that tetanus spores may be isolated from wounds in which *Cl. tetani* is not producing toxin; furthermore, spores may remain dormant in the tissues for long periods of time, germinating and producing toxin after a wound has healed. In a recent study in the United States, it was found that *Cl. tetani* could not be isolated from 7 percent of the cases studied, even though the symptoms of tetanus were undeniable.

The wounds that lead to tetanus may be very small and seemingly insignificant because the toxin is effective in extremely small amounts. The primary requirement for production of the disease is that the wound provide an anaerobic environment, which is necessary for toxin production.

Tetanus toxin acts on nerve cells; the mechanism of its action, however, is not fully understood. It is a protein which diffuses away from the infected wound and may enter the circulation; it can also travel along the nerves to the spinal cord. In the central nervous system it interferes with control of certain reflex activities, resulting in violent spasmodic contractions of muscles that counteract each other. These contractions can be triggered by any minor stimulus. The muscles of the jaw are particularly susceptible, hence the common name "lockjaw" has been given to tetanus.

The incubation time of tetanus may be as short as four days after infection, or as long as many weeks if dormant spores, which must first germinate, are present in the tissues. Tetanus begins with restlessness, irritability, stiffness of the

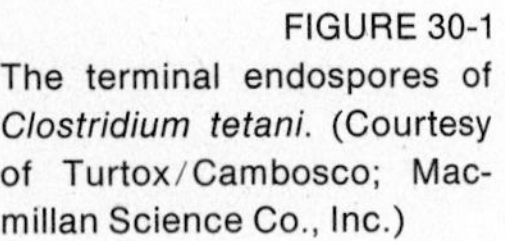

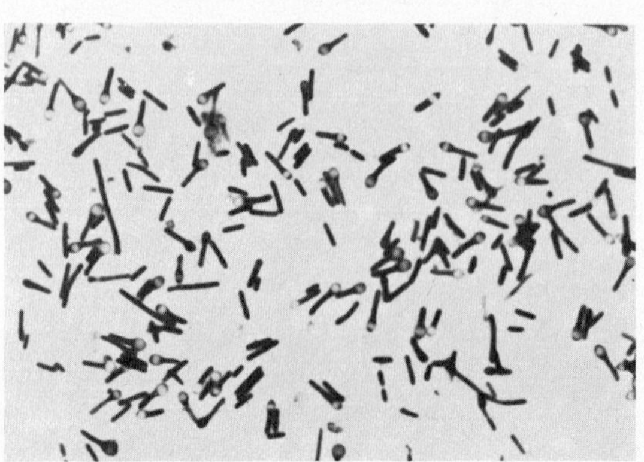

FIGURE 30-1
The terminal endospores of *Clostridium tetani.* (Courtesy of Turtox/Cambosco; Macmillan Science Co., Inc.)

neck, a tight jaw, and sometimes convulsions, particularly in children. As more muscles tense, the pain grows more severe, breathing becomes labored, and after a period of almost unbearable pain, the patient often dies of respiratory failure.

Tetanus is treated by administering antitoxin, and, because there is only one antigenic type of toxin, only one type of antitoxin is necessary. Gamma globulin from human beings immunized with tetanus toxoid is the treatment of choice. This antitoxin does not cause severe hypersensitivity reactions in humans and is much more effective than the horse antitoxin formerly used. The antitoxin cannot affect toxin that is already bound to nervous tissue, however. This may explain why tetanus antitoxin often has only limited effectiveness. In addition to antitoxin treatment, a wound must be cleansed and all dead tissue and other material that would provide anaerobic conditions removed. Penicillin can kill any actively multiplying clostridia it reaches, and help to prevent the formation of more exotoxin.

Prophylatic immunization with tetanus toxoid is by far the best weapon against tetanus. Three injections of the toxoid are given, usually beginning during the first year of life. A "booster" dose is given after a year, and another when the child enters school. Once immunity has been established by this regime, "booster" doses at about 10-year intervals will maintain an adequate level of protection. More frequent doses of toxoid are not recommended because of the danger of an allergic reaction following intensive immunization. This sort of hypersensitivity reaction can result when antigen (the toxoid) is introduced into an individual who already has large quantities of humoral antibodies against the antigen.

Gas Gangrene

In addition to tetanus, bacteria of the gas-gangrene-producing group of clostridia can cause serious wound infections. This group of gas-forming organisms includes *Cl. perfringens* (also known as *Cl. welchii*) and several other species. At present gas gangrene is rare except in battlefield situations where wounds cannot be promptly treated; however, it occurs occasionally as a result of surgery, illegal abortion, and accidents.

NONCLOSTRIDIAL ANAEROBIC INFECTIONS OF WOUNDS

It has become apparent that many wound infections are caused by anaerobes other than clostridia. This perception is due to improved and more convenient methods for isolating anaerobic organisms. Although much remains to be learned about these anaerobic infections, they are caused principally by certain organisms of the normal flora, namely, Gram-negative rods of the genera *Bacteroides* and *Fusobacterium;* and by Gram-positive cocci, principally streptococci.

In addition to these organisms, certain funguslike bacteria can cause wound infections. Among these is *Actinomyces israelii,* Gram-positive, anaerobic,

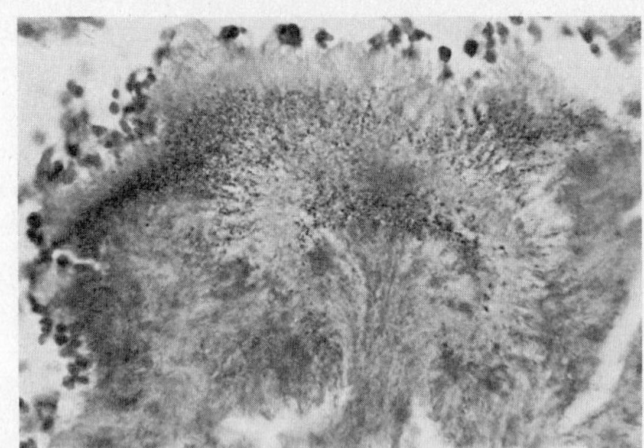

FIGURE 30-2 *Actinomyces israelii.* These funguslike bacteria are part of the normal flora of the oral cavity of man, but under unusual circumstances they can invade and cause an infection. This specimen is part of a colony of *A. israelii* in pus from an infected patient. The colony appears as a yellow mass; therefore it is called a "sulfur granule." Filaments of bacteria can be seen radiating from the edge of the colony. (Courtesy of Turtox/Cambosco; Macmillan Science Co., Inc.)

branching bacteria which are occasional members of the normal flora of the alimentary and upper respiratory tracts. Within pus from infected areas these actinomycetes form microcolonies known as "sulfur granules" because of their yellow color. When examined microscopically, a "sulfur granule" is seen to consist of filaments of the bacteria arranged in radial fashion around the periphery of each granule (Figure 30-2). As might be expected, wounds caused by dental procedures sometimes become infected with actinomyces or other anaerobic bacteria that are part of the normal flora of the mouth. Actinomycetes can also invade the lung or gastrointestinal tract and cause a variety of conditions other than wound infections.

ANIMAL BITES

Both normal flora and pathogens of animals can be transmitted to man via animal bites. The virus disease, rabies, is one of the most feared infections transmitted in this manner, but less serious bacterial infections also occur. For example, the bites of cats, dogs, and some other mammals are frequently contaminated with *Pasteurella multocida.* This small Gram-negative rod is similar in appearance to the causative agent of the plague, *Yersinia pestis,* and also resembles the bacteria that cause tularemia, *Francisella tularensis. Pasteurella multocida* organisms are facultatively anaerobic and fastidious, but grow readily on rich media. Inhabitants of the normal flora of the mouth and upper respiratory tract of their natural animal hosts, they can cause severe infection when body defenses are suppressed or stressed (as by a viral infection). Different strains of *P. multocida* show differences in virulence and host preference. In human beings, infections with *P. multocida* occur principally following animal bites and can be severe; however, they usually respond to appropriate antibiotic therapy.

Wounds from cat scratches and bites occasionally lead to a disease known as "cat-scratch fever." This disease affects lymph nodes draining the area of injury.

Typically, it is a mild illness lasting from a few weeks to a few months. The causative agents have not been definitely identified, but chlamydia are probably responsible for some cases. The disease affects man only, and cats that transmit it give no evidence of illness. Rat bites can also lead to infections with bacteria carried in the oral cavity of rats, notably *Spirillum minus* and *Streptobacillus moniliformis.*

FUNGAL INFECTIONS OF WOUNDS

Over the 20-year period from about 1925 to 1945, more than 3,000 cases of an unusual type of wound infection were reported in South African mine workers. Typically, wounds caused by splinters from mine timbers became ulcerated, often without causing pain. After about a week, nodules developed in a chain along the lymphatic vessel draining the wound and these nodules then developed into ulcers. The lymph node draining the area usually became enlarged, and the lesions persisted for long periods of time even though the patients were usually not very ill. These miners were infected with a common fungus, *Sporothrix schenckii* (also known as *Sporotrichum schenckii*), which grows well on timbers in mines where both the temperature and the humidity are consistently high. It is also common on vegetation, and can infect florists, gardeners, and other workers who suffer wounds from contaminated thorns or splinters. The disease rarely becomes generalized but usually follows the chronic, persistent course described above. The fungus may also spread in the skin, causing scaly, flat plaques or wart-like lesions, but inoculation by a puncture wound more often produces lymphatic involvement.

The causative fungus of this disease, *S. schenckii,* is dimorphic. In nature, or in cultures incubated at room temperature, it forms a fluffy mold mycelium, tan to brown in color. In vivo and in vitro at 37°C, it grows as a budding yeast. In the mold form, the hyphae of *S. schenckii* are much thinner than those of most fungi and spores are formed along the hairlike hyphae. The spores (conidia) are oval or rounded, about 2 to 5 μm in length, and occur in clusters on branches (conidiophores) at right angles to the hyphae, producing a flowerlike appearance (Figure 30-3a). The yeast phase at 37°C and in tissues is characterized by

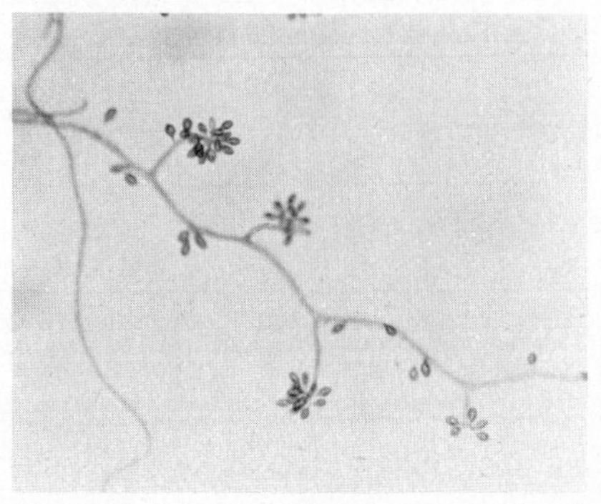
(a)

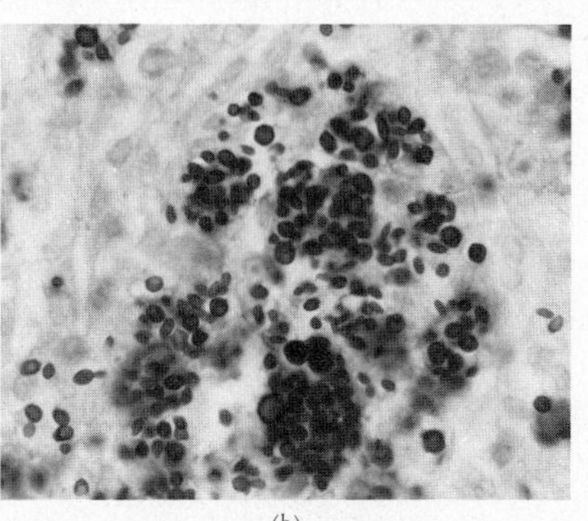
(b)

FIGURE 30-3
Sporothrix schenckii, a dimorphic fungus. (a) The mold phase exhibits slender hyphae with elongated ovoid conidia in a flowerlike arrangement. (b) The yeast phase from infected tissue. Some budding cells can be seen.

elongated cigar-shaped cells (up to 5 μm long) with from one to three buds at both ends of the cell (Figure 30-3b).

Other fungi can invade and infect wounds, but none produces such a characteristic, chronic disease as that resulting from *S. schenckii.*

SUMMARY

Wounds interfere with normal defenses and therefore predispose to infection. *Staphylococcus aureus* is the most frequent cause of wound infections. Other aerobic microorganisms often responsible for wound infection include streptococci, enterobacteria, and pseudomonas. Wounding often leads to the development of anaerobic conditions in the tissues and subsequent invasion by anaerobic bacteria such as clostridia, anaerobic streptococci, bacteroides, fusobacteria, and actinomycetes.

Surgical wounds may become infected with *S. aureus* that are resistant to penicillin. The presence of sutures in a wound predisposes to staphylococcal infection. *Streptococcus pyogenes* and enteric bacteria are also frequent causes of surgical wound infections.

Burns are particularly susceptible to infections. *Pseudomonas aeruginosa* is one of the most prevalent agents in burn infections and a frequent cause of death in burn patients because of its resistance to antibiotics.

Tetanus is an important disease resulting from anaerobic infections of wounds with *Clostridium tetani.* The disease is caused by the potent tetanus nerve toxin; it is therefore treated with tetanus antitoxin and prevented by immunization with tetanus toxoid. Gas gangrene is also produced by members of the genus *Clostridium.*

Animal bites and scratches may transmit microorganisms from animals to humans, resulting in serious disease. Among the infectious agents transmitted in this manner are rabies virus, *Pasteurella multocida,* and others.

Various fungi can infect wounds. One of the most readily recognized fungus infections is caused by *Sporothrix schenckii,* which invades small wounds and causes a characteristic lymph node involvement and skin ulceration.

QUESTIONS

REVIEW

1. List some of the aerobic or facultative bacteria that most often cause wound infections.
2. Why is *Pseudomonas aeruginosa* likely to infect severe burn wounds? Since this organism is often considered "nonpathogenic," is a pseudomonas wound infection cause for concern? Why?
3. In theory, tetanus can be eradicated, yet it still occurs. Why is this and what can be done to eliminate tetanus?
4. Can tetanus be treated effectively. How?

5. List a few kinds of infection that can be transmitted by animal bites or scratches.

THOUGHT

1. Why might burn wounds be even more susceptible to infection than surgical wounds, and burn patients more likely to develop serious or fatal infections?
2. Before abortion was legalized, illegal abortions frequently resulted in severe infections, or even in death of the patient from infection. Speculate about the reasons for this, and about which organisms probably caused most of the infections.

FURTHER READING

ARBUTHNOTT, J. P., "Staphylococcal Toxins." *Microbiology—1975*. Ed. D. Schlessinger. Washington, D.C.: American Society for Microbiology, 1975, pp. 267–271. A summary of the toxins that may contribute to the pathogenicity of staphylococci.

DUNCAN, C. L., "Role of Clostridial Toxins in Pathogenesis." *Microbiology—1975*. Ed. D. Schlessinger. Washington, D.C.: American Society for Microbiology, 1975, pp. 283–291. A brief summary of information concerning the nature and mode of action of the clostridial toxins, including among others the toxins that cause tetanus and gas gangrene.

FINEGOLD, S. M., J. E. ROSENBLATT, V. L. SUTTER, and H. R. ATTEBERY, Scope Monograph on Anaerobic Infections. Kalamazoo, Mich.: The Upjohn Company, 1972. A succinct and well-illustrated summary of anaerobic infections.

FISHER, M. W., "Polyvalent Vaccine and Human Globulin for Controlling *Pseudomonas aeruginosa* Infections." *Microbiology—1975*. Ed. D. Schlessinger. Washington, D.C.: American Society for Microbiology, 1975, p. 416.

PAVLOVSKIS, O. R., L. T. CALLAHAN III, and M. POLLACK, "*Pseudomonas aeruginosa* Exotoxin." *Microbiology—1975*. Ed. D. Schlessinger. Washington, D.C.: American Society for Microbiology, 1975, pp. 252–256. An in-depth look at an important exotoxin of pseudomonas, one that may play a significant role in the pathogenesis of infections caused by *P. aeruginosa*.

CHAPTER 31 INFECTIONS INVOLVING THE BLOOD VASCULAR AND LYMPHATIC SYSTEMS

ANATOMICAL AND PHYSIOLOGICAL CONSIDERATIONS

Almost everyone today is familiar with the general structure and function of the body's blood vessel system, although few people had ever considered the possibility of blood circulating before William Harvey's convincing argument in 1628. The heart, a muscular pump enclosed in a sac called the *pericardium,* supplies the force that moves the blood. As shown schematically in Figure 31-1, the heart is divided into a right and a left side, separated by a wall of tissue through which blood normally does not pass after birth. The right and left sides of the heart are each divided into two chambers, one of which receives blood (*atrium*) and the other which discharges it (*ventricle*). Although not common, infections of the heart valves, muscle, and pericardium may be disastrous because of their effect on the vital functions of the circulatory system. Blood from the right ventricle flows through the lungs and into the atrium of the left side of the heart. This blood then passes into the left ventricle and is pumped through the aorta to the arteries and capillaries that supply almost all of the tissues of the body.

A system of veins collects the blood from the capillaries in the tissues and carries it back to the atrium of the right heart. Because pressure in the veins is low, one-way valves help keep the blood flowing in the right direction. Thus, as the blood flows around the circuit, it alternately passes through the lung and through the tissue capillaries. Also during each circuit, a portion of the blood passes through other organs, such as the spleen, liver, and lymph nodes, all of which, like the lung, contain fixed phagocytic cells of the mononuclear phagocyte system. Foreign material, such as microbes, may be removed by these phagocytes on passage of the blood through such tissues.

Mononuclear phagocyte system, page 371

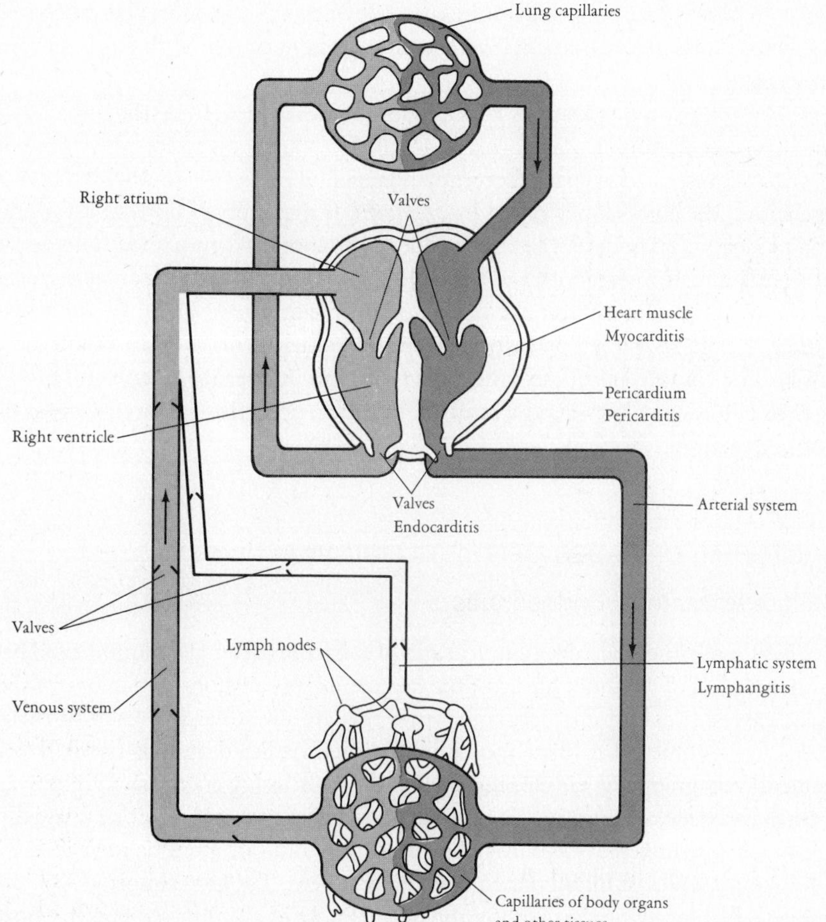

FIGURE 31-1
The blood vascular and lymphatic systems. The lymphatic drainage from the lung is omitted for clarity.

LYMPHATIC SYSTEM

The system of lymphatic vessels begins in tissues as tiny tubes that differ from blood capillaries in having closed distal ends and being somewhat larger. Inside these lymphatic vessels is *lymph,* an almost colorless fluid. This fluid comes from plasma that has oozed through the blood capillaries to bathe the tissue cells and which is then taken up by the lymphatics. Unlike the blood capillaries, the lymphatics readily take up foreign material such as invading microbes or their products. The tiny lymphatic capillaries join progressively larger vessels in a manner similar to the roots of a plant. Many one-way valves in the lymphatic vessels keep the flow of lymph moving in a general direction away from the lymphatic capillaries. Both contraction of the vessel walls and compression by the movements of muscles force the lymph along.

At many points in the system, lymphatic vessels drain into small, roughly bean-shaped bodies called *lymph nodes.* These nodes are so constructed that foreign material, such as bacteria, is trapped; the nodes also contain cell types that are involved in phagocytosis and antibody production. Lymph flows out of the nodes

Lymphoid tissues, page 386

through vessels that eventually unite into one or more large tubes that discharge the lymph into the blood vessel system through connections with large veins in the chest.

An infection of a hand or a foot is sometimes evidenced by the spread of a red streak up the limb from the infection site. This streak represents the course of lymphatic vessels that have become inflamed in response to the infectious agent, and the condition is called *lymphangitis.* It may stop abruptly at a swollen and tender lymph node and later continue to yet another lymph node. This arrest of an inflammation demonstrates the ability (even though sometimes temporary) of the lymph nodes to clear the lymph of an inflammatory agent. As indicated earlier, the blood and lymph carry such antimicrobial agents as white blood cells, antibodies, complement, lysozyme, beta lysin, and interferon. Lymph and blood may clot in regional vessels as a result of inflammation arising from infection or antibody-antigen reactions.

BACTERIAL DISEASES OF THE BLOOD VASCULAR SYSTEM

Subacute Bacterial Endocarditis

Table 31-1 outlines the important features of subacute bacterial endocarditis. This is an infection of the inner lining of the heart and is usually localized to one of the heart valves. It commonly occurs in hearts that are abnormal as a result of rheumatic fever, some other disease, or a birth defect. Afflicted persons typically become ill very gradually, developing fever and slowly losing energy and vigor over a period of weeks or months. The causative organisms of endocarditis are usually shed from the infected valve into the circulation and can often be identified by culturing samples of blood drawn from an arm vein. They are almost always organisms frequently present among the normal flora of the infected person, most commonly oral alpha hemolytic streptococci or other streptococci of low virulence, or *Staphylococcus epidermidis.*

Alpha hemolysis, page 429

As indicated in Chapter 27, a few organisms of the normal flora frequently gain entrance to the bloodstream during dental procedures, brushing of teeth, or

TABLE 31-1
Features of Subacute Bacterial Endocarditis

Host Factors	*Microbial Cause*	*Pathogenesis*
Structural abnormality of heart: birth defect, rheumatic fever, other diseases	Normal flora: *Staphylococcus epidermidis*, viridans streptococci, other	Turbulent blood flow Clot formation Colonization of clot Enlargement of infected clot Release of fragments: plugging of vessels; damage due to immune complexes

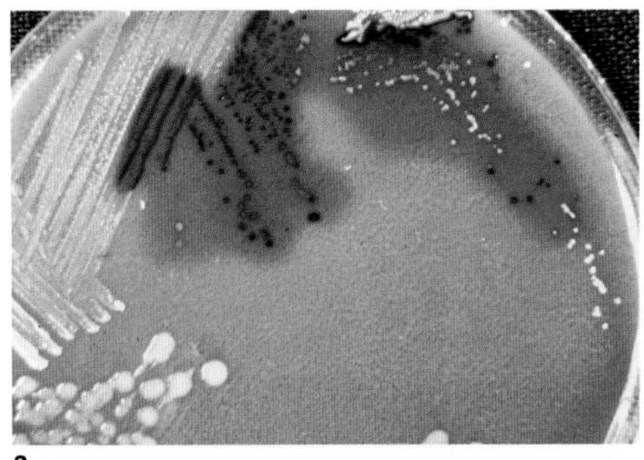

a

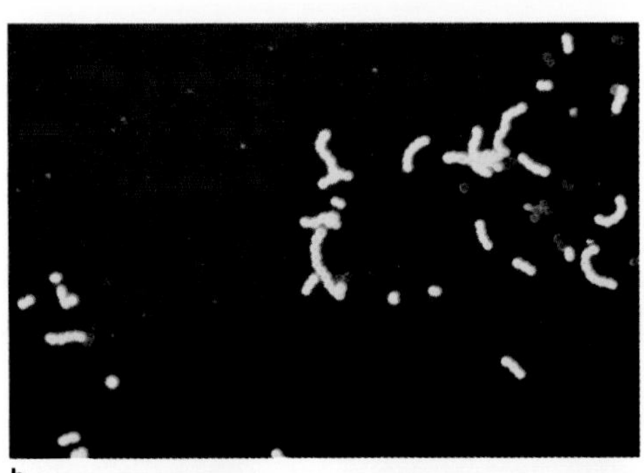

b

c

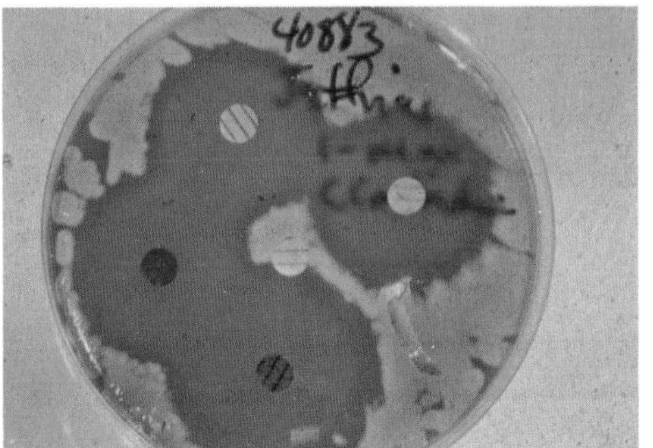

d

(a) Oxidase test. The dark-colored colonies on this plate are showing a positive oxidase reaction. **(b)** *Streptococcus pyogenes* from a fluid medium culture, stained with specific fluorescent-labeled antibodies. **(c)** Test for diphtheria toxin. The top horizontal streak is an avirulent *Corynebacterium;* the bottom streak is *Corynebacterium diphtheriae.* The vertical band is filter paper soaked with diphtheria antitoxin. Note the fine lines of precipitate at a 45° angle to the bottom streak. **(d)** Antibiotic susceptibility of *Clostridium perfringens* on sheep blood agar medium. Note the zones of inhibited growth and hemolysis about the antibiotic-containing discs. (a courtesy of J. Portman; b, d courtesy of C.E. Roberts; c courtesy of M.F. Lampe.)

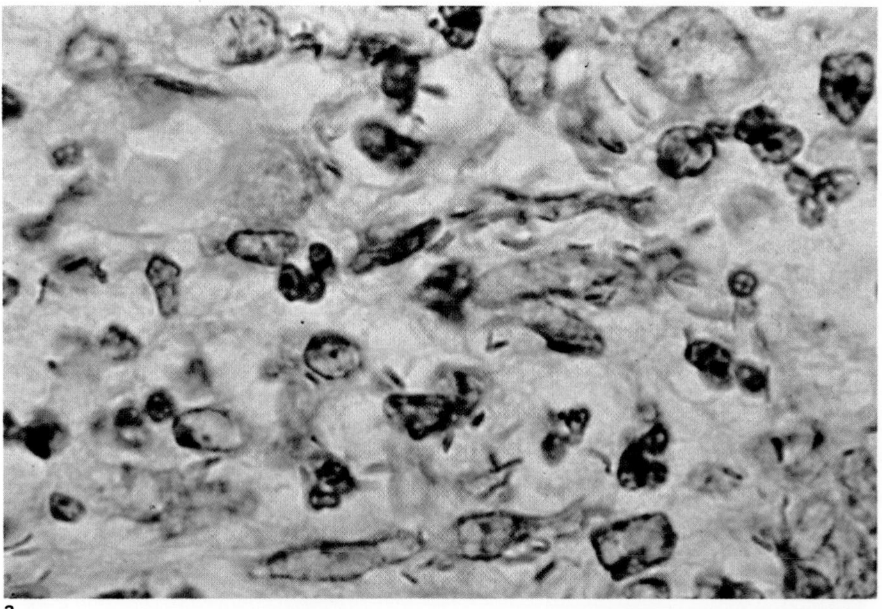

a

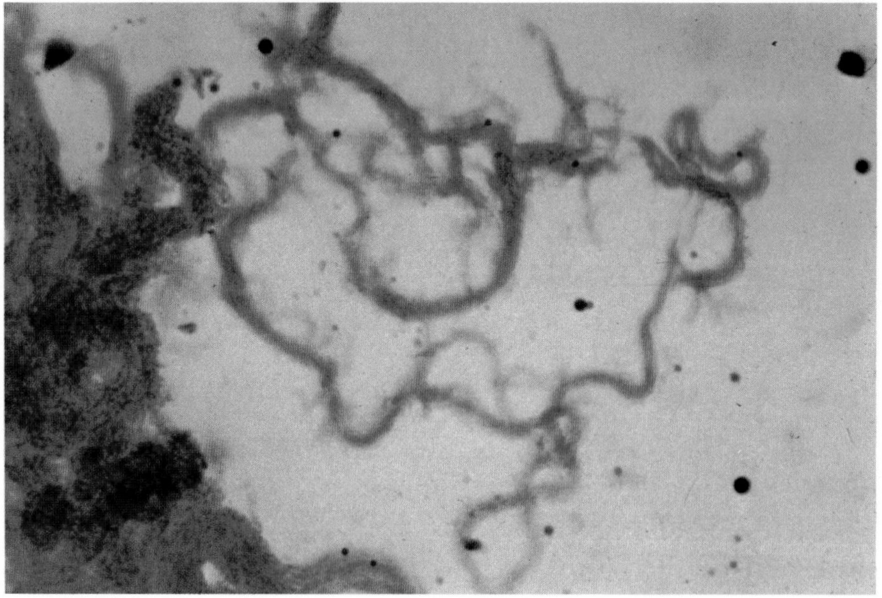

b

(a) Tuberculous infection of the small intestine (acid-fast stain), now a rare disease because of milk pasteurization and tuberculin testing of dairy cows. **(b)** Cording of *Mycobacterium tuberculosis* (acid-fast stain). Characteristic growth of virulent strains on moist media. (Courtesy of C.E. Roberts.)

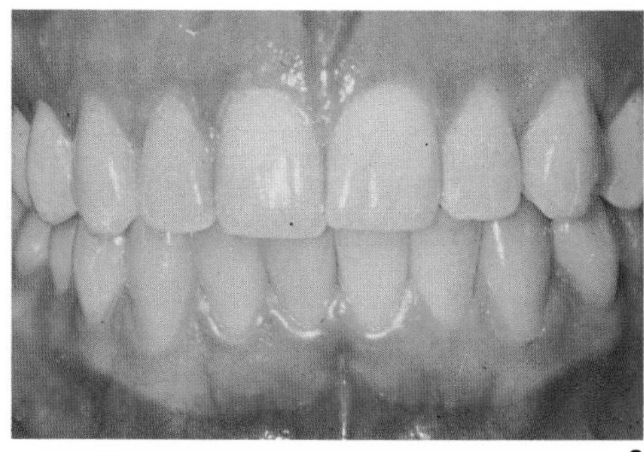

a

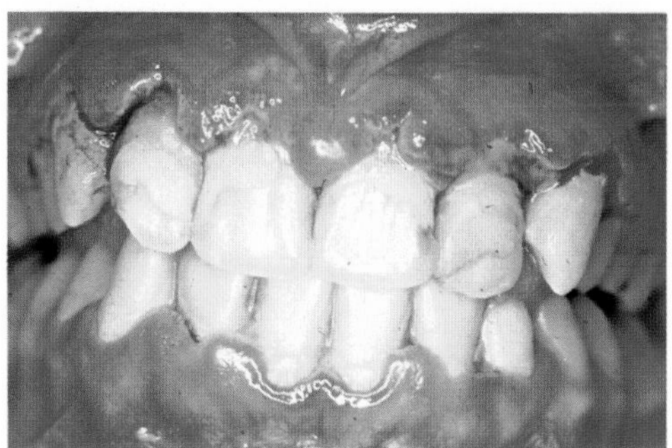

b

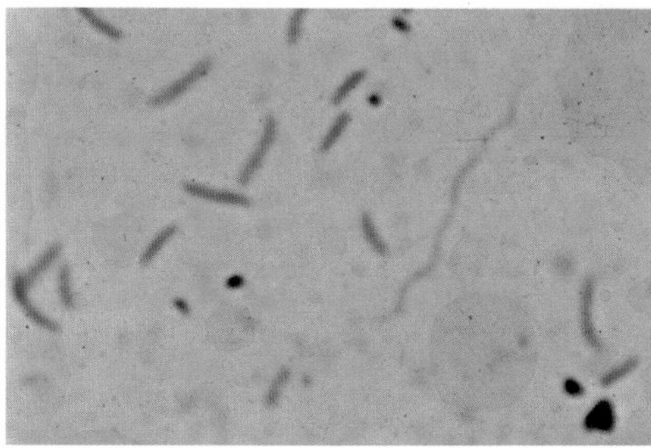

c

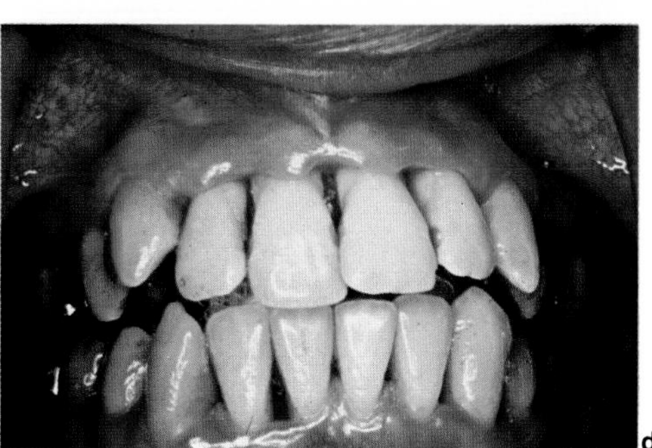

d

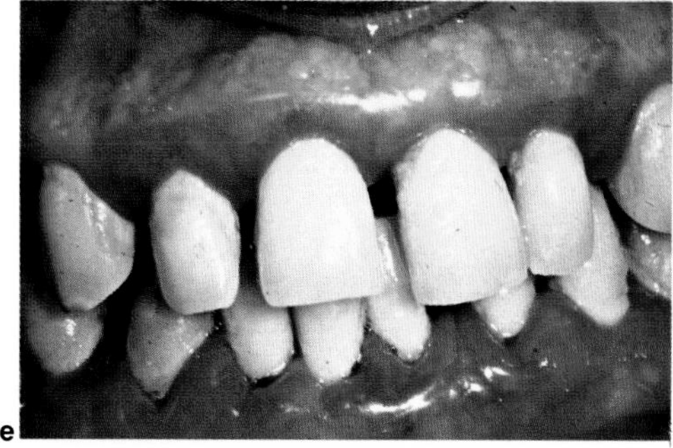

e

Oral diseases. **(a)** Normal. **(b)** Acute ulcerative necrotizing gingivitis. **(c)** Gram-stained smear from b. **(d)** Chronic stage of b. **(e)** Periodontal disease marked by subgingival plaque and intense inflammation. (a, b, d, e courtesy of R.C. Page; from S. Schluger, R.A. Yuodelis, and R.C. Page, *Periodontal Disease*. Philadelphia: Lea and Febiger, 1977; c courtesy of C.E. Roberts.)

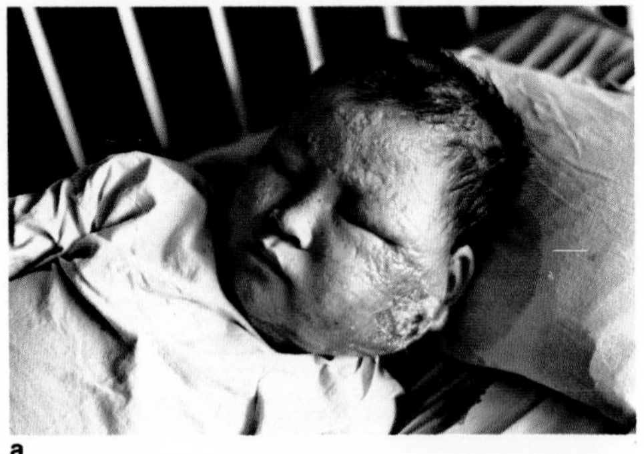

a

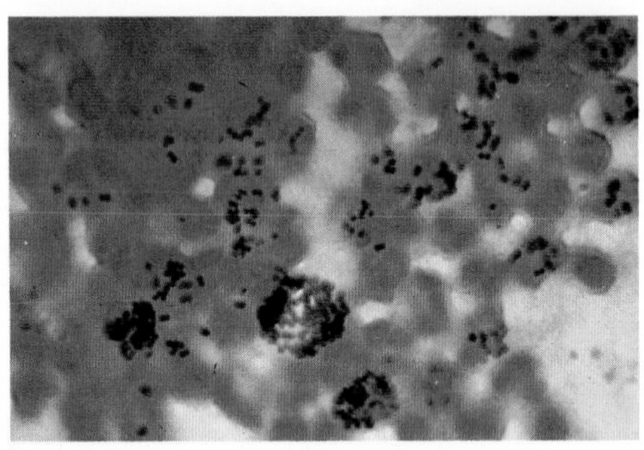

b

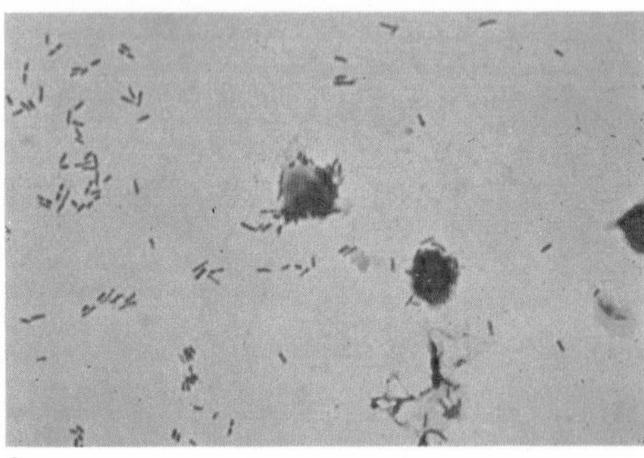

c

(a) Severe staphylococcal infection of a baby's face. **(b)** Staphylococcal abscess (Gram stain). **(c)** Bacteria in urine (Gram stain). (Courtesy of C.E. Roberts.)

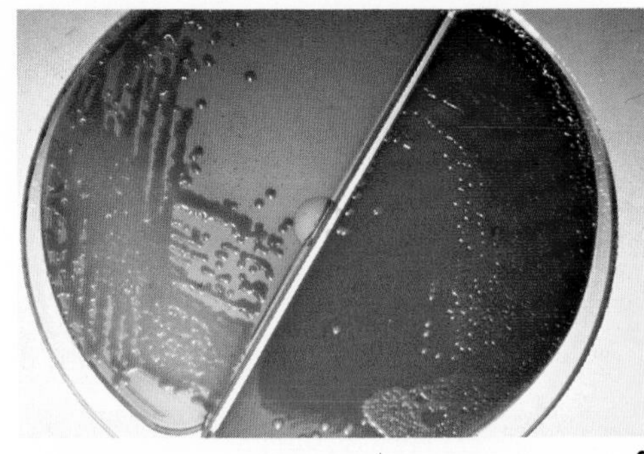

d

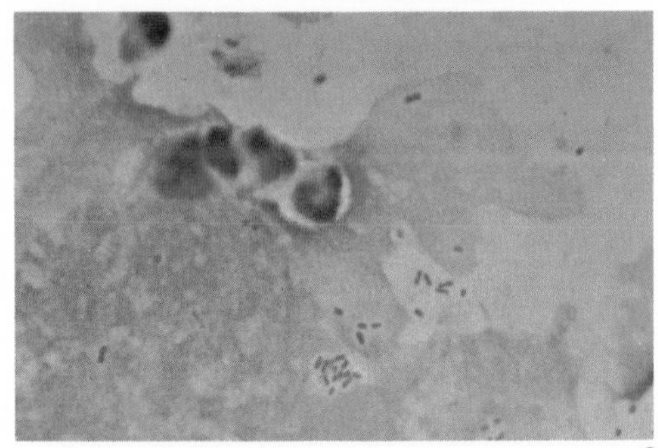

e

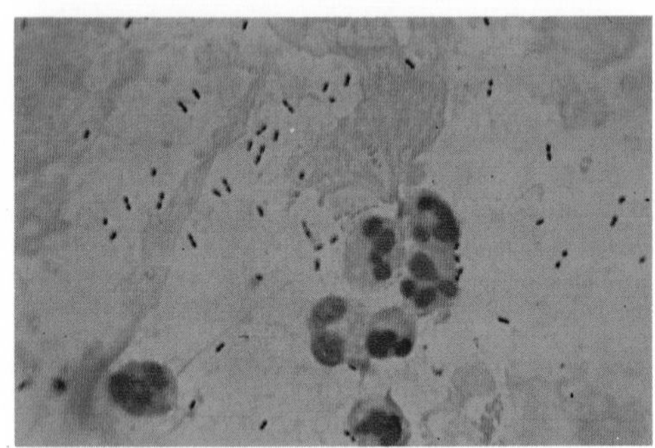

f

(d) Bacteria cultured from urine. **(e)** Sputum from a patient with *Klebsiella* pneumonia (Gram stain). **(f)** Sputum from a patient with pneumococcal pneumonia (Gram stain). (Courtesy of C.E. Roberts.)

other trauma. These are normally eliminated quickly by body defense mechanisms. On the other hand, in an abnormal heart, turbulent blood flow fosters the formation of a film of blood clot and traps such organisms in areas where phagocytes have difficulty acting. Thus the high levels of antibody often present in bacterial endocarditis are of little value and may even foster the progress of the infection by clumping the causative bacteria and depositing them on the heart muscle. Once a clot has been formed, the organisms multiply extensively and more clot may be progressively deposited around them, gradually building up a fragile mass. Bacteria and pieces of infected clot continually break off and, if the latter are large enough, may block important blood vessels and lead to tissue death, or to the weakening and ballooning out of larger vessels. Circulating immune complexes may lodge in the kidney, producing a form of glomerulonephritis. Even though the organisms normally have little invasive ability, great masses of them growing in the heart are sometimes able to burrow into the heart or valve tissue, resulting in valve rupture.

Immune complex reaction, pages 407–408

In some instances of infective endocarditis, culturing the blood from an arm vein fails to yield the causative agent. This is especially likely to occur when the infection is on the right side of the heart and blood from the infected site must pass through both the lung and the tissue capillaries (with their phagocytic cells) before reaching the arm vein. In other instances the organism may be too fastidious to grow in the usual bacteriologic media and may even be rickettsiae (obligate intracellular parasites).

Infections of the heart can also be caused by more virulent microbes such as *Staphylococcus aureus* or *Streptococcus pneumoniae,* usually producing a rapidly progressing illness called *acute bacterial endocarditis.* Such organisms are much more likely to invade the heart and its valves and produce permanent tissue destruction than those causing subacute bacterial endocarditis.

Staphylococci, pages 426–427
Streptococci, pages 427–429

Septicemia

"Blood poisoning" (septicemia) is an illness associated with microbes or their products circulating in the bloodstream. (Bacteria sometimes enter the bloodstream without causing illness, and this is referred to as *bacteremia*). The main findings in septicemia are fever and a drop in blood pressure, probably due at first to impaired strength of the heart and the walls of blood vessels. In severe septicemia the blood vessel damage may be irreversible. If the drop in pressure is marked and prolonged, there is insufficient flow of blood to supply adequate oxygen to the organs and other tissues, and shock results. Septicemia is almost always the result of an infection somewhere else in the body and is thus commonly caused by one of the usual bacterial pathogens. In addition, alterations in normal body defenses as the result of medical treatment have in recent years resulted in many septicemias from microbes that normally have little invasive ability. These infections commonly originate from the normal flora of the body.

Gram-negative rod shaped bacteria are most often responsible for septicemia, although Gram-positive bacteria and fungi can also be causative. Septicemia

Enterobacteria, pages 435–437

Pseudomonads, pages 433–434

caused by Gram-negative rods can be devastating. Shock is common and only about half of all persons afflicted with this kind of infection survive. Cultures of the blood in these cases usually reveal facultatively anaerobic Gram-negative rods such as *Escherichia coli, Enterobacter aerogenes, Serratia marcescens,* or *Proteus mirabilis.* Among the aerobic Gram-negative rod-shaped bacteria encountered in septicemia are organisms commonly found in nature, such as *Pseudomonas aeruginosa.* This organism has extraordinary biochemical capabilities, is very resistant to antibiotics, and is able to grow under a variety of conditions unfavorable to the usual bacterial pathogens.

Bacteroides, pages 438–439

Not all of the Gram-negative organisms causing septicemia will grow in the presence of air. As mentioned in Chapter 27, anaerobic Gram-negative rods of the genus *Bacteroides* make up a major percentage of the normal flora of the large intestine, and these not infrequently cause septicemia. Formerly this group of organisms was poorly classified, and many strains that were isolated failed to fit into existing species. Today, better procedures and equipment have evolved for studying the strict anaerobes. Gas chromatographic examination is an especially useful tool for identifying this group of bacteria. This technique readily determines the characteristic metabolic products of the bacteria.

Endotoxin, Table 19-3, page 374

How Gram-negative bacteria produce septicemia and shock is still not completely understood, but the probable sequence of events is shown in Table 31-2. Gram-negative bacteria contain endotoxin, a poisonous material making up part of their outer cell walls. This type of toxic substance has been discussed in Chapter 19. When injected into the bloodstream of laboratory animals in milligram doses, it produces fever and shock, often leading to death.

Despite these effects of endotoxins defined under laboratory conditions, there is as yet no agreement about the extent to which endotoxins are released in infections, nor about the exact role they play in specific diseases caused by Gram-negative bacteria. Nevertheless, it is very likely that endotoxins cause the principal damage to the host in many septicemias, especially those resulting from relatively avirulent bacteria of the normal flora and the environment. Other bacterial products, including protein toxins, probably also play a role in some cases of septicemia.

TABLE 31-2
Probable Sequence of Events in Septicemia Caused by Gram-negative Rod Bacteria

1. Bacteria enter the bloodstream from a focus of infection.
2. Endotoxin is released.
3. Endotoxin acts on blood vessels causing impaired contractile strength, which becomes irreversible if prolonged.
4. Dilatation of blood vessels results in a drop in blood pressure.
5. Insufficient pressure and maldistribution of blood result in an inadequate supply of oxygen to vital organs.
6. Death results in many cases, even though the infection is controlled, because of damage to the circulatory system from endotoxin.

BACTERIAL DISEASES OF THE MONONUCLEAR PHAGOCYTE SYSTEM

Tularemia

Not long ago there appeared in Vermont an outbreak of disease characterized by fever and enlargement of lymph nodes. Altogether 72 cases were identified, and all occurred among persons who had been in contact with fresh muskrat pelts or skinned muskrats. These people proved to have tularemia, caused by an agent widespread among wild animals and arthropods—an agent which was often fatal in the days before antibiotics were available for treatment.

The causative microbe of tularemia, *Francisella tularensis*[1] is a pleomorphic, nonmotile, aerobic Gram-negative rod that derives its name from Tulare County, California, where it was first studied. The organism can be cultured from blood and other materials taken from infected persons and animals. Unfortunately, contaminating microbes from the environment or the normal flora, if present in the inoculum, may sometimes obscure the growth of *F. tularensis.* If this occurs, a portion of the material is injected into a laboratory animal such as a guinea pig or white mouse. The defense mechanisms of the animal quickly destroy the contaminating flora, while the pathogen invades the animal's tissues. Fluorescent antibody or other serological techniques can then be used to identify *F. tularensis* from the infected tissues. *Francisella tularensis* also produces a soluble antigen which can be extracted from tissue and identified with specific antiserum (*Ascoli test*).

Francisella tularensis, page 434

Precipitin reaction, pages 390–391

Tularemia occurs in many areas of the Northern Hemisphere, including all the states of the United States except Hawaii. In the eastern United States, human infections usually occur in the winter months, as a result of skinning rabbits (thus the common name "rabbit fever"). In the West, infections result primarily from the bites of ticks and flies, and thus usually occur during the summer. *Francisella tularensis* appears to infect through unbroken skin, but it more likely enters the body through small cuts or scratches that may not be visible, or through the mucous membranes of the eye or mouth. It typically causes an ulcer at the site of entry, and enlargement of the lymph nodes in the surrounding area of the body, and these may become filled with pus. The organisms spread to other parts of the body via the lymphatic and blood vessels. *F. tularensis* are of interest because (like *M. tuberculosis*) they are ingested but not killed by phagocytic cells, and they readily grow within them. This may explain why the tularemia persists in some persons even though they have high titers of antibody in their blood. The mechanisms of cell-mediated immunity are probably responsible for ridding the host of this and other organisms that tend to persist intracellularly. Both delayed-type hypersensitivity (demonstrated by injecting antigen from the bacterium into the skin of an infected person) and humoral antibody quickly arise during infection, and their demonstration can be used to help identify tularemia when positive cultures of *F. tularensis* cannot be obtained.

Delayed hypersensitivity, pages 411–413

[1] Formerly known as *Pasteurella tularensis.*

Brucellosis

Brucella abortus, page 434

Brucellosis ("undulant fever," and "Bang's disease") is contracted primarily from diseased cattle, pigs and goats. In animals the responsible bacteria cause chronic infections involving the mammary glands and the uterus, thereby contaminating milk and causing abortions. Human beings develop a disease characterized by fever, body aches, and weight loss, which characteristically subsides without treatment but may then relapse one or more times during a period of weeks or months. *Brucella abortus* is the species most commonly transmitted by cattle; but there are five additional *Brucella* species which have other host preferences. In the United States 60 percent of the cases of brucellosis occur in workers in the meat packing industry; less than 10 percent arise from ingestion of raw milk or other unpasteurized dairy products. As with tularemia the organisms responsible for brucellosis penetrate mucous membranes or breaks in the skin and are disseminated via the lymphatic and blood vessels to the heart, kidneys and other parts of the body. Like *F. tularensis, Brucella* species are resistant to phagocytic killing and can grow intracellularly. Antibiotic treatment is effective, and without treatment the disease is debilitating but rarely fatal. The most important control measures against brucellosis are pasteurization of dairy products and inspection of domestic animals for evidence of the disease.

A VIRAL INFECTION OF THE HEART MUSCLE

Myocarditis, or inflammatory disease of the heart muscle, may be caused by bacteria, fungi, or protozoa but is more commonly the result of a viral infection. Sometimes this inflammation primarily involves the sac around the heart (pericardium) and is called *pericarditis.* Involvement of the heart muscle may produce only pain in the chest and abnormalities of the electrical impulses of the heart, but if the infection is extensive, the heart is unable to contract effectively, and heart failure may result. This problem is most likely to occur in infants but can arise at any age.

Picornaviruses, Table 17-1, page 341

Appendix III, pages 711–713

Identification of the causative microbe of myocarditis can sometimes be made from cultures of the fluid accumulating in the pericardial sac or from cultures of feces or throat secretions plus demonstration of a rise in neutralizing antibody during the course of the illness. In most of the instances where an agent has been recovered in myocarditis, it has been a Coxsackie virus.

The virus responsible for myocarditis presumably reaches the heart by way of the bloodstream or lymphatic system after infecting the respiratory or gastrointestinal tract. However, the reason for only a few infected persons manifesting myocarditis during any given Coxsackie virus epidemic is unknown. Once in the heart the virus replicates in and extensively damages the muscle cells, and some of these cells die and are replaced by scar tissue. In some terminal cases of myocarditis, after an illness of many months, viral antigen can still be demonstrated in the heart muscle by using fluorescent antibody against Coxsackie virus. Mysteriously, the virus cannot usually be recovered in cultures even though large amounts of viral antigen are demonstrable. This suggests the possibility of a persistent infection with a defective form of the virus, and progressive heart damage from an immune response to the infected muscle cells.

Coxsackie viruses are able to grow and produce disease in mice, but susceptibility of these mice to the virus is markedly age-dependent. In fact, only newborn mice are consistently affected. Moreover, Coxsackie viruses differ from each other in the types of mouse tissue cells they will attack. Group A viruses characteristically damage the muscles, whereas group B attack the central nervous system, liver, pancreas, and fatty tissues. Different Coxsackie viruses also show differing tissue affinities in human beings. Most persons with viral myocarditis have Group B Coxsackie infections.

Because only a small fraction of persons infected with Coxsackie viruses develop serious illness, and very few of these die, there has not been much effort to establish control measures against these organisms. No vaccines are available, and as with other enteroviruses, sanitary practices are unlikely to limit spread of these agents, although they should help decrease the infecting dose of virus received by its victims. Excretion of the virus in the feces of an infected person may continue for weeks, even after mild, unnoticed infections. The presence of healthy disseminators in a community undoubtedly aids in the spread of Coxsackie viruses.

A VIRAL DISEASE OF THE LYMPH NODES: INFECTIOUS MONONUCLEOSIS

Infectious mononucleosis ("mono") is a disease familiar to many students because of its high incidence among economically privileged persons between the ages of 15 and 24 years. The term *mononucleosis* refers to the fact that persons infected with this condition have an increased number of mononuclear white blood cells (white blood cells with unsegmented nuclei) in their bloodstreams. The most dramatic symptoms of the disease are fever, sore throat, and enlargement of the lymph nodes and spleen. In occasional instances complications such as myocarditis, meningitis, hepatitis, or paralysis develop. One very interesting aspect of infectious mononucleosis is the development of an antibody in the bloodstreams of sick persons that will react with an antigen on the surface of red blood cells taken from sheep. In fact, for most laboratories this is the only practical way to separate patients with infectious mononucleosis from those with other infectious conditions closely resembling it.

See Figure 20-6, page 387

In 1968 reports appeared with some very interesting developments relating to infectious mononucleosis. These derived from work extending back to 1964 on a peculiar virus seen in cell cultures prepared from a malignant tumor of the lymphatic tissues common in parts of Africa. The virus was shown to be unrelated to any known virus, although its appearance was similar to that of the herpes viruses. The virus is now popularly called the Epstein-Barr virus (or EB virus) after its discoverers. Using EB virus grown in tissue cell cultures as antigen, it was

Herpes virus, Figure 17-1, page 341

Burkitt's lymphoma, page 362

possible to detect and quantify anti-EB virus antibody in the bloodstreams of various persons. The highest titers of this antibody were found in persons with the type of malignant lymphatic tumor which had originally yielded the virus, but many other persons were also found to have the antibody. Of special interest, however, was the finding that persons who had contracted infectious mononucleosis consistently lacked the antibody to EB virus before their disease started but had developed high titers of this antibody during their illness. The antibody has been found to persist for years after infection. Moreover, a close relative of EB virus could be cultured from the mononuclear blood cells of persons infected with mononucleosis in a manner similar to its culture from the malignant tumor. People with infectious mononucleosis do not develop the lymphoma, although EB virus probably causes the African tumor, other factors are of crucial importance in the etiology of the malignancy.

Present evidence indicates that the EB-like virus of infectious mononucleosis is widespread and infects most economically disadvantaged groups at an early age without producing significant illness. In such populations infectious mononucleosis is quite rare. On the other hand, more than 50 percent of students entering some United States colleges have shown no antibody to EB virus and presumably have escaped past infection with the agent. Infectious mononucleosis occurs almost exclusively in such antibody-negative subjects. Epidemiological evidence indicates that kissing may be an important mode of transmission of infectious mononucleosis in young adults. The virus is not highly contagious, even its intrahousehold spread is rare.

TOXOPLASMOSIS: A PROTOZOAN DISEASE

Toxoplasmosis is a disease caused by the small crescent-shaped protozoan, *Toxoplasma gondii* (Figure 31-2). This microbe can infect a broad range of hosts, including both birds and mammals, and disastrous epidemics among commercially valuable animals have occasionally occurred. Overt disease in human beings as a

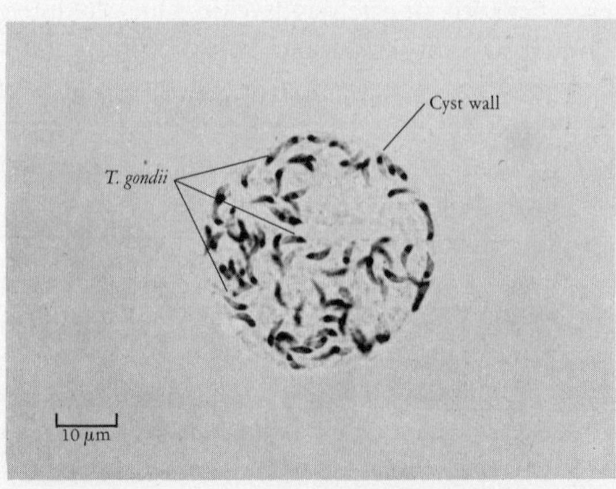

FIGURE 31-2
Toxoplasma gondii cyst. (Courtesy of J. Remington.)

result of this organism is unusual, although infection is common. Three main illness patterns are seen: (1) infection of the fetus resulting from the asymptomatic infections of the expectant mother—in this instance the baby may be born mentally defective or merely suffer slight damage to the retina of its eye; (2) infections of adults, especially those with serious underlying diseases, may be manifest as myocarditis or generalized systemic involvement, affecting most body tissues; and (3) a mild illness resembling infectious mononucleosis and usually seen in young adults. Studies of students at the University of California and at Harvard College have shown about one out of five were infected with *Toxoplasma.* Infection is thought to occur mainly from eating improperly cooked meat, and an outbreak of five cases once occurred among a group of New York medical students who had eaten inadequately cooked hamburger. Their illness resembled infectious mononucleosis, and they had lymph node enlargement lasting about three months.

The life cycle of *T. gondii* has long been a mystery, but it is now known that the organism is a coccidium of cats, infecting the cells lining the cat intestine (coccidia are sporozoa—distant relatives of malaria parasites). In one American city about half the cats tested were infected, although they showed no overt signs of illness. Other animals become infected when they ingest food or water contaminated by feces from infected cats. The organisms then enter the tissues of the new host—perhaps a cow—where they develop into a cystic form containing many infectious progeny. These persist alive for an indefinite period and can be transmitted when another animal, such as a human, ingests the inadequately cooked flesh of the cow. However, the complete cycle of growth of *T. gondii* can only develop with infection of the intestinal-lining cells of a cat or closely related animal species. It has been established that persons in close contact with cats have a higher toxoplasma infection rate than do those who do not have cats.

Sporozoa, Table 15-1, pages 305, 312–313

Toxoplasma gondii is of general interest because of its wide host range. The organisms are infectious for almost all cells of warm-blooded animals except nonnucleated red blood cells. *T. gondii* was also one of the first infectious agents shown to produce general enhancement of resistance to intracellular infections by other species of microorganisms. The practical significance of this observation is as yet undefined, but in view of the widespread and benign nature of most human infections, it is interesting to speculate on the possible role of this coccidium of cats in ameliorating human disease. Nevertheless, expectant mothers would do well to avoid cats and rare meat.

SUMMARY

There are two networks of vessels in the body, the blood vascular system and the lymphatic system. As the heart pumps blood around the blood vascular circuit, any microbes present are likely to be removed by the phagocytic cells of the spleen, lung, and other organs. Unlike the blood vessels, the lymphatic vessels begin as blind-ended capillaries which are readily permeable to microbes. Lymph is subject to the decontaminating action of phagocytes in the lymph nodes. Failure of body defense mechanisms may result in infection of the heart valves and lining (endocarditis), the muscle and outside covering of the heart (myocarditis and pericarditis), and the lymphatic vessels (lymphangitis). Microorganisms circulating in the bloodstream may release toxins and produce septicemia. Tularemia and infectious mononucleosis are diseases in which the lymph nodes are often prominently involved. Toxoplasmosis is an unusual manifestation of the widely spread infections caused by *Toxoplasma gondii,* a protozoan parasite of cats.

QUESTIONS

REVIEW

1. How do avirulent bacteria of the normal flora cause the serious disease bacterial endocarditis?
2. Give the genera of three Gram-negative, rod-shaped bacteria that can cause septicemia. Name the toxic substance common to all such bacteria.
3. Name the virus responsible for infectious mononucleosis.
4. Tularemia can be identified as the cause of death of an animal in less time than it takes to recover and identify *Francisella tularensis* in culture. Explain how.
5. What is the host range of *Toxoplasma gondii?*

THOUGHT

1. Why is it that infections by Coxsackie viruses are manifest in different ways, sometimes by heart disease and sometimes by disease of other parts of the body?
2. How would the elimination of cats from the world jeopardize the existence of *Toxoplasma gondii* (if at all)?

FURTHER READING

FRENKEL, J. K., "Pursuing Toxoplasma," *Journal of Infectious Disease* 122:553–559 (December 1970).

HALL, J. G., "The Flow of Lymph," *New England Journal of Medicine* 201:720–722 (September 1969).

MAYERSON, H. S., "The Lymphatic System," *Scientific American* (June 1963).

NIEDERMAN, J. C., G. MILLER, H. A. PEARSON, J. S. PAGANO, and J. M. DOWALIBY, "Infectious Mononucleosis. Epstein-Barr–virus Shedding in Saliva and the Oropharynx," *New England Journal of Medicine* 294:1355–1359 (June 1976).

WIGGER, C J., "The Heart," *Scientific American* (May 1957).

CHAPTER 32

MULTICELLULAR PARASITES

The animal subkingdom Metazoa includes a number of organisms parasitic for human beings. The majority of these fall into three phyla: Platyhelminthes, Nematoda, and Arthropoda. Organisms of the first two phyla are primitive wormlike creatures, but those of the third phylum, Arthropoda, including insects, arachnids, and crustaceans, are highly advanced on the evolutionary scale. A few members of the phyla Platyhelminthes and Nematoda carry microbes that infect other animals but not human beings. By contrast, Arthropoda includes some extremely important vectors of human pathogens.

ARTHROPODS AND ARTHROPOD-BORNE INFECTIONS

Examples of some arthropod-borne infections are shown in Table 32-1 along with their causative agents and principal vectors.

Mosquitoes

The mosquito (Figure 32-1) is one of the most medically important of the insects. The mouth parts of the mosquito consist of sharp stylets which are forced through the skin to the subcutaneous capillaries. One of these needlelike stylets is hollow, and salivary secretions of the mosquito are pumped through it. These secretions help keep the victim's blood from clotting as it is drawn into a tube formed by the other mouth parts of the insect, and they may also act on the victim's blood vessels to increase blood flow. The injected mosquito saliva contains antigens to which some individuals are hypersensitive, and it may also contain

TABLE 32-1
Examples of Arthropod-borne Diseases

Disease	*Causative Agent*	*Vector*
Malaria	*Plasmodium* species	Mosquitoes (*Anopheles* species)
Equine encephalitis	Arboviruses	Mosquitoes (*Culex* species)
Plague	*Yersinia pestis*	Flea (*Xenopsylla cheopis*)
Typhus	*Rickettsia prowazekii*	Louse (*Pediculus humanus*)
Rocky Mountain spotted fever	*Rickettsia rickettsii*	Ticks (*Dermacenter* species)

agents causing serious infectious diseases. Only in the female mosquito are the mouth parts adapted to piercing animal skin, however; the male mosquito feeds instead on plant juices. Ovarian development and egg production in mosquitoes are quite markedly enhanced by a blood meal, indicating that blood is the source of important hormones or nutrients for this insect which contribute to its reproductive functions. The muscles of the throat of the mosquito suck blood into the digestive system and into large storage areas. The latter allow the mosquito to take in as much as twice its body weight in blood, giving it a relatively good chance of picking up microbes circulating in the host's capillaries. Other parts of the mosquito digestive tract are the stomach, in the wall of which the malaria parasite develops, and the midgut, through which the parasite penetrates into the mosquito's body cavity to infect the insect's salivary glands.

IMPORTANCE OF MICROSCOPIC IDENTIFICATION

The medically important mosquitoes are largely included in the genera *Aedes, Culex,* and *Anopheles.* Precise identification of species and subspecies of these genera is extremely important and depends largely on microscopic examination of antennae, wings, claws, mating apparatus, and other features. Correct identification is often essential to control of medically important vectors because different species and subspecies of mosquitoes differ greatly in their breeding areas, time of feeding, and choice of host.

Francisella tularensis, page 434

Although mosquitoes may act as vectors for certain worms, and for bacterial agents such as *Francisella tularensis,* the protozoan and viral agents transmitted by

FIGURE 32-1
Mosquito internal anatomy.

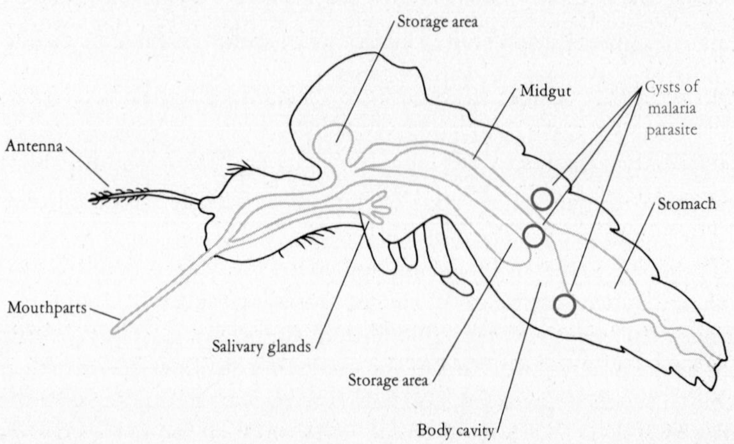

mosquitoes are numerically far more important. A few of these mosquito infections that are important to human beings are considered in the following sections.

Malaria Malaria, an ancient scourge of mankind, was suspected of being mosquito-borne at least as long ago as the second decade of the eighteenth century. This suspicion was dramatically confirmed in the late 1800s when mosquito control measures in Havana led to a drop in the incidence of malaria from about 9 of 10 persons to 1 in 50. In the early 1900s effective mosquito control reduced the incidence of malaria and yellow fever and allowed completion of the Panama Canal.

Human malaria is caused by species of *Plasmodium,* protozoa of the class Sporozoa. Four species are involved, *P. vivax, P. falciparum, P. malariae,* and *P. ovale,* which differ in morphology and, in some instances, life cycle, type of disease produced, and treatment.

Sporozoa, Table 15-1, pages 305, 312–313

The most characteristic feature of human malaria is the occurrence of paroxysms of fever at 1- to 3-day intervals, with periods of relative well-being in between. The protozoan parasite of the infection, one of the four species of *Plasmodium,* grows and divides in the red blood cells of the host (Figure 32-2). After a time the offspring of this division rupture the red blood cells in which they are growing and are released into the plasma. The victim's intermittent fevers result from this periodic release of the plasmodia. Most of the young plasmodia then enter new red blood cells, multiply, and the cycle is repeated. Some of them, however, develop into specialized sexual forms (*gametocytes*), different from the other parasites in both their appearance and susceptibility to antimalarial medicines. These sexual forms of plasmodium cannot develop further in the human host and are not important in causing the symptoms of malaria. They are, however, infectious for certain species of mosquitoes (especially certain species of *Anopheles*) and are thus ultimately responsible for the transmission of malaria from one person to another.

Infection of a mosquito begins when it ingests the sexual forms of the malarial parasite. Shortly after entering the intestine of the mosquito, the male gametocyte produces tiny whiplike bodies which unite with the female gametocyte in much the same way as do the sperm and ovum of higher animals. The resulting zygote enters the wall of the midgut of the mosquito and forms a cyst, which enlarges as the zygote undergoes meiosis and division into numerous asexual organisms. The cyst then ruptures into the body cavity of the mosquito and the released parasites find their way to the salivary glands where they may be injected into a new human host. It is of interest that the plasmodial infection of the mosquito fails to impair the insect's normal function or longevity. This may imply a long evolutionary association of the parasite and the mosquito host, with selection of those plasmodial variants most favorable to coexistence with the host mosquito. Unhappily the relationship between the plasmodium and man is not so advanced, and malaria is often a severe and sometimes fatal illness in human beings.

Injected into a new human host by the mosquito, malarial parasites are car-

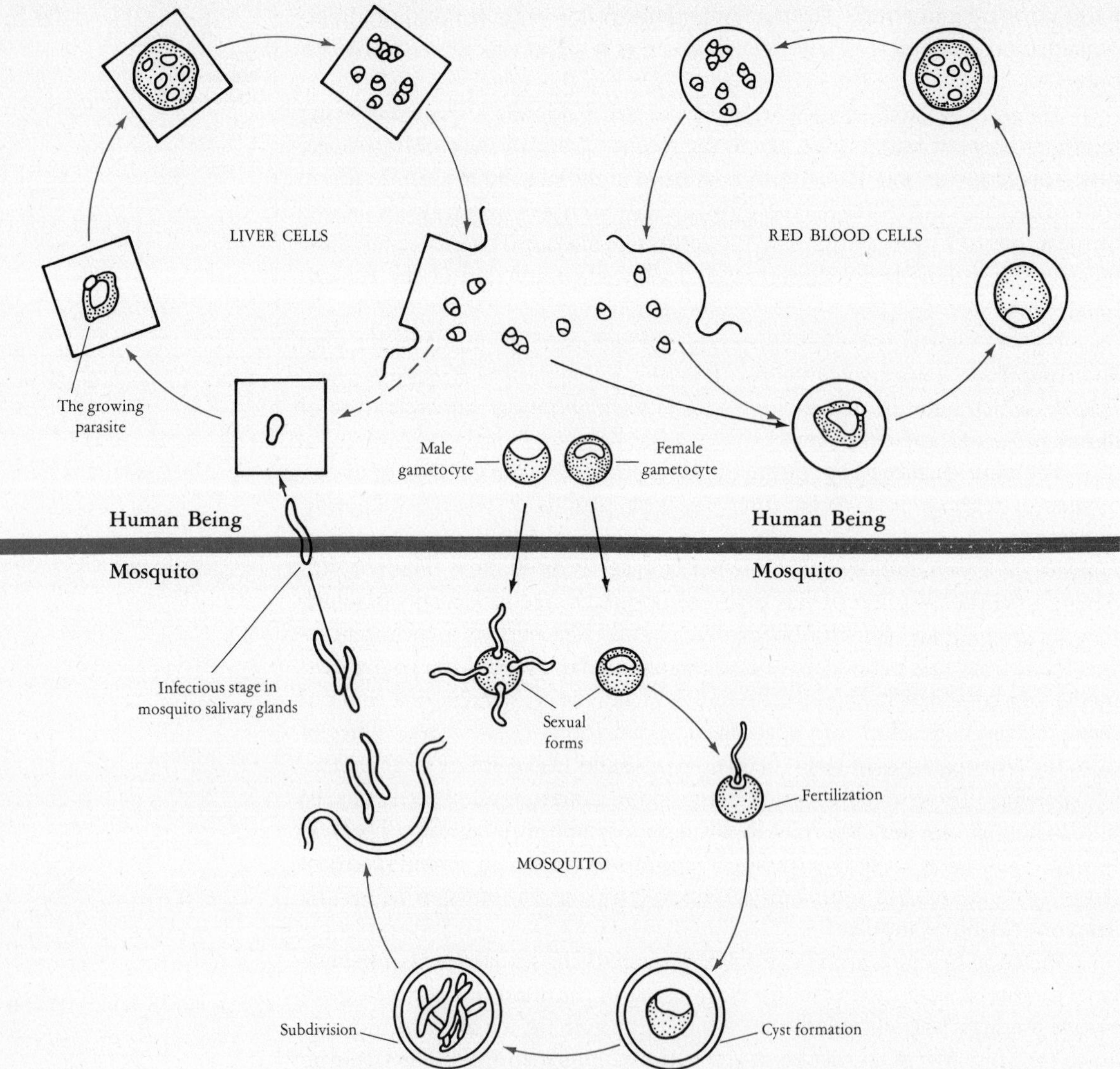

FIGURE 32-2 Life cycle of *Plasmodium* species.

ried by the bloodstream to the liver where they leave the circulation to infect liver cells. In these cells each parasite enlarges and subdivides, producing thousands of daughter cells. These are released into the human bloodstream and establish the cycle of infection described above, involving red blood cells. With some species of *Plasmodium,* organisms released from the liver cells may also infect new liver cells, thus establishing a cycle of infection in the liver. A sustained liver infection may result, which may persist despite effective treatment of the bloodstream infection, and malaria may recur later, when treatment is discontinued.

In most areas of the world, visitors can avoid contracting malaria by taking chloroquine once weekly for 1 week before and during their visit, and for 6 weeks after leaving the area. However, strains of *Plasmodium* resistant to chloroquine are present in some areas, making prophylaxis uncertain. Malaria is controlled by using insecticides and eliminating breeding areas for the principal mosquito vectors, and in giving treatment to infected patients. Patients are identified by preparing stained smears of their blood and examining these microscopically for the malarial parasites. Intensive efforts, often with assistance from the World Health Organization, have markedly reduced or eliminated mosquito-borne malaria from many countries. Complete eradication of human malaria is theoretically possible, but has proved impossible in practice. Ceylon, for example, was formerly subject to serious malarial epidemics, with one in 1935 causing an estimated 80,000 deaths. With intensive eradication efforts the incidence of malaria in Ceylon had dropped to only 17 cases in 1963, at which time control measures were largely discontinued and surveillance efforts relaxed. Subsequently, the incidence of the disease promptly rose once again, so that in the single month of February 1968 over 42,000 cases of malaria were diagnosed on Ceylon. Formerly common in the United States, malaria became very rare with the establishment of control measures. The dramatic reappearance of malaria in the United States, associated with the return of military personnel from Southeast Asia (each of the 50 states reported cases in 1970, with a total of almost 4000 cases), was largely due to malaria acquired outside the country. In some instances transmission of the disease to civilians has resulted from blood transfusions. Transmission can also occur when heroin addicts share syringes; one such episode in 1971 led to 48 infections.

Yellow fever. Yellow fever is caused by a mosquito-borne arbovirus. This disease is characterized by fever, aches and vomiting, often associated with bleeding from the nose and gastrointestinal tract, and jaundice. The mortality rates from yellow fever often reach 40 percent. The reservoir of the disease is now mainly infected mosquitoes and primates living in tropical jungles of Central and South America, and Africa, but the disease was formerly common in the United States. Control of yellow fever is achieved by control of its principal mosquito vector, *Aedes aegypti,* and by using a live vaccine for immunizing people who might become exposed.

Arboviruses, Appendix III, pages 711–713

Arthropod-borne encephalitis. In addition to carrying the arbovirus responsible for yellow fever, mosquitoes may also bear viruses that cause *encephalitis,* an inflammatory process of the brain. Human and domestic-animal encephalitis caused by arboviruses occurs in endemic or epidemic fashion every year in the United States. During 1975, for example, cases of this disease were identified in 32 of the contiguous United States. Among the causative viruses are the "equine encephalitis viruses," which often infect horses but which can also cause death, mental disorder, or motor nerve impairment in human beings.

The encephalitis viruses can pass from one mosquito to another only through common exposure to another host, normally one of the birds living in a swamp. Although the viruses remain in the infected mosquito for the life of the mosquito, they cannot spread directly from one mosquito to another nor from the mosquito to its egg (transovarially). The fact that under proper circumstances a mosquito may be infected with the virus for its entire lifespan indicates how well adapted the parasitic agent is to its host. Every now and then, when the host mosquitoes are unusually plentiful and the number of infected birds is high, the encephalitis virus may be spread to other areas and to less adapted hosts. Mosquitoes may then transmit the virus to domestic fowl, horses, or man. The transfer of virus from mosquito to swamp bird to mosquito to bird, and so on, is often referred to as a "cycle," although it is more like a never-ending chain with repeating links. The spread of the virus to other hosts is usually a dead end because of the high mortality among the new hosts and because the species of mosquitoes that usually bite these hosts are less able to sustain the infection and transmit the causative virus. It is likely, however, that on occasion mutants among infecting encephalitis viruses may multiply and persist in their new vector and cause less serious disease in the new host, so that a new cycle develops.

Control of equine encephalitis is achieved chiefly through mosquito control, although vaccines given to animals may be helpful in some instances.

Fleas

Fleas are wingless insects that depend on their powerful hind legs to jump quickly from place to place.

The anatomy of a flea is diagrammed in Figure 32-3. Points of importance in the identification of flea species include the spines or "combs" about the head and first chest segment of this insect; the muscular pharynx (for sucking blood from the host) and the long esophagus, followed by the spiny valve composed of rows of teethlike cells, are also noteworthy.

Yersinia pestis, page 436

Plague. Plague is primarily a disease of rats and other rodents, and of the fleas they carry. The flea contracts the plague bacillus, *Yersinia pestis*[1] by feeding

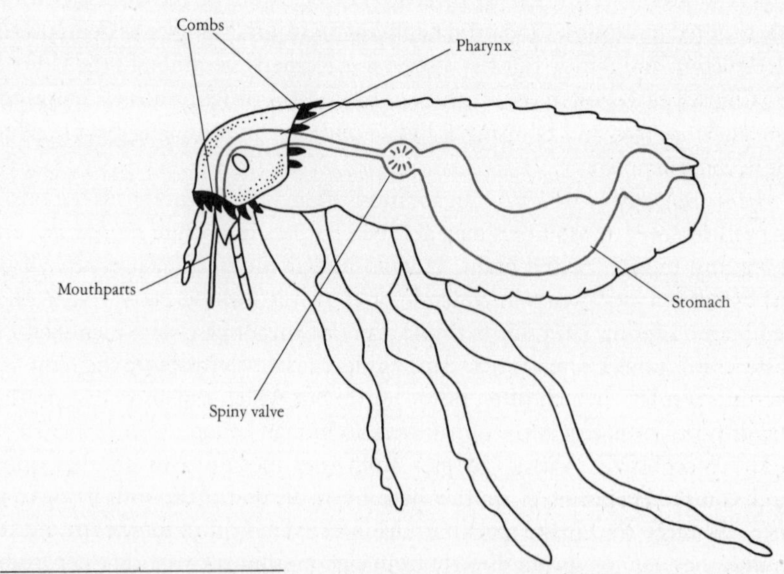

FIGURE 32-3
Anatomy of a flea.

[1] Formerly called *Pasteurella pestis.*

on an infected rodent. So many of the bacilli may be present in the blood of an infected rat that a flea may ingest 25,000 of these microbes in the course of a single feeding! The organisms pass into the stomach of the flea and multiply there, producing sticky masses of bacteria which can cause obstruction by adhering to the spiny valve. With obstruction, blood sucked into the flea esophagus cannot pass into the stomach but is regurgitated into the puncture wound in the host's skin, carrying the plague bacilli with it. Moreover, the hungry flea tries again and again to feed and may infect several hosts with *Y. pestis.* Infected fleas may live for long periods, because the obstruction in the spiny valve often resolves spontaneously, but the infection persists, and *Y. pestis* is then excreted with the flea feces. This fecal material may also infect human beings and other animal hosts when rubbed or scratched into flea bites.

Even when many susceptible rodents are present, the mortality rate from *Y. pestis* infection may be so high that the flea vectors of the disease find a rapidly diminishing supply of their natural hosts. In this case they will turn to human beings, who thereby become infected with *Y. pestis* and develop bubonic plague. Carried by tissue fluids to the lymph nodes near the site of a flea bite, the bacteria cause enlargement of the nodes ("bubo"—thus "bubonic" plague). Large numbers of the bacilli are then released into the bloodstream and distributed to all areas of the human body. In some cases bleeding beneath the skin is a prominent feature of bubonic plague; the darkened blood and the dusky appearance of dying patients may have suggested the name "Black Death" for the European epidemic of the fourteenth century.

Yersinia pestis bacilli in the blood are carried to the lungs, where they may produce a form of the disease which spreads easily from person to person without the mediation of fleas. The bacilli enter the bronchial secretions; with coughing they are sprayed into the air and can thus be inhaled by others. The disease resulting from inhalation rather than flea bite is called "pneumonic plague" and accounts for most epidemic spread among human beings.

In both bubonic and pneumonic plague the causative bacteria can be seen in infected material and readily grown on laboratory media. *Y. pestis* are Gram-negative enterobacteria which stain more intensely at each end, looking somewhat like a safety pin.

In early 1968, 90 cases of plague with 36 deaths occurred in villages of central Java. Indonesian and American health officers imposed strict quarantines to prevent person-to-person spread of the disease and gave plague vaccine to persons living in areas adjacent to the epidemic. Rats were trapped, and the most frequent parasite found on them was the rat flea, *Xenopsylla cheopis.* Villages were heavily sprayed with DDT to reduce the numbers of fleas. The epidemic was quickly controlled, and cases ceased to appear within about 2 months. Plague in human beings is an uncommon disease in the United States today, although it has been identified in the wild rodent population of at least 15 states and thus still remains a threat. Indeed, 15 human cases of plague occurred during 1976, scattered among four southwestern states. In 1977 several human cases and one death were reported in Oregon.

Lice

Like fleas, lice are small wingless insects that prey on warm-blooded animals by piercing their skin and sucking blood. However, the legs and claws of lice are adapted for holding onto body hair and clothing, rather than for jumping from place to place. Like fleas and mosquitoes, lice are susceptible to microbial infections by agents which can be transmitted to man.

The most notorious of the lice, *Pediculus humanus,* or "cootie," is from 2 to 3 mm long, with a characteristically small head and thorax and large abdomen (Figure 32-4). This louse has a membranelike lip with tiny teeth which anchor it firmly to the skin of the host. Within the floor of the mouth is a piercing apparatus somewhat similar to that of the fleas and mosquitoes. *Pediculus humanus* has only one host, a human being, but easily spreads by direct contact or by contact with personal items, especially in situations of crowding and poor sanitation. A similar organism, *Phthirus pubis* ("crab louse"), is not uncommon among young adults; it is most frequently transmitted during sexual intercourse. It is not known to be a vector of infectious disease.

Salmonella typhi, page 436

Rickettsiae, page 439

Epidemic typhus. Epidemic typhus (not to be confused with typhoid, caused by *Salmonella typhi*) is caused by *Rickettsia prowazekii,* a tiny obligate intracellular bacterium transmitted by *P. humanus.* This louse-borne disease tends to occur under crowded and unsanitary living conditions, especially in areas of water shortage and during the social disruption of wartime. In fact, typhus has been responsible for many more deaths than have the armaments of war. Patients with this disease develop fever abruptly, followed in a few days by rash and confusion (typhus means "hazy"). Damage to small blood vessels then becomes prominent, resulting in gangrene and hemorrhages beneath the skin. Involvement of the heart, brain, and kidneys may contribute to the illness, and mortality from typhus ranges from about 10 to 40 percent in different epidemics.

Lice feeding on typhus-infected persons ingest the infectious rickettsiae. The microorganisms are then taken up by the cells that line the gut of the louse, and multiply within them. The infected cells die and liberate the rickettsiae, which then pass through the louse intestine and are excreted in the feces. Scratching the itchy louse bites causes a new victim to rub some of the contaminated louse excreta into the bite wound, thereby inoculating the victim with

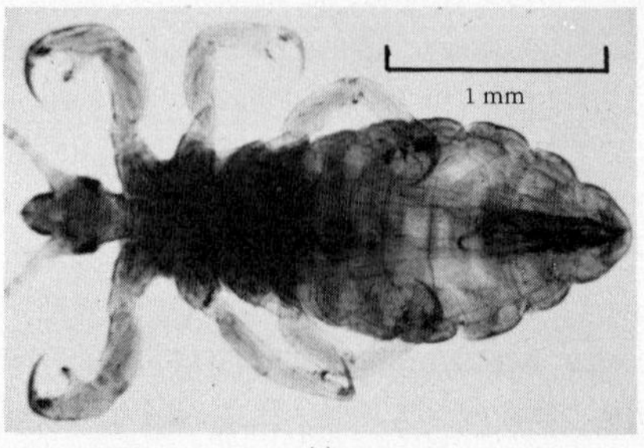

(a)

(b)

FIGURE 32-4
Lice. (a) *Pediculus humanus.* (b) *Phthirus pubis.* (Courtesy of F. Schoenknecht.)

R. prowazekii. Infection may at times also occur by inhalation of dried louse feces or by accidentally rubbing it into the eyes.

Like other rickettsiae, *R. prowazekii* does not grow on cell-free media, but only in susceptible living cells. The organisms can be recovered from human patients or lice by inoculating infected blood or other material into laboratory animals, embryonated eggs, or tissue cell cultures. However, it is easier and safer for laboratory workers to look for a rise in the titer of antibody to typhus rickettsiae in a patient's blood. This is done by making use of a dramatic example of antigenic sharing by members of two greatly different genera of bacteria, *Rickettsia* and *Proteus*. Certain strains of *Proteus vulgaris* and *P. mirabilis* have cell wall antigens that react with the antibodies formed against *R. prowazekii* and some other rickettsiae. The resulting antibody-antigen reaction produces agglutination of the *Proteus*. This procedure, employing strains of *Proteus* to diagnose rickettsial disease, is known as the *Weil-Felix* test.

Epidemic typhus can readily be controlled by using insecticides to kill lice. For all practical purposes transmission of the disease from person to person can result only from *P. humanus* infestation, and human beings are the only reservoir for *R. prowazekii*. Unfortunately, complete eradication of typhus is complicated by the fact that persons who recover from the disease may continue to have living rickettsiae in their body tissues for many years without apparent ill effects. But under conditions of stress, fading immunity, and other unknown factors, these organisms may multiply and again cause typhus. If *P. humanus* is present, the recrudescent disease may be transmitted to others, producing a new wave of epidemic typhus. *Rickettsia prowazekii* is thought to persist in tissues through a delicate balance between defensive factors of the host and attributes of the microbe which protect it against these factors, but the exact mechanisms involved are unknown.

Ticks

Ticks are not insects but arachnids. Arachnids differ from insects in their lack of wings and antennae, and in the fact that their thorax and abdomen are fused. Furthermore, adult arachnids have four pairs of legs as compared to the three pairs which insects have. *Dermacentor andersoni* (Figure 32-5), a common American tick, has a broad host range—more than 25 animals. Most of the ticks that attack humans are intermittent parasites, attaching from time to time to feed on human blood, then dropping off to rest, grow, or deposit eggs. Their bite is usually completely undetected and one discovers the tick with surprise, firmly attached to the skin and getting swollen with ingested blood.

The attachment of ticks to the skin is very secure. In fact, if one tries to pull off a tick, the mouth parts may separate from the tick rather than tearing from the skin of the host. How the tick attaches itself without causing pain to its host is unclear. Some ticks produce a toxin powerful enough to cause paralysis. Paralyzed human beings and animals usually recover rapidly following removal of the tick.

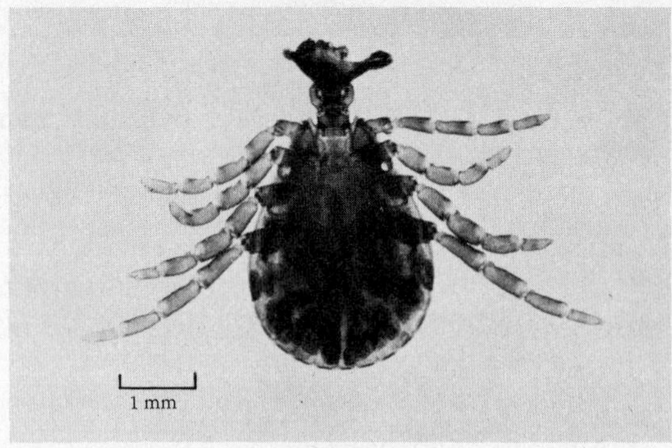

FIGURE 32-5
The tick, *Dermacentor andersoni.* The mouth parts are imbedded in a piece of the host's skin torn away when the tick was removed. Note that four pairs of legs are present as opposed to the three pairs of insects.

Rocky Mountain spotted fever. Ticks are subject to microbial infections, including those caused by protozoan, viral, and bacterial agents—some of which are infectious for humans and domestic animals. For example, *D. andersoni* may be infected with *R. rickettsii* which may be passed between ticks and wild rodents without producing serious disease in either the vector or the reservoir host. In human beings, however, the rickettsia produces *Rocky Mountain spotted fever,* a serious illness somewhat resembling typhus. The observation that rickettsiae can pass from generation to generation of tick through tick eggs (transovarial passage) is of great interest, demonstrating the ability of yet another microbial species to coexist with its arthropod host for long periods without harming it significantly.

Other tick-borne diseases. In addition to carrying *R. rickettsii, Dermacentor andersoni* may also harbor the pathogenic rickettsia *Coxiella burnetii,* the cause of *Q fever,* usually contracted by people through inhalation of infected tick feces or by the drinking of milk from an infected cow. The disease typically results in a mild pneumonia. Another tick-borne agent of human importance is the arbovirus of *Colorado tick fever.* This virus is endemic in wild rodents of the Rocky Mountain region of the United States. In human beings it typically produces illness with fever, but no rash. Finally, ticks may be infected with the bacterium *Francisella tularensis,* the Gram-negative rod which causes tularemia. Like *R. rickettsii, F. tularensis* can pass transovarially. Members of another genus of ticks transmit species of *Borrelia,* the spirochetes responsible for relapsing fever.

Tularemia, page 565
Borrelia, page 441

NEMATODA (ROUNDWORMS)

Pinworms

The roundworms are a very large group of parasites, some of which, like *Enterobius vermicularis,* the pinworm of children, are cosmopolitan and widely known. *Enterobius* worms live in the large intestine, migrating to the anus where they discharge their eggs. The inflammation and itching which result often produce sleeplessness and behavioral disorders. There are effective medications for the treatment of pinworms.

Trichinosis

Trichinosis is characterized by fever, muscle pain, swelling around the eyes and sometimes a rash. Occasional cases are fatal because of damage to the heart or brain. The disease occurs worldwide.

The cause of trichinosis is *Trichinella spiralis,* a tiny roundworm 1 to 4 mm long which lives in the small intestine of meat-eating animals, especially rats, pigs, bears, dogs, and human beings. The female worm discharges living young into the lymph and blood vessels of the host's intestine without an intervening egg stage, and these larvae are carried to all parts of the body. They cause an inflammatory reaction in which many of them die and are dissolved by body defense mechanisms. Those that arrive alive in the muscles of the host often lodge there. The muscle tissues react to encase these organisms with scar tissue and inflammatory cells, and the worms then stay alive for months or years within the muscle. If the flesh of the host is eaten by human beings or another animal, the digestive juices of the new host release the worm from its case, permitting it to burrow into the new host's intestinal lining, where it matures, and where the mature female worms begin producing larvae in the new host to complete the life cycle of the worm. Each female *Trichinella* may live 4 months or more and produce 1500 young. It is the penetration of these young worms into the tissues of the host's body that is responsible for the symptoms of trichinosis.

There is no satisfactory treatment for severe trichinosis. Prevention of trichinosis in human beings depends on cooking meat adequately and throughout, so that all parts reach at least 150°F. Pork has been the chief offender, presumably because of the practice of feeding uncooked garbage containing meat scraps to pigs. However, beef ground in a machine previously used for pork has also resulted in cases of trichinosis, and bear meat is a notorious cause. Government inspection of meats does not detect *Trichinella*-infested meats, and adequate destruction of the worm larvae by freezing is often impractical because of the low temperatures and long time required.

PLATYHELMINTHES (THE FLATWORMS)

Flatworm parasites of human beings fall into two major groups: flukes, which are relatively short, flat, bilaterally symmetrical worms that generally attach by one or more sucking discs, and tapeworms, usually longer than flukes and ribbonlike in appearance.

Schistosomiasis

Schistosomiasis is a disease caused by flukes of the genus *Schistosoma.* It is often a chronic, slowly progressive illness resulting in damage and loss of function of the liver, with resulting malnutrition, weakness, and accumulation of excess fluid in the abdominal cavity. The worldwide incidence of schistosomiasis has been estimated at 100 million cases. In the Western Hemisphere the disease occurs chiefly in the Caribbean area and in South America, and many thousands of cases have been seen in cities of the United States to which Latin Americans have emigrated.

Schistosoma mansoni, a common cause of schistosomiasis, is about 10 mm long and lives in the small veins of the human intestinal wall. Its life cycle is depicted in Figure 32-6. The female worm discharges eggs, some of which are swept into the liver by the flow of blood, while others rupture through the blood vessels and adjacent intestinal lining to enter the feces. The major symptoms of the disease are caused by the eggs, the many of which lodge in the liver producing inflammation and scarring and eventual loss of liver function.

The cercarial form of *Schistosoma mansoni* infects human beings by burrow-

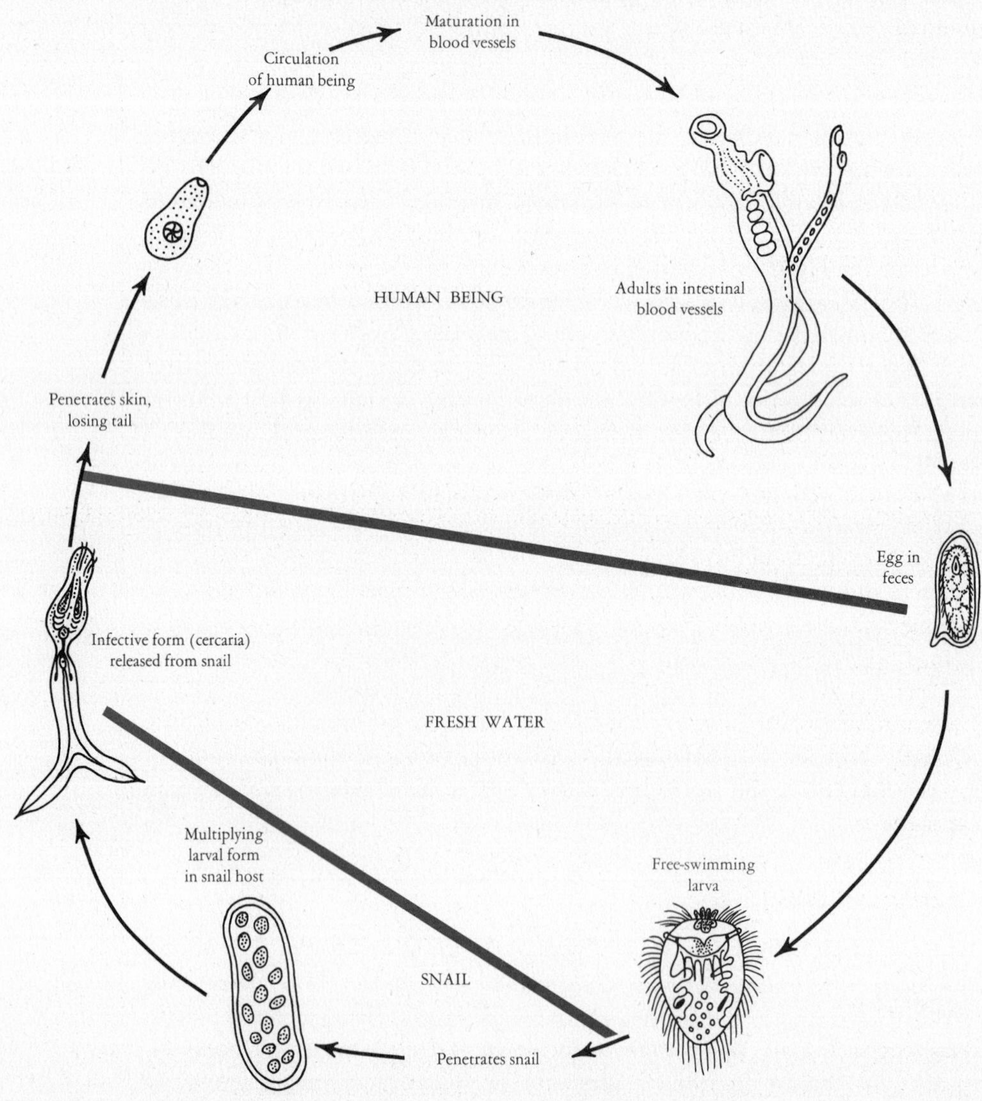

FIGURE 32-6 Life cycle of *Schistosoma mansoni* (developmental forms not to scale).

ing through the skin of persons wading in infested waters. The worms then penetrate to the blood vessels beneath the skin, leaving their tails at the body surface, and are carried by the blood flow to the intestinal veins where they become mature and complete their life cycle. The worms can live for more than 25 years and continue to produce eggs and consequent liver damage.

Measures used in the control of schistosomiasis include treatment of infected cases, sanitary disposal of feces, chemical and biological control of snail hosts, and anticercarial measures. The expanding use of irrigation rice farming, as well as the existence of animal hosts for some species of schistosomes, has made control difficult.

"SWIMMERS' ITCH"

The disease "swimmers' itch," common in parts of North America and other areas of the world, is caused by the cercariae of schistosomes of birds or other animals. These cercariae penetrate the skin and then die; they are unable to complete their life cycle in human beings.

Tapeworm Infestation

People contract tapeworms mainly by eating infested, inadequately cooked beef, pork, or freshwater fish. The adult worms commonly have a commensal relationship with the human host; they live by absorbing juices from the human intestine and cause few or no symptoms. However, some large tapeworm species sometimes absorb enough vitamin B_{12} to cause anemia, and in unusual circumstances, tapeworm developmental stages can invade human tissue and cause serious disease.

IMPORTANCE OF INTERMEDIATE HOSTS

Figure 32-7 gives the life cycle of the fish tapeworm, *Diphyllobothrium latum.* Note that, as with many other multicellular parasites, the earlier stages of development of tapeworms occur in intermediate hosts (for example, the snail and fish with *D. latum*), while adulthood with egg production occurs in the definitive host (for example, a human being). The developing stages of *D. latum* are invasive and grow within the tissues of the intermediate host, while the adult normally simply attaches to the intestinal lining of the definitive host. Human beings can sometimes become infected by the intermediate developmental forms of tapeworms, producing serious disease. This usually occurs with accidental ingestion of the tapeworm eggs, but may also come about with the practice of applying raw meat to open wounds in the mistaken belief the meat will aid recovery. If tapeworm larvae are present in the meat, they may penetrate the tissues via the wound.

Control of tapeworms depends on adequate cooking of meat and fish, and on proper disposal of human feces.

SUMMARY

The importance of arthropod parasites lies mainly in their susceptibility to microbes which are pathogenic for human beings. Mosquitoes may be infected with such agents as *Plasmodium* species (causes of malaria) and the viruses responsible

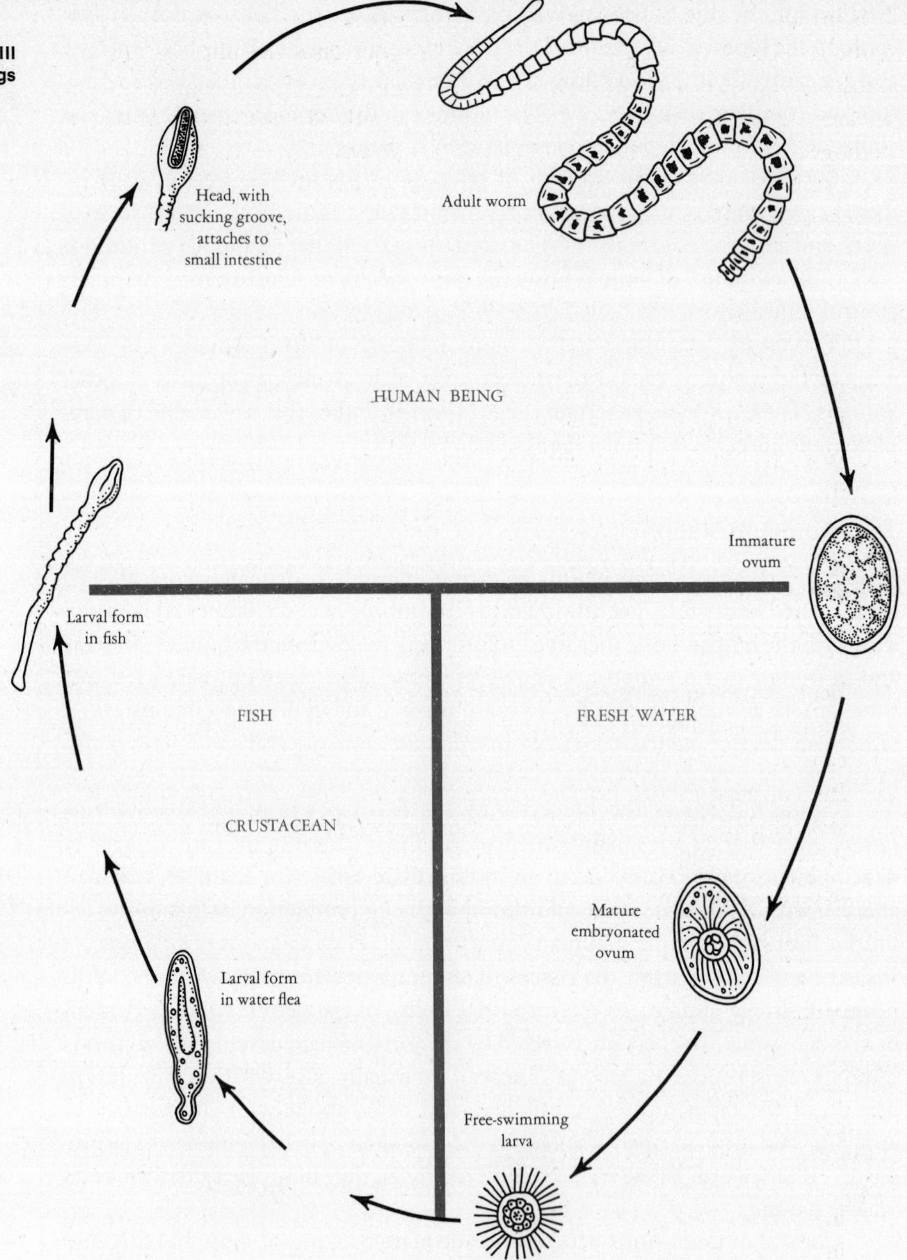

FIGURE 32-7 Life cycle of *Diphyllobothrium latum* (developmental forms not to scale).

for yellow fever and equine encephalitis. Fleas may carry the enterobacterium *Yersinia pestis,* the cause of plague, and lice, *Rickettsia prowazekii,* the cause of typhus. Arachnids such as ticks may also be hosts for microbial pathogens including *Rickettsia rickettsii* (the cause of Rocky Mountain spotted fever), *Coxiella burnetii* (the cause of Q fever), and *Francisella tularensis* (the cause of tularemia).

It is interesting that in some instances maintenance of an arthropod-borne microbe requires a vertebrate host, whereas in other cases the microbe can pass transovarially among insects. Some arthropods tolerate their infection without apparent harm. The wormlike multicellular parasites are not vectors of human pathogens, but can themselves cause serious disease. Some examples of such pathogenic organisms are *Enterobius* and *Trichinella* (roundworms), and *Schistosoma* and *Diphyllobothrium* (flatworms).

QUESTIONS

REVIEW

1. Describe the life cycle of *Plasmodium* species in the human host.
2. List the methods employed in control of malaria.
3. What is arbovirus encephalitis? Give the natural history of one of the causative agents.
4. In what ways might precise laboratory identification of the arthropod responsible for transmitting an epidemic disease help to influence control strategy?
5. Give the cause and mode of transmission of bubonic plague.

THOUGHT

1. Because mosquitoes feed on both human beings and plants, what do you think about the possibility of mosquitoes transmitting agents of human diseases to plants and vice versa?
2. Because parasitic arthropods may feed on more than one mammalian host, it seems likely that simultaneous infection with agents of human and animal diseases sometimes occurs. From knowledge of earlier chapters, would genetic interchange among these agents be possible?

FURTHER READING

BROWN, H. W., *Basic Clinical Parasitology.* 4th ed. New York: Appleton-Century-Crofts, 1975. A readable text covering most protozoan and multicellular parasites of medical importance.

HAWKINS, F., "The Clock of the Malaria Parasite," *Scientific American* (June 1970).

JAMES, M. T., and R. HARWOOD, *Herms's Medical Entomology.* 6th ed. New York: Macmillan, 1969. Describes the identification, habits and habitat of medically important insects.

JONES, J. C., "The Sexual Life of a Mosquito," *Scientific American* (April 1968).

KADIS, S., T. C. MONTIE, and S. J. AJL, "Plague Toxin," *Scientific American* (March 1969).

LANGER, W. L., "The Black Death," *Scientific American* (February 1964).

ROUECHE, B., "Alerting Mr. Pomerantz." *Eleven Blue Men and Other Narratives of Medical Detection.* New York: Berkley Medallion, 1965. About rickettsial pox.

YOELI, M., "Animal Infections and Human Disease," *Scientific American* (May 1960).

ZINSSER, H., *Rats, Lice and History.* Boston: Little, Brown, 1935. About typhus.

CHAPTER 33

ANTIMICROBIAL DRUGS

Biochemical unity, pages 8–9

Similarities in the biosynthetic pathways of all living cells have been discussed previously. Fortunately there are also differences, and substances can be found that take advantage of these differences and interfere with the vital biochemical pathways of microbial cells at concentrations that do little or no harm to mammalian cells. Such substances are said to have *selective toxicity;* they poison one type of cell but spare others. The search for such substances began in the early part of the twentieth century with the observation that some dyes stained one kind of bacterial cell very well but had little affinity for another. It was reasoned that dyes that had selective affinity for a group of cells would selectively kill them as well. Although this idea was only partly true, it led to the discovery of sulfa drugs, the first of a long series of medicines demonstrating substantial activity against pathogenic microbes when given to people in doses producing little or no toxic effect. The ratio of the maximum tolerated dose to the minimum dose required to cure infections is called the *therapeutic index.* The quest for medicines of high therapeutic index represents one of the greatest success stories of modern science.

THERAPEUTIC INDEX

As shown in Table 33-1, antimicrobial medicines can be grouped according to their effectiveness against different kinds of microbes. For example, the *macrolides* are chiefly active against Gram-positive bacteria, whereas the *polymyxins* are useful against Gram-negative bacteria. Antimicrobial medications may also be classified according to whether their principal action results in the killing of microbes or merely inhibiting them. Thus penicillin is classed as *bactericidal,* while tetracycline is *bacteriostatic.* Antimicrobial agents generally exert their effect at concentrations measured in micrograms per milliliter (one microgram equals one-thousandth of a milligram).

TABLE 33-1
Spectra of Activity of Antimicrobial Medicines

Primarily active against Gram-positive bacteria	
Penicillins	Vancomycin
Bacitracin	Macrolides
Primarily active against Gram-negative bacteria	
Polymyxins	
Activity against many Gram-positive and Gram-negative bacteria	
Tetracyclines	Cephalosporins
Chloramphenicol	Sulfa drugs
Aminoglycosides	
Active against mycobacteria	
Aminoglycosides	Isonicotinic acid hydrazide (INH)
Cycloserine	Para-aminosalicylic acid (PASA)
Rifampin	Pyrazinamide
	Ethionamide
Active against protozoa	
Emetine	Sulfa drugs
Quinine	Primaquine
Chloroquine	Chloroguanide
Active against fungi	
Amphotericin	Nystatin
Griseofulvin	

ORIGIN AND DEVELOPMENT OF ANTIMICROBIALS

Antimicrobial medications derive from two sources: plant extracts or other chemicals, and products of microbial cells. Antimicrobials originally derived from microbial sources are called *antibiotics.* Some antibiotics are now extensively modified in the chemistry laboratory or even synthesized without the help of living microbes.

Most of the major groups of antibiotics were discovered before 1955, and most of the significant antibiotic advances since then have come as a result of modifying the older forms. Indeed, as shown in Table 33-2, only three groups of microbes have yielded useful antibiotics: actinomycetes (filamentous branching bacteria), bacteria of the genus *Bacillus,* and saprophytic molds. Furthermore, almost all useful antibiotics have come from only four genera: *Streptomyces, Bacil-*

Actinomycetes, page 260

Molds, pages 289–290

TABLE 33-2
Sources of Antibiotics used for Systemic[a] Administration

Actinomycetes	Bacillus *Species*	*Molds*
Tetracyclines	Polymyxins	Penicillins
Chloramphenicol	Bacitracin	Cephalosporins
Macrolides		Griseofulvin
Aminoglycosides		
Polyenes		
Vancomycin		
Cycloserine		

[a] Systemic means given by injection or by mouth for absorption into the circulatory system, as opposed to use on body surfaces.

lus, Penicillium, and *Cephalosporium.* To obtain a desired antibiotic a carefully selected strain of the appropriate producing species is inoculated into medium and incubated. As soon as the maximum antibiotic concentration is reached, the drug is extracted from the medium and extensively purified. Its activity is then standardized against a reference sample of pure antibiotic, and it is put into a form suitable for administration (flavored, buffered, coated to protect against stomach acid, put into capsules, pressed into tablets, or mixed with substances to prolong absorption). New antibiotics must receive extensive evaluation of their properties: their spectrum of activity against pathogens; their toxicity (to define permissible doses); the identification of pathways for their metabolism and excretion by the host; the measurement of their concentrations obtained in tissues, blood, and other host body fluids; the effect of protein-binding and pH on their activity against microbes; and their stability under various conditions. In many instances the microorganism that produces an antibiotic also yields several substances closely related to the antibiotic. Mutants of the microbial strain which give the best production of the most desirable of these substances are then selected; a mutant with superior ability to produce a certain type of penicillin, for example, was obtained when mold cultures were exposed to X-rays to induce mutations.

IMPORTANT GROUPS OF ANTIMICROBIALS

Sulfonamides (Sulfa Drugs)

The discovery of sulfa drugs resulted from a systematic and tedious examination of hundreds of dyes carried out at the I. G. Farben Company in Germany during the early 1930s. These dyes were tested for activity against streptococcal infections, and one, prontosil, was found to be effective. Mysteriously the dye was effective in vivo but not in vitro. British and French scientists shortly solved this puzzle by showing that the activity of the dye actually resulted from sulfanilamide which was formed in the animal body by the splitting off of a chemical side chain of the dye. Sulfanilamide was found to be readily absorbed and to have a high therapeutic index, although it was insoluble at acid pH and therefore prone to precipitate in the kidney, producing damage. Because it is a small molecule not greatly bound by serum protein, sulfanilamide diffuses readily into mammalian tissues and spinal fluid. It is easily modified chemically so that a whole family of sulfa drugs with different properties is now available. Unfortunately, the widespread and indiscriminate use of sulfa drugs has led to the selection, in many genera of pathogenic organisms, of mutants that resist sulfa drugs so that the drugs are limited in their usefulness today. In addition, certain forms of sulfa result in severe allergic reactions sufficiently often to make other antimicrobial medications preferable.

Resistance of *Neisseria meningitidis,* page 542

Other Chemical Agents

In addition to sulfa drugs, several other small molecular weight chemicals (Figure 33-1) selectively inhibit metabolic processes of certain microbes. *Trimethoprim*

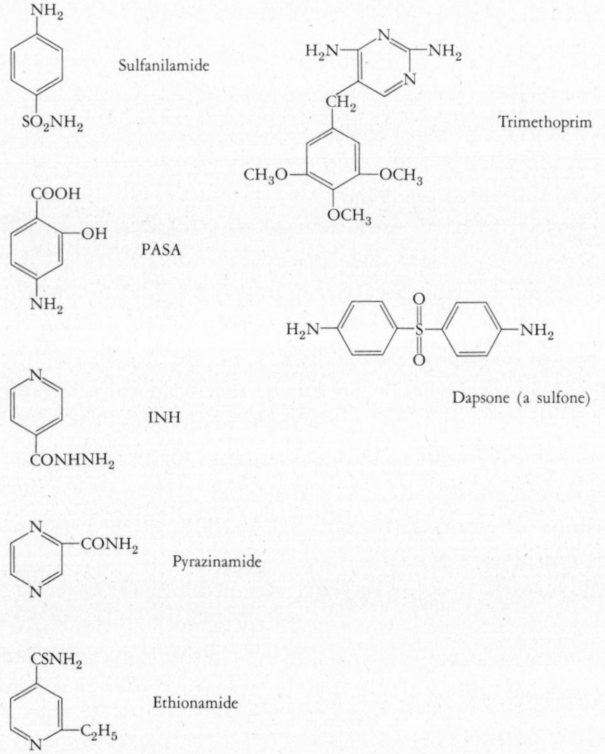

Sulfanilamide

Trimethoprim

PASA

Dapsone (a sulfone)

INH

Pyrazinamide

Ethionamide

FIGURE 33-1
Synthetic antimicrobials.

markedly potentiates the antimicrobial action of sulfonamides, and the two types of medication are often given together. Sulfones are useful in treating leprosy, and para-amino-salicylic-acid (PASA) is a drug useful only in treating tuberculosis. However, some other simple cyclic compounds, including *isonicotinic acid hydrazide* (INH), are also useful in treating tuberculosis.

Leprosy, pages 543–545
Tuberculosis, pages 489–491

Ancient remedies sometimes remain useful. For example, over 300 years ago Spanish monks in Peru learned that the cinchona bark used medically by the native people was effective against malaria. It is now known that *quinine* (Figure 33-2) is the chief antimalarial ingredient in cinchona bark, and this substance is still useful in treating persons infected with *Plasmodium falciparum* strains resistant to more modern medications. Little is known of the mode of action of quinine except that it has diverse effects on specific microbial and mammalian cells. Only certain developmental forms of the malarial parasite are susceptible to it. Other agents useful in treating diseases caused by protozoa are given in Figure 33-1 and Table 33-1. Although some, like emetine, are ancient remedies, their modes of action are generally still poorly defined.

Malaria, pages 573–575

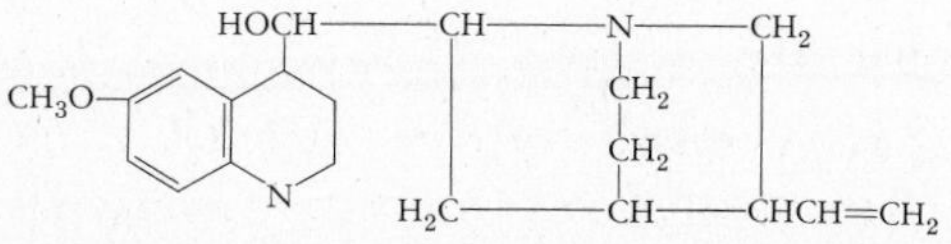

FIGURE 33-2
Quinine, an ancient antimalarial medication still lifesaving in some cases.

Penicillins

Penicillum, Table 14-2, page 292

Penicillin was discovered accidentally in 1928 when Sir Alexander Fleming, working at St. Mary's Hospital in London, noticed that a mold contaminant on a culture plate containing staphylococci had resulted in a zone of killing of the bacteria. The mold was a species of *Penicillium*—and so was derived the name of the antibacterial substance, *penicillin,* that it produced. Filtrates of broth cultures of the mold proved to have insufficient activity for treating infected wounds, however, and it was not until a decade later that sufficiently potent preparations of penicillin became available for treating infections. The intervention of World War II spurred British and American workers to develop means for large-scale production and to determine the chemical structure of penicillin. Several different penicillins were found to be present in the mold cultures, and these were designated F, G, X, K, O, and so on. Of these, penicillin G (or benzyl penicillin; see Figure 33-3), was found to be most suitable for treating infections, being relatively stable and possessing high antibacterial activity. It was then discovered that the addition of phenylacetic acid to the mold growth medium markedly increased the synthesis of penicillin G.

Penicillin was the first safe and effective antibiotic for systemic therapy, and it remains today the drug with the greatest therapeutic index. However, penicillin G has the drawback that it is unstable in acid solutions and therefore cannot consistently survive stomach acid and be absorbed after oral administration.

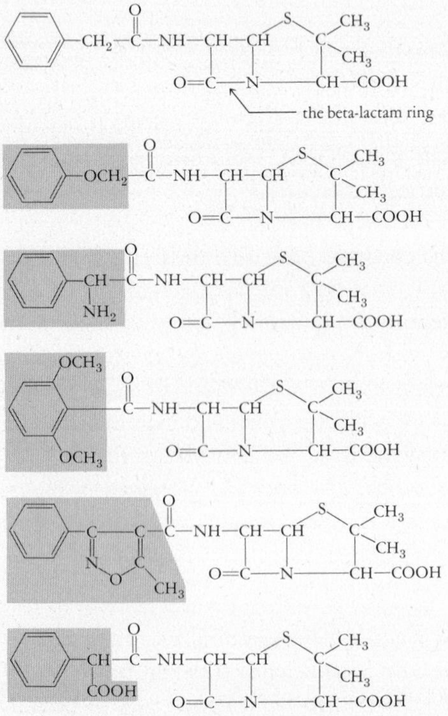

FIGURE 33-3
Some members of the penicillin family. Shaded areas indicate the modifications responsible for the changes in properties.

the beta-lactam ring

Penicillin G

Penicillin V
(acid-resistant)

Ampicillin
(broadened spectrum acid-resistant)

Methicillin
(penicillinase-resistant)

Oxacilin
(acid- and penicillinase-resistant)

Carbenicillin
(broadened activity against Gram-negative rods)

Moreover, its spectrum of activity is limited to Gram-positive bacteria (such as *Streptococcus pyogenes*) and a few Gram-negative organisms (such as *Neisseria meningitidis*). It is completely ineffective against bacteria which produce penicillinase, the enzyme that destroys the drug. The development of penicillin V, a relatively acid-stable compound, only partially overcame these difficulties. Penicillin V was synthesized by the *Penicillium* when phenoxyacetic acid was added to the mold cultures.

In 1959 English scientists discovered that the main portion of the penicillin molecule—6-amino penicillanic acid—could easily be produced naturally and then modified chemically by the addition of different chemical side chains. From the many resulting compounds, several were selected for use in treating infections (Figure 33-3). Further alterations in 6-amino penicillanic acid have yielded a large family of penicillins with differing properties for use in various kinds of infections. Carbenicillin and ticarcillin, for example, are effective against many strains of *Pseudomonas aeruginosa,* a species generally resistant to other penicillins.

Pseudomonas aeruginosa, pages 433–434

Although nontoxic, penicillins occasionally cause death when administered to people who are allergic to them (an estimated 300 to 500 deaths per year in the United States); this happens because, as mentioned in Chapter 21, the penicillins themselves are nonantigenic but can act as haptens when they or their breakdown products attach to body proteins. For this reason, doctors and nurses always ask patients about penicillin allergy before giving them the drugs. In the past some people probably became hypersensitive to penicillin because of unknowing exposure to it in vaccines, or in milk from cows under treatment with the drug, but strict controls now make this possibility less likely.

Haptens, page 385

Another drawback of the penicillins is their instability in aqueous solutions, so that they must be used promptly after preparation to avoid loss of antibacterial activity. Even water condensing from the air may decrease the activity of penicillin powders that have been refrigerated. Also, widespread and often inappropriate use of penicillins has resulted in a selective increase in penicillin-resistant strains of bacteria as the penicillin-sensitive strains were killed. Some penicillin-resistant bacteria even grow in penicillin solutions and can cause infections when such contaminated medicines are administered. Finally, as is the case for most medicines taken orally, food may interfere greatly with the absorption of penicillins.

Cephalosporins

A strain of mold of the genus *Cephalosporium,* obtained from the sea near a sewage outlet, was found to produce antibiotic substances somewhat similar in structure to penicillin (Figure 33-4). One of these substances, cephalosporin C, could be modified chemically to produce useful antibiotics. Some of the resulting cephalosporins are susceptible to stomach acid and must be given by injection; others can be given orally. Like the penicillins, the cephalosporins are bactericidal and have a high therapeutic index. They are active against many Gram-positive and Gram-negative pathogens. Their chemical structure is sufficiently different from that of the penicillins to make them relatively resistant to penicillinase and to permit

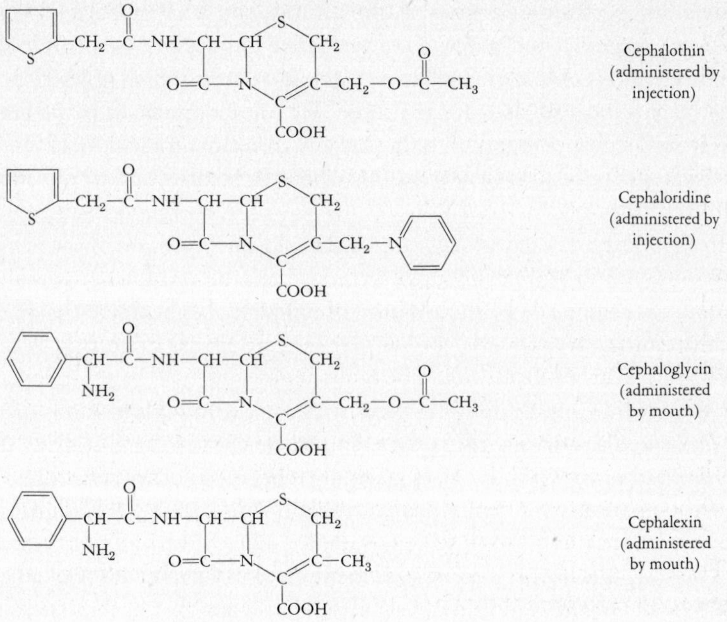

Cephalothin (administered by injection)

Cephaloridine (administered by injection)

Cephaloglycin (administered by mouth)

Cephalexin (administered by mouth)

FIGURE 33-4
Cephalosporin derivatives.

their administration to some people who are allergic to the penicillins. The cephalosporins are, however, susceptible to cephalosporinases produced by a variety of bacteria.

Tetracyclines

Unlike the penicillins and cephalosporins, which are of fungal origin, the tetracyclines (Figure 33-5) are produced by certain strains of the actinomycete, *Streptomyces.* A family of useful tetracyclines has been produced through selection of appropriate producing strains of the organism and by chemical alteration. This family includes oxytetracycline, chlortetracycline, demeclocycline, minocycline, and tetracycline. All have a very similar spectrum of activity, but vary in their stability, toxicity, and affinity for blood proteins. They are often referred to as "broad spectrum" antibiotics because they show activity against strains from many different genera of bacteria, both Gram-positive and Gram-negative. The term broad spectrum, however, may be misleading because of the high incidence of tetracycline resistance among various genera of bacterial pathogens. The tetracyclines are usually given orally and achieve relatively low levels of antimicrobial

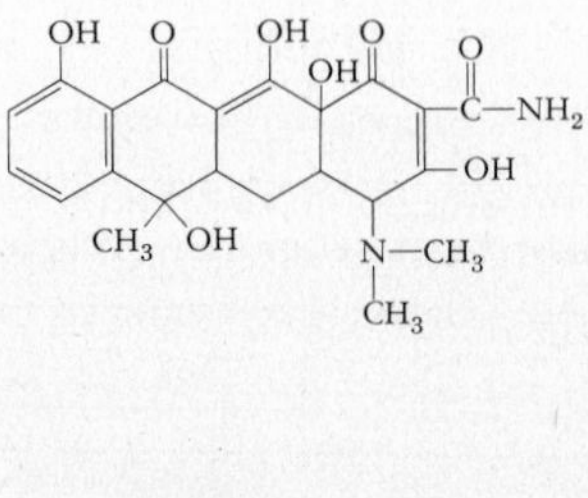

FIGURE 33-5
Tetracycline. Modifications of this structure yield oxytetracycline, chlortetracycline, demeclocycline, minocycline, and others.

activity in body fluids. They are bacteriostatic and are easily bound and inactivated by foods and metallic ions.

Tetracyclines, like many other drugs, bind reversibly to serum proteins. The binding is an equilibrium reaction, so that a fraction of the antibiotic remains free to react with microbes and to diffuse into tissue fluids. Because the fraction of free antibiotic is small, the concentrations entering body fluids are relatively low. Excretion is delayed because protein-bound antibiotic does not normally enter the urine.

In the past, tetracyclines have been widely and indiscriminately used in animal feeds, in infection of human beings when the etiologic agent could not be readily determined, and in efforts to prevent secondary bacterial infections in persons suffering from colds, influenza, measles, heart failure, or surgical wounds. It is now known that the use of tetracyclines to prevent secondary infection commonly *increases* the chance of infection, probably because the tetracyclines interfere with the protective effect of the normal bacterial flora. Thus tetracyclines may foster colonization or overgrowth by resistant microbes, including *Staphylococcus aureus, Candida albicans, Pseudomonas aeruginosa,* and species of *Proteus* and *Klebsiella.* The once extensive use of tetracyclines has resulted in an increasing percentage of resistant strains of *Streptococcus pyogenes, Neisseria gonorrhoeae,* and *Bacteroides* species, all of which were previously susceptible to the drug. The therapeutic index of the tetracyclines is low. The toxic effects of these drugs range from diarrhea and discoloration of teeth to growth retardation of infants. Severe malfunction of the liver can result from large doses, and tetracyclines administered after prolonged storage (such as in the family medicine chest) can seriously derange kidney function through a deterioration product of the drug.

Staphylococci, pages 426–427
Candida albicans, Table 14-3, page 298

Chloramphenicol

Originally isolated from *Streptomyces venezuelae,* chloramphenicol (Figure 33-6) is now synthesized more cheaply by chemical methods without the need for biosynthetic steps. Like the tetracyclines, it has a broad spectrum of antimicrobial activity and is generally bacteriostatic in the concentrations it reaches in body fluids. It has a higher therapeutic index than the tetracyclines, is less bound to serum protein, and readily diffuses into cells and spinal fluid to act on bacterial pathogens beyond the reach of other antibiotics. However, it is toxic at high doses—mainly due to its interference with the normal development of red blood cells. Much more serious, however, is a reaction occurring in about 1 of every 40,000 patients who receive chloramphenicol. This complication—aplastic anemia—is characterized by the inability of the body to form white and red blood

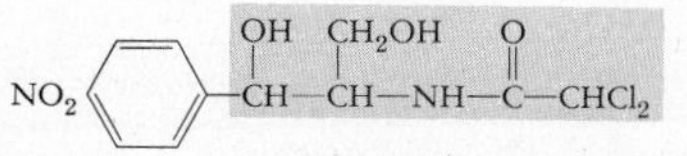

$$NO_2\text{-}C_6H_4\text{-}\underset{\displaystyle OH}{CH}\text{-}\underset{\displaystyle CH_2OH}{CH}\text{-}NH\text{-}\overset{\displaystyle O}{\overset{\|}{C}}\text{-}CHCl_2$$

FIGURE 33-6 Chloramphenicol. The structure of the side chain (shaded portion of the molecule) is of critical importance to its antimicrobial activity, but the part of the molecule responsible for toxic effects is not known.

cells and often ends in death from infection or from the development of leukemia. For this reason, use of chloramphenicol is generally restricted to life-threatening infections where equally effective alternative treatment is not available. Nevertheless, chloramphenicol is extensively used in economically poor countries because of its low price. Furthermore, during the years of peak chloramphenicol use in the United States, fewer than half the persons developing aplastic anemia had received the antibiotic, showing that there are causes of aplastic anemia other than chloramphenicol.

Aminoglycosides

Glycosidic bonds, Figure 2-19, page 43

The aminoglycosides are classified as a group because all have two or three chemical components linked by glycosidic bonds (Figure 33-7). These antibiotics are bactericidal and share many properties, but they do not necessarily cover the same spectrum of antimicrobial activity. Streptomycin, the first of the aminoglycosides, was found after screening 10,000 cultures of soil bacteria; the producing strain, *Streptomyces griseus,* was originally isolated from the throat of a chicken! Streptomycin is a bactericidal antibiotic, active against many Gram-positive and Gram-negative bacteria, and one of the few antibiotics that is active against *Mycobacterium tuberculosis.* Streptomycin has little activity against anaerobic bacteria or under acid conditions. Because it is poorly absorbed from the gastrointestinal tract, it must be given by injection. Unfortunately, the usefulness of this promising drug was impaired soon after its discovery because of the high frequency of very resistant mutants among many bacterial pathogens. The toxicity of strepto-

FIGURE 33-7
Aminoglycoside antibiotics. The shaded areas indicate the glycosidic linkages.

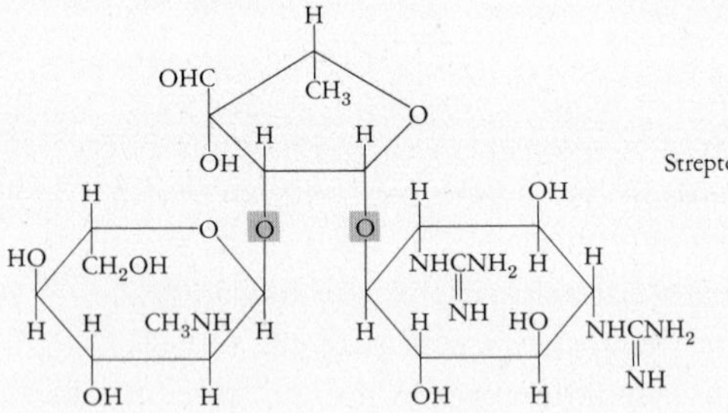

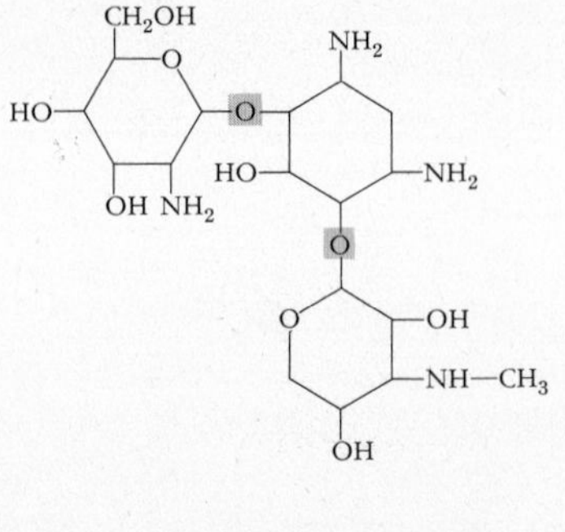

mycin involves damage to the kidney and the nervous apparatus for body equilibrium, but the most feared toxic effect of the drug is irreversible deafness. The toxic effects of streptomycin result when prolonged high concentrations of the drug occur in the bloodstream. This situation is especially likely to happen in persons with kidney diseases that interfere with normal excretion in the urine.

The aminoglycosides neomycin and kanamycin have been widely used orally to reduce intestinal enterobacteria prior to surgery, but this practice may lead to superinfection with resistant organisms, especially anaerobic bacteria such as *Bacteroides* species. Gentamicin, tobramycin, and amikacin, three newer aminoglycosides, are active against many Gram-negative bacteria that have acquired resistance to the older medications. However, strains resistant to gentamicin and tobramycin have already appeared in several areas of the United States.

Miscellaneous Antibiotics of Medical Importance

Macrolides. *Erythromycin* and other macrolide antibiotics (Figure 33-8) share the properties of being active primarily against Gram-positive bacteria (a spectrum similar to penicillin G); these drugs are usually bacteriostatic, and readily absorbed after oral administration. They have a high therapeutic index. The major use of the macrolide antibiotics is in the treatment of infections in persons allergic to penicillins. Unfortunately, resistant variants sometimes arise during the macrolide treatment of staphylococcal infections.

Lincomycins. *Lincomycin* is chemically distinct from other antibiotics but has an antimicrobial spectrum similar to that of the macrolides. The therapeutic index of the drug is high, and its action is generally bacteriostatic in the concentrations it reaches in body fluids. Lincomycin has been modified chemically and some of its derivatives are better absorbed and show greater in vitro activity than the parent compound. Although staphylococci resistant to the macrolides may be initially sensitive to lincomycins, resistant forms quickly arise on exposure to the latter drugs.

Polypeptide antibiotics. *Polypeptide antibiotics* (Figure 33-9), such as bacitracin and polymyxin, generally are poorly absorbed after oral administration. Both bacitracin and polymyxin can be administered by injection, but because of toxic-

Polypeptides, page 35

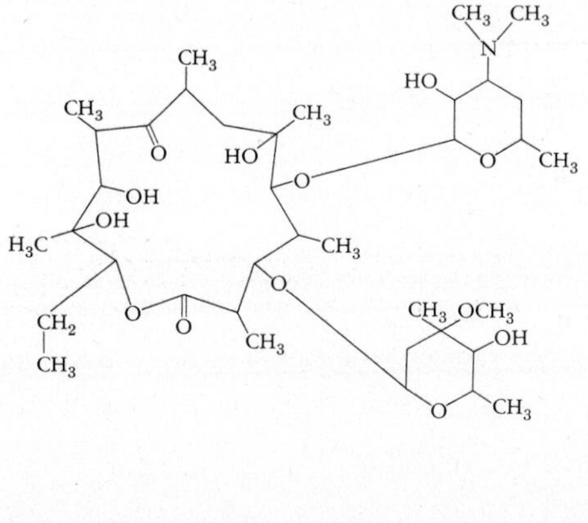

FIGURE 33-8
Erythromycin.

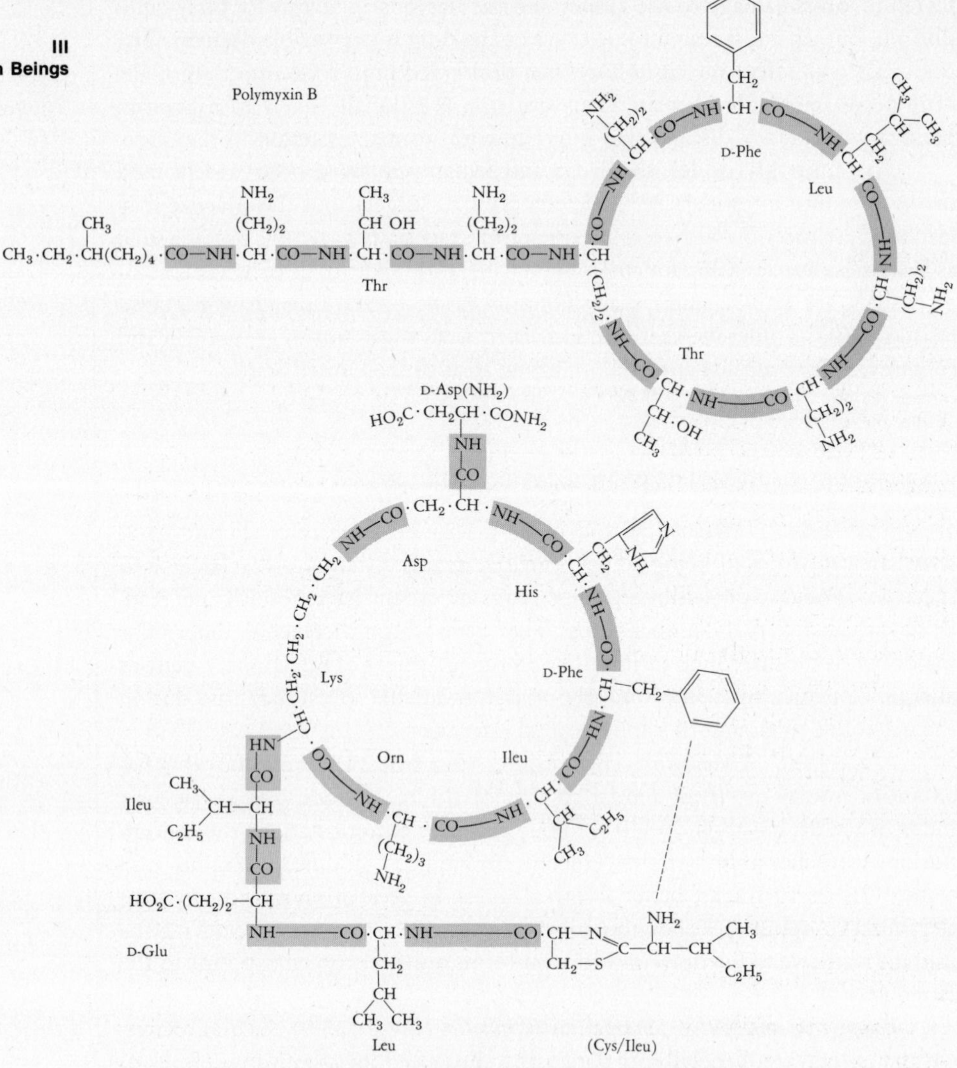

FIGURE 33-9 Polypeptide antibiotics. Shaded areas indicate peptide bonds.

ity this usage is limited to the treatment of bacteria resistant to other antimicrobial drugs. More often bacitracin and polymyxin are commonly applied to body surfaces for treating very superficial infections. Bacitracin is a bactericidal antibiotic produced by a strain of *Bacillus subtilis* originally obtained from dirt in a wound from a girl named Tracy; hence the name *"bacitracin."* It is effective in treating infections by Gram-positive bacteria such as *Staphylococcus aureus.* The polymyxins are produced by another species of *Bacillus* and, in contrast to bacitracin, are active against Gram-negative organisms. They are useful primarily in treating serious infections caused by *Pseudomonas aeruginosa* but share with bacitracin the property of poor diffusion into infected tissue. In addition, the poly-

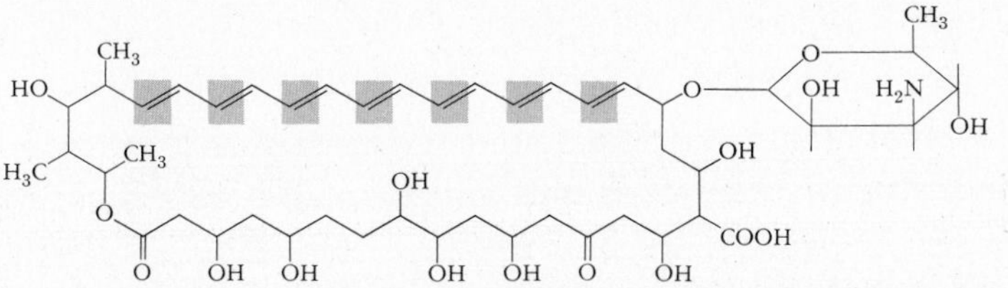

FIGURE 33-10
Amphotericin B, a polyene antibiotic. Shaded areas show unsaturated (ene) groups.

myxins are readily inactivated by pus and dead tissue, probably because they react with free nucleic acids.

Polyene antibiotics. *Polyene antibiotics* (Figure 33-10), such as nystatin and amphotericin, have activity against some fungi and protozoa and, although toxic, are among the few available agents for the treatment of infections caused by eucaryotic microorganisms.

Rifamycins. The *rifamycins,* discovered in 1957 in cultures of *Streptomyces mediterranei,* were found unsuitable for therapeutic use, but a derivative, rifampin (Figure 33-11) has shown great activity against many Gram-positive and Gram-negative organisms as well as mycobacteria. Rifampin can be administered orally and little toxicity has been noted so far. Its main defect has been that mutants highly resistant to it are readily selected from several bacterial genera. The rifamycins are chemically unrelated to other antibiotic families.

Antitumor agents. Finally, a few antibiotics have selective activity against the cells of malignant tumors. Unfortunately, the therapeutic index of these drugs is very low, and some toxic effects on the host are the rule. *Actinomycin D* (Figure 33-12) and *mitomycin C* are examples of such agents. They have given temporary arrest, lasting several years, of some tumors.

MECHANISMS OF ANTIMICROBIAL ACTION

Competitive Inhibition

In competitive inhibition the enzymatic activity of an organism is blocked by a substance that closely resembles the normal substrate for that enzyme. This is the mechanism of action of the sulfa drugs and has been considered in Chapter 7.

The vitamin folic acid, to which para-amino-benzoic acid (PABA) is con-

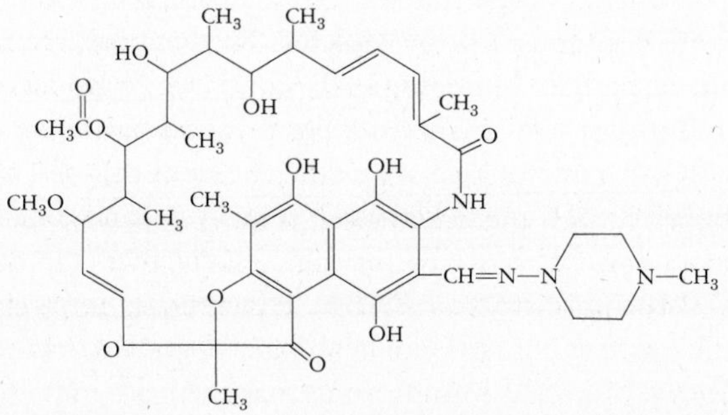

FIGURE 33-11
Rifampin, which shows both antibacterial and antiviral activity in vitro.

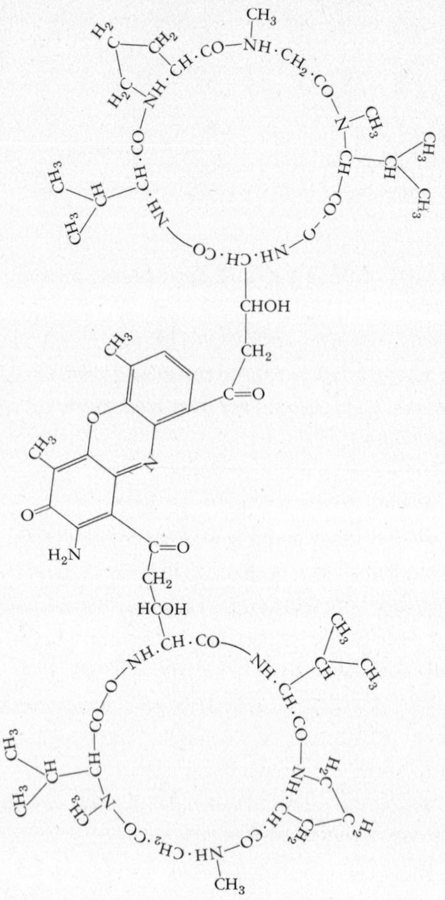

FIGURE 33-12
Actinomycin D, an antitumor antibiotic.

verted, is essential for cell growth because it is required for the production of purine and pyrimidine bases, as well as some amino acids. Mammalian cells are not affected by sulfa drugs because they cannot use PABA and similar substances to synthesize folic acid; instead, they must obtain this vitamin from dietary sources. On the other hand, many bacteria, and some protozoa and fungi, do synthesize folic acid from PABA and are therefore inhibited by sulfa drugs.

The effect of sulfa drugs on susceptible microbes is not seen until several cell divisions have occurred, because the folic acid that has been formed prior to the addition of a sulfa drug must first be depleted. Furthermore, the action of sulfa drugs can readily be reversed by adding PABA or the end products dependent on PABA metabolism to a culture that has been treated by the sulfa drug. Sulfa drugs may act on some microbes in ways different from that described above. In fact, the pathogenicity of some rickettsiae is actually enhanced by sulfonamides.

The sulfones and PASA act by the same biochemical mechanism as sulfa drugs and are also competitors of PABA. However, sulfones and PASA are uniquely effective against mycobacterial pathogens (*Mycobacterium leprae* for sulfones, *M. tuberculosis* for PASA) indicating that there are important unknown factors associated with the antimicrobial action of this group of competitive inhibitors. On the other hand, the antitubercular drug INH is a competitor of pyridoxine (vitamin B_6) in a different biochemical pathway. INH is taken up by an enzyme in place of pyridoxine (the normal substrate of the enzyme) and is converted to abnormal products. One of these products is lethal for the cell. INH is therefore bactericidal, unlike the sulfa drugs, sulfones and PASA, all of which are bacteriostatic.

Impairment of Cell Wall Synthesis

A number of antimicrobial agents act by interfering with bacterial cell wall synthesis. The unique nature of the bacterial cell wall ensures that antimicrobials acting principally at this site will have a high therapeutic index. For example, low concentrations of penicillin interfere with one of the final enzymatic reactions—the joining of adjacent layers through their amino acids—in the synthesis of peptidoglycan. Cell wall precursors also appear in the medium, probably due to the activation of a lysozyme-like bacterial enzyme that normally functions to modify the cell wall during growth of the bacterium. As a result, the bacterial cell wall is greatly weakened and the cell lyses unless it is suspended in a hypertonic medium, in which case spherical protoplasts of the organism will form. Only growing cells are affected by penicillin, because only growing cells are synthesizing their cell wall. Furthermore, in many Gram-negative organisms, penicillin has a difficult time reaching its site of action because of the presence of additional wall layers covering the peptidoglycan layer. This helps to explain why many Gram-negative bacteria are resistant to penicillin. Additionally, all eucaryotic cells are completely resistant to penicillin because only procaryotic cells have peptidoglycan in their cell wall.

Peptidoglycan, page 60

Lysozyme, page 63

The cephalosporins act in a manner similar to that of the penicillins. Bacitracin, cycloserine, and vancomycin act at other stages in peptidoglycan synthesis, but their action is not restricted to the cell wall and their therapeutic index is low. The action of cycloserine stems from its structural resemblance to the amino acid alanine; bacitracin and vancomycin act at the level of the cytoplasmic membrane, interfering with the transfer of cell wall components.

Inhibitors of Protein Synthesis

Several antimicrobial drugs useful against bacterial pathogens exert their effect by interfering with steps in protein synthesis. Only a few of these substances are selective enough to be administered safely to patients, and this is not surprising because protein synthesis is a feature of all cells, not just those of pathogenic microorganisms. Fortunately, 80S eucaryotic ribosomes are different enough in their structure that some selectivity in the action of these antimicrobial medications is possible. Even so, bacterial-type (70S) ribosomes are present in some eucaryotic cellular organelles, such as mitochondria. This may account in part for the observation that side effects are often observed in patients undergoing treatment with high doses of the antimicrobial agents that selectively affect procaryotic protein synthesis.

Ribosomes, pages 71–72

The tetracyclines and the aminoglycosides bind to the smaller of the two ribosomal subunits. Tetracycline interferes with the attachment of charged tRNA to the ribosome and therefore prevents formation of peptide bonds between amino acids. Because this effect is reversible, the tetracyclines are only bacteriostatic. The aminoglycosides bind irreversibly to the ribosome and are bactericidal. They are potent inhibitors of the initial steps in the formation of protein and can also alter the shape of the ribosome and cause the genetic code to be incorrectly read.

Translation, pages 175–179

Chloramphenicol, erythromycin, and lincomycin all bind reversibly to the larger ribosomal subunit. Chloramphenicol interferes with peptide bond formation in procaryotic cells but, unfortunately, also interferes significantly, in high doses, with protein synthesis in eucaryotic cells. Erythromycin and lincomycin also inhibit protein synthesis by binding to the ribosome, thereby interfering with formation of the amino acid chain. The sites of attachment of these two antimicrobials are closely adjacent. Lincomycin acts by inhibiting the ribosomal enzyme responsible for peptide bond formation. This enzyme, called *peptidyltransferase,* is an integral part of the larger ribosomal subunit.

Agents Active against Cytoplasmic Membranes

The polypeptide antibiotics polymyxin and amphotericin damage cytoplasmic membranes. The polymyxins bind to phospholipid components of procaryotic, and to a somewhat lesser extent, of eucaryotic cells. The polyene antibiotics, such as amphotericin, bind to sterols (lacking in most procaryotes) and thus are almost exclusively active against eucaryotic pathogens, such as *Histoplasma capsulatum* and *Coccidioides immitis.* Both groups of drugs produce major toxic effects in mammalian hosts.

Coccidioidomycosis, histoplasmosis, pages 493–497

Inhibitors of Nucleic Acid Synthesis

Rifampin acts by binding strongly to bacterial DNA-dependent RNA polymerase, and thereby inhibits mRNA synthesis. Rifampin and related antibiotics show some activity against the polymerases of both DNA and RNA viruses, including reverse transcriptase, but this has not been sufficiently great to be of therapeutic value. Actinomycin D also inhibits DNA-dependent RNA synthesis but is highly active against mammalian cells. Because the principal effect of this drug is against rapidly dividing cells, it has been useful against certain cancers.

Transcription, pages 174–175

IN VITRO DETERMINATION OF SUSCEPTIBILITY OF MICROORGANISMS TO ANTIMICROBIAL AGENTS

The Need for Susceptibility Testing

For some pathogenic species, susceptibility to therapeutic agents is predictable (for example, penicillin G always kills pneumococci, and chloroquine is generally active against the malarial agent *P. vivax*). With other organisms (such as

S. aureus), however, there is no way of knowing reliably which antimicrobial agent is likely to be effective in a given case. In treating infections it has often been the practice to try one antimicrobial agent after another on the patient until a favorable response is observed, or, if the infection is very serious, to give several antimicrobial agents together. Both approaches are undesirable, because with each unnecessary drug there are unnecessary risks of toxic or allergic effects, as well as undesirable alterations of the normal flora. *The cornerstone of rational treatment of infectious diseases thus rests with choosing the antimicrobial agent most likely to act against the offending pathogen but against as few other cells as possible.* Practical methods exist for determining the susceptibility to antimicrobial agents of the more rapidly growing bacterial pathogens. Some of the basic concepts involved in the in vitro determinations of such bacterial susceptibility are given below.

Quantification of Susceptibility

In principle, in vitro testing of susceptibility relies on preparing a series of decreasing concentrations of the antimicrobial agent in a suitable growth medium, and adding a suspension of the infecting organism to each concentration. After a period of incubation the concentrations in which visible growth of the organism has not appeared are determined (Figure 33-13). The lowest concentration capable of preventing growth is called the *minimal inhibitory concentration* (MIC). Cultures containing concentrations of the antimicrobial agent which show no growth of the microbe can be subcultured to a medium lacking antibiotics to see whether viable organisms remain. In this way the *minimal bactericidal concentration* (MBC) can also be determined. The organism is said to be "sensitive" to the lowest concentration that inhibits growth. If the concentration required to *kill* the organisms is only two to four times as much as the inhibitory concentration, the antimicrobial agent is said to be bactericidal; if higher concentrations are required, it is bacteriostatic.

Assays of Antimicrobial Agents

Mere knowledge of the MIC and MBC of a drug is not enough to tell whether it will be effective in treating an infection. It is also necessary to know whether these concentrations are likely to occur in infected tissues. To determine this, each new drug is given to human subjects in doses known to be safe and nontoxic. Samples of the blood, urine, and other body fluids are then collected at different time intervals following administration of the drug. A very sensitive organism can then be used to measure the amount of antimicrobial drug present in these body fluids. A culture of the test organism is spread over the surface of an agar medium. Holes are then punched out of the agar; some of these are filled with fluid containing known concentrations of the drug to be assayed, while others are filled with the body fluid being tested. Following incubation, zones of inhibition are formed around the agar wells, their diameter depending on the concentration of drug present in each well. Be measuring the zone sizes and plotting them against the corresponding concentrations, a curve relating inhibition-zone size to

(a) A series of tubes containing decreasing concentrations of antimicrobial agent.

8 μg/ml 4 2 1 0

(b) Addition of an invisible inoculum doubles the volume.

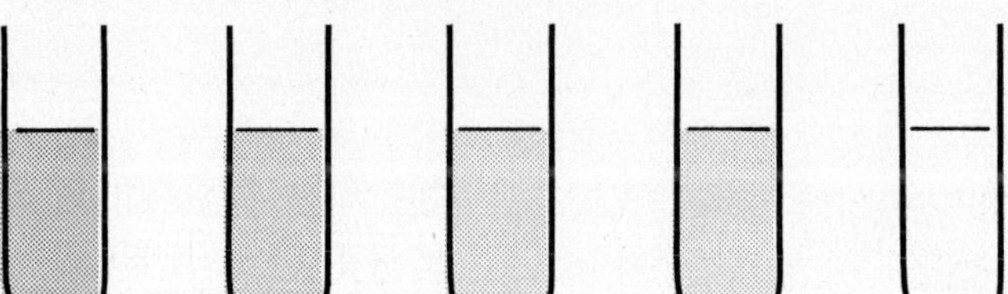

4 μg/ml 2 1 0.5 0

(c) Appearance of growth in the more dilute solutions following incubation.

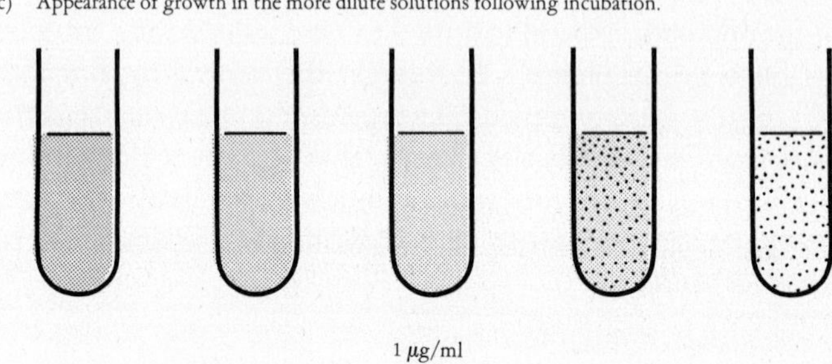

1 μg/ml
MIC

FIGURE 33-13
Minimal inhibitory concentration (MIC). In this example the MIC is 1.0 μg/ml. The tube containing medium without antibiotic is a growth control. In actual practice, an antibiotic control is also included (organism of known MIC).

concentration is obtained, from which the concentration of antimicrobial agent in the body fluid can be read (Figure 33-14). Thus concentrations present at different times following administration of a drug can be determined, as in Figure 33-15, for serum levels of a penicillin.

The Meaning of "Sensitive"

Although often used in the quantitative sense, the word "sensitive" is also commonly used in a qualitative sense to describe organisms susceptible to concentrations of antimicrobial drug known to occur in the blood of patients under treatment. The word "resistant" is used for microorganisms requiring substantially higher concentrations. Thus an organism with an MIC of only 20 μg per milliliter of polymyxin would nevertheless be called "resistant" because the blood

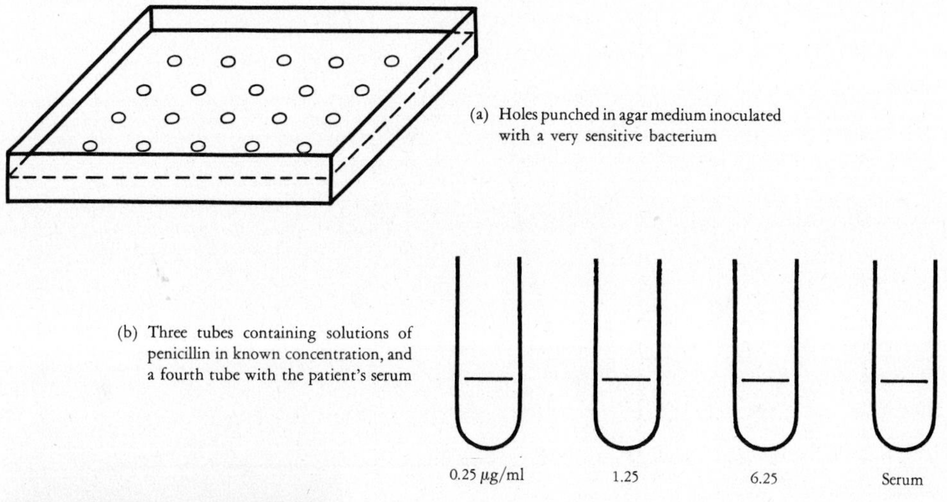

(a) Holes punched in agar medium inoculated with a very sensitive bacterium

(b) Three tubes containing solutions of penicillin in known concentration, and a fourth tube with the patient's serum

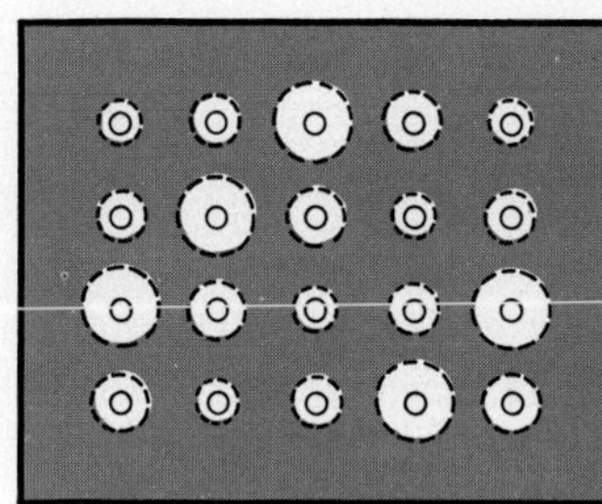

(c) The holes have been filled with fluid from the tubes and the culture plate incubated overnight; the sensitive bacterium grew everywhere except for areas around the holes in the agar

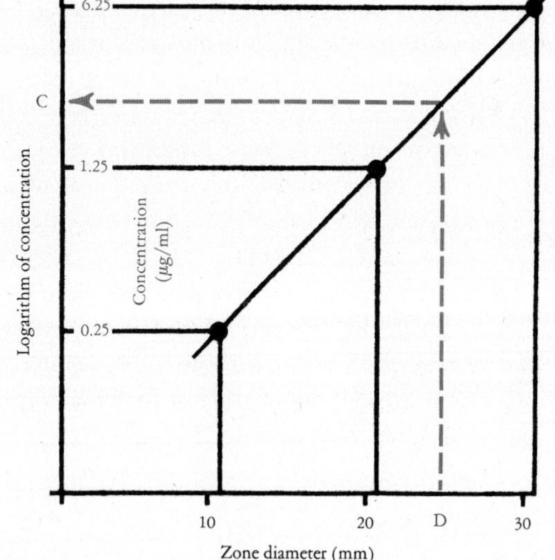

(d) The diameters of the areas of inhibited growth are averaged for each solution and plotted semilogarithmically; the diameter D of the zone of inhibition produced by the serum is used to determine the concentration C of antibiotic in the serum

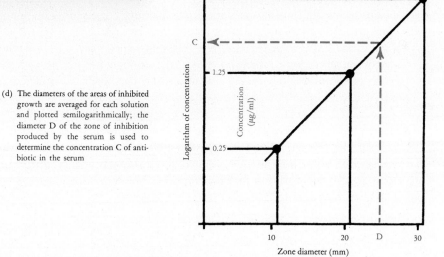

FIGURE 33-14 Assay of penicillin in a patient's serum.

levels of this drug are usually lower than 5 mg per milliliter. Microorganisms requiring inhibitory concentrations that lie on the borderline between sensitive and resistant are often called "intermediate."

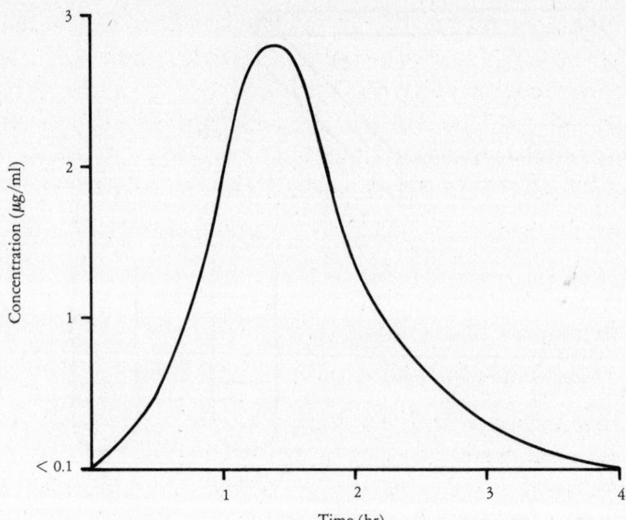

FIGURE 33-15
Concentration of penicillin in a patient's blood at different times after an oral dose.

Distinguishing Sensitive and Resistant Organisms

Because, in practice, it is technically difficult, expensive, and time-consuming to determine MICs, the qualitative susceptibilities of bacterial pathogens to antimicrobial agents are determined using filter paper discs impregnated with the agents. Thus, to determine whether the *S. aureus* strain isolated from a boil is sensitive to penicillin or not, a culture of the organism is spread over the surface of an agar medium and an antibiotic disc containing a precise amount of penicillin is placed on top of it. Following incubation, the zone of inhibition of growth is measured and, if large enough, the organism is called *sensitive* to penicillin. To determine whether the inhibition zone is large enough, it is necessary to refer to tables of values for each drug. These tables have been prepared by determining the MICs and inhibition zone sizes of the drugs simultaneously on bacteria of a wide range of susceptibility. In this way the inhibition zone diameters were correlated with the MICs, and, knowing the expected blood levels of each drug, correlations between inhibition-zone diameters and sensitivity could also be determined. Because antimicrobial agents diffuse through agar at different rates, there are different standards for each drug.

Standardization of Sensitivity Determinations

The medium used for determining sensitivity should be roughly comparable, in certain constituents, to the tissue fluids of the body; otherwise, erroneous information may result. For example, if the growth medium contains significant amounts of sulfonamide inhibitors (as many media do), organisms will appear to be resistant to sulfa drugs even though they are actually quite susceptible to these drugs in the blood of a patient, where there is little PABA and folic acid end product. Other drugs may also show altered activity in certain media; for example, tetracyclines are inactivated by di- and trivalent metallic ions and show enhanced activity at low pH, while aminoglycosides have decreased activity at low pH and increased activity at high pH. Because of such variations, a committee of the World Health Organization has developed detailed recommendations for standardization of the methods used in determining bacterial susceptibility testing. Little work has yet been done in standardizing the procedures used for determining fungal, protozoan, viral, or metazoan drug susceptibilities.

LIMITATIONS OF THE VALUE OF ANTIMICROBIAL AGENTS

Selection of Resistant Variants

During the first few years after the introduction of the sulfonamides and penicillin, there was great hope that such agents would soon do away with most infectious diseases. Yet today, microbial resistance limits the usefulness of all known antimicrobial drugs. For some organisms (such as *S. aureus*), penicillin-resistant strains were well established in nature before the introduction of penicillin, although they occurred at a low frequency. But the heavy use of penicillin (now measured in tons per year) soon fostered their selection, so that by 1950 more than 50 percent of the strains of staphylococci responsible for infections were penicillin-resistant. Furthermore, such strains do not readily disappear when usage of an antibiotic is discontinued, because they are well adapted for survival even without the competitive advantage of widespread drug usage.

Clones of drug-sensitive bacteria also contain mutants, ranging in frequency from about 10^{-3} to 10^{-9}, that are resistant to most antimicrobial agents. Under certain conditions of sustained heavy antimicrobial use, selection of these mutants may also give rise to drug-resistant strains of an organism. Generally these mutants are not well adapted for survival under ordinary conditions and are only able to maintain their dominance of a flora with the help of heavy antimicrobial drug usage. As soon as the antimicrobial drug is withdrawn, however, the organisms lose their competitive advantage and once again become only a minor part of the total population of their species. Such selection occurs with chloramphenicol-resistant mutants of *S. aureus*.

Transfer of genetic information from resistant to sensitive strains of bacteria by transduction and conjugation is another process which can markedly increase the incidence of resistance to antimicrobial medications. Transfer of the genetic information necessary for resistance occurs infrequently, but the presence of an appropriate antimicrobial drug can give a marked selective advantage to the recipient bacteria. Moreover, resistance factors (R factors), transferred by conjugation, often contain genes for resistance to several medications. Thus a person receiving a single antibiotic runs the risk of having a sensitive pathogen acquire resistance to several antibiotics simultaneously.

R factors, pages 209–211

Finally, the establishment of resistant microbial strains in a community is markedly influenced by conditions which allow these organisms to spread from person to person and to colonize. Thus, overcrowding and mobility of the human population, ventilation of houses and offices, sanitation, antibiotic usage, and the virulence of pathogens all may play major roles in establishment of resistance. The administration of an antibiotic to a person markedly increases the probability that that person will be colonized by extraneous antibiotic-resistant strains of an organism, presumably because of suppression of sensitive strains, which would otherwise compete with the resistant forms, in the person's normal flora.

Some of the mechanisms by which microbes resist the action of antibiotics and chemotherapeutic agents are now known, and many studies of these mechanisms are being carried out in the hope that means can be devised to circumvent them. Some microbes resist the action of antimicrobial drugs by degrading them: references have been made earlier to the penicillinases, microbial enzymes that can attack the penicillin molecule and destroy its activity. A number of penicillinases are known, differing in their inducibility, kinetics, and site of action against penicillins. Most act by opening the beta lactam ring of the drug (Figure 33-3), but some penicillinases attack other sites on the penicillin molecule. Resistance to still other antibiotics occurs through microbial enzymes that add a chemical group (such as an acetate or phosphate group) and thus cover up those sites on the antibiotic molecule that would normally react with vital structures in the microbes. Resistance to other drugs is achieved by changes in permeability of the cell wall or cell membrane to prevent entry of the antimicrobial agent. Finally, some sulfa drug-resistant bacteria have been shown to have increased production of the metabolic analog (PABA) of the sulfa drugs and thus to prevent the competitive biosynthetic blockade produced by the sulfa drugs in sensitive cells.

Other Limitations to Antimicrobial Effectiveness

Drugs that act against microbes either prevent their growth or increase the rate of their death. These drugs have no effect on the microbial toxins that may have already been released into a host and, in some instances, by killing microbes the drugs actually enhance the release of endotoxins, with life-threatening consequences. Moreover, the effectiveness of antimicrobial drugs is influenced by their ability to diffuse to the site of an infection. Aminoglycosides (such as streptomycin), for example, may fail to cure meningitis because they cross the blood-brain barrier poorly. In some instances, conditions at the immediate site of an infection may not be favorable for antibiotic action. Penicillin kills streptococci in a few hours in broth medium, but many days are required for the same concentration to kill these organisms if the streptococci are sequestered in a blood clot. The acid of urine also interferes with the activity of streptomycin, while the polymyxins are neutralized by the nucleic acids in pus. Furthermore, most antimicrobial agents are of little value in eliminating infectious organisms from abscesses or other sites where the organisms are not actively dividing; therefore, abscesses must be drained in order for antimicrobial drugs to be maximally effective. Antimicrobial drugs also often fail to eliminate infections caused by sensitive organisms in persons with impaired host defenses, particularly those with impaired phagocytic function. Additionally, antibiotics may prevent the development of natural immunity to an infecting agent, presumably by allowing elimination of the organism before an adequate primary immunological response can occur. Finally, good treatment is almost totally lacking for some microbes, particularly the protozoa that cause African sleeping sickness and American trypanosomiasis. There is also increasing pessimism whether continued discovery of new antimicrobials can keep pace with the development of resistance to antimicrobial medications.

Trypanosoma, pages 310–311

SUMMARY

If they have the property of *selective toxicity,* antimicrobial agents are of value in treating infections. The ratio of the maximum dose of a drug tolerated by patients to its minimum curative dose is referred to as the *therapeutic index* of the drug. Antibiotics are derived from only a few genera of microbes, and exhaustive searches of many strains are required to find new antibiotics suitable for use. The major groups of antibiotics consist of families of chemically related substances with varying properties, some of which result from the natural manipulations of producing microbes and others from chemical alterations of the products of biosynthesis. The sites of action of various antimicrobial medications include the cell wall, protein synthetic mechanisms of the cell, the cytoplasmic membrane, or nucleic acid synthesis. Sulfa drugs and some other low molecular weight chemicals act by interfering with synthesis of folic acid, a substance necessary for coenzymes responsible for the synthesis of purines, pyrimidines, and some amino acids.

The value of antimicrobial medications has been markedly impaired by the emergence and dissemination of drug-resistant microbes. Studies of the mechanisms of drug action and the mechanisms of microbial resistance have aided in finding new antimicrobial medicines. Antibiotics and other antimicrobials have been of enormous value in controlling many serious infections, but their continued value depends largely on their judicious use, to minimize the incidence of organisms resistant to them.

QUESTIONS

REVIEW

1. Give two procaryotic and two eucaryotic genera of microorganisms that contain antibiotic-producing species.
2. What are the required properties of a medically useful antibiotic?
3. How is it that bacteriostatic antimicrobial agents can cure infectious diseases?
4. Why might it be inadvisable to use antimicrobial agents to treat minor infections which would anyhow be overcome by body defenses?
5. What is meant by sensitivity of a microbe to an antimicrobial agent?

THOUGHT

1. How do you interpret the fact that so few strains of microbes produce antibiotics?
2. Why is it much harder to find an antibiotic useful against viral disease than one useful against diseases caused by bacteria?

FURTHER READING

BALDRY, P. E., *The Battle Against Bacteria.* New York: Cambridge University Press, 1965.

FRANKLIN, T. J., and G. A. SNOW, *Biochemistry of Antimicrobial Action.* 2d ed. New York: John Wiley & Sons, 1975. Brief, lucid book giving much information on modes of action and mechanisms of resistance.

GAUSE, G. F., *The Search for New Antibiotics—Problems and Perspectives.* New Haven, Conn.: Yale University Press, 1960.

GOODMAN, L. S., and A. GILMAN. eds., *The Pharmacological Basis of Medical Therapeutics.* 5th ed. New York: Macmillan, 1975.

HARE, R., *The Birth of Penicillin and the Disarming of Microbes.* London: George Allen and Unwin, 1970. Fascinating account of the fluke of luck that gave us penicillin.

ROUECHE, B., "Something Extraordinary." *Eleven Blue Men and Other Narratives of Medical Detection.* New York: Berkley Medallion, 1965.

WATANABE, I., "Infections and Drug Resistance," *Scientific American* (December 1967).

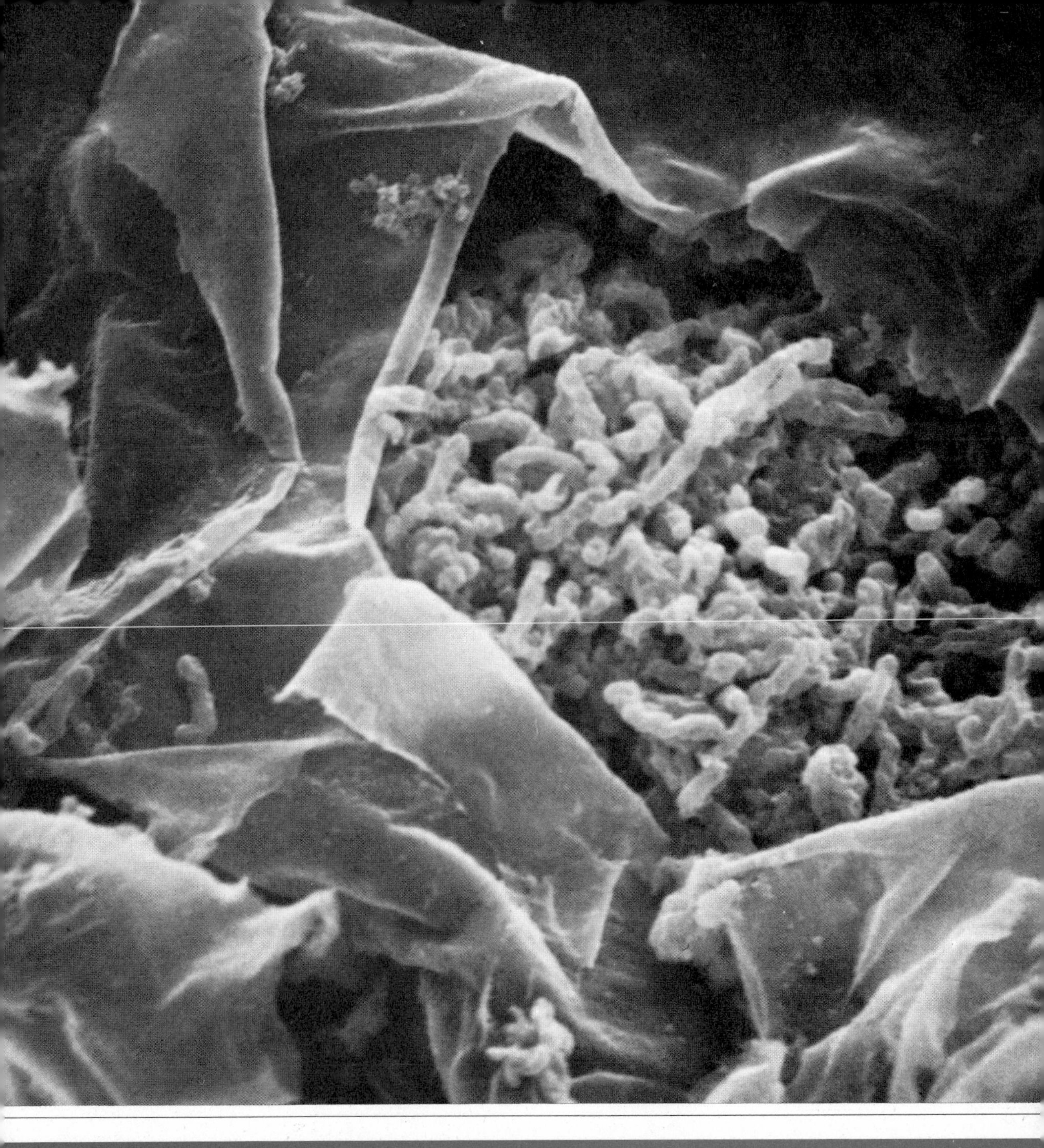

IV MICROBES AND THE ENVIRONMENT

TERRESTRIAL MICROBIOLOGY

The microorganisms that live in the soil play an indispensable role in maintaining life on this planet. The soil contains myriad microorganisms able to degrade or to chemically modify organic and inorganic molecules. It also contains organisms that are indispensable to life on earth because they transform chemical substances that cannot be readily utilized into others that serve as nutrients for a variety of organisms. For example, although human beings live in a virtual ocean of air that is 79 percent nitrogen, this element cannot be utilized by most living creatures unless it is first converted into forms of nitrogen such as those that occur in amino acids. In ways such as this, microorganisms contribute directly to the fertility of the soil.

Considerable interest in soil organisms stems from the fact that many of them produce antibiotics and other useful products. Immense screening programs have been undertaken by the drug industry with the express goal of isolating organisms that produce antibiotics. The search to find organisms capable of synthesizing compounds of industrial importance also continues. Indeed, in no other habitat can one hope to find the tremendous range of synthetic capabilities that are represented by soil organisms.

Besides its beneficial denizens, the soil is also the habitat of a number of fungi and bacteria that are pathogenic for plants and some that are pathogenic for animals and humans.

SOIL

Components of the Soil

Soil may be divided into four major components: mineral fractions, organic material, soil moisture, and soil atmosphere. The detailed composition of a soil with respect to these four components will determine the quality and quantity of plant and animal life growing in that particular soil.

Biological Content of Soil

The soil teems with life, from the submicroscopic viruses of bacteria and plants to the macroscopic earthworms and other large forms. Bacteria are the most numerous soil inhabitants, ranging from a thousand or less to several million per gram of soil. In terms of biomass they are not as important as the fungi, but they probably exceed the biomass of the algae and protozoa combined. The total mass of microbes on this planet far exceeds the total mass of animals.

The most abundant soil bacteria are small coccoid rods that are highly pleomorphic. Depending on the locality of a soil, from 5 to 35 percent of its bacteria are of the genus *Arthrobacter.* The vast majority of the remaining bacteria are included in the following genera: *Pseudomonas, Clostridium, Bacillus, Micrococcus,* and *Flavobacterium. Chromobacterium, Sarcina,* and *Mycobacterium* are less common. Members of the actinomycetes, such as the antibiotic-producing *Streptomyces* and *Nocardia,* are widespread and abundant; indeed, the characteristic odor of soil results from the presence of members of the *Streptomyces.*

Arthrobacter, page 259

Streptomyces and *Nocardia,* page 260

The most extensive microbial growth in soil takes place on soil crumbs in the form of microcolonies (Figure 34-1). Bacteria are not uniformly distributed throughout the soil, but their distribution usually follows closely that of organic matter, predominantly in the upper soil layers or top soil.

The composition of the fungal flora of the soil has been difficult to determine because the media routinely used in microbiology laboratories select for specific nutritional types of fungi and only certain fungi appear. Furthermore, the

FIGURE 34-1
A microcolony of rod-shaped bacteria separated out of the soil and prepared by the freeze-etching technique. There are approximately 10 cells, which is the average number of cells in a microcolony. Note that one cell is undergoing division. (Courtesy of L. E. Casida, Jr., and D. L. Balkwill.)

generic composition and size of the fungal flora of a soil vary with the type of soil and with its physical and chemical characteristics. The class Fungi Imperfecti consistently contributes the greatest number of fungal genera in the soil. These genera commonly include *Aspergillus, Trichoderma,* and *Penicillium.* Members of the class Phycomycetes are not as numerous, although members of the Mucorales are very common. Only a few genera of Ascomycetes are found, and although the large Basidiomycetes, the mushrooms, are present, their quantity is difficult to assess because their large size is not compatible with their being easily cultured on agar medium. Larger forms of life are present in soil in vast numbers. They range in size from the large moles and gophers through the smaller earthworms down to the barely visible nematodes and tiny insects. The numbers of these barely visible forms can be staggering. A single acre of soil may contain half a billion insect eggs and larvae and up to a billion mites.

Fungi Imperfecti, pages 296–297

RELATIONSHIPS BETWEEN MEMBERS OF SOIL FLORA

There are numerous food chains interacting in the soil, and one organism that preys upon another is itself preyed upon by still other organisms. Several unique situations within the soil food chain deserve mention because of their involvement with microorganisms. The soil nematodes are a group of roundworms, about the size of the letter S, which resemble tiny eels. Many prey on bacteria, fungi, and protozoa as well as on other nematodes. Their major importance to agriculture is based on their mouth parts, which form an essentially hollow spear that can penetrate the toughest root and draw the juices out of the corresponding plant. Furthermore, the soil nematodes serve as very important vectors for transmitting disease-causing viruses to plants. It is estimated that they cause over half a billion dollars worth of crop damage in this country each year.

An interesting and important soil relationship involves the termites and the protozoan that inhabits their stomachs. Termites are major agents of wood decay in the warmer parts of the world. In the soil, however, the termite is only half the team. The gut of this insect is tightly packed with a certain type of protozoa which is able to digest the cellulose of the wood that the termite eats and which the termite can only ingest, not digest. Neither the insect nor its protozoan partner can exist without the other.

A commonly observed important plant-fungus symbiosis occurs with the mycorrhiza (plural, mycorrhizae). These are discussed in Chapter 14.

Mycorrhiza, page 298

ENVIRONMENTAL INFLUENCES ON THE BACTERIAL AND FUNGAL FLORA

Environmental conditions affect the density and composition of the bacterial and fungal flora of the soil. The primary environmental influences include moisture, temperature, acidity, organic matter, and inorganic nutrient supply.

Moisture and Oxygen Supply

Moisture and the availability of oxygen are intimately interrelated. Wet soils are unfavorable for most bacteria simply because the filling up of pore spaces in soil with water diminishes soil aeration. The major effect of wetness is therefore a reduction in the supply of soil oxygen. Continued waterlogging of the soil changes the flora from one consisting primarily of aerobes to one consisting mostly of anaerobic species. On the other hand, when the moisture content of soil drops to a low level, as in a desert environment, the metabolic activity and number of soil bacteria are markedly reduced. Many soil nonspore-forming bacteria are resistant to drying and representatives of many major bacterial groups survive for years.

Fungi and oxygen, page 292

Because the filamentous fungi are strict aerobes, they are found close to the surface of the soil. In waterlogged soils the numbers of these fungi are also markedly reduced. Their spores, however, often persist and, once the water in the soil diminishes and the level of soil oxygen increases, the molds recover and become an important part of the flora.

Acidity

Highly acid or alkaline conditions tend to inhibit the growth of many common bacteria. Many agricultural practices that adjust soils to neutrality, such as the liming of an acid soil, promote the growth of most bacteria. If fertilizers containing ammonium are added in excess, they can inhibit bacterial growth. This results from the fact that microorganisms can oxidize ammonium to nitric acid, thereby creating a highly acid environment.

Many species of molds can multiply over a broad pH range—in the alkaline region up to a pH above 9, and in the acidic region to pH values almost as low as 2. Because most bacteria do not multiply in highly acidic environments, the fungi dominate the soil population in areas of low pH. For this reason adding lime to a soil reduces the abundance of fungi, and treating the soil with acid-forming fertilizers, such as those containing ammonium, increases the fungal abundance. This can be attested to by anybody who has seen mushrooms growing in abundance in a lawn fertilized with an acid fertilizer. The increased abundance of fungi at low pH results primarily from the fact that the bacteria are not present and competing for food.

Temperature

Temperature and enzymes, page 147

Through its effect on the rates of enzyme reactions, temperature governs the rates of biochemical processes. A warming trend favors the biochemical changes in soil that are brought about by its microbial population. Mesophiles constitute the bulk of the soil bacteria, and true psychrophiles are rare or absent. In winter the bacteria in soil are cold-tolerant mesophiles, rather than psychrophiles. Thermophiles are ubiquitous.

Temperature and growth, pages 92–94

Most soil fungi are mesophilic; thermophilic fungi are uncommon. However, the thermophiles that have been studied will grow at 50°C but not at 65°C. A few thermophilic fungal strains can be demonstrated in soil, but they are abundant only in compost heaps, which do not normally reach high temperatures.

Nutritionally, the great majority of soil bacteria are heterotrophic. The organic matter which serves as their source of energy is produced largely by the photosynthetic activity of higher plants. However, although a tremendous amount of organic material is added to soil every year, there is generally not enough to supply the energy requirements of soil bacteria; and the size of the bacterial population is therefore limited by the organic matter available. The addition of organic material, such as occurs when manure or crop residues are plowed into the soil, results in a dramatic increase in the number of bacteria and also of fungi, because the latter are also heterotrophic.

In soils in which plants are growing a variety of organic materials exudes from the plant roots. The zone which surrounds the roots and contains these exudates is termed the *rhizosphere.* The exudates provide an abundant source of material for energy and bacteria therefore tend to concentrate on or near the root surfaces of plants. Other soil organisms, such as fungi and protozoa, are also abundant in the rhizosphere.

Nutrition of fungi, page 292

Growth of vegetation depends on the cycling of the major chemical elements. These elements reach the soil, as well as aquatic environments, in organic material. This material is then degraded and the elements are restored to those forms in which they can be reutilized by plants. Oxygen and hydrogen are required elements and are usually abundant in the biosphere. But carbon and nitrogen must be continuously recycled, or plant growth would soon be limited. Phosphorus and sulfur, although needed in smaller amounts, are also essential and also pass through complex cycles.

CARBON CYCLE

The carbon cycle revolves around CO_2, its fixation into organic compounds by green plants, and its regeneration, primarily by microorganisms. The overall cycle was considered in Chapter 5. The discussion here focuses on the decomposition of organic material.

Carbon cycle, pages 115–117

One of the most important decomposition products of both animal and plant material is the heterogeneous organic *humus.* Humus is composed of very complex molecules whose exact structures are not known. Because humus contains many very complex ring molecules, it is not readily decomposed but tends to persist for long periods. Little is known about the organisms involved and the pathway by which humus is decomposed, but it appears that actinomycetes may be largely responsible for its degradation to CO_2.

Actinomycetes, page 260

The decomposition of organic matter is carried out by "decomposers," chiefly bacteria, fungi, protozoa, and small animals. The decomposition of plant residues, which represent a large source of the organic debris in terrestrial environments, involves the activity of a large number of organisms. Fungi and typical bacteria utilize the more readily decomposable organic substances of plants, such as sugars, amino acids, and proteins. The cytophagas are capable of degrad-

Cytophagas, pages 262–263

ing cellulose, a polymer highly resistant to degradation by most bacteria. Other major plant constituents, such as lignin and pectins, are generally decomposed initially by fungi. Some of the breakdown products of these constituents can then be further metabolized by bacteria. Lignin, however, is very inert, and it has proven difficult to isolate organisms able to attack this three-benzene-ring structure.

As a general rule, bacteria appear to play the dominant role in the decomposition of animal flesh; fungi appear to be the most important in the initial degradation of wood. Protozoa, mites, and nematodes may play a more important role in decomposition than was previously suspected. Therefore, if these organisms in forest litter are killed by treatments that have no effect on bacteria and fungi, the decomposition of fallen leaves and dead twigs is greatly slowed.

Methane bacteria, page 271

The main gas evolved in the aerobic decomposition of organic matter is CO_2. However, when the level of oxygen is low, as is the case with wet rice-paddy soil, marshes, swamps, and manure piles, methane (CH_4) production may be great. The biosynthesis of methane is limited to a specialized group of bacteria, the methane bacteria, all of which are strict anaerobes that live in poorly aerated environments. Three genera of methane bacteria have been described: *Methanobacterium, Methanococcus,* and *Methanosarcina.* All three genera have the ability to gain energy from the oxidation of hydrogen gas and the reduction of CO_2, according to the following reaction:

$$4H_2 + CO_2 \longrightarrow CH_4 + 2H_2O$$

Anaerobic respiration, page 159

This reaction is an excellent example of an anaerobic respiration in which CO_2 is reduced. The methane-forming organisms cannot utilize common organic compounds, as can most heterotrophs. The possible role of these organisms in energy production is considered in Chapter 39.

Microbiological Decomposition of Synthetic Chemicals

Most organic compounds of natural origin are capable of being degraded by one or more species of soil organism. But this is not the case with a large number of synthetic chemicals. In recent years people have added to the number of slowly degraded or nonbiodegradable compounds in the soil by adding thousands of tons of pesticides in many parts of the world, as well as by adding nonbiodegradable detergents and many plastics. Other compounds which have been added in large quantities, and which may also be toxic for a variety of birds and fish, are the polychlorinated biphenyls, compounds used in many manufacturing operations.

AROMATIC COMPOUNDS CONTAINING CHLORINE ATOMS OFTEN ARE NONBIODEGRADABLE

These aromatic molecules, having chlorine atoms attached to their benzene ring, are not biodegradable, and their concentration in the environment is steadily increasing. The chemical nature of *herbicides* and *insecticides* also covers an extremely broad range of organic compounds—organic acids, nitrophenols, chlorinated organic acids, and other organic substances. Many of these substances have chemical structures similar to humus, which is also degraded very slowly. And, although certain of these chemically synthesized compounds disappear from the soil within a reasonably short time—a few days or weeks—many remain undegraded for years. Chemically synthesized compounds are most likely to be biodegradable if they have a chemical composition similar to that of naturally occurring compounds. If a synthetic chemical compound is totally different from any that occurs in nature, then no organism will have the enzymes for degrading it rapidly and the compound is likely to persist for long periods of time.

It is highly desirable that toxic compounds be degradable, because it is becoming apparent that most herbicides and insecticides are not only toxic to their target weeds or insects but may also have far-ranging deleterious effects on birds and other animals. There is well-documented evidence that the nonbiodegradable pesticide DDT accumulates in the fat of predatory birds. The data in Table 34-1 illustrate the phenomenon of *biological magnification.* The continuing ingestion and reingestion of DDT, which accumulates in fat, results in a greater and greater concentration of the DDT as the food chain through which DDT is passed approaches its highest members. DDT, as well as other chlorinated hydrocarbon insecticides, also interfere with the reproductive process of the birds by interfering with egg shell formation. Fragile eggs, which break before the young can hatch, are produced. Furthermore, DDT and other chlorinated hydrocarbons may be carcinogenic. Thus, although a biodegradable compound must be continually applied to maintain its effectiveness, its biodegradability recommends its use.

COMPOUNDS DEGRADED IN LABORATORY MAY NOT BE DEGRADED IN NATURE

It is now apparent that DDT is biodegradable in the laboratory, despite its extreme persistence in nature. Why this is true is not yet understood. It may relate to the fact that the initial degradative reactions on the DDT molecule do not yield energy, and the organism therefore gains little by degrading the compound. In the laboratory, however, energy sources are provided in addition to the DDT; this enables laboratory organisms to multiply and attack the DDT. These observations highlight an important caution in the study of waste disposal by microorganisms: that biodegradability of a material under laboratory conditions does not ensure its destruction by these same organisms in their natural environment.

TABLE 34-1
Food Chain Concentration of DDT

	Parts per Million DDT Residues
Water	0.00005
Plankton	0.04
Silverside minnow	0.23
Heron (feeds on small animals)	3.57
Herring gull (scavenger)	6.00
Fish hawk (Osprey) egg	13.8
Merganser (fish-eating duck)	22.8

Source: From E. P. Odum, *Fundamentals of Ecology* (Philadelphia: W. B. Saunders, 1971), p. 74.

FIGURE 34-2
Formula of 2,4,-D (left) and 2,4,5-T (right). Note that 2,4,5-T has an additional chlorine atom attached to carbon atom 5.

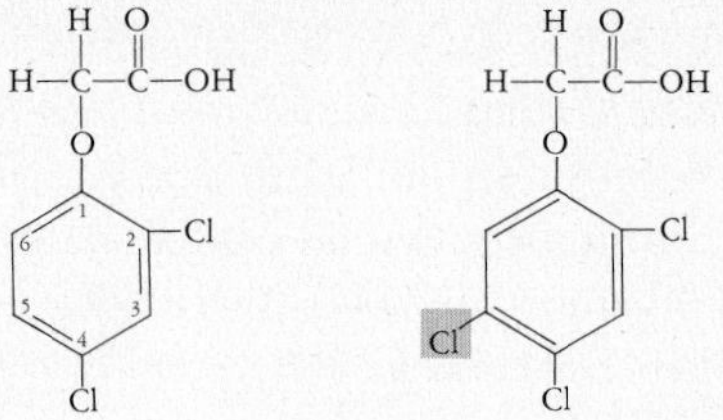

SLIGHT DIFFERENCES IN STRUCTURE AFFECT DEGRADABILITY

Relatively slight molecular changes markedly alter the biodegradability of a compound. Perhaps the best studied example involves the herbicides 2,4-dichlorophenoxyacetic acid (2,4-D) and 2,4,5-trichlorophenoxyacetic acid (2,4,5-T). The only difference between these two compounds is the additional Chlorine atom on 2,4,5-T (Figure 34-2). When 2,4-D is applied to the soil, it completely disappears within a period of several weeks, as a result of its degradation by a segment of the microbial population in the soil. However, when 2,4,5-T is applied, it is often still present for more than a year (Figure 34-3). Presumably this latter compound is much less similar to naturally occurring compounds than is 2,4-D. The persistence of a variety of herbicides is shown in Table 34-2. The estimates in this table, however, are only approximations, since many factors influence the rate of degradation of compounds by the soil flora. As a general rule, any practice which favors the multiplication of microorganisms will increase the rate of degradation. Thus raising the temperature, maintaining the pH near neutrality, and providing an optimal amount of moisture are all likely to increase the rate of degradation of most materials added to the soil.

The disappearance of a compound from the soil may result from a variety of other factors besides microbial activity; chemical reactions, destruction by light, volatilization, and leaching from the soil may all contribute.

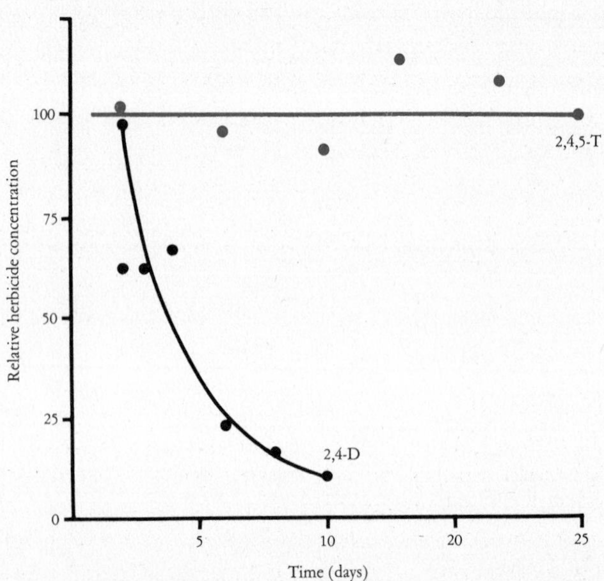

FIGURE 34-3
A comparison of the rates of disappearance of two structurally related herbicides, 2,4-D and 2,4,5-T. (From M. Alexander, *Introduction to Soil Microbiology,* John Wiley & Sons, New York, 1961.)

TABLE 34-2
Decomposition and Period of Persistence of Several Herbicides

Name of Compound	*Abbreviation*	*Persistence in Soil*	*Organisms Which Degrade*
3-(p-chlorophenyl)-1, 1-dimethyl urea	Monuron	4–12 months	*Pseudomonas*
2,4-dichlorophenoxyacetic acid	2,4-D	2–8 weeks	*Achromobacter* *Corynebacterium* *Flavobacterium*
2,2-dichloropropionic acid	Dalapon	2–4 weeks	*Agrobacterium* *Pseudomonas*
2,3,6-trichlorobenzoic acid	2,3,6-TBA	Greater than 2 years	
2,4,5-trichlorophenoxyacetic acid	2,4,5-T	Greater than 1 year	

Source: From M. Alexander, *Introduction to Soil Microbiology* (New York: John Wiley & Sons, 1961), p. 241.

Some toxic compounds may be degraded to other compounds which are just as toxic as their sources. The insecticide dieldrin, for example, can be degraded by both sunlight and by microorganisms to another compound, photodieldrin. This compound appears to be more toxic to many biological systems than is dieldrin. Photodieldrin is termed a *stable terminal residue* because it is apparently not readily degraded and accumulates in the environment.

Some cases are known in which inactive compounds can be converted to active herbicides by soil organisms. For example, 2,4–D is synthesized in soil by *Bacillus cereus.*

The concern over the nonspecific effects that insecticides and herbicides exert on the environment has encouraged efforts to develop agents that are highly specific for their target organisms, but harmless to everything else. It seems unlikely that organic chemical compounds can be developed which will display such specificity. Therefore, the efforts of many laboratories are being directed to studying natural biological agents which will affect only certain groups of plants or insects. Some success has already been achieved with these agents. These efforts and their future will be considered in Chapter 39. Great emphasis is also being put on developing and using chemically synthesized compounds, including pesticides, detergents and plastics, which can be broken down in the natural environment.

THE NITROGEN CYCLE

The element which plants require in greatest quantity is nitrogen, an important element in the makeup of proteins and nucleic acids. Although the atmosphere consists of 80 percent nitrogen, relatively few organisms can utilize this gaseous form of the element. Inorganic nitrogen is assimilated almost entirely either as nitrate (NO_3^-) or ammonium (NH_4^+). However, the bulk of the nitrogen added to soil is in the form of organic materials, largely as crop residues. Thus the

Proteins, pages 35–37

Nucleic acids, pages 37–39

Anaerobic respiration, page 159

organic nitrogen in these materials must be converted to ammonium ions (*ammonification*) and then to nitrate (*nitrification*) before it can be used. Some microorganisms convert nitrogen gas into ammonium ion (*nitrogen fixation*), which is in turn converted into the amino group of the amino acids in plant proteins. On the other hand, other bacteria can also convert nitrate to gaseous nitrogen by using nitrate as a metabolic electron acceptor in place of oxygen (*denitrification*).

All forms of nitrogen undergo a number of transformations. The entire series of reactions involving nitrogen comprises the nitrogen cycle (Figure 34-4), and microorganisms are essential at each stage of this cycle. The cycle consists of several individual transformations of nitrogen, in which the nitrogen passes through several oxidation states from its highly reduced form (NH_4^+) to its highly oxidized form(NO_3^-). The salient features of each individual step are now considered.

Ammonification

Amino acids, pages 32–35

Virtually all of the nitrogen found in the upper surface of the soil exists in organic molecules, primarily in the amino groups of amino acids in protein molecules, and in some in covalent linkages in purines and pyrimidines.

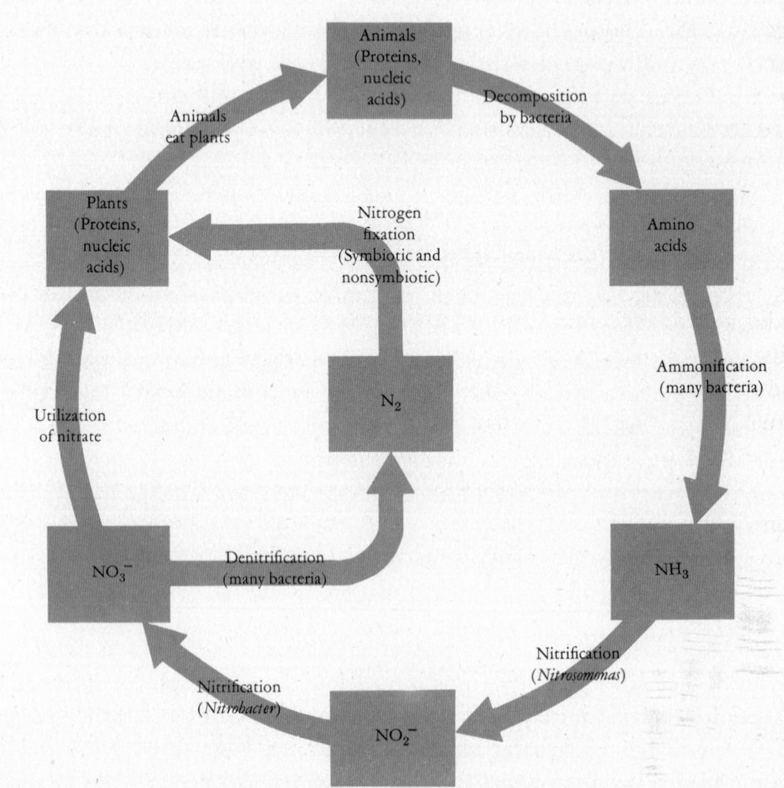

FIGURE 34-4
Nitrogen cycle.

A wide variety of microorganisms, including aerobic and anaerobic bacteria, and fungi, are capable of decomposing protein. They do this initially through the action of extracellular proteolytic enzymes, that convert protein into long chains of amino acids of varying lengths. Other enzymes then proceed to degrade these long chains into smaller chains and finally into amino acids, which are in turn degraded by a series of enzymes into CO_2, ammonium, and sulfate (derived from the sulfur-containing amino acids), and water.

Nitrification

During the process of ammonification, ammonia may be released directly from ammonium compounds into the atmosphere, but this is likely to happen only in alkaline soils in which very large amounts of organic nitrogen are being metabolized. The more usual sequence of events is the conversion of ammonium to nitrite, which is then converted to nitrate. These processes are called *nitrification.* The nitrate formed is readily utilized by plants. It will be helpful to consider the conversion of ammonium to nitrate in terms of the oxidation levels of the various elements involved. The ammonium ion (NH_4^+) is the most reduced form of nitrogen. As such it can be oxidized by bacteria with the release of energy. This oxidation occurs in two steps: the first of these is the oxidation of NH_4^+ to NO_2^-, and the second is the further oxidation of NO_2^- to NO_3^-. One group of organisms carries out the first oxidation and another group carries out the second. The organisms which actually function in nitrification in nature represent only a few genera. The bacteria oxidizing NH_4^+ to NO_2^- are members of the genus *Nitrosomonas;* the conversion of NO_2^- to NO_3^- is carried out by *Nitrobacter.*

Oxidation levels, page 151

Nitrosomonas, page 269

The reactions involved can be written as follows:

$$NH_4^+ + 1\tfrac{1}{2}O_2 \longrightarrow NO_2^- + H_2O + 2H^+ + \text{energy}$$
$$NO_2^- + \tfrac{1}{2}O_2 \longrightarrow NO_3^- + \text{energy}$$

As one would expect by looking at these reactions, they will occur only under aerobic conditions, and therefore no nitrification occurs in waterlogged soils. Both *Nitrosomonas* and *Nitrobacter* are autotrophs, utilizing CO_2 as their sole source of carbon. The energy required for the reduction of CO_2 to the constituents of cytoplasm is obtained by the oxidation of NH_4^+ to NO_2^- or of NO_2^- to NO_3^-.

Autotrophs, page 96

Nitrite is highly toxic to plants, and it is therefore important that the organisms that oxidize NO_2^- to NO_3^- be in the same environment as those that convert NH_4^+ to NO_2^-. Almost invariably, they are.

Denitrification

Certain normally aerobic bacteria that are active in proteolysis and ammonification will reduce nitrate under anaerobic conditions (anaerobic respiration). The ultimate results of this are the liberation of nitrogen gas and the net loss of nitrogen from the soil.

Anaerobic respiration, page 159

Immobilization

Utilization of ammonium in synthesis, pages 164–166

Nitrogen may be temporarily lost to plants by a process termed *immobilization.* This results from the assimilation of inorganic nutrients into the organic macromolecules of microorganisms inhabiting the soil. As bacteria and fungi grow they also utilize ammonium and nitrate salts for the synthesis of cell material, thereby converting nitrogen from a utilizable form into forms unusable by plants. The nitrogen is "sequestered" in the microorganisms and is unavailable until they die and are lysed. Only then can the nitrogen be changed into forms that can be used by plants.

Nitrogen Fixation

Nitrogen is continually being removed from the soil through the leaching action of water, which removes nitrate, through the incorporation of nitrogen into plants which are harvested, and by denitrification. Usable nitrogen must therefore be continually added to the soil to maintain its nitrogen status.

The natural supply of fixed nitrogen is limited and this limits the capacity of world agriculture. None of the processes described thus far explain how combined or *fixed nitrogen* (nitrogen in nongaseous form) is added to the soil. One source is the small amounts of fixed nitrogen, in the form of ammonium and nitrate ions, that occur in rainwater. Fertilization of soil also adds fixed nitrogen. Yet the combination of these two processes may often not compensate for the amount of nitrogen lost from the soil by denitrification. Nitrogen must therefore be fixed in other ways, and microorganisms represent the most important means for this fixation.

Nitrogen fixation is brought about by two types of organisms: those that are free-living and those that live in symbiotic association with various plants. A variety of these bacteria are capable of reducing nitrogen gas to the amino group of the amino acids that make up proteins. This process apparently occurs in several steps: in these, the nitrogen gas is first activated and then reduced to the ammonium ion. The overall reaction requires a tremendous expenditure of energy, about 15 molecules of ATP for every molecule of nitrogen fixed. The ammonium formed inside the cell is then incorporated into amino acids by transamination.

Azotobacter, page 254

Free-living nitrogen fixers. A variety of free-living microorganisms are capable of fixing nitrogen. The best known of these among the true bacteria, although not usually the most abundant, are members of the genus *Azotobacter.* These heterotrophic, aerobic, Gram-negative rods may be the chief suppliers of fixed nitrogen in grasslands and other ecosystems lacking plants with nitrogen-fixing symbionts. These organisms are rare in soils with a pH below 6.0, and members of the genus *Beijerinckia,* another heterotrophic nitrogen fixer, are found in highly acid soil. For reasons that are not clear, members of the genus *Beijerinckia* occur principally in the tropics and rarely in temperate zones.

Cyanobacteria, pages 272–273

The dominant anaerobic nitrogen-fixing organisms of the soil are members of the genus *Clostridium.* Cyanobacteria also fix nitrogen anaerobically, especially in flooded soils; indeed, rice has been cultivated successfully for centuries without the addition of nitrogen-containing fertilizer, probably because of the cyanobacteria developing in the water of paddy fields. Other nitrogen-fixing bacteria are found in the genera *Enterobacter, Chlorobium,* and *Rhodospirillum,* the latter two being photosynthetic.

Symbiotic nitrogen fixers. Although free-living bacteria are potentially capable of adding a considerable amount of fixed nitrogen to the soil, symbiotic nitrogen fixing organisms are far more significant in benefiting plant growth and crop production. The input of soil nitrogen from the microbial symbionts of leguminous plants such as alfalfa may be roughly 10 times the annual rate of nitrogen fixation attainable by nonsymbiotic organisms in a natural ecosystem.

Members of the genus *Rhizobium* are the most important symbiotic nitrogen fixing bacteria. The plants with which these organisms associate are all leguminous plants, and include alfalfa, clover, peas, beans, peanuts, and vetch. The symbiotic association of the bacteria with the plant results in the formation of nodules on the roots of the plant. Both the leguminous plants and the bacteria can grow perfectly well without each other, although the bacteria may disappear from the soil if leguminous plants are not grown in it for long periods of time. Furthermore, the bacteria are generally capable of fixing nitrogen only when they are growing in the root nodule of the plant.

The stages in the infection process of legumes by *Rhizobium* is understood at the descriptive level, but relatively little is understood of it at the biochemical level. The growth of *Rhizobium* in the rhizosphere is probably stimulated by the excretion of a number of organic compounds. The organisms multiply until they have reached a high cell density. This growth takes place primarily within a membranous layer around the outside of the root of the legume, and probably prevents its root secretions from escaping from the vicinity. The primary invasion site of Rhizobium is the root hairs of the plant (Figure 34-5). Several steps occur in the course of the invasion. One of the first of these is breakdown in the cellulose fibers which surround the root hairs; this may allow the bacteria to reach the hairs. The bacteria then excrete enzymes which dissolve the cement holding the cellulose fibers of the hairs together, and this may be responsible for the root-hair breakdown. The rhizobia then change their morphology from rod-shaped cells to spherical, highly flagellated cells called *swarmer cells.* The rhizobia may also synthesize the plant growth hormone indoleacetic acid from tryptophan excreted by the plant root, and this hormone may induce a deformation or curling of some of the root hairs.

Swarmer cells enter the curled root hairs of the legumes, proliferate, and a hyphalike infection thread is synthesized by the plant. The bacteria move within this thread, which branches into a number of root cells. In this manner the bacteria gain entrance to the root cells. If a root cell happens to be a normal diploid cell, it is usually destroyed by the infection and degenerates. If, however, the root cell is a rare tetraploid cell (a cell having four sets of chromosomes) it may develop into a nodule. Only about 5 percent of the rhizobial infections of legumes ultimately result in nodule formation. By some unexplained mechanism the in-

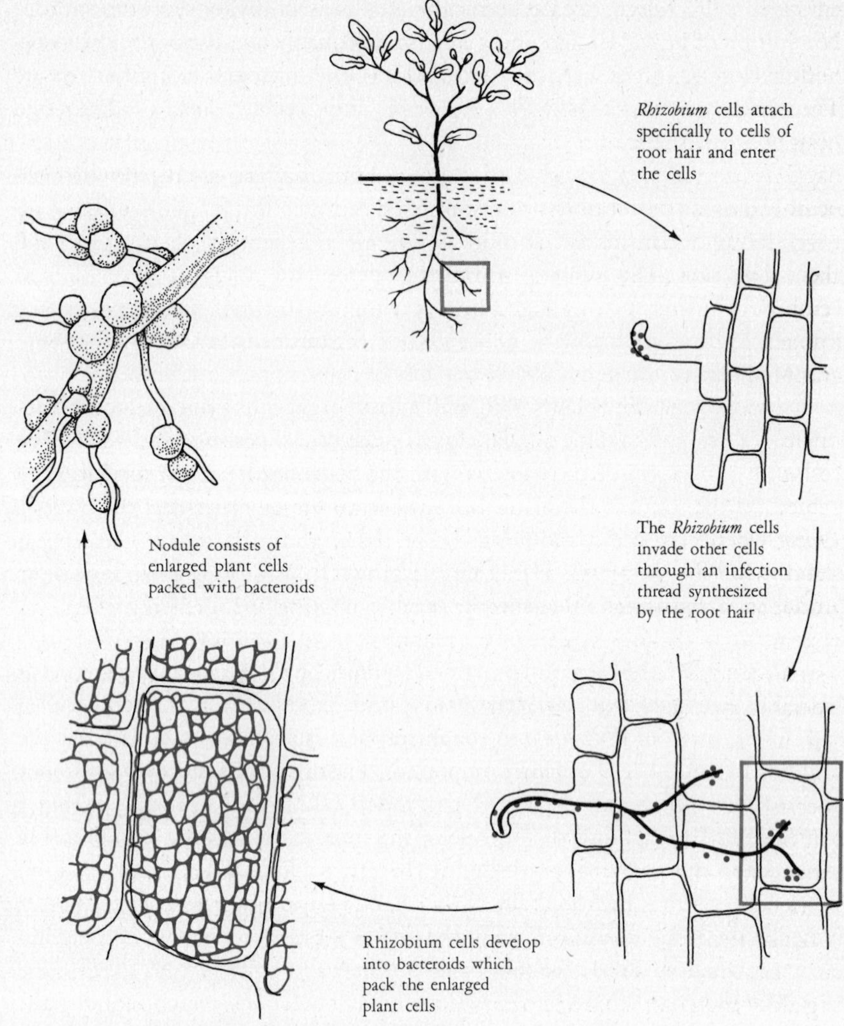

FIGURE 34-5
Symbiotic nitrogen fixation. The major steps leading to the formation of a root nodule in leguminous plants by *Rhizobium*.

fection of a tetraploid cell stimulates it to divide, and the continued division of these cells results in the formation of a nodule (Figure 34-5). The bacteria multiply rapidly within the tetraploid cells and become surrounded by a membrane synthesized by the host. They then undergo further morphological changes into swollen and irregular shaped organisms (Figure 34-5). These oddly shaped bacteria are called *bacteroids,* and their formation is apparently induced by several chemicals in the plant tissue.

One of the most intriguing questions regarding symbiotic nitrogen fixation is: What do the bacterium and the plant each contribute to the process? Within the past several years the answer has become much clearer with the observation that several species of *Rhizobium* can fix nitrogen in the complete absence of a leguminous plant. Obviously, *Rhizobium* has the genes for synthesizing the key enzyme in nitrogen fixation, *nitrogenase,* which carries out the reaction $N_2 \longrightarrow 2NH_3$. The enzyme is inactivated by oxygen, and a leguminous plant nodule provides the proper environment to protect the enzyme from oxygen. The protection is provided by a hemoglobin, a red pigment synthesized by healthy nodules containing bacteroids. Association of a bacterium with a legume happens because part of the genetic information for synthesizing the necessary hemoglobin is present in the bacterium, and part is coded by genes in the plant.

PLANT SUPPLIES ENERGY AND PROPER ANAEROBIC ENVIRONMENT IN NODULES FOR N_2 FIXATION

Different nitrogen fixing bacteria have different mechanisms for protecting their nitrogenase enzymes from oxygen. *Azotobacter vinelandii* is an obligate aerobe, yet it fixes nitrogen. Its secret is that it uses up all of the oxygen in its immediate environment through respiration, thereby maintaining an essentially anaerobic environment around its nitrogenase enzyme.

Nodules are not found on all species of legumes, and the reason for this is unclear. Perhaps those legumes that lack nodules have not developed the capacity for symbiosis, on the other hand, it is possible that the proper conditions for achieving nodulation have not yet been discovered. Likewise, not all species of *Rhizobium* are capable of invading leguminous plants. Moreover, the ability to develop a symbiotic relationship that results in nitrogen fixation varies markedly among infective strains of *Rhizobium.* Once established, the capacity of a symbiotic system to fix nitrogen is referred to as its *effectiveness,* and it is entirely possible for a strain of Rhizobium to be infective but not effective. Furthermore, there is considerable specificity between *Rhizobium* species and the particular legumes they will infect. Indeed, the speciation within the genus *Rhizobium* is based on the ability of an organism to infect certain plants (Table 34-3). A single infective species of the organism is generally able to infect several different plant species, but the basis for this specificity is not known. It appears to reside in a single bacterial gene or in a number of closely linked genes, because the ability to infect a particular plant group can be transferred from one species to *Rhizobium* to another by the process of DNA-mediated transformation.

Transformation, pages 200–203

In addition to the legumes, several genera of nonleguminous trees, including alder and ginkgo, possess nitrogen-fixing root nodules at some stages in their life cycle. The most intensively studied member of this nonleguminous nitrogen-fixing group is the alder tree, considered to be a pioneer tree because it is able to grow in nitrogen-poor soils. This ability stems from the tree's ability to fix

TABLE 34-3
The Legumes Infected by Various Species of *Rhizobium*

Legume	*Species of Rhizobium*
Alfalfa group	*R. meliloti*
Clover group	*R. trifolii*
Pea group	*R. leguminosarum*
Bean group	*R. phaseoli*
Lupini group	*R. lupini*
Soybean group	*R. japonicum*

TABLE 34-4
Quantities of Nitrogen Fixed by Microorganisms

Group	*Species or Habitat*	N_2 *Fixed per Acre per Year (lb)*
Nodulated legumes	Alfalfa	113–297
	Soybean	57–105
	Red clover	75–171
Nodulated nonlegumes	Alder tree	200
Cyanobacteria	Arid soil in Australia	3
	Paddy field in India	30
Free-living heterotrophs	Soil under wheat	14
	Soil under grass	22
	Rain forest in Nigeria	65

Source: From M. Alexander, *Microbial Ecology.* (New York: John Wiley & Sons. 1971), p. 428.

nitrogen. Although many attempts have been made to isolate the nitrogen-fixing microorganism(s) (probably an actinomycete) involved in symbiosis with the alder tree, the organisms have yet to be isolated in pure culture. This suggests that there may be an obligatory symbiotic relationship between the plant and the microorganism. No known species of *Rhizobium* will nodulate a nonleguminous plant.

Very recently, however, great excitement was generated when it was reported that a microorganism, *Spirillum lipoferum,* can fix nitrogen when associated, perhaps in a symbiotic relationship, with cereal grasses, such as corn. These bacteria also fix nitrogen when grown under microaerophilic conditions in the absence of the grain. Many field studies must be done to determine whether crop yields are greater in the presence rather than the absence of the organism.

The amount of nitrogen fixed by nitrogen fixing bacteria depends primarily on two factors, the organism and its habitat. Some idea of the amount of nitrogen fixed by the major groups of organisms can be gained from Table 34-4.

In the late nineteenth century there was concern that denitrifying bacteria were exhausting the nitrogen in the earth's soil. Today, however, there are some who think that the amount of fixed nitrogen being added to the soil is excessive. Nitrogen is fixed by industrial processes as well as by microorganisms, and the amount of nitrogen fixed industrially has been doubling about every 6 years. It has been estimated that when this amount is added to that fixed by legumes, the total exceeds the amount that is lost through denitrification. The problem which arises from this is that fixed nitrogen is being carried by water into lakes and rivers, resulting in an increase in algal growth in some areas. Furthermore, in many cases, the nitrogen in the organic wastes that come from an ever-increasing human and animal population contributes even more to the fixed nitrogen being added to our waters.

RECOMBINANT DNA AND NITROGEN FIXATION

Recombinant DNA, Figure 9-22, page 214

Many scientists are optimistic about the possibility of introducing the genes for nitrogen fixation onto plasmids and then into other bacteria in order to greatly extend the host range of symbiotic nitrogen fixation. As a beginning, the nitrogen-fixing genes from *Klebsiella pneumoniae* have been incorporated into a plasmid which was then transferred to *Escherichia coli* and other nonnitrogen-fixing species of bacteria. (Figure 34-6). These organisms were then able to fix nitrogen under anaerobic conditions. An even more futuristic idea is to introduce these genes directly into the roots or seeds of a plant. Whether the genes concerned with nitrogen fixation can be transferred and function in eucaryotes, however, remains unanswered. It may be necessary to modify the structure of the plant in order to protect the nitrogenase from inactivation by oxygen. Practical results from this approach are probably a long way in the future.

EVEN THOUGH NIF GENES ARE IN A PLANT, N_2 FIXATION REQUIRES ENERGY AND ANAEROBIC CONDITIONS

TRANSFORMATIONS: PHOSPHORUS

Phosphorus occurs in the soil in both organic and inorganic forms. Organic phosphorus-containing compounds occur almost exclusively near the surface. They are derived from surface vegetation and are accordingly found as constitu-

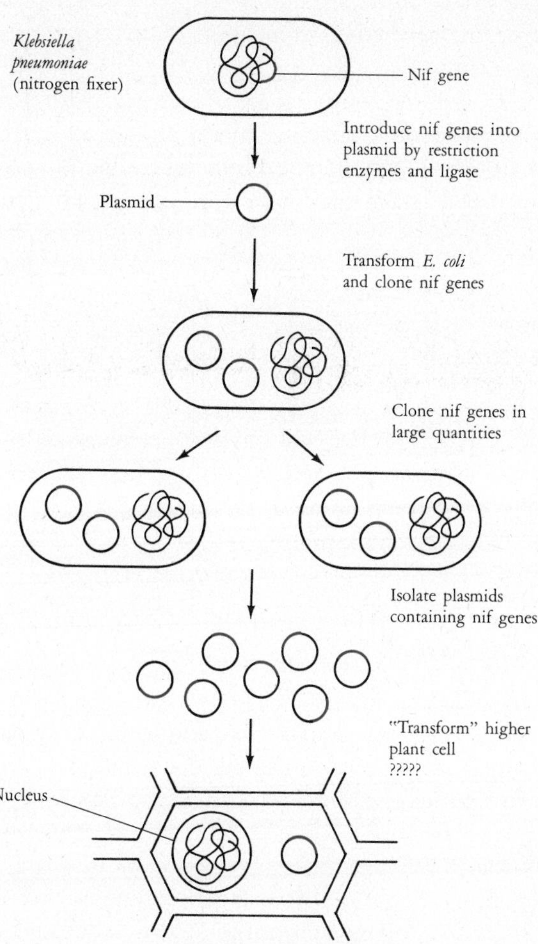
Klebsiella pneumoniae
(nitrogen fixer)
Nif gene
Introduce nif genes into plasmid by restriction enzymes and ligase
Plasmid
Transform *E. coli* and clone nif genes
Clone nif genes in large quantities
Isolate plasmids containing nif genes
"Transform" higher plant cell
?????
Nucleus

FIGURE 34-6
A highly speculative scheme of how the genes of nitrogen fixation might be cloned in *E. coli* and then introduced into nonleguminous plants.

ents of plant tissue, animals, and microorganisms. The inorganic forms of phosphorus occur largely as insoluble salts of calcium, iron, and aluminum. Plants cannot use these inorganic and organic forms readily, but orthophosphate (PO_4^{-3}) is readily used by most plants and microorganisms. Microorganisms play three major roles in phosphorus transformation: they mineralize organic phosphorus, they convert insoluble forms of inorganic phosphorus to soluble forms, and they immobilize inorganic phosphorus.

Overall Transformations of Phosphorus

Organic phosphate is first converted to orthophosphate, which can then be utilized by plants or immobilized by bacteria into their own organic forms of phosphorus. Some minerals containing phosphate may also be converted into orthophosphate by microorganisms.

Mineralization of Organic Phosphates

Bacteria and fungi are largely responsible for rendering organic phosphorus available to succeeding generations of microorganisms as well as to plants. Most microorganisms do this by synthesizing an enzyme, phosphatase, which cleaves the organic phosphate from a wide variety of compounds in which it is found. The breakdown of phosphate-containing nucleic acids is initiated by the action of nucleases which degrade DNA and RNA into their individual subunits (nucleotides). The phosphate of these nucleotides is then cleaved by the enzyme phosphatase to form nucleosides:

Structures of nucleotides, page 40

$$\text{nucleo}\textit{tide} \longrightarrow \text{nucleo}\textit{side} + PO_4^{-3}$$

Dissolution of Minerals

A wide variety of commonly occurring soil organisms are capable of dissolving calcium phosphate [$Ca_2(HPO_4)_2$ and $CaHPO_4$, respectively], which is much more soluble than the tribasic phosphate salt [$Ca_3(PO_4)_2$]. The amount of phosphate which dissolves depends primarily on the amount of acid that these organisms produce in their metabolic processes.

MINERAL CYCLE: SULFUR

Sulfur occurs in all living matter chiefly as a component of certain amino acids—methionine, cystine, and cysteine. The sulfur cycle bears many similarities to the nitrogen cycle (Figure 34-7). Like nitrogen, sulfur is present in the soil chiefly as a part of proteins, and it is taken up by plants in its oxidized form, sulfate (SO_4^{-2}). The sulfur-containing proteins are first degraded into their constituent amino acids by proteolytic enzymes excreted by a large variety of soil organisms. The sulfur of the amino acids is generally converted to hydrogen sulfide (H_2S) by a variety of soil microorganisms, a process analogous to NH_4^+ for-

Amino acids formulas, page 33

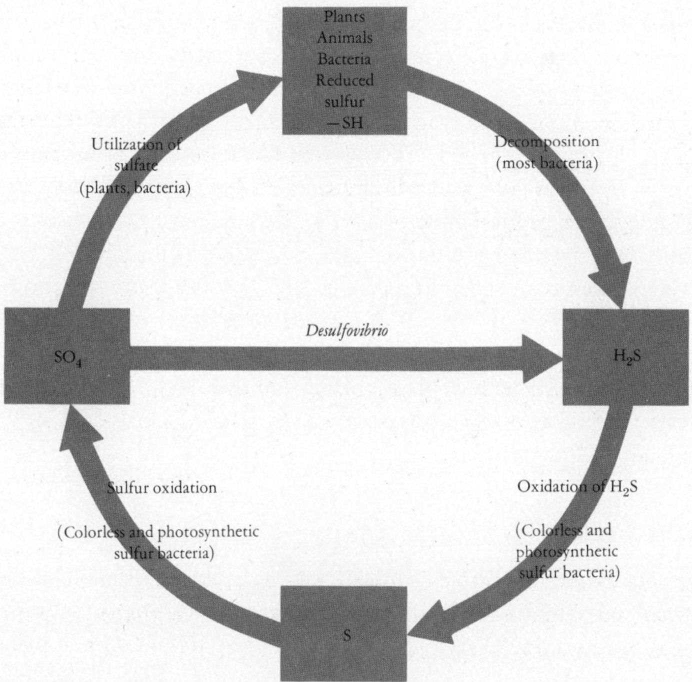

FIGURE 34-7
Sulfur cycle.

mation. The H_2S formed is highly toxic to biological systems and generally does not accumulate. Under aerobic conditions H_2S is oxidized spontaneously to sulfur and then to its most readily utilized form, sulfate, by sulfur bacteria. This process is analogous to nitrification. Under anaerobic conditions the counterparts of the denitrifying organisms operate: these sulfate-reducing bacteria reduce SO_4^{-2} to H_2S.

The microbial flora concerned with the sulfur cycle have a number of unique features. The oxidation of H_2S to SO_4^{-2} is carried out principally by two major groups of organisms, the nonphotosynthetic autotrophs, members of the genera *Thiobacillus* and *Beggiatoa* and, less commonly, by the photosynthetic autotrophs (green and purple sulfur bacteria).

Thiobacillus, page 271

Beggiatoa, page 271

The reduction of sulfate to sulfide (sulfate reduction) is carried out by a small group of anaerobic bacteria which are capable of utilizing sulfate as the electron acceptor in their anaerobic respiration. These anaerobes include organisms of the genus *Desulfovibrio* as well as a few anaerobic spore-forming rods. Hydrogen sulfide is thus likely to accumulate in waterlogged soils. Since hydrogen sulfide is toxic to higher plants, sulfate reduction is undesirable from an agricultural point of view.

BACTERIAL FERTILIZERS

In addition to the vital role they play in the recycling of elements, bacteria seem to be capable of increasing plant growth if seeds or seedling roots are immersed in

Gibberellins, page 685

a suspension of bacteria. This effect is most likely a result of plant-growth promoting hormones, such as gibberellin and indole acetic acid, which are produced by bacteria as they grow around the roots of plants in the soil. There are other possible ways in which bacteria may improve plant growth. These include the recycling of elements discussed above as well as the control of root diseases. A number of different bacteria, including species of *Azotobacter, Bacillus, Pseudomonas,* and *Clostridium,* promote plant growth. The field testing of bacterial fertilizers has not always yielded positive and reproducible results, and the increase in yield is never more than about 10 percent. Yet there are many situations in the world where an increase of this magnitude has far reaching consequences; therefore, studies on bacterial fertilizers will certainly continue in an effort to understand the role of the bacteria in plant growth and the conditions for maximizing their effect.

PLANT PATHOGENS

Although plants are susceptible to infection by a variety of agents including bacteria, viruses, and nematodes, the fungi are by far the most important plant pathogens. One recent estimate states that in Ohio alone 100 diseases of plants were incited by fungi, about 50 by bacteria, and 100 by viruses. Over 10 percent of the total cash yield of crops has been estimated to be lost through plant diseases.

Fungal Diseases

The species of fungi that cause plant diseases are found among all fungal groups: Phycomycetes, Ascomycetes, Basidiomycetes, and Fungi Imperfecti. Some of the more common diseases caused by members of each group are listed in Table 34-5.

The most important groups of fungal pathogens, as far as their economic importance to man is concerned, are the rusts and smuts. These fungi,

TABLE 34-5
Some Fungal Diseases of Plants

Disease	*Host*	*Group of Fungus Responsible*
Brown spot	Corn	Phycomycete
Soft rot	Sweet potato	Phycomycete
Powdery scab	Potato	Phycomycete
Chestnut blight	American chestnut	Ascomycete
Ergot	Rye	Ascomycete
Scab	Apple	Ascomycete
Powdery mildew	Rose	Ascomycete
Peach leaf curl	Peaches	Ascomycete
Dutch elm	Dutch elm	Ascomycete
Scab	Peach	Deuteromycete
Late blight	Celery	Deuteromycete
White pine blister rust	White pine	Basidiomycete
Leaf rust	Wheat	Basidiomycete

Basidiomycetes that lack a conspicuous fruiting body, are represented by over 20,000 species of rust and over 1,000 species of smuts. They parasitize most of the important crops, forest and ornamental plants. They are named after the appearance of the spores that appear on the host plant. Rusts, in particular, have an extremely complex life cycle, in some cases requiring two unrelated hosts for the completion of their normal life cycle.

Fungal diseases have virtually destroyed several plant species in the United States. An Ascomycete native to the Orient, accidentally introduced into the United States, has completely wiped out the American chestnut tree in parts of the country. It has now invaded Europe and threatens the European chestnut with the same fate. Another Ascomycete, accidentally introduced into the United States from Europe, is slowly but relentlessly exterminating the American elm. The fungus is carried by a bark beetle which is susceptible to DDT. However, the use of this agent has now been banned because of its harmful effects on birds and perhaps on other forms of life. Fortunately, a new antifungal agent has recently been developed which appears to hold promise in curing elm disease in trees in which the disease has not progressed very far.

Bacterial Plant Pathogens

Members of the genus *Erwinia,* Gram-negative facultative anaerobes, represent a heterogeneous collection of organisms, often characterized by their association with plants. *Erwinia amylovora* causes a soft rot of the fleshy parts of a number of plants. The organism produces an extracellular enzyme that hydrolyzes pectin, a "cement" that holds plant cells together. *Erwinia amylovora* also causes a wilting disease of plants.

Agrobacterium tumefaciens, a Gram-negative, flagellated, rod-shaped organism, is closely related to some members of the genus *Rhizobium.* The interesting feature of this relationship is that whereas *Rhizobium* infects plants and causes the formation of root nodules, (Figure 34-8a) *A. tumefaciens* causes a tumorlike growth in plants called crown gall tumor (Figure 34-8b). In crown gall tumors, destruction of the bacteria does not stop the development of the tumor once it has been initiated, suggesting that living bacteria are not needed for tumor development. Recent experiments have shown that a small fragment of a plasmid which is present in all virulent *A. tumefaciens* strains is transferred from the bacterial cell to the plant, and these few genes are responsible for the plant tumor. This points to a fundamental similarity between these plant tumors and animal tumors caused by the integration of viral genes into the chromosomes of animals. Significantly, if the plasmid of *Agrobacterium* is transferred to *Rhizobium,* this organism attains the ability to cause crown gall tumors.

Other bacterial diseases of plants are listed in Table 34-6.

Plant viruses,
pages 347–349

Viral and Viroid Diseases of Plants

A large number of plant diseases are caused by viruses that are often found in soil (Table 34-7). Plant viruses have been discussed in Chapter 17.

Viroids,
pages 18, 349

(a)

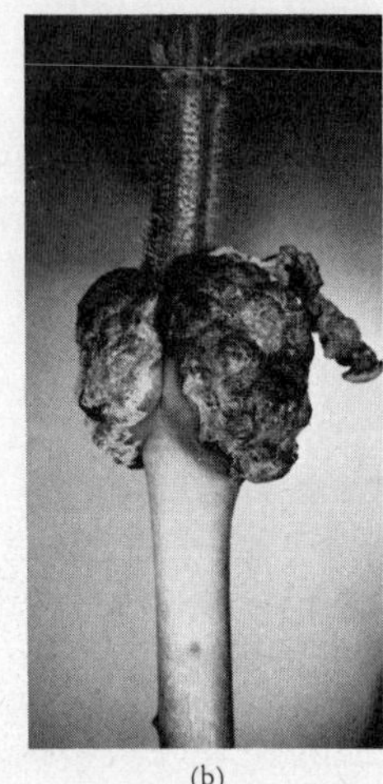

(b)

FIGURE 34-8
(a) Root nodules resulting from infection of roots of a leguminous plant by *Rhizobium.* The host plant must be a legume. (The scale represents 1 cm.) (Courtesy of F. J. Bergersey.) (b) Crown gall tumor induced by *Agrobacterium tumefaciens* on the stem of a tomato plant.

TABLE 34-6
Some Bacterial Diseases of Plants

Disease	*Host*	*Species Responsible*
Crown gall tumor	Most deciduous plants	*Agrobacterium tumefaciens*
Bacterial wilt	Tobacco Tomato	*Pseudomonas solanacearum*
Bacterial spot	Tomato	*Xanthomonas vesticatoria*
Bacterial canker	Tomato	*Corynebacterium michigense*
Fire blight	Apples, pears	*Erwinia amylovora*

TABLE 34-7
Some Viral and Viroid Diseases of Plants

	Disease	*Natural Hosts*
Viral	Tobacco mosaic disease	Tobacco Tomato
	Cucumber mosaic disease	Cucumber Delphiniums Lupins Dahlias
	Wound tumors	Clover
Viroid	Potato spindle tuber disease	Potato Tomato
	Citrus exocortis disease	Citrus fruits

SUMMARY

The soil is the natural habitat of a variety of organisms: viruses, bacteria, algae, fungi, protozoa, mites, and nematodes, to name the most prominent. These organisms play both useful and destructive roles. Their useful roles include the decomposition of organic matter as part of the cycling process of various chemical elements, such as nitrogen, carbon, oxygen, phosphorus, and sulfur.

Most compounds generated by biological processes are degradable by living organisms. The more complex molecules, such as certain aromatic compounds, are degraded by only a very few species of organisms. Simpler structures are generally capable of being decomposed by a large number of different organisms. In general, animal flesh is degraded by bacteria and plant material by fungi. A large number of complex hydrocarbons, which often contain chlorine atoms and are chemically unrelated to naturally occurring compounds, are nonbiodegradable. Many of these compounds, commonly found in pesticides, are fat soluble and accumulate in the fat of organisms in the food chain, with harmful effects in some cases.

The elements comprising living matter must be continuously recycled. This recycling process involves a variety of microorganisms that convert the elements in the organic molecules in decaying matter into inorganic forms which are utilized by plants for the synthesis of their own organic molecules. Fixed nitrogen is being continually lost from the soil through the action of denitrifying organisms and so must be returned. Nitrogen-fixing organisms, both symbiotic and free-living, are able to convert nitrogen gas into ammonium, which is then converted into plant protein.

The most important plant pathogens are the fungi, and several species of trees in the United States have been almost completely eradicated by fungal disease. Viruses are the next most important plant pathogens, whereas bacteria are the least important.

QUESTIONS

REVIEW

1. Consider the oxygen and nitrogen cycles. Which steps in these cycles require energy; which provide energy to the cell?
2. In the symbiotic fixation of nitrogen, what does the plant supply to the bacteria, and what do the bacteria supply to the plant?
3. What are the major groups of organisms in the soil?
4. What major groups of microorganisms are pathogenic for plants? Give an example of a disease caused by each group.
5. List some of the genera of nonsymbiotic nitrogen-fixing microorganisms.

THOUGHT

1. A farmer is not pleased if heavy rains occur soon after he adds nitrate fertilizer to his soil. Why?
2. The major advantage in using nonbiodegradable pesticides has led to the disuse of these substances in many cases. Explain.

FURTHER READING

Brill, W., "Biological Nitrogen Fixation," *Scientific American* (March 1977).

Cloud, P., and A. Gibor, "The Oxygen Cycle," *Scientific American* (September 1970).

Deevey, E. S., Jr., "Mineral Cycles," *Scientific American* (September 1970).

Scientific American (September 1976). The entire issue is devoted to food and agriculture. An article of special interest is "The Cycles of Plant and Animal Nutrition" by J. Janick, C. Noller, and C. Rhykerd.

Tsutomu, H., *Microbial Life in the Soil: An Introduction.* New York: Marcel Dekker, 1973. The most up-to-date introductory text, covering the full range of topics associated with microbial life in the soil. It is a rather advanced, concisely written book which refers to numerous original research papers.

AQUATIC MICROBIOLOGY

And if from man's vile arts I flee
And drink pure water from the pump;
I gulp down infusoria,
And quarts of raw bacteria,
And hideous rotatorae,
And wriggling polygastricae,
And slimy diatomacae,
And various animalculae
Of middle, high and low degree.

—WILLIAM JUNIPER, *The True Drunkard's Delight*

Virtually all water contains living organisms, although not always the varieties described in the quotation above. Some are accidental contaminants, but others are aquatic organisms that normally live in predominantly aqueous environments. These unseen, often unsuspected creatures play vital roles in the earth's aquatic ecosystems and in the geochemical cycles of the earth at large. In fact, human life could not survive for very long without the benefits derived from these organisms' activities. For example, those aquatic microorganisms that are photosynthetic transform energy from the sun into chemical energy, and most of them also produce oxygen, essential for animal life. Others that are heterotrophic decompose wastes and participate in cycling of the elements essential for life. This chapter is concerned with the nature of aquatic microorganisms and their environments, the complex interactions that occur within aquatic ecosystems, and the effects of water pollution, both on these ecosystems and on the larger environment.

THE NATURE OF AQUATIC ENVIRONMENTS

Water has several unique properties that help to make it a necessary part of the environment for many microorganisms and that also contribute to its essential functions within all living cells. For example, water can adsorb to colloids and particles by forming hydrogen bonds; it can absorb light of certain wavelengths, an important property for photosynthetic aquatic microbes; and it is an extremely efficient solvent, a property that is of paramount importance for all organisms, whether they are terrestrial or aquatic.

Because water is such an excellent solvent, natural water supplies are actually dilute solutions of many substances. There is probably no truly pure water, even in laboratories where multiple distillations yield highly purified samples, because minute amounts of the components of glass or other containers dissolve in the distillates. The expression "pure water" is generally used to refer to water that is safe to drink but far from pure in the chemical or microbiological sense.

Rain has fewer impurities than water from other natural sources (such as rivers or lakes) because it is a distillate which becomes contaminated only by the material suspended in air. As indicated in Table 35-1, samples of rainwater have been found to contain very small quantities of dissolved nutrients. As compared with rain from a nonindustrial area, rainwater collected in industrial regions may be considerably richer in dissolved materials, especially sulfate, ammonium ions, and other materials that are potential nutrients for microorganisms. It may also contain more toxic substances. When rainwater reaches the ground, it becomes further enriched with dissolved materials.

The composition of fresh water reflects its source, which may be either surface waters (such as streams, rivers, and lakes) or underground water (frequently simply called *groundwater*). Gases such as CO_2 or methane may be dissolved in groundwater. The presence of various gases dissolved in water helps to determine the microbial flora that can grow in it.

The course of water through underground strata changes its properties in ways other than the addition of dissolved gases and other nutrients. Hot springs owe their increased temperatures to passage of groundwater through deep strata of the earth's crust. Ordinarily, water temperature increases about 1°F for every

TABLE 35-1
Approximate Amounts (mg/liter) of Various Substances in Natural Water Supplies

Water Source	CO_3^{2-}	SO_4^{2-}	Cl^-	NO_3^-	NH_3	Mg^{2+}	Ca^{2+}	Na^+	K^+	*P(Total)*
Rain	—	2.0	0.5	0.2	0.5	$\geq$0.1	0.1–10.0	$\geq$0.4	$\geq$0.03	—
Fresh water										
Wisconsin										
soft waters	69.6	20.5	9.9	64.0	158.0	37.7	46.9	10.9	4.8	23.0
River (mean)	73.9	16.0	10.1	—	—	17.4	63.5	15.7	3.4	—
Seawater	73	2,712	19,353	0	0	1,294	413	10,760	387	0.03

Source: Data compiled from G. E. Hutchinson, *A Treatise on Limnology*, vol. 1 (New York: John Wiley & Sons, 1957); D. R. Kester, I. W. Duedall, D. N. Conners, and R. M. Pytkowicz, *Limnology and Oceanography* 12:176–179 (1967); and T. D. Brock, *Principles of Microbial Ecology* (Englewood Cliffs, N. J.: Prentice-Hall, 1966).

50 ft of depth. Hot spring waters often contain large quantities of dissolved substances, including sulfur compounds such as H_2S.

The temperature of natural waters can range from approximately 0°C to nearly 100°C. It is amazing to find that various species of microorganisms can survive at any temperature within this range. Even in boiling geyser basins where the water temperature is greater than 90°C, species of *thermophilic* bacteria adapted to grow in this environment have been selected. At the other extreme, certain fungi and bacteria (*psychrophiles*) can grow near or even below freezing temperatures.

The pH of natural waters, determined by their content of dissolved substances, can also vary greatly. For example, the acid springs in Yellowstone National Park generally have a pH between 2 and 4; nevertheless, they support the growth of microorganisms adapted to a very low pH. Other natural waters are extremely alkaline, with a pH greater than 9; these are also the natural habitat of some microorganisms. No matter how inimical the aquatic environment may seem, some kind of microbe is able to survive or even thrive in it as long as nutrients sufficient for multiplication are present.

Freshwater rivers and lakes are commonly rich in nutrients (Table 35-1). Of course, there is considerable variation from one body of fresh water to another, depending on factors such as the geochemical character of an area and the nature and amount of effluents added to its waters. Thus in areas where calcium salts are plentiful in the soil, the calcium content of lakes may be considerably higher than the average figure stated in Table 35-1.

As water follows its course to the sea, it becomes more and more enriched with salts, leading to differences in the salinity of various bodies of water. Seawater is relatively high in salt content (containing on the average about 3.5 percent salt as compared with less than 0.05 percent for fresh water) but poor in certain other nutrients, such as phosphate and nitrate, which are found in more abundance in fresh water. Seawater therefore supports the growth of *halophilic microorganisms* but usually contains fewer organisms than fresh water. The medium described in Table 35-2, used to grow marine microbes in the laboratory, was designed to approximate seawater. Its formula indicates the proportions of the various nutrients and salts commonly found in seawater.

HALOPHILES

An unusual situation occurs in the salt lakes found in hot and dry regions of the earth. Here inflowing water with its dissolved ions is trapped because there are no outlets. As water flowing into the lake evaporates, more enters, keeping the water level fairly constant but increasing the content of dissolved salts. A familiar example is the Great Salt Lake in Utah, where the salinity is considerably greater than that of seawater. This sort of environment restricts the growth of all except extremely halophilic organisms.

Other examples of specialized aquatic habitats in which microorganisms have become adapted to an unusual milieu include sulfur springs and mineral springs, which contain unusually high concentrations of iron, magnesium, or other minerals.

Within the wide continuum of aqueous environments ranging from fresh

TABLE 35-2
Formula for Artificial Seawater Used for Cultivating Marine Organisms

Salts	*Grams per Liter*
NaCl	19.40
$MgCl_2$	8.78
Na_2SO_4	3.25
$CaCl_2$	1.23
KCl	0.55
$NaHCO_3$	0.16
KBr	0.08
H_3BO_3	0.02
$SrCl_2$	0.02
NaF	0.0025

Source: Adapted from D. R. Kester, I. W. Duedall, D. N. Conners, and R. M. Ptykowicz. *Limnology and Oceanography* 12:176–179 (1967).

water to salt lakes, tremendous varieties of microbial species exist and interact. Any body of water represents a complex ecosystem. Energy can enter the system in the form of light, being converted to chemical energy by photosynthetic primary producers, either algae or cyanobacteria (in aerobic conditions) or bacteria (in anaerobic areas). The primary producers require inorganic nutrients, utilizing CO_2 as their carbon source. They serve as the first step in the food chain, providing food for the protozoa and small invertebrates that in turn serve as sustenance for fish.

In addition to dissolved inorganic nutrients, preformed organic substances also find their way into water in the form of discarded wastes, dead leaves, animal corpses, and the like. These are transformed into inorganic materials by the action of decomposers, including both bacteria and fungi. The decomposers are also part of the food chain, serving as nutrients for heterotrophic protozoa and invertebrates. A small part of the organic material in water is not decomposed but sinks to the bottom to become part of mud sediments or to be transformed eventually into coal or oil.

Oxygen is one of the most important limiting elements within aquatic environments because of its low solubility in water. Decomposers, such as the pseudomonads and species of *Cytophaga, Caulobacter, Hyphomicrobium,* and other genera, use the available dissolved O_2 for oxidizing organic materials. The aerobic metabolism of many microbes can therefore lead to a depletion of O_2 and consequently to conditions sufficient for anaerobic growth. Rivers are continuously aerated by the flow of water and are not as readily depleted of O_2 as are lakes. During warm weather, lakes can actually become stratified into aerobic and anaerobic layers. Colder, denser water sinks to the bottom and is rapidly depleted of O_2, while warmer water layers at the top are replenished with O_2 from the air. Water samples taken from various levels of such a stratified lake contain quite different populations of microorganisms. Aquatic environments can also become

EUTROPHICATION

anaerobic following *nutrient enrichment* (*eutrophication*).

Many aquatic microorganisms have unique capacities for exploiting their environments. The concentration of dissolved nutrients in water is frequently low compared with that in solid media; thus organisms which have special abilities have been selected for aquatic survival. Many algae and aquatic bacteria, for example, can attach to stones, plants, or other solid surfaces and continually withdraw nutrients from the water flowing past them. Others float in the plankton and are moved about so that they are able to reach more nutrients than they would if they were attached.

The Microbial Flora of Fresh Water

The microbial flora of fresh water varies considerably with the kind of water being considered. For instance, samples of tropical rainwater have been found to contain relatively few microorganisms which had been airborne in dusts and subsequently washed from the air by rain. A majority of the organisms found were actinomycetes and other soil bacteria, with fewer fungi and very few algae. After rainwater falls to the ground, it soon acquires more microorganisms of various kinds, depending on the nature of its new environment.

Algae, cyanobacteria, and protozoa make up the bulk of the microbial mass of fresh water. The species of algae and cyanobacteria in lakes, and their abundance at different seasons, varies considerably. Algae and protozoa have been discussed in some detail in Chapters 13 and 15, respectively, and are not further considered here, except to stress that they are abundant in aquatic environments and are a predominant and an integral part of aquatic ecosystems.

Bacteria are also essential to aquatic ecosystems. In addition to animal pathogens and other terrestrial species that are passively carried into rivers and lakes, bodies of fresh water contain species of bacteria that grow in nature only in such an environment. The population of the freshwater lake described in Figure 35-1 provides a good example of some bacteria native to freshwater habitats. Lakes can be compared with cities, and their bacterial flora with city dwellers; various groups settle where the environment is suitable for their existence, and there is thus a constant shifting as the environment changes. To carry the analogy one step further, just as city dwellers often cluster around the industrial plants that employ them, bacteria also tend to live in areas of a lake where their particular capabilities can be used to support their growth.

Thus aerobic bacteria that can degrade organic materials are usually found near the surface of lakes, where the oxygen content is highest, or in other aerated portions of the lakes; here they can act on animal wastes and plant or algal remains in the water. For example, pectins and cellulose from dead algae are broken down by the metabolic activities of heterotrophic aerobic bacteria, such as some of the pseudomonads. Cellulose from dead plants and algae is also often degraded by species of *Cytophaga.* These bacteria produce a cellulose-degrading enzyme that is associated with their cell surfaces so that the bacterial cells must actually make contact with the cellulose in order to act on it.

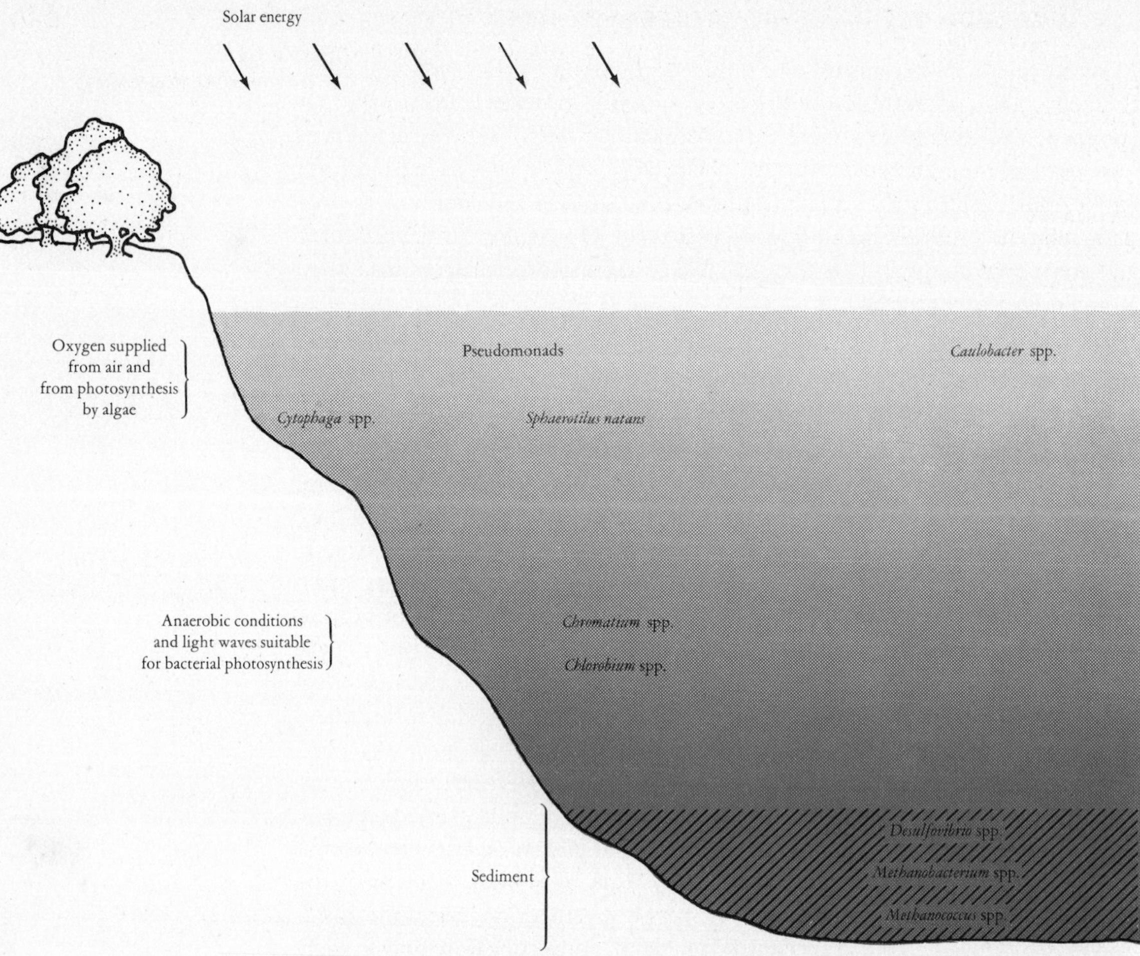

FIGURE 35-1 Representative bacteria in a freshwater lake. The species and their relative numbers varies considerably from one lake to another, and in a single lake from one time to another.

Deeper in lake waters, where oxygen is depleted and anaerobic organisms can grow, the microbial flora is quite different from that of the aerobic upper layers. In areas where light of the proper wavelength penetrates, photosynthetic bacteria prosper within the limited stratum where conditions are favorable to them. Here the purple and green sulfur bacteria, such as *Chromatium* and *Chlorobium,* oxidize H_2S gas dissolved in the water, using photosynthesis as their system for generating energy. Elemental sulfur or sulfate ions are produced as a result of this process.

Among the many anaerobic bacteria that are active in the sediment of lakes are members of the genus *Desulfovibrio,* which reduce sulfate ions to sulfide ions. Some of the sulfides remain in the mud, but others become dissolved in the lake water as H_2S, which can be oxidized by other microbial species, leading to a cycling of sulfur from one form to another.

Other members of the flora of mud in lake sediments are active in the anaerobic decomposition of settled organic materials. Species of *Methanobacterium* and *Methanococcus,* for example, produce methane from organic substances, and some of the clostridia decompose nitrogen-containing organic substrates to yield amines and subsequently ammonia.

The prosthecate bacteria, also part of the normal flora of lakes and ponds, are uniquely adapted to exploit their environment. For example, species of stalked bacteria of the genus *Caulobacter* are able to live in water that is low in nutrients. It is thought that their stalk, an outgrowth of the bacterial cytoplasm and cell wall, serves as an area of absorption which permits the cell to take in nutrients more efficiently than can nonstalked cells. The appendages of *Hyphomicrobium* species may have the same function. Some bacteria possess gas vacuoles, special organelles that allow the bacteria to stratify at certain levels in a lake.

Prosthecate bacteria, page 264

Unusual concentrations of various nutrients in natural waters leads to selection of a normal flora capable of utilizing those nutrients.Therefore iron springs that contain large quantities of ferrous ions support the growth of species of *Gallionella* and *Leptothrix.* Similarly, sulfur springs are the natural habitat of both the photosynthetic and nonphotosynthetic sulfur-metabolizing bacteria.

Although only a small portion of the total microbial activity in a lake community has been mentioned here, it is apparent that the activities of freshwater bacteria are many and complex.

The Microbial Flora of Seawater

The numbers and kinds of microorganisms in seawater are restricted by several factors. In addition to the salinity of the water and the small amounts of certain nutrients present in it, the limiting factors of growth include a high pH (around 8.0), a limited supply of vitamins or other growth factors, and the intensity and wavelength of light that can penetrate the water to serve as an energy source. Thus, although the coastal zone is rich in marine microorganisms, the deeper waters of the ocean are not so well populated and the varieties of microbial species that thrive in them are fewer than are found in fresh waters.

Attempts to make even so gross a measurement as the total mass of living organisms (or biomass) per liter of seawater have not been very successful because of technical difficulties. One method that has been used is direct counting of organisms under the microscope, but this is tedious. Another disadvantage of this technique is that seawater contains particulate debris which sometimes cannot be distinguished visually from microorganisms. Then too, not all microbes survive long enough to be counted. Attempts to grow the organisms in the laboratory lead to even less accurate results because of variations in the growth requirements of the many species present. Probably the most accurate method used to determine marine biomass has been the measurement of ATP by the luciferin-luciferase enzyme system. This technique gives an estimate of the number of viable organisms in a given volume of water, based on a carbon : ATP ratio of approximately 250 for most marine microbes. The measured amounts of ATP, in micrograms per liter, have ranged from 0.3 or more near the surface of the ocean to 0.05 at a depth of a few hundred meters, and to 0.005 at depths below 1000 m. Thus, in general, the *biomass of seawater decreases with increasing depth.*

BAROPHILES

The deep sea contains remarkably few bacteria. This may result partly from lack of energy sources and from the water pressure, which increases about one atmosphere per 10-m of depth, or to as much as 1000 atmospheres (15,000 lb/sq in) in the deepest parts of the ocean. Many microorganisms cannot survive greatly increased pressures. Those which can, called *barophiles,* are the only ones found deep in the sea.

Much remains to be learned about the various species of microorganisms that make up the biomass of seawater. Certainly the algae and protozoa of plankton account for a large proportion of the mass of organisms near the surface. Bacteria are found in smaller proportions, along with a few fungi and an undetermined number of viruses. Some organisms among the marine flora are probably there only temporarily as a result of having been washed from soil or air or discarded with wastes into the sea, especially in waters of the coastal zone.

Algae, see Chapter 13

Marine algae include much of the phytoplankton of the sea, the plantlike seaweeds, part of the flora of coral reefs, and other species. It has been estimated that the mass of phytoplankton in seawater is about 10 times that of the zooplankton. Some of the genera of algae and cyanobacteria commonly found in the sea are given in Table 35-3.

Protozoa, see Chapter 15

Protozoa make up a large part of the zooplankton. They feed on bacteria in surface films and on submerged structures, such as masses of algae. Actively motile protozoa exist in open water where they can move to food supplies, whereas some of the ameboid or attached ciliated species feed on bacteria and other organic materials at the bottom of relatively shallow areas of the sea.

One reason it is difficult, if not impossible, to determine the bacterial content of seawater accurately is that most marine bacteria are found associated with particles, often less than 0.1 mm in size, in the water near the surface of the sea. These bacteria, which utilize nutrients from the particles and from the water, are usually eaten by snails and clams.

Even though it has been estimated that bacteria account for only about from 1 to 5 percent of the total biomass of the sea, these organisms are of great importance in contributing to the growth of other marine flora. Several genera of bacteria commonly found in seawater are listed in Table 35-3. Some of these organisms oxidize organic materials to CO_2 which can be used by algae, resulting in the production of O_2 through photosynthesis. The O_2 produced can in turn be utilized for additional bacterial oxidations. Equally important is the production by bacteria of vitamins (especially B_{12}) which are essential for the growth of many algae.

A number of bacterial species found in the sea are pathogenic or nonpathogenic parasites of marine animals and plants. For example, a group of vibrios (typified by *Vibrio anguillarum*) are of major importance as fish pathogens. These halophilic bacteria have been found in about 50 species of fish from at least 14 countries. Marine vibrios are not always pathogenic; many fish contain nonpath-

TABLE 35-3
Some Representative Genera of Microorganisms Commonly Found in the Marine Environment

Genus	*Characteristics*
Algae	
Asterionella	Diatom
Fragilaria	Diatom
Navicula	Diatom
Gonyaulax	Dinoflagellate
Gymnodinium	Dinoflagellate
Bacteria	
Trichodesmium	Cyanobacteria
Oscillatoria	Cyanobacteria
Nostoc	Cyanobacteria
Mycoplana	Gram-negative rods; polar flagella
Pseudomonas	Gram-negative rods; polar flagella
Vibrio	Gram-negative rods; polar flagella
Spirillum	Gram-negative rods; polar flagella

ogenic, commensal vibrio species among their normal gastrointestinal flora. It has been postulated that some of these commensal vibrios can act as opportunists causing disease following injury or stress to the fish that carry them.

Some infectious diseases of fish closely parallel human diseases in both their nature and their causative bacteria. One fish ailment resembling the human disease tularemia is caused by *Pasteurella piscicida,* a bacterial species quite similar to *Francisella tularensis,* which causes human tularemia. Mycobacteria are also pathogens of some marine and freshwater fish as well as of man; however, the species of mycobacteria infecting fish are generally distinct from those causing similar human disease and vice versa. *Mycobacterium marinum* and other mycobacteria have been implicated in fish tuberculosis. Although *M. marinum* can cause infections in human beings, the human disease they produce is quite different from human tuberculosis, which is caused by *M. tuberculosis.* It is probable that pathogenic *Pasteurella* and *Mycobacterium* species are transmitted among fish as a result of the fish eating portions of other, infected, fish. There is no evidence that either of these fish pathogens can grow in seawater outside their hosts.

Marine environments, even though they are somewhat restrictive, support complex populations of microorganisms that are essential links in food chains. In addition to their food value, marine algae are responsible for supplying a large proportion of the O_2 produced by photosynthesis. These and other functions of marine microorganisms make both these and freshwater microbes a vital part of the complicated web of interacting living creatures.

THE CONSEQUENCES OF WATER POLLUTION

Pollution of bodies of water often interferes with the functions of aquatic microorganisms. Trash-littered beaches, dead fish, unsightly and foul-smelling lakes, and epidemics of diseases caused by waterborne bacteria and viruses are only a few of the consequences of water pollution, and have been recognized for a long time. During recent years many others have become apparent.

Water pollution is a complex problem which cannot be discussed comprehensively here. Therefore, only two facets of water pollution, important in microbiology, are considered: pollution caused by pathogenic organisms and pollution which leads to eutrophication.

Pollution with Pathogenic Organisms

The idea that certain diseases can be transmitted via water was not generally recognized during ancient times. Moses and the Israelites, for example, had strict rules and laws of hygiene that must have prevented the spread of many diseases transmitted by direct contact or by means of animal vectors; however, little or nothing concerning water supplies is included in these laws. There is reason to believe that Alexander the Great, about 1000 years later, realized the dangers of contaminated water, because it is reported that he had his troops boil their drinking water.

As the years passed more and more people began to associate certain diseases

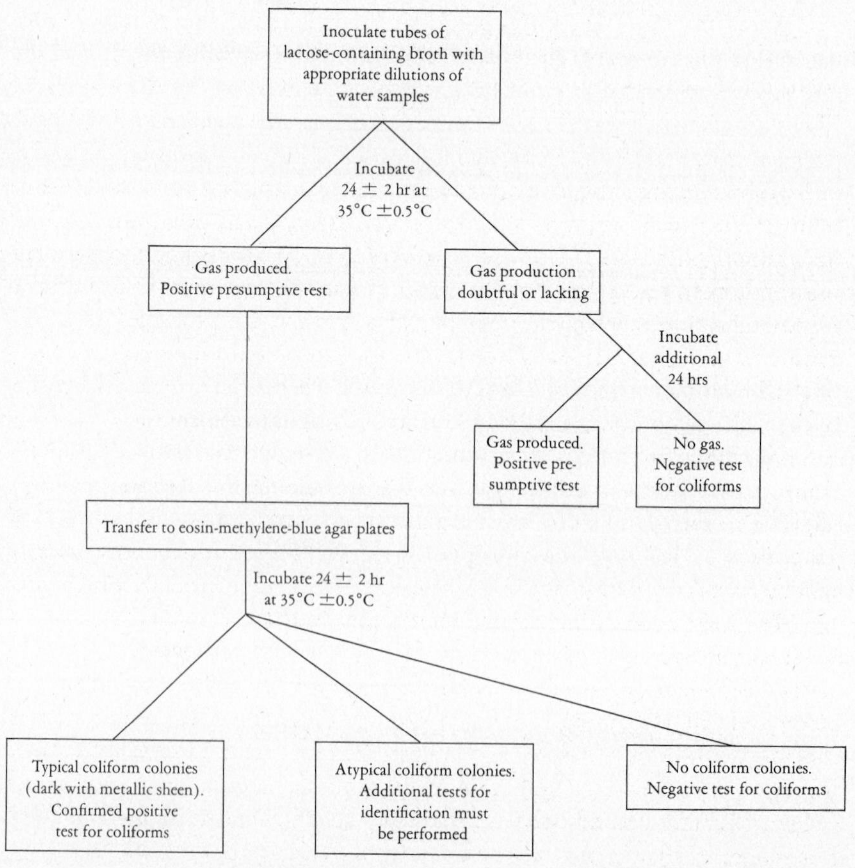

FIGURE 35-2(a)
Analysis of water for fecal contamination. Method recommended by the United States Public Health Service for testing for coliforms in drinking water. Coliform organisms are all aerobic and facultative, Gram-negative, nonspore-forming, rod-shaped bacteria that ferment lactose with gas formation within 48 hr at 35°C. (Adapted from *Standard Methods for Examination of Water and Waste Water,* 13th ed., American Public Health Association, 1971.)

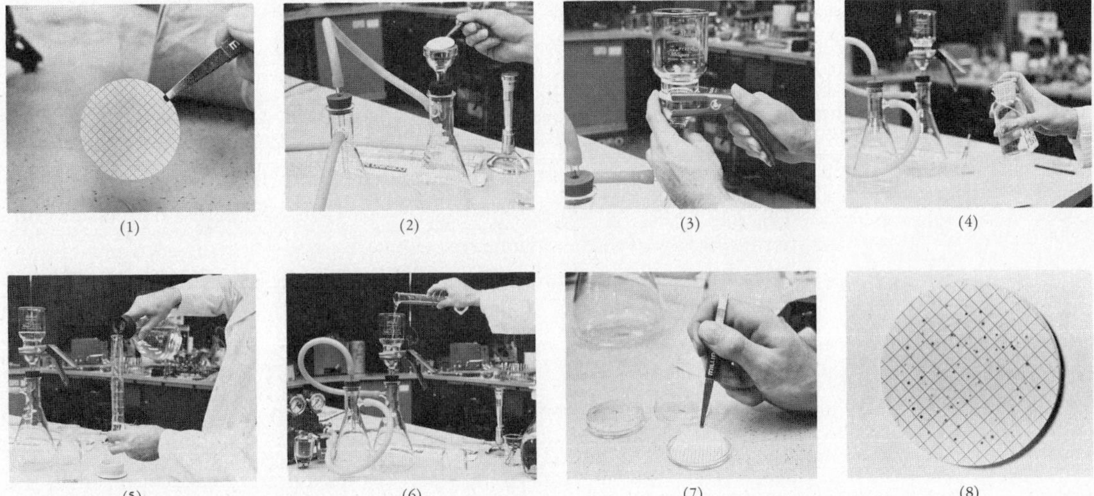

1. Sterile filter with grids for counting is handled with sterile forceps.
2. The membrane is placed on the filtering apparatus.
3. The apparatus is assembled.
4. A water sample is mixed well.
5. A portion of the sample is measured, and if necessary it is diluted with sterile diluent.
6. The portion is filtered by vacuum. Any bacteria that are present are retained on the filter.
7. The filter is placed on a dish of medium designed to select for coliforms. Nutrients can diffuse through the filter and colonies are formed on the gridded membrane where they can be counted readily.
8. The number of colonies on the filter reflects the number of coliform bacteria present in the original sample.

FIGURE 35-2(b) Analysis of water for fecal contamination. Cellulose acetate membrane (Millipore filter) method. (Courtesy of the Millipore Corporation.)

with drinking water from particular sources. In London in 1854, about 30 years before cholera vibrios were discovered, John Snow traced the source of a cholera outbreak to water from a single pump on Broad Street, and suggested that the disease had been caused by self-replicating agents in the water. As soon as methods for culturing bacteria became available, it was easy to prove that water can transmit a variety of pathogens. Thus for many years a high priority has been assigned to avoiding the pollution of water with human or animal wastes that may contain pathogenic organisms. Sewage disposal or treatment to eliminate or minimize this type of pollution has become prevalent in most economically advanced countries, and great progress toward sanitary water is being made in all parts of the world. Nevertheless large waterborne epidemics occur from time to time, even in cities with water supplies that are well monitored for fecal contamination, the usual source of pathogens in water.

The 1965 epidemic of salmonellosis in Riverside, California, illustrates several important points about the pollution of water with pathogenic microorganisms. This small city obtained most of its water from wells. The supply was monitored carefully, using a standard procedure (Figure 35-2) to detect fecal contamination. This procedure is based on the fact that fecal *Escherichia coli* is able to survive in water and is usually much more numerous than the pathogens in feces. Therefore, it is more practical to test for *E. coli* as an indicator of fecal contamination than to try to isolate the relatively few pathogens which may be present in the water. The water supply in Riverside was chlorinated only when tests indicated a significant degree of coliform contamination. At the time of the city's epidemic, coliform counts were within acceptable limits; nevertheless, more than 16,000 cases of gastroenteritis occurred. Of these, more than 70 were severe enough to require hospitalization, and three patients died. The causative organism, *Salmonella typhimurium,* was shown to be 10 times as numerous as *E. coli* in the water supply, accounting for the failure to detect dangerous contamination by using the standard tests.

The Riverside epidemic was unusual in that the causative pathogen outnumbered *E. coli* in the water, and that, consequently, fecal contamination was not detected. Waterborne epidemics are, however, by no means unique. Accidents involving city water supplies are reported each year; for example, typhoid epidemics involving hundreds of cases have occurred in recent years at Aberdeen, Scotland, and at Zermatt, Switzerland. Often the source of such an epidemic can be traced to a break in a pipe which permits fecal contamination to enter the water supply. Controlled experiments have shown that bacterial pollutants can travel more than 60 m in groundwater. In actual outbreaks of disease the pathogens of dysentery and typhoid have been shown to spread in groundwater laterally for as far as 250 m. Hence, constant vigilance is required for even the best designed and operated water supplies, in order to minimize the dangers of their contamination by pathogens.

Eutrophication

Another important microbiological aspect of water pollution is the process of nutrient enrichment known as *eutrophication.* This process leads to an increased productivity that commonly upsets the normal ecological balance and creates various kinds of problems, including unsightly algal scums, obnoxious odors, and the death of fish caused by oxygen depletion.

Algae can become a nuisance in lakes for natural reasons, but humans are frequently the culprits responsible for adding the algal nutrients that consequently cause lakes to become overproductive. Probably the material most frequently added is treated sewage which, in spite of treatment, is rich in nutrients produced by the bacterial degradation of human wastes. In recent years sewage has become an even more effective fertilizer for algae because of the large quantities of phosphate-containing detergents in sewage wastes. Because limited supplies of phosphates and nitrates are often limiting factors for the growth of algae in unenriched lakes, increased amounts of phosphates can lead to tremendous increases in the numbers of algae present, resulting in algal blooms. When these algae then die, they may create odor nuisances. As they are degraded by bacteria, the oxygen supply in deep water may be depleted so that fish cannot survive. In lakes that have excess phosphate either because of enrichment or for natural reasons, addition of other substances, such as nitrates, can cause further increases in algae.

Algal blooms, page 276

Eutrophication and other forms of pollution have been in large part responsible for causing severe deterioration of rivers and lakes such as the Hudson

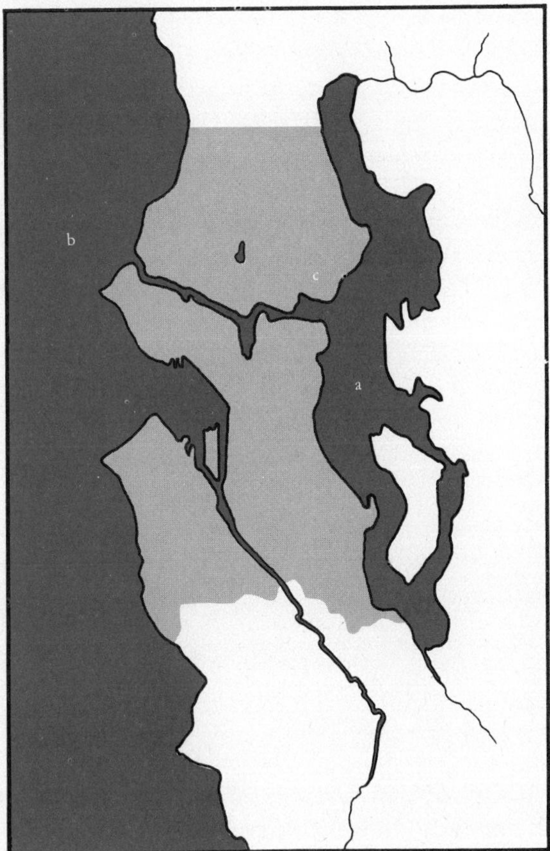

FIGURE 35-3
Lake Washington (a) and Puget Sound (b) are connected by a ship canal that intersects the City of Seattle. The University of Washington (c) is located adjacent to Lake Washington. The lake, formerly eutrophic, is now clear as a result of improved sewage disposal plans. All sewage from the city is treated and discarded into the strong currents of Puget Sound.

River and Lake Erie. Concentrated efforts are being made at present to reverse this process in many areas. One of the most extensive studies of eutrophication and its successful reversal has been made by scientists at the University of Washington who studied Lake Washington, a large freshwater lake. Figure 35-3 indicates that the city of Seattle, including the University of Washington, is situated between the saltwater Puget Sound and Lake Washington. For many years raw sewage from the city was emptied into the large lake. But during the 1930s the algal growth and odor of the lake became objectionable, and untreated sewage was diverted at sea level into Puget Sound, the very large body of saltwater on the opposite side of the city. From 1940 to 1952, however, the population around Lake Washington increased tremendously and during these years 10 sewage treatment plants were built, all of which discharged treated sewage into the lake. Each year the eutrophication of the lake increased and the once-clear lake waters became less transparent as the algae in the lake thrived. High concentrations of nitrates and phosphates in the lake's water led to rapid and abundant growth of cyanobacteria of the genus *Oscillatoria.*

Growing public concern about the deterioration of Lake Washington resulted in a vote in 1958 to reorganize the Seattle sewage disposal system. Over $120 million was allocated for the project, and by the early 1970s Lake Washington, which no longer received treated sewage effluent, had regained much of its clarity and the number of fish in the lake had increased. In addition, raw sewage from the city of Seattle was no longer discharged into the Sound; plants treated the sewage before discharging it deep into Puget Sound. The sound is large enough and has enough exchange of water with the Pacific Ocean that eutrophication is not likely to result. It is, of course, necessary to maintain vigilance to be sure that Puget Sound does not become overloaded with other kinds of pollution, such as toxic metals and oil from industrial wastes or tankers.

The Lake Washington study shows that eutrophication can be successfully reversed. It also indicates the need for informed voters who are aware of the problems involved and are willing to support the efforts necessary for the correction of this process.

MICROORGANISMS AND POLLUTION

The interaction of aquatic microorganisms which participate in biological cycles is of the utmost importance in combating water pollution. There is a kernel of truth in the misstatement that a stream or an ocean is a natural waste-disposal unit; it is true that most natural substances can be recycled by aquatic ecosystems. However, these systems cannot be overwhelmed and remain effective. There is a limit to their capacities—a limit that is often exceeded.

Microbial degradation of pollutants is not always as effective in the deep sea as it is in certain other environments. In one study it was observed that food recovered from a sunken submarine after 10 months at a depth of 1540 m was remarkably well preserved (Figure 35-4); subsequent experiments showed that the rates of degradation by microorganisms were 10 to 100 times slower in the ocean depths than at the same temperature in the laboratory. The high pressure, lack of oxygen, and low temperature of water at great depths have a tremendous influence on microbial activities. Findings such as these must be kept in mind when materials are discarded into deep seas.

Even when aquatic microorganisms are metabolically active, they do not always operate in ways that benefit man. In the case of heavy metals and other

(a)

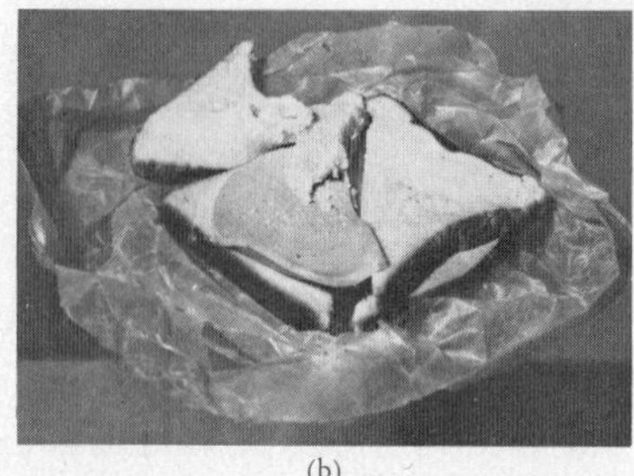
(b)

(c)

FIGURE 35-4 Lack of microbial degradation of organic materials in the deep sea. Foods from the sunken and recovered research submarine *Alvin* were strikingly well preserved after more than 10 months at a depth of 1540 m in the ocean 135 miles southeast of Massachusetts. (Courtesy of H. W. Jannasch; H. W. Jannasch, K. Eimhjellen, C. O. Wirsen, and A. Farmanfarmaian, *Science* 171:672–675, 1971.)

poisons, microbial activity may serve to transform or concentrate these substances to such an extent that they become harmful for human beings. For example, metallic mercury or mercury salts are often present in industrial wastes discarded into bodies of water. In this environment the mercury compounds are converted by bacterial action into methyl mercury, a powerful human nerve toxin. Fish concentrate the methyl mercury in their tissues to such an extent that severe disease or death can occur in human beings who eat the fish. In Japan almost 100 people have died from eating fish caught in Minamata Bay, where an industrial plant discarded its mercury-containing wastes. Many other persons have suffered severe mercury poisoning, both in Japan and in other parts of the world. For this reason fishing has been banned in some areas, such as Lake St. Clair in the Great Lakes area, and swordfish have been banned as food because of their high content of mercury.

In order to combat water pollution, people must, first of all, understand the various roles played by aquatic microbes and, equally important, cooperate with the microorganisms and not overwhelm their capacities.

SUMMARY

Water containing dissolved nutrients serves as the natural habitat of a variety of aquatic microorganisms. Water from various sources differs considerably in its content of nutrients, offering a continuum of aqueous environments, from rainwater to fresh water to salt water. Any body of water represents an ecosystem in which the microbial flora depends on the particular environmental conditions present. Therefore the microbial flora of a freshwater lake differs considerably from the flora of rainwater or from marine flora. Within a body of water tremendous differences in microbial flora are also observed, as, for example, between the ocean surface and the ocean depths.

Water pollution is a complex problem. Two important facets of this problem are pollution caused by pathogenic organisms and that which leads to eutrophication. Water pollution with pathogenic organisms is monitored by examining the water for *Escherichia coli* which serves as an indicator of fecal contamination. This method is usually, but not always, adequate; nevertheless constant vigilance is required for even the best water systems to minimize the dangers of contamination with pathogens. Eutrophication, by pollution of water with microbial nutrients, leads to increased production of large numbers of algae and cyanobacteria. When these organisms die and are degraded by bacteria, oxygen is depleted by bacterial metabolism. As a result fish and other oxygen-requiring organisms cannot survive, and the polluted body of water deteriorates. Eutrophication can be successfully reversed, but the reversal usually requires expensive and prolonged efforts.

Aquatic microorganisms which participate in biological cycles are of the utmost importance in combating water pollution. However, it is necessary for people to understand the various roles played by microbes to ensure cooperation with the microorganisms so that their capacities are not overwhelmed.

QUESTIONS

REVIEW

1. In Yellowstone National Park and other parts of the West, the following habitats are part of the scenery. Match the different kinds of organisms with their appropriate habitats (more than one kind of organism may be appropriate for a single habitat).

Habitat	*Kind of Organism*
A. Acid springs	1. Thermophiles
B. Boiling geyser basins	2. Acidophiles
C. Salt lakes	3. Psychrophiles
D. Mountain streams	4. Mesophiles
E. Freshwater lakes	5. Halophiles
F. Glaciers	6. Algae
	7. Bacteria

2. Describe the food chain in freshwater lakes, starting with the primary producers. Include the decomposers.
3. Arrange the following water samples in the order of their increasing biomass: deep ocean, surface of ocean near shore, salt lake, rain, stream polluted with sewage.
4. What is eutrophication and what can be done about it?

THOUGHT

1. It has been said that "the solution to pollution is dilution." Discuss the reasons why this is not a valid statement.
2. Water supplies are vulnerable to deliberate contamination by saboteurs. How might enemy agents use microorganisms or their products to poison water supplies, and what measures could be taken to prevent or counteract such sabotage?

FURTHER READING

ALEXANDER, M., *Microbial Ecology.* New York: John Wiley & Sons, 1971. A broad view of microbial ecology, including the oceans and fresh water as natural ecosystems.

BROCK, T. D., *Principles of Microbial Ecology.* Englewood Cliffs, N.J.: Prentice Hall, 1966. An excellent text on microbial ecology.

HEUKELEKIAN, H., and N. C. DONDERO, eds., *Principles and Applications in Aquatic Microbiology.* New York: John Wiley & Sons, 1964.

ODUM, E. P., *Fundamentals of Ecology.* 3d ed., Philadelphia: W. B. Saunders, 1971. Includes chapters on fresh water ecology and marine ecology.

SNIESZKO, S. F., ed., *A Symposium on Diseases of Fishes and Shellfishes.* Special Publication no. 5. Washington, D.C.: American Fisheries Society, 1970.

Standard Methods for the Examination of Water and Wastewater. 13th ed. Washington, D.C.: American Public Health Association, 1971.

WOOD, E. J. F., *Marine Microbial Ecology.* New York: Reinhold, 1965.

CHAPTER 36

THE MICROBIOLOGY OF WASTE TREATMENT

During the early development of civilization when communities were small and separated, disposal of wastes was not a major concern. Raw wastes could be emptied into rivers or lakes, or buried in the ground, where they were either degraded by microbial activity or vastly diluted before the contaminated waters reached other human beings. As men began to congregate to form large cities, however, the problems of waste disposal and of obtaining pure water increased; contamination of water by fecal wastes led to major epidemics of diseases caused by waterborne pathogens. These problems have been compounded during the last few centuries by increased industrialization and the consequent addition of industrial wastes to water supplies.

At present a large majority of the population of the United States is concentrated in cities that cover a very small percentage of the land. To understand what this means in terms of waste treatment, consider that every day the average American uses about 150 gal of water, 4 lb of food, and 19 lb of fossil fuel, which is converted into some 120 gal of sewage, 4 lb of trash or rubbish, and almost 2 lb of air pollutants! This means that a city with only 1 million inhabitants is faced with the disposal of 120 million gal of domestic sewage daily, to say nothing of industrial and other wastes.

In large American cities the collection of domestic sewage is usually carefully controlled. Treatment of the sewage wastes, however, is often far from adequate, leading many to wonder how long large cities can continue to exist. In many smaller towns and communities the problems of waste treatment are even less efficiently handled. As late as 1972 it was estimated that only from 30 to 50 percent of Americans lived in areas with sewage treatment, while the rest relied on cesspools or septic tanks, which are often inadequate and lead to contaminated water supplies and other forms of pollution. Even today, it is clear that sewage treatment represents major problems for both the small town and the metropolis. Nevertheless, there is increasing recognition that this tremendous liability is also a potential asset in terms of energy. In order to realize this potential, it is necessary that wastes be recycled and transformed into useful materials that are no longer a hazard to health and to the environment. In this way, the abundant resources represented by wastes can contribute to adequate energy supplies in the future.

Microbial activity is extremely important in the recycling of waste materials. The sections that follow consider the nature of sewage and other wastes and some of the major methods of waste treatment currently being used. It will become apparent that most of these methods depend on the conversion, by microorganisms, of organic materials to inorganic forms—the process of *mineralization* or *stabilization.* In general the *primary treatment* of sewage involves the physical processes of screening and sedimentation to remove from it large objects and some of its particulate material. *Secondary treatments* involve chemical and biological processes for converting as much as possible of the materials remaining in sewage into odorless inorganic substances that can be reutilized. *Tertiary treatment* involves the removal from sewage of phosphates and nitrogen compounds that could cause eutrophication.

MINERALIZATION

THE NATURE OF SEWAGE

Domestic sewage contains human wastes, which represent a source of both pathogenic and nonpathogenic microorganisms, as well as a variety of other organic materials such as food wastes and cleaning compounds. The use of garbage disposal units adds even more organic material to domestic sewage. Even so, sewage consists of only approximately from 1 to 2 percent organic solids; the rest is water. Thus the microorganisms of human wastes are greatly diluted by the water in raw sewage.

Large cities frequently produce sewage containing considerable amounts of industrial wastes, which vary widely depending on their source. Sometimes these materials are toxic for microorganisms and for other forms of life. For example, studies have shown that sewage effluent from the Los Angeles area has added to the coastal waters of California considerable quantities of such nonbiodegradable compounds as DDT and the polychlorinated biphenyl compounds, which threaten marine food sources. Heavy metals, such as mercury, cadmium, copper, nickel, and zinc, are also frequent industrial by-products. A comparison of parts per million of a few heavy metals in sewage sludge from the highly industrialized Muskegon County, Michigan, and from the relatively nonindustrial city of San Diego, California, is shown in Table 36-1. Unless special measures are taken to remove such toxic industrial wastes, they may kill or inhibit waste-degrading microorganisms, consequently destroying these organisms' ability to stabilize sewage during treatment.

Different kinds of problems arise when industries add huge quantities of

TABLE 36-1
Heavy Metals in Sewage Sludge from Industrialized and Nonindustrial Areas

	Concentration (ppm) in sludge from	
Elements	*Highly Industrialized Muskegon County, Mich.*	*Nonindustrial San Diego, Calif.*
Cadmium	1–110	0–2.5
Chromium	26–580	0
Copper	24–690	20–33
Lead	6–510	3–11
Nickel	Trace–150	0–0.75
Zinc	90–2280	67–200

organic food wastes to sewage. For example, it has been estimated that a single brewery may discard as much degradable organic material as a medium-sized city. Even though such discarded material is nontoxic, it greatly increases the load of solids that must be stabilized (or mineralized) by sewage treatment. Recent studies on waste treatment in the potato processing industry further illustrate this problem. Whereas in the past potatoes were consumed without processing (and the peelings were, perhaps, fed to livestock), between one-third to one-half of the 15 million tons of potatoes produced annually in the United States today are processed to frozen French fries, potato chips, instant potato flakes, and other products. In recent years the untreated wastes produced from such processing have accounted for from 20 percent to more than 50 percent of the potato tonnage used, and contributed a biologically degradable waste load equivalent to that of 5.5 million people. This problem was compounded by passage of the Clean Water Restoration Act of 1972, which provides that discharge of pollutants into navigable waters be eliminated by 1985. Consequently, food processors have been developing better methods for peeling potatoes and preparing other foods so as to minimize the wastes produced and to utilize those wastes efficiently.

Clearly the increase in convenience and comfort that results from industrialization has all too often produced tremendous increases in both degradable and nonbiodegradable waste materials and a consequent demand for improved methods of adequate waste disposal. The billions of dollars now being allocated by public and private sources for research and development in waste disposal have come late; nevertheless, considerable advances in this area are being made.

PRINCIPLES OF MICROBIAL DEGRADATION

During the *aerobic treatment of sewage* microbial oxidations of organic compounds yield carbon dioxide and inorganic nitrogen-containing nutrients for plants. This cycle is completed when the plants are eaten, transformed into animal products, and eventually converted into human or animal organic wastes. During the *anaerobic decomposition of sewage* similar changes occur, except that anaerobic bacteria ferment organic compounds. The products of this fermentation are subsequently utilized through anaerobic respiration. The methane bacteria are important in anaerobic degradation, converting the small breakdown products formed by other bacteria from organic carbon compounds into CO_2 and CH_4 (methane or natural gas). In the presence of hydrogen, methane bacteria can further convert CO_2 into methane as well. The remaining CO_2 can be metabolized by photosynthetic organisms and plants and the CH_4 generated by the methane bacteria can either be discarded, conserved for fuel, or oxidized to CO_2 by the activities of a few genera of aerobic bacteria, *Methanomonas* in particular.

SEWAGE TREATMENT

BOD

The term *biochemical oxygen demand* (BOD) designates the oxygen-consuming property of a waste-water sample; it is roughly proportional to the amount of degradable organic material present in the water sample. To measure the BOD, a well-aerated sample of water is incubated in a sealed container under standard conditions of time and temperature (usually 5 days at 20°C), and the amount of oxygen consumed by the microorganisms growing in the sample is measured. The higher the BOD the more oxygen is being used in biological degradation processes during the incubation. Thus high BOD values reflect large amounts of degradable organic materials in a sample of waste water or other material. For example, rich nutrient media used for laboratory cultivation of bacteria have BOD values of approximately from 2000 to 7000 mg of oxygen per liter of solution, as compared with 100 to 300 mg/liter for unpolluted natural waters. The decrease in BOD of waste water during treatment reflects the effectiveness of the treatment in converting organic wastes in the water to inorganic materials.

Effective treatment should decrease the BOD of sewage as much as possible, and should also remove toxic and other objectionable materials from the sewage. The methods of treatment chosen for sewage depend on a number of factors, including the amount of sewage material, its BOD, the presence of toxic materials, and the nature of the receiving waters (that is, the bodies of water into which the sewage treatment products are emptied).

Primary treatment of sewage is designed to remove materials which will settle, or sediment out. During this step sewage is passed through a series of screens to remove large objects such as sticks, rags, and trash (Figure 36-1a) and then allowed to settle for a period of from 90 min to 2 hr. Sometimes aluminum sulfate, ferrous sulfate, or other chemicals are added to coagulate or flocculate particles so that they settle more rapidly. After the settling period the remaining fluid is given secondary treatment, and the sedimented material from the primary tanks is usually either sent to a large tank called a *digester* for further treatment or is incinerated.

Secondary treatment of sewage (Figure 36-1b) is designed to stabilize most of the organic materials and reduce the BOD of the sewage. The mode of stabilization of sewage by populations of aerobic organisms, most often used as second-

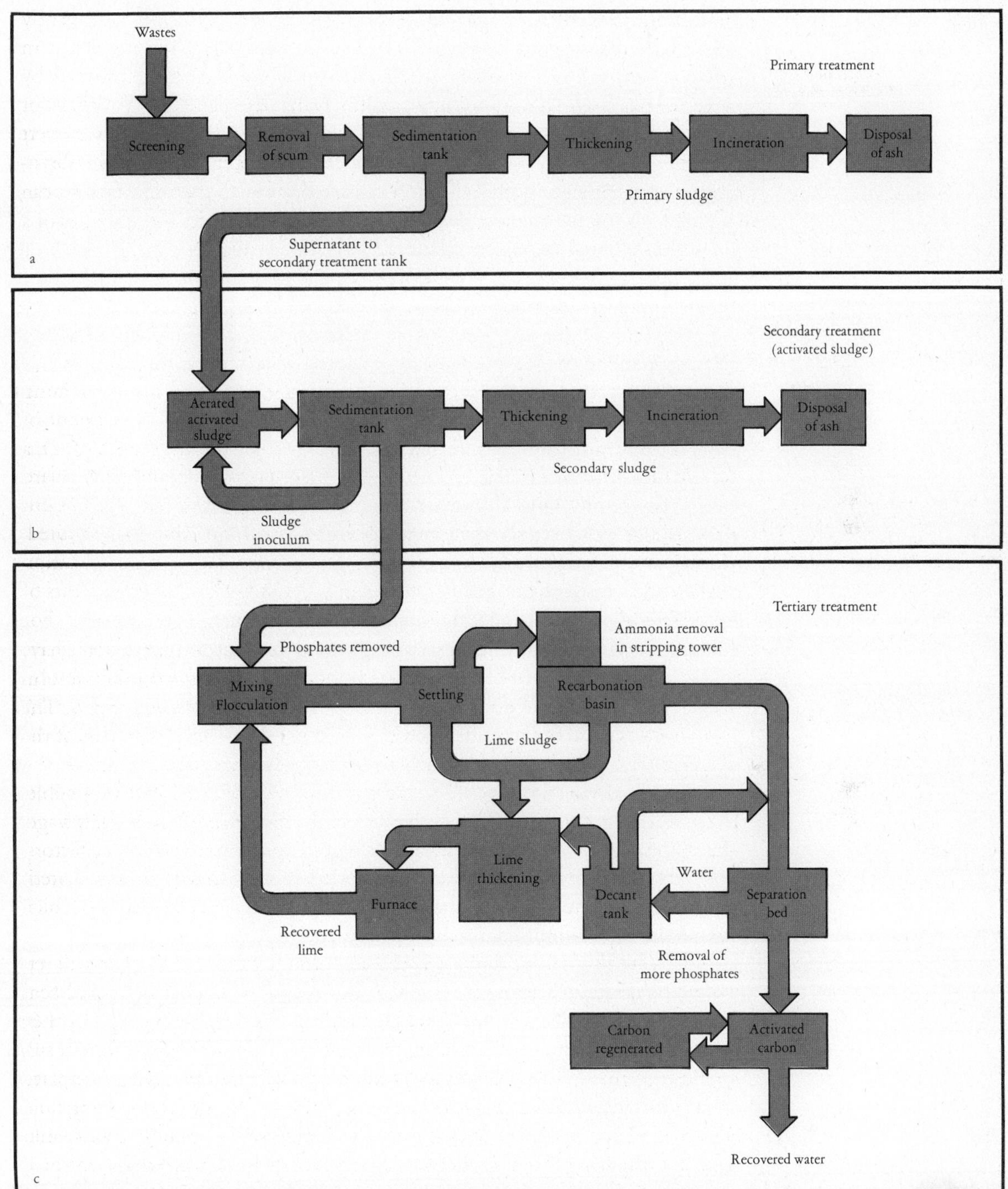

FIGURE 36-1 Metropolitan waste treatment. The most advanced waste treatment schemes utilize tertiary as well as primary and secondary treatments. (a) Primary treatment. (b) Secondary treatment. (c) Tertiary treatment.

ary treatment in larger sewage disposal plants, is called the *activated sludge method.* During this treatment the sewage serves as a nutrient source for mixed populations of aerobic organisms adapted to grow in it. An abundance of oxygen is supplied by mixing the sewage in an aerator. Most of the biologically degradable organic material is converted into gases or oxidized products, and a very small percentage is incorporated into the cell material of the organisms growing in the sewage. This process is often the most satisfactory way to treat domestic sewage, but it can be ruined by the presence of toxic industrial wastes.

Complicated ecological relationships occur during activated sludge treatment. The ecosystem in the treatment is analogous to that in natural waters, where animal corpses and other organic materials are degraded by microbial action. Many of the same microbial interactions also occur in both sewage treatment and in the self-purification of natural waters, but sewage treatment is designed to speed microbial activities under controlled conditions and in a confined space. Different microbial populations act in sequence, some degrading large organic molecules into smaller ones, and these are in turn further oxidized by still other organisms. Bacteria make up a major proportion of the microorganisms present in sewage treatment. Among those isolated in one study of activated sludge were bacteria of the genera *Acinetobacter, Alcaligenes, Brevibacterium, Caulobacter, Cytophaga, Flavobacterium, Hyphomicrobium, Pseudomonas,* and *Sphaerotilus,* and yeasts of the genus *Debaromyces.*

After the bacteria and fungi used in sewage treatment have utilized certain nutrients, these microorganisms serve as the food for ciliates, rotifers, nematodes, and other larger forms of life. The end result is a *small increase in the mass* of organisms present in a treated sewage and a *large decrease in the amount of degradable organic materials.* The increased microbial mass in the sewage is then removed to the digester, a portion being left behind as the inoculum to act on a new load of waste materials.

Within a sewage digester anaerobic organisms act on the solids remaining in sewage after its aerobic treatment. The digester provides *anaerobic stabilization* and removes water from the sewage so that a minimum of solid matter remains in it. Here again various populations act sequentially. It is here also that the anaerobic methane-forming bacteria can perform their role of converting the simple organic acids in a sewage into the useful end products methane (CH_4) and CO_2.

The greater efficiency of aerobic over anaerobic metabolism is reflected in sewage treatment by the more efficient removal of biologically degradable organic materials in the aerobic-activated sludge method of sewage treatment than in methods that utilize anaerobic organisms. Strictly anaerobic sewage treatment may reduce the BOD of a sewage by from 35 to 75 percent, while the aerobic-activated sludge treatment usually removes about 95 percent of the BOD. However, the aeration procedure necessary for the activated sludge treatment of sewage requires a considerable input of energy. Furthermore, BODs from 10 to 20 times as great as those occurring in the activated sludge method can be handled by anaerobic methods, with the result that the anaerobic treatment method can handle a greater BOD per acre per day.

Pathogenic bacteria are generally eliminated from sewage during secondary treatment; however, disease-producing viruses may survive. Pathogens account for only a very small proportion of the total number of bacteria in feces, and they are greatly diluted by the water in sewage. Most of them soon die because they are poorly adapted for growth relative to other organisms present in sewage. During secondary treatment of sewage pathogens must compete for nutrients with the huge mass of bacteria that have been adapted to grow best at the temperature and conditions provided. As a result most pathogenic bacteria are rapidly overgrown and eliminated by their competitors. Animal viruses, on the other hand, lack appropriate host cells in sewage and cannot replicate there, although they may survive for long periods. If, therefore, large quantities of virus particles are present in raw sewage, some may be recovered after secondary treatment. Furthermore, although sewage effluents are often chlorinated before being discharged into receiving waters, chlorine treatment at this stage is of little value for killing viruses, because the virus particles are commonly protected from this chemical by their enclosure within small aggregates of effluent materials.

At present many sewage treatment plants discard the residue from secondary treatment into receiving waters, often causing *eutrophication.* As noted in Chapter 35 the large quantities of phosphates or nitrates remaining in sewage after its stabilization often increase the growth of microorganisms that gradually deplete oxygen, threatening other forms of aquatic life. The tertiary treatment of sewage, to remove nitrates and phosphates, can greatly alleviate this problem.

Tertiary treatment of sewage to remove materials that remain after its secondary treatment depends on either biological or chemical steps, or on both. At South Lake Tahoe, California, over 7 million gal of sewage can be converted within one day to water pure enough to drink. In the sophisticated, three-stage treatment system used at that location, both physical and chemical methods are used for the tertiary treatment, with part of the by-products being reclaimed and solid wastes burned (Figure 36-1c).

The chemical and physical removal of sewage materials during tertiary treatment at Lake Tahoe involves a number of steps. Following secondary treatment by the activated-sludge method, lime is added to the treated water to coagulate and precipitate phosphate-containing particles. The resulting precipitates are allowed to settle in a clarification tank. Gaseous ammonia in the water, derived from the mineralization of organic compounds, is removed by passing the water from the clarification tank through a "stripping tower." In the tower rough-textured hemlock slats break the falling water into droplets; NH_3 gas escapes and is dispersed by a fan. The water is then neutralized by adding CO_2 and is passed through a series of temporary holding ponds before being filtered through a sequence of cylinders. Some of the filters, made of coal, sand, and garnet, remove remaining particulates; others, made of activated charcoal, remove detergents, pesticides, and other toxic materials. Chlorine is added to the purified water as a final step in its treatment to kill or inhibit any remaining microorganisms and to oxidize any odor-producing substances that may remain.

In some designs for the tertiary treatment of sewage, chemical precipitation of phosphates has been combined with biological removal of nitrates. Certain bacteria (particularly species of *Pseudomonas* and *Bacillus*) can reduce nitrates (NO_3^-) completely to N_2 (denitrification). The N_2 gas that is formed is inert, nontoxic, and easily removed.

Treatment of Small Quantities of Sewage

Considerable progress has been made in recent years in developing better methods for the disposal of urban wastes. At present, however, these methods are not practical for small communities or for isolated dwellings. In such settings, other means of waste disposal—which employ much the same basic principles of microbial degradation but in different ways—must be used.

Small towns sometimes depend for their waste disposal on a process called *lagooning,* in which sewage is channeled into shallow ponds, or lagoons, where it remains for from several days to a month or more depending on the design of the lagoon. During this period settling occurs and sewage materials are stabilized by anaerobic or aerobic organisms or both. Pathogenic bacteria are usually eliminated by competition, as described previously.

Trickling filters (Figure 36-2) are frequently used for smaller sewage treatment plants; however, they are also sometimes used in place of activated sludge digesters for secondary treatment. The trickling filters spray controlled amounts of sewage over beds of coarse gravel and rocks. The pebbles and rocks in the beds then become coated with a film of organisms that aerobically degrade the sewage. The film, about 2 mm thick, consists of an outer layer in which fungi predominate, a middle layer of fungi, algae, and cyanobacteria, and an inner layer composed largely of bacteria, fungi, and algae. Protozoa, especially some of the ciliates, are also present. The trickling filter is used to adjust the rate of sewage flow over the rocks so that waste materials can be maximally degraded. As is the case with the activated sludge method, different populations act in turn to degrade various compounds. Nematodes, rotifers, ciliates, bacteria and other organisms cooperate during sewage stabilization, both with trickling filters and in lagoons.

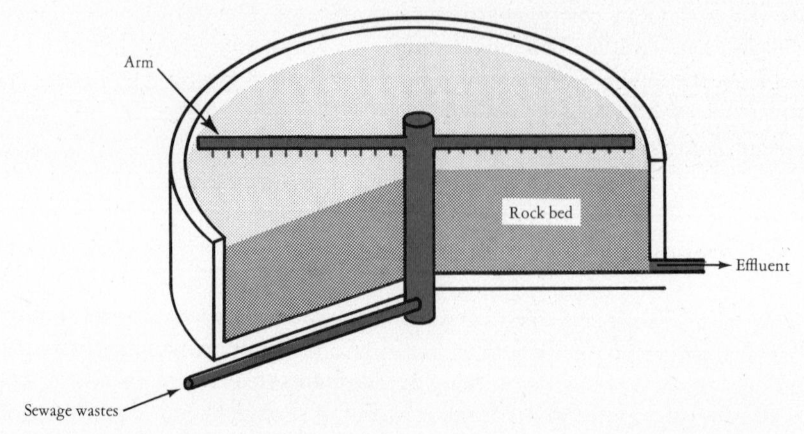

FIGURE 36-2
The trickling filter. Sewage wastes are channeled into the revolving arm and trickle through holes in the bottom of the arm onto a gravel and rock bed. The rocks are coated with microorganisms that stabilize the sewage as it trickles through the bed so that the effluent has a greatly reduced load of degradable organic materials.

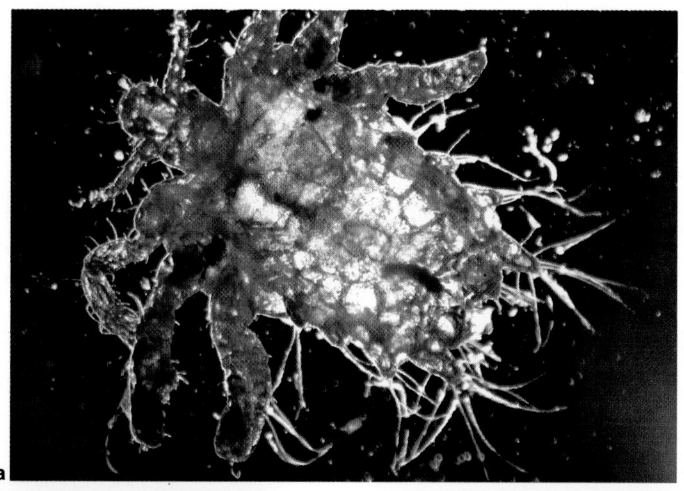
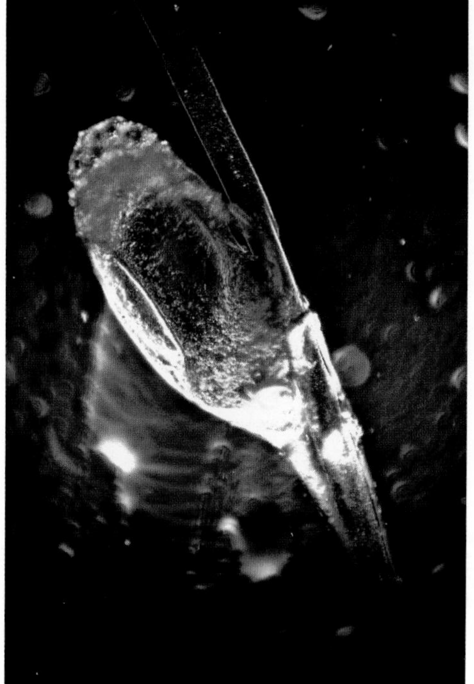
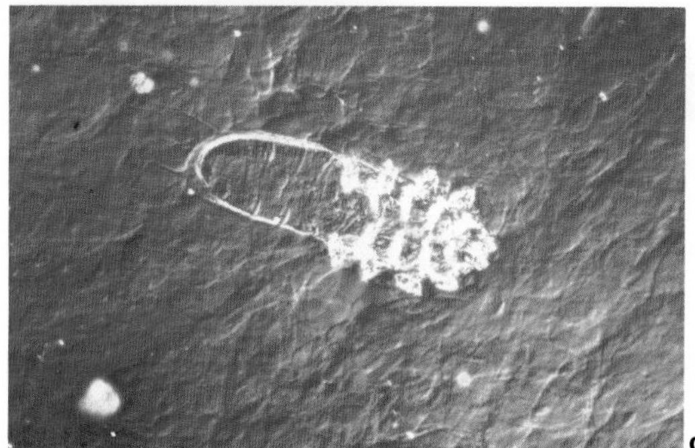
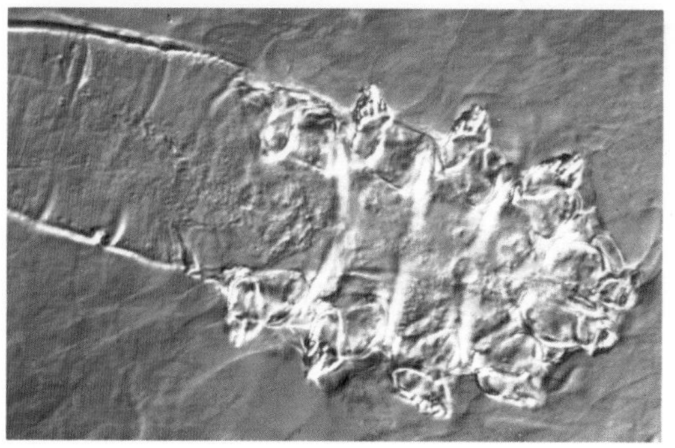

Common multicellular parasites of human beings. **(a, b)** *Phthirus pubis* (crab louse): adult and egg. **(c, d)** *Demodex folliculorum*, a tiny mite residing in hair follicles. (Courtesy of F. Schoenknecht.)

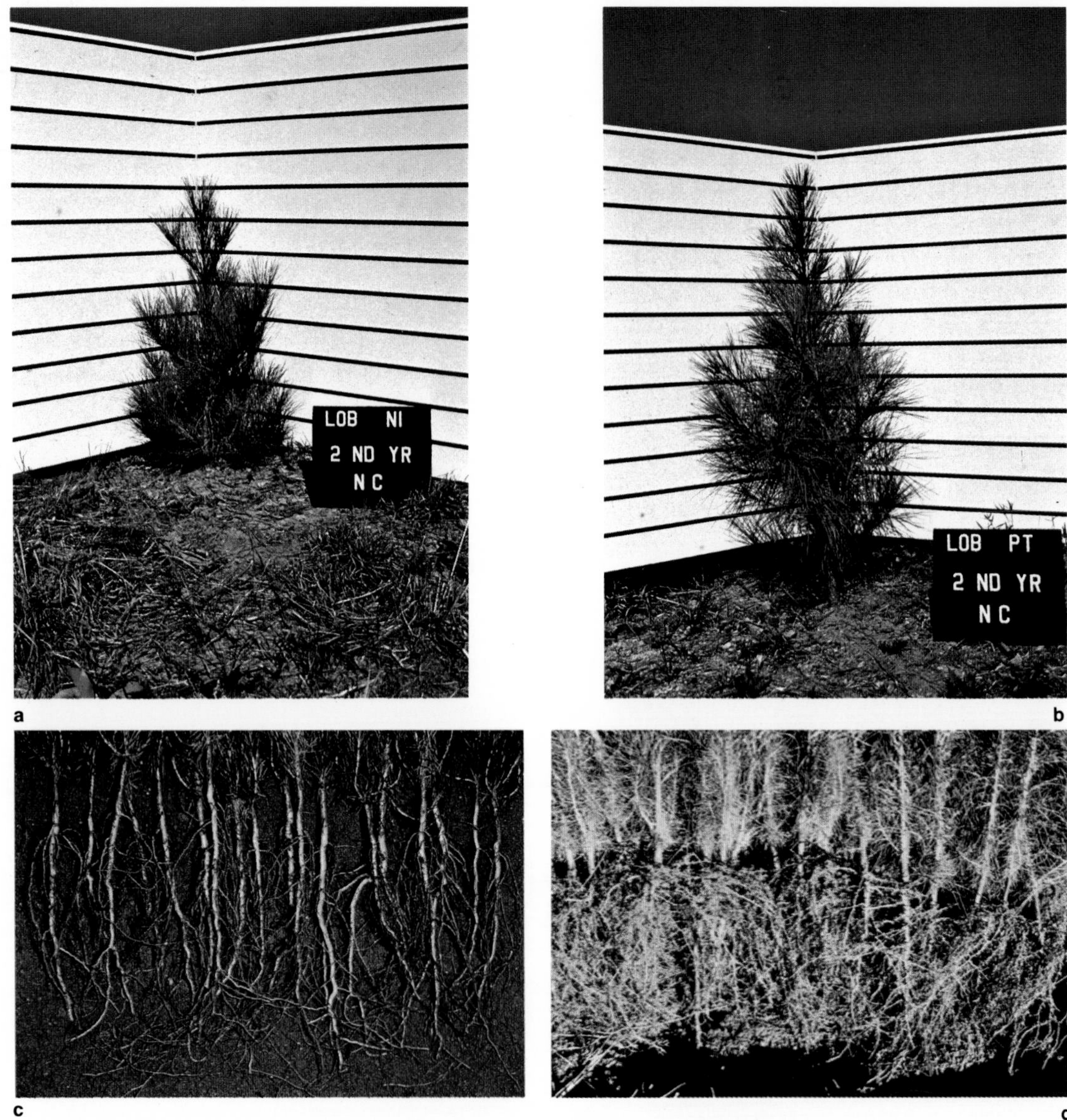

Two-year-old loblolly pines. **(a)** Not inoculated. **(b)** Inoculated with the mycorrhizal fungus *Pisolithus tinctorius*. **(c)** Roots of trees not inoculated. **(d)** Roots of trees inoculated with *P. tinctorius* showing mycorrhizal colonization. (Courtesy of D.H. Marx/USDA.)

Microbial growth in fresh water. **(a)** A sulfur spring showing a whitish growth of *Thiothrix* species. **(b)** Photomicrograph of a wet mount showing the *Thiothrix*. **(c)** A heavily polluted stream. Water draining from a landfill of municipal garbage enters the stream from the left. **(d)** Closer view of the brownish scum containing myriads of microorganisms. (Courtesy of J.T. Staley.)

a

b

c

Studying marine waters. **(a)** A Nansen bottle, a device for taking samples at different depths. **(b)** A bathythermograph for measuring temperature differences. **(c)** A plankton-sampling device. (Courtesy of J.T. Staley.)

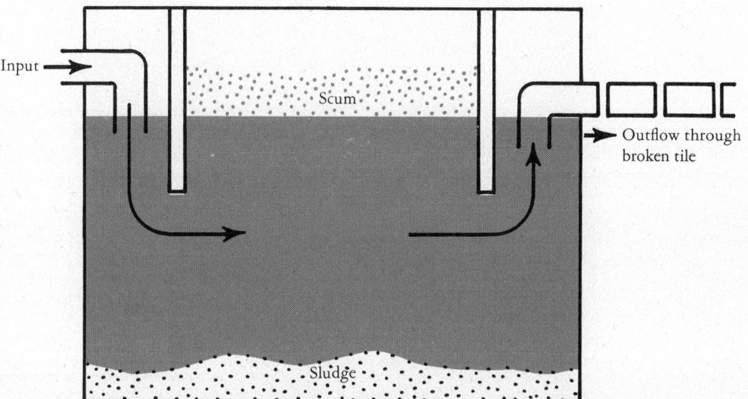

FIGURE 36-3 The septic tank. Sewage wastes enter the tank through the input pipe. Within the tank, solid materials settle and undergo anaerobic stabilization. Materials that do not settle exit through the outflow pipe through broken tile which permits seepage into the drainage field. Conditions in the drainage field must be aerobic so that materials remaining can be degraded by the activities of aerobic microorganisms. If the drainage area is not properly designed, contaminated materials readily enter adjacent surface waters.

Isolated dwellings or very small communities customarily rely on septic tanks for sewage disposal. In theory the septic tank is satisfactory; in practice, however, it often fails. Figure 36-3 illustrates the design of a septic tank. Sewage is collected in a large tank in which much of the solid material settles and is degraded by anerobic microorganisms. The fluid overflow from the tank has a high BOD and must be passed through a drainage field of sand and gravel. Theoretically stabilization should occur in the drainage field in the same manner described for the trickling filter. However, stabilization depends on adequate aeration and sufficient action by aerobic organisms associated with sand and gravel in the drainage field—conditions which are often not met. For example, clay soil under a drainage field may prevent adequate drainage, allowing anaerobic conditions to develop, or toxic materials may inhibit microbial activity in the drainage field. There is also always the possibility that the drainage from a septic tank contains pathogens; the tank must therefore never be allowed to drain where it can contaminate water supplies.

Research is presently being directed toward finding better ways of dealing with small quantities of sewage. One method that has proven satisfactory in some small communities is the *integrated pond treatment.* Good results have been obtained from this in handling sewage, especially in suburban subdivisions where land is divided into lots. By using an integrated pond system, the lot sizes of a small community can be much smaller than where septic tanks are used because individual septic tanks and drainage fields are not needed. Instead, the land can be subdivided into small lots with a common portion set aside for a series of ponds (Figure 36-4). Sewage is channeled into successive ponds, which are carefully designed to carry out both aerobic and anaerobic stabilization. By the time the water reaches the final pond, it is suitable for decorative or recreational use. This type of system depends on both proper design and adequate maintenance, and

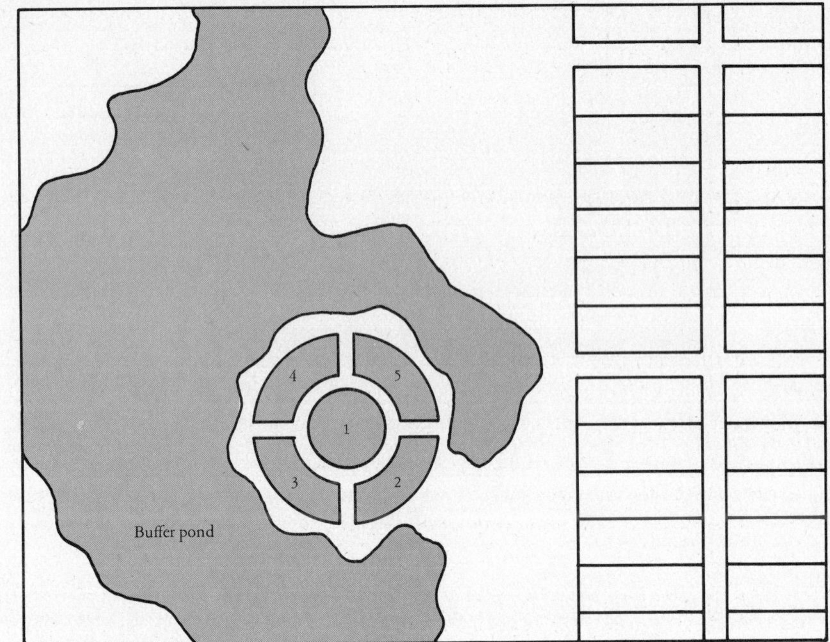

FIGURE 36-4
An integrated pond design for sewage disposal. Sewage wastes from all the lots in the subdivision (right) are pumped into the bottom of the primary pond (1) where they undergo anaerobic stabilization. Grease rises to the top and the effluent removed from near the bottom is transferred to the next pond (2). Further (aerobic) stabilization occurs in each of the other ponds. Cleaner water near the surface is transferred to the next pond in sequence (3,4,5, and the buffer pond). The buffer pond, surrounded by plantings, provides a parklike area for the subdivision. In addition, lots can be made smaller and easier to maintain because individual septic tanks and drainage fields are not necessary.

requires careful monitoring. However, in communities where it has been used it has provided safe and efficient waste disposal with the added bonus of a park area around the pond system.

Progress is also being made in many other areas of waste treatment. For example, soil filters are being developed to remove gases (such as H_2S) that are formed during sewage treatment. If the soil is kept moist so that bacteria can grow, the malodorous and toxic H_2S is removed during passage through the soil and converted by bacteria (such as *Thiobacillus* sp.) to sulfate ions.

UTILIZATION OF TREATED WASTE RESIDUES

One area of waste treatment in which research and development are greatly needed is in the utilization of treated waste residues. Clearly, receiving waters deteriorate following the addition of treated wastes, as a result of the nutrients added to them, as well as from toxic materials, and changes in their pH and temperature. Even when the best available methods of tertiary sewage treatment are used, the quality of receiving waters is decreased. For instance, although the recovered water at Lake Tahoe is pure enough to drink, it still contains some microbial nutrients. For this reason the purified water is collected in reservoirs, rather than being rechanneled into the lake.

Wastes as precursors of single-cell proteins, page 692

Increasing efforts are being made to utilize waste residues in the production of crops and single-cell protein. Some of these efforts and some of the problems involved in the utilization of treated waste residues are discussed in Chapter 39.

It is apparent that many problems in waste disposal remain to be solved.

The treatment of wastes and the utilization of waste residues currently offers one of the most challenging areas of research for microbiologists and other scientists.

SUMMARY

Microbial activity is extremely important in the recycling of waste materials and in reclaiming water from domestic and industrial wastes. Most methods of waste treatment involve the mineralization of organic wastes to inorganic materials by microbial action. Toxic products in the wastes may interfere with treatment procedures.

Primary treatment of sewage wastes involves screening and sedimentation to remove large objects and larger particulate material; secondary and tertiary treatments involve a number of methods of mineralization. The activated sludge method of secondary treatment is currently the most satisfactory means of treating large quantities of domestic wastes. Tertiary treatment consists of either physical, chemical, or microbial methods for removing and sometimes reclaiming the materials left in water after secondary treatment of sewage.

Small quantities of sewage are sometimes treated by lagooning, by use of trickling filters, in septic tanks, or in a system of integrated ponds. Better methods for utilizing treated waste residues are being sought.

QUESTIONS

REVIEW

1. What is the nature of domestic sewage? Of industrial wastes?
2. How have modern "convenience products" such as prepared foods, detergents, and insecticides affected the problems of waste treatment?
3. Outline an ideal sewage treatment system, including primary, secondary, and tertiary treatments.

THOUGHT

1. Suppose you own 150 acres of ocean-front property that you want to develop into a community of vacation homes, retaining as much as possible of the natural setting. Of course, safe and effective sewage treatment must be a part of the plan. What are the advantages and disadvantages of each of the following for such a development?

   Individual septic tanks for each building
   Trickling filter
   Integrated pond system

2. Plan a system for waste disposal that would be feasible for space travel.

FURTHER READING

BASCOM, W., "The Disposal of Waste in the Ocean," *Scientific American* (August 1974). This article presents rational approaches to the safe disposal of wastes into the ocean.

GAUDY, A. F., Jr., and E. T. GAUDY, "Microbiology of Waste Waters," *Annual Review of Microbiology* 20:319–336 (1966). A review of waste-water microbiology prior to 1966.

HOOVER, S. R., "Prevention of Food-processing Wastes," *Science* 183:824–829 (1974). A look at new ways to decrease the BOD that results from food processing.

MITCHELL, R., *Introduction to Environmental Microbiology.* Englewood Cliffs, N.J.: Prentice-Hall, 1974. Written for students of environmental engineering and the basic sciences.

ODUM, E. P., *Fundamentals of Ecology.* 3d ed. Philadelphia: W. B. Saunders, 1971.

TABER, W. A., "Wastewater Microbiology," *Annual Review of Microbiology* 30: 263–277 (1976). An updated review of wastewater microbiology.

THE MICROBIOLOGY OF FOODS AND BEVERAGES

Foods are part of an ecosystem in which microorganisms, insects, and a variety of other living creatures play important roles. This chapter is concerned in part with the activities of microorganisms during food spoilage and with the methods of preventing such microbial spoilage. Although foods are spoiled also by oxidation, enzymatic action, insects, rodents, and a multitude of other mechanisms, these are beyond the scope of this discussion. This chapter also deals with microbial activities in the preparation of foods and beverages.

SPOILAGE OF FOODS

Fruits, vegetables, dairy products, and meats all contain a variety of microorganisms on their surfaces. This is to be expected, because microbes are ubiquitous and are especially plentiful in the soil and around animals. Anyone who prepares food has the responsibility of making sure that the numbers and kinds of organisms present at the time of its consumption are such that the food is both safe to eat and appetizing.

Microorganisms Responsible for Food Spoilage

Microbial food spoilage usually results from microbial activity following proliferation of the organisms present in and on foods. For example, potentially destructive fungi are found on virtually all apples; thus the preservation of fresh apples depends in part on controlling the number of fungi present by culling rotten apples and adjusting the environment so as to discourage fungal growth.

Food microbiologists have made extensive studies of the numbers of various kinds of bacteria and fungi in food samples. Representative microbial counts for meats, fruits, vegetables, and other foods may be found in most of the references at the end of this chapter. These counts vary considerably, depending on the kind of food and on its general condition. For instance, a comparison of frozen breaded shrimp samples revealed a high of 178 million organisms per gram of shrimp in poor condition and a high of only 7 million per gram for shrimp samples in good condition. In other studies ground beef was found to contain as many as 33 million bacteria per gram. Although no standard for ground beef has been accepted, the suggested upper limits range from 1 to 10 million organisms per gram.

The microorganisms that spoil foods are usually nonpathogenic for man. This is not surprising when the conditions of their growth are considered. Foods are most often kept well below the normal body temperature of 37°C. Pathogenic bacteria, which usually grow best near 37°C, thus prefer a warmer environment than is provided under food storage conditions. Similarly, the nutrients available in fruits, vegetables, and other foods are often not suitable for the optimum growth of human pathogens, and the pH of foods is often more acid or alkaline than that of body fluids.

A number of genera of Gram-negative bacteria are important agents of food spoilage; *Pseudomonas* species, for example, are often involved in spoilage of a wide variety of foods. The pseudomonads have extensive metabolic capabilities and are hence able to grow on many different kinds of foods. Some of these organisms can utilize proteins, lipids, and carbohydrates, enabling them to grow readily on meats as well as on foods higher in carbohydrates. Psychrophilic pseudomonads that can multiply at refrigerator temperatures are especially notorious for spoiling meats and many other foods. Certain species of *Achromobacter* and *Alcaligenes* are other psychrophilic agents of food spoilage.

Members of the genus *Acetobacter* can transform ethyl alcohol to acetic acid (vinegar). Although this property is very beneficial for commercial producers of vinegar, it can be a great detriment to wine producers. In fact *Acetobacter sp.* are one of the principal spoilage bacteria of wines. Milk products are sometimes spoiled by Gram-negative rods of the genus *Alcaligenes* which cause "ropiness" of milk and make cottage cheese slimy.

Gram-positive bacteria frequently also spoil foods. In addition to some streptococci, Gram-positive food-spoilage bacteria include species of *Leuconostoc, Lactobacillus, Micrococcus, Staphylococcus, Bacillus,* and *Clostridium. Streptococcus, Leuconostoc,* and *Lactobacillus* are all lactic acid-producing bacteria. Certain strains of *Staphylococcus aureus* grow readily in high-carbohydrate foods and secrete enterotoxins that cause staphylococcal food poisoning, discussed in Chapter 27. Gram-positive rods, both aerobic of the genus *Bacillus* and anaerobic of the genus *Clostridium,* are particularly troublesome during the high temperature preparations of foods because their spores are heat resistant. *Bacillus coagulans* and *Bacillus stearothermophilus* cause sour spoiling of corn and certain other canned vegetables. Many of the clostridia putrefy foods held under anaerobic conditions; *Clostridium*

Clostridial food poisonings, pages 518–520

botulinum, however, is of primary importance because of its ability to secrete an extremely potent lethal exotoxin into anaerobically prepared or stored food (Chapter 27).

Requirements for Growth of Spoilage Organisms in Foods

All microorganisms require water in order to grow; however, the amount of moisture that permits growth varies with different kinds of microbes. Food microbiologists use the term *water activity* to designate the amount of moisture available in foods. Salted or dried foods have low water activity, or low amounts of water available for microbial growth, as compared with most fresh foods. Consequently, many bacteria are inhibited in salted or dried foods but fungi may be able to grow on these foods.

The pH of a food is also important in determining which organisms can survive and thrive in or on it. The pH may also determine whether toxins are produced by contaminating microorganisms, as is the case with *C. botulinum.*

Any medium, including food, represents a complex interaction of factors affecting microbial growth. In addition to water activity and pH, such other factors as osmotic pressure, amount of oxygen available, and temperature have a great influence on the numbers and kinds of microorganisms that will propagate. Also, the nutrients present in a food have a great deal to do with the kinds of organisms that will grow in or on it. For example, if a certain food is deficient in a given vitamin, any organism requiring that vitamin cannot normally grow in the food. The hardy pseudomonads or other bacteria capable of synthesizing the vitamin will then be able to take over and grow, if other conditions are favorable. Thus, pseudomonads often spoil foods because they have extensive capabilities for synthesizing essential nutrients, and because they can grow under a variety of conditions.

Rinds, shells, and similar coverings aid in protecting some foods from the invasion of spoilage organisms. Even so, sooner or later a microorganism with just the right properties for living on that food will break these defenses and cause spoilage.

Antimicrobial Substances in Foods

Some foods contain antimicrobial substances which may help to prevent spoilage and are sometimes useful in protecting human beings and other animals against disease. For example, egg white is rich in lysozyme. Thus, even if lysozyme-susceptible bacteria breach the protective shell of an egg, they are destroyed by lysozyme before they can cause spoilage. Other examples of naturally antimicrobial foods include cranberries, which are rich in antimicrobial benzoic acid, and fresh milk, which contains several antimicrobial substances. In fact cows' milk contains an antibacterial peroxidase system analogous to that found in human phagocytes. Such peroxidase systems provide important defense mechanisms in both cattle and human beings. Onions, garlic, radishes, and many other foods also contain substances that kill or inhibit microorganisms. There is evidence that the Egyptians who built the pyramids consumed large quantities of garlic and radishes, and it has been postulated that these foods were used to prevent or combat infection among them; however, there is no scientific evidence that the antimicrobial substances garlic and radishes contain are active in the human body.

FOOD PRESERVATION

Some methods of food preservation, such as drying and salting, have been known throughout the ages, whereas others have been discovered or developed more recently. The major methods of preserving foods are by high-temperature treatment, low-temperature storage, desiccation, chemical treatment, and irradiation. Each of these methods is discussed separately in the following pages.

High-Temperature Treatment

In 1809 a Frenchman named Appert gained fame and fortune by developing the method for canning food that, with little modification, is still used today. Although he had never heard of bacteria and other microorganisms, Appert empirically worked out a procedure for preserving foods in stoppered jars that were heated for hours in a boiling water bath.

Several general rules hold true for high-temperature food-preservation methods. First, as the temperature is increased, the time that the food must be heated in order to preserve it decreases. This fact reflects the thermal death times (TDT) of contaminating microorganisms. Table 37-1 indicates how the TDT of bacterial spores decreases as the temperature increases. Second, the concentration of microorganisms affects the conditions necessary for killing; the higher the concentration of bacteria, the more intensive the heat treatment required, if all other factors remain constant. For example, in tests carried out under exactly the same conditions, the spores in a given volume of medium containing 72 billion spores of *C. botulinum* had a TDT at 100°C of 240 min, whereas with 650,000 spores in the same volume the TDT was 85 min; but when 328 spores were heated at 100°C, the TDT was only 40 min.

A third general rule concerning high-temperature preservation of food is

TABLE 37-1
Some Representative Thermal Death Times of Bacterial Spores[a]

Bacterial Species	*Thermal Death Time at 100°C (min)*	*Thermal Death Time at 120°C (min)*
Bacillus subtilis	15 to 20	—
Clostridium botulinum	360	5
Bacillus coagulans	1140	17

[a]Times required to kill a given number of spores *under conditions of testing.* It should be emphasized that the times may vary under different conditions.

that *the condition of the vegetative cells or spores influences their heat resistance.* Even within a given strain of bacteria, heat resistance varies depending on the way in which a culture has been grown, the phase of growth of its cells, the age of spores if these are present, and other factors. For example, spores become more heat resistant for a time as they age, a factor which could significantly alter the TDT for a bacterial strain. A fourth general rule of high temperature food preservation is that *the environment or material in which microorganisms are heated has a major influence on their heat resistance.* Moist heat kills much more efficiently than dry heat, as evidenced by the efficacy of autoclaving as a means of sterilization. Generally, also, organisms are most resistant to heat at the pH which supports their optimum growth; therefore extremes in pH often cause increased killing. Organisms are killed by heat more readily in high concentrations of salt, and they are less heat sensitive in media containing considerable quantities of proteins and fats (protein or fat protection).

Importance of "wet heat," page 128

Table 37-2 presents data illustrating the influences of environment on heat resistance and, consequently, on the times required for processing of foods. Baked beans are alkaline, and their low water content does not permit heat to penetrate them quickly; they must therefore be heated for a long time to kill any spores that may be present. Organisms in cream-style corn are more resistant to heat than those in corn packed in brine, because of the high salt content of the brine and the protection of fats and proteins in the cream sauce. Tomatoes are highly acid and will not readily support the germination of spores; also, any spore-forming clostridia that may be present in tomatoes do not readily form toxin at this low pH. Thus tomatoes can be processed safely at a lower temperature and for a shorter time than the other foods listed.

In addition to canning, another means of high-temperature processing of foods is *pasteurization.* In this process food preservation is usually not the primary aim; rather, pasteurization is a heat treatment which destroys all nonspore-forming pathogenic organisms without significantly altering the quality of a food. Thus pasteurization of milk or other foods does not sterilize them, but only kills all nonspore-forming, disease-producing organisms. Vinegar and a few other foods, on the other hand, are pasteurized not primarily to kill pathogens (which are not likely to survive and grow at the low pH of this food) but to destroy certain spoilage organisms. The factors influencing high-temperature canning,

TABLE 37-2
Recommended Time and Temperature Necessary for Processing Several Foods[a]

Food	*Temperature (°F)*	*Time (min)*
Baked beans	240	105
Corn, cream style	240	90
Corn, in brine	240	50
Tomatoes	212	34

[a] Recommendations based on bulletins issued by the National Canners Association.

Pasteurization, page 130

discussed previously, are also applicable to pasteurization. The two methods of pasteurization in general use are the high-temperature–short-time method (HTST) and the low-temperature–long-time (LTLT) or holding method. For example, with milk the minimum requirements for the HTST method are 71.7°C (161°F) for 15 sec and those for the holding method are 61.7°C (145°F) for 30 min. On the other hand, ice cream mix, which is richer in fats than milk, requires 82.2°C (180°F) for about 20 sec, or 71.1°C (160°F) for 30 min. As is the case with canning, heat treatment for pasteurization must be adjusted to the individual situation.

Low-Temperature Treatment

The growth of microorganisms is temperature-dependent, and low temperatures are useful in food preservation.

The cold storage of some fruits and vegetables (for example, apples and potatoes) can preserve these foods for months. In this method the food is stored at temperatures near 10°C in the dark and at the appropriate humidity.

Refrigeration can be adjusted to various temperatures, depending on the food in question. Most household refrigerators maintain a temperature around 4°C, although the temperature may vary from 0 to 10°C in different parts of the refrigerator. The proper relative humidity and ventilation in the refrigerator are important in maintaining good quality of the foods. Sometimes, with chilling, other methods of preservation, such as ultraviolet irradiation, are used for better results.

During this century freezing has become a major means of preserving foods. This method permits foods to be stored for many months, or even longer, with little loss of color, flavor, and other criteria of quality. Various methods of freezing and a wide range of freezer temperatures are used, depending on the food being processed.

Desiccation

For many centuries drying has been used to preserve food. This process decreases the water activity of a food below the limits required for microbial growth. Often, drying by natural means (sun drying), or by artificial means is supplemented by other methods, such as salting or adding high concentrations of sugar or small amounts of preservative chemicals to a food. Lyophilization (freeze drying) has also become rather widely used for preserving some foods. In this process the frozen food is dried in a vacuum, and the quality of the reconstituted product is much better than with ordinary freezing or drying methods.

Although drying stops growth, it does not kill all bacteria in or on foods. For example, a number of cases of salmonellosis have been traced to dried eggs. This occurs because egg shells may be heavily contaminated with species of pathogenic salmonellae from the gastrointestinal tract of the hen. When the eggs are cracked, many of the bacteria enter and subsequently survive drying; these bacteria can cause disease in susceptible individuals who eat the inadequately cooked egg product. Some states now have laws stating that only pasteurized dried eggs can be sold to prevent the transmission of such pathogens.

Chemical Preservatives

Reference has been made to methods of adding chemicals to preserve foods. Salting involves the addition of large amounts of salt that increase the osmotic strength—and actually result in decreasing the water content—of the food. High sugar content of foods (as in jams, jellies, honey, preserves, and sweetened condensed milk) leads to a similar result. Pickling combines salt and a very acid medium, and is effective in controlling the growth of most microorganisms. Sodium benzoate, sorbic acid, propionate, and other chemicals are often added to certain foods to retard microbial growth. Nitrites and nitrates in foods inhibit spore germination; their presence in processed meat therefore helps to protect against botulism, but there is concern about their potential *carcinogenicity*.

Radiation

Irradiation can be lethal to microorganisms and it is thus useful as a means of preserving foods. Adequate irradiation has been obtained from ultraviolet rays, gamma rays, and X-rays. Ultraviolet rays are restricted to surface applications, whereas gamma rays and X-rays penetrate foods. Microwave energy is limited in use for food preservation because of the heat it generates.

Methods have been developed for irradiating foods to give essentially the same results given by heat treatments such as sterilization (as in canning) and pasteurization. It is probable that irradiation will become more widely used in coming years. However, because enzymes in a food are not destroyed by irradiation, enzyme degradation of the food can continue. Therefore, irradiated foods may spoil more rapidly or not be as appetizing as heat-treated foods unless these natural enzymes are destroyed by other means, such as mild heating.

CONTROL OF MICROBIAL HAZARDS IN FOODS

Standards for Control

Food production is carefully controlled in the United States. Federal, state, and local agencies cooperate in inspections to see that protective laws are enforced. The Federal Food, Drug, and Cosmetic Acts set standards for foods that are shipped between states or territories, and require accurate labeling. Other federal laws provide for the inspection of meats and poultry produced for interstate shipment.

State and local agencies enforce many of the laws that protect food consumers. Of course, food manufacturers also usually act as their own watchdogs.

As a result, the microbiological quality of most foods sold in the United States is very good, compared with that in many parts of the world.

Microbial Food Poisoning

Bacterial food poisonings, pages 517, 518

The United States has not been free of microbial food poisoning. Many cases occur each year from foods prepared either commercially, at home, or in large quantities in institutions. The major kinds of bacterial food poisonings, caused by salmonellae, staphylococci, and clostridia, are discussed in Chapter 27.

Additional Microbial Hazards of Foods

In addition to bacterial food poisonings, certain other hazards may result from ingesting foods. Among these are diseases sometimes transmitted via contaminated foods. Examples are brucellosis, listeriosis, leptospirosis, tularemia, tuberculosis, Q-fever, streptococcal infections, poliomyelitis, and hepatitis. Adequate cooking and pasteurization will prevent transmission of pathogens in this manner.

Algal toxins, pages 284, 285
Fungal toxins, page 518

Very dangerous forms of food contamination can be caused by algal and fungal toxins, as mentioned in previous chapters.

MICROBES AND THE PRODUCTION OF FOODS AND BEVERAGES

Ancient methods of preparing foods and beverages by utilizing microbial activities have often been improved by scientific research. However, some of these processes, including the production of fine cheeses and wines, are far from being fully understood.

Milk Products

MILK

Lactic acid bacteria, page 253

Milk is normally sterile until secreted; by the time cow's milk reaches the market, however, it contains a multitude of microorganisms introduced from the udder and during handling. Along with other ubiquitous contaminants, the lactic-acid bacteria *Streptococcus lactis* and *Lactobacillus* species are commonly present. These Gram-positive bacteria, described in Chapter 12, grow especially well in milk, which is rich in lactose. *Streptococcus lactis* causes the souring of unpasteurized milk by rapidly producing enough lactic acid to lower the pH of milk to 4.8 or below, which coagulates the milk proteins. Pasteurization kills most spoilage bacteria as well as vegetative bacterial pathogens. Therefore pasteurized milk does not sour as rapidly as unpasteurized milk and remains palatable for many days at refrigerator temperature.

In the United States commercially marketed milk is usually pasteurized and its bacterial content is monitored. The United States Public Health Service has developed standards for grading milk as Grade A or Grade B (Table 37-3) on the basis of bacterial counts. Direct microscopic counts can be made of the numbers of visible organisms (both living and dead) in samples of raw milk; plate counts

TABLE 37-3
Recommended Standards[a] for Dairy Products

Product	*Maximum Number of Microorganisms per Milliliter or Gram*
Grade A raw milk prior to commingling with other milk	100,000
Grade A raw milk commingled, prior to pasteurization	200,000–300,000
Grade A pasteurized milk and milk products (except cultured products)	20,000 (5–10 coliforms)
Grade A pasteurized cultured products	10 coliforms
Ice cream and other frozen desserts	50,000 (10 coliforms)
Liquid ice cream mix for machines	25,000 (5 coliforms)
Grade B raw milk	600,000
Grade B pasteurized milk	40,000

[a] The recommended standards vary slightly from one community to another

of the number of colonies grown from samples enumerate only living organisms capable of growing under the conditions of culture. Both counts are used in the grading of milk. Grade B milk, which has a higher bacterial count, is usually made into cheese, ice cream, or other products.

YOGURT

Although many of the lactic acid bacteria can ferment the lactose in milk, the end products of their fermentation vary considerably with the species or even the strain used. Whereas the growth of *S. lactis* causes souring and spoilage, certain lactobacilli (for example, *Lactobacillus bulgaricus*), often with the aid of some streptococci and yeasts, can ferment milk to yogurt.

BUTTER

Fresh pasteurized milk and yogurt account for about 45 percent of all the milk consumed in the United States; approximately 20 percent of the milk output is used for butter production. Butter can be kept much longer than fresh milk because between 80 and 85 percent of butter consists of the fats of milk—butterfat—emulsified with 12 to 16 percent water, along with small amounts of salt and milk proteins. Butterfat is a unique combination of glycerides, compounds that contain some of the fatty acids. Most of the bacteria and fungi present in butter do not thrive on such a high-lipid diet. In addition the low water content of butter discourages the growth of most microorganisms. *Pseudomonas putrefaciens* or other pseudomonads are the most common spoilage organisms of butter. They cause rancidity by breaking down the glycerides in butterfat to yield free fatty acids. Butyric acid is particularly prominent in rancid butter.

CHEESE

The manufacture of cheese accounts for another 10 to 15 percent of the nation's output of milk. In general, cheese-making involves coagulation of milk proteins to form solid curds which can be separated from the milk liquid (whey) and treated in various ways to yield characteristic cheeses. Lactic acid bacteria may be used to produce the proper acidity to clot milk into a firm curd, enzymes (rennet or pepsin) may be added to clot the milk proteins, or both may be used together.

Cottage cheese is the fresh curd of whole or skim milk (soured with

S. lactis) removed from the whey without further treatment. Cream cheese is a similar curd made from cream. Sometimes lactobacilli are added to slightly decompose the protein and increase the flavor of these cheeses, but usually they are simply concentrated milk solids. Cottage cheese is high in protein, and if skim milk is used rather than whole milk, the resulting product is low in fat content and in calories. Cream cheese, on the other hand, is rich in fats and high in calories. Neither keeps well because fresh curds retain a high content of water and readily support growth of spoilage organisms. These cheeses have little flavor because they have not been ripened by microbial action, which results in the production of substances responsible for the flavors and aromas of other cheeses.

Hard cheeses are made by heating fresh curds at low temperature for an hour or two and then removing the whey. During the curing or ripening period, which may last months or years, the action of nonproteolytic bacteria or molds transforms the curds to hard cheese. The distinctive flavors of various cheeses result from the activities of different microorganisms. For example, very hard Parmesan and Romano cheeses are ripened by bacteria, as are the less hard Cheddar and Swiss cheeses. If the bacteria produce gas during the ripening process, holes are formed in the cheese (in much the same way holes are formed in bread when CO_2 is produced by the action of yeasts).

Semisoft and soft cheeses are ripened by bacteria or molds or both. The microbial degradation of fats and proteins in the milk curds used results in the soft final product. Limburger and Liederkrantz are soft cheeses that are ripened by bacteria; Camembert and Brie are ripened by molds. *Penicillium camemberti,* for example, grows on the surface of Camembert cheese and degradation of the curds proceeds inward. The more extensive the activities of a mold, the riper and softer the cheese will be. Roquefort cheese is also ripened by mold.

The infinite variety of microorganisms that can ripen curds and the complexity of their metabolic processes account for the fact that the production of fine cheeses remains an art, even when it is scientifically controlled.

Wine

A discussion of cheeses leads naturally to a consideration of wines, for the two have been partners for many centuries. It is not too surprising that early man learned to make wine, because grapes and other fruits naturally contain the yeasts needed to transform their juice to wine. These yeasts and other microbial flora collect in the waxy film, or bloom, that covers the grape and gives it a dull, matted appearance. As many as 10 million yeast cells may be found on the surface of a single grape; perhaps 1 percent of these are *Saccharomyces cerevisiae,* variety *ellipsoideus,* the strain largely responsible for the fermentation of grape sugars to wine. However, some of the other organisms present on grapes may represent a disadvantage for winemakers by producing unwanted products.

Wine production begins with the selection of the fruit to be used (usually grapes), because the quality of the final product depends, to a large extent, on the initial content of the fruit's sugars, acids, water, and other chemicals. Selection of the appropriate starting materials is of the utmost importance to the wine producer, but the microbiologist is primarily concerned with selection of the proper yeasts to act on the raw materials.

Commercially, wine is made by crushing carefully selected grapes in a machine which removes the stems and collects the resulting solids and juices, called *must.* For red wine, the entire must of red grapes is put into the fermentation vat, the red color and rich flavors of red wines being derived from components of the grape skin and seeds. For the production of white wines, only the juice of either red or white grapes is fermented. Rosé wines get their light pink color from the entire crushed red grape being fermented for about one day, after which the juice is removed and fermented alone. In the fermentation vat, sulfur dioxide (SO_2) is often added to the must to inhibit growth of the natural microbial flora of the grape, and cultures of specially selected strains of *S. cerevisiae* are then added. These strains are usually more resistant than most yeasts to the antimicrobial action of SO_2. Fermentation proceeds at a carefully controlled temperature, which varies with the type of wine, for a period ranging from a few days to several weeks. During this time most of the sugar is fermented into ethyl alcohol, resulting in a final alcohol content of less than 14 percent (except under special conditions). At the proper time during fermentation of red wines the juice is separated from the pulp and skins by a wine press and fermentation is allowed to continue to completion. The particulate debris settles and the wine is cleared by filtration prior to aging. The production of wines is similar throughout the world, but certain steps vary. For example, sulfur dioxide is usually not added during the production of French wines.

An amazingly complex variety of microbial transformations occurs during the fermentation and aging of wines. Enzymes of yeasts and other microorganisms catalyze the conversion of sugars (glucose, fructose, and others), organic acids, amino acids, pigments, and many other components of fruit into numerous different substances. Only a few of these reactions have been identified. The fermentation of glucose to ethyl alcohol, which has been most extensively studied, is discussed in Chapter 7.

Alcoholic fermentation, pages 150, 154, 155

The presence of lactobacilli in wine is not always advantageous. These bacteria, along with other acid-tolerant lactic acid and acetic acid bacteria, are the most prevalent cause of wine spoilage. They are usually controlled by adding SO_2 or inhibitory chemicals to the final product.

Every mole of glucose or fructose that is fermented to alcohol yields 2 moles each of ethyl alcohol and CO_2. This release of CO_2 causes a bubbling or "working" in the wine vat. Usually all the gas is released before the wine is bottled, resulting in a still (noncarbonated) wine. Other processes are used to prepare carbonated wines such as champagne.

A number of the beverages produced by *Saccharomyces* activity are listed in Table 37-4. Note that almost anything that can be fermented by yeasts has been used to make alcoholic beverages, including fruit juices, molasses, and grain mashes. Beverages with an alcohol content greater than about 15 percent are prepared by distilling or adding alcohol to naturally fermented products. Some-

TABLE 37-4
Some Alcoholic Beverages Produced by *Saccharomyces*

Beverage	*Starting Material*	*Procedure*
Beer	Germinated grain (malt)	Natural fermentation
Table wine	Fruit juice	Natural fermentation
Sake	Rice	Amylase from mold (*Aspergillus oryzae*) converts starch to sugar which is naturally fermented
Fortified wine (sherry, port)	Fruit juice	Natural fermentation plus addition of brandy to increase alcohol content to about 20%
Brandy	Fruit juice	Natural fermentation followed by distillation to increase alcohol content to about 40 to 50%
Whiskey	Grain mash	
Rum	Molasses	
Vodka	Potatoes	

times special flavors are obtained by using more than one kind of fermentation reaction. For example, lactic acid bacteria are used to produce lactic acid in grain mash for making sour-mash whiskey; *S. cerevisiae* subsequently ferments the sour-mash to produce alcohol, and distillation results in whiskey with a characteristic flavor.

The Japanese wine, sake (Table 37-4), is of special interest because its production depends on several microbial fermentation reactions. First, cooked rice is inoculated with the mold *Aspergillus oryzae* which produces amylase and thereby degrades the rice starch to sugar. Then *Saccharomyces* converts the sugar to alcohol, and lactic acid bacteria add to the flavor by producing lactic acid and other metabolites.

Beer

Beer is made by the fermentation of germinated grain (malt), usually barley, by species of *Saccharomyces*. Some 5000 years ago underbaked bread made from germinated barley was used to make beer. Small pieces of bread were soaked in water; after a few days the resulting mash fermented and was sieved or filtered and sometimes flavored with herbs or honey. Later hops came into use, being added to give a characteristic bitter flavor to the beer.

The carbohydrates of grains exist largely in the form of starch and must be converted to simple sugars before they can serve as substrates for yeast fermentation. This conversion of starch to sugars is carried out by enzymes which develop in the grain soon after germination. The germinated grain, or malt, is soaked in water to yield an extract called malt wort that contains the sugars and other nutrients needed for growth of the yeast. The wort is boiled before the yeast is added, partly to destroy the enzymes of the grain as well as most microorganisms that may be present, but mainly to concentrate the wort and improve the flavor. The flowers of hops are added primarily for their bitterness, but they also contain antibacterial substances that inhibit the growth of unwanted bacteria. Special brewers' yeasts are used for the fermentation, most often *S. carlsbergensis,* a yeast especially selected for making beer. The yeast ferments the wort mixture for about a week. The beer is then allowed to settle and filtered to remove microorganisms and other particulate materials before aging, carbonating, and bottling, canning, or other packaging. The process is shown in Figure 37-1. It differs from

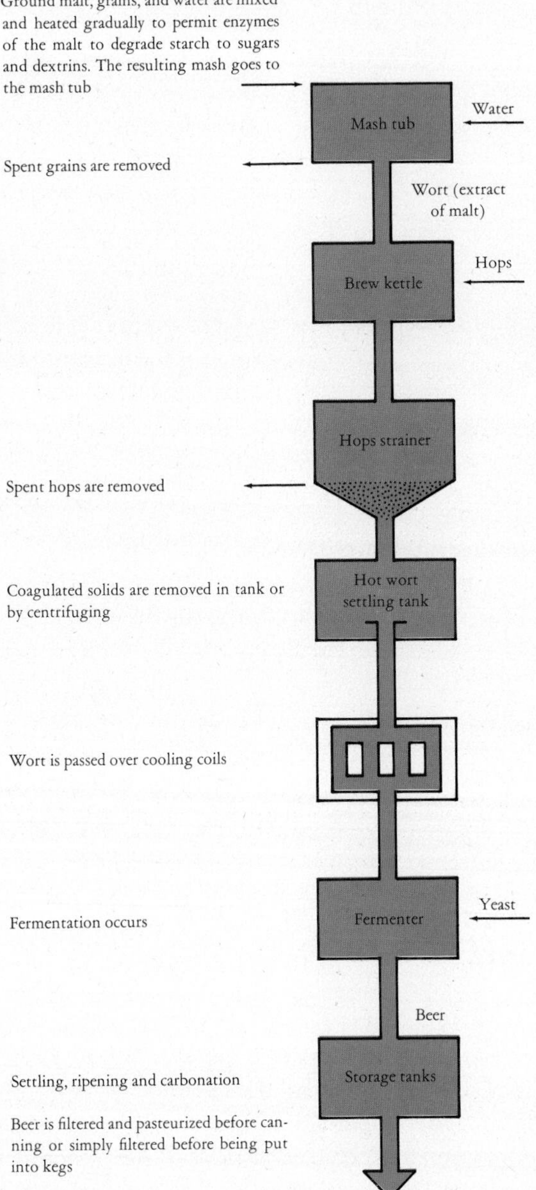
Ground malt, grains, and water are mixed and heated gradually to permit enzymes of the malt to degrade starch to sugars and dextrins. The resulting mash goes to the mash tub
Mash tub
Water
Spent grains are removed
Wort (extract of malt)
Brew kettle
Hops
Hops strainer
Spent hops are removed
Coagulated solids are removed in tank or by centrifuging
Hot wort settling tank
Wort is passed over cooling coils
Fermentation occurs
Fermenter
Yeast
Beer
Settling, ripening and carbonation
Storage tanks
Beer is filtered and pasteurized before canning or simply filtered before being put into kegs

FIGURE 37-1
The commercial production of beer.

wine-making in a number of ways, principally in that the sugars for fermentation of beer are derived from the breakdown of the starch in grain rather than being natural sugars. The flavor of beer results from the various combinations of grain and hops used, and the alcohol content varies from 3.2 to 6 percent.

Bread

In addition to their use in making wine and beer, strains of *S. cerevisiae* have also been selected especially for making bread. The yeasts utilize the sugar in bread dough and produce alcohol and CO_2; because this fermentation proceeds for just a few hours, however, only small quantities of alcohol are made. During baking, the volatile alcohol escapes and CO_2 bubbles are trapped in the bread, producing holes and resulting in the "lightness" of raised breads.

"Active dry yeast" (lyophilized) has become more popular than yeast cakes, especially for home baking, because it can be kept much longer and does not have to be refrigerated. The dried yeast cells retain their viability better when they are sealed under vacuum, or when CO_2 or nitrogen is substituted for air in the package.

Microorganisms other than yeasts can also contribute to better bread products. For example, amylase produced by some of the filamentous molds is sometimes added to degrade the starch in bread flour, resulting in more gas production during bread-making.

Vinegar

Vinegar is an aqueous solution, containing at least 4 percent acetic acid, obtained from the oxidation of ethyl alcohol by species of acetic acid bacteria. The alcohol of apple cider is the substrate for cider vinegar; wine vinegar is prepared by the action of acetic acid bacteria on the alcohol in wine. Vinegar can be made from many alcoholic beverages; however, wine, malt, and cider vinegars are most important industrially.

Acetic acid bacteria are strictly aerobic, Gram-negative rods, characterized by their ability to carry out a number of incomplete oxidations. They can tolerate high concentrations of acid and can thus grow in acidic wines or ciders, oxidizing alcohol to acetic acid, by the following reactions:

$$\underset{\text{ethyl alcohol}}{2CH_3CH_2OH} + \underset{\text{oxygen}}{O_2} \longrightarrow \underset{\text{acetaldehyde}}{2CH_3CHO} + \underset{\text{water}}{2H_2O}$$

$$\underset{\text{acetaldehyde}}{2CH_3CHO} + \underset{\text{oxygen}}{O_2} \longrightarrow \underset{\text{acetic acid}}{2CH_3COOH}$$

The intermediate product, acetaldehyde, is quickly converted into acetic acid and does not accumulate during the process. *Acetobacter aceti, A. orleanse,* and *A. schuezenbachi* are species important in the manufacture of vinegar.

The transformation of alcohol to acetic acid (acetification) may be slow or rapid. The best quality (and most expensive) vinegars are obtained by the old French or Orleans process of slow acetification. In this process oak barrels half full of wine are placed on their side; the barrels have holes drilled in their upper parts to provide the aerobic environment necessary for the growth of *Acetobacter.* A "starter" of vinegar containing acetic acid bacteria is added to the wine in the barrels and the bacteria develop as a film on the surface of the liquid. Approximately 1 to 3 months are required for the maximum oxidation of ethyl alcohol to acetic acid in this method.

Rapid acetification in a vinegar generator is widely used to produce most vinegars. The generator consists of a tank (usually wooden, to resist the highly acid vinegar) containing wood shavings or other materials that provide a large surface area. Acetic acid bacteria develop aerobically as a film on the wood shavings. The alcoholic substrate is sprayed on the shavings and trickles through, being oxidized en route by the abundance of *Acetobacter* organisms. In principle the vinegar generator operates much as the trickling filter does during degradation of wastes to provide a large surface area for aerobic metabolism.

Trickling filter, page 658

Sauerkraut

Sauerkraut is the product of a natural lactic acid fermentation carried out by lactobacilli and certain other bacteria that are normally present on cabbage. The cabbage is shredded and layered, with salt between the layers. The salt content, 2.0 to 2.5 percent of the total weight, inhibits the growth of many bacteria and also dehydrates the cabbage. The layers are firmly packed and pressed down to provide an anaerobic environment. The fermentation results in acid formation which soon inhibits the growth of most microorganisms but favors the growth of *Leuconostoc, Lactobacillus plantarum, Lactobacillus brevis,* and similar species which grow and produce large amounts of lactic and acetic acids, alcohol, and CO_2, and smaller amounts of other compounds. When the desired flavor has been attained, usually after 2 to 4 weeks at warm room temperatures, the sauerkraut is canned.

PICKLES

Lactic acid fermentations are also important in making pickles. The same species of lactic acid bacteria that ferment sauerkraut are also used to produce pickles.

SUMMARY

Microbial activity plays a large role in food spoilage. Important conditions that regulate the growth of spoilage organisms on foods are the foods' water activity (or water content), pH, osmotic pressure, amount of available oxygen, temperature, and nutrient supplies. Some foods contain antimicrobial substances that discourage the growth of certain organisms.

Food preservation can be achieved by high-temperature treatment, low-temperature storage, drying, chemical treatment, and irradiation.

Microbial hazards of foods include salmonella, staphylococcal, clostridial, and other food poisonings, the transmission of various pathogens by contaminated foods, and poisonings by certain algal or fungal toxins.

Microbes are important in the production of a wide variety of foods, including butter, cheeses and other milk products, wines, beer, bread, vinegar, sauerkraut, pickles, and others.

QUESTIONS

REVIEW

1. Why are species of *Pseudomonas* frequently implicated in food spoilage?
2. What is meant by the water activity of a food? Which would you expect to spoil faster, a food with a high water activity or a food with a low water activity?
3. Which is more likely to be recovered from a jar of spoiled pickles, yeast or bacteria? Why?
4. What are some of the foods produced by lactic acid fermentations? What organisms are involved in preparing each of these foods?
5. Outline the uses of *Saccharomyces* species in the production of foods and beverages.

THOUGHT

1. If you were to own a large food-processing company, outline some of the general plans for quality control or for microbiological research and development that you might follow.
2. What kinds of microbiological hazards might be encountered in the following foods and how can they be overcome?

    Oysters grown in fecally polluted waters
    Eggs with shells that were cracked at the time of purchase
    Frozen turkey
    Fresh raw milk

3. Strains of *Penicillium* that produce penicillin can be isolated from fresh fruits. It might therefore be supposed that penicillin is frequently ingested in fresh foods, but in fact it is not. Why is this the case?

FURTHER READING

AMERINE, M. A., "Wine," *Scientific American* (August 1964).

FORD, B. J., *Microbiology and Food.* London: Northwood Publications, 1970. A small volume integrating principles of hygiene, food spoilage, and food poisoning.

FRAZIER, W. C., *Food Microbiology.* New York: McGraw-Hill, 1958. Presents the basic principles of food microbiology.

JAY, J. M., *Modern Food Microbiology.* New York: Van Nostrand Reinhold, 1970. A standard textbook.

KLEYEN, J. and J. HOUGH, "The Microbiology of Brewing," *Annual Review of Microbiology* 25: 583–608 (1971).

NICKERSON, J. T., and A. J. SINSKEY, *Microbiology of Foods and Food Processing.* New York: American Elsevier, 1972. A book presenting information on the microbiology of foods and of food processing and the microbiology of food-borne diseases, suitable for professional food handlers.

COMMERCIAL APPLICATIONS OF MICROBIOLOGY

Human beings have used microorganisms throughout history, but only during the last century has much been learned about the organisms involved. With increasing knowledge, it has been possible to improve ancient methods that had been discovered empirically, such as bread- and wine-making, so as to make them more productive commercially. Besides these older applications, new practical and commercial applications of microbial activites are continually being devised. As an example, the discovery that some organisms produce substances which kill or inhibit other microbes opened the vital area of antibiotic drug production, and new antibiotic drugs are being constantly sought and frequently discovered.

Continued research and increased understanding of microbial activities will undoubtedly lead to other benefits that can be developed commercially. Environmental improvement, food supply, energy production, and space exploration are only a few of the areas that are expected to employ microorganisms to an even greater degree in the future than they do today (Chapter 39). In this chapter a few of the important commercial applications of microbiology other than food production are considered.

AMINO ACIDS AND OTHER ORGANIC ACIDS

Human beings can synthesize 12 of the 20 major amino acids; the other 8, the essential amino acids, must be supplied. Some microorganisms can readily synthesize the essential amino acids, as well as others, and a number of derepressed mutant strains of bacteria have been selected which produce unusually large quantities of a particular amino acid.

Amino acids, page 35

In the past the essential amino acid L-lysine had been produced commercially by the joint activities of selected strains of two common species of bacteria, *Escherichia coli* and *Enterobacter aerogenes.* In the first part of this commercial process, an *E. coli* mutant that cannot produce lysine supplies enzymes which catalyze the incorporation of nitrogen from inorganic ammonium ions into the amino groups of the organic compound diaminopimelic acid. In the second part of the process, enzymes of *E. aerogenes* catalyze the removal of CO_2 from diaminopimelic acid, yielding L-lysine:

$$\text{Ammonium ions} + \text{glycerol} + \text{corn-steep liquor} \xrightarrow{\text{enzymes of } E.\ coli} \text{diaminopimelic acid}$$

Diaminopimelic acid and L-lysine, page 61

$$\underset{\text{diaminopimelic acid}}{COOH{-}CHNH_2{-}CH_2{-}CH_2{-}CH_2{-}CHNH_2{-}COOH}$$

$$\xrightarrow{\text{enzymes of } E.\ aerogenes} CO_2 + \underset{\text{L-lysine}}{CH_2{-}NH_2{-}(CH_2)_3{-}CH{-}NH_2{-}COOH}$$

Today, lysine is produced commercially by a simpler method, using mutants of *Micrococcus glutamicus* that can synthesize large quantities of lysine directly. Other strains of *M. glutamicus,* which is a Gram-positive coccus, are used to produce the nonessential amino acid L-glutamic acid from molasses. L-glutamic acid is also produced from corn-steep liquor by the Gram-positive rod *Brevibacterium flavum.*

Organic acids (including lactic, citric, gluconic, and butyric acids) are also produced industrially by microbial activity. Lactic acid, for example, can be synthesized by *Lactobacillus bulgaricus* from the lactose in the whey that is a waste product of cheese manufacture. More often, glucose or other sugars are used as substrates for fermentation by *L. delbruckii.* The lactic acid produced is used in the manufacture of food products, for treating hides in leather making, in pharmaceuticals, paints, and plastics. Butyric acid, produced by species of *Clostridium,* is an important organic solvent often used in paints and plastics.

Molds are used commercially to produce citric, oxalic, and gluconic acids. Special strains of *Aspergillus niger* ferment the sugar in beet molasses (as well as sugars from other sources) to citric acid. Gluconic acid can also be made commercially by *A. niger,* but other strains which produce gluconic acid, and not other organic acids, must be selected. A simple glucose-salts medium provides the substrate for gluconic acid production by *A. niger.*

A recent experimental study has shown that *A. niger* is also capable of converting almost all of the sugar in brewery waste to fungal mass, accompanied by the considerable production of citric acid. This conversion can reduce the biochemical oxygen demand of brewery wastes by 96 percent, as well as producing citric acid and single cell protein in the fungal mass. Such microbial activities are not only of economic importance, but also contribute to improvement of the environment.

Vitamins

Since yeasts synthesize vitamins very efficiently, certain strains of yeast are often used as food supplements. An acomycete (*Ashbya gossypii*) is widely used industrially to produce riboflavin. A particularly important commercial process is the use of actinomycetes and some other bacteria, including species of *Streptomyces* and also certain strains of *Bacillus megaterium,* to produce vitamin B_{12}.

Steroids

Although it is difficult to recover large quantities of *steroid hormones* from animal sources, it is possible to manufacture these hormones microbially from sterols or related compounds obtained from animals or plants. Special strains of the mold *Rhizopus* and of the bacteria *Corynebacterium* and *Streptomyces* are used to change particular chemical groups on the sterol and similar molecules by enzyme activity, thereby producing the steroids (Figure 38-1).

Sterols, page 44

ANTIBIOTICS

The commercial manufacture of antibiotics, begun more than 30 years ago with the growth of *Penicillium notatum* in flasks, has developed into today's multibillion dollar antibiotic industry. The tremendous development of the industry in the United States is illustrated by statistics showing that in 1945 only about 12,000 lb of antibiotics were produced commercially, compared with well over 3 million lb produced in 1961, since which time the annual production has continued to increase.

In the American antibiotic industry, penicillin is currently produced by special strains of *P. chrysogenum.* The natural penicillin molecule that is produced can subsequently be altered chemically to give a variety of semisynthetic penicillins with special properties (see Figure 33-3).

See Figure 33-3, page 590

Antibiotics represent only a small proportion of the metabolic products of the microorganisms which synthesize them. In fact they seem to be minor byproducts of microbial metabolism, not essential to the cells which produce them, and it is not clear how or why they are made. In order to harvest sufficient quanti-

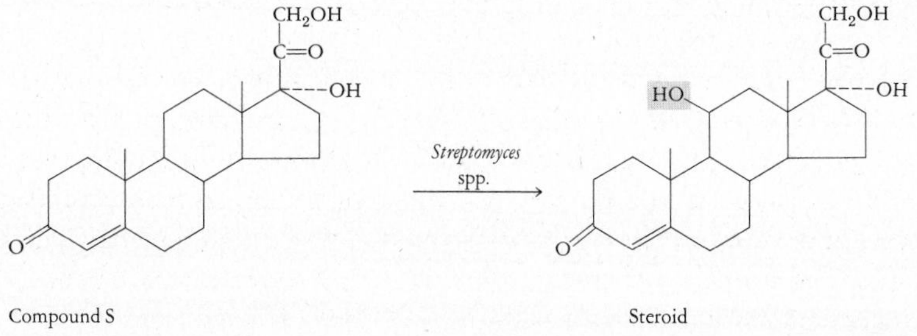

FIGURE 38-1
Steroid transformation. Special strains of bacteria or molds can enzymatically convert compound S, an inactive substance widely distributed in nature, to valuable steroids such as hydrocortisone.

ties of antibiotics, mutant strains of microorganisms, which produce large amounts of a particular antibiotic, have been selected. Even so, the large residues of microorganisms left after antibiotics have been produced present tremendous disposal problems. In the production of penicillin, for example, only a tiny fraction of the final culture fluid is recovered as penicillin, while huge amounts of mold mycelium and expended culture medium must be discarded. Unless properly treated, these wastes are an environmental hazard because they are rich in nutrients. Perhaps in the future they will be efficiently recycled as sources of energy or food.

VACCINES, TOXOIDS, AND ANTITOXINS

Vaccines, toxoids, and other materials used to immunize humans and animals against disease must be prepared under the most rigid conditions. Microbial exotoxins, such as the diphtheria and tetanus toxins that are prepared commercially, must be transformed into innocuous toxoids and carefully tested before being used. Similarly, vaccines made from infecious organisms, whether attenuated and living (as, for example, the Sabin polio vaccine) or inactivated (such as the influenza vaccines), must be produced industrially and yet scrupulously monitored to ensure their safety. There are several recorded instances where mistakes have been made in the preparation of vaccines. In an incident some years ago a batch of Salk polio vaccine was not completely inactivated; as a result, a number of children who received the vaccine developed paralytic poliomyelitis. Many commercially made immunizing agents are grown in chick embryos or cells; they thus present a real hazard to those few persons who are allergic to eggs. Even the tiny amount of egg antigens present in an immunizing dose of a vaccine can cause a potentially very dangerous anaphylactic reaction in an individual who is highly sensitive to eggs.

Some of the commercially produced immunizing agents of major importance are listed in the table in Appendix IV.

Killed or inactivated infectious agents (such as bacteria and viruses) may be used in vaccines to stimulate the production of antibody without causing disease. Alternatively, living, attenuated agents may be administered in order to immunize an individual. For example, the pertussis (whooping cough) vaccine is made of inactivated bacteria, whereas the Sabin polio vaccine, in common use today, consists of living mutants of the polioviruses which can infect, but these mutants are not able to invade the central nervous system nor do they lead to severe disease.

In 1976 the "swine flu" became a threat, and the United States government initiated a national mass immunization campaign against this influenza. The vaccine used was produced by four large pharmaceutical companies from a hybrid virus prepared for this purpose. The hybrid was produced in the laboratory by genetic recombination between a "swine flu" virus (isolated from the swine flu outbreak early in 1976 at Fort Dix, New Jersey and a standard influenza virus known to grow well in eggs. Genetic analysis of the RNA of the hybrid showed that only the two genes coding for hemagglutinin and neuraminidase antigens came from the "swine" influenza virus, whereas six other genes were derived from the standard influenza virus strain. In addition to using the vaccine made by inactivating the hybrid swine influenza virus, the 1976 inoculation campaign used another vaccine—containing inactivated viruses of both the hybrid influenza virus strain and the Victoria influenza strain that had been a major cause of influenza since 1968.

Manufacturers are now developing influenza vaccines which employ only the isolated, soluble forms of important immunogens of the influenza virus. These immunogens, the hemagglutinin and the neuraminidase antigens, can be removed from the virus particles in a relatively purified but still highly antigenic form. When used to immunize, they reportedly induce immunity very effectively with fewer unwanted side effects than those produced by the whole virus vaccines.

Obviously, the variability of influenza viruses and other microorganisms presents a challenge to the commercial producer of vaccines as well as to the microbiologist. Use of sophisticated microbiological techniques is essential in the production of safe and efficient vaccines.

In commercial toxoid preparation from exotoxins, fairly mild treatment with formalin or heat can be employed to destroy exotoxin toxicity while retaining an exotoxin's ability to incite antibody formation. Antitoxins are sometimes necessary to prevent or treat diseases caused by powerful exotoxins. For the commercial preparation of antitoxins, horses or other large animals are immunized with toxoid, and the gamma globulin portion of their blood serum is separated. When possible, gamma globulin from immunized human volunteers, instead of animals, is prepared, in order to decrease the risk of hypersensitivity reactions (serum sickness and others) in people who receive an antitoxin. Human antitoxin is also preferable because it is not inactivated and removed from the circulation as rapidly as are foreign antisera. Antitoxins are used primarily in treating botulism, tetanus, diphtheria, snake venom, and certain other kinds of intoxications.

Vaccines and antitoxins must be standardized during their preparation to ensure that each dose has the proper amount of activity. Reference standards for these substances have been established by federal health agencies and are carefully enforced; in a number of cases, international standards, established by the World Health Organization, are used.

ENZYMES

Earlier discussions have shown that yeasts, molds, and bacteria produce an immense number of enzymes. Fungi are used commercially in the manufacture of

Enzymes, pages 143–147

TABLE 38-1
Some Enzymes That Are Produced Industrially by Microorganisms

Enzyme	*Reaction Catalyzed*	*Organisms Frequently Used for Production*
Amylase	Starch to maltose	*Aspergillus* sp.
Maltase	Maltose to glucose	*Rhizopus* sp.
Proteases	Proteins to polypeptides or amino acids	*Bacillus subtilis* *Aspergillus* sp.
Pectin-degrading	Degradation of pectin	*Aspergillus* sp. *Rhizopus* sp.
Invertase	Sucrose to glucose and fructose	*Saccharomyces cerevisiae*
Penicillinase	Degradation of penicillin	*B. subtilis* *B. cereus*
Streptodornase	Degrades DNA	*Streptococcus pyogenes*
Streptokinase	Degrades fibrin clots	*S. pyogenes*

invertase (an enzyme which converts sucrose to glucose and fructose), proteases (enzymes which degrade proteins to their constituent peptides and amino acids), and a large variety of other enzymes, a few of which are outlined in Table 38-1.

In recent years detergents containing certain enzymes of *Bacillus subtilis* had gained domestic and commercial popularity, since these enzymes can remove many kinds of stains from fabrics. Thermophilic strains of *Bacillus* were used in the detergents because their enzymes are often more heat resistant than those of mesophilic strains. The detergents were often sold in powder form and generated dusts which, when inhaled, sometimes resulted in allergic sensitization to the bacterial enzyme proteins. This kind of sensitization (probably a mixture of IgE-mediated and immune-complex-mediated allergic reactions, see Chapter 21) led to a number of serious respiratory problems in industrial workers exposed to large amounts of the detergent dusts. As a result, the enzyme-containing detergents have lost popularity in the United States, although they are still used in Europe.

MICROORGANISMS IN SEA FARMING

Sea farming, or *mariculture,* is becoming increasingly important on a commercial scale in many developed countries of the world. Anyone who has recently priced shrimp, oysters, and other marine delicacies can appreciate the strong commercial interest in their cultivation.

Shellfish hatcheries culture a number of species of algae as food for mollusk larvae. These algae include various flagellates and diatoms representing the phyla Chlorophyta and Chrysophyta. Shrimp larvae are also fed algae, the small flagellates being used for shrimp larvae that are just hatched, and the larger diatoms for older and larger shrimp larvae.

Except for the bivalve mollusks and the early stages of shrimps, commercially grown marine animals do not feed directly on microorganisms; however, they are often fed brine shrimp or rotifers, and these in turn are grown by feeding on microbes. The single-cell proteins produced by microorganisms are also used in mariculture as fish food.

GIBBERELLINS

The gibberellins comprise a group of substances which have remarkable capacities for stimulating plant growth and blossoming. These substances include gibberellic acid and closely related compounds with a complex cyclic organic acid structure. The fungi *Gibberella* and *Fusarium* are used commercially to produce gibberellins. Figure 38-2 illustrates the effects of gibberellins on plant growth.

DEXTRANS

Some species of *Leuconostoc* can convert sucrose into both fructose and the large polymers of glucose called dextrans. Solutions of dextran isotonic with blood are used to supplement plasma transfusions for patients who have had extensive blood loss. The dextrans have a number of other uses both in medicine and in the scientific laboratory.

Dextran, page 43

OTHER APPLICATIONS

In addition to those chemicals already mentioned, alcohols and acetones can be commercially manufactured from microbial activity. Industrial ethyl alcohol may be produced from molasses by *S. cerevisiae* and other yeasts, and distilled. The lactose in whey, the starch in grains, and the sugars in molasses are all readily converted by yeasts into ethyl alcohol suitable for industrial use. Butyl alcohol and acetone are produced by the anaerobic metabolism of some species of *Clos-*

(a)

(b)

FIGURE 38-2
The effects of gibberellins on plant growth. Untreated plants are pictured on the left, and plants treated with gibberellins are on the right. (a) Bush beans. (b) Head lettuce. (Courtesy of M. J. Bukovac.)

tridium; such fermentation, however, is not currently used commercially because it is now cheaper to produce acetone and butanol synthetically.

Linen, hemp, and other fibers are made by retting, a process in which microbial enzymes break down the pectin in plant stems, releasing the natural fibrous materials in the stems. In preparing linen, for example, bundles of flax are submerged in water; anaerobic conditions then develop, and these permit clostridia to grow in the material. During a period lasting about two weeks clostridial enzymes attack the pectin in the flax and free the fibers, which are subsequently cleaned and made into linen cloth. Other bacteria can carry out an aerobic retting process.

MICROORGANISMS IN BIOASSAYS

Microorganisms are frequently used to assay the amounts of certain substances present in a mixture. Some organisms can detect traces of materials too small to be detected by chemical assays. Microbes can also distinguish between the D- and L-isomers of amino acids when these are not susceptible to distinction by ordinary chemical analyses.

Vitamins are usually measured by microbiological assay. In this procedure, strains of bacteria which require a vitamin are grown on a medium lacking it; vitamins are present only in the material added to the medium. Growth of the bacteria occurs only if the required vitamin is present and growth is proportional to the amount of vitamin present.

SUMMARY

Microbial activities have many commercial applications. Apart from the production of foods, discussed in a previous chapter, other commercial applications of microorganisms include the manufacture of amino acids and a number of other organic acids, vitamins, antibiotics, vaccines, toxoids, antitoxins, enzymes, and a host of other useful products. Frequently microorganisms are also used to assay the amount of certain substances, such as vitamins, present in a sample.

QUESTIONS

REVIEW

1. List some of the molds that are commercially important in enzyme production.
2. What genera of microorganisms are able to convert plant sterols to valuable steroid hormones?
3. What are the principles involved in producing toxoids, antitoxins, and vaccines?
4. What are the hazards of using "enzyme detergents"? What is the source of the enzymes in the detergents?
5. How can the essential amino acid L-lysine be produced commercially by microbial activity?

1. Suggest some commercial processes that could be developed to utilize waste products. (Chapter 39 comments on some of these possibilities.)
2. How would you go about screening many microbial cultures for the possible production of antimicrobial agents?

FURTHER READING

HANG, Y. D., D. F. SPLITTSTOESSER, and E. E. WOODAMS, "Utilization of Brewery Spent Grain Liquor by *Aspergillus niger*," *Applied Microbiology* 30:879–880 (1975).

MILLER, B. M., and W. LITSKY, *Industrial Microbiology*. New York: McGraw-Hill, 1976. A multiauthored series of chapters on many aspects of commercial microbiology, food and petroleum microbiology, microbiological deterioration, and biological aspects of water quality.

PRESCOTT, S. C., and C. G. DUNN, *Industrial Microbiology*. 3d ed. New York: McGraw-Hill, 1959.

Progress in Industrial Microbiology. Vols. 1–13, 1959–1974. Despite the success of these review volumes in presenting current information about industrial microbiology, the series ceased being published in 1974. The last volume (Vol. 13, 1974) contains chapters on single-cell protein, biodeterioration, and vaccine production.

RYTHER, J. H., and J. C. GOLDMAN, "Microbes as Food in Mariculture," *Annual Review of Microbiology* 29:429–443 (1975). A review restricted to the role of microorganisms as food, directly or indirectly, for commercially produced marine organisms.

UNDERKOFLER, L. A., and R. J. HICKEY, eds. *Industrial Fermentations*. Vols. I and II. New York: Chemical Publishing, 1954.

UNDERKOFLER, L. A., ed. *Developments in Industrial Microbiology*. Vol. 16. (1976).

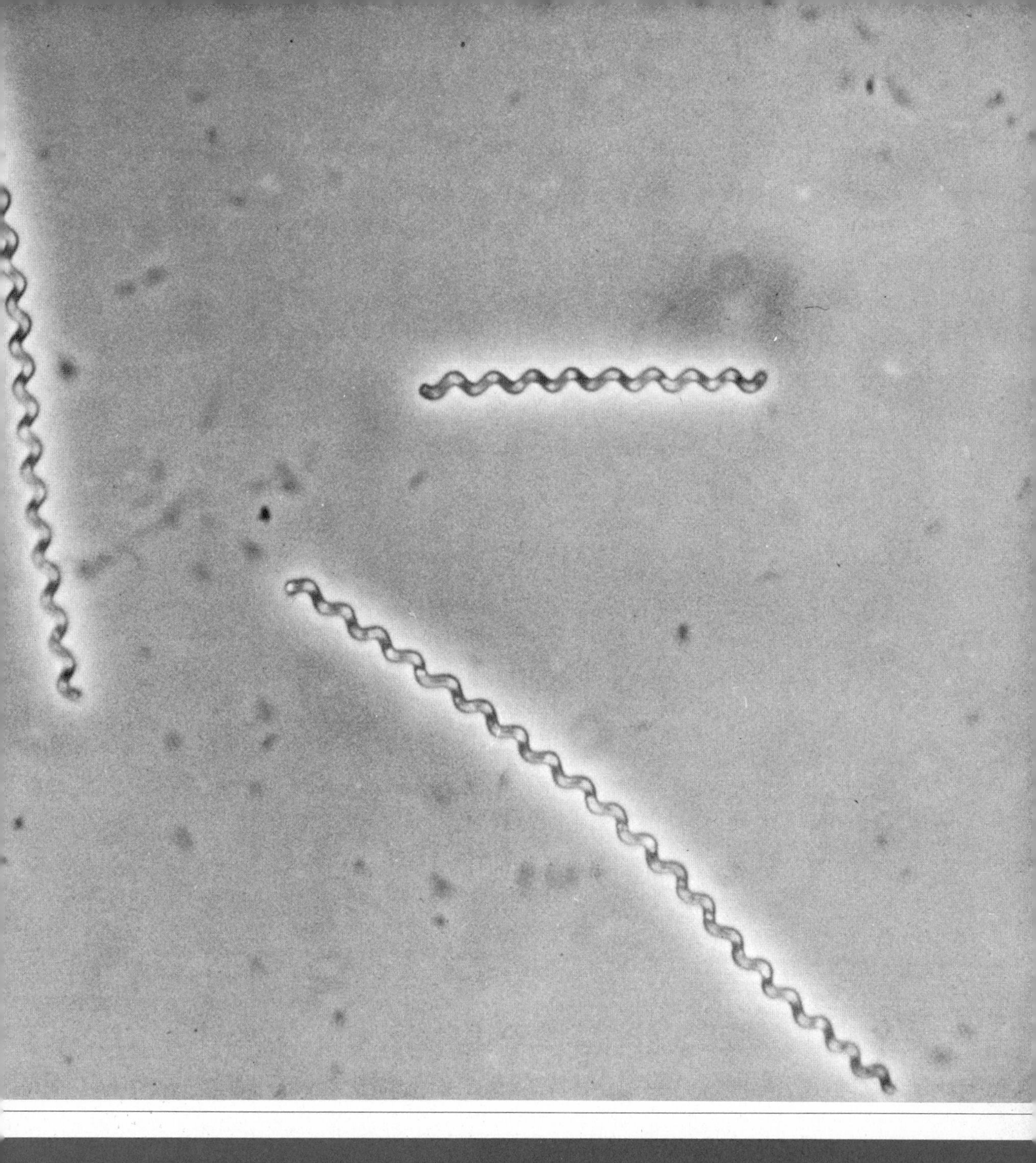

V THE CONTINUING FRONTIER

CHAPTER 39

LOOKING AHEAD

History teaches us that men and nations behave wisely once they have exhausted all other alternatives.

—ABBA EBAN

We are all aware that the human population has been increasing rapidly. Two thousand years ago the earth had between 100 and 200 million people; this number increased to 500 million by the seventeenth century, and today, only four centuries later, there are about 4 billion persons on this earth. Even this mind-boggling figure is expected to double within our lifetime. The earth's huge population thus threatens to outstrip the capacity of the earth to feed it, and it certainly will soon exhaust the natural resources on which our present lifestyles depend. Nonrenewable energy sources are particularly threatened. Moreover, the availability of energy and food production are inextricably intertwined, as witnessed by the need for irrigation, production and application of fertilizers, farm machinery and food processing.

SINGLE-CELL PROTEIN

In 1966 a conference concerned with the production of food from microorganisms was held at the Massachusetts Institute of Technology. It was there that the term *single-cell protein* (SCP), representing the protein in microbes, was coined in order to overcome any possible unpleasant connotation associated with the eating of bacteria, seaweeds, or fungi. Although the term is not always strictly correct, because some of the organisms that produce SCP are multicellular, it is useful in designating a potential source of food protein that is already of some importance and may become overwhelmingly so in the years to come.

Growth in the world's population, along with an alarming decrease in available energy sources, point to the inescapable conclusion that we may have to turn to SCP as a major food source. Another potential specter—one which has been with us since biblical times—is the threat of flood and drought. This was dramatically illustrated during the summer of 1976, when record droughts afflicted major grain-producing areas in most of Europe and in parts of every continent, while floods inundated other areas. Both flood and drought result in extensive crop loss. Beyond these threats, meteorologists speculate that the earth's climate may be undergoing profound changes, perhaps of a cyclic nature. If the average worldwide temperature were to drop just 1°C, the growing seasons in the temperate zones would be shortened enough to cause critical reductions in the earth's food supplies. Thus the development of SCP is especially attractive in that it enables food to be produced independent of weather or climate. Moreover, with the earth's population expected to double in a mere 30 years, serious food shortages seem inevitable, regardless of climate.

Another advantage of SCP is that it can be produced rapidly. For example, it has been estimated that half a ton of growing yeast can produce 50 tons of protein per day. Because from 30 to as much as 70 percent of a cell, on a dry-weight basis, consists of protein, the efficiency of microbial conversion of nutrients to protein is great and the yield of protein is high. In addition, the land, water, and personnel needed to produce SCP are much less than are needed in current agricultural practice. For example, a single yeast-producing plant in Minnesota is now expected to produce as much protein in one year as 400,000 acres of land devoted to raising beef cattle; the expenditure of energy and labor will also be far smaller.

Composition of bacterial cell, page 12

Diverse substrates, including some wastes that currently present disposal problems, can serve as single-cell protein precursors. Each year, in the United States alone, some 3.5 billion tons of waste material must be handled at a cost of $4.5 billion. Half of this waste consists of cellulose and materials related to it. At Lousiana State University a method for converting cellulose-containing sugar cane wastes to SCP has already been developed on an experimental basis. Under carefully controlled conditions, the wastes are chopped and partially degraded chemically before being inoculated with bacteria of the genus *Cellulomonas,* which further degrade the materials to the sugars and other metabolic products that can be used by other bacteria. The bacteria so grown contain about 60 percent protein, which can be separated, dried and used for animal feed.

Cellulose, page 42

Very simple substrates can be used by some algae and cyanobacteria. For example, the photosynthetic cyanobacterium *Spirulina maxima* grows on organic waste materials, bicarbonate, and sunlight; diluted sewage sludge serves as a good source of organic nutrients. For centuries *S. maxima* has been utilized as food by the people of the central African country of Chad. Currently, *S. maxima* is being grown in the alkaline waters of Lake Texcoco in Mexico on a commercial scale. Eucaryotic algae, such as kelp, are actually farmed in the ocean even now, and plans are being made to expand this farming operation.

Cyanobacteria, pages 272–273

In addition to algae and cyanobacteria, some bacteria also can convert simple substrates into protein. *Hydrogenomonas* species, for example, utilize CO_2 and H_2 to produce high protein-containing cells, and some species of *Pseudomonas* can convert methanol to SCP. Other pseudomonads and some yeasts as well use more complex substrates, such as kerosene or fuel oil, as substrates for protein synthesis. One source estimates that 3 percent of the world's current annual oil production is enough to meet the protein needs of everyone on earth for one year.

Pseudomonas, pages 252–254

The various SCP products discussed so far have been used as animal feeds and have not generally been proposed as food for direct human consumption. However, various kinds of yeasts have long been accepted as food for human beings or have been used in food preparation. Perhaps the most familiar is *Saccharomyces cerevisiae,* the yeast used for making bread, wine, and beer (see Chapters 7 and 37). At present, yeasts grown commercially on wastes from corn processing are being used primarily for animal feed, but torula yeast is being "farmed" in an indoor tank where it grows on ethyl alcohol. The crop of this yeast is now used as a supplement to salad dressing, and there is the possibility that it may be used in the very near future as a supplement in such processed foods as meat patties, pasta, frozen pizzas, and sauces.

Saccharomyces, page 673

Yeasts also are being used to reclaim nutrients from highly complex starting materials. For example, two different yeasts are now being used to produce animal feed from potato starch-water. One of these yeasts produces enzymes which degrade carbohydrates to sugars, while the other species utilizes the resulting sugars to reproduce itself, with the production of protein. However, in the United States, SCP is not being used to any great extent, even in animal feeds.

A major disadvantage of using SCP for human consumption is its widespread lack of acceptance; SCP may not be pleasing in taste, and it may, furthermore, cause gastrointestinal upset. This intestinal response may relate to toxic properties of cell walls and to the high nucleic acid content of SCP. Thus research is now being conducted on the possibility of using naturally occurring wall-less organisms, such as the alga *Dunaliella,* as a source of single-cell protein. Another objection to SCP is that many SCP products lack one or more of the essential amino acids, very often sulfur-containing cysteine and methionine, which must then be added as a supplement.

Oddly enough, little is known about the hazards of using SCP as human food. Until more is learned about the possible toxicity of, or adverse reactions to SCP, this material will probably continue to be primarily important as an animal feed. When more is known and world conditions change, it is conceivable that SCP will assume a central role in the feeding of people throughout the world, while at the same time alleviating problems of waste disposal.

ENERGY PRODUCTION

Few of our natural resources have been as instrumental in shaping modern society as the fossil fuels. This tremendous store of energy was delivered to the earth in the form of sunlight trapped by photosynthesis and stored, over hundreds of millions of years, in the chemical bonds of organic material and coal. Since the dawn of the industrial era, we have consumed this energy store at an increasingly rapid rate. In the United States alone, total energy usage has doubled every 16 years since World War II. Much energy is wasted, particularly in the conversion of one form of energy to another. For example, over three-quarters of the United States' electrical power is produced from fossil fuels, with almost 70 percent of the energy from these fuels lost in power production and transmission. Automobiles, representing one of the least efficient modes of transportation, consume 20 percent of the nation's energy. Presently, the average per capita consumption of oil alone is more than 1000 gal per year (including gasoline, plastics, fertilizer, and other petroleum products).

What role can microorganisms play in our food-energy dilemma? They offer no immediate solutions, but exciting possibilities lie in their high growth rates and in their collective ability to carry out a wide range of energy conversions. Microorganisms can, for example, trap solar energy in photosynthesis and convert useless organic materials to useful ones such as food and fuel. Moreover, our growing understanding of genetics and genetic engineering have the potential for significantly broadening the usefulness of microbial energy conversions.

Photosynthetic Organisms in Energy Conversions

Photosynthesis, pages 160–162

The photosynthetic activity of living cells provides a way of concentrating and storing solar energy. For example, a slowly growing tree produces firewood; aquatic plant and microorganisms concentrate and store energy even more rapidly. In experimental ponds algae growing on the nutrients in processed sewage have produced as much as 500 lb of algae per acre per day. This mass of organic material is a possible source of fuel, either for burning directly or by conversion to another form. One method of such conversion is the application of heat under pressure, causing pyrolysis of an organic material, a process similar to that which produces crude oil in nature. A chemistry teacher in Michigan uses this method to produce gasoline from garbage and lawn clippings with an apparatus he has built in his back yard. A similar process obtains 1.3 barrels of oil from each ton of dry organic waste fed into it. Organic material can also be burned at a high temperature with limiting oxygen to make producer gas (mainly a mixture of hydrogen and carbon monoxide); this gas was used to power trucks and tractors in Europe and China during World War II.

Ethanol

Ethanol formation, page 155

Microbial action on organic material can also be employed to obtain fuel. Ethanol can be produced by fermentation and is an excellent high energy, clean-burning fuel. The use of ethanol has not been economical in the past because energy was required to distill ethanol from its fermentation reaction mixtures. But the use of less expensive starting materials for fermentative ethanol production promises to compete in cost with the present chemical synthesis of this substance. For example, in converting waste paper to ethanol, the cellulose degrading enzyme of a mold converts the cellulose to glucose; a second organism then ferments the glucose to ethanol. A recently developed process of this type promises to produce ethanol at less than 10 cents per liter. In the United States, ethanol production from waste paper alone could replace 10 percent of our current gasoline needs.

Methane

Methane is a clean-burning, odorless gas with a heating value about two-thirds that of natural gas. It has been used for many years for cooking, heating, and running stationary internal combustion engines. In Germany before World War II, 22,000 automobiles were adapted to run on methane; the powering of automobiles, however, is not the best use of methane since it takes about 150 cu ft of methane (the yield from digesting a single day's manure from five cows) to equal one gallon of gasoline. Furthermore, methane must obviously be drastically compressed for use in transportation vehicles. Methane can, however, be converted to methanol, a much better fuel for automobiles of current design.

Organic waste can be digested anaerobically to produce methane ("marsh gas"), and methane production also goes on in nature, in lake and marine sediments and in the intestines of human beings and other animals. Indeed, mine disasters have resulted from the accumulation of natural methane, because the gas is explosive when mixed with air.

The methanogenic bacteria (the organisms which produce methane naturally) are a diverse group of strict anaerobes presently classified into four distinct genera. Some of these organisms are thermophilic. The methanogenic bacteria produce methane as the last product of a complex series of reactions that involve a number of different organisms. In this process, when organic material is sealed in a tank, oxygen is quickly depleted by facultative anaerobes. Complex polysaccharides, such as cellulose, are then degraded, first to monosaccharides such as glucose, and then to such organic acids as acetic acid, as well as to hydrogen and carbon dioxide.

Methane formation
page 616

$$\text{Organic material} \xrightarrow[\text{(several steps)}]{\text{anaerobic digestion}} CH_3COOH + H_2 + CO_2$$

$$CO_2 + 4H_2 \longrightarrow \underset{\text{methane}}{CH_4} + 2H_2O$$

The total yield of this reaction is 3 moles of methane from each mole of glucose, but the energy value of the methane produced is only slightly less than that from the 2 moles of ethanol obtainable from a mole of glucose. Unlike ethanol, however, methane is poorly soluble in water and only about half as dense as air. With a modest temperature increase and agitation of the reaction mixture, it quickly separates and collects in the top of the vessel, along with the carbon dioxide produced in the reaction. It is important to note that the production of methane causes no loss of nitrogen from the reaction mixture. Thus, harvesting methane from organic wastes does not impair their potential value as fertilizer for the growing of algae or plants.

Recent studies employing thermophilic methanogenic bacteria and continuous flow digesters indicate that in the United States the conversion of agricultural organic wastes to methane could replace over 20 percent of the natural gas currently being used. Conversion of organic material to methane obviously has great potential. However, much remains to be done to understand the biochemical mechanisms of, to obtain the best combinations of bacteria for, and to design efficient digesters for commercial methane production.

In discussing the commercial production of methane, it is important to recognize that if the organic material used for methane production were not converted to methane, it might be used for other purposes connected with energy and food production. For example, if this organic material were wood or had a high carbon content, it might be burned directly to provide heat. It is also important to take into account that the costs in money and expenditures of energy required for transporting organic material to a fermentation plant large enough to enable the methane generating process to be run efficiently might be prohibitive.

Hydrogen

Hydrogen is a clean-burning fuel much like methane in its general characteristics, but it contains less energy per equivalent volume. Hydrogen gas is produced by numerous bacteria. For example, the pathogen *Clostridium perfringens* produces large quantities of hydrogen gas in its fermentation of carbohydrates. Again, a mixed bacterial system might be necessary to degrade complex organic materials before hydrogen could be generated efficiently, and hydrogen-utilizing bacteria would need to be excluded from this system.

Clostridium perfringens, pages 430, 555

Another interesting approach to hydrogen production employs photosynthetic procaryotes such as cyanobacteria. In the absence of nitrogen gas the nitrogen-fixing enzyme systems (nitrogenases) of these organisms can reduce the hydrogen of water to hydrogen gas; this reaction requires a large amount of energy in the form of ATP generated by photosynthesis. Thus hydrogen production by cyanobacteria occurs only in light and in an atmosphere which excludes oxygen and nitrogen. In theory 10 percent of the solar energy absorbed by cyanobacteria could be recovered in hydrogen gas, but so far the yields have been low and have not been sustained.

Nitrogenase, page 625

Hydrogen has great appeal as an energy source because of the variety of microbial production methods for it and the availability of the necessary substrate (water). However, on the basis of our present knowledge, microbial production of hydrogen is an inherently inefficient process and is unlikely to make a major impact on our energy needs. Despite this, the intriguing possibility of developing a microbial hydrogen generator will undoubtedly continue to spur scientific investigations in this area.

BIOLOGICAL PEST CONTROL

General Considerations

From the previous discussions of food production and energy, it should be clear that the availability of energy and the production of food are inextricably interrelated. Without the gasoline to run its tractors, modern agriculture could not exist. Without the energy to produce the hydrogen needed to make ammonia fertilizers, crop production would be severely curtailed. One approach to increasing the output of food production, without drastically increasing energy consumption, is to ensure that insect pests do not destroy crops either in the field before they are harvested or while they are being stored.

Insects destroy approximately 10 percent of all crops grown in the United States, resulting in a loss of from \$5 to \$6 billion annually. In many parts of Africa the food losses caused by insect pests may reach 75 percent. In addition to the loss of lives which results indirectly from their destruction of food crops, insects are a major direct health problem in many parts of the world. Malaria, transmitted by mosquitoes, affects 100 million people a year in Africa and kills 800,000. Over 700,000 people in Africa are blinded each year by a species of black fly. The tsetse fly which carries the protozoa that causes African sleeping sickness now prevails in over 25 percent of Africa.

People have waged chemical warfare against insects for many centuries. Chemical insecticides were first used in Europe in the 1600s, and in the 1800s pyrethrum, a compound obtained from the chrysanthemum and still in use today, was introduced to kill fleas. However, the single most important development in insect control was the synthesis of DDT (*d*ichloro*d*iphenyl*t*richloroethane). In 1943 this substance so effectively stemmed a typhus epidemic in Naples that it soon came into general use against a wide variety of insects responsible for spreading malaria, typhus, and encephalitis. So successful was DDT that a number of other chlorine-containing insecticides such as aldrin, chlordane, and heptachlor were manufactured, beginning shortly after World War II. Through use of these insecticides alone, farmers have increased their crop yields by 10 percent.

DDT, page 617

Insects have been on the earth some 400 million years, and probably not a single species has ever been completely eradicated through human efforts. In part, this reflects the presence of insects in very large numbers, and that they can mutate to become resistant to deleterious agents, and can move away from danger. The belief is now widespread that we can realistically only hope to keep insects under such reasonable control that their damage is economically acceptable. One view is that this goal must involve an integrated approach. Potent chemical insecticides will probably always be needed to cope with any acute insect problem, but the major responsibility for keeping insects under control year in and year out will probably rest on biological methods. These can be aimed at specific insect targets without adversely affecting either human beings or the environment. Furthermore such techniques may move with the insect and multiply at its expense, and can therefore be long-lived. In the following discussion, we shall emphasize one type of biological insect control—control by microorganisms.

Microbial Control Agent

That insects suffer from infectious diseases has been known since the third century A.D. Aristotle recognized that bees suffered disease and in 1870 Pasteur correctly diagnosed an infectious disease of silkworms as being caused by a protozoan. Insect diseases are known to be caused by virtually all members of the microbial world, including typical bacteria, mycoplasma, viruses, rickettsia, protozoa, and fungi. However, relatively few microbial agents are currently being used commercially to combat insects. This probably results from the fact that chemical insecticides have been so successful that there has until recently been no need to develop more expensive and less predictable biological agents. However, this situation is now changing, and many companies are actively developing and marketing biological insect control agents.

The desirable attributes of these control agents are shown in Table 39-1. Unfortunately, no single organism has all of the features shown. As the table indicates, microbial control agents can be applied in one of two ways, either as insecticides or as agents which will establish themselves and persist in the environment. In the first case, the agent is added in large amounts and at regular intervals, much as a chemical insecticide is applied. In the second, the agent is added as a relatively small inoculum which will persist in an insect population, allowing nature to take its course. It is desirable to achieve a balance of the two methods.

Bacillus thuringiensis and Other Bacilli in Pest Control

B. thuringiensis is the most widely recognized and most intensively investigated of all bacterial insect pathogens. The organism is a Gram-positive, spore-form-

TABLE 39-1
Desirable Attributes of Two Types of Microbial Control Agents

Persistent (Established in Population)	*Microbial Insecticides*
Spreads rapidly	Easily applied
Able to find susceptible host	Inexpensive
Persistent	Virulent
Safe	Easily produced
Able to control to levels of minimal economic importance	Stores well
Predictable control	Safe
	Able to control to levels of minimal economic importance
	Predictable control

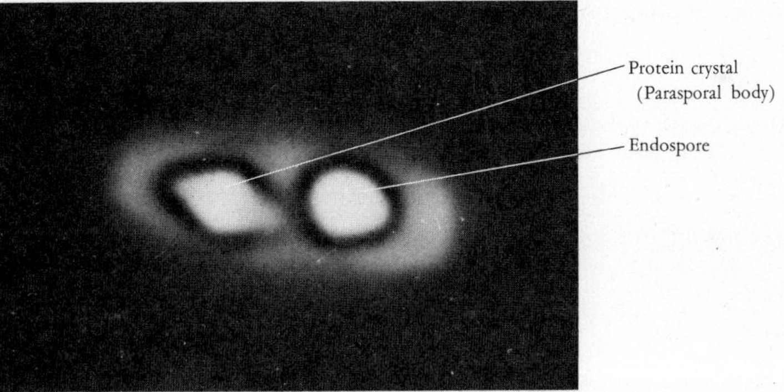

FIGURE 39-1
Protein crystal (parasporal body) and endospore in a cell of *Bacillus thuringiensis*. (Courtesy of C. L. Hannay; Reproduced by permission of the National Research Council of Canada. From *Canad. J. Microbiol.* 1:694, 1955.)

ing rod, with one unusual characteristic that separates it from other bacilli: it synthesizes a diamond-shaped crystal of toxin, the *parasporal body,* at the time of sporulation (Figure 39-1). The properties of this crystalline toxin are the basis of the insecticidal properties of the bacterium. The crystal is soluble only under alkaline conditions, such as are found in the midgut of a large group of damage-causing insects. Following ingestion by a susceptible insect, the parasporal body of the microbe dissolves in the insect midgut and the products of the crystal act as an endotoxin which affects integrity of the insect gut membrane. The membrane ruptures, the contents of the gut spill out, and general paralysis ensues. Most economically important insects are killed by the toxin and even by ingestion of the spores of *B. thuringiensis* alone. The pathology of the latter killing effect is unknown.

More than a dozen companies are now producing this insecticide commercially. The bacilli are conveniently grown in 40,000-liter fermentation tanks and their spores and crystals collected. The mixture is spread using the same equipment that is used for the application of chemical insecticides. The *B. thuringiensis* insecticide has been used to protect vegetables, field crops, ornamental plants and fruit, as well as forest and shade trees. Its host spectrum is relatively large and it effectively controls such insect pests as the cabbage looper, cabbage worm, horn worm, alfalfa caterpillar, tobacco budworm, green clover worm, fruit tree leaf roller, grape leaffolder, cutworm, gypsy moth, Western hemlock looper, and Western tent caterpillar. The *B. thuringiensis* insecticide also controls several pests of stored flour, cornmeal, almonds, and tobacco, and its spores and crystals can effectively reduce the louse population on chickens without harming the animals.

Other members of the genus *Bacillus* have also been used commercially as insect control agents with marked success. These include *B. popillae* and *B. lentimorbus,* both of which persist in the environment. Both organisms cause "milky disease," so called because the hemolymph of the insects they infect becomes whitish. In contrast to *B. thuringiensis,* these organisms can sporulate only in an infected insect. Commercial preparations of spores of the two bacilli are obtained from diseased Japanese beetle larvae. However, once a susceptible insect has ingested the spores, they germinate and cause disease. The vegetative cells of the organism then sporulate and these are disseminated by the dying insect host to other susceptible insects. Thus, one dusting application to a heavily infested soil may keep an area free of insects for periods of up to 10 years. The Japanese beetle and European chafer are two of the principal pests attacked by milky disease. These two insects alone feed on over 200 species of plants, causing millions of dollars of crop damage each year.

Extensive field testing of the three species of *Bacilli* discussed above has been carried on for a long time. None of the three organisms appears to pose any threat to humans or animals, and no pollution of the environment has been detected from their use. In addition, insects have not developed any apparent resistance to these microorganisms or to their products.

Viruses in Pest Control

More than 300 viruses representing seven major virus groups have been isolated from insects. However, because of the potential risk of virus cross-infections with higher animals, the World Health Organization has recommended that only members of the *baculovirus group* should be considered as potential pesticidal agents (Figure 39-2). So far as is known, this group has no chemical, physical, or biological properties in common with any of the viruses found in vertebrates or plants. Within the group of baculoviruses, the *nuclear polyhedrosis viruses* seem to be the most promising insect-control agents. They are named for the large inclusion bodies called *polyhedra* which they form within the nuclei of the cells they infect. These polyhedra consist of a lattice of protein molecules which enclose the virions. When ingested by a susceptible insect, the polyhedra dissolve in the insect gut and the virions are released. For most of the virus incubation period, diseased insects show no signs of infection. One of the first signs of infection is that the hemolymph, normally clear, turns to a milky white fluid. At this stage the insect stops feeding and becomes very sluggish. Following death the insect's tissues disintegrate, its fragile skin breaks and its body fluids, containing large numbers of newly formed infectious polyhedra, are released.

Ordinarily the baculoviruses are quite virulent and their introduction into

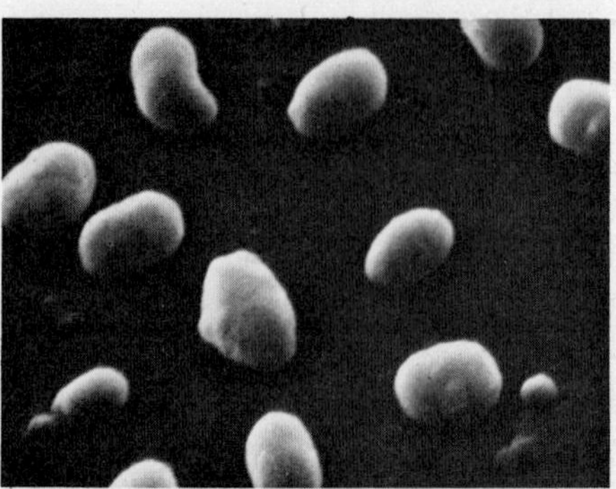

FIGURE 39-2
A baculovirus of the salt marsh caterpillar as viewed through a scanning electron microscope. (×37,500). (Courtesy of D. J. Witt; D. J. Witt, S. E. Milligan, and G. R. Stairs, *Virology*, 69:357, 1976.)

a susceptible insect population quickly decimates it. This group of viruses has been isolated from several major orders of insects, all of which show the same signs of disease. Eastern tent caterpillars and the tussock moth, which is responsible for major defoliation of Douglas fir trees, are particularly susceptible. The world's most serious grassland pest, the East African army worm, which has been known to denude 300 square miles of grassland in a few weeks, is also susceptible to baculovirus infection.

The future of microbial pest control appears promising. The numbers of microbes that infect insects is great and very few have been exploited. Furthermore, it should be possible to modify the properties of the infectious agents by selecting mutants which may have altered virulence or host range patterns. One can envision the use of mixtures of microbial insect control agents to increase their effectiveness. The costs of achieving an effective environmentally acceptable method of insect control will not be cheap, but there is good reason to hope that people will be able to maintain a period of détente with the insects for many years to come.

THE FIELD OF MEDICINE

The great successes of the twentieth century in the control of infectious diseases have rested primarily on such public health measures as sanitary sewage disposal, good nutrition, and immunization. The development of effective antimicrobial medications has played an important, yet relatively minor role in these gains. Although the gains against infectious disease have been made in the face of a rapid world population increase, shortages of cheap power and food and the increasing population density threaten these gains. Moreover, the rapid spread of resistance to antimicrobial agents is already causing some alarm, as well as renewed interest in the development of new and more effective vaccines.

Microbiological research will continue to focus on those diseases that appear to be infectious but for which a cause has not yet been identified. Recent advances include identification of rotaviruses as causes of winter diarrhea, and the linking of a bacterium with the "Legionnaires' disease" that occurred in Philadelphia in 1976. However, the emphasis of such research will not so much be on identifying a causative organism for each disease, but on the mechanisms by which the responsible organisms cause the disease. Historically, preoccupation with finding a single cause for each disease has hindered progress because some diseases can have several causes, and also because certain strains of a given microbial species can cause disease while others cannot. For example, diarrhea resulting from a buildup of cyclic AMP in cells of the small intestine can be caused by both *Vibrio cholerae* and *E. coli*. However, only those *E. coli* strains possessing the plasmid which codes for heat labile enterotoxin can cause the buildup of cyclic AMP; this plasmid can be passed to common bacterial species other than *E. coli*. The powerful tools of molecular biology are being applied to study the factors influencing distribution of such plasmids and their expression among groups of bacteria.

Vibrio cholerae, pages 437, 514

Cyclic AMP, pages 231–232

Plasmids, pages 209–211

The fruits of discoveries in molecular biology are also appearing in other areas of medicine. For example, the most intriguing recent developments in medicine have been in the field of virology, especially as it relates to such diseases of unknown cause as leukemia, rheumatoid arthritis, systemic lupus erythematosus, multiple sclerosis, myasthenia gravis and many others. It now seems possible that these diseases are caused by viruses, but it would be premature to say that the causes of any of these diseases are known, or even to predict that effective treatment for them will be forthcoming shortly. However, it is clear from studies of these diseases that we are entering a new era in the understanding of how viruses and human beings can interact.

Slow viruses, pages 349–351

Endogenous viruses, pages 359–360

One important discovery concerning viral infections was made more than a decade ago when Dr. Robert Rustigian of Tufts University infected tissue cell cultures with measles virus and then cultivated the cells in the presence of measles antibody. When measles virus was added to susceptible cells in the *absence* of antibody, the virus rapidly multiplied and destroyed the cells with the release of large numbers of infectious virus particles. However, Dr. Rustigian found that when the infected cells were incubated *with* antibody, a new type of infection was established in which the cells were not killed, although they produced abundant viral protein and very few infectious viral particles. This implied that a *persistent infection* with a defective mutant of the measles virus had been established.

The type of persistent infection established *in vitro* by Dr. Rustigian undoubtedly occurs occasionally under natural conditions and gives rise to diseases both because of the persistent infection and because of the immune response of the body against viral proteins. Persistent diseases of this nature are likely to have the following characteristics:

1. They are rare complications of common viral infections.
2. They have a slow onset and progression.
3. They involve mainly a single organ or body system.
4. Viral antigen is readily detected, while the infectious virus itself is difficult or impossible to demonstrate.

Several diseases mentioned earlier in this book will probably be shown to fall into this category, including subacute sclerosing panencephalitis (SSPE) and multiple sclerosis following measles and other virus infections, progressive hepatitis after hepatitis B viral infections, and progressive myocarditis or generalized myopathy after Coxsackie B virus infections. A number of other diseases are probably examples of this kind of persistent infection.

Measles, page 468

Hepatitis, pages 508–510

Myocarditis, pages 566–567

A second type of persistent infection occurs with certain DNA and RNA viruses that possess the enzyme reverse transcriptase. Such viruses have the potential of establishing persistent infections by integrating their DNA into the chromosome of their host cell. Whether or not the virus DNA becomes and remains integrated can depend on both host genetic factors and on the metabolic state of the host cell. The consequences of this type of persistent infection can include:

Reverse transcriptase, pages 357–359

1. No detectable abnormality of host cell function.
2. Production of an abnormal antigen by the host cell.
3. Recurrent episodes of host cell destruction (caused by excision of the viral genome from the host chromosome and replication of mature virus).
4. Defective host immunity (as, for example, when infected B cells are destroyed by cellular immunity).

Evidence is accumulating that a number of such human diseases as leukemia, and "autoimmune" diseases such as lupus erythematosus, will prove to be due to this type of persistent infection. Moreover, certain "hereditary" diseases, such as diabetes, will quite possibly be shown to result partly from persistent viruses passed from parents to offspring.

Better understanding of persistent viral infections will help in the quest for treatment of these diseases, or at least in avoiding treatments that are harmful. There are, however, additional potential benefits of research in this area. For example, more than 30 serious congenital human diseases result from the inability of the body to produce an essential enzyme. If the gene controlling production of the missing enzyme could be introduced into the cells of such people they would be normal. In theory, one could incorporate such a missing gene into a viral genome and then introduce the gene into the affected person by a harmless persistent infection with the virus. At present, the practical difficulties in accomplishing this feat are enormous, and its hazards are largely unknown.

The destruction of unwanted cells is another possible use of persistent viral infections that is currently under active investigation. For example, a persistent viral infection in a person receiving a kidney transplant could theoretically kill or inhibit the immune cells that attack the transplant and cause its rejection. The great specificity of viral infections has also suggested their use to infect and destroy certain cancers. These are some of the ideas spurring current research in microbiology. So far, they have led to important new findings in immunology; whether new modes of treatment will be forthcoming from them is as yet uncertain.

Transplantation, pages 413–414

SUMMARY

Microorganisms have the potential for playing important roles in the food-energy dilemma which faces the earth's expanding population. Because microbes can grow, with the synthesis of protein, on a large variety of waste materials, these organisms have the potential of serving as sources of food protein. In the United States many laboratories are experimenting with a variety of starting materials and organisms for food protein production, but although there are many promising approaches, single cell proteins are not being produced on a commercial scale. The use of microorganisms to produce metabolic products which can serve as a large-scale source of fuel is intriguing, but probably impractical for the foreseeable future. The use of certain bacteria and viruses in pest control is practical. Although chemical pesticides have dominated the pesticide market in the past, an environmentally aware public is beginning to recognize some unfortunate consequences of using long lasting chemicals that kill indiscriminately. *Bacillus thuringiensis* and the polyhedral necrosis virus are both being used commercially with good success as insecticides in many situations. Medical microbiology is moving more strongly in the direction of better prevention of infectious disease through new and improved vaccines, in understanding the epidemiology of the virulence and the antimicrobial resistance plasmids of pathogenic microorganisms, and in investigating persistent viral infections.

QUESTIONS

REVIEW

1. List some of the advantages provided by single cell protein as a source of food. Give some of its disadvantages.
2. List some of the potential fuel compounds produced in bacterial metabolism.
3. What characteristics of *B. thuringiensis* make it a useful agent for biological pest control?
4. List the characteristics of disease that result from persistent infection with agents that do not contain reverse transcriptase.

THOUGHT

1. What properties of an organism would make it ideal for the production of single cell protein?
2. Discuss the characteristics of a microbe that would make it an ideal agent for pest control.
3. How can you account for each of the disease characteristics of persistent infections produced by viruses that lack reverse transcriptase.

FURTHER READING

SINGLE-CELL PROTEIN

KIHLBERG, R., "The Microbe as a Source of Food," *Annual Review of Microbiology* 26: 427–466 (1972). An excellent review of the production and use of single-cell protein.

MATELES, R. I., and S. R. TANNENBAUM, eds., *Single Cell Protein.* Cambridge, Mass: M.I.T. Press, 1968. A multiauthored report on the production of single-cell protein.

MILLER, B. M., and W. LITSKY, *Industrial Microbiology.* New York: McGraw-Hill, 1976. This is by far the most current and well-written book in this field. Each area is covered by an authority. Chapter 13, on petroleum microbiology has an up-to-date discussion on single-cell protein.

ENERGY

CALVIN, M., "Photosynthesis as a Resource for Energy and Materials," *American Scientist,* 64:270–278 (May–June 1976). An excellent and readable discussion of potential photosynthetic applications to food and energy problems.

HEICHEL, G. H., "Agricultural Production and Energy Resources," *American Scientist* 64: 64–72 (January–February 1976). Interesting discussion of agricultural energy conversions.

BIOLOGICAL PEST CONTROL

BURGES, H. D., and N. W. HUSSEY, ed., *Microbial Control of Insects and Mites.* New York: Academic Press, 1971. An authoritative discussion, by experts in the field, on the current knowledge, achievements and future prospects for biological control of insects. Very complete but becoming a bit out of date.

CANTWELL, G. E., ed., *Insect Disease.* Vols. I and II. New York: Marcel Dekker, 1974. These two volumes are an introduction to insect pathology for advanced undergraduate students, but they serve as source material for biological insect control.

SUMMERS, M., R. ENGLER, L. FALCON, and P. VAIL, eds., *Baculoviruses for Insect Pest Control: Safety Considerations.* Washington, D.C.: American Society for Microbiology, 1975. The published proceedings of a symposium held in 1974. A very detailed and up-to-date discussion, by authorities in the field, on the most important group of viruses for insect control. There are 35 individual chapters and a question-and-answer section after each chapter.

INFECTIOUS DISEASES

DATTA, S. K., and R. S. SCHWARTZ, "Infectious(?) Myasthenia," *New England Journal of Medicine* 291:1304–1305 (December 1974). Short article discussing the role of persistent viral infections and heredity in autoimmune diseases.

HOLLAND, J. J., "Slow Inapparent and Recurrent Viruses," *Scientific American* (February 1974).

KLEIN, G., "The Epstein-Barr Virus and Neoplasia," *New England Journal of Medicine* 293:1353–1357 (December 1975). Illustrates the cooperative interaction of host cell genetics and viral transformation.

APPENDIX I

PATHWAY OF GLUCOSE DEGRADATION THROUGH THE GLYCOLYTIC AND TCA CYCLES

See figure on opposite page.

Appendix I
Pathway of Glucose Degradation through the Glycolytic and TCA Cycles

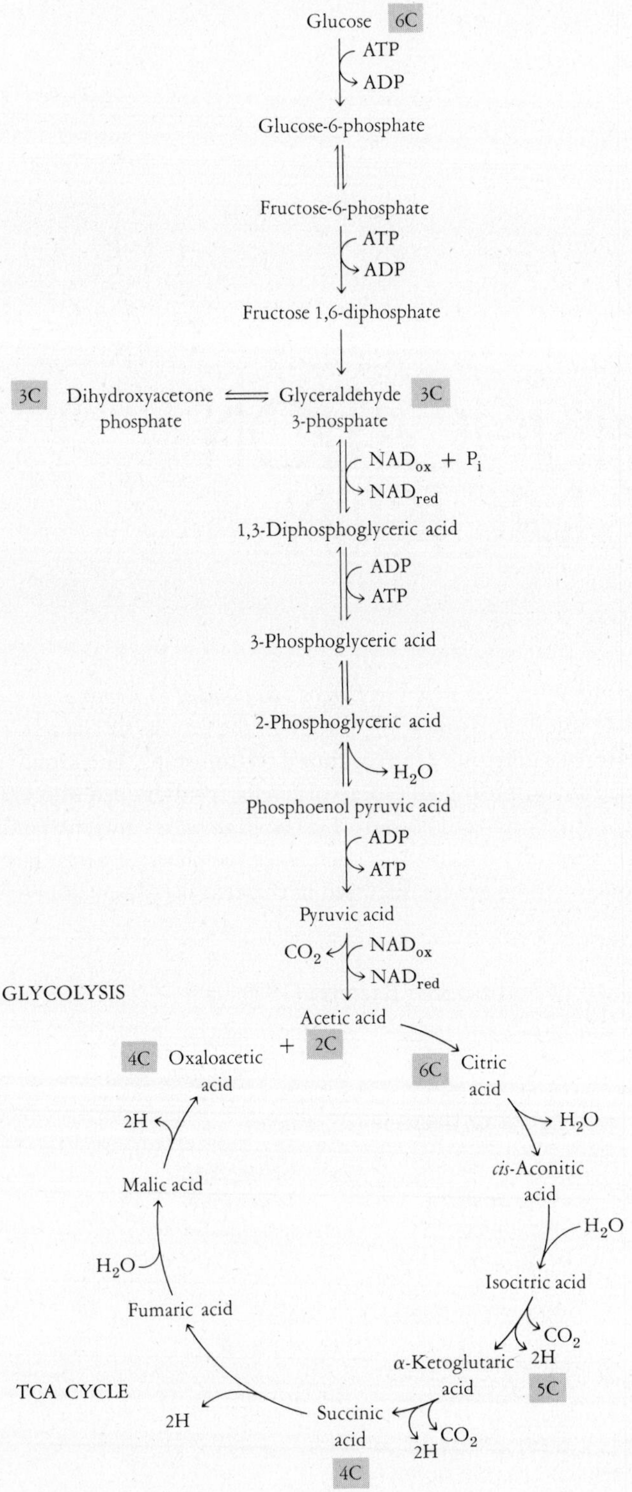

Glucose 6C
ATP → ADP
Glucose-6-phosphate
⇅
Fructose-6-phosphate
ATP → ADP
Fructose 1,6-diphosphate
↓
3C Dihydroxyacetone phosphate ⇌ Glyceraldehyde 3-phosphate 3C
NAD_{ox} + P_i → NAD_{red}
1,3-Diphosphoglyceric acid
ADP → ATP
3-Phosphoglyceric acid
⇅
2-Phosphoglyceric acid
→ H_2O
Phosphoenol pyruvic acid
ADP → ATP
Pyruvic acid
CO_2 ← NAD_{ox} → NAD_{red}

GLYCOLYSIS

Acetic acid 2C
+
4C Oxaloacetic acid
6C Citric acid
→ H_2O
cis-Aconitic acid
H_2O →
Isocitric acid
→ CO_2, 2H
α-Ketoglutaric acid 5C
→ CO_2, 2H
Succinic acid 4C
→ 2H
Fumaric acid
H_2O →
Malic acid
→ 2H
Oxaloacetic acid

TCA CYCLE

APPENDIX II

CLASSIFICATION OF BACTERIA

The grouping of bacteria given in *Bergey's Manual of Determinative Bacteriology*[1] represents an interim classification already in the process of changing as new information reveals previously unrecognized relationships. The kingdom Procaryotae, defined as comprising all procaryotic cells, is subdivided into cyanobacteria (Division I) and bacteria (Division II). The *Manual* is concerned only with further classification of Division II, which is composed of 19 parts. These parts are given below, with the genera discussed in this text listed beneath each part.

Part 1: **Phototrophic Bacteria** (18 genera)

Rhodomicrobium	*Chlorobium*
Rhodospirillum	*Chromatium*

Part 2: **Gliding Bacteria** (27 genera)

Stigmatella	*Cytophaga*
Chondromyces	*Leucothrix*
Beggiatoa	*Thiothrix*
Simonsiella	

Part 3: **Sheathed Bacteria** (7 genera)

Sphaerotilus
Leptothrix

[1] R. E. Buchanan and N. E. Gibbons, eds., *Bergey's Manual of Determinative Bacteriology,* 8th ed. (Baltimore: Williams and Wilkins, 1974).

Part 4: **Budding and/or Appendaged** (17 genera)

Hyphomicrobium
Caulobacter
Gallionella
Ancalomicrobium

Part 5: **Spirochetes** (5 genera)

Spirochaeta
Treponema[2]
Borrelia[2]
Leptospira[2]

Part 6: **Spiral and Curved Bacteria** (6 genera)

Spirillum[2]
Campylobacter[2]
Bdellovibrio[2]

Part 7: **Gram-negative Aerobic Rods and Cocci** (20 genera)

Pseudomonas[2]
Azotobacter
Rhizobium
Agrobacterium
Halobacterium
Beijerinckia
Alcaligenes
Brucella[2]
Bordetella[2]
Francisella[2]

Part 8: **Gram-negative Facultatively Anaerobic Rods** (26 genera)

Escherichia[2]
Salmonella[2]
Shigella[2]
Enterobacter[2]
Serratia[2]
Proteus[2]
Yersinia[2]
Vibrio[2]
Erwinia
Klebsiella[2]
Chromobacterium
Streptobacillus[2]
Photobacterium
Flavobacterium[2]
Haemophilus[2]
Pasteurella[2]

Part 9: **Gram-negative Anaerobic Bacteria** (9 genera)

Bacteroides[2]
Fusobacterium[2]
Desulfovibrio

Part 10: **Gram-negative Cocci and Coccobacilli** (6 genera)

Neisseria[2]
Acinetobacter[2]
Branhamella

Part 11: **Gram-negative Anaerobic Cocci** (3 genera)

Veillonella[2]

[2] Medically important genera.

Part 12: **Gram-negative Chemolithotrophic Bacteria** (17 genera)

Nitrobacter
Nitrococcus
Nitrosomonas
Thiobacillus
Siderocapsa

Part 13: **Methane-producing Bacteria** (3 genera)

Methanobacterium
Methanosarcina
Methanococcus

Part 14: **Gram-positive Cocci** (12 genera)

Micrococcus[2]
Staphylococcus[2]
Streptococcus[2]

Part 15: **Endospore-forming Rods and Cocci** (6 genera)

Bacillus[2]
Clostridium[2]

Part 16: **Gram-positive Asporogenous Rod-shaped Bacteria** (4 genera)

Lactobacillus
Listeria[2]

Part 17: **Actinomycetes and Related Organisms** (39 genera)

Corynebacterium[2]
Arthrobacter
Brevibacterium
Propionibacterium
Actinomyces[2]
Mycobacterium[2]
Nocardia[2]
Streptomyces[2]
Thermoactinomyces[2]

Part 18: **Rickettsias** (18 genera)

Rickettsia[2]
Coxiella[2]
Chlamydia[2]

Part 19: **Mycoplasmas** (4 genera)

Mycoplasma[2]
Spiroplasma

[2] Medically important genera.

GROUPS OF ANIMAL VIRUSES

No universal classification of animal viruses has been generally agreed upon, but most of the various groups of viruses infecting human beings and other animals can be divided into thirteen categories, 5 consisting of DNA viruses and 8 of RNA viruses. The viruses containing DNA are shown in Table III-1.

TABLE III-1
DNA Viruses

Group	*Important Characteristics*	*Examples*
1. Pox viruses	Very large ovoid viruses of complex structure; unlike other DNA viruses, they replicate in the cytoplasm	Cause smallpox, cowpox, rabbit myxomatosis, and economically important diseases of domestic fowl
2. Herpes viruses	Medium to large-sized, enveloped viruses that frequently cause latent infections; some are oncogenic	Human oral and genital herpes simplex, cytomegalovirus infections, varicella-zoster, and EB virus infections of human beings; infections of numerous warm- and cold-blooded vertebrates
3. Adenoviruses	Medium-sized, naked viruses; some cause persistent infections of tonsil tissue; some are oncogenic in heterologous hosts; base ratios vary widely although the appearance is uniform within the group	About 40 types known to infect the respiratory and intestinal tracts of human beings; frequent cause of conjunctivitis or sore throat; many additional types infect other animals
4. Papovaviruses	Small viruses, often oncogenic in animals other than human beings. Name derives from *pa*pilloma, *po*lyoma, *va*cuolating viruses	Viruses that cause human warts and certain degenerative brain diseases
5. Parvoviruses	Very small viruses; contain single-stranded DNA in some cases; some require a helper virus in order to propagate; also called picodnaviruses (pico = small, dna = DNA)	Infect rodents, swine, and arthropods; found in human beings in association with adenovirus helpers, but their pathologic significance is uncertain

Characteristics of RNA virus groups are presented in Table III-2.

TABLE III-2
RNA Viruses

Group	*Important Characteristics*	*Examples*
1. Picornaviruses	Very diverse group of small, naked viruses; their name derives from "pico," meaning small, and RNA; one subgroup, the enteroviruses, is acid-resistant; the second subgroup, the rhinoviruses, is sensitive to acid	Enteroviruses infect the intestine and a variety of body tissues; about 70 types infect human beings, including polio viruses, coxsackieviruses and echoviruses; rhinoviruses principally infect the respiratory tract and are the major cause of human colds
2. Togaviruses	Large, diverse group of medium-sized, enveloped viruses, many of which are transmitted by arthopods which they also often infect	Cause human infections such as rubella, yellow fever, equine encephalitis, dengue and others; numerous other vertebrates and invertebrates are susceptible to togavirus infections
3. Myxoviruses	Medium-sized, enveloped viruses, usually showing projecting spikes; the viral RNA is arranged in six segments; the three subgroups of myxoviruses are distinguished by antigenic differences in their nucleoproteins	The influenza viruses of human beings and other animals
4. Paramyxoviruses	Similar to myxoviruses but somewhat larger and with unsegmented RNA	Human rubeola and mumps; Newcastle disease of chickens; distemper of dogs
5. Reoviruses	Unlike other animal viruses, reoviruses contain double-stranded RNA; strongly resemble certain plant viruses	One group, the rotaviruses, is the principal cause of vomiting and diarrhea in children; Colorado tick fever virus falls into another reovirus subgroup; numerous other animals are susceptible to reoviruses
6. Rhabdoviruses	Bullet-shaped, enveloped viruses	Includes rabies virus and other viruses of mammals, fish, and insects
7. Retroviruses	Generally resemble myxoviruses in size and shape but possess the enzyme reverse transcriptase; one subgroup, the oncorna (*onco*genic RNA) viruses, transforms cells and causes tumors; a second subgroup causes certain slow virus infections of animals	Rous sarcoma virus of chickens; mouse mammary tumor virus; cat leukemia virus; visna virus of sheep
8. Arenaviruses	Medium-sized, enveloped viruses with a granular appearance	Lassa fever virus of human beings; lymphocytic choriomeningitis virus of mice

Other groups of animal viruses include the coronaviruses (which resemble myxoviruses but have petallike surface projections), and numerous arboviruses (arthropod-borne viruses) that fail to fit into existing groups. Also certain viruses which have only recently been characterized (such as the RNA virus of hepatitis A, the DNA virus of hepatitis B, and some of the "slow viruses") have not yet been precisely classified.

APPENDIX IV

SOME IMMUNIZING AGENTS OF MAJOR IMPORTANCE

Agent (*Active against*)	*Production Method*	*Other Information*
Diphtheria toxoid (exotoxin of *Corynebacterium diphtheriae,* the cause of diphtheria)	Exotoxin is converted to toxoid by treatment with formalin	Given in DPT[1], (see schedule); booster of diphtheria and tetanus toxoids recommended at 12–14 years and as needed during adulthood; immunity is usually long-lasting (10 years or longer)
Tetanus toxoid (exotoxin of *Clostridium tetani,* the cause of tetanus)	Exotoxin is converted to toxoid by treatment with formalin	Given in DPT[1] (see schedule); booster of diphtheria and tetanus toxoids recommended at 12–14 years and tetanus toxoid booster every 7–10 years thereafter
Pertussis vaccine (*Bordetella pertussis,* the cause of whooping cough)	Killed *B. pertussis*	Given in DPT[1] (see schedule)
Poliovirus vaccine (polioviruses, causes of paralytic poliomyelitis)	Living, attenuated viruses of the three types grown in monkey kidney or human cell cultures	Given orally on the same schedule as DPT[1] (see schedule); further boosters not recommended unless there is increased risk of exposure to the virus (via travel, patient care)
Rubeola vaccine (rubeola virus, the cause of measles)	Living, attenuated virus grown in chick embryo cell cultures	Given intramuscularly at 1 year of age or later
Mumps vaccine (mumps virus)	Living, attenuated virus grown in chick embryo cell cultures	Given intramuscularly after 1 year of age or to adolescents and adults who have not had mumps
Rubella vaccine (rubella virus, the cause of German measles)	Living, attenuated virus grown in duck embryo, rabbit, or human cell cultures	Given subcutaneously, usually to females aged 1 year to puberty (often at age 8–10 years)

Agent (*Active against*)	*Production Method*	*Other Information*
Influenza vaccine (influenza viruses)	Prevalent types of influenza viruses, grown in egg allantoic fluid and killed with formalin or ultraviolet irradiation	Given subcutaneously or intradermally to high-risk individuals annually.
Bacille Calmette-Guerin, or BCG (*Mycobacterium tuberculosis*)	Living, attenuated tubercle bacilli of the BCG strain	Widely used in other countries but in the United States recommended only for certain high-risk individuals (now being used also as a means of immunotherapy to increase nonspecific resistance to some tumors)
Rabies vaccine (rabies virus)	Rabies virus grown in duck embryo and inactivated with propiolactone	Given only to those with high risk of exposure to rabies, such as veterinarians
Typhoid vaccine (*Salmonella typhi,* the cause of typhoid fever)	Killed suspension of *S. typhi*	Gives limited short-term protection; given to travelers to endemic areas and high-risk individuals, such as laboratory workers
Cholera vaccine (*Vibrio cholerae,* the cause of cholera)	Killed suspension of *V. cholerae*	As for typhoid vaccine
Yellow fever vaccine (yellow fever virus)	Living, attenuated virus, grown in eggs or cell cultures	Gives good degree of protection for about 10 years; given to travelers to endemic areas

[1] DPT stands for diphtheria, pertussis, tetanus immunizing mixture, which is used for young children. Doses are given intramuscularly at 2–3 months, 4–5 months, 6–7 months, 15–19 months, and 4–6 years.

For further reading see *Vaccines: An Update,* Department of Health, Education and Welfare, Publication no. (FDA) 74-3013. Available from Consumer Information, Public Documents Distribution Center, Pueblo, Colo. 81009, and from U.S. Government Printing Office, No. 1974-544-19157, at no charge. Also see *Microbiology—1975,* ed. D. Schlessinger. Washington, D.C.: American Society for Microbiology, 1975, pp. 399–416. The report of a symposium on the development of new vaccines against *Haemophilus influenzae* type b, meningococci, respiratory viruses, and pseudomonas.

GLOSSARY

In addition to terms and their definitions, the following list includes the names of organisms referred to in the text. Where the pronunciation of a term or name is not obvious, the authors have attempted to show the commonly used pronunciation within parentheses following the term.

abscess (ab′-sess) A localized collection of pus surrounded by inflamed tissue.

Acetobacter (a-see′-toe-bak′-ter)

Acinetobacter (a-sin-et′oh-bak-ter)

Actinomyces (ak-tin-oh-my′-seez)

Actinoplanes (ak-tin-oh-plane′-eez)

activated sludge method A method of sewage treatment in which wastes are aerated and degraded by complex populations of aerobic microorganisms.

activation energy The energy required to elevate molecules from one energy level where they are nonreactive to a higher energy level where they can react spontaneously.

active site The sequence of amino acids on an enzyme molecule which is concerned with binding the substrate. Also called *catalytic site.*

active transport The energy-requiring process by which molecules are carried across cell boundaries. Often the molecules are moved from a lower to a higher concentration.

adenosine triphosphate (ATP) (a-den′-o-seen tri-fos′-fate) A chemical compound that is the major storage form of energy in cells. The chemical energy is stored in the form of two high-energy phosphate bonds, which release energy when hydrolyzed.

adenoviruses (ad′-e-no viruses)

adsorption (viral) The process of the interaction of a virus with specific receptors on host cells, resulting in retention of the virus on the cell surface.

Aedes (ah-ee′-deez)

aerobic organisms (aerobes) Organisms that can utilize oxygen as the final electron acceptor during metabolism.

Aeromonas (air-oh-moan′-az)

aerotolerant Able to survive and perhaps grow in the presence of 20 percent oxygen, although not utilizing it for metabolism.

agar (ah′-grr) A polysaccharide obtained from various species of seaweeds, used to gel materials (especially microbiological media). It cannot be broken down by most bacteria.

Agrobacterium (ag-roh-bak-teer′-ee-um)

akinetes (ack′kineats) Nonmotile resting cells produced by some algae; characterized by a thickened cell wall and food reserves.

alginates (al′-jin-ates) Polysaccharides derived from some of the brown algae; used to aid in emulsifying materials (such as causing suspension of small globules of one liquid in a second liquid with which the first will not mix).

allergy (hypersensitivity) An immunological response that results in damage to the responder, usually manifested by impaired function of a tissue.

Allomyces (al-oh-my′-seez)

allosteric protein (al-o-steer′-ic pro′-teen) A protein, commonly an enzyme, which changes its shape or conformation as a result of binding small molecules. The change in shape affects its ability to carry out its catalytic or other function.

allosteric site The sequence of amino acids on an allosteric protein which binds the small molecules. It is distinct from the active or catalytic site.

ameba, pl. amebae A unicellular organism with an indefinite changeable form. Also spelled amoeba.

amino acid (ah-mean′-o acid) An organic compound containing both a carboxyl (COOH) and an amino (—NH_2) group, bonded to the same carbon atom. The 20 amino acids which are the subunits of proteins vary in the structure of their side chains.

aminoglycoside (ah-mee-no-gly′-ko-side) A group of chemical compounds derived from sugars such as glucose and containing amino groups. Several antibiotics (including streptomycin) are aminoglycosides.

ammonification (a-moan′-i-fi-ka-shun) The reactions which result in the release of ammonia (NH_3) from organic nitrogen-containing molecules.

anaerobic jar (an-air-ro′-bic) A glass jar from which oxygen can be evacuated to provide an oxygen-free environment. Used in the growth of anaerobic organisms.

anaerobic organisms (anaerobes) Organisms that do not use oxygen as the final electron acceptor during metabolism; organisms that grow in the absence of air.

anaerobic respiration The metabolic process in which electrons are transferred from one compound to an inorganic acceptor molecule other than oxygen. The most common acceptor molecules are carbonate, sulfate, and nitrate.

anamnestic (memory) response (an′-am-nes′-tick) The immunological response to a second or subsequent dose of antigen; characterized by a more rapid, efficient and long-lasting response than resulted following the first (primary) dose of antigen.

Ancylostoma duodenale (an-kil-lahst′-toh-mah duo-deen-ah′-lee)

anemia (an-eem′-ee-a) A condition characterized by having less than the normal amount of hemoglobin.

anopheline (a-noff′-fa-leen) Refers to mosquitos of the genus *Anopheles.*

antibiotic A chemical substance produced by certain molds and bacteria which inhibits the growth or kills other microorganisms. Some antibiotics can now be synthesized chemically and others can be altered by chemical methods so as to enhance their usefulness.

antibody (an′-ti-body) Protein produced by the body in response to the presence of an antigen; can combine specifically with that antigen.

antibody reaction site The part of an antibody molecule that combines with the specific antigen.

anticodon (anti-ko′-don) The three nucleotides in transfer RNA which are complementary to a sequence of three nucleotides in messenger RNA (a codon).

antigen (an′ti-jen) A substance that can incite the production of specific antibodies and can combine with those antibodies.

antigenic determinant The part of an antigen molecule that combines with the specific antibody.

antiseptic A substance that will inhibit or kill microbes. As often used, this term implies that the substance is sufficiently nontoxic to apply to skin or other tissue.

antiserum A serum useful in treatment or in identification of antigens because of its content of specific antibody or group of antibodies.

Arthrobacter (are-throh-bak′-ter)

arthrospores (arth′-ro-spores) Fungal spores formed by breaking apart of the cells composing septate hyphae.

Ascaris lumbricoides (ass′-kah-riss lum-brik-koid′-eez)

ascus, pl. asci (ask′-us, ass-sigh) Saclike structure in which fungal spores (ascospores) are borne.

Aspergillus (ass-per-jill′-us)

asporogenous (ay-spore-ah′-jen-us) Not producing spores.

asymptomatic (ay-sim-to-mat′-ik) Without symptoms. When used to describe infections, the infected person notices no abnormalities.

atopic (ay-tow′-pick or ay-topic) Allergic; reactions that are caused by antibodies of the class IgE.

attenuated Modified so as to be incapable of causing disease under ordinary circumstances.

attenuator region (a-tĕn′-ū-ā-tor) A control region of DNA which must be traversed by RNA polymerase molecules before the polymerase can begin to transcribe structural genes.

autoclave A device employing steam under pressure and used for sterilizing materials stable to heat and moisture.

autoimmune Immune or hypersensitive to some constituents of one's own body.

autolyze To break up or destroy a cell, usually by means of its own enzymes.

autotroph (aw′-tow-trof) An organism that can utilize CO_2 as its main source of carbon.

auxotroph (ox′-o-trof) Term applied to laboratory-derived mutants of a microorganism that requires an organic supplement, other than a carbon and energy source, for growth; for example, an amino acid, purine, pyrimidine, or vitamin.

avirulent (ay-veer′-u-lent) Lacking virulence. When used to describe microbes, it means properties that normally promote the ability of an agent to cause disease are lacking.

axial filament The structure responsible for motility of spirochetes. It resembles a flagellum that lies within the layers of the cell wall.

axilla (ak-sil′-a) Armpit.

Azotobacter (ah-zoh-toh-bak′-ter)

Bacillus subtilis (bah-sil′-us sut′-till-lis)

B. megaterium (B. meg-a-teer′-e-um)

B. polymyxa (B. poly-mix′-a)

bacteriocin (back-tear′-e-o-sin) Protein molecules, produced by bacteria, which kill other strains of the same or closely related species. Bacteriocins are named after the organisms that produce them, that is, enteric bacteria such as *E. coli* produce colicins; *B. megaterium* synthesizes megacins. The chemical structure and mode of action differ for different bacteriocins.

bacteriocin factor (back-tear′-e-o-sin factor) A DNA plasmid or episome that carries the genetic information for the synthesis of a bacteriocin.

bacteriophage (back-tear′-e-o-fayge) A virus that infects bacteria; often abbreviated "phage."

Bacteroides fragilis (back-ter-oid′-eez frah-jil′-us)

bacteroids (back′tear-oids) Bizarre, irregularly shaped forms which certain bacteria can assume under certain environments; for example, *Rhizobium* in root nodules.

Balantidium coli (bal-an-tid′-ee-um coh′-lee)

barophiles (bear′-oh-files) Organisms that grow only under conditions of high pressure, such as those adapted to grow in ocean depths (where the pressure is as high as 1000 times the normal at sea level).

Bartonella bacilliformis (bar-ton-ell′-ah bah-sil-ah-for′-miss)

basal body The part of the bacterial flagellum which attaches the flagellum to the cell.

base analog A chemical whose structure is similar (analogous) to one of the purine or pyrimidine bases found in DNA or RNA.

base pair A purine with its complementary pyrimidine in DNA and RNA; adenine pairs with thymine, and guanine with cytosine, for example.

basidium, pl. basidia (bah-sid′-e-um, bah-sid′-e-ah) A clublike structure on which certain fungal spores (basidiospores) are borne.

Bdellovibrio bacteriovorus (dell-oh-vib′-ree-oh back-ter-oh-vor′-us)

Beggiatoa (bej-ee-a-toe′-ah)

Beijerinckia (by-yer-ink′-ee-ah)

benign (be-nine′) Nonmalignant. Favorable or neutral rather than serious.

Bifidobacterium (by-fid-oh-back-teer′-ee-um)

binary fission (bye′-ner-e′ fish′-en) An asexual reproductive process in which one cell splits into two independent daughter cells.

biochemical mutants A general term applied to mutants which have defects in biochemical pathways; generally applied to mutants defective in the synthesis or degradation of small molecules.

Biochemical Oxygen Demand (BOD) The oxygen-consuming property of water.

biological magnification The phenomenon of the increasing concentration of a chemical substance, such as a pesticide, in organisms the higher they are in the food chain.

biomass (bye′-o-mass) The total mass of living organisms in a given volume (for example, in seawater it is the total mass of living organisms per liter). The total weight of all organisms in any particular environment.

biopsy (bye′-opsee) The surgical removal of small pieces of tissue for microscopic examination to aid in diagnosis.

biosphere (bye′-o-sphere) The portion of the earth in which life exists.

biotype Strain of microorganism characterized by its pattern of biochemical reactions with different substances.

blastogenic response The division of cells in response to certain stimuli. The cells enlarge (to blast forms) in the course of division. For example, lymphocyte transformation in response to specific antigen is a blastogenic response.

Blastomyces dermatitidis (blast-oh-mye′seez der-mah-tit′-id-us)

blastospore (blast′-oh-spore) Reproductive cell formed by budding, as in the yeasts.

bloom (of algae) Overabundant growth of algae, often as a result of enrichment with a nutrient that would otherwise be scarce enough to limit growth.

Bordetella pertussis (bor-deh-tell′-ah per-tuss′-siss)

Borrelia recurrentis (bor-rell′-e-ah ree-cur-ren′-tiss)

Brucella (bru-sell′-ah)

budding An asexual form of cell multiplication in which protuberances (buds) form at various sites on the parent, enlarge, and develop into new individuals.

bursa equivalent The lymphoid tissue of mammals that is analogous to the Bursa of Fabricius of birds; the lymphoid organ that influences the development of B lymphocytes. The location of the bursa equivalent in mammals is not known.

calcareous (kal-care′-e-ous) Containing calcium oxide (lime).

Candida albicans (can′-did-ah al′-bik-anz)

capsid (kap′-sid) Protein coat that surrounds and protects the nucleic acid of a virus.

capsomere (kap′-so-mere) A subunit of the protein capsid of a virus.

carbohydrate An organic compound consisting of many hydroxyl (—OH) groups and containing either a ketone ($-\overset{\overset{\large O}{\|}}{C}-$) or aldehyde ($-\overset{\overset{\large O}{\|}}{C}-H$) group. Examples include sugars, cellulose, glycogen, starch.

carcinogen A cancer-causing agent.

carcinogenesis (car-sin-oh-jen′uh-sis) The induction of cancer.

carcinogenicity Cancer-causing capacity of an agent.

carrier An individual who has pathogenic microbes in or on his or her body without showing any signs of illness. The carrier state occurs during incubation and convalescence of infectious disease and with asymptomatic infection, colonization, or contamination. As usually used, the term implies that the microbes have access to the exterior of the body and thus potentially to other people.

catabolism (ka-tab′-o-lism) The metabolic degradation of organic compounds.

catabolite repression (ka-tab′-o-lite repression) The inhibition of genes governing a pathway of degradation by substances that can be efficiently utilized as sources of energy.

catalase (kat′-a-laze) An enzyme, found in human beings and many microorganisms, which degrades hydrogen peroxide to oxygen and water.

Caulobacter (caw-loh-back′-ter)

cell-bound antibodies Antibodies that are attached to cells, either because they were not released from the cells that formed them or because they have attached to other cells.

cellulose A polysaccharide, composed of glucose subunits. The most abundant organic compound in the world.

central dogma The colloquial phrase given to the major theme of molecular biology that the flow of information follows the pattern: DNA to RNA to protein.

Cephalosporium (sef-alloh-spore′-ee-um)

cercaria (ser-care′-ee-ah) The last developmental stage of the free larvae of flukes, distinguished by possessing a tail.

chelate (key′-late) To remove metallic ions from a solution by combining them with certain organic compounds.

chelicera (kel-iss′-serah) Appendages close to the mouth of arachnids; evolutionary equivalent of pinchers.

chemostat (keme′-o-stat) A growth chamber with an inlet and outlet which allows a controlled continuous addition of nutrients and outflow of cells.

chert A dull-colored, flintlike quartz often found in limestone.

chitin (kite′-in) A polysaccharide found in the cell walls of fungi and in the external covering of arthropods and crustaceans.

Chlamydia (klah-mid′-ee-ah)

Chlamydomonas (klah-mid′-o-moan′-as)

chlamydospore (klah-mid′-oh-spore) A form of asexual spore of some fungi, characterized by a thickened wall and by resistance to adverse environmental conditions, but not to the same degree as the bacterial endospore.

chloramphenicol (klor-am-fen′-i-kol) A bacteriostatic antibiotic produced by a species of *Streptomyces* which inhibits protein synthesis in a variety of Gram-positive and Gram-negative organisms.

chlorophylls (klor′-o-fills) Green pigments necessary for photosynthesis.

chocolate agar Blood agar which has been heated under carefully controlled conditions so as to release hemin and destroy an enzyme which breaks down NAD. The heating causes the blood to assume a chocolate color.

Chondrococcus columnaris (kon-droh-kock′-us cahl-um-nare′-iss)

chorioallantoic membrane (kore′-e-oh-allan-tow′-ick membrane) The outer (or) one of the membranes surrounding an embryo, such as in an embryonated chicken egg, often used to grow viruses.

chromatographic column (crow-mat′-oh-graf′-ick column) An absorbent material packed into glass columns through which mixtures of substances are passed. Separation of components of the mixture may be achieved by selective absorption to different parts of the absorbent material or by differences in the ease by which the substances are removed (eluted) from the absorbent material.

ciliates (sill′-e-ates) Protozoa that bear multiple cilia; for example, the paramecia.

cilium, pl. cilia (sill′-e-a) A short, projecting hairlike organelle of locomotion, similar to a flagellum but thicker and having different internal structure.

clone (klohne) A family of cells derived from a single parental cell by repeated divisions; the offspring of a single cell.

Clostridium botulinum (Kloss-trid′-ee-um bot-u-line′-um)

Cl. perfringens (Cl. per-fringe′-enz)

Cl. tetani (Cl. tet′-an-ee)

coccidia (cock-sid′-e-ah) An order of protozoa in the class Sporozoa.

Coccidioides immitis (Kock-sid-ee-oid′-eez im′-mit-iss)

coenocytic (seen′-oh-sit′-ick) Having many nuclei within a single-celled organism.

coenzyme (co-en′-zyme) A small, organic molecule that transfers small molecules from one enzyme to another.

colicin (kole′-i-sin) One kind of bacteriocin; a protein produced by one strain of enteric bacteria which kills other strains of enteric bacteria.

coliform (cole′-e-form) Gram-negative rods, including *Escherichia coli* and similar species, which normally inhabit the colon. Commonly included in the coliform group are *Enterobacter aerogenes, Klebsiella* sp., and other related bacteria.

colloid (koll′-oid) A suspension of finely divided particles, approximately 5 to 5000Å in size, which do not settle rapidly and are not readily removed by filtration.

colonization Establishment of a site of reproduction of microbes on a material, animal, or person without necessarily resulting in tissue invasion or damage.

combined therapy The term applied to the simultaneous use of more than one technique in the treatment of a disease; often refers to the use of several antibiotics simultaneously.

commensalism (koh-men′-sa-lism) State of living on or in an organism without causing harm or giving benefit to the organism.

community All of the organisms inhabiting a common habitat and interacting with one another.

competent cell A bacterial cell which is capable of taking up and integrating high molecular weight DNA into its own chromosome in DNA-mediated transformation.

competitive inhibition The inhibition of enzyme activity caused by the competition of the inhibitor with the substrate for the active (catalytic) site on the enzyme. Impairment of function of an enzyme is due to its reaction with a substance chemically related to its normal substrate.

complement (C) A system of at least 11 serum proteins that act in sequence, producing certain biological effects concerned with inflammation and the immune response.

complex lipids Lipids that contain other elements in addition to hydrogen, oxygen, and carbon.

compost heap A pile of organic plant materials, commonly leaves and grass, which undergoes decomposition by microorganisms.

congener (kon′-jen-er) A thing closely related to another; in a chemical sense, a structure closely related in form to another chemical structure.

congenital Existing at birth.

conidiospores (cone-id′-e-oh-spores) Fungal spores borne on specialized structures called conidiophores.

conjugation A mechanism of gene transfer in bacteria which involves cell-to-cell contact.

conjunctiva (kon-junk-tye′-vah) The mucous membrane that lines the inner surface of the eyelids and exposed surfaces of the eyeball.

constitutive enzyme (kun-stit′-u-tive) An enzyme whose synthesis is not altered in response to changes in the environment.

contagious (kun-tay′-jeeus) Infections transmissible by contact; catching.

coralline (kor′-rahl-lean) Corallike. Refers to a class of corallike algae with hard, calcareous cell walls (the class Corallinaceae).

corals (kor′-rahls) Primitive, calcareous plants that grow principally in tropical waters.

core enzyme of RNA polymerase The portion of the enzyme which is specifically concerned with catalyzing the formation of ester bonds between nucleotides. It has no specificity as to where it initiates or stops transcribing.

corepressor (ko′-ree-press-sore) The small molecule which activates the repressor (regulator) protein to regulate gene function.

corticosteroid (kor-tik-oh-stee′-roid) A medication or hormone which resembles in structure and action corticosterone, a hormone produced by the outer portion of the adrenal gland.

Corynebacterium diphtheriae (koh-ryne-nee-bak-teer′-ee-um dif-theer′-ee-ee)

C. hoffmani (C. hoff′-man-ee)

C. xerosis (C. zeer-oh′-siss)

covalent bond (ko-va′-lent) A strong chemical bond formed by the sharing of electrons.

Coxiella burnetii (cox-see-ell′-ah bur-net′-ee-ee)

crustacean (krus-tay′-sheean) A member of the class Crustaceae, which includes water fleas, shrimp, crabs, and similar forms.

Culex (ku′-lex)

cysts (sists) Dormant resting cells characterized by a thickened cell wall.

cytochromes (site′-o-chromes) Iron-containing substances that help convey electrons to molecular oxygen in the electron transport chain. They function in the generation of ATP.

cytopathic effects (CPE) (site-oh-path′-ick) Observable changes in cells in vitro produced by viral action; for example, lysis of cells or fusion of cells.

Cytophaga (sye-toff′-ah-gah)

cytoplasmic membrane The flexible structure immediately surrounding the cytoplasm in all cells.

dalton (doll′-ton) A unit of weight equal to the weight of a single hydrogen atom.

darkfield microscopy A microscopic technique in which no light is visible except that which is reflected from an object. This results in the object being light and the background dark.

DDT An insecticide, dichlorodiphenyl trichloroethane, that is toxic to human beings and other animals when swallowed or absorbed through the skin.

death phase The stage in which the number of viable bacteria in a population decreases at an exponential rate.

deciduous Having parts, such as leaves, that fall off at a certain season.

decomposers Heterotrophic microorganisms that decompose organic substances, with the production of inorganic materials.

decontamination Removing or inactivating pathogenic microbes and their toxic products.

defective virus A virus that lacks a part of its nucleic acid which consequently cannot direct the replication and release of complete, infective virions.

Demodex folliculorum (dem′-oh-dex foh-lik-u-lore′-um)

denature To destroy the native configuration of a macromolecule, such as protein or nucleic acid.

denitrification (de-nite′-trif-fi-ka-shun) The process by which nitrate is reduced to nitrogen gas.

density The ratio of the mass of a substance to its volume.

deoxyadenosine monophosphate (dAMP) One of the subunits of DNA, a mononucleotide in which the sugar is deoxyribose. The other subunits of DNA are abbreviated dTMP, dGMP, and dCMP.

deoxynucleoside triphosphates (de-ox′-e-new-klee-o-side try-fos′-fates) Purine and pyrimidine bases, bonded to the sugar deoxyribose, which in turn is bonded to a string of three phosphate molecules.

deoxyribose (de-ox′-e-rye-bose) A 5-carbon sugar found in DNA.

Dermacentor andersoni (der-mah-sen′-ter an-der-soh′-nee)

dermis The portion of the skin lying just beneath the surface layers and containing nerves, blood vessels, and connective tissue.

Desulfovibrio (dee-sull-foh-vib′-ree-oh)

dextran (dex′-tran) Branched polysaccharide of D-glucose subunits; often found as a storage product in bacteria and yeast.

diabetes Condition characterized by excessive urine production. Sugar diabetes is the colloquial term for diabetes mellitus, a condition marked by deficiency of the hormone insulin and consequent elevation of blood and urine glucose.

diaminopimelic acid (dye′-uh-mean′-oh-pie-meal′-ick acid) A chemical found in nature only in procaryotic organisms, particularly in the cell-wall mucocomplex of bacteria.

diatomaceous earth (dye-uh-tow-may′-see-ous) Deposits of empty cell walls of diatoms; this extremely hard silica-containing material has many practical uses. Also called *diatomite* and *kieselguhr*.

diatoms (dye′-uh-toms) Algae with hard silica-containing cell walls; members of the class Bacillariophyceae, phylum Chrysophyta.

diauxic growth (dye-ox′-ik) The two-stage growth curve displayed by certain bacteria when grown in a medium containing two different compounds which can serve as sources of energy.

dicotyledenous (die-kot-ti-lee′-den-ous) Pertaining to a group of plants (dicots) that produce two-seed leaves when the seed germinates. Most flowering plants are dicots; grasses are not.

diffusion The movement of molecules from a region of high concentration to a region of low concentration.

digester A large tank in which sedimented wastes are degraded by anaerobic microorganisms.

dikaryon A cell which contains two different kinds of nuclei resulting from the fusion of two cells. During sexual reproduction these nuclei fuse to produce the diploid state; meiosis follows and haploid spores are produced.

dimer (dye′-mer) The molecule resulting from the association of two identical molecules.

dimorphic (dye-more′-fick) Having two forms, as in the fungi that may exist either as molds or in yeastlike forms.

dinoflagellates (dye-no-fla′-jell-ates) Algae of the class Pyrrhophyta, primarily unicellular marine organisms.

Diphyllobothrium latum (dye-fil-oh-bah′-three-um lah′-tum)

diploid (dip′-loyd) Refers to a cell in which there are two copies of each type of chromosome (2N). Each chromosome of a pair originates from a chromosome of one of the parent sexual cells that fused to begin the line of diploid cells.

direct selection The technique of selecting mutants by plating organisms on a medium on which the desired mutants will grow but the parent will not.

disaccharide (dye-sack′-a-ride) A carbohydrate molecule consisting of two monosaccharide molecules joined together.

disease A process resulting in tissue damage or alteration of function, producing symptoms, or noticeable by laboratory or physical examination.

disinfection Killing pathogenic microbes on or in a material without necessarily sterilizing it. Use of this term usually implies that a liquid or gaseous chemical agent is employed for the microbial killing.

distal Farthest from the middle, opposite of proximal.

DNA hybridization A technique for assessing the relatedness between DNA molecules. The strands of DNA are pulled apart (denatured) and their ability to associate (renature) with the single-stranded DNA from another organism is measured. Only DNA strands with complementary base sequences will associate.

DNA polymerase (DNA poll-lim′-er-race) The enzyme which catalyzes the formation of DNA. A number of different DNA polymerase enzymes exist, and the function of each in vivo is not known.

donor cell The bacterial cell which transfers its chromosome, or a portion of its DNA, to another cell (the recipient cell).

dorsoventrally In the direction from back to front; for example, in vertebrates, from vertebral column to the abdomen.

doubling time The time required for the number of cells in a population to double in number; essentially the same as the generation time.

eclipse The period during which virus exists within the host cell as free nucleic acid.

ecological niche The particular role that a species plays in the activities of the ecosystem as well as the physical space occupied by an organism.

ecology The study of the relationships of organisms to one another and to their environment.

ecosystem (ee′-ko-system) An environment and the organisms which inhabit it; for example, a lake with all its inhabitants (from bacteria and algae to fish) make up a particular ecosystem. All of the organisms in a habitat plus all of the factors in the environment with which the organisms interact.

ectoparasite A parasite that lives or feeds on the outer surface of the host's body.

eczema A condition characterized by a blistery skin rash, with weeping of fluid and formation of crusts, usually due to an allergy. Eczema vaccinatum is the disease resulting from infection of a person with eczema by vaccinia virus, the agent used to immunize against small pox.

eczema vaccinatum (eks-e′-mah vak′-sin-ah-tum) A sometimes fatal disease characterized by a generalized pox eruption that results from vaccinia virus infection of a person with eczema. A common consequence of exposure of a patient with eczema to someone recently vaccinated against smallpox.

electron A subatomic particle of negative electrical charge that orbits the positively charged nucleus of an atom. For maximum stability an atom must have a certain number of electrons in its outermost orbit.

electrophoresis (e-lek′-tro-for-ee′-sis) A technique for separating proteins based on the fact that proteins differ in their electrical charges and so will move at different rates in an electrical field.

encephalitis (en-sef-ah-lye′-tis) Inflammation of the brain.

endemic Referring to the usual prevalence of a disease in a given geographic area, as opposed to epidemic.

endospore (en′-dough-spore) A very resistant form of a resting bacterial cell which develops from a vegetative cell by a series of biochemical reactions called *sporulation.*

end product The chemical compound which is the final product in the sequence of a metabolic pathway.

end product inhibition Also called *allosteric inhibition* or feedback inhibition. The inhibition of the first enzyme of a biosynthetic pathway by the end product which combines with the allosteric site of the enzyme, thereby changing its shape and altering its enzymatic activity.

end product repression The regulation of gene activity by a process in which the end product of a biosynthetic pathway activates the repressor molecule which then combines with the operator region and prevents the functioning of RNA polymerase.

energy The ability to do work.

enrichment culture A technique used for isolating an organism from a mixed culture by manipulating conditions so as to favor growth of the organism sought while minimizing growth of the other organisms present.

Entamoeba histolytica (en-tah-mee′-bah his-toh-lit′-ik-ah)

Enterobacter aerogenes (enter-oh-bak′-ter air-ah′-jen-eez)

Enterobius vermicularis (enter-oh′-bee-us ver-mik-u-lah′-ris)

envelope (of bacterial cell) The portion of the cell that encloses the cytoplasm, including the cytoplasmic membrane, cell wall, and capsule if present.

envelope (viral) The outer lipid-containing layer possessed by some virions. Obtained from modified host cell membranes when the virus leaves the cell.

enzyme An organic catalyst. A protein molecule which lowers the activation energy of substrates allowing them to react at temperatures compatible with life.

epidemic The presence of many cases of a disease within a region.

epidermis (epee-der′-miss) The outermost skin layers.

Epidermophyton (epee-der-moff′-fit-ton)

episome (ep′-pi-soam) A particle of functional DNA which can exist either covalently bonded to the bacterial chromosome or in an extrachromosomal state.

equine Pertaining to horses.

Erwinia (er-win′-ee-a)

Erysipelothrix (air-ree-sip′-ell-oh-thrix)

erythrasma (air-ree-thraz′-mah) A painless condition involving moist areas of the skin, characterized by redness or brown discoloration and large numbers of the bacterium *Corynebacterium minutissimum.*

erythrocytes (e-rith′-row-sites) Red blood cells.

erythromycin (e-rit-throw-my′-sin) An antibiotic, produced by a strain of *Streptomyces,* which inhibits protein synthesis primarily of Gram-positive organisms.

Escherichia coli (esh-er-ik′-ee-a coh′-lee)

ester bond A bond, found in lipids and nucleic acids, in which a hydroxyl group (—OH) is bonded to a carboxyl group (—COOH) with the removal of HOH.

eucaryotic cell (you-carry′-aw′-tick) A complex cell type, characterized by having a nuclear membrane, mitochondria, and numerous chromosomes. All living cells except bacteria and blue greens are eucaryotic.

eutrophication (you′-trow-fi-kay′-shun) Nutrient enrichment leading to over-production of algae.

excise To cut out.

extrachromosomal Not associated with the chromosome. Applied to episomes and plasmids in bacteria, and to DNA in eucaryotic organelles.

exudate A material such as pus composed of fluid from the vascular system, cells, and sometimes products of tissue breakdown.

facultative organisms Those organisms able to carry out both aerobic and anaerobic metabolism. May also be used in terms of other properties such as photosynthesis (for example, facultatively photosynthetic organisms can use either photosynthesis or other means of gaining energy).

fastidious Exacting; used to refer to organisms that require many growth factors.

fats Simple lipids consisting of esters of glycerol with fatty acids.

feedback inhibition Another term for allosteric or end product inhibition. The inhibition of the first enzyme of a biosynthetic pathway by the end product of the pathway.

fermentation The metabolic process in which the final electron acceptor is an organic compound.

fission (fish′-en) An asexual reproductive process in which one cell splits into two or more independent daughter cells; in binary fission two daughter cells are produced.

fixation The combination of an element with other atoms in covalent linkage; for example, the combination of nitrogen gas with hydrogen to form NH_3, or of carbon into organic compounds.

fixed nitrogen The form of nitrogen when it is combined with other elements, that is, NH_3, $—NO_2^-$, NO_3^-.

flagellates (fla′-jell-ates) Organisms that bear flagella. The term is usually applied to protozoa or algae.

flagellum, pl. flagella (fla-jell′-um) A long, whiplike organelle of locomotion made up of intertwined molecules of protein.

Flavobacterium meningosepticum (flay-voh-bak-teer′-ee-um men-injoh-sep′-tik-um)

flavoproteins (flay-vo-pro′-teens) Enzymes that form a part of the electron transport scheme. The flavin portion of the molecule is a coenzyme which is a derivative of the vitamin riboflavin.

fluke A short, flattened parasitic worm of the class *Trematoda.*

F^- cell A bacterial cell that does not contain an F particle, but can act as a recipient and receive one from an F^+ cell.

follicle A very small excretory sac, such as the ones from which body hair protrudes.

fomite (foh′-mite) An inanimate object, such as a book, tool or towel, which can act as a transmitter of pathogenic microbes even though not supporting their growth.

forespore One of the more readily identified stages in the process of sporulation, identified as a refractile body which is not yet resistant to heat.

F^+ cell A bacterial cell that can transfer the extrachromosomal F (fertility) particle to a recipient (F^-) cell.

F′ cell A bacterial cell which carries an extrachromosomal F particle to which is attached a fragment of chromosomal DNA.

Francisella tularensis (fran-siss-sell′-ah tu-lah-ren′-siss)

free energy (*G*) Energy that is available to do work, particularly in causing chemical reactions.

fruiting body Specialized reproductive structures containing many spores; examples are mushrooms (the fruiting bodies of certain basidiomycetes) and the fruiting bodies of slime molds and some myxobacteria.

frustules (frus′-tewles) Hard, silica-containing cell walls of diatoms.

Fusobacterium (fuzoh-bak-teer′-ee-um)

galactose (ga-lak′-toes) A 6-carbon sugar. Together with glucose it forms the disaccharide lactose (milk sugar).

gametes (gam′-eats) Haploid cells which fuse with other gametes to form the diploid zygote in sexual reproduction.

gametocyte (gam-ee′-tow-site) Cell that produces or becomes a mature sexual cell.

gas chromatography A technique of separating and identifying gaseous components of a substance.

gastroenteritis (gas′-trow-en-ter-ite′-is) Inflammation of the gastrointestinal tract, usually characterized by diarrhea, nausea, and vomiting.

gene A portion of a chromosome which carries the genetic information for the synthesis of one polypeptide chain.

generation time The time required for the population of cells to double in number. Also called *division time.*

genetic code The composition and sequence of all of the sets of three nucleotides which code for the amino acids. The sequence of three nucleotides which codes for one amino acid is called a *code word.*

genetic engineering The deliberate modification of the genetic properties of an organism either through selection of desirable traits, the introduction of new information on DNA, or both.

genetic recombination The process by which a segment of DNA of one chromosome is integrated into the DNA of another chromosome.

genome (gene′-ome) The total genetic information needed to reproduce a cell or a virus. In bacteria, it is equivalent to one chromosome.

genotype (gene′-o-type) The sum total of the genetic constitution of an organism. Much of the genotype is not expressed at any one time.

genus (gene′-us) A category of related organisms, usually containing several species; the first name of an organism in the Binomial System of Classification.

germicide An agent that kills microbes.

germinate To begin to grow.

germination The sum total of the biochemical and morphological changes that an endospore or other resting cell undergoes prior to its becoming a vegetative cell.

Giardia lamblia (jee-are′-dee-ah lamb′-lee-ah)

glomerulonephritis (glom-er-ulo-nef-rye′-tiss) A kidney disease characterized by inflammation of the capillary loops through which portions of blood plasma are normally filtered.

glucose A 6-carbon sugar. Also called dextrose.

glycoproteins Proteins that contain sugars as part of the molecule.

glycosidic bond (gly-ko-sid′-ik) A bond in carbohydrates which links a hydroxyl group (—OH) of one sugar with the aldehyde group ($-\overset{\overset{\displaystyle O}{\|}}{C}-H$) of another.

Gonyaulax (gone-ee-aw′-lax)

granuloma (gran-you-lome′-ah) A localized collection of macrophages and other cells that is formed around a foreign body or an infectious agent in the tissues.

growing point The site on the DNA molecule at which replication is taking place.

growth curve The graphic representation of the change in numbers of organisms in a culture with time.

growth factor An organic compound other than the carbon and energy source which an organism requires and cannot synthesize; for example, a vitamin, amino acid, FAD, and so on.

Haemophilus influenzae (hee-moff′-ill-us in-flew-en′-zee)

half-life ($T_{1/2}$) Time required for the disappearance of half of a sample. With reference to immunoglobulins, the time required for the disappearance from the blood of half of the molecules in a sample injected into the bloodstream.

Halobacterium (hah-loh-bak-teer′-ee-um)

halophile (hale-oh-file)

halophilic organism Literally means "salt-loving" organism which will grow in a medium to which has been added a concentration of NaCl to inhibit the growth of other bacteria. Obligate facultative halophiles will grow in NaCl but do not require it; obligate halophiles require NaCl; extreme halophiles will grow only in media where the NaCl concentration is very high.

haploid (hap′-loyd) The chromosome state in which each type of chromosome is represented only once (1N), although there may be several copies of the same chromosome. All procaryotic cells are haploid; the sexual cells of eucaryotic forms are also haploid.

hapten (hap′-ten or hap′-teen) Substance that can combine with specific antibodies but cannot incite the production of those antibodies unless it is attached to a large carrier molecule.

haustorium, pl. haustoria (house-tore′-ee-um) Tiny rootlike projections of fungi that function to increase the surface area and aid in uptake of nutrients by absorption.

heat labile Destroyed by heat.

heat shock The short heat treatment, usually 60 to 70°, given to bacterial endospores to induce the process of germination.

hemadsorption (heem-ad-sorp′-shun) The adsorption of red blood cells to host cells infected with certain viruses. Hemadsorption results from incorporation into the host cell membrane of viral proteins which combine with the red blood cells.

hemagglutination (heem′-uh-gloot-uh-nay′shun) Agglutination, or clumping together, of red blood cells.

hemolymph The circulating nutritive fluid of certain invertebrates.

hemolysin (hee-mah′-liss-in) A substance which disrupts the membrane of red blood cells, causing them to release their hemoglobin.

herbicide A chemical which is used to kill weeds. However, many herbicides are not selectively toxic, and they will kill most plants and animals if not carefully applied.

hermaphroditic (her-maff-roh-dit′-ik) Possessing both male and female sex organs.

herpes simplex (her′-peez sim′-pleks) A blistery localized skin rash caused by herpes simplex virus, usually located on the lip (cold sore, fever blister) or the genitalia.

herpes viruses (herp′-eez viruses)

herpes zoster (her′-peez zos′-ter) A localized blistery infection of the skin supplied by one or a group of sensory nerves, sensory nerve ganglia, and sometimes other tissues, often painful, by the varicella-zoster virus.

heterotroph (organotroph); heterotrophic organism (het′-uh-row-trowf) An organism that obtains energy from organic compounds. An organism that utilizes an organic compound as its main source of carbon.

hexose The general term for a 6-carbon sugar.

Hfr cell (high frequency of recombination) A strain of bacterium that has the ability to transfer its chromosome to an F^- cell. The F particle is integrated into the chromosome of the Hfr cell.

high-energy bond A bond which yields a large amount of energy when the bond is broken. This energy can be harnessed to do work.

hippurate Salt of hippuric acid, $C_9H_9NO_3$, normally excreted in the urine of herbivorous animals.

histamine (hiss′-ta-meen) A low molecular weight substance found in an inactive form in certain animal and plant tissues that, upon release, may cause relaxation of the walls of blood vessels, increased permeability of the vessels, passages of the lung, and other effects. It is one of the substances playing a role in the inflammatory process.

Histoplasma capsulatum (hiss-toh-plaz′-mah cap-su-lah′-tum)

homologous Corresponding in structure, position and origin. Refers to DNA molecules which have identical base sequences and therefore arise from the same species.

hook One of the structural components of the bacterial flagellum.

horizons The layers of soil, generally three, that are distinguishable when soil is examined in cross section.

hormogones or hormogonia (horm-oh-gohnes) Short rows of filamentous cells that function as reproductive elements; the hormogones break off from old filaments and glide away to form new filaments.

host An organism on or in which smaller organisms or viruses live, feed, or reproduce. When dealing with parasites having complex life cycles, the host in which the adult lives, or the one in which sexual reproduction takes place, is called the *definitive host.* A host harboring larval or asexually reproducing forms is called an *intermediate host.*

humoral antibodies (hume′-uh-ral) Antibodies in body fluids, such as blood.

humus (hew′-mus) A dark colored, highly complex organic material in soil that is not readily degraded by microorganisms.

hyaluronic acid (hye-al-you-ron′-ik) A mucopolysaccharide gel which fills intercellular spaces in tissue and also occurs as a capsular substance of some bacteria.

hyaluronidase An enzyme that hydrolyses hyaluronic acid, an intercellular mucopolysaccharide of body tissues.

hydrocarbon Any compound composed of only hydrogen and carbon.

hydrogen atom This atom consists of one proton and one electron which moves in orbit around the electron. Weight = 1 dalton.

hydrogen bond A weak attraction (bond) between an atom which has a strong attraction for electrons and a hydrogen atom which is covalently bonded to another atom that attracts the electron of the hydrogen atom.

hydrolytic reaction A reaction in which a molecule reacts with water and is broken down into two or more smaller molecules. $AB + HOH \rightarrow AOH + HB$.

hydrophobic A substance which lacks an affinity for water.

hydrophobic bonds Bonds which arise as a result of the tendency of polar water molecules to exclude nonpolar molecules. The nonpolar molecules associate with one another (by van der Waals forces).

hypersensitivity reaction (allergic reaction) An immunological reaction that causes impaired functioning of the host making the immunological response.

hypertonic (hye-per-ton′-ik) A fluid having an osmotic pressure greater than another fluid with which it is compared.

hypha, pl. hyphae (high′-fah, high′-fee) Filament of fungal cells. A single filament of a fungus. A large number of hyphal filaments (hyphae) constitute a mycelium.

icosahedral symmetry (eye-coh-suh-heed′-ral) A symmetry characterized by a structure which has 20 sides, and each side is an equilateral triangle.

imino The group N—H; the amino acid proline has this group and is an imino acid.

"immune" (lysogenic bacteria) Bacteria which are not susceptible to a particular temperate phage because they carry that phage in prophage form. The "immunity" depends on the presence of a repressor whose synthesis is directed by the prophage.

immune cells Lymphoid cells (usually lymphocytes) capable of participating in cell-mediated immunity.

immune response Specific response to an antigen by production of humoral antibodies or immune cells. Primary—the response which occurs after first exposure to an antigen. Secondary—the response which occurs after second or subsequent exposure to an antigen (see anamnestic response).

immunity State of protection; for example, state of protection against the mumps virus. Natural or innate immunity—protection that results from the genetic make up of the host; for example, domestic animals have an innate immunity to mumps virus. Acquired immunity—immunity gained as a result of exposure to an agent; for example, immunity to mumps virus is usually acquired (actively) by response to infection with the virus, but it may also be acquired (passively) by the administration of specific antibodies formed by another host.

immunofluorescence (im-mune′-oh-flewo-res′-ence) Fluorescence resulting from a reaction between a substance and specific antibodies that are bound to a fluorescent dye.

immunoglobulins (im-mune-oh-glob′-yew-lins) Antibodies or antibodylike protein molecules, usually part of the gamma globulin portion of blood.

immunological enhancement (im-mune-oh-lodge′-e-kul) Enhanced survival of incompatible grafts (tumors or normal tissue) caused by specific humoral antibodies.

immunological tolerance (immunological unresponsiveness) Failure to respond to a potential antigen; lack of response to a specific substance that would normally cause a response.

immunosuppression Nonspecific inhibition (suppression) of the ability to make an immune response, resulting in an overall depression of the response to all or most antigens.

impetigo (im-pa-tye′-go) A bacterial disease of the skin, often highly contagious, characterized by small blisters, weeping fluid, and formation of crusts.

inclusion body Round, oval, and sometimes irregular body occurring in the cytoplasm or nuclei of cells infected with certain viruses or other intracellular parasites.

indicator agar (differential medium) Agar containing components that can be modified in distinctive and readily recognized ways by specific microorganisms. The medium serves to differentiate between various organisms.

indirect selection The selection of mutants by determining which organisms do not grow on a certain medium; the mutants do not grow, the parents do grow.

inducer A substance responsible for activating certain genes, thereby resulting in the synthesis of new proteins.

inducible enzyme An enzyme that is synthesized only in response to a particular substance in the environment, the inducer.

induction (viral) The process by which a prophage is excised from the host cell DNA.

induration (in′-dew-ray′-shun) Thickening.

inert (noble) elements An element whose outer electron shell is filled; therefore it does not form bonds with other elements.

infection Invasion of tissues (including skin or mucous membranes) by microbes with or without the production of disease.

infective dose (ID_{50}) The dilution of virus at which 50 percent of inoculated hosts are infected.

inflammatory response A nonspecific response to injury characterized by redness, heat, swelling, and pain in the affected area.

infusion The liquid extract of any material soaked in water.

insecticide Any substance that kills insects. If applied improperly, it may kill other organisms as well.

inspissation (in-spiss-say′-shun) Making a fluid thicker by evaporating volatile components.

integration With reference to DNA, the covalent bonding of DNA from a donor organism with the recipient chromosome.

interferons (in-ter-fear′-ohns) Proteins that are produced by virus-infected cells (or by cells in response to certain other stimuli) that prevent the replication of viruses in other cells.

in vitro Literally means "in glass"; in culture; outside of the body.

in vivo In the body.

iodophore (aye-o′-do-fore) A complex of iodine with a carrier molecule that liberates free iodine in solution. Such compounds are used as antiseptics.

ion An atom or group of atoms that has gained or lost an electron and therefore carries a negative or positive charge.

isomers Two compounds which have identical chemical compositions but differ in configuration in that one is the mirror image of the other.

Isospora belli (eye-soss′-pore-ah bell′-ee)

isotope Two forms of the same element which have the same atomic number but different atomic weights. One of the forms is unstable and decomposes spontaneously, so that its presence can be readily detected.

jaundice (jawn′-diss) A condition characterized by abnormally large amounts of certain hemoglobin breakdown products in the blood, skin, and mucous membranes, giving them a yellow color.

kinetic energy Energy which results from motion.

kinetics (kin-et′-icks) Used to describe the dynamics of ongoing processes, such as growth of bacteria or enzyme reactions.

kinin (kye′-nin) A group of peptide substances found in blood and tissue in inactive form, which upon activation play a role in causing inflammation, blood clotting and other effects.

Klebsiella (kleb-see-ell′-ah)

label A marker, commonly a radioactive atom which allows the location of the molecule to be monitored.

Lactobacillus bulgaricus (lack-toe-ba-sil′-lus bull-gar′-e-kus)

L. casei (L. case′-e-eye)

lactose A sugar (disaccharide) consisting of one molecule of glucose and one of galactose.

lag phase The stage in the growth of a bacterial culture characterized by extensive synthesis of nucleic acid and protein but no increase in the number of viable cells.

lamella (la-mel′-la) A layer of internal cell membranes as organelles of photosynthesis.

larva (lar′-va) An immature form of certain animals which differs morphologically from the adult form.

larynx (la′-rinks) Area between the upper end of the windpipe and the throat and containing the vocal cords; "voice box."

lecithin (less′-sith-in) A phospholipid found in animal tissues.

legume (leg′-ume) A group of plants (including peas, beans, and clover) with pods that split in two when mature. Some can develop a symbiotic relationship with nitrogen-fixing bacteria.

Leptospira (lep-toh-spire′-ah)

lethal dose (LD_{50}) The dilution of virus or other test agent which kills 50 percent of inoculated hosts.

leukocytes (lou′-ko-sites) White blood cells. There are several morphological types: polymorphonuclears (polys), lymphocytes, eosinophiles, monocytes, and basophiles.

lichen (like′-en) A fungus and an alga or cyanobacterium growing together to form a single organism; an alga and a fungus living symbiotically.

lignin (lig′-nin) A hard, complex carbohydrate which is found in plant cell walls.

lime A basic material, the oxide of calcium.

lincomycin (link-o-my′-sin) A bacteriostatic antibiotic, produced by *Streptomyces,* which inhibits protein synthesis.

lipid Any of a diverse group of organic substances which are relatively insoluble in water but soluble in alcohol, ether, chloroform, or other fat solvents.

lipopolysaccharide (lip′-o-poly-sak′-a-ride) The macromolecule formed by the covalent attachment of lipid with polysaccharide.

lipoprotein (lip-o-pro′-teens) A weak combination of lipid and protein.

Listeria (liss-teer′-ee-ah)

log phase The stage of growth of a bacterial culture when the cells are multiplying exponentially.

lumen (lou′-men) The cavity within a tubular structure such as the intestine, blood vessel, or windpipe.

lymphocyte (limf′-oh-site) Small, round, or oval lymphoid cell with a large nucleus and scanty cytoplasm. Lymphocytes are inactive cells, but when stimulated they can transform into active, dividing cells. They are found in blood and lymph and also in many body tissues.

lymphoid (lim′-foid) Pertaining to lymphocytes or the lymphatic system, as lymphoid tissue is tissue composed largely of lymphocytes, and lymphoid cells are cells characteristic of lymphoid tissue.

lymphoma (lim-fo′-ma) A tumor or neoplastic disorder of lymphoid tissue.

lyse To disrupt or dissolve. Often used with respect to disruption of cells, or *cell lysis.*

lysogenic (bacteria) (lye-so-jen′-ick) Bacteria which are carrying a prophage of a temperate virus.

lysogenic conversion The change in the properties of bacteria as a result of their carrying a prophage; for example, the conversion of nontoxin-producing bacteria to toxin-producing bacteria by the action of a prophage.

lytic (virus) (lit′-ick) A virus that replicates within a host cell and causes lysis (death and disruption) of the host cell.

macrolides (mak′-roh-lydes) A group of antibiotics such as erythromycin, composed of compounds having a large cyclical structure.

macromolecule (mak-roh-mol′-e-kule) A large molecule made up of subunits bonded together in a characteristic fashion.

macrophage (mak′-roh-fayge or mack′-row-fahge) A large mononuclear phagocyte. Macrophages are found in virtually all parts of the body.

malaise (ma-layze′) A vague feeling of being ill.

malignancy (ma-lig′-nan-cy) A progressive increase in virulence or seriousness of any pathologic condition leading eventually to death.

mannose (man′-nose) A 6-carbon sugar.

maturation (viral) The assembly of complete, infective virions from viral components replicated separately within a host cell.

meiosis (my-o′-sis) The process by which the chromosome number is reduced from diploid (2N) to haploid (1N). Segregation and reassortment of the genes occur in this process, and gametes or spores may be produced as the end product of meiosis.

meningitis (men-in-jye′-tiss) Inflammation of the meninges, membranes that cover the brain and spinal cord.

merozoite (mare-o-zo′-ite) The youngest and smallest of the developmental forms of malarial parasites growing in the mammalian host.

messenger RNA (mRNA) A type of RNA which transcribes the genetic information of the DNA of the cell; mRNA serves as a template for protein synthesis.

metachromatic Staining different colors with the same dye. For example, *C. diphtheriae* may stain blue with red inclusions when the dye methylene blue is applied.

metachromatic granules Granules inside certain bacteria which stain a different color than the dye used for staining, as red with a blue dye. Some are known to consist of long chains of phosphate molecules termed volutin.

metazoan Refers to members of the metazoa, multicellular animals which show tissue differentiation.

Methanobacterium (meth-ain-oh-bak-teer′-ee-um)

Methanococcus (meth-ain-oh-kok′-us)

methanogenic Producing methane.

Methanosarcina (meth-ain-oh-sahr′-sin-ah)

microaerophile, microaerophilic organism (my-kroh-air′-oh-file) Organism that requires low concentrations of oxygen for growth (lower than the concentration in air). Sometimes used to indicate organisms that can grow in air but grow much better under anaerobic conditions.

Micrococcus (my-kroh-kock′-us)

microgram One millionth of a gram (one thousandth of a milligram). Abbreviated μg.

Microsporum (my-kros′-pore-um)

mineralization (stabilization) Conversion of organic to inorganic materials.

mineralize To convert organic substances into inorganic molecules.

miracidium (meer-a-sid′-ee-um) The free-swimming larva which emerges from the eggs of flukes.

mitosis (my-toe′-sis) A process of chromosome duplication. The chromosomes divide longitudinally, and the daughter chromosomes then separate to form two identical daughter nuclei.

molt Shedding of the outer covering of arthropods and other forms.

monolayer cell culture An in vitro culture in which the cells form a single layer on the container.

mononuclear phagocyte system See reticuloendothelial system.

monosaccharide (mono-sak′-kar-ride) A sugar, a simple carbohydrate, generally having the formula $C_nH_{2n}O_n$ where n can vary from 3 to 8. The most common are 5 and 6.

Moraxella (more-ax-ell′-ah)

motor nerve Nerve responsible for controlling muscles (as opposed to nerves transmitting sensations).

mucocomplex The rigid backbone of the cell wall, consisting of two major subunits, N-acetyl muramic acid and N-acetyl glucosamine, and a number of amino acids. Also called peptidoglycan and murein.

mucoprotein (mu-ko-pro′-teen) A protein associated with a carbohydrate.

mucosa A mucous membrane; the lining of body cavities that communicate to the exterior.

mutagen (mutagenic agent) (mu′-ta-jen) Any agent which increases the frequency at which DNA is altered, thereby resulting in an increase in the frequency of mutations.

mutant (mu′-tent) An organism whose base sequence in DNA has been modified, resulting in an altered protein that gives the cell properties which differ from the parent.

mutation (mu-ta′-shon) A modification in the base sequence of DNA in a gene resulting in an alteration in the protein coded by the gene.

mutualism Both host and organism benefit. A living together of two organisms whereby both benefit from the association.

mycelium, pl. mycelia (my-seal′-e-um, my-seal′-e-ah) A tangled, matlike mass of fungal hyphae, exemplified by the cottony growth of molds. The collection of hyphae composing a fungus.

Mycobacterium leprae (my-koh-bak-teer′-ee-um lepp′-ree)

M. tuberculosis (M. tuberk-you-loh′-siss)

Mycoplasma pneumoniae (my-koh-plazz′-mah nu-moan′-ee-ee)

mycorrhiza (my-koh-rize′-ah) A symbiotic relationship between certain fungi and the roots of plants.

myxoviruses (mix′-oh-viruses)

nares (nair′-eez) Nostrils.

nasopharynx (nay-zo-fair′-inks) Area where the nasal passages join the throat.

natural amino acid The L-isomer of the amino acid. Only L amino acids are found in proteins.

necrotic (nee-krot′-ic) Dead. Used when referring to dead cells or tissues that are in contact with living cells, as necrotic (dead) tissue in wounds.

Neisseria catarrhalis (nye-seer′-ee-ah cat-arr-al′-iss)

N. gonorrheae (N. gahn-oh-ree′-ee)

N. meningitidis (N. men-in-jit′-id-iss)

neoplasm (nee′-o-plasim) See tumor.

niacin (nye′-a-sin) A B vitamin. It can be transformed into the coenzyme nicotinamide adenine dinucleotide (NAD).

nicotinamide adenine dinucleotide NAD (nik-o-tin′-a-mid add′-den-neen di-new′-klee-o-tide) A coenzyme which transfers H atoms. It is a derivative of the vitamin niacin. When NAD is carrying the H atom, it is indicated as NAD_{red}.

nitrification (nye-tri-fi-kay′-shun) The conversion of NH_3 or NH_4^+ to nitrite (NO_2^-) and nitrite to nitrate (NO_3^-).

Nitrobacter (nye-troh-bak′-ter)

nitrogen fixation The combination of nitrogen gas with other atoms to form compounds such as NO_2^-, NO_3^-, NH_3.

Nitrosomonas (nye-troh-soh-moan′-ass)

Nocardia asteroides (noh-kar′-dee-ah ass-ter-oid′-eez)

nonpolar bond A bond in which the charge is evenly distributed on each of the atoms which form the bond.

nonsense codon A sequence of three purines and pyrimidines in messenger RNA which does not have a corresponding complementary sequence of three purines and pyrimidines in tRNA. Therefore the presence of a nonsense codon terminates the synthesis of a polypeptide chain.

nucleotides The basic subunits of ribonucleic acid, consisting of a purine or pyrimidine, covalently bonded to ribose which is covalently bound to a phosphate molecule. The subunits of DNA are called deoxyribonucleotides.

numerical taxonomy A technique for determining the relationship between organisms by determining the number of characteristics that the organisms have in common.

oncogenesis (on-ko-jen′-i-sis) The process of tumor formation.

oncogenic (on-ko-jen′-ick) Tumor-inducing.

operator region The gene which combines with an active repressor molecule.

operon A group of linked genes controlled as a single unit.

opsonins (op′-soh-nins) Substances that cause increased phagocytosis; for example, antibodies and complement.

opsonize Increase the susceptibility of microorganisms to phagocytosis.

organ A structure composed of different tissues, coordinated to perform a special function. Absent in members of the Protista.

organelle (or-gan-ell′) Any structure which performs a specific function in a cell.

orifice (or′-ih-fiss) Entrance or outlet of a body cavity.

origin Refers to the site on the DNA molecule at which DNA synthesis is initiated in the process of chromosome duplication.

orthophosphate (or-tho-fos′-fate) The inorganic form of phosphate; sometimes abbreviated P_i.

osmosis (os-moh′-sis) The passage of a solvent through a membrane from a dilute solution into a more concentrated one.

osteomyelitis (oss′-tee-oh-my-e-light′-iss) Inflammation of the bone.

otitis (oh-tye′-tiss) Inflammation of the ear. Inflammation of the chamber just beneath the eardrum is called *otitis media.*

outgrowth The stage in the development of endospores into vegetative cells which follows endospore germination.

oxidative phosphorylation (oxidate′-tiv fos-for-i-lation) The generation of energy in the form of ATP which results from the passage of electrons through the electron transport chain to a final electron acceptor.

papovaviruses (pap-oh′-vah-viruses)

para-amino-benzoic acid (PABA) A growth factor for certain bacteria; a precursor of the vitamin folic acid.

parasite An organism that lives in or on another organism (the host) and gains benefit at the expense of the host.

parvoviruses (par′-voh-viruses)

Pasteurella multocida (pass-ture-ell′-ah mull-toss′-id-ah)

pasteurization (pass′-ture-yze-ay-shun) The processes of heating food or other substances under controlled conditions of time and temperature (for example, 63°C for 30 min) to kill pathogens and reduce the numbers of other microbes.

pathogenesis (path-oh-jen′-isiss) The sequence of changes which culminates in disease.

pathogenic Causing disease. "Pathogenic" is also used to designate microbes which commonly cause infectious diseases, as opposed to those which do so uncommonly or never.

peat Partially decayed plant material found in swamps.

pectins (pek′-tins) Polysaccharide substances found in the cell walls of some algae; also found in ripe fruits and used to make jellies. A polysaccharide which serves as an important component of many plant cell walls.

Pediculus humanus (ped-ik′-u-luss hu-man′-us)

pellicle (pell′-e-kull) Thickened, elastic cell covering. Differs from the cell wall in that the pellicle is not rigid.

penetration (viral) The process of introducing viral nucleic acid into a host cell.

Penicillium (pen-e-sill′-e-um)

pentose A general term for any 5-carbon sugar.

peptide bond The covalent bond which joins adjacent amino acids together in proteins: an amino group of one amino acid is bonded to the carboxyl group of the next amino acid, with the formation of water.

peptidoglycan The rigid backbone of the bacterial cell wall, consisting of two major subunits, N-acetyl muramic acid and N-acetyl glucosamine, and a number of amino acids. Also called mucocomplex and murein.

Peptococcus (pep-toh-kok′-us)

Peptostreptococcus (pep-toh-strep-toh-kok′-us)

peristalsis (pay-riss-tall′-siss) Wavelike contractions of the digestive tract responsible for propulsion of contents through the tract.

permeability The capacity of a membrane or other structure to allow passage of substances through it.

permease (per′-me-ace) A system of enzymes which are concerned with the transport of nutrients into the cell.

pesticide A general term for any agent that kills pests, either plant or animal. If applied indiscriminately, it may kill other forms of life.

petrolatum (pet-tro-lay′-tum) A greasy, jellylike substance made from petroleum, often referred to by the trade name vaseline.

phagocytosis (fag′-oh-sigh-toe′-sis) Cellular ingestion of particulate materials within membrane-bound vesicles. The particles are engulfed by pseudopods of the cell.

phenotype (feen′-o-type) The sum total of the observable properties of an organism, resulting from the interaction of the environment with the genotype.

phospholipid (fos-fo-lip′-id) A lipid which has a phosphate molecule as part of its structure.

Photobacterium (foh-toh-bak-teer′-ee-um)

photooxidation A chemical reaction occurring as a result of absorption of light energy in the presence of oxygen. It is sometimes responsible for the death of microbes.

photosynthesis (foh-toh-sin′-the-sis) The sum total of the metabolic processes by which light energy is utilized to convert CO_2 and a reduced inorganic compound to cytoplasm. $6H_2X + 6CO_2 \rightarrow C_6H_{12}O_6 + 6X$.

phototrophic Utilizing sunlight as an energy source.

Phthirus pubis (theer′-us pu′-biss)

phylum, pl. phyla (fie′-lum, fie′-lah) A large division of related families in classifying living organisms. The classification is subdivided progressively from kingdom → subkingdoms → phyla → classes → orders → families → genera → species.

phytoplankton (fie-tow-plank′-ton) The floating and swimming algae and procaryotic organisms of lakes and oceans.

picornaviruses (pick-orn′-ah-viruses)

pigments Coloring substances. In algae these are substances such as chlorophylls, carotenoids, and phycobilins that absorb light energy and give color to the algae.

pilosebaceous gland (pile-oh-seh-bay′-see-us) The tiny gland (sebaceous gland) that secretes an oily substance into the hair follicle.

pinocytosis (pine′-oh-sigh-toe′-sis) Cellular ingestion of liquids; a process essentially similar to phagocytosis.

Pityrosporum orbiculare (pit-ee-rah′-spore-um orb-ik-u-lah′-ree)

P. ovale (P. oh-val′-lee)

plankton (plank′-ton) Organisms that float in water.

plaque (plack) A clear area in a lawn or a monolayer of cells. Viral plaques are created by viral lysis of infected cells within the clear area.

plaque (dental) A collection of bacteria tightly adhering to a tooth surface and responsible for dental caries.

plasma (plaz′-ma) The fluid portion of blood before clotting.

plasma cells Lymphoid cells specialized to produce immunoglobulins in large quantities.

plasmodium, pl. plasmodia (plaz-mode-e-um, plaz-mod-e-ah) A mass of cytoplasm, enclosed by only a cytoplasmic membrane (but lacking a cell wall) and therefore free to ooze over a substrate. An example is found in the acellular slime molds.

Plasmodium vivax (plazz-moh′-dee-um vee′-vax)

pleomorphic (plee-oh-more′-fik) Occurring in various shapes.

poly-β-hydroxybutyric acid granules A storage reserve of carbon and energy which occurs as long chains of individual subunits forming readily visible granules.

polychlorinated biphenyl compounds (PCBS) Nonbiodegradable, toxic organic chemicals that are widely used in plastics and paint manufacture. (When discarded as wastes into bodies of water, these compounds are toxic for marine organisms.)

polyene (polly-een′) A chemical compound containing alternating single and double bonds; a group of antibiotics having such a property.

polymer A high molecular weight compound which results from the chemical combination of large numbers of subunits.

polymorphic (polly-more′-fick) Having different distinct forms at various stages of the life cycle.

polymorphonuclear (PMN) leukocyte (polly-morf-oh-new′-kle-er lew′-koh-site) A phagocyte with a lobed nucleus; also called neutrophilic granulocyte, poly, or neutrophil.

polypeptide A chain of amino acids bonded together by peptide bonds. The length varies from several amino acids to the length coded for by one gene.

polysaccharide (polly-sak′-karride) Long chains, branched or unbranched, of monosaccharide subunits.

polysome (polly′-soam) A number of ribosomes attached to the same molecule of messenger RNA.

population A group of organisms of the same species; that is, an interbreeding group.

potential energy The energy that is available in a substance as a result of its position in relation to its surroundings.

primary producers In the food chains, the organisms that carry out photosynthesis and thus produce organic materials from CO_2 and H_2O, using sunlight as an energy source.

primary structure The sequence of amino acids in a protein molecule.

primed cell A cell that has been specifically stimulated by a given antigen and can either produce more cells capable of responding to that antigen, synthesize immunoglobulin, or mediate reactions of cell-mediated immunity.

procaryotic cell (pro-karry-ot-tik) The simple cell type, characterized by the lack of a nuclear membrane, the absence of mitochondria, and a haploid state at all times.

progeny Offspring; descendants.

promoter region The region on the DNA at which the enzyme RNA polymerase binds and initiates the process of gene transcription.

prophage (pro-fayge or pro-fahge) The nucleic acid of a temperate phage when it is integrated with the host cell DNA.

Propionibacterium acnes (proh-pee-ah-ee-bak-teer′-ee-um ak′-neez)

prostaglandin (pross-tah-gland′-in) A group of substances found in various tissues and body fluids which cause relaxation of the walls of certain blood vessels and other effects. Like kinins and histamine, it probably plays a role in inflammation.

protease (pro′-tee-ace) An enzyme that breaks down proteins. Proteases differ markedly in their activity at different pH. Some proteases degrade proteins to amino acids, others degrade proteins only to polypeptides.

protein (pro′-teen) A macromolecule containing one or more polypeptide chains.

Proteus mirabilis (proh′-tee-us mee-rab′-il-us)

Protista (pro-tis′-ta) The kingdom in which the members of the microbial world are classified; lack of extensive tissue differentiation characterizes its members.

protoplast (pro′-tow-plast) A cell which has its rigid cell wall removed.

prototroph (pro′-toh-trof) An organism which has no organic growth requirements other than a source of carbon and energy.

pseudohyphae (sue-dough-high′-fee) A cellular filament resembling hyphae but formed by adhering elongated yeast cells which have formed by budding.

Pseudomonas aeruginosa (sue-dough-moan′-ass a-rue′-gin-o′-sa)

pseudomycelium (sue-dough-mye-see′-lee-um) A mass of fungal filaments resembling hyphae but composed of loosely united elongated cells formed by budding.

pseudopod, pl. pseudopods or pseudopodia (sue′-dough-pod, sue-dough-pohd′e-uh) Extension of cytoplasm that aids in engulfing particles and functions in the motility of ameboid cells.

pupa (pew′-pah) A stage in the development of many insects in which most of the changes from wormlike form to adult occur.

pure culture A culture which contains only a single strain of organism.

pus cells Cells in pus, principally dead PMNs.

putrefaction (pew-tre-fak′-shun) The decomposition of proteins with the production of foul-smelling products.

pyogenic (pye′-oh-jen′-ik) Producing pus.

pyrophosphate (pie-row-fos′-fate) Two phosphate molecules bonded together.

quaternary ammonium compound (quat-ter′-nary) A substance with an N atom bonded to four organic groups, thereby giving a positive charge to the N atom. A chloride or bromide ion commonly makes up the rest of the molecule.

radioactive label A substance emitting an ionizing radiation which can be detected and thereby used to locate the radioactive compound.

raphe (ray′-fee) Open, elongated slit in the frustule of some diatoms, through which mucilage can extrude; involved in the motility of diatoms.

receiving waters Bodies of water into which waste treatment products are emptied.

recipient cell The cell that receives DNA from a donor cell.

recombinant DNA DNA that results from the combining of DNA molecules from two or more sources. More recently it connotes DNA formed by combining DNA molecules in vitro.

recrudescent (ree-kroo-dess′-ent) Return of symptoms of a disease after a period of improvement.

red tide Ocean waters colored by the proliferation of red single-celled algae in sufficient numbers to kill fish.

refractile (ree-frak′-tile) A body that bends a ray of light which passes through it so that the body often appears brighter than the surrounding objects.

regulatory genes Genes whose function is to control the rate of synthesis of other gene products. Regulatory genes often code for repressor proteins.

replica plating A technique for the simultaneous transfer of organisms of a large number of separated colonies from one medium to others. Used in the indirect selection of bacterial mutants.

replication (viral) The process of synthesis of viral components within a host cell, culminating in the formation of complete virions.

replication fork The Y-shaped structure in the replicating DNA molecule; the site of DNA replication.

replicative form A double-stranded form of nucleic acid, produced during multiplication of single-stranded viruses and consisting of the single strand of nucleic acid and a complementary strand. The replicative form serves at the intermediate in the synthesis of new single strands of viral nucleic acid.

repressor A regulatory substance (usually a protein) that prevents the function of a certain gene or genes.

repressor protein (regulatory protein) A protein capable of regulating the function of a gene or group of genes.

reproduction The initial process through which propagation of species is carried on: asexual, in which a single cell divides, or sexual, in which two cells unite before daughter cells are formed.

reservoir The sum of all the sources of an infectious agent.

resistance transfer factor (RTF) A set of genes, associated with R factors that carry the genetic information for the transfer of R factors.

resolving power The distance two objects must be from one another before they can be distinguished with the microscope as two separate objects.

respiration (res-spir-ray′-shun) The sum total of metabolic steps in the degradation of foodstuffs when the electron (hydrogen) acceptor is an inorganic compound.

restriction enzyme An endonuclease that cleaves DNA at specific purine and pyrimidine sites.

reticuloendothelial system (re-tick′-yew-low-end-oh-theal-e-al) The system of mononuclear phagocytes distributed throughout the body in such a way as to clear foreign materials from the circulation; particularly abundant in liver, spleen, lung, lymph nodes, and bone marrow. Certain mononuclear phagocytic cells of the body located especially in liver, spleen, lung, lymph nodes and bone marrow.

reverse mutation or reversion The modification of the base sequence in a mutant gene to its original (parent) sequence.

reverse transcriptase An enzyme that synthesizes DNA complementary to an RNA template.

R factor A transferable plasmid, found in many enteric bacteria, which carries genetic information for resistance to one or more chemotherapeutic agents. Often associated with a transfer factor that is responsible for conjugation with another bacterium and transfer of the R factor.

R genes The genes on the R factor which are specifically concerned with resistance to chemotherapeutic agents.

rhabdoviruses (rab′-do-viruses)

Rhizobium (rise-oh′-bee-um)

rhizopodial (rise-oh-poh′-dee-al) Algae that are motile by means of amebalike pseudopods.

Rhizopus (rise′-oh-puss)

rhizosphere (rise′-o-sphere) The region immediately surrounding the roots of plants.

Rhodomicrobium vannielii (road-oh-my-kroh′-bee-um van-neel′-ee-ee)

ribose (rye′-bose) A 5-carbon sugar found in ribonucleic acid.

ribosomal RNA (rye-bow-soam′-mal) A type of RNA which is found in ribosomes.

Rickettsia prowazekii (rik-ett′-see-ah prow-ah-zek′-ee-ee)

RNA polymerase (poh-lim′-mer-ace) An enzyme that catalyzes the synthesis of RNA from ribonucleoside triphosphates; the enzyme that links together nucleotide subunits during the transcription of DNA to form RNA. Its action represents the first step in protein synthesis.

RNA synthetase An enzyme that polymerizes mononucleotides into RNA, using RNA as a template.

Saccharomyces cerevisiae (sac′-a-row-my′-seas sara-vis′-i-ee)

Salmonella typhi (sall-moh-nell′-ah tye′-fee)

salmonellosis (sall-moh-nell-oh′-sis) Infection with pathogenic Gram-negative rods of the genus *Salmonella.* Well-known forms of salmonellosis include the common gastroenteritis caused by many strains of *S. enteritidis* and the less common typhoid fever caused by *S. typhi.*

saprophyte (sap′-roh-fyte) A microorganism which lives on dead organic material.

sarcoma (sar-ko′-ma) A tumor of connective tissue.

Sarcoptes scabiei (sar-kop′-teez scabe′-ee-ee)

schistosomiasis (shiss-toh-soh-mye′-assiss) The disease resulting from infection with flukes of the genus *Schistosoma.*

Schistosoma mansoni (shiss-toh-soh′-mah man-soh′-nee)

secondary bonding forces Weak chemical bonds, such as hydrogen bonds, van der Waals forces.

selective enrichment A technique for specifically encouraging the growth of a particular group or even a single species of organism.

selective medium A medium that has components which restrict growth to organisms of a particular type.

semiconservative replication The type of replication of DNA in which new daughter strands complementary to each of the original strands are synthesized in each round of replication.

sensitized lymphoid cells Lymphoid cells, usually lymphocytes, that have the capacity to respond to specific antigen in such a way as to cause reactions of delayed-type hypersensitivity and cellular immunity or production of antibody in the anamnestic response.

septate Divided by membrane; for example, fungal filaments may be septate or may lack septa (aseptate or nonseptate).

septum (sep′-tum) A membrane separating two cavities or two cells in a filament of cells.

serological Having to do with serum.

serotype Strain of microorganism distinguished by its reaction with specific antiserum.

Serratia marcescens (ser-ray′-shee-ah mar-sess′-senz)

serum (see′-rum) The fluid portion of blood which remains after the blood clots.

sessile (sess′-ile) Fixed or attached.

sexual recombination The exchange of portions of DNA molecules, from each parent, in a cell. Provides for the cell having some properties of each of the parental cells.

Shigella dysenteriae (shigg-ell′-ah diss-en-tair′-ee-ee)

Siderocapsa (sid-er-oh-cap′-sah)

Siderococcus (sid-er-oh-kok′-us)

sigma (σ) factor Subunit of RNA polymerase that recognizes specific sites on DNA for initiation of RNA synthesis; a subunit of the enzyme RNA polymerase which is responsible for determining where the enzyme begins transcribing.

Simonsiella (see-moan-see-ell′-ah)

simple lipid A lipid that contains only the elements C, H, and O.

single-cell protein (SCP) The protein contained in members of the microbial world which may serve as a source of food.

soap A substance resulting from reaction of a fatty acid and an alkali; a salt of a fatty acid.

somatic cells (so-mat′-ic) All of the cells of the body except the germ cells (eggs and sperm).

specificity A term defining selective reactivity between an antigen and its corresponding antibodies or primed lymphocytes.

Sphaerotilus (sfeer-aht′-il-us)

spherule (sphere′-ule) Large, round, thick-walled structure containing many fungal spores; characteristic of the tissue phase of *Coccidioides immitis.*

sphincter (sfingk′-ter) A ringlike muscle that normally constricts a passage or orifice of the body and relaxes as required for normal function.
spicule (spick′-ule) A needlelike structure.
Spirillum (spye-rill′-um)
Spirochaeta (spyro-kee′-tah)
spontaneous mutants Mutants which arise without the addition of recognized mutagenic agents.
sporangium (spore-ran′-gee-um) Saclike structure within which asexual spores are formed.
spores Specialized reproductive cells; asexual spores germinate without uniting with other cells, whereas sexual spores of opposite mating types unite to form a zygote before germination occurs.
sporocyst (spore′-oh-sist) A stage in the life cycle of certain protozoa in which two or more of the parasites are enclosed within a common wall.
Sporothrix (*Sporotrichum*) *schenckii* (spore-oh-trick′-um shenk-ee-ee)
sporozoan (spore′-oh-zoe′-uhn) A protozoan of the class Sporozoa.
sporozoite (spore-oh-zoh′-yte) A developmental form in the life cycle of certain protozoan parasites such as *Plasmodium* species. The sporozoite is the form in which the malarial parasite is transmitted to human beings by the infected mosquito.
sporulation Spore formation. Also used to mean a process of multiple fission by which some protozoa divide.
stabilization (mineralization) Conversion of organic materials to inorganic substances.
stable terminal residue The last in a series of soil degradation products which is not capable of being further degraded by natural means, and therefore remains.
Staphylococcus aureus (staff-il-oh-kok′-us aw′-ree-us)
S. epidermidis (S. epi-der′-mid-iss)
starch A polysaccharide consisting of repeating glucose molecules.
stasis (stay′-sis) A staying in one place, or a stoppage of flow, for example, of blood.
stationary phase The stage of growth of a culture in which the number of viable cells remains constant.
sterilization Rendering an object or substance free of all viable microbes. Practically speaking, to sterilize an object is to make it extremely improbable that a single living organism or virus remains.
Streptococcus pyogenes (strepto-kok′-us pye-ah′-jen-eez)
Streptomyces griseus (strepto-mye′-seez gree′-see-us)
S. mediterranei (S. med-it-ter-an′-ee-ee)
S. venezuelae (S. ven-ez-u-ayl′-ee)
substrate The substance on which an enzyme acts to form the product.
succession (ecological) The predictable and orderly sequence of change in a community.
sucrose A disaccharide consisting of a molecule of glucose and one of fructose.
sugar A common term for the low molecular weight carbohydrates. Common table sugar is sucrose. There are D- and L-forms of sugars; the D-forms are found in nature.
sulfanilamide (sul-fa-nil′-a-mide) A drug whose chemical structure is very similar to that of para-amino benzoic acid. It competes with PABA for the enzyme which converts PABA to folic acid.
superinfection An infection added to one already present.
suture (su′-chore) Stitch used by a surgeon to close a wound.

swarmer cells The cells of *Rhizobium* which penetrate the root hairs to set up the symbiotic association with legumes. They are more nearly spherical than the rod-shaped cells of *Rhizobium.*

symbiosis (sim-bye-oh′-siss) Living together.

synapsis (sin-nap′-sis) The pairing of identical chromosomes which occurs prior to the first division in meiosis.

synthetic medium A medium in which every component has been identified with respect to its chemical composition and quantity.

taxonomy The science of the classification of organisms.

temperate phage A phage which can either become integrated into the host cell DNA as a prophage, or replicate outside the host chromosome and lyse the cell.

template The molecule that serves as the mold for the synthesis of another molecule. Macromolecular synthesis involves templates.

termites A type of insect which harbors cellulose digesting protozoa in its gut.

tetracycline (tet-ra-sigh′-clean) An antibiotic produced by a member of the genus *Streptomyces;* acts by inhibition of protein synthesis.

therapeutic index The ratio of minimum toxic dose to the effective dose of a medication.

Thermoactinomyces (ther-moh-ak-tin-oh-mye′-seez)

thermophile Preferring or requiring a higher temperature than most (for example, above 40°C with bacteria).

thermostable Resistant to heat.

Thiobacillus (thy-oh-bah-sill′-us)

Thiothrix (thy′-oh-thrix)

thorax The chest.

tincture An alcoholic solution of a medication or other chemical substance.

tissue A group of cells, of one or a few similar types, which associate into a structural subunit and perform a specific function.

titer (tight′-er) The concentration of a substance in solution; for example, the amount of a specific antibody in serum, usually measured as the highest dilution of serum which will give a positive test for that antibody. The titer is often expressed as the reciprocal of dilution; thus a serum which gives a positive test when diluted 1:256, but not at 1:512, is said to have a titer of 256.

toxin A poisonous substance originating from microbes, plants, or animals.

toxoid A substance derived from a toxin in which toxicity has been destroyed while maintaining its ability to induce immunity to the toxin.

Toxoplasma gondii (tox-oh-plazz′-mah gahn′-dee-ee)

transamination (trans-am-in-a′-shun) A reaction in which keto $acid_1$ and amino $acid_2$ react to form keto $acid_2$ and amino $acid_1$. There is an exchange of the keto and amino groups.

transcription The process of transferring genetic information coded in DNA into informational messenger RNA; the copying of the message of DNA into RNA by a process which involves base pairing of complementary RNA with the DNA and polymerization of the bases so ordered.

transduction Transfer of genetic information by temperate phages. *Generalized transduction* means transduction of any part of a bacterial chromosome. *Restricted or specialized transduction* means transduction of a particular portion of the bacterial chromosome adjacent to the site of prophage integration.

transfer RNA (tRNA) A type of RNA which binds and carries amino acids to the ribosome.

transformation A conversion from one form to another. In the context of animal virology, this is the process by which viruses change normal cells to malignant cells. In bacterial genetics, this is the process of transfer of DNA as "naked" DNA. Lymphocyte transformation refers to blastogenesis in lymphocytes.

translation The process by which genetic information in the messenger RNA directs the order of the amino acids in protein during protein synthesis.

transovarial Passing from one generation to another through infection of the egg by the parent.

Treponema pallidum (trep-oh-nee′-mah pal′-ee-dum)

Trichinella spiralis (trick-in-ell′-ah speer-ahl′-iss)

Trichomonas vaginalis (trick-oh-moan′-us vaj′-in-alice)

Trichophyton (trick-oh-phye′-ton)

trophozoites (trowf′-oh-zoe′-ites) The vegetative forms of some protozoa, as opposed to the resting forms called cysts.

tumor A swelling involving a mass of new tissue which grows independently of surrounding tissue and may invade it. Neoplasm is a synonym.

tyndallization (tin-dal-i-za′-shun) The technique of eliminating endospores by treating a material with heat sufficient to kill vegetative organisms, incubating to allow any spores to germinate and develop into vegetative cells, and then reheating to kill the new vegetative cells. Tyndallization requires that the substance allow germination of endospores if they are present.

typhus An infectious disease caused by *Rickettsia prowazekii.*

ultracentrifuge A high-velocity centrifuge used to separate colloidal or submicroscopic particles, such as proteins or viruses.

ultraviolet (UV) light Electromagnetic radiation with a wavelength between 175 and 350 nm (shorter than visible light). Certain wavelengths, absorbed by nucleic acids result in mutation and death.

urethritis Inflammation of the urethra, often infectious in origin.

valency Capacity to combine with something else; for example, of atoms to combine in specific proportions with other atoms and of antibodies to combine with a certain number of antigenic determinants.

van der Waals interactions (van-der-walls) Weak chemical bonds which result from the nonspecific attraction of atoms when they come near each other.

vascular Pertaining to blood or lymphatic vessels.

vector An animal, often an insect, which carries an infectious agent from one host to another.

vegetative cell The growing or feeding form of a microbial cell, as opposed to resistant resting forms.

vegetative phage Phage that is replicating within a host cell and producing mature virions.

vehicle An inanimate carrier of an infectious agent from one host to another.

Veillonella (vah′-yon-ell-ah)

vesicle A small sac.

vestigial (ves-tijh′-e′-al) Occurring as a visible trace or sign of something that no longer exists.

Vibrio cholerae (vib′-ree-oh kahl-er-ee)

viridans streptococci Alpha hemolytic streptococci that do not fall into recognized serotypes.

virion (veer′-e-ohn) A viral particle consisting of nucleic acid surrounded by a protein capsid and, in some viruses, an outer envelope.

viroids (vye′ roids) Single-stranded molecules of RNA that have been shown to cause several plant diseases. Their mechanisms of action are unknown.

virulence The sum of properties of an organism which enhances pathogenicity.

virulent phage A bacteriophage that causes lysis of the host cell following phage replication in the cell. As opposed to temperate phage it cannot be integrated into the chromosome of the host.

virus An obligate intracellular parasite consisting of a bit of nucleic acid, surrounded by a protein coat and sometimes enclosed by an envelope. Different viruses are capable of infecting animals, plants, and bacteria.

vitamin A substance, required in very small quantities, which is converted into a coenzyme.

volutin (vol′-u-tin) Another name for polyphosphate compound; occurs as metachromatic granules; long chains of phosphate molecules joined together to form a storage material in certain cells.

Xenopsylla cheopis (zeen-oh-sill′-ah kee-op′-iss)

Yersinia pestis (yer-sin′-ee-ah pess′-tiss)

zooplankton (zoe-oh-plank′-ton) The floating and swimming small animals and protozoa of open waters.

zoospores (zoe′-oh-spores) Motile, asexual spores.

zygote (zigh′-goat) Diploid cell formed by the sexual fusion of two haploid gametes; a cell formed by the sexual union of a male and female gamete.

INDEX

Major references are indicated by boldface page numbers.

D

E

F

G

J

K

L

M

Q

R

U

V